River Discharge to the Coastal Ocean
A Global Synthesis

Rivers provide the primary link between land and sea, historically discharging annually about 36 000 km^3 of freshwater and more than 20 billion tons of solid and dissolved sediments to the global ocean. Together with tides, winds, waves, currents, and geology, rivers play a major role in determining the estuarine and coastal environment. The movement of freshwater and the distribution of river-derived sediments to the ocean have fundamental impacts on a wide variety of coastal environments, ranging from the Mississippi and Nile deltas, to coastal Siberia, to the Indonesian archipelago.

Utilizing the world's largest database – 1534 rivers that drain more than 85% of the landmass discharging into the global ocean – this book presents a detailed analysis and synthesis of the processes affecting the fluvial discharge of water, sediment, and dissolved solids. The ways in which climatic variation, episodic events, and anthropogenic activities – past, present, and future – affect the quantity and quality of river discharge are discussed in the final two chapters. The book contains more than 165 figures – many in full color – including global and regional maps. An extensive appendix presents the 1534-river database as a series of 44 tables that provide quantitative data regarding the discharge of water, sediment and dissolved solids. The appendix's 140 maps portray the morphologic, geologic, and climatic character of the watersheds. The complete database is also presented within a GIS-based package available online at www.cambridge.org/milliman.

River Discharge to the Coastal Ocean: A Global Synthesis provides an invaluable resource for researchers, professionals, and graduate students in hydrology, oceanography, geology, geomorphology, and environmental policy.

JOHN D. MILLIMAN is Chancellor Professor of Marine Science at the Virginia Institute of Marine Science at the College of William and Mary, Virginia, and has been at the forefront of studies of river runoff to the coastal ocean since the 1970s. He is an Honorary Professor at the Institute of Oceanology, Chinese Academy of Sciences, and recipient of the 1992 Francis P. Shepard Medal for excellence in marine geology, and has published more than 180 research papers and 10 books.

KATHERINE L. FARNSWORTH is an Assistant Professor in the Department of Geoscience at the Indiana University of Pennsylvania. Her research examines the interaction between fluvial and marine systems including the flux and fate of sediments to the coastal oceans.

River Discharge to the Coastal Ocean

A Global Synthesis

JOHN D. MILLIMAN
Virginia Institute of Marine Science
College of William and Mary
Gloucester Pt., Virginia USA

KATHERINE L. FARNSWORTH
Department of Geoscience
Indiana University of Pennsylvania
Indiana, Pennsylvania USA

CAMBRIDGE
UNIVERSITY PRESS

University Printing House, Cambridge CB2 8BS, United Kingdom

Cambridge University Press is part of the University of Cambridge.

It furthers the University's mission by disseminating knowledge in the pursuit of education, learning and research at the highest international levels of excellence.

www.cambridge.org
Information on this title: www.cambridge.org/9781107612181

First published 2011
First time paperback 2013

A catalogue record for this publication is available from the British Library

Library of Congress Cataloguing in Publication data
Milliman, John D.
River discharge to the coastal ocean : a global synthesis / John D. Milliman, Katherine L. Farnsworth.
p. cm.
Includes bibliographical references and index.
ISBN 978-0-521-87987-3
1. Stream measurements. 2. Rivers. 3. Fluvial geomorphology.
I. Farnsworth, Katherine L. II. Title.
GB1203.2.M55 2011
551.48'3–dc22 2010034896

ISBN 978-0-521-87987-3 Hardback
ISBN 978-1-107-61218-1 Paperback

Contents

Foreword

We began this book primarily to collate into a single database relevant environmental data for rivers that discharge directly into the global ocean. As the book evolved, however, it began to take on a life of its own, and now contains, following a brief introductory chapter, three rather long chapters that attempt to give an overview to the environmental controls on fluvial discharge, short-term and longer-term temporal variations, and the impact of human activities on global rivers and their watersheds.

As geological oceanographers, we tend to view rivers differently than most hydrologists, geochemists or geomorphologists. While our data, analyses and interpretations of these data touch upon fundamental processes that govern fluvial runoff, physical, and chemical weathering, transport, dispersal and sedimentation, we do not explain in detail how fluvial systems work. We do, however, include a rather extensive bibliography that can help direct the interested reader to relevant literature.

In presenting such a large database – more than 1500 rivers – some of the data almost invariably will prove to be erroneous. We may have transposed numbers, and almost surely we have missed some key data sources. Where errors or omissions occur, we request the reader notify us – kindly, if possible – so that we can correct our mistakes and set the record straight.

There are many people whose help made this book possible. Because of the long time – more than three decades for JDM – over which we acquired and analyzed these data, we almost certainly have inadvertently omitted several (we hope not many) people, to whom we apologize. The Global River Data Centre (GRDC) and the US Geological Survey National Water Information System (NWIS) provided a large number of data used throughout the book. Charlie Vörösmarty's RIVDIS and Arctic RIMS databases were also very helpful, particularly in creating their various user-friendly internet databases. We also made liberal use of Michel Meybeck's global river database (Meybeck and Ragu, 1996), without which our discussion of chemical weathering in Chapter 2 would have been impossible.

Maria Michailova deserves special mention for the way in which she substantially clarified the extensive database for Russian rivers, many of which we either did not know or got wrong in our earlier attempt at a collation of global data (Milliman *et al.*, 1995). We also thank Yoshi Saito for help with data from Japan and China; Steve Smith for data from Mexico and southeast Asia; Yang Zuosheng, Wang Ying, Wang Houjie, S. L. Yang and Kevin Xu (China); Jim Wilson (Ireland); Kristinn Einarsson (Iceland); Lea Kauppi and Pirkko Kauppila (Finland); Des Walling (England); Wolfgang Ludwig, Maria Snoussi and Albert Kettner (Mediterranean and Black Sea rivers); Guadalupe de la Lanza Espino (Mexico); Bastiaan Knoppers (Brazil); Pedro Depetris (Argentina); Juan Restrepo (Colombia); John Largier (South Africa); Pham Van Ninh (Vietnam); Murray Hicks and Berry Lyons (New Zealand); Kao Shuh-ji (Taiwan); Harish Gupta (India); Peter Harris (Australia); and Professor da Silva (Sri Lanka). Jeff Mount and Jon Warrick opened our eyes regarding California rivers.

There are many other colleagues and friends without whom parts of this book would have been difficult, if not impossible, to write. Phil Jones (Climate Research Unit, University of East Anglia) provided us with a 100-year precipitation database from which Kehui (Kevin) Xu was able to synthesize several maps used in Chapter 2. Kevin also provided us with a number of other maps that we used throughout Chapters 3 and 4. Jon Warrick, James Syvitski, Kao Shuh-ji, Noel Trustrum, Nate Mantua and Bob Gammish were extremely helpful in supplying us with photographs used in Chapters 3 and 4, as was Juan Restrepo in sharing with us his land-use maps of the Magdalena watershed. Mike Page's and Noel Trustrum's photo of a dammed lake (Fig. 3.55) is also gratefully acknowledged. The time spent with Leal Mertes on the Santa Clara provided much insight and many memories, and she would have almost certainly provided us with a startling cover for this book; we mourn her passing.

Additional conversations with and insights from James Syvitski, Des Walling, Michel Meybeck, and Jean-Luc Probst over the past 15 years also were extremely helpful. We thank them all for their continued friendship.

Bob Meade has been an admirable mentor and long-time steadfast friend. Our many conversations and insights over the years have proved particularly useful in helping us understand better the whys and wherefores of, as Bob would term it, potamology. Bob's critical reading of the first draft of this book proved invaluable; any omissions or errors in this book almost certainly occurred after his careful editing.

A particular heart-felt thanks goes to Marilyn Lewis, at the SMS/VIMS library, who, through the wonders of Interlibrary Loan, obtained for us many of the books and reprints that facilitated our access to the river literature.

Harold Burrell's graphic skills, coupled with his patience as we continually reconsidered and re-thought figures for the book, are gratefully acknowledged.

Many of the data and insights in this book stem directly or indirectly from research grants awarded by the US National Science Foundation (NSF), the US Office of Naval Research (ONR), and the US Naval Oceanographic Office (NAVO). We thank in particular Peggy Schexnayder (NAVO) for her unfailing interest and encouragement over the 12 years that we spent working on this book.

Finally, JDM thanks Ann Milliman for enduring those many evenings when thoughts of rivers and this book took precedence over some of the more domestic and romantic aspects of our life. KLF thanks her family and friends for their patience and encouragement throughout the process.

1 Introduction

"Data! Data! Data!" (Holmes) cried impatiently.
"I cannot make bricks without clay."
Dr. J. Watson as transmitted to A. Conan Doyle

"Give me the facts, Ashley,
and I will twist them the way I want to suit my argument"
(statement attributed to W. Churchill)

Rivers and the coastal ocean

Rivers provide the primary link between land and sea, annually discharging about 36 thousand km^3 of freshwater and more than 20 billion tons (Bt) of solid and dissolved material to the world ocean. These fluxes, together with physiography and oceanographic setting, help determine the character of the estuarine and coastal environment. Although discharged water and sediments are generally confined to the coastal zone, if a flood is sufficiently large (e.g. Amazon River) or the shelf sufficiently narrow (e.g. southern California or eastern Taiwan) fluvial-driven plumes can extend to or beyond the shelf edge. In addition to their link to the coastal ocean, rivers historically have played key roles in human habitation and history, providing water, nutrients, transportation, and protection, among other things, for people living within their drainage basins.

Because of the wide range of physical and societal functions that a river can serve, one appealing – yet also daunting – aspect in the study of rivers and their watersheds is the diversity of perspectives and approaches used in their study. Geochemists and geologists often view a river in terms of landscape denudation or sediment transport, whereas geomorphologists may be more concerned with landscape character and its evolution. Engineers design and plan human adaptations to a watershed, while planners and policy makers may focus on the societal implications of these anthropogenic changes. Oceanographers tend to view rivers in terms of their discharge – and the fate of that discharge – to the coastal ocean. One outcome of this diversity of approaches is an ever-expanding database and an even wider range of published and unpublished literature. This is seen by the variety of journals and books referenced in this book. It may be unusual, for example, for an oceanographer to read *Catena*, a journal devoted to soil science, or a geomorphologist to read *Marine Geology*, but for someone interested in rivers, these journals – and a great many more like them – become almost required reading.

Superimposed on the interest in local, regional and global watersheds and their discharge to the marine environment has been a growing concern about the impacts of global climate change and human perturbations on present and future water resources (e.g. Shiklomanov and Rodda, 2003; Vörösmarty and Meybeck, 2004). An increasing number of international programs, such as those under the auspices of the International Geosphere–Biosphere Programme (IGBP), focus on the connective links between rivers, their watersheds, and the coastal ocean. The number of organizations (and, sadly, their acronyms) involved in water-related issues has literally exploded over the past 15–20 years into a veritable smorgasbord of alphabetical constructions. Gleick (2002), for example, used more than 10 pages to list water-related websites, and the number has increased substantially since then. Of particular relevance to this book are the programs within IGBP, most notably LOICZ (Land–Ocean Interactions at the Coastal Zone), PAGES (Past Global Changes), GAIM (Global Analysis, Integration and Modeling), ILEAPS (Integrated Land Ecosystem–Atmosphere Processes Study), WCRP (World Climate Research Programme), and GWSP (Global Water System Project). These IGBP efforts have led to a series of valuable and timely books dealing with global change. Of particular relevance to rivers are *Global Change and the Earth System* (Steffen *et al.*, 2003), *Paleoclimate, Global Change and the Future* (Alverson *et al.*, 2003), *Vegetation, Water, Humans and the Climate* (Kabat *et al.*, 2003), and *Coastal Fluxes in the Anthropocene* (Crossland *et al.*, 2005).

About this book

In this book we attempt to document and provide an overview and understanding of river fluxes to the coastal ocean. Our global database of more than 1500 rivers (see Fig. 1.1) includes only rivers that discharge directly to the ocean, not tributaries. The Mississippi and Amazon, for example, are included but not the Ohio or the Negro.

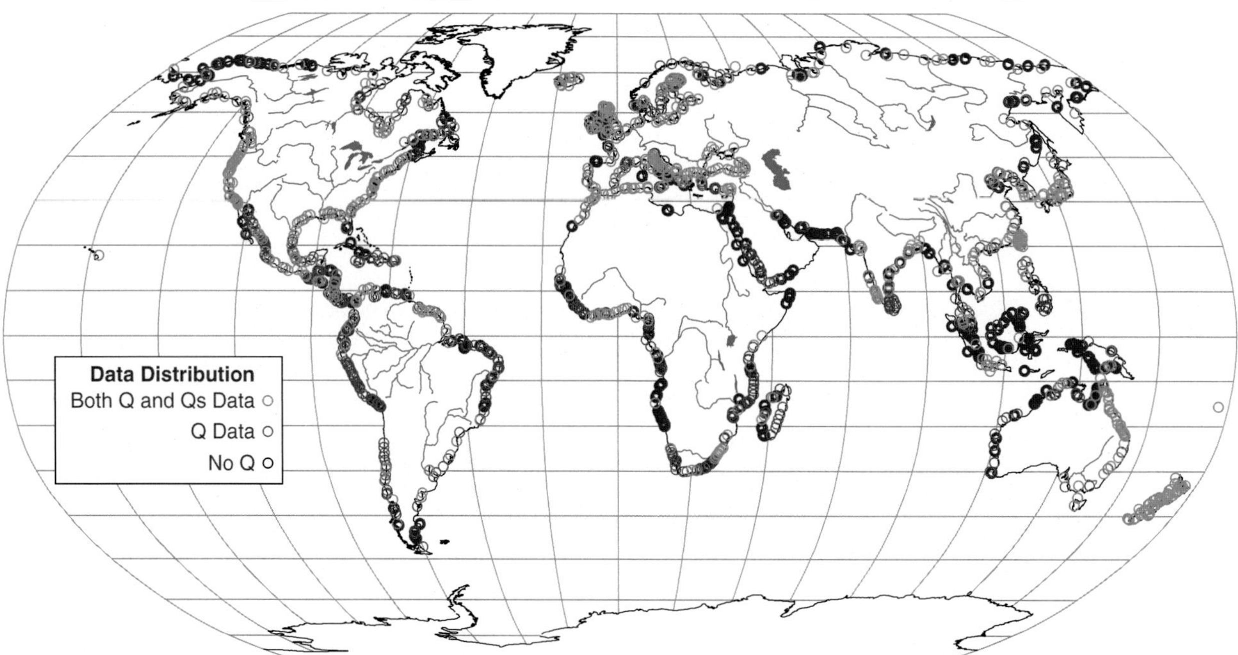

Figure 1.1. Locations of the 1534 rivers represented in our global database. Green circles represent rivers for which mean annual discharge and sediment and/or dissolved solid discharge are available; red circles represent those rivers for which only discharge is available; and black circles rivers for which no discharge values have been reported (or at least not which we could find in the literature).

Early attempts to collate a global database suffered from the lack of data, particularly for rivers that drain developing countries and for smaller rivers, even though they play critical roles in the global delivery of fluvial sediment to the coastal ocean (Milliman and Syvitski, 1992). A 1994 LOICZ-sponsored GLORI (Global River Index) meeting in Strassbourg (France), hosted by Jean-Luc Probst, provided the initiative to expand the global river database, and a subsequent LOICZ–GLORI compilation of more than 600 rivers (Milliman *et al.*, 1995) provided a template on which future additions or corrections could be added. This was followed by an expanded GLORI database collated by Meybeck and Ragu (1996). Based in large part on the international response to the LOICZ report, the database has grown to the 1500+ rivers presented in the appendix of this book.

We define a river as a linear depression that drains to progressively lower elevations – in this book, ultimately to the ocean. By this definition, frequency and quantity of discharge are not factors in delineating a river, although they certainly help define the character of a river. Thus a wadi in Sudan or an ephemeral stream in Mexico is geomorphically, if not hydrologically, as much a river as the Amazon, even though its flow may be infrequent or, in the case of Libyan rivers, presently non-existent.

Our database contains entries from more than 100 countries, the most entries (128) being from the USA. But other countries have a surprising number of entries; for example, 60 from Mexico, 30 from South Africa, and even 11 from Yemen. The rivers included in the database in part depends on the size of the country. For larger countries, such as USA, Russia or China, we generally list only rivers larger than 3000 km^2 in area, except where annual suspended-sediment or dissolved-solid data are available. For smaller countries, such as Italy or Cuba, we include some rivers with basin areas smaller than 1000 km^2.

The 1534 rivers in our database collectively drain 86 600 000 km^2 of watershed. The cumulative area of the 34 rivers with watersheds larger than 500 000 km^2 (Fig. 1.2) accounts for 52 900 000 km^2, half of the ~105 000 000 km^2 that drains into the global ocean (Fig. 1.3b). Assuming that the number of global rivers is inversely proportional to their basin areas (Fig. 1.3a) and that our database includes essentially all rivers with drainage basins larger than 30 000 km^2 (292 rivers), we derived an algorithm (# rivers = 16.92*(basin area in millions of km^2)$^{-0.7903}$; r^2 = 0.997) that allows us to calculate the number of global rivers relative to their basin areas. Our algorithm, for example, predicts that there are 644 global rivers with basin areas greater than 10 000 km^2. Our database, in fact, shows 643 rivers (Fig. 1.3a) in this size range, suggesting that that our database contains all or nearly all rivers with drainage basins larger than 10 000 km^2. Cumulative basin area for these 643 rivers is 83 200 000 km^2.

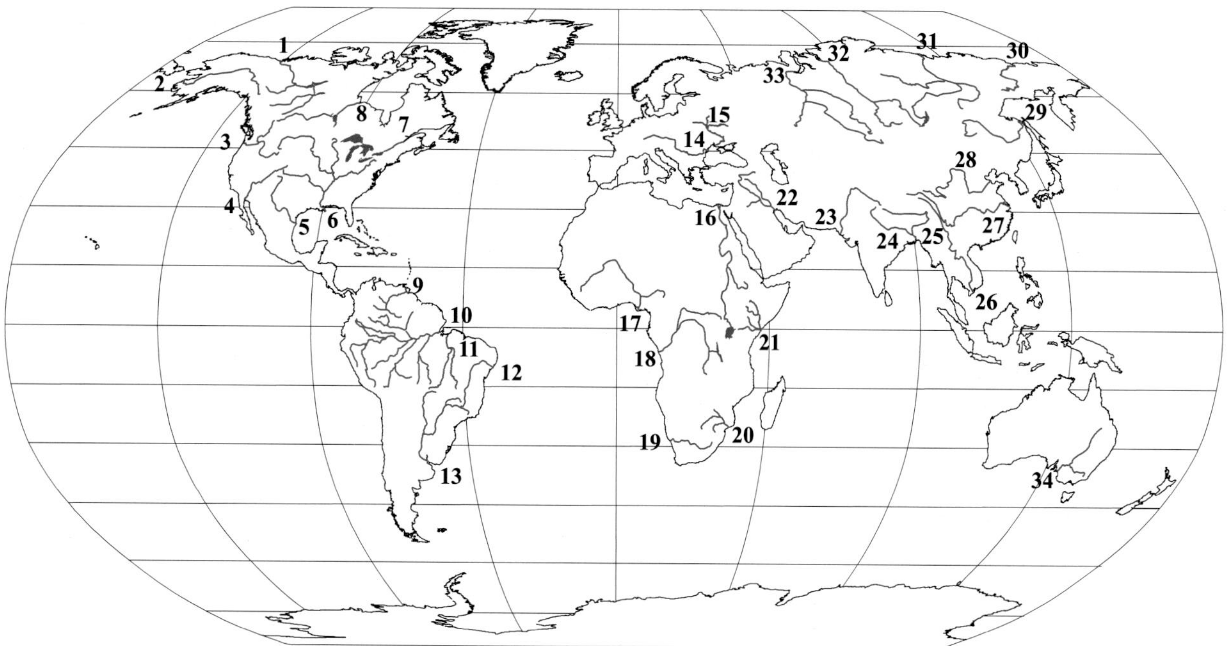

Figure 1.2. Locations of the 34 rivers with basin areas greater than 500 000 km^2; collectively these drainage basins account for half of the land area draining to the global ocean. 1, MacKenzie; 2, Yukon; 3, Columbia; 4, Colorado; 5, Rio Grande; 6, Mississippi; 7, St. Lawrence; 8, Nelson; 9, Orinoco; 10, Amazon; 11, Tocantins; 12, Sao Francisco; 13, Parana; 14, Danube; 15, Dniepr; 16, Nile; 17, Niger; 18, Congo; 19, Orange; 20, Limpopo; 21, Shebelle-Juba; 22, Shatt al Arab; 23, Indus; 24, Ganges; 25, Brahmaputra; 26, Mekong; 27, Changjiang; 28, Huanghe; 29, Amur; 30, Kolyma; 31, Lena; 32, Yenisei; 33, Ob; 34, Murray.

Because our database is oriented towards rivers larger than 3000 km^2, it is less inclusive for smaller rivers. Of the approximately 24 500 global rivers with basin areas larger than 100 km^2, we calculate that there are ~23 000 rivers having watersheds between 100 km^2 and 3000 km^2 in area; collectively they drain about 10 000 000 km^2. Our database includes only 450 of these rivers, draining a cumulative area of 700 000 km^2 (Fig. 1.3b). In spite of their relative paucity, these 450 rivers nonetheless represent perhaps the most extensive small-river database yet published.

At the end of the book we present our GIS-based database – in both printed form and as a online at www.cambridge.org/milliman – which provides an environmental characterization of the 1534 rivers. For ease of presentation, we divide the world into 44 regions, for each of which we present three maps that identify river location and drainage basin morphology, drainage basin geology and average runoff (both annual and monthly), and drainage basin geology. The database lists the body of water into which the river discharges and important climatic and geomorphic characteristics, such as basin area, maximum elevation, geology, and discharge volumes of water, sediment and dissolved sediments; see pages 165–169 for a more complete discussion of the database and maps.

Chapter 2 discusses the discharges of water, suspended and dissolved solids to the global ocean, as well as the environmental factors that control these fluxes. Chapter 3 addresses temporal variations and changes, ranging from climatic cycles to the impact of episodic events (e.g. floods, or volcanic eruptions). Few rivers and their watersheds, however, are immune from human activities and their environmental impacts, and in Chapter 4 we discuss some of the impacts of human-induced change on rivers, culminating in a short discussion of probable impact(s) of present and future use and climate change. Said in another way, Chapter 2 describes how rivers work and the final two chapters throw up numerous caveats and cautions to any synthetic interpretations based on long-term means. The discussion and our data presented herein should not be considered complete, but rather as a moving target that will evolve as new data become available and as future shifts in climate and watershed character, both natural and anthropogenic, make themselves felt.

Other global databases

It was only in the early nineteenth century that river discharge was systematically monitored, first in northern Europe, most notably the Gøta (Sweden), Nemanus (Lithuania), and Rhine (Germany) rivers, and 50 years later in North America (St. Lawrence). Few non-European discharge measurements pre-date the twentieth century. As

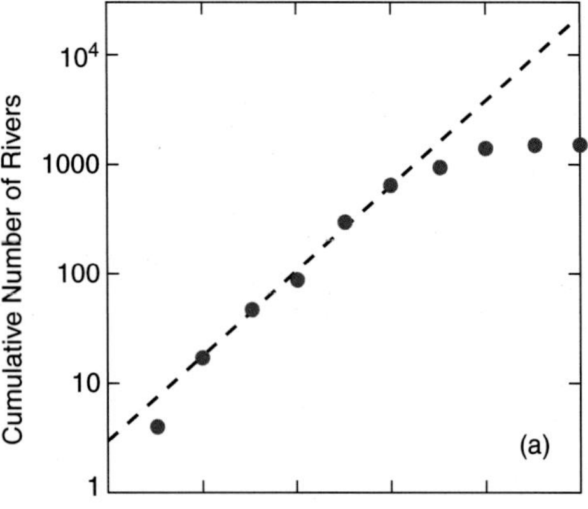

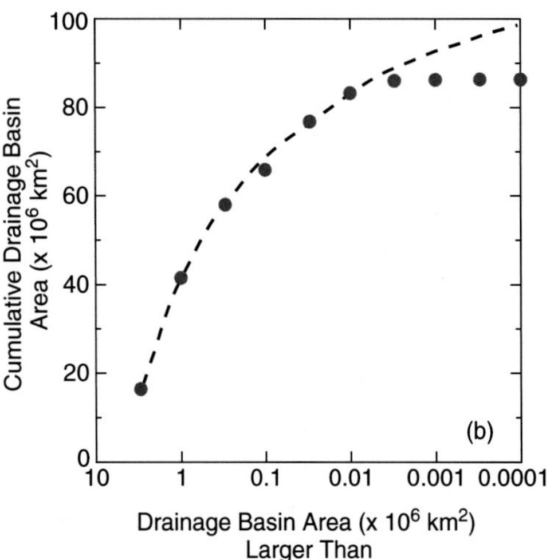

Figure 1.3. (a) Calculated number of global rivers (dashed line) and number of rivers in our database (red dots) vs. drainage basin size. The algorithm (# rivers = 16.92*(basin area in millions of km^2)$^{-0.7903}$) on which the calculation is based was derived from rivers in our database larger than 1 000 000 km^2, 300 000 km^2, 100 000 km^2 and 30 000 km^2. This plot suggests that our database effectively captures all or nearly all rivers larger than 10 000 km^2 that discharge into the global ocean. (b) Calculated cumulative basin area vs. basin size (dashed line) closely follows cumulative basin areas from our database (red dots) for rivers larger than 10 000 km^2. Given our emphasis on rivers larger than 3000 km^2, our database includes relatively few smaller rivers. The calculated global drainage basin areas for rivers larger than 10 km^2 in area – 103 000 000 km^2 – closely approximates the total land area draining into the global ocean, 105 000 000 km^2, lending confidence to our calculations.

such, the global database – or least that which is accessible – is rather thin both spatially and temporally.

In terms of both coverage and time-series, the UNESCO compilation of river discharge data remains a singularly valuable contribution. Beginning with *Discharge of Selected Rivers of the World* (1969), UNESCO ultimately published monthly and annual discharge records of 1000 rivers, a number of records extending back into the nineteenth century. Publication of an African river database (UNESCO, 1995) was particularly useful, since discharges for many of these rivers would otherwise have been difficult if not impossible to access. UNESCO World River reports ceased publication in 1992 (data entries extending only to 1984), but fortunately Charles Vörösmarty and his colleagues at the University of New Hampshire (www.watsys.unh.edu) maintained the UNESCO database on a GIS-based web page (www.rivdis.sr.unh.edu). The New Hampshire group also compiled valuable data sets for the Arctic and Latin American rivers. The ArcticRIMS (http://rims.unh.edu/data.shtml) webpage, another contribution from the Vörösmarty group, has proved particularly useful in accessing up-to-date discharge records for most pan-Arctic rivers.

The Global Runoff Data Centre (GRDC; http://www.grdc.bafg.de) in Koblenz, Germany, has the largest and most active global database. As of July 2008, it had captured water discharge data from more than 7300 stations, many records presented as daily, monthly, and yearly discharges. As many of the data are from tributaries or rivers that drain to inland basins (e.g. central Asia), we find that only 611 of the GRDC rivers discharge directly to the sea, and only 21 have records longer than 100 years; the Gøta's discharge record extending back to 1807 being the longest. Cumulative basin area upstream of the 611 river-gauging stations is ~61 000 000 km^2 (Table 1.1), about 60% of the total land area draining to the global ocean, and collectively these 611 rivers discharge ~65% of the global fluvial water. Only ~30% of the GRDC rivers, however, have discharge records longer than 50 years (Table 1.1; Fig. 1.4a), which is barely long enough to encompass short-term climatic cycles such as El Niño–Southern Oscillation or North Atlantic Oscillation. To capture longer cycles, such as the Pacific Decadal Oscillation or the Atlantic Multidecadal Oscillation (see Chapter 3), more than 50 years of data are needed. Moreover, as of 2008, the records for less than half of the 611 GRDC ocean-discharging rivers extended beyond 2000 (Fig. 1.3b); of these, only 124 rivers had records longer than 50 years (Fig. 1.4a). Cumulative basin area upstream of gauging stations for these 124 rivers is only 16 000 000 km^2, and collectively they account for only ~12% (4400 km^3/yr) of the average global discharge.

It is not surprising that European and North American rivers rank high in terms of number of rivers in the GRDC database, as well as their lengths of record (Fig. 1.4b). In contrast, African rivers are woefully underrepresented: the GRDC lists only four African rivers with >50 years of data (Nile, Congo, Senegal, Orange), and data entries for most African rivers end by the mid 1980s.

Of the world's 12 highest-discharge rivers (see Table 2.3), one (Irrawaddy) is not found in the GRDC database, two

Table 1.1. *Summary of Global River Data Center (GRDC) database as of July 2008. Of the 611 rivers in the GRDC database that discharge directly to the ocean, 179 (<30%) are represented by more than 50 years of discharge data, about 2/3 of them (135) from Europe or North America. Collectively, South American and Asian rivers account for nearly 2/3 of the cumulative global discharge, but only 30 of these rivers are represented by more than 50 years of data.*

	# Rivers	Avg. yrs data	> 50 yrs data	ΣArea (× 10^6 km^3)	ΣQ (km^3/yr)
Europe	144	50	65	5.7	1550
N. America	173	47	70	13.6	3150
S. America	69	24	10	11.2	8450
Africa	71	25	4	8.3	2100
Asia	86	33	20	20	5600
Oceania	68	37	11	2.2	570
Totals:	**610**	**40**	**179**	**61**	**21 400**

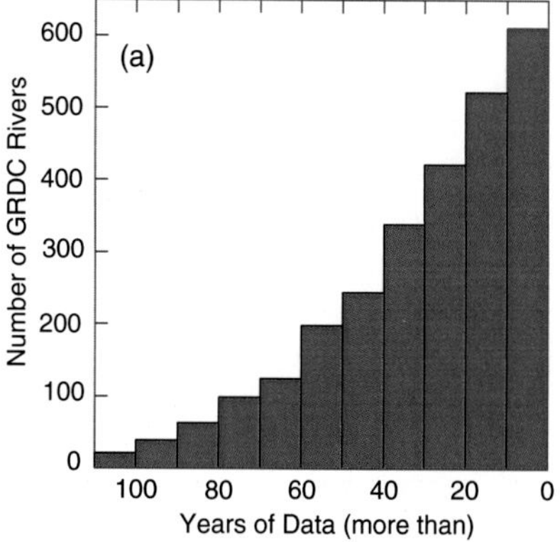

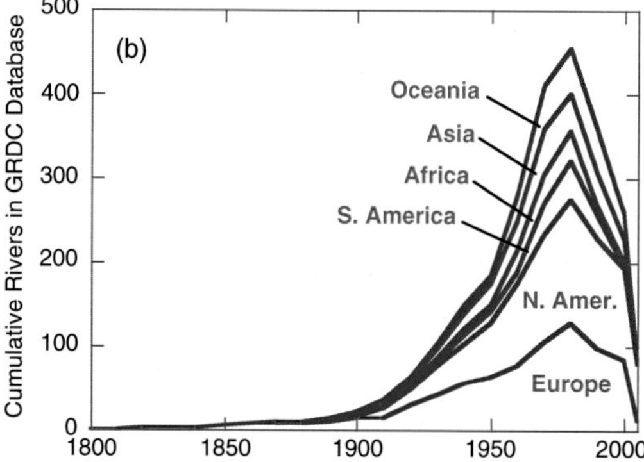

Figure 1.4. (a) Number of years of data for the 611 rivers in the GRDC database that discharge directly into the global ocean. Relatively few rivers are represented by more than 50 years of data. (b) GRDC discharge data, 1800–2005. Much of the steep decline of between 1980 and 2000 reflects a decrease in the reporting of discharge data, particularly from Africa and South America, and in part reflects a decline in global river monitoring (Vörösmarty *et al*., 2001). The decrease in post-2000 data is partly the result of the lag time needed for some countries to forward their data to GRDC.

(Brahmaputra, Mekong) are represented by fewer than 15 years of post-1950 data, the Ganges has no post-1973 data, the Amazon's data begin at 1968, and the Congo's (at Kinshasa), Orinoco's and Parana's GRDC data end in 1983, 1989, and 1994, respectively. In fact, of the 12 rivers, the 2008 GRDC database lists post-1995 discharge for only the Changjiang, Lena, Ob, Yenisei, and Mississippi. Viewed another way, of the world's 50 largest rivers in terms of discharge (accounting for ~55% of the total global discharge), 14 are either not listed (e.g. Salween, Meghna, Fly, San Juan) in the GRDC's meta-database or only upstream data are listed (e.g. Niger, Zambezi, Khatanga). For the 50-yr period between 1951 and 2000, more than half of the collective ~20 000 km^3/yr discharge in the GRDC rivers is represented by less than 30 years of data (Fig. 1.5). If one were to rely solely on GRDC data, meaningful trends over this 50-yr period would be difficult to detect.

Despite the GRDC's laudable effort to collate a global discharge database, the above paragraphs suggest that some of GRDC's accomplishments have fallen short of their goals. The problem lies not with GRDC but rather with those countries who either have not measured river discharge or have been reluctant to share their data with the global community. India, Indonesia, Iran, Iraq, Italy, and Ivory Coast, to mention only the "I" countries, apparently have ceased (or have severely limited) submitting river data to GRDC. This has only increased the data disparity between countries (Table 1.2). Compounding the problem, in recent years many gauging stations have been closed (Vörösmarty *et al*., 2001). This problem is particularly acute for rivers draining the higher latitudes, where longer records are needed to help delineate short- and longer-term effects of global climate change (see Chapter 4).

We have made particular use of a database compiled by Meybeck and Ragu (1996), which, unfortunately, may prove difficult for some readers to obtain. The 545 rivers listed by Meybeck and Ragu are principally confined to those rivers

Table 1.2. *GRDC database (2008) for selected countries whose rivers discharge into the global ocean.*

Country	# Rivers in our database	# Rivers with GRDC data	# Rivers with post-1997 GRDC data
USA	124	71	62
Canada	85	56	33
Australia	82	35	31
Japan	25	22	22
Russia	67	31	15
New Zealand	69	15	14
Mexico	62	25	18
India	43	14	0
Italy	45	4	1
Indonesia	88	2	0

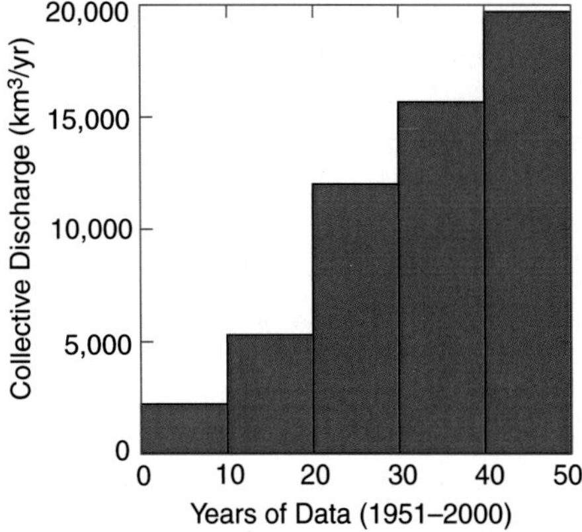

Figure 1.5. Length of record, 1951–2000, for the world's 50 largest rivers in terms of annual discharge; the collective 19 700 km^3/yr represent ~55% of the global total. Of this total, ~60% (12 000 km^3/yr) is represented by <30 years of GRDC-accessible data.

with drainage basins >10 000 km^2, annual water discharges >10 km^3/yr, or annual suspended loads >5 million tons (Mt)/yr. Some smaller rivers that drain polluted watersheds are also included. Of particular importance is Meybeck's and Ragu's documentation of reported concentrations, loads and yields of dissolved solids and nutrients discharged from many of these rivers.

Fierro and Nyer (2007) recently published their third edition of *The Water Encyclopedia*, which contains more than 1100 tables and 500 figures. For US rivers, Fierro and Nyer provide a reasonable access to primary data although, regrettably, many of the fluvial data are presented in English, not metric, units. While the subtitle of the book promises a guide to internet resources, as the authors state in their Preface, the internet historically has been unable to provide adequate data. The internet landscape, however, is rapidly changing and improving. Some countries, USA, Australia, Taiwan, and China, to mention a few, have initiated accessible internet web pages. The extensive US Geological Survey (USGS) dataset (www.nwis/usgs.gov) allows one to access discharge, sedimentological, and geochemical data, all of which were extremely useful in the preparation of this book. In early 2008, GWSP produced an on-line water atlas (http://atlas.gwsp.org) that provides maps (and their databases) for a number of environmental and socio-economic aspects of the Global Water System. But for many global rivers one is forced to rely on available published literature or personal assistance from international colleagues.

In the 1990s, the International Hydrological Programme (IHP), under UNESCO, began FRIEND ("Flow Regimes from International Experimental and Network Data" – we wonder how long it took to create *that* acronym?), which has developed a number of regional working groups. The southern Asia group, under the leadership of K. Takeuchi, released several comprehensive reports on various rivers in the region (Takeuchi *et al.*, 1995; Jayawardena *et al.*, 1997).

The United Nations Global Environment Monitoring System (GEMS) river database, distributed by the Canada Centre for Inland Waters in Burlington, Canada (http://www.cciw.ca/gems/intro.html), has centered its attention on major fluvial watersheds as a measure of regional and global water quality. Sixty-six countries have submitted data to GEMS; the total number of stations in the GEMS data archives exceeds 700. The GEMS menu offers a wide range of parameters including inorganic and organic constituents as well as pH and Biological Oxygen Demand (BOD). Meybeck *et al.* (1989) and Fraser *et al.* (1995) presented overviews of the GEMS program as well as listing data for 124 rivers. In recent years, however, the GEMS effort appears to have flagged. For instance, on-line data for the Rhone River ends in 1994, the Acheloos (Greece) in 1995, and the Guayas/Duale (Ecuador) in 1983, whereas Godavari (India) data begins in 1996. Moreover, the types of data submitted by each country vary, some countries submitting many measurements for each river, others only a few. Creating a regional or global time series based on GEMS data alone seems unlikely.

In recent years a number of working groups have attempted to collate available river discharge data. One of the more comprehensive collections is by the Woods Hole World Rivers Group (http://www.whoi.edu/page.do?pid=19735). As of the writing of this book, there were more than 1500 rivers in their on-line *Land2Sea* database, many of the rivers with watersheds smaller than 1000 km^2. The oft-cited FAO/AGL database (http://www.fao.org/landwater/aglw/sediment/default.asp), last updated in 2005, contains sediment yields for 872 rivers. These entries, however, include yields from upstream stations, tributaries of larger rivers (the Mekong River within Thailand, for instance, has 49 entries), and rivers draining inland countries (Ethiopia and Lesotho, for example); sediment yields for rivers relative to our discussion probably number less than 100. On a much smaller scale, Eurosion (http://eurosion.org/database/index.html/) has collated water and sediment discharges for a number of larger European rivers. Similar types of projects have increased in recent years.

Dams, a topic discussed in Chapter 4, have been documented intensively by the International Commission on Large Dams (ICOLD), as witnessed by their comprehensive register published in 1988. Since then, however, ICOLD's data have been restricted to ICOLD members. The World Commission on Dams (WCD; http://www.dams.org) was formed in 1998 to review and assess the design, construction and decommissioning of dams, often, it seems, at odds with ICOLD. Their first definitive report, mentioned further in Chapter 4, was issued in 2000.

Problems with existing data

Given the wide variety of sources for the data listed in our tables and discussed in the following chapters, one must acknowledge the many potential problems and pitfalls that can – and often do – affect the veracity of the data: bad measurements, unreliable rating curves, inadequate monitoring, watershed modification, erroneous transcription of the data, etc. Of the 77 rivers in our database that have reported or assumed pre-dam sediment loads greater than 20 Mt/yr, only 19 are considered to have adequate up-to-date data (Table 1.3), 11 of which are in China and Taiwan. The calculated annual sediment load for the Changjiang at the Datong gauging station, for example, is based on 30–60 daily to bi-weekly (depending on river stage) surface, subsurface and near-bottom suspended sediment samples taken at 10 to 12 cross-river stations, altogether thousands of samples annually.

The reported or assumed sediment loads of 41 rivers in (Table 1.3), by contrast, are based either on uncertain or out-of-date data (23 rivers) or are rivers for which we can find no reported measurements (18 rivers) (Table 1.3). The reported sediment load for the Susitna River (Alaska), for instance, is derived from measurements taken in the 1950s; because of ensuing human changes to the landscape (e.g. logging, mining) as well as climate change over the past 50 years, we judge these data to be marginal in terms of representing the present-day Susitna. Sadly, the Mississippi may be the only sediment-rich US river that is adequately monitored (by the US Army Corps of Engineers).

To compound the problem further, many of the rivers in our database have been dammed or irrigated such that reported water and sediment discharge may over-estimate (sometimes greatly) the actual present-day discharges to the coastal ocean. Some examples of erroneous sediment data are given in Table 1.4; other examples are discussed in greater detail in subsequent chapters.

Uneven geographic distribution

Although our river database attempts to represent a uniform geographical distribution of river data, the global distribution of the data unfortunately remains uneven. At first glance at Table 1.5, the distribution in our database looks reasonably well balanced – 199 rivers for Africa, 244 for Europe, etc. But on closer inspection we see that there are only 22 African rivers for which we have found dissolved-solid data, compared with 65 European rivers. Moreover, we can find no reported data for more than 20% of the African, Central American/Caribbean, Eurasian, and Oceania rivers listed in our appendices. Of the 70+ rivers draining western South America, for instance, only five have reported sediment data and only five have dissolved-solid data. Any estimates of suspended or dissolved deliveries from western South America are therefore clearly precarious. Likewise, there are few sediment or dissolved data for the rivers draining southern Africa or Australia, and essentially none for Central America or Caribbean rivers (see Fig. 1.1). The lack of monitored data for Philippine and (particularly) Indonesian rivers is particularly frustrating, since the few available data suggest that collectively these numerous small rivers represent major sources for both suspended-sediment and dissolved-solid discharges to the global ocean (see Chapter 2).

Uneven data quality

In our tables and in the ensuing discussion we have compiled data reported by many scientists and engineers who used a variety of measuring techniques over different periods of time. Suspended-sediment samples, for instance, may have been collected from a bucket lowered into the side of a river, by depth-integrating samplers lowered from a bridge, or water samples taken from a moving boat. Some reported data may represent a single measurement, others long-term averages.

The problem of data reliability becomes clearer if we compare reported basin areas, the one fluvial parameter that should be relatively easy to quantify and easy to replicate. Of eight published estimates of the areas of the world's 10 largest rivers listed in Table 1.6, only four rivers (the Congo, Mississippi, Yenisei, and Lena) have listed areas that are reasonably consistent. The Niger's reported basin area, in contrast, ranges from 1.2 to 2.2 × 10^6 km^2, and the Nile's from 1.8 to 3.8 × 10^6 km^2. Some of these discrepancies can be explained by the different methods used to estimate basin area. Using digital elevation models with a 30′ resolution, for example, Vörösmarty *et al.* (2000) equated potential flow pathways as a measure of basin area. Their estimate of the Nile basin area (3 800 000 km^2) is 30% higher than other published estimates, but it does include dry drainage basins that may well have discharged into the Nile in the recent geological past (Chapter 3). The problem with using digital elevation databases can be seen in comparing some of the basin

Table 1.3. *Subjective appraisal of the quality of suspended sediment data for rivers whose estimated pre-dam sediment loads are reported to or assumed to have exceeded 20 Mt/yr. Data quality is based on the rigor with which measurements were made, length of record, and date of last reported measurements. The Copper River's (Alaska) load, for example, is based on measurements last taken in 1965 from a gauging station that is no longer maintained. Until recently, the Irrawaddy's sediment discharge and sediment load were based on measurements taken in the nineteenth century (Gordon, 1885), re-evaluated in 2007 by Robinson* et al.*, and judged to be questionable in quality and present-day relevance. Publication of recent discharge data (Furuichi* et al.*, 2009), however, changed our appraisal of the Irrawaddy's database from "poor" to "fair".*

Good		Fair		Poor	
		ALB	Semani	AUS	Ord
CAN	Fraser	ALB	Vijose	BAN	Ganges
CHI	Daling	ARG	Parana	BRA	Amazon
CHI	Hanjiang	BUR	Irrawaddy	BRA	Tocantins
CHI	Huanghe	CAN	MacKenzie	BUR	Kaladen
CHI	Liaohe	CAN	Skeena	BUR	Salween
CHI	Luanhe	COL	Magdalena	CGO	Congo
CHI	Pearl	IDA	Godavari	COL	Patia
CHI	Yangtze	IDA	Krishna	EC	Guayas
FRA	Rhone	JAP	Tenryu	EGT	Nile
MEX	Colorado	NZ	Waiapu	IDA	Damodar
MOR	Sebou	PNG	Fly	IDA	Mahandi
PAK	Indus	PNG	Purari	IDA	Narmada
ROM	Danube	RUS	Amur	INO	Barito
RUS	Lena	SA	Orange	INO	Barum
TW	Beinan	TH	Chao Phrya	INO	Brantas
TW	Choshui	VN	Song Hong	INO	Cimanuk
TW	Hualien			INO	Digul
TW	Kaoping			INO	Hari
USA	Mississippi			INO	Kajan
				INO	Kampar
				INO	Kapuas
				INO	Mahakam
				INO	Membarano
				INO	Musi
				INO	Pulau
				IRQ	Shatt Arab
				KEN	Tana
				MEX	Grijalva
				MOZ	Limpopo
				NIG	Niger
				PNG	Kikori
				PNG	Sepik
				SOM	Juba
				TAN	Rufiji
				USA	Alsek
				USA	Copper
				USA	Susitna
				USA	Yukon
				VEN	Orinoco
				VN	Mekong

Table 1.4. *Previous and current estimates of sediment loads transported by several of the rivers listed in our database. (1) Gibbs (1967); (2) Dunne* et al. *(1998); (3) Inman* et al. *(1998); (4) this book; (5) NEDECO (1973); (6) Restrepo and Kjerve (2000 a,b); (7) Qian and Dai (1980); (8) Wang* et al. *(2006); (9) Milliman and Meade (1983); (10) Wasson* et al. *(1996); (11) Xu* et al. *(2007).*

River	Previous estimate (Mt/yr)		Current estimate (Mt/yr)		Reason for "error"
Amazon	500	(1)	1200	(2)	Bad sampling
S. Clara (1968–85)	9.3	(3)	3	(4)	Bad rating curve
Magdalena	240	(5)	140	(6)	Inadequate data
Huanghe	1100	(7)	<100	(8)	Water consumption + drought
Murray	30	(9)	1	(10)	Error in transcription
Changjiang	500	(9)	120	(11)	50 000 dams

Table 1.5. *Regional distribution of rivers that appear in the appendices from which many of the plots and much of the synthesis in Chapter 2 are based. Oceania (primarily Australia, New Zealand, and Indonesia) and Africa account for more than half the rivers (161) for which we can find no data. (NB. Central America includes the Caribbean islands.)*

Region	# Rivers	Discharge data	Sediment load data	Dissolved load data	No data
N. America	272	247	137	102	25
C. America	66	50	10	7	16
S. America	170	155	46	24	15
Europe	249	225	192	64	24
Africa	195	142	66	22	53
Eurasia	72	55	27	15	17
Asia	263	219	162	105	44
Oceania	244	136	91	26	108
Totals	1531	1229	731	365	302

Table 1.6. *Reported areas (× 10^3 km^2) of 10 largest river drainage basins as cited in the literature. GEMS data come from Fraser* et al. *(1995) and are also utilized by Meybeck and Ragu (1996). L & P = Ludwig and Probst (1998); Times = The* Times World Atlas *(1999); V* et al. *= Vörösmarty* et al. *(2000); Oki (1999), R & K = Renssen and Knoop (2000); D & T = Dai and Trenberth (2002).*

River	L & P	GEMS	Times	V *et al.*	Oki	R & K	D & T	This book
Amazon	5903	6112	7050	5854	6140	6400	6356	6300
Congo	3704	3690	3700	3699	3730	3820	3699	3800
Mississippi	3246	3270	3250	3203	3250	3240	3203	3300
Ob	3109	2550	2990	2570	3000	2750	2570	3000
Nile	1874	2960	3349	3826	2960	2830	3826	2900
Parana	2868	2600	3100	2661	2970	2760	2661	2800
Yenisei	2567	2550	2580	2582	2610	2600	2582	2600
Lena	2465	2440	2490	2418	2350	2460	2418	2500
Niger	1540	1240	1890	2240	2110	1640	2240	2200
Amur	1926	1920	1855	2903	1870	1880	2903	1900

areas calculated using the USGS Hydro1k; the Huanghe, for example, is listed as 990 000 km^2, compared with the generally accepted value of 780 000 km^2, the Orinoco as 950 000 km^2 vs. 1 100 000 km^2, and the Pyasina (Russia) as 64 000 km^2 vs. 180 000 km^2. Another source of discrepancy is that some reported basin areas are based on the area upstream of the seaward-most gauging station. For example the Niger's basin area reported by GEMS (Fraser *et al.*, 1995) apparently does not include the Benue River, which drains about 1 000 000 km^2. Other discrepancies may result from typographical errors that subsequently have been passed on in the published record (see below).

If it is so difficult to achieve consistency for a relatively straight-forward parameter such as basin area, what hope do we have in finding accuracy and consistency for more difficult parameters such as suspended and dissolved concentrations and transport? This quandary is discussed at greater length in Chapter 2.

Analytical and reporting errors

Poor analytical techniques can lead to errors that are perpetuated in the literature. The oft-cited nutrient fluxes from Russian Arctic rivers (Gordeev *et al.*, 1996), for instance, appear extremely high for pristine rivers. Reported NH_4–N values, for example, may be 2–3 orders of magnitude inflated due to inaccurate analyses (Holmes *et al.*, 2000; 2001).

Unfortunately, it is often difficult to vouch for the accuracy of many of the reported data, which in some cases may simply represent errors in data transcription. Once reported, these transcription errors may be recycled in other papers, thus perpetuating the error. A transcription error may explain the 2 900 000 km^2 basin area for the Amur River (generally agreed-upon area is 1 900 000 km^2) reported by Vörösmarty *et al.* (2000) and then cited by Dai and Trenberth (2002). Milliman and Meade (1983) listed the annual sediment load of the Murray–Darling River (in southern Australia) as 30 Mt/yr, whereas the proper number is probably 1 Mt/yr (Wasson *et al.*, 1996) (Table 1.4). Owing to a typographical error, Milliman and Syvitski (1992) listed the sediment load of the Mackenzie River as 42 Mt/yr rather than 142 Mt/yr, an error noted by Macdonald *et al.* (1998), but which continued to be cited in subsequent papers (e.g. Holmes *et al.*, 2002).

Duration of measurements and temporal change

Given the temporal fluctuations in river discharge, how long a record is needed to provide a reasonable estimate of mean discharge? Can we assume, for example, that annual sediment load for the Rio Terraba (Costa Rica; basin area 4800 km^2) is 1.9 Mt/yr (Krishnaswamy *et al.*, 2001) when this number is based on a single year (an El Niño year, at that) of gauging? Ironically, although annual and inter-annual variability are inversely related to basin size (see Chapter 3), large rivers generally are more thoroughly monitored. Capturing the impact of a storm or flood is therefore more likely to be missed on a small river than on a large one (Walling and Webb, 1988; Walling *et al.*, 1992). How, for example, does one factor in a three-day 1969 flood of the Santa Clara River (which luckily was monitored; see Chapter 3), one that discharged more sediment than the previous 20 years combined (Inman and Jenkins, 1999; Warrick and Milliman, 2003) (Fig. 1.6)? The timing of the sampling also can be important. Holmes *et al.* (2002), for example, suggest that the differences in reported sediment loads for rivers draining the Russian Arctic (e.g. 12–26 Mt/yr for the Lena, 4.7–16 Mt/yr for the Kolyma) may largely reflect different monitoring schedules.

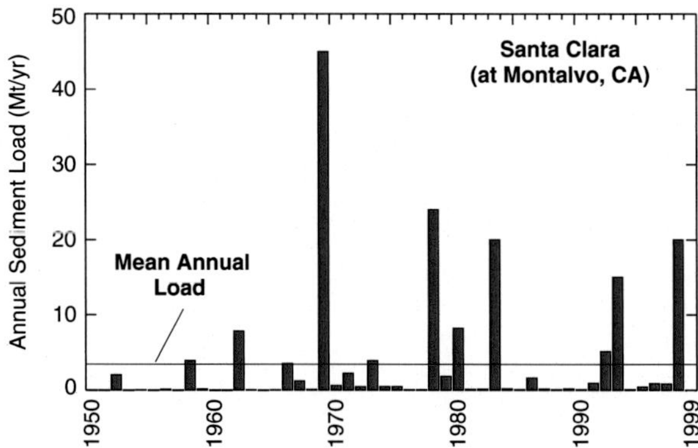

Figure 1.6. Annual sediment discharge from the Santa Clara River, as measured at the Montalvo gauging station, southern California. Note that the calculated mean load (3 Mt/yr) stems largely from flood-derived discharges in 1969, 78, 83, 93 and 98. Ignoring these one- to three-day events, collectively representing only about 20 days in a total of 50 years, the mean sediment load of the Santa Clara would be only ~0.5 Mt/yr.

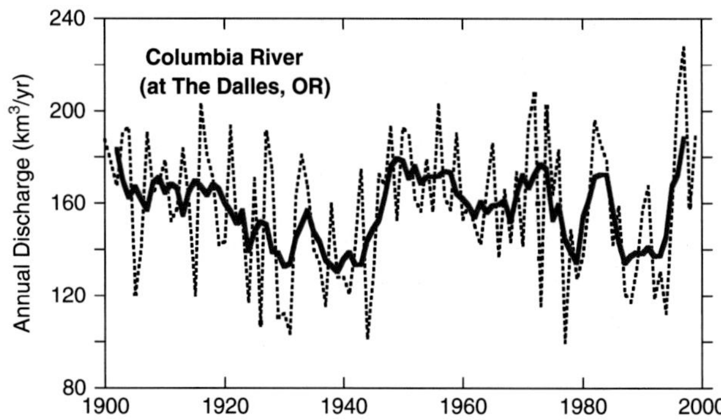

Figure 1.7. Annual discharge of the Columbia River (at The Dalles, Oregon), 1900–2000. The black dashed line shows annual discharge, the blue solid line shows the five-year running mean. Although long-term record shows no significant trend in annual discharge, interannually it deviated by more than 50% (100–227 km^3/yr) from the long-term mean of 158 km^3/yr.

The mean discharge of the Columbia River (670 000 km^2 basin area) in the Pacific Northwest is 5200 m^3/s, but during the 1930s and again in the late 1980s and early 1990s average annual discharge was ~4000 m^3/s, whereas in the 1950s it was ~6000 m^3/s (Fig. 1.7). Factoring in cyclic discharge, discussed further in Chapter 3, is particularly problematic in rivers that have been inadequately monitored.

Of even greater concern is the changing discharge of rivers whose flow has been affected by climate change and/or human impact. Over the past 60 years, for instance, discharges from the many large rivers (e.g. Indus, Yellow, Krishna, Don, Murray–Darling, etc.) have changed dramatically in response to both climatic variability and (more importantly) to increased water use and consumption. See Chapter 4 for a more thorough discussion of this problem.

Considering all these potential problems, it is no wonder that data for many rivers only can be considered approximations, which helps explain why we report values only to their second digit (see appendix).

A shrinking database

Increased demographic pressures and predicted climate change (see Chapter 4) require increased river monitoring – water discharge but also suspended sediments and dissolved solids. In fact, in recent years the exact opposite has occurred, particularly in such climatically important areas as the Arctic (Vörösmarty *et al.*, 2001) and southern Asia. Despite our best efforts to compile a functional global database, discharges reported in this book may be based on measurements made 20–40 (or more) years ago (especially in India and much of Africa) or they may come from second- or third-hand sources (e.g. data reported for Indonesian rivers). In the years since suspended- or dissolved-solid sampling ended, the watershed may have been altered (e.g. deforestation, reforestation, mining, etc.), climate may have shifted, the river may have been diverted, or the gauging station relocated.

More importantly, many river gauging stations have been decommissioned. Vörösmarty *et al.* (2001) cited a 25% reduction in the number of Canadian gauging stations since 1990, and in recent years more than 100 long-term gauging stations have been closed annually in the United States alone (cf. Lanfear and Hirsch, 1999). The Santa Clara River gauging station at Montalvo, a station at which a number of major floods were monitored (see Fig. 3.50), for instance, was decommissioned in 2004. Of greater concern is the decommissioning of pan-Arctic gauging stations, where future climate change may first be noted. The number of pan-Arctic rivers presently monitored approximates the number occupied in the early 1960s, but is only about 60% the number monitored in the early 1980s; the most severely affected have been gauging stations that had been occupied for 15 years or less (Shiklomanov *et al.*, 2002). Pilot Station on the Yukon River, one of the largest (830 000 km^2) relatively unspoiled rivers in North America, for instance, was deactivated in 1996, leaving the Stevens Village station (with an upstream drainage area ~60% that of Pilot Station) the most downstream station (Brabets *et al.*, 2000). Pilot Station was reactivated in 2001, but the five-year data-gap remains.

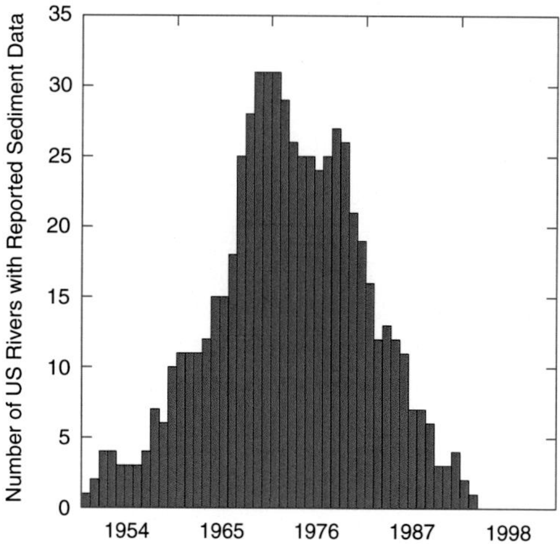

Figure 1.8. Daily suspended-sediment measurements taken in USA rivers discharging directly into the coastal ocean, as listed in the USGS NWIS (National Water Information System) database (http://co.water.usgs.gov/sediment/seddatabase.cfm). The number of rivers monitored daily for suspended sediments peaked in the late 1960s and began declining in the early 1980s. The last entries in the USGS NWIS database are for 1995. Suspended sediment is still being monitored in some rivers, but the data are not being included in the online database.

The problem of decreasing dissolved- and suspended-sediment monitoring is even more acute. The USGS web page (http://co.water.usgs.gov/sediment/) lists suspended-sediment data for 59 US rivers that appear in our database, most of which were measured between 1965 and 1980, peaking in the early 1970s (Fig. 1.8). By the mid 1990s the only rivers effectively monitored were those linked to the NAWQA (National Water Quality Assessment) program, and at present only the Missouri–Mississippi is thoroughly monitored, although by the US Corps of Army Engineers, not by USGS. To calculate sediment discharge for US rivers, one must therefore increasingly rely on sediment-rating curves based on data that are 30–40 years old that, considering landuse change, become less relevant with the passage of time. Now compare the poor USA database with Taiwan, which has probably the best monitored rivers in the world. A sufficient number of suspended sediment samples are taken from most Taiwan rivers each year, for instance, to create annual stratified rating curves (Kao *et al.*, 2005; Kao and Milliman, 2008).

The reason for the shrinking global database is not difficult to understand: establishing a sediment-sampling station is expensive, and its long-term maintenance is sufficiently costly that, given a choice, many governments would rather use those moneys to address other priorities. Unfortunately the declining database increasingly limits our ability to quantify and predict the discharge of fluvial water and its constituents to the ocean, particularly in response to present and future changes in landuse and climate (see Milly *et al.*, 2008).

Some final cautions

We cannot emphasize too strongly that the values (particularly for suspended sediment) cited in tables throughout this book mostly represent measurements taken at a gauging station upstream from the river mouth. In some rivers, the station is located a considerable distance upstream to avoid any tidal influence. The Obidos gauging station, for example, 900 km upstream from the mouth of the Amazon River, represents *only* 4 600 000 km^2 of upstream basin area, compared with a total drainage basin area of 6 300 000 km^2. As much of the downstream area has an annual runoff greater than 1 m/yr (or more), more water is clearly discharged at the mouth than is reported at Obidos. Sediment discharge at the river mouth, however, may be less than that calculated at Obidos, depending on the amount of downstream floodplain deposition. The Eel River gauging station at Scotia in northern California (upstream basin area of 8640 km^2), in contrast, is located only 25 km from the head of the river's small estuary. The Eel's flood plain is small (<150 km^2) and the river discharge is event-driven (see Chapter 3), so that sediment discharge measured at Scotia probably approximates what actually escapes to the ocean (Sommerfield and Nittrouer, 1999).

Finally, one must always take into account the uniqueness of each river. A book about the Deschutes River, a tributary to the lower Columbia in western Oregon, edited and largely written by O'Connor and Grant (2003), is entitled *A Peculiar River.* While the Deschutes may be "peculiar," so are most other rivers. A change in one or two watershed parameters, even slightly, can alter the character of the river, sometimes dramatically. Creating algorithms from one or a group of rivers to predict present or future values for neighboring rivers – as we do here and many others have done before us – may be only slightly less masochistic than tilting at windmills.

With these various caveats and cautions in mind, we trust that the reader will bear with us through the ensuing chapters.

2 Runoff, erosion, and delivery to the coastal ocean

Ev'ry valley shall be exalted,
And ev'ry mountain and hill made low
G. F. Handel (after Book of Isaiah 40:4)

Introduction

Much of the scientific interest in rivers revolves around attempting to quantify the flux and fate of fluvial discharge and to understand the processes that dictate these fluxes. No matter the motivation, a comprehensive understanding of fluvial processes and fluxes requires a synthetic approach, one that covers a wide range of spatial and temporal scales – local to global, hours to millennia – over which these processes occur and vary. In this chapter we discuss fluvial runoff and erosion and the transfer of their products to the coastal zone. We attempt to delineate the environmental factors that control these fluxes by utilizing both published literature and the database that we have collated in the book's appendix and GIS-based materials on the accompanying website, www.cambridge.org/milliman.

This exercise, however, must be viewed within the context of numerous previous efforts that collectively have laid the foundation for much of what is said here. To mention just a few previous studies that have dealt with suspended and dissolved solid transfer: Fournier (1949), Livingstone (1963), Holeman (1968), Lisitzin (1972), Baumgartner and Reichel (1975), Meybeck (1979, 1988, 1994), Milliman and Meade (1983), Walling and Webb (1983, 1996), Berner and Berner (1987), Meade *et al.* (1990), Milliman and Syvitski (1992), Summerfield and Hulton (1994), Stallard (1995a, b), Meade (1996), Edmond and Huh (1997), Ludwig and Probst (1998), Syvitski and Milliman (2007), and de Vente *et al.* (2007). The total number of relevant studies stretches well into the hundreds.

The following discussion primarily concerns long-term means of the erosion and weathering processes and transfer of water and solids, ignoring for the moment temporal variations (discussed in Chapter 3) and impact(s) of anthropogenic activities (Chapter 4).

The hydrologic cycle and water discharge

River discharge (Q), often reported as m^3/s or km^3/yr, primarily reflects the climate and size of the drainage basin. A plot of mean annual discharge from 1100 rivers in our database (Fig. 2.1) shows that basin area alone explains (statistically) 68% of the variance in Q. Big rivers as a rule tend to have greater discharge than smaller rivers, but for any given basin area, discharge can vary by two to three orders of magnitude. The Fortescue River in Australia, for instance discharges <0.2% that of the Rajang River in Malaysia (0.23 km^3/yr vs. 110 km^3/yr) even though the two rivers drain similar-size areas (49 000 vs. 51 000 km^2). Similarly, the Haast River (South Island, New Zealand) discharges as much water annually (6 km^3) as the Brazos River (Texas) although draining < 1% the watershed area (930 vs. 120 000 km^2). In both cases the critical variable defining discharge is climate; specifically, the drainage basin's hydrologic budget.

The hydrologic budget

Although runoff (R) and discharge (Q) are sometimes used interchangeably, the former is defined as discharge normalized by basin area: $R = Q/A$, which allows one to distinguish the roles of basin area (A) and climate in defining discharge. Referring to the previous paragraph, runoff from the Haast River exceeds 6000 mm/yr (6 $km^3/yr/930\ km^2$) whereas Brazos runoff is 50 mm/yr (6 $km^3/yr/120\,000\ km^2$).

In meteorological terms, runoff is defined as the difference between water gain (via precipitation) and water loss:

$$R = P - \sum(ET + S + C)\cdot$$

That is, runoff (R) = precipitation (P) minus the sum of evapotranspiration (ET), water storage (S), and water consumption (C). Evapotranspiration is the water loss via physical evaporation and plant transpiration. Storage refers to the loss of water to both surface (lake) and subsurface (groundwater) reservoirs. In a natural drainage basin unaffected by anthropogenic activities, long-term storage is assumed to be zero; short-term groundwater storage can vary but is difficult to quantify. Globally groundwater plays a secondary role in freshwater discharge to the ocean. Hydrologic models suggest that groundwater may represent only 5–6% of the total freshwater discharged directly to the global ocean (Zekster and Dzhamalov, 1988; Zekster and Loaiciga, 1993), but in terms of the seaward transfer of dissolved ions

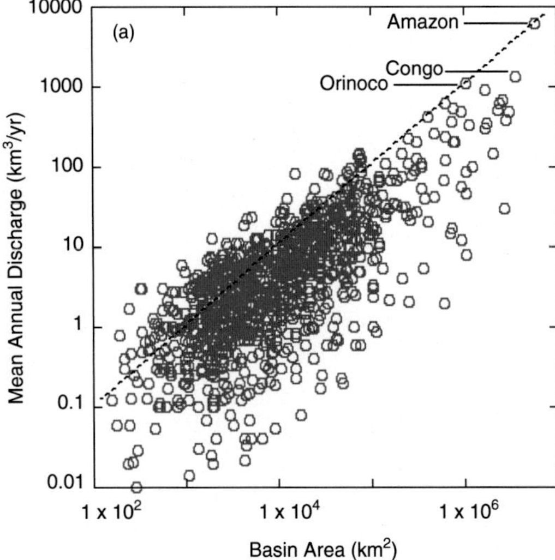

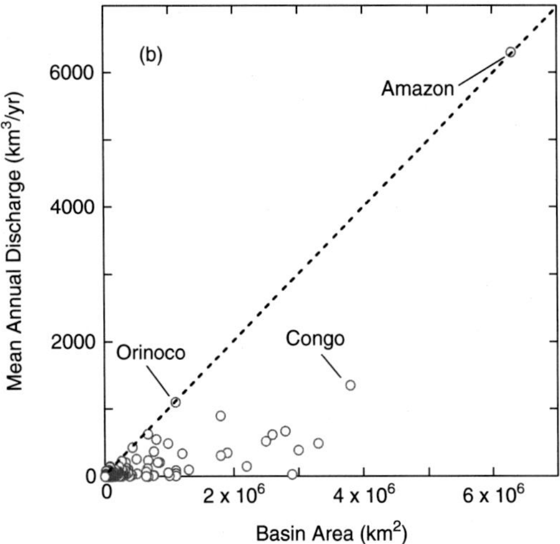

Figure 2.1. Mean annual discharge vs. basin area for more than 1100 rivers that discharge to the coastal ocean. Data come from the appendices at the back of the book. The log–log presentation (a) allows us to display data over many orders of magnitude, whereas the linear plot (b) relegates most rivers to the blob of circles in the lower left corner. But the linear plot underscores the global prominence of the Amazon River in terms of both basin area and discharge, its ~6400 km^3/yr exceeding the combined discharge of the next seven largest rivers. Diagonal dashed lines indicate 1000 mm/yr runoff.

and nutrients to the coastal ocean, groundwater plays an important regional and global role (Moore, 1996; Church, 1996). Dzhamalov and Safronova (2002) estimate 1 billion tons (Bt) of dissolved solids are discharged globally each year via groundwater, about 25% that of fluvially discharged dissolved solids (see below).

Water consumption is the loss of water by human use (e.g. evapotranspiration from irrigated cropland). It differs from "water withdrawal," which is the total amount of water withdrawn from the watershed, some of which may return to the system (e.g. via sewage outfall). Although consumption can be considerable in heavily regulated watersheds (Milliman *et al.*, 2008), in natural systems it generally represents a minor loss.

Precipitation

Precipitation occurs when water-laden air is cooled. This can occur along a front separating warm from cold air masses or through the upward convection of warm, humid air into the cooler atmosphere, resulting in localized thunderstorms. Not surprisingly, global and regional precipitation patterns reflect latitudinal circulation cells that transport momentum and energy from low to high latitudes, but also longitudinal wind patterns. The highest rates of global precipitation occur around the climatic equator, ~10N (Figs. 2.2 and 2.3), and on the windward sides of continents. Orographically controlled precipitation occurs when warm moist air cools as it rises over mountains, thus releasing water. The most commonly cited example of orographic control is the cooling of moisture-laden air as it rises over the Himalayas during the SW monsoon season, but similar processes are noted in other regions bordering high mountains, such as the California Sierra Nevada and the Colorado Rockies (Fig. 2.4). Viviroli *et al.* (2007, and references therein) state that mountains, which they term "water towers," are particularly important in arid climates, as they can produce ~five-fold greater runoff than the surrounding lowlands.

Oceanic circulation also can play a key role in regional precipitation. As warm air associated with the Gulf Stream and Kuroshio currents reaches higher latitudes, for instance, both precipitation and latent heat are released, one result being the warming of subpolar coastal Norway and Alaska. The combined effects of orography and oceanography help explain the longitudinal asymmetry in runoff between the eastern and western Americas (see Fig. 2.5).

Atmospheric circulation – and its domination by latitudinal cells – means that tropical areas tend to have high rainfall, subtropical mid-latitudes (~30N and S) low rainfall, and subpolar regions (latitudes ~60°) slightly higher precipitation (Fig. 2.3). Superimposed upon this picture is a longitudinal variation in atmospheric circulation (E to W at tropical and subpolar latitudes; W to E in mid-latitudes). In addition, topographic/orographic controls of rainfall and the oceanographic influence of warm oceanic waters transported to higher latitudes on the eastern sides of ocean basins (via the Gulf Stream and Kuroshio) help dictate both regional and global precipitation and evaporation (Figs. 2.2, 2.4a–2.6).

Combining the effects of these factors, precipitation varies globally from <10 mm/yr to >5000 mm/yr. Basin-wide

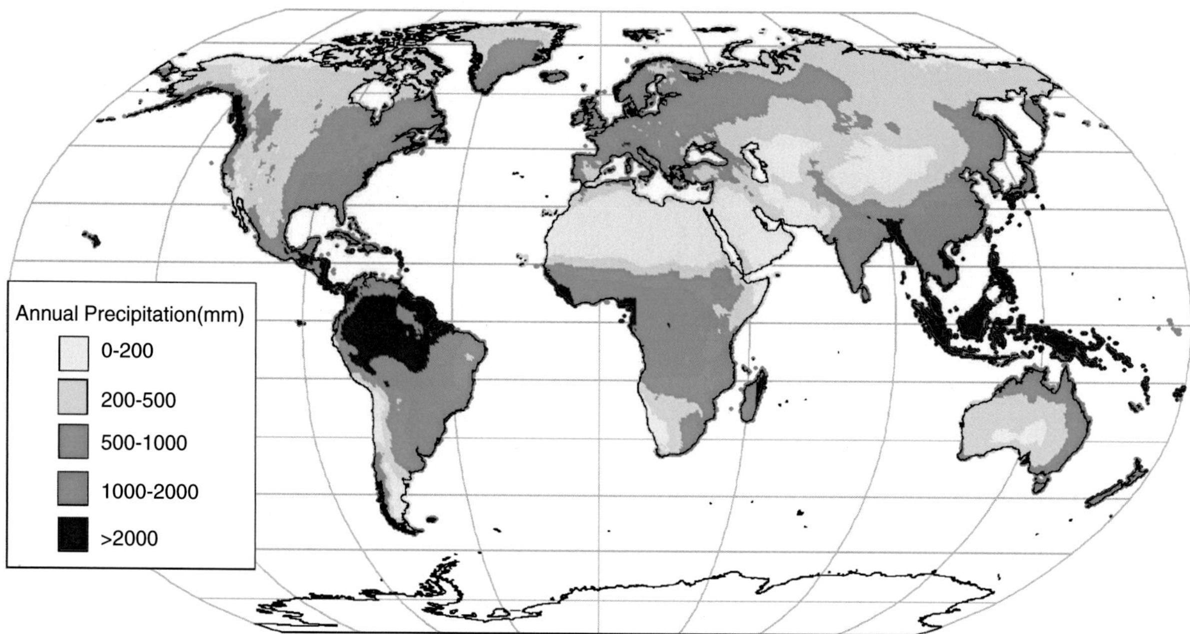

Figure 2.2. Global distribution of annual precipitation, 1901–2000, based on 0.5-degree grid data from the Climatic Research Unit (CRU), East Anglia University.

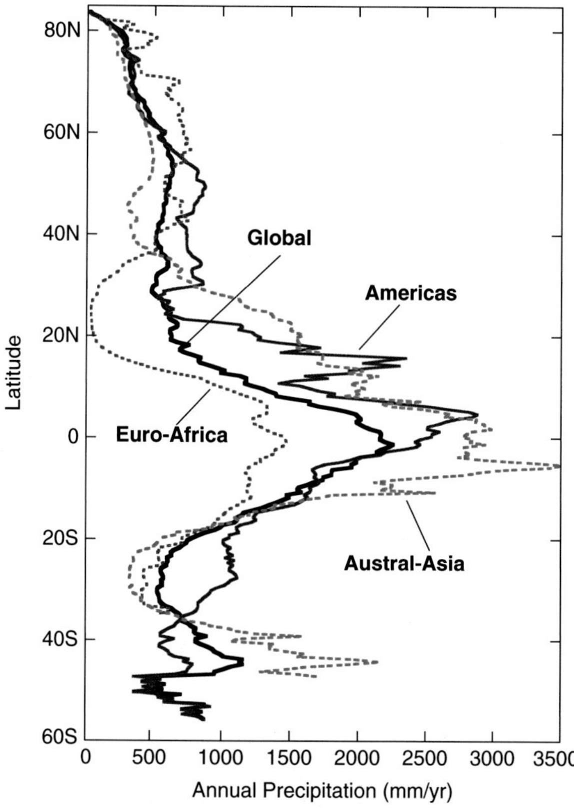

Figure 2.3. Latitudinal distribution of global precipitation (bold black line) and in the Americas (blue line), Euro-Africa (red dashed line), and Austral-Asia (green dashed line). Note the significantly lower Euro-African precipitation between ~20N and 25S. Data from CRU, East Anglia University.

precipitation in some of the New Guinean mountains can approach or exceed 10 000 mm/yr, whereas mean annual precipitation in some arid regions is nil except for rare storms (see Chapter 3).

Evapotranspiration

Unlike precipitation, which is easily measured by a rain gauge (although the gauge's trapping ability can depend on wind speed as well as the form of the precipitation – e.g. rain or snow; Bogdanova *et al.*, 2002), evapotranspiration (commonly termed *ET*) is difficult to quantify. *ET* represents the integration of two interrelated processes: (1) physical evaporation, which is a function of air temperature, humidity, and wind velocity, can be measured by pan evaporation; (2) plant transpiration also depends on the ecosystem and the plants within that ecosystem (e.g. grassland vs. climax forest; deciduous vs. conifer), thus making it difficult to measure. Because of the difficulty in quantifying evapotranspiration, *ET* is often reported as potential evapotranspiration – *PET* (Fig. 2.6) – the adjective "potential" indicating how much would be evaporated given sufficient precipitation.

Evapotranspiration tends to decrease with increasing latitude (Figs. 2.6 and 2.7) in response to decreasing temperature and changing vegetation (e.g. deciduous vs. evergreen trees). The pronounced latitudinal fluctuation in fluvial runoff (Fig. 2.5) thus reflects both variations in precipitation (Figs. 2.2 and 2.3) and evapotranspiration (Figs. 2.6 and 2.7). According to Budyko (1974), *ET* in humid regions is controlled primarily by *PET*, whereas in more arid regions *ET* is

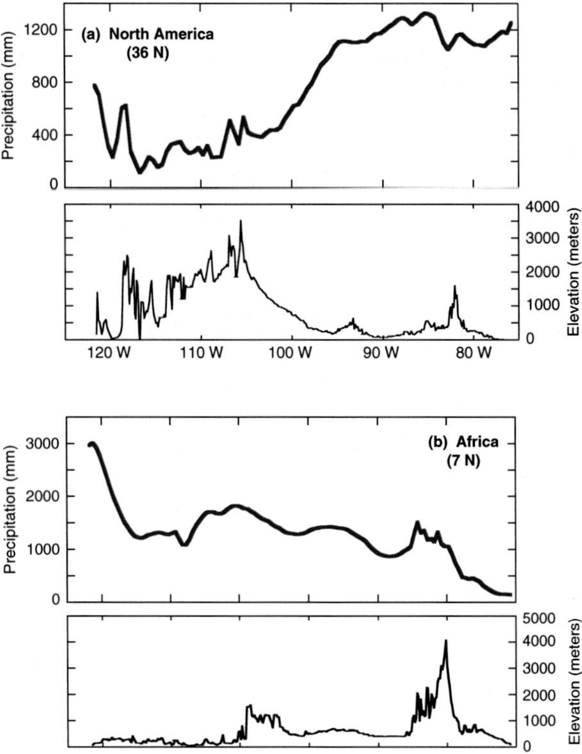

Figure 2.4. Longitudinal variation in precipitation across USA (at latitude 36N) (a) and equatorial Africa (latitude 7N) (b), illustrating the influences of orography (as shown by topography) and longitudinal atmospheric circulation on precipitation. The orographic control is particularly obvious in the western USA, where the windward (western) sides of the Sierra Nevada and Rocky Mountains receive greater precipitation than the leeward (eastern) sides. Precipitation data from CRU, University of East Anglia.

controlled more by the level of precipitation. In the Sahara Desert, for example, *PET* exceeds 2000–2500 mm/yr, whereas *P* averages only 200 mm/yr. The fact that the few scattered Saharan rivers exhibit any runoff – at least in some years – reflects the fact that precipitation often falls during short, intense storms, allowing short-term *P* to exceed *ET* (see Chapter 3).

River runoff

Unlike many climate classifications that are based on annual precipitation (e.g. Köpper–Geiger, 1954; Meigs, 1953), we categorize rivers on the basis of their annual fluvial runoff (see appendix). We recognize four classes of rivers: arid (runoff <100 mm/yr), semi-arid (100–250 mm/yr), humid (250–750 mm/yr) and wet or high-runoff rivers (>750 mm/yr). This classification differs from that used by Meybeck (2003; Meybeck and Vörösmarty, 2005): arheic (runoff <3 mm/yr), digoheric (<30 mm/yr), mesorheic (30–300 mm/yr), hyperheic (runoff >300 mm/yr). Thus, although discharge for the Amazon (basin area of 6 300 000 km^2) is three orders of magnitude higher than that of the Choshui (basin area of 3100 km^2) – (6300 km^3/yr vs. 4.3 km^3/yr) – both rivers are classified as wet (R = 1000 mm/yr and 1400 mm/yr, respectively). The Congo (R = 1300 km^3/yr per 3 800 000 km^2 = 355 mm/yr) is a humid river, whereas the adjacent Niger (R = 190 km^3/yr per 2 200 000 km^2 = 86 mm/yr) is an arid river.

Greatest runoff is noted in small rivers that drain wet, mountainous terrain (orographic effects facilitating high precipitation); seven of the ten greatest runoff rivers listed in Table 2.1 have basin areas <2000 km^2. All low-runoff rivers listed in Table 2.1, in contrast, drain arid or subarid climates. Even though the southern hemisphere accounts for a relatively small portion of the global landmass, more than half of the low-runoff rivers listed in Table 2.1 are located south of ~10S, supporting the suggestion that southern hemisphere rivers have a different runoff character than northern hemisphere rivers (Finlayson and McMahon, 1988; McMahon *et al.*, 1992). Peel *et al.* (2001) have hypothesized that this difference, particularly apparent in temperate arid rivers, may reflect the dominance of evergreen trees in southern hemisphere watersheds, which leads to greater evapotranspiration (relative to northern hemisphere deciduous-dominated watersheds), particularly during winter months.

In considering global runoff, we are impressed by the dominance of arid drainage basins – much of the Arctic, most of Africa (coastal west Africa and some Madagascar watersheds being the only areas where R >750 mm/yr), essentially all of Australia, the Middle East, and parts of western North and South America (Fig. 2.8). High-runoff watersheds are located predominantly in northern South America and southern Asia/Oceania; rivers draining SW South America, NW North America, Labrador, Iceland, and Norway also have high runoffs.

Because most large rivers (>500 000 km^2 in area) drain arid or semi-arid continental interiors (see Fig. 1.2) – with the notable exceptions of the Amazon, Orinoco and Brahmaputra rivers – many large rivers are classified as arid or subarid. Collectively these large arid to semi-arid rivers drain nearly twice as much global area as large humid and wet rivers (Table 2.2). Even though rainfall near the mouth of the Niger River exceeds 1000 mm/yr, for example, most of the river's watershed lies in arid west Africa; mean annual runoff for the entire Niger watershed is only 86 mm/yr. Similarly, much of the Mississippi drainage basin lies within the arid and semi-arid Missouri and Arkansas watersheds (see Chapter 4); its annual runoff (150 mm/yr) classifies it as a semi-arid river even though the tributary Tennessee and Ohio rivers have runoffs exceeding 500 mm/yr.

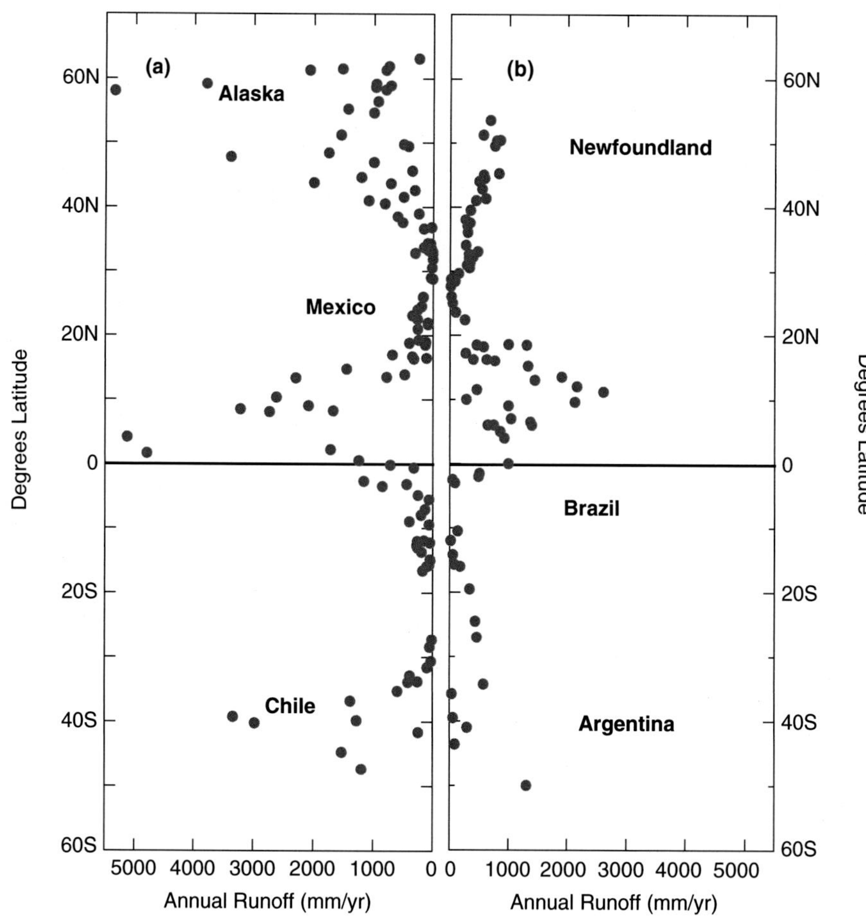

Figure 2.5. Latitudinal variation in hydrologic runoff from western (a) and eastern (b) North, Central and South America (NCSA), reflect the control that latitudinal circulation cells have on precipitation. Topography, oceanography, and evapotranspiration help explain local and regional variations in these patterns. Note that eastern NCSA runoff peaks (~2700 mm/yr) at ~10N, while along western NCSA it peaks at ~5N (reaching >5000 mm/yr). Mid-latitude runoff on both sides of NCSA is <100 mm/yr, but increases at higher latitudes. High Alaskan runoff is explained by both oceanic circulation and orographically controlled precipitation.

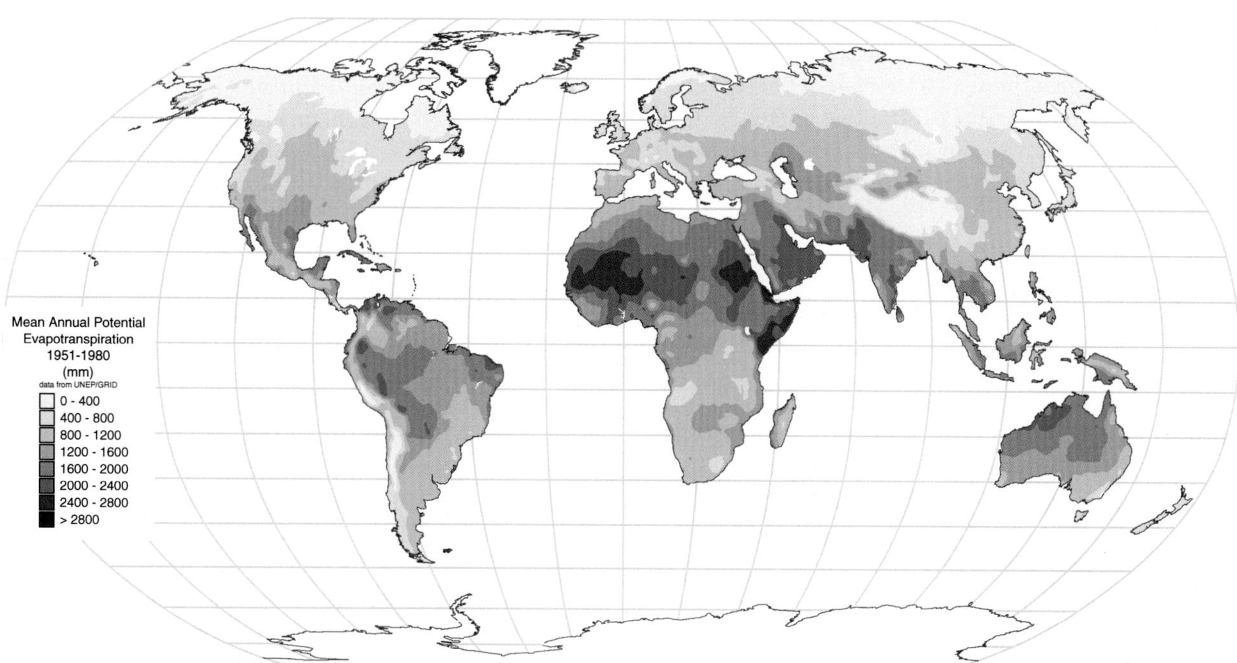

Figure 2.6. Global distribution of potential evapotranspiration (PET). Note the dependence of PET on air temperature and humidity (see Fig. 2.2).

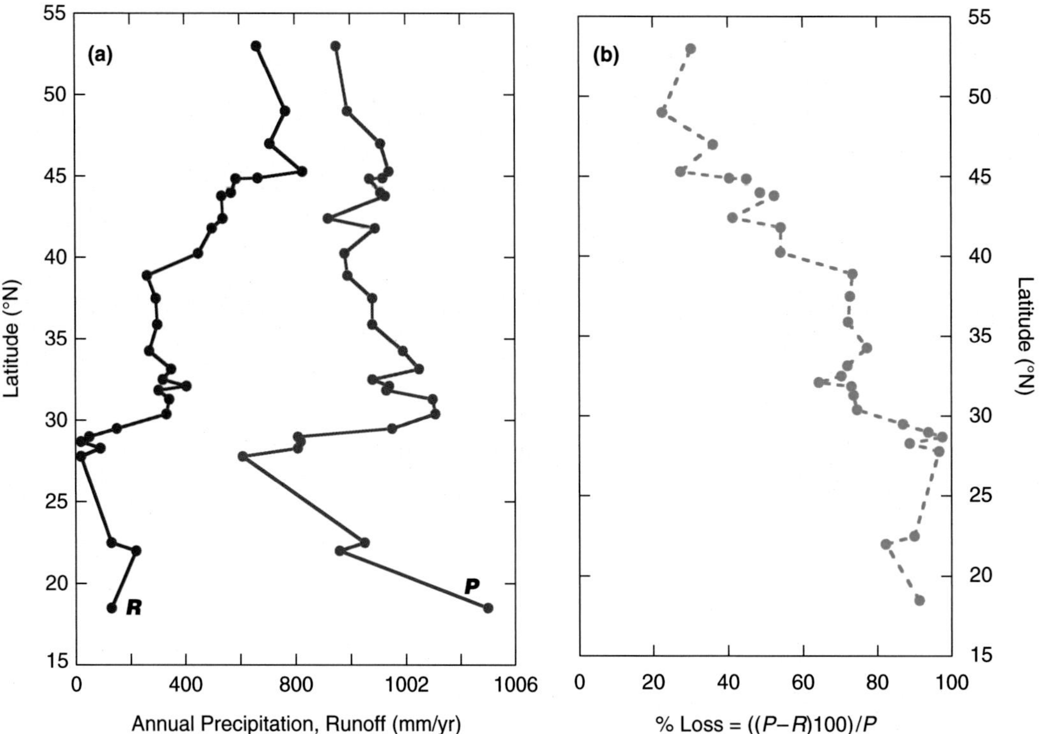

Figure 2.7. (a) Latitudinal variation in river runoff (*R*) and basin-wide precipitation (*P*), and (b) % water loss [(*P*–*R*)*(100)]/*P* in eastern Central and North American river basins. Assuming minimal storage and consumption, water loss essentially reflects evapotranspiration (*ET*).

The 12 rivers that individually discharge more than 400 km^3 annually (Table 2.3) occupy about 25% of the total land area emptying into the global ocean and collectively account for more than 35% of the total fluvial water annually discharged to the global ocean. Other than the Amazon and Orinoco, most high-discharge rivers are characterized by either high runoff offsetting relatively small drainage basins (e.g. Irrawaddy) or by large drainage basins that offset low runoff (e.g. Yenisei, Lena).

Freshwater discharge to the global ocean

We calculate that rivers discharge ~36 000 km^3 of water to the global coastal ocean (Fig. 2.9); the mean global runoff is ~350–360 mm/yr. This discharge estimate is similar to those made by Nace (1967), Baumgartner and Reichel (1975: ~37 700 km^3/yr discounting Antarctic and Greenland ice-melt runoff, the latter alone accounting for >250 km^3/yr; Hanna *et al.*, 2009; Mernild *et al.*, 2008), Berner and Berner (1987: 36 000 km^3/yr), Milliman (1990: 35 000 km^3/yr), Dai and Trenberth (2002: 37 000 km^3/yr), and Fekete *et al.* (2002: 37 500 km^3/yr). Our estimate is somewhat lower than the 38 500–41 000 km^3/yr calculated by Fekete *et al.* (1999), Oki (2006), GRDC (www.grdc.bafg.de), Shiklomaov (in Gleick, 1993) and Peucker-Ehrenbrink (2009), and substantially lower than the 43 000–47 000 km^3 discharge reported by Korzoun *et al.* (1977), L'Vovich (1974) and Shiklomanov and Rodda (2003) (see also summary in Oki, 1999). The differences in calculated global discharge are best explained by differences in the databases and differences in the methods used to extrapolate the runoff from 40% of the world's surface for which discharge data are lacking. Estimates in cumulative global discharge are likely to contain errors equivalent to at least one Congo – i.e. 1000–1500 km^3/yr (R. H. Meade, personal communication).

Given the latitudinal distribution of global precipitation (Figs. 2.2, 2.3, and 2.7), it is not surprising that rivers draining northern South America and southern Asia/Oceania account for about half of the total freshwater (~19 000 km^2/yr) discharged to the global ocean (Figs. 2.9 and 2.10a). By virtue of its extensive drainage basin and high runoff, the Amazon's discharge (6300 km^3/yr) exceeds the combined discharge of the next eight largest rivers (Table 2.3). Arctic rivers, by contrast, collectively drain about 20% of global watershed but account for "only" about 4800 km^3/yr; their mean annual runoff is less than 200 mm/yr. Of the 11 regions shown in Fig. 2.9, only NE South America and parts of western South America (see Fig. 2.10a) can be regarded as being high-runoff (mean runoff 860 mm/yr). Three other regions (western South and North Americas, southern Asia/Oceania) have runoffs greater than the global average (360 mm/yr), whereas most of Australia and Africa are semi-arid to arid (Fig. 2.10b). Rivers draining

Table 2.1. *Greatest and smallest runoffs (in bold) of rivers within our global river database; global mean is 360 mm/yr. Most high-runoff rivers (>4000 mm/yr) drain tropical or temperate mountains (see Fig. 2.5), and with exception of the San Juan River all have small watersheds. In contrast, rivers with runoffs <5 mm/yr drain a variety of elevations, but many lie in the southern hemisphere. Climate is represented by three groups of letters: first group refers to mean basin temperature: Tropical (Tr), Subtropical (STr), and Temperate (Te); the second group refers to mean annual runoff: Arid (<100 mm/yr: A) and Wet (>750 mm/yr: W); and the third group refers to season of maximum discharge: Summer (S), Autumn (Au) and Winter (W), Continuous (C), and Desert (D).*

River	Country	Basin area (× 10^3 km^2)	Climate	Discharge (km^3/yr)	Runoff (mm/yr)
Hokitika	New Zealand	0.35	Te-W-C	3.1	**8900**
Esk	New Zealand	0.25	STr-W-W	1.7	**6800**
Naya	Colombia	2	Tr-W-W	13	**6500**
Haast	New Zealand	0.93	Te-W-C	6	**6450**
Speel	USA (Alaska)	0.58	Te-W-S	3.1	**5300**
San Juan	Colombia	16	Tr-W-C	82	**5100**
Taramakau	New Zealand	1	Te-W-C	4.8	**4800**
Baudo	Colombia	5.4	Tr-W-Au	24	**4400**
Micay	Colombia	4.4	Tr-W-C	19	**4300**
Global					**360**
Besor	Israel	3.7	STr-A-W	0.01	**3**
Casamance	Senegal	37	Tr-A-S	0.1	**3**
Loa	Chile	33	STr-A-W	0.1	**3**
Holgat	South Africa	1.6	STr-A-D	0.003	**2**
Murchison	Australia	82	STr-A-S	0.2	**2**
Spoeg	South Africa	1.6	STr-A-D	0.003	**2**
Dasht	Pakistan	36	STr-A-S	0.04	**1**
Omaruru	Namibia	14	STr-A-D	0.02	**1**
Swartlinjies	South Africa	1.7	STr-A-D	0.002	**1**
Hoanib	Namibia	18	STr-A-D	0.01	**0.5**
Ugab	Namibia	33	STr-A-D	0.01	**0.3**

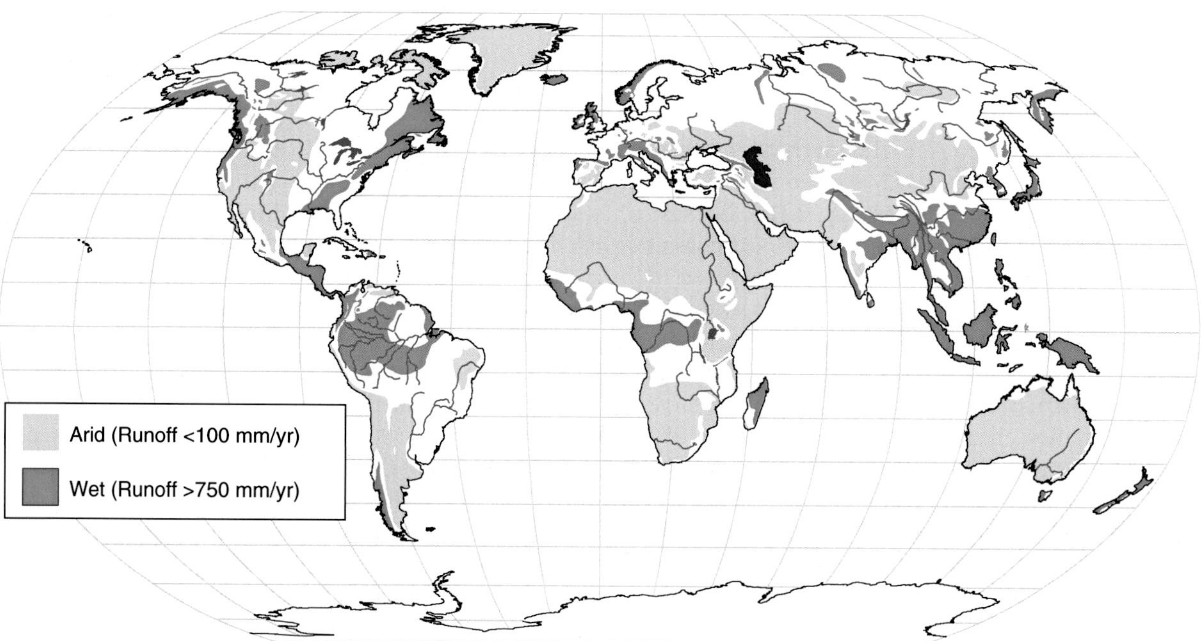

Figure 2.8. Global distribution of arid (R <100 mm/yr) and wet (R >750 mm/yr) watersheds; after Baumgartner and Riechel (1975).

Table 2.2. *Hydrologic classification, based on fluvial runoff (discharge/basin area), of world rivers with watershed areas >500 000 km². Numbers in parentheses refer to the watershed area in millions of km². The cumulative basin area of 52 900 000 km² represents slightly more than half of the total land area draining into the global ocean. See Fig. 1.2 for location of these rivers.*

Arid (<100 mm/yr)	Semi-arid (100–250 mm/yr)
Colorado (Mexico) (0.64)	Amur (1.9)
Huanghe (0.75)	Dnepier (0.58)
Indus (0.98)	Kolyma (0.66)
Murray–Darling (1.1)	Lena (2.5)
Nelson (1.1)	MacKenzie (1.8)
Nile (2.9)	Mississippi (3.3)
Niger (2.2)	Ob (3.0)
Orange (1.0)	Parana (2.8)
Rio Grande (0.67)	Sao Francisco (0.63)
Shebelle–Juba (0.8)	Yenesi (2.6)
Zambesi (1.4)	Yukon (0.85)
Total land area: 13.6 × 10⁶ km²	**Total land area: 21.7 × 10⁶ km²**
Humid (250–750 mm/yr)	**Wet (>750 mm/yr)**
Changjiang (1.8)	Amazon (6.3)
Columbia (0.67)	Brahmaputra (0.67)
Congo (3.8)	Orinoco (1.1)
Danube (0.82)	**Total land area: 8.1 × 10⁶ km²**
Ganges (0.98)	
Mekong (0.8)	
St. Lawrence (1.2)	
Tocantins (0.76)	
Total land area: 10.8 × 10⁶ km²	

Africa, in fact, discharge only 2700 km³/yr, a continent-wide runoff of 90 mm/yr (Figs. 2.9 and 2.10).

North and Central America

Runoff in North and Central American rivers ranges from <100 mm/yr in the SW USA and N Mexico to >1000–2000 mm/yr in southern Central America (Fig. 2.11a). Because of large watersheds and locally high runoff (in coastal Alaska), rivers in the James Bay–Hudson Bay region of northern Canada and central and southeastern Alaska discharge 1300 km³/yr and 960 km³/yr, respectively. Rivers discharging into the Gulf of Mexico, Gulf of St. Lawrence, and the Arctic Ocean have low runoffs, but their large drainage basins allow them to discharge 990 km³/yr, 750 km³/yr, and 660 km³/yr, respectively. We have few data for Puerto Rico, Hispaniola, Cuba, Newfoundland, and Baffin and Victoria islands, but their collective size exceeds 1000 km², necessitating that we include them in our calculations even though, at best, they are rough estimates.

South America

The fluvial character of South America is dominated by two large river systems, the high-runoff Amazon in the north and the moderate-runoff Parana/Uruguay system in the south (Fig. 2.11b). Together with the Orinoco and coastal rivers, NE South America discharges 8200 km³ of freshwater to the coastal ocean; the Magdalena and coastal rivers draining northern Colombia and Venezuela discharge an additional 350 km³. Rivers draining eastern and southern Brazil, much of Argentina and southern Peru and northern Chile stand in stark contrast, with runoffs of <100 mm/yr. Perhaps most interesting are the rivers draining western Colombia and northern Ecuador, a number of which have runoffs exceeding 3000 mm/yr (e.g. San Juan River); collectively these NW rivers discharge an estimated 400 km³ of water to the equatorial eastern Pacific. Southern Chilean rivers discharge roughly an equal amount as the NW rivers, whereas rivers along the arid central western coast contribute essentially no fresh water to the global ocean.

Table 2.3. *Global rivers that discharge more than 400 km³ of water annually; annual discharge in bold. Collectively these drainage basins account for ~25% of the total land area draining into the global ocean, and they discharge ~35% of the freshwater reaching the ocean. Most rivers are high mountain (>3000 m maximum elevation), but climates vary widely; runoffs for the 12 rivers range from <200 mm/yr (Mississippi, Parana) to 1000 mm/yr (Amazon, Orinoco, Irrawaddy).*

River	Country	Basin area (× 10^3 km^2)	Climate	Discharge (km^3/yr)	Runoff (mm/yr)
Amazon	Brazil	6300	Tr-W-S	**6300**	1000
Congo	Congo, DR	3800	Tr-H-S	**1300**	340
Orinoco	Venezuela	1100	Tr-W-S	**1100**	1000
Changjiang	China	1800	Te-H-S	**900**	500
Bramaputra	Bangladesh	670	STr-W-S	**630**	940
Yenisei	Russia	2600	A-SA-S	**620**	240
Mekong	Vietnam	800	Tr-H-S	**550**	690
Lena	Russia	2500	A-SA-S	**520**	220
Ganges	Bangladesh	980	STr-H-S	**490**	500
Mississipppi	USA	3300	Te-SA-Sp	**490**	150
Parana	Argentina	2600	STr-SA-C	**460**	180
Irrawaddy	Burma	430	Tr-W-S	**430**	1000
Totals		**27 000**		**13 800**	500

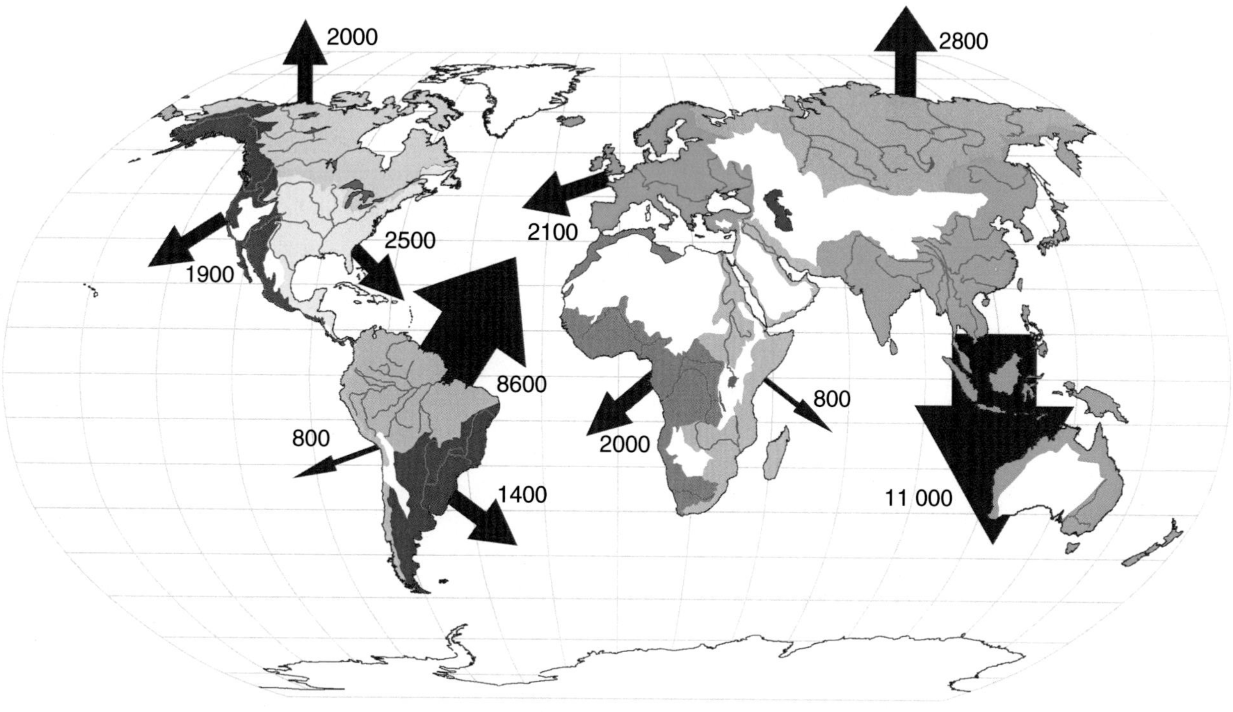

Figure 2.9. Fluvial discharge of freshwater to the global coastal ocean. Numbers are mean annual discharge (km^3/yr); the arrows are proportional to these numbers. Cumulative arrow width is the same for subsequent global maps in this chapter, thus facilitating comparison of the relative fluvial discharge of freshwater, suspended solids (Fig 2.29), dissolved solids (Fig.2.38), and dissolved silica (Fig. 2.49).

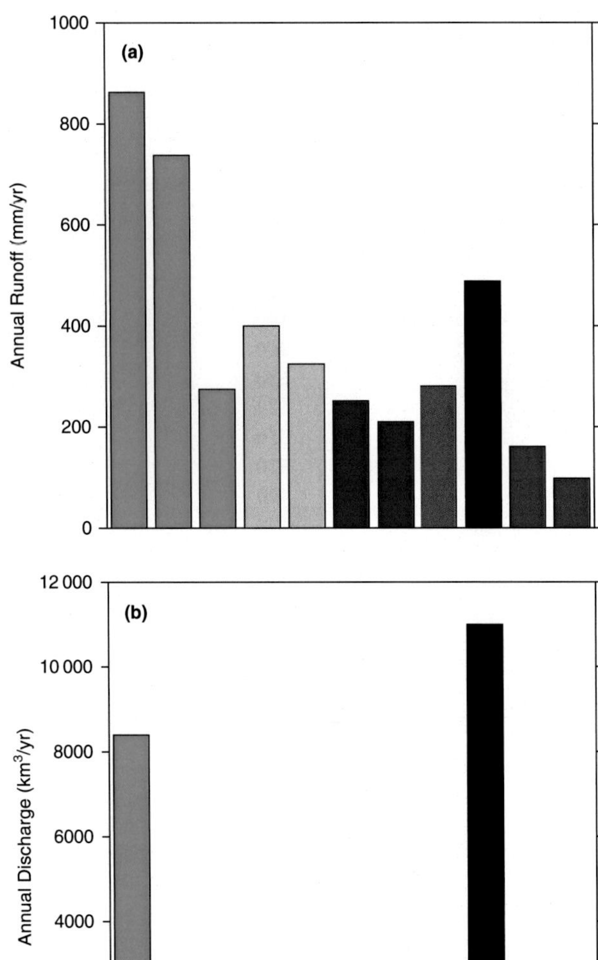

Figure 2.10. Annual runoff (a) and water discharge (b) to the global ocean from the 11 regional areas portrayed in Fig. 2.9. Note that only northern and parts of western South America can be classified as wet (>750 mm/yr), whereas the Eurasian Arctic and Africa are semi-arid or arid.

Europe

If one were to judge only from the small arrows in the panel in Figure 2.11c, one would assume that European rivers are rather uniform and, to someone used to working in the New World or Asia – to put it kindly – rather boring. While it is true that the runoff from many northern and central Europe rivers can be viewed as being monotonously semi-arid (<250 mm/yr), the contrasts, in fact, are stark. Watersheds influenced by the Gulf Stream (Iceland and Norway) have runoffs >750 mm/yr, Albanian and western Greek rivers have runoffs >1000 mm/yr; in contrast, far eastern Mediterranean rivers are semi-arid (Fig. 2.11c). Nowhere is the disparity in runoff more apparent than in the Black Sea drainage basin: northern rivers, such as the Don and Dniepr, have runoffs <100 mm/yr, whereas some Georgian rivers in the southeastern part of the region have runoffs exceeding 1500 mm/yr (Jaoshvili, 2002; Milliman *et al.*, 2010).

Africa

One reason that we have not included detailed maps of Africa or the Russian Arctic is that rivers draining both areas are rather consistent. Of the large African rivers, for example, only the Congo (runoff 340 mm/yr) is non-arid; pre-dam runoffs for the Nile, Niger, Zambesi, Orange, and Shebelle averaged only about 50 mm/yr. Many of Africa's major secondary rivers (e.g. Limpopo, Volta, and Senegal) are also arid.

Asia and Oceania

Southern Asian and Oceania rivers collectively account for about 30% of the fluvial discharge to the global ocean, but the contrasts between various areas within southern Asia and the high-standing islands are noteworthy (Fig. 2.11b). Runoff from the major Himalayan rivers, for example, ranges from 91 mm/yr for the Indus in the west to 940 mm/yr and 1000 mm/yr for Brahmaputra and Irrawaddy farther east. Highest runoff is seen in Taiwan, the Philippines, and Indonesia, runoff in some rivers exceeding 3000 mm/yr. In contrast, many Australian rivers, including the Murray–Darling, have runoffs less than 50 mm/yr.

Eurasian Arctic

The very large Russian Arctic rivers (Ob, Yenisei, Lena, and Kolyma; combined drainage basin area 8.8×10^6 km^2) are all semi-arid, with an average runoff of 185 mm/yr. Because the shorter Russian rivers (Severnaya Dvina, Khatanga, Indigirka, Pechora, etc.) drain less of the arid Eurasian interior, their runoff is somewhat higher (260 mm/yr), but only the Severnaya Dvina and Pechora (310 mm/yr and 410 mm/yr, respectively) in the far west can be considered humid rivers.

Another perspective

The Arctic Ocean accounts for less than 5% of the global ocean (17×10^6 km^2), but the watershed draining into the Arctic totals 21×10^6 km^2, about 20% of the land area draining into the global ocean. The drainage area/ocean ratio is 1.2. By contrast, the South Pacific accounts for 1/4 of the global ocean area, but combined watersheds draining into the South Pacific total only 5×10^6 km^2, giving a drainage area/ocean area ratio of ~ 0.05 (Table 2.4). The greatest freshwater input is to the North Atlantic (largely because of the Amazon and, to a lesser extent, the Orinoco and Mississippi), but the

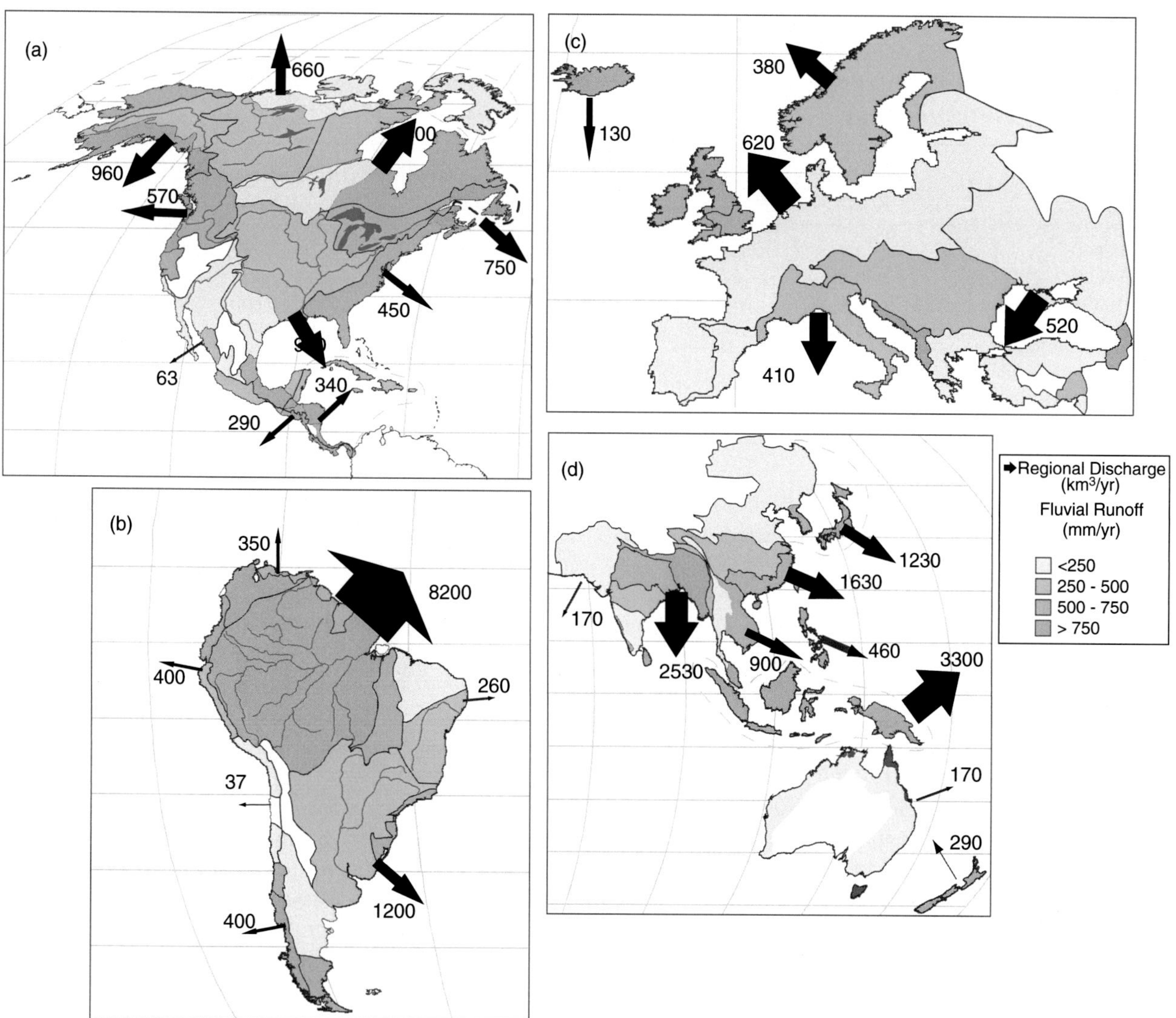

Figure 2.11. Regional discharge from (a) North America, (b) South America, (c) Europe, and (d) Austral-Asia. Runoff for the various polygons is color-coded, from arid to wet. Arrows in each panel are proportional to water discharge within that area. Where discharge data are meager, we rely on data from adjacent polygons. Areas with generally similar physiography, geology, and climate are delineated by black borders. Africa and the Eurasian Arctic are excluded from this analysis because of the general uniformity of their physiography, geology, and climate.

greatest input per unit basin area is to the Arctic (280 mm/yr if distributed evenly over the entire basin). The lowest input per unit area is the South Pacific (45 mm/yr) (Table 2.4). The timing of discharge to the various basins, however, shows stark seasonal contrasts, peak discharge to the Arctic and Pacific occurring in June, versus May for the Atlantic and August for the Indian Ocean (Dai and Trenberth, 2002).

Sediment: erosion and discharge

Introduction

Fluvial sediments can be transported in suspension or as bedload along the river bottom. The term suspended sediment normally refers to solids coarser than 0.45–0.62 μm, depending on the nominal pore diameter of the filter; finer material is considered dissolved if one disregards colloidal matter. Suspended sediment includes wash load, the fine fraction that remains more or less continuously in suspension, and bed material load, the sediment incorporated into the suspended load during higher flow. The dividing line separating bed material load from bed load varies with time and flow conditions: as river flow increases, what is normally transported as bed load can be resuspended and incorporated into the bed material load.

Many, if not most, fluvial sediment data refer to suspended sediment rather than total load, in large part because measuring bed load is, at best, problematic. Bed load measurement, for example, can alter surface roughness of the

Table 2.4. *Cumulative oceanic areas, drainage basin areas, discharge, and runoff of rivers draining into various parts of the global ocean. For this compilation, it is assumed that Sumatra and Java discharge into the Indian Ocean, and that the other high-standing islands in Indonesia discharge into the Pacific Ocean. Rivers discharging into the Black Sea and Mediterranean are assumed to be part of the North Atlantic drainage system.*

	Ocean area (× 10^6km^2)	Drainage basin (× 10^6km^2)	Discharge (km^3/yr)	Sediment flux (Mt/yr)	Dissolved flux (Mt/yr)
North Atlantic	44	30	13 200	2500	1350
South Atlantic	46	12	3400	400	240
North Pacific	83	15	6100	7200	660
South Pacific	94	5	4100	3900	650
Indian	74	14	4000	4000	520
Arctic	17	21	4800	350	480
Total	360	98	36 000	18 000	3800

riverbed, whether by installation of a recording instrument on the bottom or digging a pit to record the speed at which it is filled. Although new non-intrusive laser and acoustic Doppler techniques offer some hope for less intrusive measurement, most workers continue to assume that bed load represents a relatively small fraction of the total sediment load, perhaps only ~10% (C. Nordin, cited by Milliman and Meade, 1983). The 10% estimate is probably too high for most larger meandering rivers (e.g. 1–2% for the Yukon River, Brabets *et al.*, 2000; <5% for the lowermost Mississippi River, Nittrouer *et al.*, 2008). For smaller rivers, particularly short mountainous rivers, on the other hand, 10% may be too low. Rovira *et al.* (2004), for instance, estimate that 75% of the sediment load discharged by the Tordera River (basin area 894 km^2), in southern Spain, is as bed load, which is transported almost exclusively during floods, some of which may be bed material load.

Sediment discharge, often termed sediment load, refers to sediment transport: mass per unit time. Some workers appear to confuse discharge or load with sediment concentration, thus writing such statements as "...average sediment load is 35 mg/l..." (reference here purposefully deleted). A particularly useful – but sometimes confusing – term is sediment yield, which we define as the sediment discharge divided by drainage basin area upstream of which discharge is measured, in much the same way that runoff is water discharge normalized for basin area. Some workers use the term synonymously with sediment load; Phillips (1990), for example, defines yield as "sediment transport out of the drainage basin." For those workers equating yield and load, what we consider as sediment yield is termed specific sediment yield. As with runoff, sediment yield allows one to compare the loads of disparate-sized rivers, for example the Congo River (12 t/km^2/yr) vs. the Choshui River (13 000 t/km^2/yr). Waythomas and Williams (1988) have cautioned that plotting sediment yield vs. basin area (as we do in this chapter) is statistically invalid since area is common to both axes. Nevertheless, sediment yield helps us to delineate the key environmental factors that control sediment erosion and fluvial transport.

Although it is often equated with denudation rate (e.g. Gilluly, 1955; Judson and Ritter, 1964; Li, 1976; Harrison, 1994), sediment yield refers to sediment delivery averaged over the area of the entire drainage basin. But in reality, denudation is never uniform throughout the entire basin. In large drainage basins, in fact, sediment often is stored downstream, thereby resulting in a high up-basin denudation and negative down-basin denudation. Holeman (1980), for instance, calculated that only ~10% of the total eroded sediment in the conterminous United States actually reaches the ocean, and Wasson *et al.* (1996) state that only ~ 1% of the eroded soil in Australia is discharged to the ocean. As will be seen at the end of this chapter, sediment storage, both temporary and long-term, is critical to understanding the correlation between sediment yield and basin area. To add to the confusion, human influence on solid and dissolved denudation cannot be minimized; present-day sediment yields in the eastern USA, although relatively low, are ~four to five times greater than they were prior to European settlement (Meade, 1969). Chapter 4 deals in greater detail with anthropogenic impacts on sediment erosion and delivery.

No matter the scale – watershed-specific, regional or global – freshwater discharge can be measured in a relatively straight-forward way by measuring river height; the margin of error is relatively small – perhaps no more than 5–20% in extreme cases. Water discharge only requires frequent measuring a river's gauge height (stage) above an arbitrary datum, and referring the gauge height to a previously constructed graph (rating curve) of gauge height vs. measured discharge (Fig. 2.12a), corrected in response to changes in river channel configuration. Moreover, because most river flow is measured in similar ways, often with uncertainties of no more than 10% (Fekete *et al.*, 2002; Shiklomanov *et al.*, 2006), we

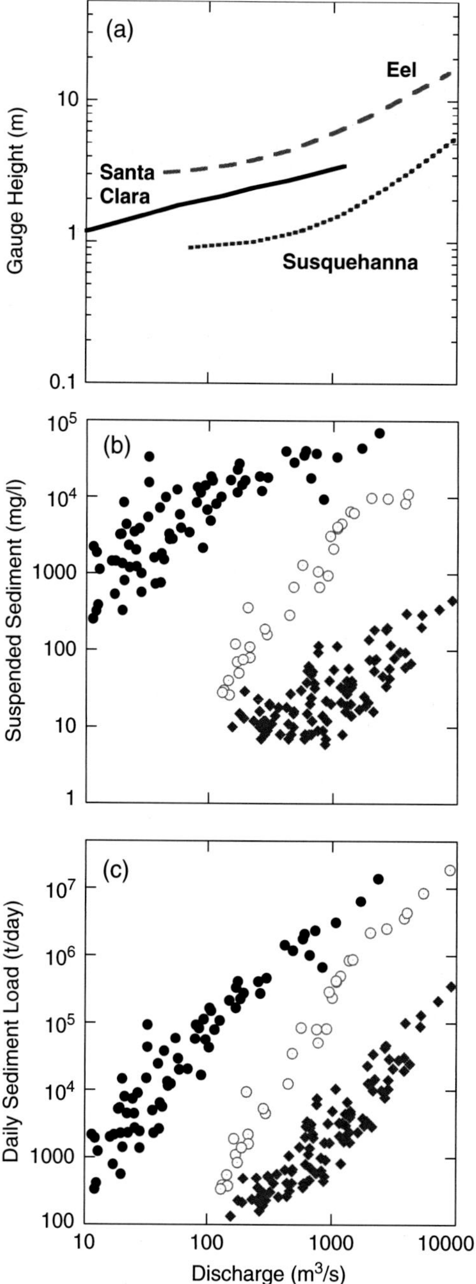

Figure 2.12. Discharge- and sediment-rating curves for the Santa Clara (at Montalvo, California: 4100 km^2 basin area, 100 mm/yr runoff; black notations), Susquehanna (at Harrisburg, Pennsylvania: 71 000 km^2, 450 mm/yr; blue notations), and the Eel (at Scotia, California: basin area 8000 km^2, 900 mm/yr; red notations) rivers. Water discharge is calculated by comparing measured gauge height to rating curves shown in (a). Measured suspended-sediment concentrations can be correlated with river discharge to derive a sediment-rating curve (b), from which sediment load (c) (i.e. concentration times water discharge) can be calculated. As water discharge appears in both terms, sediment-rating curves tend to follow a close-fitting power function. Data from the USGS.

can compare and contrast discharges of various rivers with a fair degree of certainty.

Sediment discharge is more difficult to quantify. Suspended sediment can be sampled in a diversity of ways, from long-term depth- and point-integrated cross-channel sampling used by the US Geological Survey (e.g. Edwards and Glysson, 1999; Carvalho, 2008) to a one-time dip-bucket sample, which is almost guaranteed to misrepresent both spatial and temporal variations in sediment concentration, and hence sediment discharge. The expense and difficulty in obtaining suspended sediment samples mean that even well-monitored rivers are often sampled too infrequently to catch important short-lived events, such as floods or landslides.

Historically, the most reliable monitoring of suspended sediment concentrations has combined periodic suspended sediment sampling with closely spaced sampling during major discharge events (Walling and Webb, 1988; Robertson and Roerish, 1999; Bridge, 2003). But in contrast to some larger rivers that have had long-term monitoring, many rivers have been sampled only long enough to allow workers to construct an operational sediment-rating curve, after which sampling often has been reduced or eliminated, the assumption being that the constructed curve can adequately represent the river's sediment discharge as well as the watershed's erodibility. Assuming no temporal change in watershed character, of course, is often based more on hope than on reality; the watershed's response to natural occurrences such as fire or landslides (Chapter 3) or to a change in landuse (Chapter 4) can greatly alter the erosion and/or subsequent transport of eroded sediments. In Taiwan, for example, rapidly changing annual and interannual conditions may require semi-annual revisions of the sediment rating curves (Kao *et al.*, 2005; Kao and Milliman, 2008). At the opposite end of the scale, the oft-cited 260 Mt/yr discharge of the Irrawaddy River (Burma) published by Gordon (1885) was based on measurements made in 1877–78, and almost certainly is an underestimate. Reanalysis of Gordon's collection techniques and data led Robinson *et al.* (2007) to increase Gordon's estimate by nearly 40%, which more recent measurements confirm (Furuichi *et al.*, 2009).

Published sediment loads for most rivers are derived from the algorithms produced from sediment-rating curves, but anyone who has carefully considered the multitude of inherent problems with these curves cannot help but question how accurately they represent the river's actual sediment discharge. In addition to the sampling problems mentioned above, one has to worry about how the rating curve is constructed, particularly the wide degree of scatter that often characterizes such curves (Fig. 2.12b). Hysteresis seen in many rating curves reflects the varying capacity of a river to transport more sediment during its rising stage than during its falling stage, by which time much of the available sediment has been removed. In other words, many rivers evolve seasonally from transport-limited to supply-limited (Hudson, 2003). Hysteresis loops also can reflect different timing of sediment and water contributions from various tributaries within a river system. Depending on the number of major tributaries flowing into a river, this may result in several hysteresis loops within a single year; the convoluted corkscrew pattern of Mississippi River's pre-dam rating curve in 1950 (see Fig. 4.37c) is perhaps an extreme example.

Even if one disregards hysteresis, creating a meaningful rating curve can be disconcerting. For example, does one use a stratified seasonal rating curve or not, have there been sufficient sediment measurements during peak flow, etc.? Taking into account such uncertainties, Walling (1978; Walling and Webb, 1981) and Warrick and Farnsworth (2009) suggested that a 50% error in estimated annual suspended sediment loads is not only possible but perhaps acceptable. Given these problems, one should not place too much faith in many (most?) published sediment loads. Of the 77 rivers whose mean annual sediment loads are estimated to exceed 20 Mt/y shown in Table 1.3, for example, we consider only ~25% of the estimates to be good, whereas ~50% are considered to be poor, out of date, or non-existent.

A sediment load reported to three or four digits, say 10.83 Mt/yr, should be treated with particular suspicion. One can achieve a greater sense of reality by using only the first two digits, say 11 Mt/yr, although it can be argued that only the first digit is actually meaningful, in this case, 10 Mt/yr. Depending on the sampling techniques and sampling scheme used to derive the sediment load, in fact, reporting a range between 6 and 16 Mt/yr might be closer to the truth. Throughout this book and in our database we report values to the second digit.

Of particular concern to us are the reported loads for rivers draining Indonesia and the Philippines, since, as discussed later in this chapter, the few published data suggest that these rivers collectively are a major source of fluvial sediment to the global ocean. Many of the Indonesian and Philippine data come from second- and third-hand sources, and most presumably reflect short-term measurements of unknown quality. The Sumatra river sediment data that we cite in the appendix and use in many of the following calculations, for instance, may be too high according to colleagues who have worked on eastern rivers that flow through the island's broad coastal lowlands (B. Cecil and T. Jennerjahn, personal communication). A similar problem is seen in the Mahakam River, which drains ~80 000 km^2 in eastern Borneo. Using preliminary data, Allen *et al.* (1979) estimated the Mahakam's annual load to be ~10 Mt/yr, which equates to an annual sediment yield of ~120 t/km^2/yr, somewhat lower than the global mean. Douglas (1996), in contrast, estimated that the rivers draining the Tertiary mudstones in eastern Borneo should have annual yields of ~1000 t/km^2/yr, which would give the Mahakam an annual sediment load of 50–100 Mt/yr. As new data become available, estimates for Indonesian and Philippine sediment discharge almost certainly will need to be re-evaluated.

Compounding the problem of quantifying sediment flux is that reported discharge, no matter its accuracy, represents measurements upstream from any tidal influence, which in some cases may lie hundreds of kilometers upstream from the river mouth. For example, the Yangtze River's downstream-most gauging station at Datong is 400 km upstream from the East China Sea. Almost certainly sediment is deposited downstream of Datong and perhaps new sediment introduced, so that the reported sediment load may not represent the actual amount of sediment that reaches the coastal ocean.

Environmental controls on sediment delivery to the coastal ocean

It is almost axiomatic to state that sediment erosion and subsequent transport are controlled by drainage basin size and topography/gradient, bedrock geology, climate (particularly precipitation/runoff), rainfall severity, vegetation cover, and anthropogenic activity (Ludwig and Probst, 1998; Leeder *et al.*, 1998; Meybeck, 2003; Syvitski *et al.*, 2005; Syvitski and Milliman, 2007; Kao and Milliman, 2008), the basic variables used in the Universal Soil Loss Equation. Thus, unlike river discharge, we need to look not only at precipitation and evapotranspiration (i.e. river runoff) but we also must figure in the influences of basin morphology and source-rock lithology as well as temporal change and the impact of events (Chapter 3) and human activity (Chapter 4). The continuing debate, as witnessed by the wide range of interpretative and predictive models, has been over which of these factors play primary and secondary roles in determining a river's sediment discharge.

All published interpretative and predictive models regarding fluvial sediment loads are based on factors that control sediment supply (erosion), transport (discharge), or a combination of the two. Early studies by Fournier (1949, 1960) and Langbein and Schumm (1958) concluded that precipitation is the dominant factor controlling sediment load. Douglas (1967) found a good correlation between rainfall variability and sediment load, and Jansson (1988) suggested that climate (temperature and precipitation/runoff, particularly "rainfall aggressiveness") was the determining factor in soil erosion. Using a larger database, Wilson (1973) concluded that precipitation has relatively little influence on sediment yield; rather, he argued, basin area and land use are the primary forcing functions. Ahnert (1970) agreed,

Table 2.5. *Highest and lowest average pre-dam suspended-sediment concentrations (in bold), in descending order, of the 700+ rivers in our database. Global average is 500 mg/l. Rivers with highest concentrations tend to lie in subtropical arid to semi-arid climates, the Erhjen (Taiwan) and Jaba (Bougainville) being stark exceptions. Rivers with lowest concentrations, by contrast, are located in Europe, many of them draining erosion-resistant shield-dominated terrain.*

River	Country	Area (× 10^3 km^2)	Climate	Runoff (mm/yr)	Sed. load (× 10^3 t/yr)	Sed. yield (t/km^2/yr)	Sed. concentration (mg/l)
Miliane	Tunisia	2	STr-A-D	10	0.9	450	**45 000**
Agrioun	Algeria	0.66	STr-H-W	260	4.8	7300	**28 000**
Wadi Sihan	Yemen	4.9	STr-A-D	20	2.8	570	**28 000**
Isser	Algeria	4.2	STr-A-W	86	8.3	2000	**23 000**
Erhjen	Taiwan	0.14	STr-W-S	3600	10	71000	**20 000**
Jaba	Bougainville	0.46	Tr-W-S	2800	26	56000	**20 000**
Huanghe	China	750	Te-A-S/Au	57	1100	15000	**19 000**
Draa	Morroco	110	STr-A-W	7	14	13	**17 500**
Dalinghe	China	23	Te-A-S/Au	91	36	1600	**17 000**
Santa Clara	USA	4.1	STr-A-W	44	3	7300	**17 000**
Ventura	USA	0.48	STr-SA-W	110	0.38	1700	**15 000**
Global							**500**
Dalaven	Sweden	29	SA-H-S	520	0.03	1	**2**
Fluvia	Spain	1	Te-SA-W	300	0.002	10	**2**
Ljungan	Sweden	13	SA-H-S	340	0.01	1	**2**
Lule	Sweden	25	SA-H-S	640	0.04	2	**2**
Skellefte	Sweden	12	SA-H-S	410	0.009	1	**2**
Rane	Sweden	4.1	SA-H-S	320	0.002	0.5	**1**
Rhine	Netherlands	220	Te-H-C	340	0.07	0.3	**1**
Siikajoki	Finland	4.4	SA-H-S	320	0.002	0.4	**1**

stating that "...mean precipitation has a negligible effect on denudation rate," whereas local basin relief plays a major role, a suggestion supported by Gunnell's (1998) study of denudation rates on the southern Indian shield. Catchment gradient, suggested Phillips (1990), accounts for 70% of the contribution to maximum expected variations in erosion rates, a conclusion similar to that of Summerfield and Hulton (1994).

Jansen and Painter (1974) suggested that sediment yield could be related to temperature, discharge (i.e. precipitation) or basin relief, whereas Pinet and Souriau (1988), Milliman and Syvitski (1992), Summerfield and Hulton (1994), and Syvitski and Milliman (2007) concluded that basin elevation (rather than climate) is a primary determinant. Milliman and Syvitski (1992) also confirmed Schumm and Hadley's (1961) observation that sediment yield is inversely proportional to watershed area, although there are regional contradictions to this global trend (de Vente *et al.*, 2007). Milliman and Syvitski (1992) and Montgomery and Bandon (2002) pointed out that tectonism and the rate of tectonic uplift, rather than local relief per se, may play key deterministic roles in landscape degradation. In Taiwan, for example, earthquake-generated landslides create the source of sediment that is subsequently eroded during typhoon-forced rains (Dadson *et al.*, 2004; Milliman *et al.*, 2007; Goldsmith *et al.*, 2008).

Ludwig and Probst (1996, 1998) were among the first to propose a numerical model that could extract the relative importance of the environmental factors controlling sediment discharge. Runoff intensity, rainfall variability, basin slope, and rock hardness were considered the primary factors, although the Ludwig and Probst model also suggested (wrongly, as discussed later in this chapter) that erodibility increases with aridity. One problem, which those authors readily acknowledged, was that the Ludwig and Probst model was based on a rather slim database (60 rivers) that was weighted towards large rivers, many in Europe, even though most European rivers have very low yields. The high-yield landscape in southeast Asia and Oceania was represented by only 11 rivers.

In recent years James Syvitski and his co-workers (Syvitski and Alcott, 1995; Syvitski and Morehead, 1999; Syvitski *et al.*, 2003, 2005) have been particularly active in developing dimensional analyses to predict long-term sediment loads. Syvitski and Milliman (2007) presented the BQART model, utilizing a database of 488 rivers from which they quantified the importance of river discharge,

Table 2.6. *Highest and lowest average pre-dam annual sediment loads, in descending order (in bold). 13 of the 15 highest loads are in rivers whose headwaters exceed 3000 m in elevation; 7 drain the Himalayas. Rivers with the lowest sediment loads are located in Scandinavia and the British Isles, most with headwaters <1000 m (upland rivers), many <500 m (lowland rivers).*

River	Country	Area (× 10^3 km^2)	Elevation	Runoff (mm/yr)	Sed. load (Mt/yr)	Sed. yield (t/km^2/yr)	*Q*sc (g/l)
Amazon	Brazil	6300	High Mt	6300	**1200**	190	0.19
Huanghe	China	750	High Mt	15	**1100**	1500	19
Brahmaputra	Bangladesh	670	High Mt	630	**540**	810	0.86
Ganges	Bangladesh	980	High Mt	490	**520**	530	1.1
Changjiang	China	1800	High Mt	900	**470**	260	0.52
Mississippi	USA	3300	High Mt	490	**400**	120	0.82
Irrawaddy	Burma	430	High Mt	430	**260**	600	0.6
Indus	Pakistan	980	High Mt	<10	**250**	250	2.8
Orinoco	Venezuela	1100	High Mt	1100	**210**	140	0.14
Godavari	India	310	Mountain	92	**170**	550	1.8
Mekong	Vietnam	800	High Mt	690	**150**	190	0.27
Magdalena	Colombia	260	High Mt	230	**140**	540	0.61
Fly	Papua New Guinea	76	High Mt	180	**110**	1100	0.44
Song Hong	Vietnam	160	High Mt	120	**110**	690	0.92
Skellefte	Sweden	12	Lowland	410	**0.009**	1	2
Welland	England	0.53	Lowland	210	**0.007**	13	63
Conon	Scotland	0.96	Mountain	1600	**0.006**	6	4
Slaney	Ireland	1.8	Upland	610	**0.006**	3	5
Teith	Scotland	0.52	Mountain	1400	**0.005**	10	7
Liffey	Ireland	1.4	Lowland	335	**0.004**	3	8
Karjaanjoki	Finland	2	Lowland	320	**0.002**	1	3
Rane	Sweden	4.1	Upland	320	**0.002**	0.5	1
Siikajoki	Finland	4.4	Lowland	320	**0.002**	0.4	1
Mandalselva	Norway	1.7	Upland	880	**0.001**	1	1

basin area, relief, temperature, as well as the combined effect of glacial erosion, basin-wide lithology, trapping by lakes and reservoirs, and human-induced erosion. Geological factors (basin relief, basin area, lithology, and ice/glacial erosion), they concluded, explain 65% of the variation in sediment loads; climatic and anthropogenic factors collectively explain an additional 30% of the variability.

The diversity of opinions cited above, often derived by different workers using many of the same data, clearly emphasizes the difficulty in prioritizing the factors controlling sediment erosion, transport and discharge to the ocean. As pointed out by Hovius and Leeder (1998, p. 2):

It should therefore not come as a surprise that empirical studies of modern drainage basins... have not so far produced a reliable universal relationship between sediment yield and catchment characteristics. Although strong relations between erosion rates and one or more catchment characteristics can be found locally, the search for universality, in this form, is futile. The principal reason for this is that sediment efflux of a catchment is the integrated effect of a series of tectonic, climatic and geomorphic processes...

Of 760 rivers for which we have sediment data, we find a great range in average suspended sediment concentrations (1–45,000 mg/l, with a global mean of 500 mg/l; Table 2.5), sediment loads (0.001–1200 Mt/yr; Table 2.6), and annual sediment yields (0.3–71 000 t/km^2/yr, with a global annual mean of 190 t/km^2/yr; Table 2.7). The relative plethora and range of values within our database allow us the "luxury" of sorting out the relative importance of the various environmental factors, which we attempt to do in the following pages. Because our database includes only rivers that discharge directly into the ocean, it is not compromised by data "double-dipping," in contrast to Wilson's (1973) database of 1500 rivers, which included more than 1200 tributaries to larger rivers.

Before we proceed in this discussion, we should warn the reader that we recap here many of our thought processes developed over the years as we have sought to delineate the relative importance of the various environmental factors controlling

Table 2.7. *Highest and lowest pre-dam fluvial suspended-sediment yields (in bold) in descending order; global annual mean, 190 t/km². Many high-yield rivers drain mountainous elevations (<3000 m), more than half of them in Taiwan. Maximum elevation of the highest-yield river, the Erhjen (average annual yield 71 000 t/km²), however, is only 460 m. All high-yield rivers are small, with drainage basins <3000 km², and most have runoffs greater than 2000 mm/yr. Most low-yield rivers, on the other hand, drain lower elevations or erosion-resistant shield rocks; runoff for low-yield rivers tends to be <500 mm/yr.*

River	Country	Area (× 10³ km²)	Elevation	Runoff (mm/yr)	Sed. load (× 10³ t/yr)	Sed. yield (t/km²/yr)	*Q*sc (g/l)
Erhjen	Taiwan	0.14	Upland	3600	10	**71 000**	20
Jaba	Bougainville	0.46	Mountain	2800	26	**56 000**	20
Hoping	Taiwan	0.55	High Mt	2200	16	**29 000**	13
Waiapu	New Zealand	1.7	Mountain	1600	35	**21 000**	12.5
Hokitika	New Zealand	0.35	Mountain	8900	6.2	**18 000**	2
Hualien	Taiwan	1.5	Mountain	2500	25	**17 000**	5.3
Peinan	Taiwan	1.6	High Mt	2300	20	**14 000**	4.9
Choshui	Taiwan	3.1	High Mt	2000	40	**13 000**	6.2
Waiho	New Zealand	0.29	Mountain		3.4	**12 000**	
Cikeruh	Indonesia	0.25	Mountain		2.8	**11 000**	
Tungkang	Taiwan	0.47	Mountain	2300	5.2	**11 000**	4.7
Tsengwen	Taiwan	1.2	Mountain	2000	12	**10 000**	5
Yenshui	Taiwan	0.22	Lowland	1400	2.2	**10 000**	7.3
Global						**190**	
Murray–Darling	Australia	1100	Mountain	22	1	**1**	42
Jaguaribe	Brazil	81	Upland	49	0.06	**1**	15
Guadiana	Spain	72	Upland	125	0.07	**1**	8
Kymijoki	Finland	37	Lowland	320	0.05	**1**	4
Iijoki	Finland	14	Lowland	360	0.014	**1**	3
Karjaanjoki	Finland	2	Lowland	49	0.002	**1**	3
Oulujoki	Finland	25	Lowland	320	0.024	**1**	3
Dalaven	Sweden	29	Upland	520	0.03	**1**	2
Ljungan	Sweden	13	Upland	340	0.01	**1**	2
Skellefte	Sweden	12	Lowland	410	0.009	**1**	2
Mandalselva	Norway	1.7	Upland	880	0.001	**1**	1
Rane	Sweden	4.1	Upland	320	0.002	**0.5**	1
Siikajoki	Finland	4.4	Lowland	320	0.002	**0.4**	1
Rhine	Netherlands	220	Mountain	340	0.07	**0.3**	1

sediment (and dissolved-solid) erosion (and weathering) and transport. The impatient or faint-hearted reader may prefer to skip the following pages and proceed directly to the concluding section at the end of this chapter. While many of the conclusions are understandably similar to those of Syvitski and Milliman (2007), here we assess, piece by piece, the relevant factors that control fluvial sediment discharge.

Drainage basin area and morphology

It seems reasonable to assume that sediment discharge is proportional to basin area, larger rivers having greater loads than smaller rivers. But the number of exceptions to this rule of thumb underscores the inexactitudes of such an assumption (Fig. 2.13a). The Waiapu River (North Island of New Zealand, basin area 1700 km²), for example, discharges far more sediment (26 Mt/yr) than the much larger (1 100 000 km²) Murray–Darling River (1 Mt/yr); the São Francisco (Brazil) and the Brahmaputra (Bangladesh) rivers drain similar size watersheds (640 000 and 670 000 km², respectively) but have drastically different annual sediment loads: 6 vs. 540 Mt/yr, respectively. Conversely, although sediment yield tends to increase with decreasing basin area (Fig. 2.13b), some large rivers have higher yields than smaller rivers (de Vente *et al.*, 2007); for example, the Ganges (980 000 km²; annual yield 530 t/km²/yr) vs. the Karjaanjoki (Finland, 2000 km²; annual yield 1 t/km²/yr).

We can deconvolve this relationship a bit further by factoring in drainage basin morphology. Following Milliman and Syvitski (1992), we have categorized sediment discharge and yield of 760 rivers on the basis of maximum

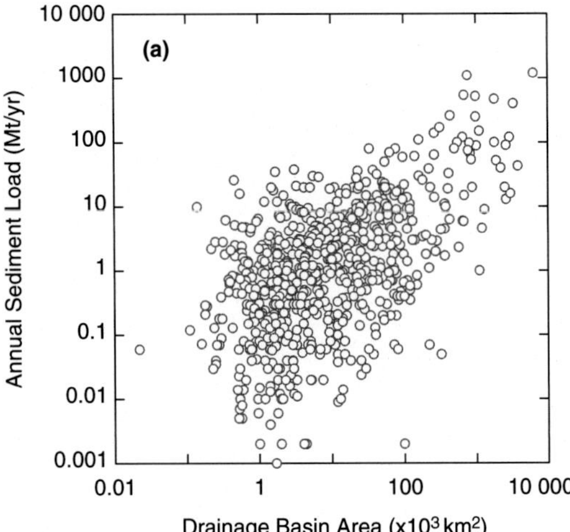

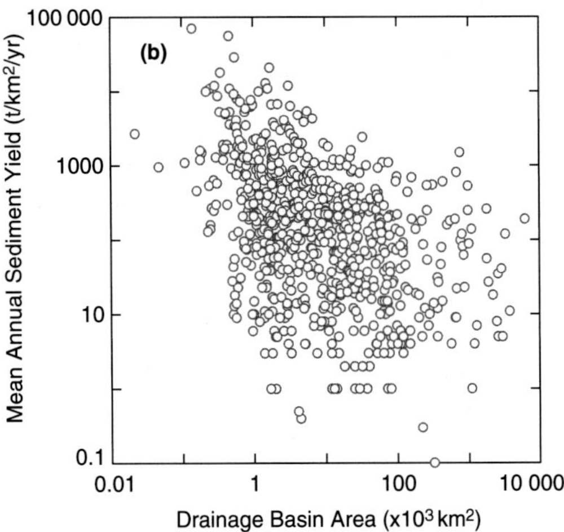

Figure 2.13. Variation of annual sediment discharges (a) and yields (b) relative to drainage basin area for 760 rivers discharging into the coastal ocean.

elevation of the drainage basin: >3000 m (high mountain), 1000–3000 m (mountain), 500–1000 m (upland), and <500 m (lowland). The resulting figures – 2.14 and 2.15 – show that sediment load and yield generally increase as maximum drainage basin elevation increases. This, of course, should be no surprise, as Ahnert (1970) and Pinet and Souriau (1988) previously suggested that relief is the main driving function of basin denudation. By categorizing rivers based on maximum elevation, Milliman and Syvitski (1992) found a reasonably workable relationship between basin area and sediment load and an inverse relation between basin area and sediment yield, much like those shown in Figs. 2.14 and 2.15. Summerfield and Hulton (1994) used a slightly expanded global database to derive the relation:

$$\text{denudation rate (mm/yr)} = 0.00721e^{(0.0015*\text{elevation})}.$$

High-mountain rivers tend to have one to three orders of magnitude greater load/yields than similar-sized lowland rivers (Figs. 2.14 and 2.15). The inverse relation between yield and basin area shown in Fig. 2.13 also becomes clearer. A high-mountain, mountain or upland river with a 1000 km² drainage basin, for example, generally has one to two orders-of-magnitude greater sediment yield than a river whose watershed is greater than 1 000 000 km² in area (Fig. 2.15). Sediment discharges and yields of coastal plain rivers (maximum elevation <100 m), in contrast, show no correlation with basin area (Milliman and Syvitski, 1992, their Fig. 6G).

There are a number of reasons for the direct relationship between sediment discharge (or yield) and elevation. Most obvious, rivers that drain higher elevations have steeper average gradients (Fig. 2.16). Elevation also serves as a tectonic surrogate: lying within orogenic belts, mountains are more likely to experience earthquakes (Inman and Nordstrom, 1971; Audley-Charles *et al.*, 1977; Potter, 1978), which can play a greater role in denudation than elevation or relief per se (Milliman and Syvitski, 1992; Montgomery and Brandon, 2002; Dadson *et al.*, 2004). But not all mountains are equally susceptible to earthquakes – either in frequency or magnitude. Mountains that lie along the edge of tectonic plates – such as the western Americas, but especially along the south Asian plate (including Taiwan, Philippines and Indonesia) – are particularly susceptible to ongoing tectonism and thus frequent and large earthquakes (Fig. 2.17). Mountains produced from past tectonic events, such as the Rockies or Urals, often are not major producers of sediment; their present high elevations, in fact, give testimony to the relatively low rates at which they shed sediment (R.H. Meade, personal communication).

The orographic control on precipitation (see above) means that mountains also are more likely to experience increased precipitation (Reiners *et al.*, 2003), and higher elevations experience greater snow and ice cover. Some high mountains may have year-round ice cover, which can greatly increase both physical and chemical weathering (Dedkov and Mozzherin, 1992; Schmidt and Montgomery, 1995; Collins, 1996; Hallet *et al.*, 1996; Anderson, 2005; Green *et al.*, 2005), particularly if the snowpack or glacier is warm-based (Tomasson, 1991; Andrews and Syvitski, 1994; Syvitski and Alcott, 1993). Vezzoli *et al.* (2004), for instance, found a direct relationship between sediment yield and glaciated area in the western Alps. The breaching of ice-dams can also produce a Jokulhlaup (see Chapter 3), another unique feature of glaciated terrain that can have a major role in both erosion and sediment transfer. Because present-day glaciers are relatively restricted (<1% of the global land area), they probably do not play a major role in present-day global erosion or sediment discharge; but they almost surely played prominent role during and following

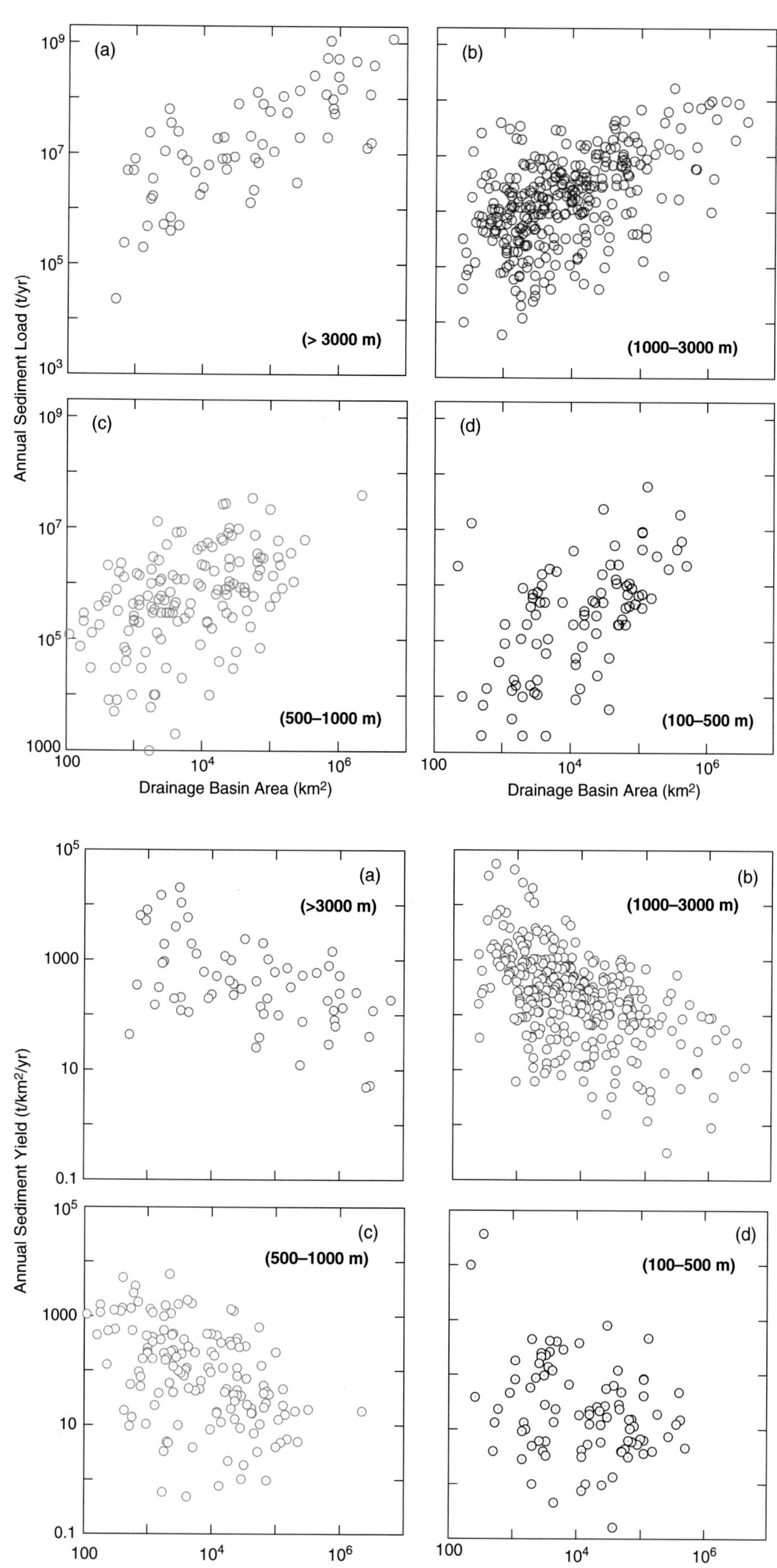

Figure 2.14. Variation of sediment load with basin area for various watersheds as defined by the maximum elevations of their headwaters: (a) high-mountain (>3000 m), (b) mountain (1000–3000 m), (c) upland (500–1000 m), and (d) lowland (100–500 m).

Figure 2.15. Variation of sediment yield vs. basin area for (a) high-mountain (>3000 m), (b) mountain (1000–3000 m), (c) upland (500–1000 m), and (d) lowland (100–500 m) rivers.

Pleistocene glacial maxima (Syvitski and Milliman, 2007, and references therein). Moreover, the "buzzsaw" erosion by frost-cracking also can play important roles in rockfall erosion (Hales and Roering, 2009). All of the above factors mean that erosion and the likelihood for mass down-slope movement (landslides, debris flows, slumps) increase with increasing watershed elevation (Hovius, 1988; Summerfield and Hulton, 1994; Wilkinson and McElroy, 2007); locally gully erosion also can increase in importance (Kelsey, 1980; Hicks *et al.*, 2000, 2004).

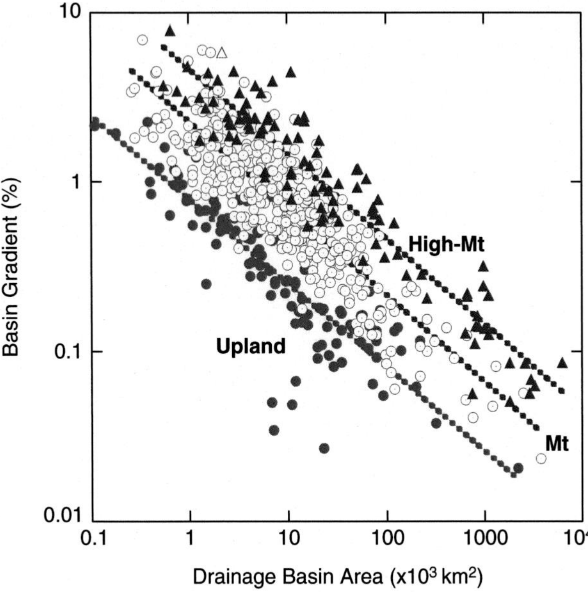

Figure 2.16. Mean gradient (maximum elevation/river length) versus basin areas for high-mountain (>3000 m; blue triangles), mountain (1000–3000 m; red dot-circles), and upland (maximum elevation 500–1000 m; green dots) rivers.

The inverse relationship between sediment yield and basin area is not dissimilar from that of mean gradient and basin area, suggesting a causal relationship. It is not that the mountains in smaller watersheds are steeper, but rather that smaller rivers tend to have relatively less floodplain area in which sediment can be stored. For example, <10% of the Amazon (basin area 6 300 000 km^2; annual sediment yield 190 t/km^2/yr) lies at elevations higher than 500 m, whereas >500 m elevations account for more than half of the Choshui River watershed (basin area 3100 km^2; annual sediment yield 13 000 t/km^2/yr). Galy and France-Lanord (2001) calculated that as much as half of the sediment eroded in the Ganges–Brahmaputra's watershed is deposited as bed load or on floodplains, not dissimilar from that calculated by Goodbred and Kuehl (1998). Said another way, one reason that large rivers have lower sediment yields is because mountains represent a lesser part of their drainage basin and floodplains a greater part.

The importance of the downstream storage of fluvial sediment has been demonstrated for the Rio Madeira basin of the Amazon, where a considerable portion of the river's sediment is deposited as it leaves the Andes Mountains

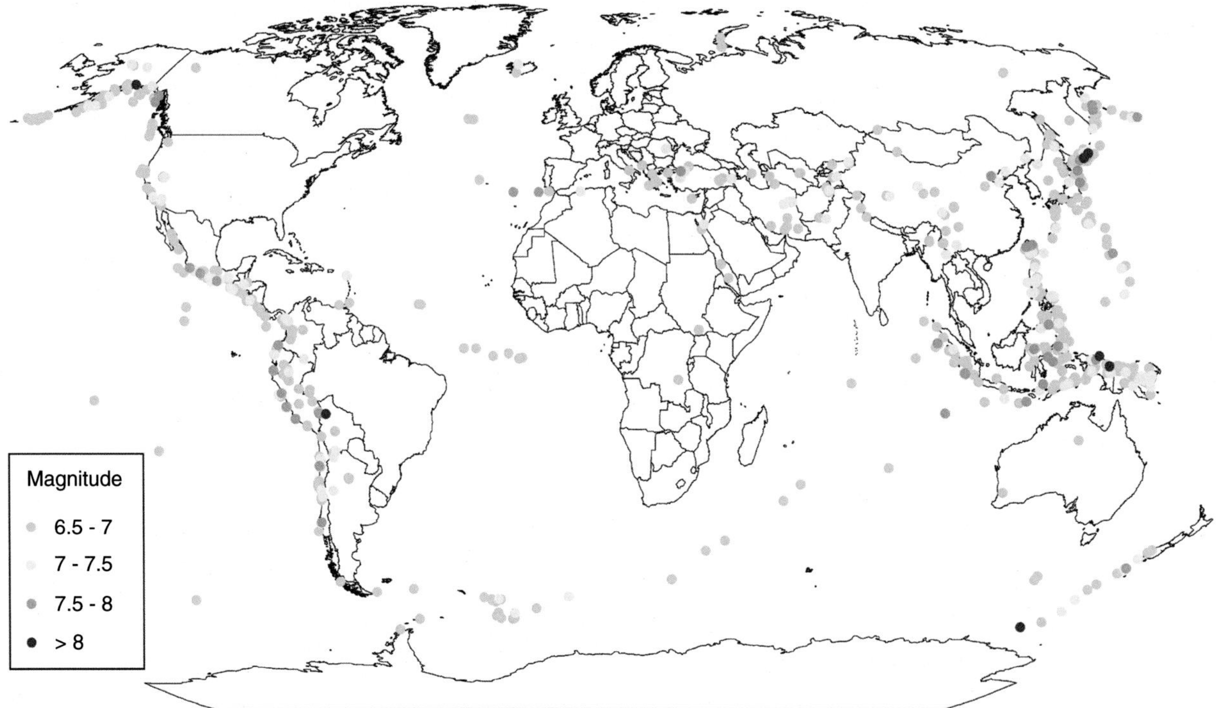

Figure 2.17. Global distribution of major earthquakes, which we define as magnitudes >6.5, 1950–2000. Note the frequency and magnitude of earthquakes in the Taiwanese–Indonesian region as well as along the west coasts of the Americas. Data from USGS.

and crosses the Amazon lowlands (Guyot *et al.*, 1996); some of it is only temporarily stored before it is remobilized and moved further downstream (Dunne *et al.*, 1998). Aalto *et al.* (2006) have estimated that as much as 50% of the sediment eroded from the Andes is deposited in the Amazon foreland basin, and Bourgoin *et al.* (2007) calculated sediment accretion of 1.6 mm/yr along a 130-km stretch of the Amazon's Curuai floodplain. Another case in point is the Yangtze River: for the 20 years prior to the closing of the Three Gorges Dam (in 2003), average sediment discharge at the downstream Datong gauging station (basin area 1 700 000 km^2) was 100 Mt/yr less than it was at the upstream Yichang gauging station (basin area 1 000 000 km^2) (see Fig. 4.44). Most of the 100 Mt/yr "lost" between Yichang and Datong was deposited on the lower river's flood plains, lakes and river channels (Xu *et al.*, 2007; see Chapter 4 for a further discussion).

Smaller rivers also are more responsive to episodic events (see Chapter 3), which, together with the above two factors (greater erosion and less downstream deposition), translates into increased down-slope mass movement and thus greater erosion. Similarly, rapid water flow means less likelihood for short-term storage, thus increased downstream transport (Hovius *et al.*, 2000; Dadson *et al.*, 2004) (see below).

Aalto *et al.* (2006) quantified the sediment loads/yields for 47 Andean rivers whose drainage basins vary from 17 000 km^2 to 81 000 km^2 in area (mean area 11 000 km^2). By limiting their study to mountainous basins in which erosion far exceeds sedimentation, they effectively minimized any impact from human activities. They concluded that basin gradient and lithology account for 90% of the variance in sediment yield. By analyzing sediment loads for nearly 500 rivers, Syvitski and Milliman (2007) concluded that basin area and relief account for ~65% of the data variance. However, the wide degree of scatter in the values seen in the previous figures underscores the importance of other factors. In fact, in some rivers relief and basin area may actually be secondary factors in controlling a river's sediment load (see Owens and Slaymaker, 1992). Restrepo and Syvitski (2006) found, for instance, that runoff and peak discharge events account for 58% of the variance in the Magdalena River's (Colombia) sediment yield. Other critical factors are discussed in the following paragraphs.

Geology and lithology

The erodibility of a rock or sediment depends, respectively, on its lithology and/or size of constituent particles, which thereby reflect its composition, mode of formation, and age. Younger rocks are often more erodible than older, harder rocks (Pinet and Souriau, 1988; Gaillardet *et al.*, 1999; Dupré *et al.*, 2003; Syvitski and Milliman, 2007 and references therein). Mudstone is more erodible than sandstone (reflecting differences in composition and/or mode of formation) or shale (reflecting a difference in age), and extrusive igneous debris is more easily eroded than intrusive igneous rocks. Hicks *et al.* (1996) showed that New Zealand rivers draining mudstone have much higher yields than rivers draining greywacke or sandstone, which in turn have higher yields than rivers draining metamorphic rocks (Fig. 2.18), not unlike that noted by Aalto *et al.* (2006) for rivers in the Bolivian Andes. Based on rock hardness, Syvitski and Milliman (2007) delineated six classes of erodibility: (1) hard, acid plutonic or metamorphic (least erodible); (2) older shield rocks; (3) basaltic and carbonate rocks; (4) slate and shale; (5) unconsolidated sediments; and (6) exceptionally weak crushed material, such as loess (most erodible). Identifying the controlling lithology of a watershed, of course, becomes problematic for larger basins where the rock types tend to be more diverse.

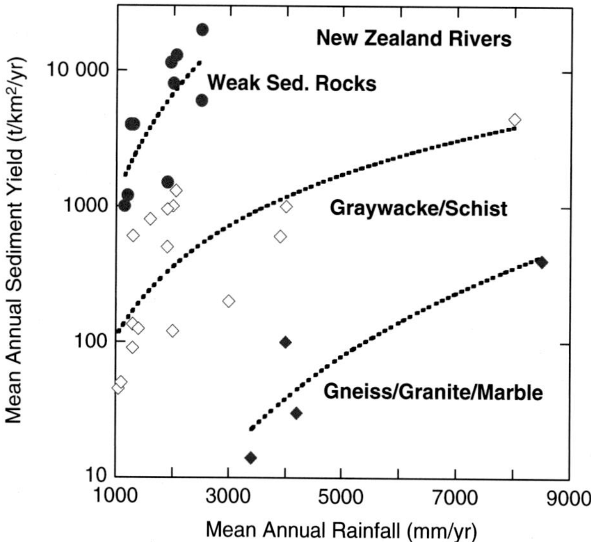

Figure 2.18. Sediment yields of New Zealand rivers as a function of annual runoff and the lithologic character of the upper drainage basins. After Hicks *et al.* (1996).

As described in the appendix, the geological character used in our database defines the geologic character of a river's upper watershed (presumably the prime source of a river's suspended sediment) on the basis of age of rocks rather than lithology, in part because of the inherent uncertainty in defining the average lithologic character of a watershed. Cenozoic and Mesozoic rocks – both sedimentary/metamorphic and igneous – are more easily eroded than older, pre-Mesozoic rocks (Fig. 2.19), in part because the pre-Mesozoic rocks that are still around give testimony to their ability to withstand erosion (R. H. Meade, personal communication). But the scatter seen in Figure 2.19 is considerable, which precludes

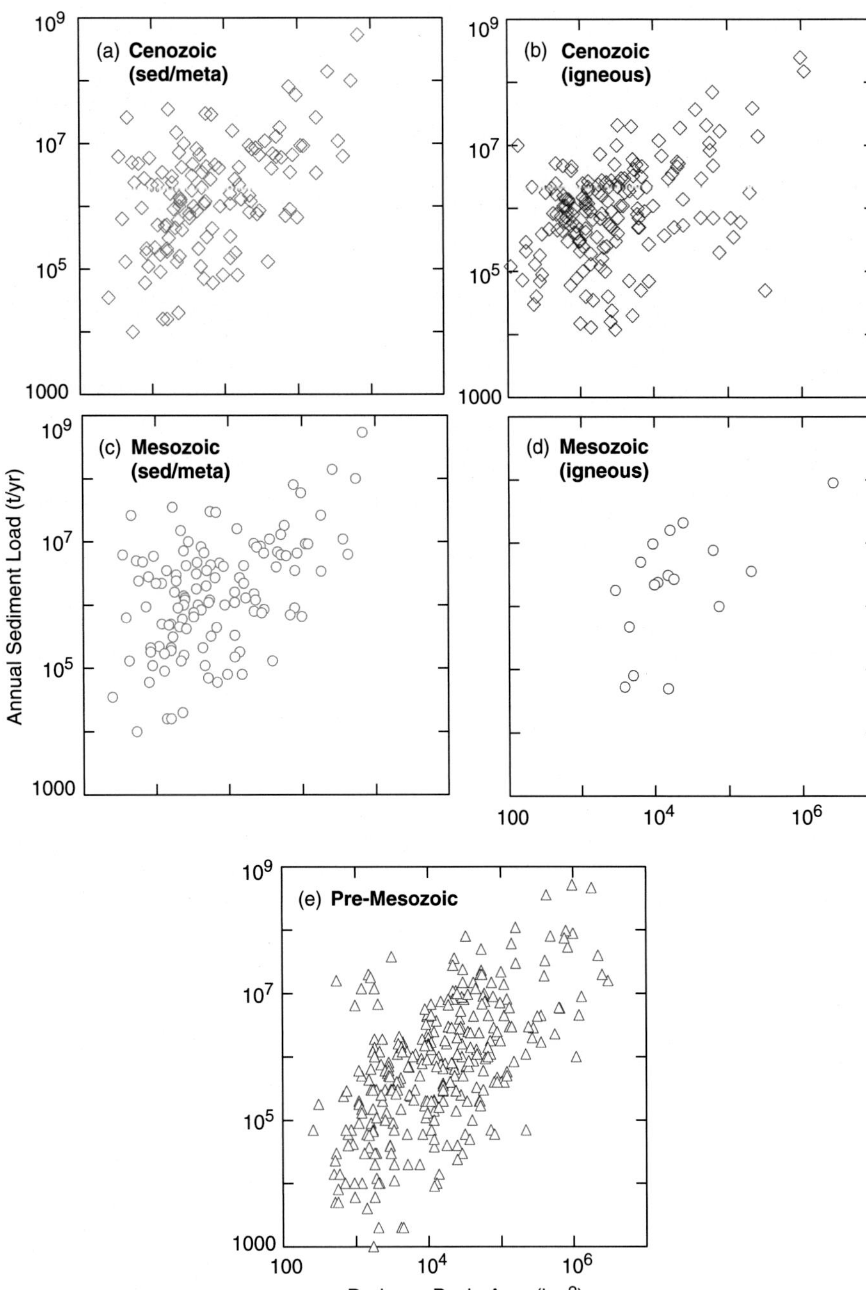

Figure 2.19. Fluvial sediment loads relative to basin area and primary geologic age/lithology of the upper watersheds. Sed/meta = sedimentary/metamorphic.

establishing statistically significant trends; some rivers draining Mesozoic igneous rocks, for example, have higher sediment loads than rivers draining Cenozoic sedimentary or metamorphic rocks.

Combining the data used in Figs. 2.14 and 2.19 affords us a look at the geologic makeup of rivers draining the four morphologic groupings (Fig. 2.20). Rivers that drain young terrain generally have greater loads than similar-sized rivers draining older rocks, but the considerable scatter precludes establishing statistically significant trends. Syvitski and Milliman (2007) calculated that lithology accounts for 13% of the variance in determining sediment load.

Summarizing to this point, we can see that drainage basin area and topography play key roles in determining the sediment load of a river, and that lithology also is important. But the scatter of data points is more than a little disconcerting. So far, however, we have ignored one key

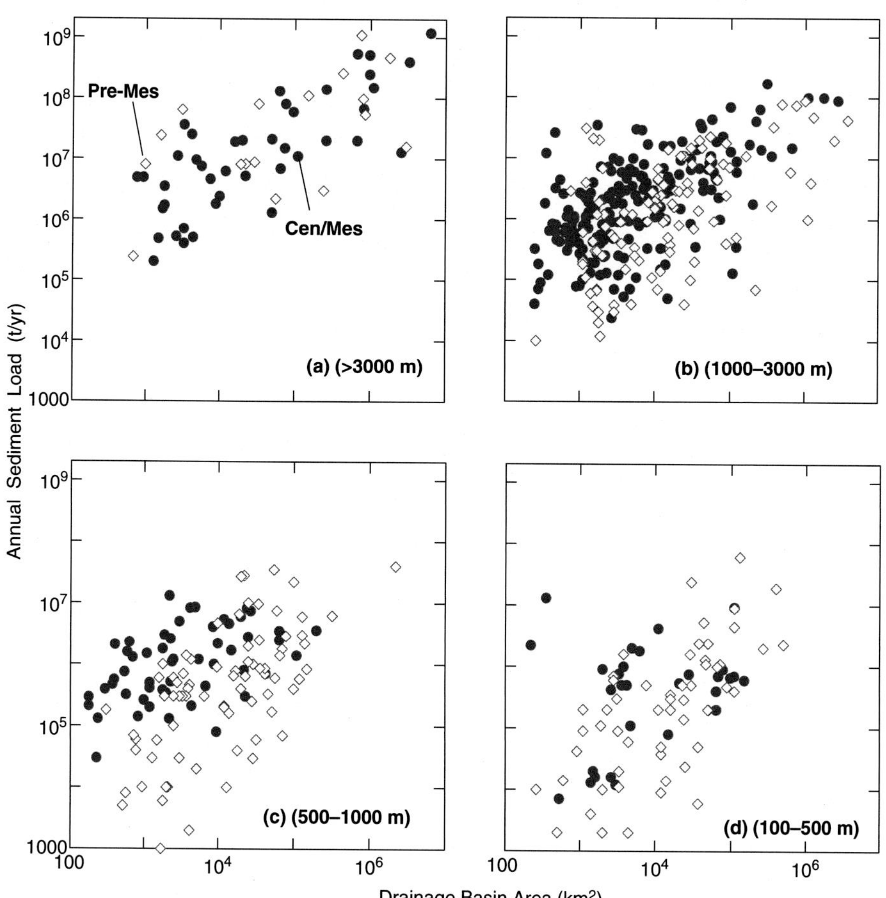

Figure 2.20. Sediment discharge vs. drainage basin area in (a) high-mountain (>3000 m), (b) mountain (1000–3000 m), (c) upland (500–1000 m), and (d) lowland (100–500 m) rivers draining younger (Cenozoic-Mesozoic – red dots) and older (pre-Mesozoic – blue diamonds) lithologies.

factor: climate – specifically precipitation and river runoff. When we combine climate with the geomorphic and geological factors discussed above, the trends become clearer and the scatter significantly reduced.

Climate: precipitation and runoff

Among the earliest – and at the time most comprehensive – studies on sediment erosion were those by Fournier (1949, 1960) and Langbein and Schumm (1958), both of which concluded that precipitation played *the* key environmental role in erosion and sediment transport. Even though restricted by a lack of data, Fournier was able to incorporate 96 sediment loads into his database, although some of the data were double entries. From this small and imperfect database, Fournier speculated that rivers whose basins receive less than 500 mm of annual precipitation have much higher sediment yields than basins with higher precipitation (Fig. 2.21a). The sparse vegetation in desert climates, Fournier argued, was not able to retard the impact of periodically heavy rains, thereby resulting in increased (albeit episodic) rates of erosion. Wetter climates, so went the argument, support luxuriant vegetation, which tends to retard the erosive impact of the raindrops, increases water infiltration into the soil, binds the soil, reduces overland runoff, and maintains soil roughness. But when annual rainfall exceeds a certain level, Fournier argued, both vegetation and soils could be eroded, resulting in the U-shaped precipitation-sediment yield curve shown in Fig. 2.21a.

Relying principally on sedimentation rates in manmade reservoirs (upstream drainage areas generally ranging between 25 km^2 and 125 km^2) throughout the central and Rocky Mountain sections of the western United States, Langbein and Schumm (1958) reached a similar conclusion as Fournier for low-rainfall rivers, highest erosion occurring at ~400 mm/yr precipitation and decreasing steadily as rainfall (and vegetative cover) increased. Perhaps because they lacked data from high-runoff watersheds, Langbein and Schumm reported no evidence of increased erosion at high levels of rainfall (Fig. 2.21a).

While the Fournier and Langbein–Schumm curves have been widely cited as having important implications in paleohydrologic studies (see Schumm, 1977), proving their applicability on either regional or global levels has produced

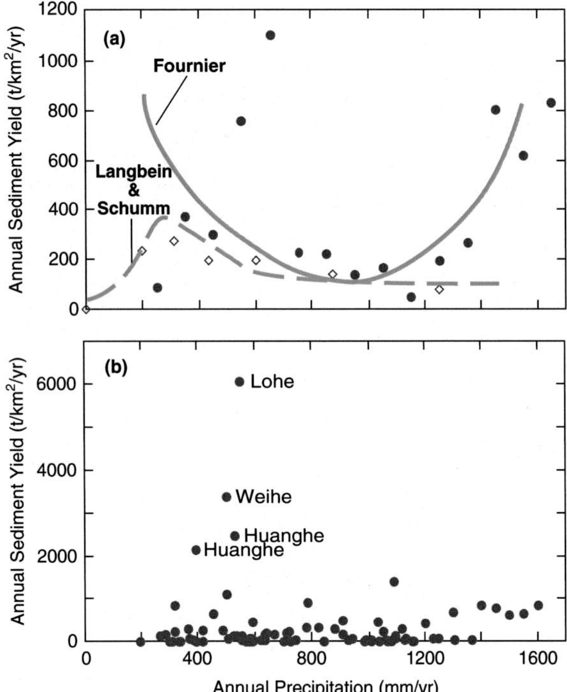

Figure 2.21. (a) Variation of sediment yield with precipitation as proposed by Fournier (red dots; 1949, 1960) and Langbein and Schumm (blue diamonds; 1958). Both trends are based on group-averaged data. (b) Fournier's complete data set from which he calculated group averages shown in (a). Note the three-order-of-magnitude spread in sediment yields for any level of precipitation.

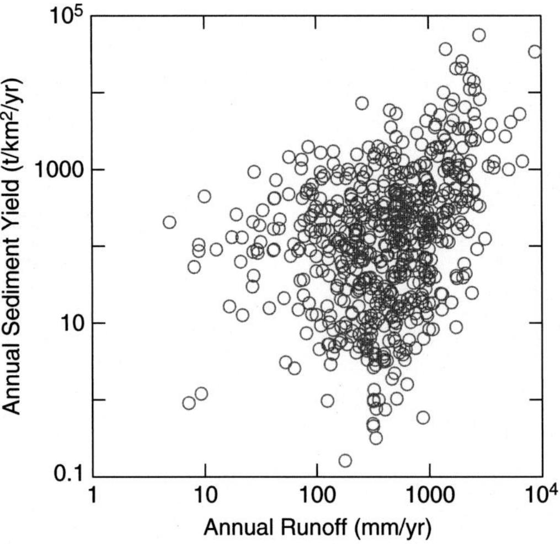

Figure 2.22. Relation between annual sediment yield and annual river runoff. Although this plot hints at increased sediment yield with increasing runoff, the yield at any runoff can vary by as much as five orders of magnitude. In this plot, the roles of basin area, morphology and geology are not considered.

mixed results. At first glance, as pointed out by Walling and Webb (1983), the Fournier/Landbein–Schumm curves appear to be intuitively reasonable – the lack of vegetation leads to accelerated erosion in arid environments, and sufficiently heavy rains, no matter what the vegetative cover, have the capacity to erode: hence Fournier's U-shaped curve. Although he cited neither Fournier nor Langbein–Schumm, Molnar (2001) suggested that late Cenozoic aridity may have led to increased mountain erosion and removal, primarily through bedload transport during high-discharge events. Modeling by Istanbulluoglu and Bras (2006) also concluded that sediment transport decreases (their Fig. 7) when annual precipitation exceeds ~250–350 mm/yr, in response to the capacity of vegetation to reduce runoff (their Fig. 8).

The inverse relationship between precipitation and erosion, however, has not been universally accepted. Utilizing 1500 individual data points, Wilson (1973) found no relation between precipitation and sediment yield, although his curve, as well as those proposed by Douglas (1967) and Ohmori (1983), suggested a sharp increase in sediment yield at precipitation levels greater than 800–1000 mm/yr. Ahnert (1970) similarly concluded that "...mean annual precipitation has no noticeable effect on denudation rate."

The lack of confirmation between runoff and sediment load led Walling and Webb (1983) and Meade *et al.* (1990) to conclude that the Langbein–Schumm curve may only apply to certain areas, such as continental climates in the USA. Inbar (1992) suggested that the Langbein–Schumm curve might be valid only in Mediterranean-type climates, a point supported by the high levels of erosion noted in the upper Tafna River in northwest Algeria (Megnounif *et al.*, 2007). Janda and Nolan (1979) and Riebe *et al.* (2001b), however, could find only minimal climatic control on erosion rates in the Sierra Nevada, even though there was an order-of-magnitude range in annual precipitation in their study area. Similarly, Renwick (1996) found little support for the assumption that climate (or, it should be noted, basin relief) necessarily dictated sediment yield. Walling and Webb (1983) concluded by stating, "Current evidence concerning the relationship between climate and sediment yield emphasizes that no simple relationship exists."

Thankfully, Fournier (1960) provided us with his database, which illustrates some of the inherent problems with his derived precipitation–sediment relationship Fig. 2.21b). Drainage basin areas utilized in his study varied from 1 000 000 km^2 (Indus, Yangtze) down to <3000 km^2 (Upper Iowa and Raritan rivers), and from mountain (Indus, Ganges) to coastal plain (Raritan, Passaic) rivers. Given the variation of sediment yield relative to basin area and basin

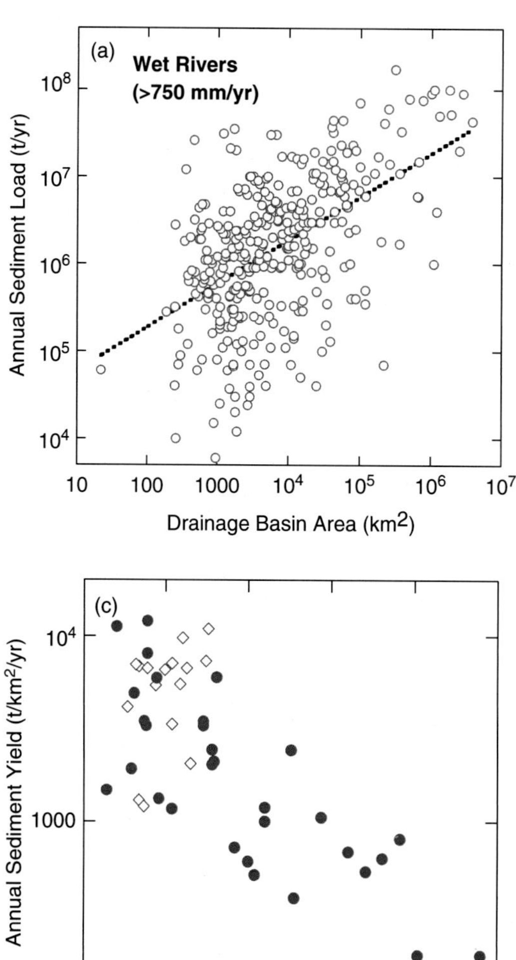

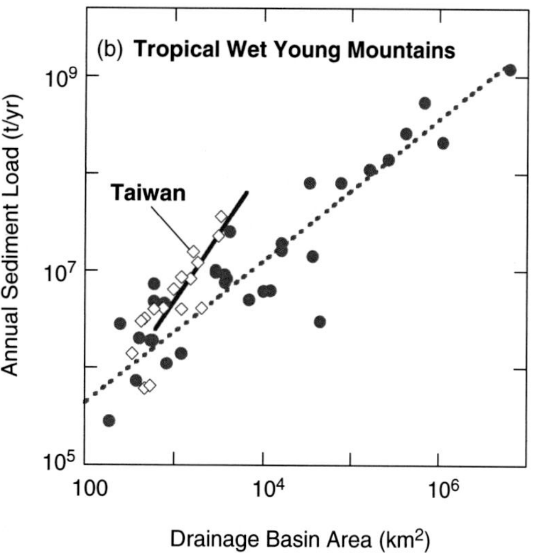

Figure 2.23. (a) Variation of sediment loads and yields vs. drainage basin area for all wet rivers (annual runoff >750 mm/yr). Sediment loads (b) and sediment yields (c) vs. drainage basin area for high-runoff rivers draining "young" tropical mountains, primarily rivers in southern Asia and Oceania, and several from northern South America, all areas with active tectonism. Taiwan rivers are highlighted because of their greater loads and yields, which may be related in part to the present-day tectonic activity on the island (Dorsey, 1988; Lundberg and Dorsey, 1990).

topography noted above, the wide range of sediment yields listed by Fournier (1960) should not be surprising. Perhaps the biggest problem with the Fournier dataset, however, is that to maximize his few data, he group-averaged the data, presenting average yields in 100-mm/yr-precipitation bins. But group-averaging can lead to potentially dangerous extrapolations, and if, as Walling and Webb (1983) pointed out, one river were added or removed, the resulting curve would have been much different. For example, the high sediment yields at 400–6000 mm/yr precipitation (Fig. 2.21b) in large part reflect the very high sediment yield of several low-runoff northern Chinese rivers, most notably the Huanghe (Yellow River). Equally important, Fournier did not take into account the roles of basin area, topography, and geology on sediment erosion, which, as we have shown above, collectively play significant roles in determining sediment load and yield. When Fournier's individual data points are plotted (Fig. 2.21b), in fact, the scatter appears to be almost random.

The lack of a simple relationship between annual precipitation and sediment yield is underscored by the large amount of scatter seen in a comparison of runoff and sediment yield (Fig. 2.22), although closer inspection does suggest a possible increase in sediment yield with increasing runoff: all rivers with annual sediment yields greater than ~10 000 t/km²/yr have mean annual runoffs >1000 mm/yr. Interestingly, however, there is no evidence of high yields at very low runoff, counter to the Fournier and Langbein/Schumm curves, even though many of the highest mean suspended-sediment concentrations are found in low-runoff rivers (Table 2.5). Runoff alone, therefore, appears to play only a relatively minor role in determining

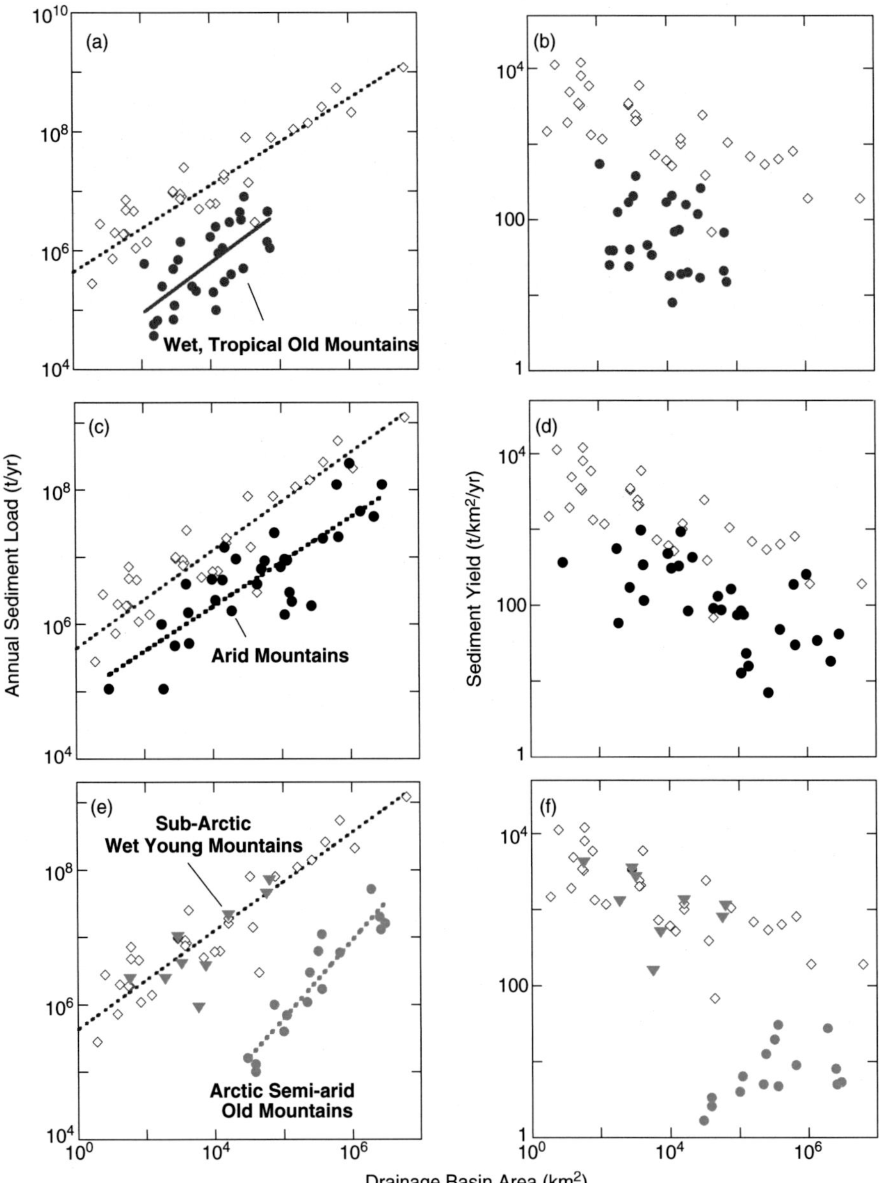

Figure 2.24. Comparison of sediment loads (left) and yields (right) of high-runoff tropical mountain rivers (open blue diamonds; see Fig. 2.23) compared to rivers draining high-runoff (wet) old tropical mountains (red dots; a, b), arid mountains (black dots; c, d), and semi-arid arctic (green dots) and high-runoff sub-arctic mountains(green triangles) (e, f).

the sediment yield of a river, 3% according to Syvitski and Milliman (2007). When taken in tandem with the other environmental drivers, however, runoff can become a key factor.

Collective impact of environmental controls

We can now view the cumulative roles of runoff, basin area, topography, and lithology on sediment erosion and transport by using two slightly different approaches: (1) by factoring in runoff to correlations we established above, and (2) by factoring in basin area, topography, and geology to runoff.

For the first approach, the sediment yields of high-runoff rivers (runoff >750 mm/yr) show a direct – but poorly defined – correlation with basin area (Fig. 2.23a). If we consider only tropical wet rivers that drain young (dominant lithology being Cenozoic sedimentary rocks) mountains in southeastern Asia and Oceania, as well as rivers draining northern South America, however, we see a two- to four-order-of-magnitude increase in sediment loads yields over a four-order-of-magnitude range in drainage basin areas (Fig. 2.23b, c). Although the R^2 of the power-function correlation between annual sediment load and basin area is 0.93 (Fig. 2.23b), there is some scatter among smaller rivers. Some of this scatter reflects a greater range in runoff (see Chapter 3) as well as greater impacts from human activity. Taiwanese rivers, whose watersheds periodically have very high typhoon-induced runoff and are impacted by human activity, for example, tend to have greater loads and yields relative to other tropical wet young mountain rivers (Fig. 2.23b, c) (see Kao and Milliman, 2008).

We can now compare the relation between sediment load–basin area for high-runoff mountain rivers whose lithologies are predominantly metamorphic or igneous; we term these wet, tropical old mountains. For this comparison we have chosen southern Asian (e.g. Maylasia), west Africa, and northeastern South America high-runoff rivers that drain primarily Paleozoic or Precambrian rocks. These wet, tropical old mountain rivers have sediment loads and yields one-to two-orders-of-magnitude smaller than similar-sized wet tropical rivers draining young mountains (Fig. 2.24a, b).

The role of precipitation can be seen by comparing arid (<100 mm /yr) rivers that drain young mountains with similar-sized wet tropical young mountain rivers. Arid rivers include those located in southern California and northwestern Africa and Eurasia; drainage areas range from 100 (San Juan Capistrano, CA) to 1 000 000 km^2 (Indus). While the scatter is considerable, arid mountain rivers tend to have roughly 5–10 times lower yield (Fig. 2.24c, d) than do wet young mountain rivers. The reason for the great range of values for arid rivers in part reflects the role of episodic events on arid rivers and the degree to which these events have been monitored, as discussed further in Chapter 3.

Extending this comparison, we now look at the combined influence of climate, morphology, and geology on sediment load and yield on high-latitude rivers – those draining the Eurasian Arctic (arctic/subarctic, semi-arid rivers eroding primarily pre-Mesozoic mountains) and rivers draining southeast Alaska and Iceland (subarctic, high-runoff rivers eroding primarily post-Paleozoic mountains). Cold high-latitude rivers should have lower loads than tropical rivers, as they flow more intermittently and some of their basins may be underlain by permafrost, which tends to retard erosion (Syvitski *et al.*, 2000; Syvitski and Milliman, 2007). Interestingly, however, southeast Alaska and Icelandic rivers have very similar sediment loads as similar-sized wet tropical rivers, whereas Eurasian Arctic rivers have two- to three- orders-of- magnitude smaller loads and yields relative to basin area. The reasons for this divergence seem obvious: topography (high, glaciated young mountains in Alaska vs. old mountains in the Eurasian Arctic), lithology (young sedimentary rocks vs. old crystalline rocks), and runoff

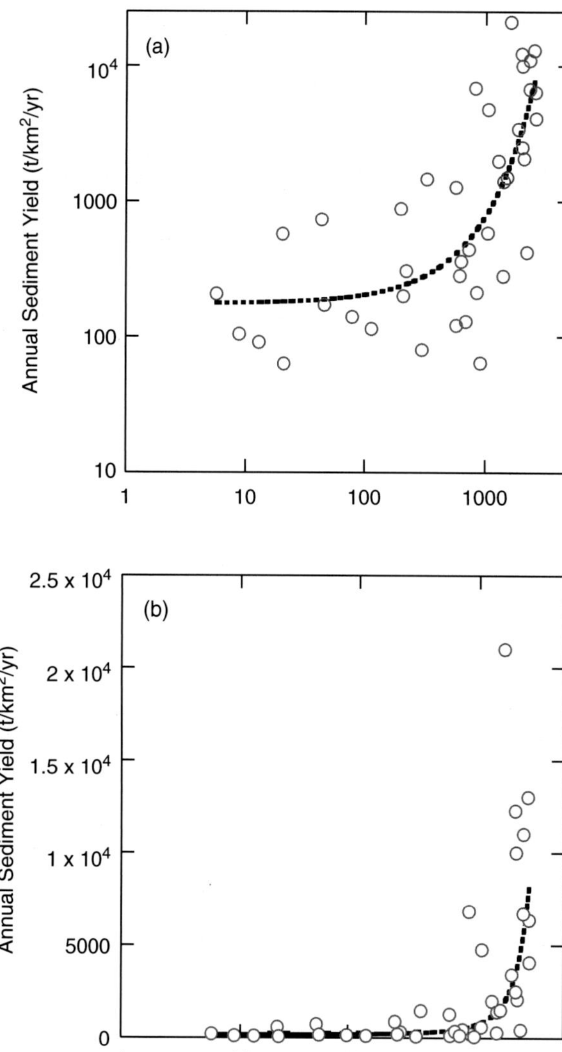

Figure 2.25. (a) Variation of sediment yields with fluvial runoff in small (1000–5000 km^2) tropical and semi-tropical young-mountain rivers. As most of the data come from rivers draining southern Asia and Oceania, the lithologies of the upper reaches in most of the drainage basins are presumably dominated by Cenozoic sedimentary rocks (Peucker-Ehrenbrink and Miller, 2003). (b) Sediment yields presented on a linear scale dramatizes the effect of runoffs greater than 1000 mm/yr.

(high-runoff Alaskan rivers vs. semi-arid Eurasian arctic rivers). Given the closeness of fit between Alaskan and tropical rivers, temperature appears to play a minor role in controlling sediment load.

While the plots in Figs. 2.23 and 2.24 show considerable promise in estimating sediment load, the amount of scatter in these various plots is not inconsiderable. Alternatively, if algorithms to calculate/estimate sediment loads are to be derived, they must include variables such as temporal and spatial patterns of precipitation, local geology (e.g. tectonic activity), and human activity (Syvitski and Milliman, 2007). As a first attempt, we look at rivers draining small mountainous watersheds (1000–5000 km^2), assuming that the differences in sediment storage in similar-sized drainage basins are relatively small. One reason to begin with small rivers is that they exhibit a wide range of fluvial runoff, from ~10 mm/yr to >4000 mm/yr. Also, owing to their small basin areas, geology/lithology tends to be more consistent and thus better defined. Most rivers shown in Fig. 2.24 drain young mountains in southeast Asia, Oceania and southern California, where the lithology is dominated by young (Mesozoic and Cenozoic) sedimentary rocks (Peucker-Ehrenbrink and Miller, 2003).

Even a casual glance at Fig. 2.25 shows the positive relation between yield and runoff. With the possible exception of the Santa Clara River (southern California), there is no evidence of a low-runoff river having a high sediment yield; we suspect that the Santa Clara's high yield (750 t/km^2/yr) may be partly in response to it draining the highly erosive Transverse Range (Inman and Jenkins, 1999; Warrick *et al.*, 2008) as well as the impact of episodic winter storms (Warrick and Milliman, 2003; see Chapter 3). Sediment yield for these small rivers increases as a power function to about 800–1000 mm/yr runoff; at higher runoffs it appears to increase exponentially. At a runoff of 1500 mm/yr, for example, yields can exceed 1000 t/km^2/yr, and beyond ~2000 mm/yr they can reach or exceed 5000 t/km^2/yr (Fig. 2.25b). Many of these watersheds are exposed to earthquakes and periodically intense rainfall events – typhoons in Taiwan, cyclones in New Zealand – both of which can accelerate landslide activity and thus landscape denudation (Hovius *et al.*, 2000; see Chapter 3).

Medium-size rivers with 20 000–80 000 km^2 drainage basin areas that discharge from tropical young mountains show somewhat similar runoff–yield relationships as small rivers, although some rivers (e.g. Sai-gon and Ma rivers in Vietnam), seem to have much lower yields than smaller rivers with similar runoffs (Fig. 2.26a). As with small rivers shown in Fig. 2.25, there is no indication of increasing sediment yield at lower runoff.

Owing to the general lack of data for many types of small rivers, we can use medium-size tropical young-mountain rivers as an internal standard to compare suspended sediment yield vs. runoff for other types of rivers. Medium-size tropical old-mountain rivers, for example, generally have lower yields, although a number of rivers (e.g. Damodar, Narmada, Ord) have yields similar to medium-size young-mountain rivers with similar runoffs (Fig. 2.26a). Most medium-size temperate young-mountain rivers have yields not dissimilar from tropical young-mountain rivers, whereas many temperate rivers draining older mountains have substantially lower yields (Fig. 2.26b). Medium-size subarctic and arctic rivers draining young and old mountains show markedly different runoff trends. The few medium-size subarctic and arctic young-mountain rivers (in Alaska) for which we have data have yields very similar those in tropical young-mountain rivers, whereas arctic old-mountain rivers with runoff <500 mm/yr have one- to two-order-of-magnitude lower yields (Fig. 2.26c).

Runoff thus appears to have a far greater influence on fluvial sediment yield than assumed by some workers. Topography and lithology clearly play critical roles, but drainage basin temperature may play a less defining role when precipitation and runoff are high, as witnessed by the close similarity between the high sediment yields of southeast Alaskan and Icelandic rivers and those draining tropical young mountains.

Global sediment flux

As mentioned at the beginning of this discussion, estimating the global flux of particulate solids to the global ocean is a daunting task. Not only are sediment data not as abundant nor as accurate as we would like (or need), but the impacts of on-going change – both natural and anthropogenic – hinder meaningful correlations. Moreover, the inverse relation between sediment yield and basin area (plus the fact that the sediment loads for most small rivers remain undocumented) means that – unlike water discharge and runoff – we cannot easily calculate a regional or, therefore, global sediment discharge and yields. Finally, it must be emphasized that for many rivers, particularly larger rivers with extensive floodplains and deltas, the sediment load measured upstream almost certainly is greater than the actual amount of sediment actually reaching the coastal ocean (Meade, 1996; Meade *et al.*, 2000). Goodbred and Kuehl (2000), for instance, have estimated that about 1/3 of the Brahmaputra measured sediment load is deposited along the lower reaches of the Bengal delta floodplain, and Xu *et al.* (2004) have calculated that about 10% of the Yellow River sediment load measured at the Lijin gauging station is deposited on the lower delta; 80% is deposited on or near the delta front.

Because small mountainous rivers generally have small floodplains and are more susceptible to floods, we assume that less sediment is sequestered in smaller drainage basins, but data that might confirm or deny this assumption remain elusive. Sediment transport also can be interrupted by lakes or (as seen in Chapter 4) man-made reservoirs (Vörösmarty *et al.*, 2003; Svyitski *et al.*, 2005). The Great Lakes in eastern North America, for example, trap most of the sediment load that ultimately would be transported by the St. Lawrence River. As a result, most St. Lawrence sediment comes from rivers that enter the St. Lawrence

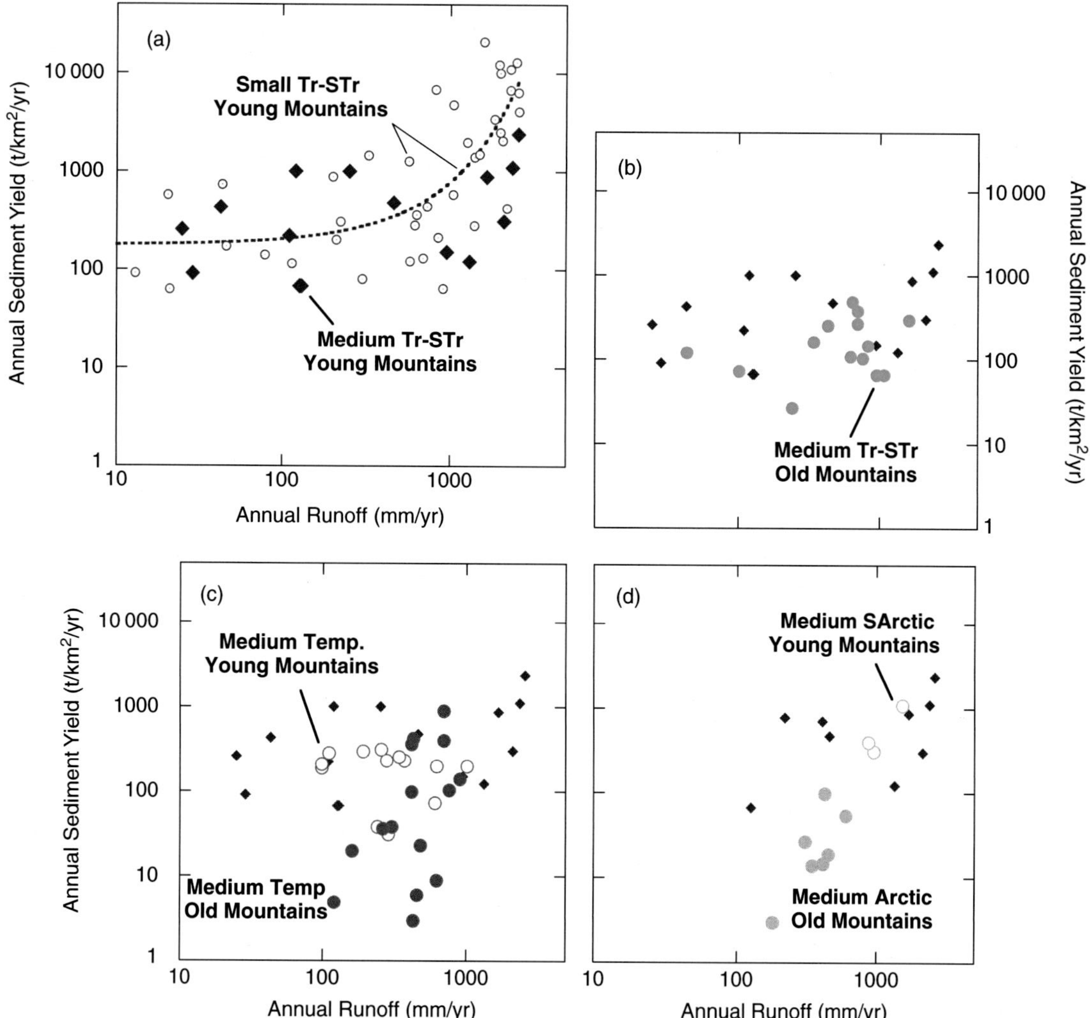

Figure 2.26. (a) Variation of sediment yield with fluvial runoff for tropical and sub-tropical young mountain rivers with small (1000–5000 km^2, blue dotted-circles) and medium-size (20 000–100 000 km^2, blue diamonds) drainage basins. The figures to the right and below show sediment yield vs. fluvial runoff of medium-size tropical old-mountain rivers (green dots; b); temperate young- and old-mountain rivers (red circles and dots, respectively, c); and arctic old-mountain and subarctic young-mountain rivers (orange dots and circles, respectively, d) compared with medium-size tropical young-mountain rivers (again portrayed as blue diamonds).

downstream of the seaward-most gauging station at Cornwall, Ontario (Rondeau *et al.*, 2000). Moreover, natural lakes and constructed reservoirs can provide ideal conditions by which nutrients and silicates are taken up by primary production, thus affecting dissolved as well as suspended loads (Humborg *et al.*, 1997; Conley *et al.*, 2000; see Chapter 4).

Despite these inherent problems, our analysis clearly shows the dependence of sediment yield on lithology (Fig. 2.28) and basin elevation (Fig. 2.27), as well as earthquake activity (Fig. 2.17), precipitation (Fig. 2.2), and basin area. Because small rivers can have far greater sediment yields than large rivers, a disproportionate amount of sediment enters the ocean from small mountain rivers (Milliman and Syvitski, 1992), which, we regret to say, recently have been relegated by some workers to the acronym SMR.

Although our discussion here centers on inorganic fluxes, it should be noted that the yield and character of fluvially transported particulate organic carbon (POC) also varies greatly with basin size (Lyons *et al.*, 2002) and sediment yield. Blair *et al.* (2003) and Leithold *et al.* (2006) have found that POC in small low-yield rivers tends to be dominated by modern plant material whereas the POC in small high-yield mountainous rivers tends to be older, reflecting derivation from eroded soils and rocks rather than living plants. This idea, however, has been recently challenged by Drenzek *et al.* (2009), who suggest that pre-aged terrestrial

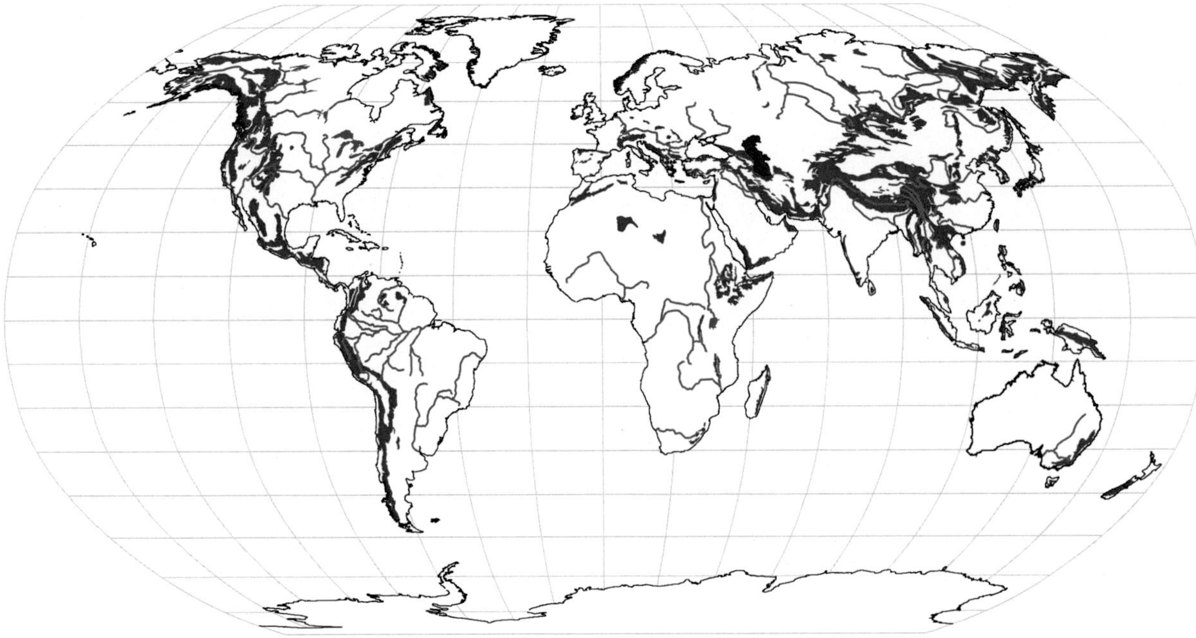

Figure 2.27. Global distribution of mountainous terrain (elevations >1000 m).

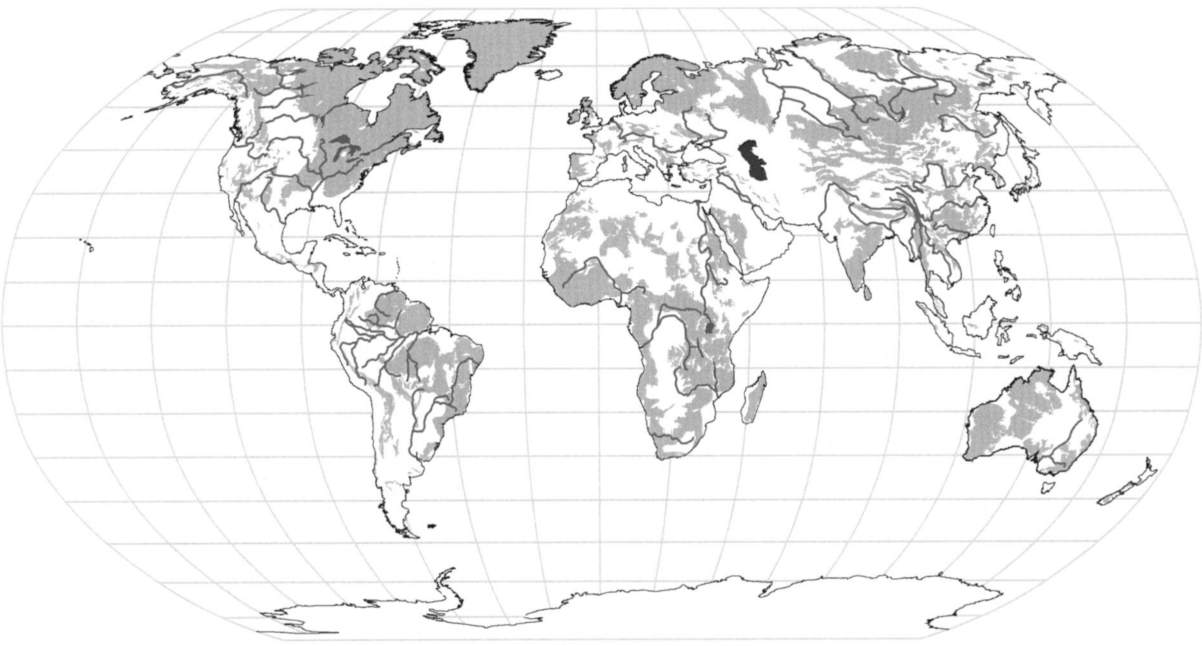

Figure 2.28. Global distribution of pre-Mesozoic surficial rocks.

carbon rather than petrogenic carbon may account for a considerable portion of total organic matter in active margin sediments.

One way to calculate sediment flux to the coastal ocean is to compute the basin areas of rivers draining a specific area in which geological and environmental conditions (especially precipitation) are more or less constant. Using an algorithm that relates measured sediment loads for rivers in that region to basin area, one can estimate collective sediment discharge from that area. Using this technique, Milliman (1995) calculated that the rivers draining the island of New Guinea collectively discharge

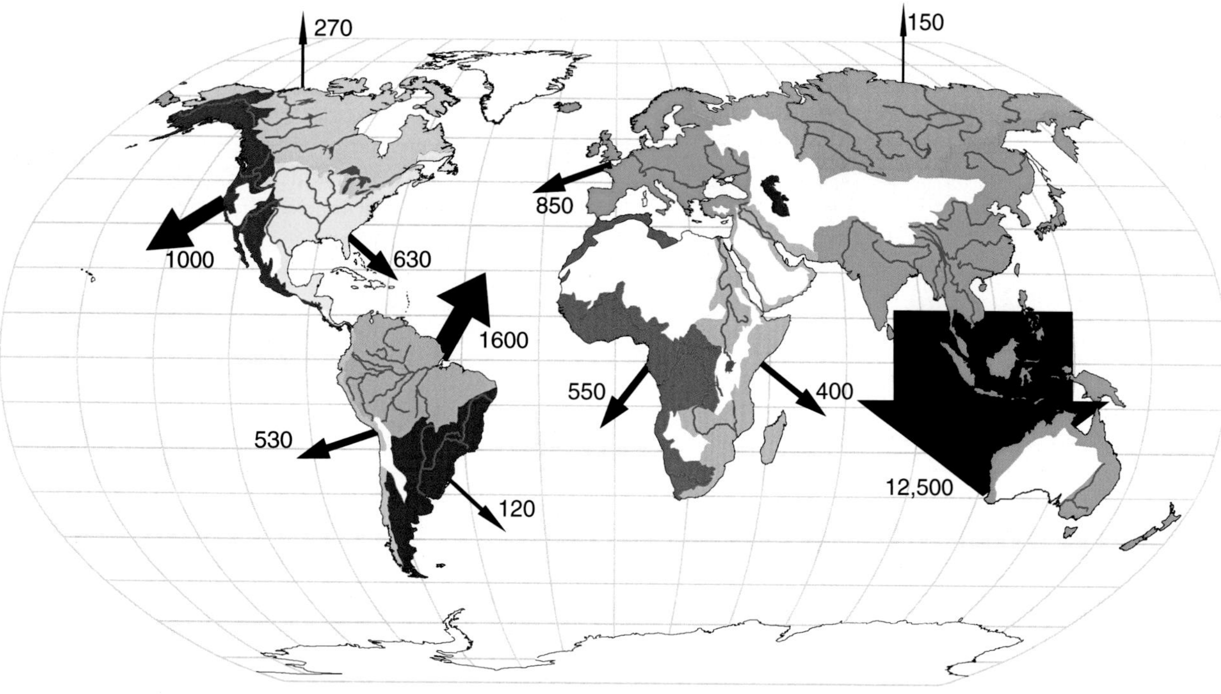

Figure 2.29. Annual discharge of fluvial sediment to the global coastal ocean.

1.7 Bt of sediment annually, about the same amount as the cumulative loads discharged by all North American rivers. Milliman *et al.* (1999) used a similar technique to show that the 465 largest rivers draining the six large islands in the East Indies (including New Guinea) collectively deliver 4.2 Bt/yr. We have used here a similar method to calculate regional sediment discharge, from which global numbers in Fig. 2.29 were derived. It should be pointed out that, as in previous figures, we use, where possible, pre-dam sediment loads.

Summing the numbers, we calculate that total sediment discharge to the global ocean is 19 Bt/yr (Fig. 2.29). This value is larger than the 16 Bt reported by Milliman and Meade (1983) and the 14 Bt/yr estimated by Syvitski *et al.* (2005). This number reflects the very high yields from southeast Asia/Oceania as well as western North and South America (Fig. 2.30). Sediment derived from southeast Asia and Oceania account for about 2/3 of the global sediment delivery (although Indonesian river data are notoriously suspect). In contrast, rivers draining pre-Mesozoic terrain have a mean annual sediment yield of 75 t/km^2/yr, less than half the average global annual yield of 190 t/km^2/yr. While Figs. 2.29 and 2.30 show the uneven global distribution of sediment erosion and delivery to the coastal ocean, regional differences can be even more stark, as discussed in the following sections.

Regional fluxes of fluvial sediment

North and Central America

Because of the mountainous terrain, young erodibile rocks, and continuous heavy precipitation, rivers draining southeastern Alaska, with annual sediment yields approaching 1000 t/km^2/yr, account for nearly 25% (650 Mt/yr) of the sediment discharged from North and Central American rivers (Fig. 2.31a). Rivers draining to the Gulf of Mexico contribute almost as much sediment (500 Mt/yr), but they drain more than seven times the area as southeastern Alaskan rivers. The relatively large sediment flux to the Canadian arctic (190 Mt/yr) mostly comes from the MacKenzie (120 Mt/yr), second largest North American river. The low sediment flux from eastern North America reflects the relatively low-elevation, old mountainous landscape that they drain. Rivers draining much of northeastern Canada have annual sediment yields <10 t/km^2/yr.

South America

Not surprisingly, watersheds draining the northern Andes, which are characterized by mountainous topography, relatively young, erodible rocks, and heavy precipitation, are a major source to the global sediment budget. Although the Amazon, Orinoco, Magdalena have relatively low annual yields (<200 t/km^2/yr; Fig. 2.31b), in part owing to their extensive floodplains, collectively

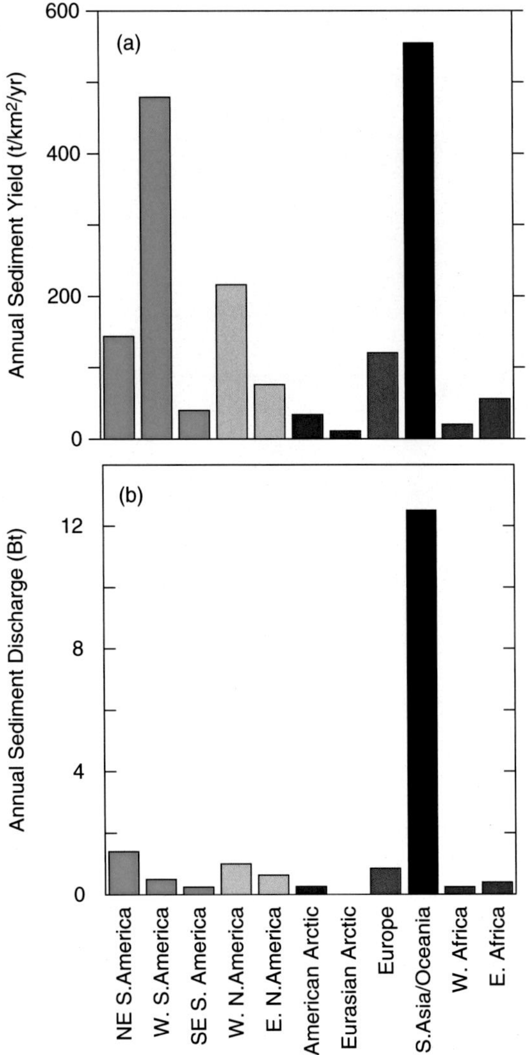

Figure 2.30. Fluvial sediment yields (a) and sediment discharges (b) to the global ocean from the various regions shown in Fig. 2. 29.

they discharge 1.4 Bt/yr of sediment. Rivers draining the Guiana Precambrian shield, in contrast, have low annual yields (<100 t/km^2/yr) and low sediment loads. Northwestern rivers have much higher annual yields, locally >1000 t/km^2/yr, the result of very high precipitation (Fig. 2.2) and steep relief (Fig. 2.28); collectively northwest rivers discharge an estimated 280 Mt/yr of sediment to the equatorial eastern Pacific. Southern Chilean rivers also have high runoff, but our estimate of 200 Mt/yr of discharge sediment is based more on a guess than the few available data. Rivers in eastern and southern Brazil, much of Argentina, and southern Peru and northern Chile, with their low runoffs and generally low-lying watersheds, offer a stark contrast to the northern and northwest rivers; annual sediment yields are generally < 100 t/km^2/yr and their cumulative sediment load is < 200 Mt/yr (Fig. 2.31b).

Europe

In terms of rivers and the terrain that they drain, Europe can best be viewed as two separate continents. Most northern European rivers flow primarily through low-lying terrain, and nearly every river has been dammed, channelized or otherwise "managed" by human intervention. The annual yield from most of these rivers is <10 t/km^2/yr (Fig. 2.31c), some Swedish and Finnish rivers having annual yields of ~1–3 t/km^2/yr (Table 2.5). Collectively, northern European rivers discharge ~45 Mt of sediment annually, equivalent to the cumulative load of Icelandic rivers, whose relatively high annual yields (>300 t/km^2/yr) reflect small drainage basins, erodible young mountains, and high runoff.

The major sediment flux from Europe comes from the rivers draining into the Mediterranean Sea off the southern slopes of the Alps (Fig. 2.31c). Runoff tends to be high, characterized by brief but intense precipitation events. Moreover, rocks are easily eroded, intensified by poor land conservation over the centuries (see Chapter 4). Drainage basins also are generally small, thus minimizing floodplain deposition. An extreme example of the southern European rivers is Albania's Semani River (a name perhaps recognized only by the most fervent fluviophile). Prior to being dammed, this small river (5600 km^2) discharged 30 Mt/yr of sediment; in fact, pre-dam sediment annual yields from Albanian rivers averaged 2800 t/km^2/yr, two- to three-orders-of-magnitude greater than for northern European rivers. Rivers draining into the southeastern Black Sea have annual sediment yields an order-of-magnitude or more higher (400–500 t/km^2/yr vs. 4–40 t/km^2/yr) than the much larger Don and Dniepr rivers to the north (Fig. 2.31c; Milliman *et al.*, 2010).

Africa

Although the mean elevation of Africa is higher than any other continent, and the names of many of its rivers are easily recognised – the Nile, Congo, Niger, and Limpopo – the sediment yield from African rivers is far lower than for any other continent save Australia (and, presumably, Antarctica). The collective pre-dam sediment yield of the Congo, Nile, Niger, Zambesi, and Orange, which together drain > 40% of the African continent, was only 25 t/km^2/yr. Pre-dam yields from large secondary rivers (Volta, Sanaga, Rufiji, Limpopo, and Senegal) were only slightly higher, ~60 t/km^2/yr. Comparing the Congo and Nile with the previously mentioned southern European rivers illustrates the point: pre-dam loads of the Vijose and Semani rivers (59 Mt/yr) in Albania equaled the Congo's annual load (~60 Mt/yr) and was half that of the pre-dam Nile (120 Mt/yr) even though they drain < 1% of the areas drained by the Congo or the Nile. The low yields from

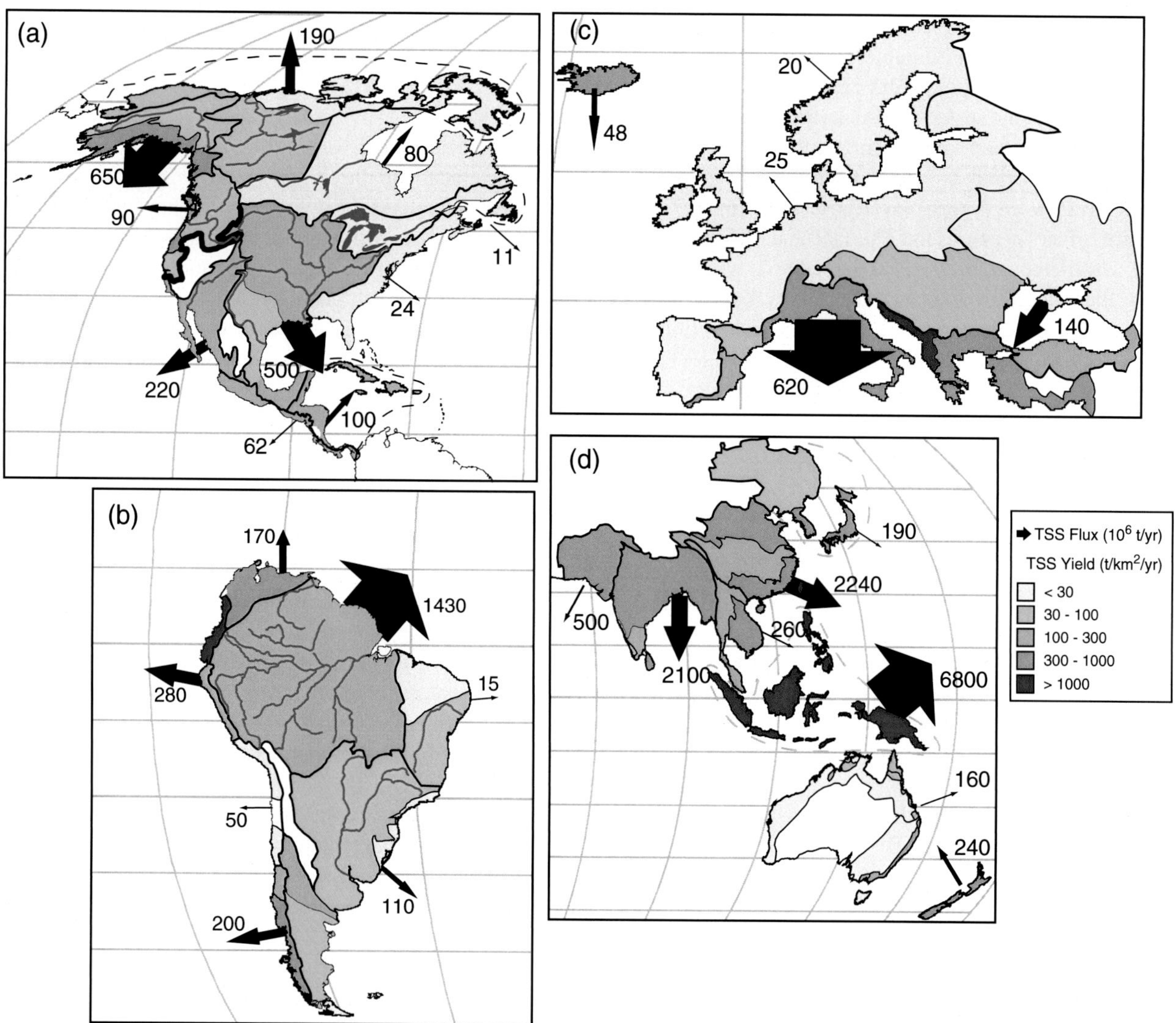

Figure 2.31. Calculated regional yields and fluvial deliveries of sediment from (a) North American, (b) South American, (c) European, and (d) Austral-Asian rivers to the coastal ocean. Annual sediment yields for the various polygons are color-coded, from <30 to >1000 t/km². Arrows in each panel are proportional to sediment discharge within that area. Where sediment data are meager, we have extrapolated yields and loads from adjacent polygons. Large areas with similar physiographies, geology, and climates are bordered by black lines. Africa and the Eurasian Arctic are not shown because of the general uniformity of their physiography, geology, and climate as well as low sediment discharge (see Figs. 2.29 and 2.30).

Africa stem primarily from the generally gentle landscape, arid climate, and old cratonic lithologies that dominate many of the watersheds.

A disproportionate amount of sediment from Africa comes from the high-yield rivers in Morocco and Algeria and in Madagascar (regional annual yields >300 t/km²/yr), the result of steep, high mountains, relatively young, erodible rocks, periodically heavy rains, and, long-term human impact (McNeill, 1992). Probst and Amiotte-Suchet (1992) estimated that pre-dam northwest African rivers discharged about 500 Mt/yr, nearly 20% of the sediment discharge by all African rivers. Madagascar rivers might discharge a similar amount of sediment, but the data are too scarce and problematic to allow us to even guess at their actual sediment delivery.

Southern Asia and Oceania

Given the discussion in the preceding pages, it should come as no surprise that the rivers draining Indonesia and the Philippines are particularly important in terms of basin erosion and sediment discharge, accounting for more than half of the sediment discharged from Asian/Oceania rivers (Fig. 2.31d). Runoff from mountains dominated by Cenozoic sedimentary and volcanic bedrock (Peucker-Ehrenbrink and Miller, 2003) result in sediment yields higher than 1000 t/km^2/yr throughout the Philippines, Indonesia, and Taiwan. In Taiwan, where Cenozoic sedimentary rocks account for more than 85% of the surface bedrock (Peucker-Ehrenbrink and Miller, 2003; Kao and Milliman, 2008), the average annual sediment yield is ~9000 t/km^2/yr, and locally it exceeds 50 000 t/km^2/yr per year (Kao and Milliman, 2008). In contrast, rivers in peninsular Thailand and Malaysia drain deeply weathered granites (Douglas, 1996), which helps explain their relatively low annual yields (<300 t/km^2/yr). Collectively, the high-standing islands of Oceania (including New Guinea) discharge an estimated 6.8 Bt of sediment annually to the coastal ocean (~40% of the global total) even though these islands account for less than 2.5% of the global land area draining into the ocean.

On a more local scale, however, yields and loads can vary greatly, as seen in Taiwan, where yields for various rivers vary by nearly two orders of magnitude, reflecting local variations in gradients, earthquake frequency and intensity, lithology, and precipitation (Kao and Milliman, 2008). By contrast, the low annual sediment yields (<100 t/km^2/yr) from northeastern Asian and Australian rivers are a function of low watershed relief, old rocks, and low runoff. Wasson *et al.* (1996) calculated that although sheet, rill, and gully/channel erosion in Australia accounts for 28 Bt sediment removal annually, sediment delivery to the ocean is only ~0.3 Bt/yr, reflecting the transport-limited nature of Australian rivers (Olive and Rieger, 1986). New Zealand rivers, although much smaller in size, discharge about 50% more sediment than Australian rivers, the result of a wet climate and tectonically active mountains.

Himalayan rivers, stretching from the Indus River in the west to the Yellow River in the East, discharge an additional 5.1 Bt/yr of sediment (pre-dam values), 1.1 Bt of this total by the pre-1990 Yellow River; see Chapter 4 for a discussion of the Yellow River and its dwindling sediment load. Because of their much larger watersheds – and thus more extensive floodplains – Himalayan rivers have much lower sediment yields than Indonesian rivers; pre-dam yields for the Indus and Yangtze were ~300 t/km^2/yr.

Eurasian Arctic

Having three of the largest rivers in the world – Ob, Yenesei, and Lena – Eurasian Arctic rivers drain 14.4 × 10^9 km^2 of land area, but they discharge only ~150 Mt/yr of sediment (Figs. 2.29 and 2.30). Gordeev (2006) reports an even smaller flux: 100 Mt/yr, equating to an annual sediment yield of 7.5 t/km^2. Yields of the Ob, Yenesei, Lena, and Kolyma (total drainage basin area 8.8 × 10^6 km^2) average only 6 t/km^2/yr. Yields for medium–large Russian Arctic rivers (individual drainage basins >100 000 km^2) are slightly higher, 13 t/km^2/yr. The collective effect of low rainfall, relatively gentle topography, and old crystalline rocks, as well large watersheds and extensive floodplains, helps explain these low yields.

Summary

We calculate that rivers discharge about 19 billion tons (Bt) of sediment annually to the coastal ocean (Figs. 2.29), equating to a global annual sediment yield of 190 t/km^2/yr. This is the same as the global sediment load calculated by Beusen *et al.* (2005) using a multiple linear regression model, but about 1/3 higher than that estimated by Syvitski *et al.* (2005). Our estimate, however, comes with several disclaimers. As stated previously, our estimates are based on sediment fluxes measured at seaward-most gauging stations, which in some cases may lie hundreds of kilometers upstream from the ocean. Floodplains between the gauging station and the ocean provide likely sinks for downstream sediments. Secondly, to obtain an understanding of geomorphic and environmental factors responsible for erosion and discharge of sediment, we have used, where possible, pre-dam values, which do not represent present-day conditions. These yields, however, also reflect human-induced erosion, induced by such activities as deforestation, farming, mining, etc. (see Chapter 4). Thirdly, the values represent mean annual values; depending on climatic drivers and events, such as typhoons, droughts or earthquakes, regional discharge can vary greatly from year to year, as shown for the Santa Clara River in Fig. 1.6 and discussed further in Chapter 3.

Although our calculated global sediment delivery is almost exactly that computed by Holeman (1968) 40 years earlier and recently by Peucker-Ehrenbrink (2009) using a similar database as ours, and somewhat higher than the 15 Bt/yr suggested by Milliman and Meade (1983) (Table 2.8), the general similarity in these estimates hides several important differences. Because of his limited database, Holeman calculated the average sediment yield of Asian rivers from south Asian rivers, and then estrapolated these yields over the entire Asian continent, apparently unaware of the very low sediment yields from the very large Arctic rivers. As a result, Holeman's estimated sediment flux from mainland Asia (14.5 Bt/yr) to be about three times greater than our estimate (5.3 Bt/yr). Counterbalancing these high discharges from mainland Asia, Holeman lacked data from rivers draining high-standing islands in Oceania and thus did not include them in his budget (Table 2.8). Finally, the lack of data from western North and South America and from rivers draining the southern Alps led both Holeman and Milliman and Meade (who concentrated their analysis on rivers delivering >15 Mt/yr) to underestimate fluxes from these regions. Peucker-Ehrenbrink (2009), on the other hand, calculates South America to have greater (3600 Mt/yr) and Africa lower (1000 Mt/yr) sediment discharges than we have calculated (2300 Mt/yr and 1500 Mt/yr, respectively).

Table 2.8. *Comparison of estimated fluvial sediment fluxes by Holeman (1968), Milliman and Meade (1983), and this book. Loads expressed in Mt/yr.*

	Holeman	Milliman and Meade	This book
North/Central America	1800	1500	1900
South America	1100	1800	2300
Europe	290	230	850
Africa	490	530	1500
Eurasian Arctic	–	84	150
Asia	14 500	6300	5300
Oceania	–	3100	7100
Totals:	18 300	13 500	19 100

As far as we know, no global sediment budget has taken into account the impact of glacial erosion in high-latitude landmasses, particularly Greenland and Antarctica. Annual sediment delivery from wet-base glaciers, as noted by Gurnell *et al.* (1996), can range from 100 to 10 000 t/km^2/yr, depending in part on catchment area. A review of the scattered data, mostly representing only a few months of measurements, suggests that Greenland may be a main contributor of sediment to the North Atlantic Ocean (Hasholt, 1996). In contrast, sediment yields from rivers draining cold-base glaciers (e.g. Bayelva River in Svalbard) appear to be significantly lower than from rivers draining temperate glaciers, presumably because of the absence of meltwater access to the glacier bed (Hodson *et al.*, 1998).

Perhaps the greatest difference between many earlier estimates and our present understanding of global sediment flux is the amount of fluvial sediment discharged to active margins (Milliman and Syvitski, 1992). Rivers draining western North and Central America, for instance, with watersheds representing <30% of the their respective continental landmasses, deliver >50% of the sediment leaving those continents (Fig. 2.31a). This difference has important ramifications regarding the fate of discharged sediment. Passive margins tend to have wide continental shelves (e.g. Yellow and East China seas), often bordered by large estuaries (e.g. Baltic Sea and Chesapeake Bay). As a result, during sea-level high stands, much of the sediment delivered by passive-margin rivers, such as the Amazon, Orinoco, and Yangtze, can be deposited in estuaries or the inner and middle continental shelf. Sediment may only escape to deeper water during falling sea level or if the sediment wedge progrades seaward to the shelf edge (e.g. present-day Mississippi). In contrast, many rivers draining active-margins tend to discharge directly onto narrow shelves, often by-passing their small or non-existent estuaries. Moreover, because smaller active-margin rivers tend to be more event-driven (see Chapter 3), flood-generated fluvial discharge can escape to or beyond the outer shelf.

Dissolved solids

Introduction

The database for dissolved solids and their chemistry is more sparse than that for suspended sediments. For most intents and purposes, Clarke's seminal compilations and discussions (Clarke, 1924a, b) remained the most accessible and widely used database until Livingstone's (1963) updated summary in 1963. But the database remained geographically uneven. Livingstone, for example, presented two tables with dissolved solid concentrations for Japanese rivers and lakes, but no data for the Ganges, Brahmaputra, Mekong, Red, Yangtze, Yellow, or Irrawaddy rivers (even now we have few data for the Irrawaddy).

In the past 40 years published dissolved data have increased almost exponentially, reflecting technological and analytical advances, increased access to previously remote study sites, and heightened societal relevance (e.g. chemical pollution). Clarke's and Livingstone's efforts have been relegated primarily to the historical, although still valuable, summaries. The level of debate regarding chemical weathering has been heightened in recent years because of the obvious role of CO_2 (and its central role in the global-warming debate; see Molnar and England, 1990; Raymo and Ruddiman, 1992; Krishnaswami and Singh, 2005) in the weathering of Ca and Mg silicates and carbonates, the so-called "Urey Reaction" (Urey, 1952):

$$CaSiO_3 + CO_2 \rightleftharpoons CaCO_3 + SiO_2$$
$$MgSiO_3 + CO_2 \rightleftharpoons MgCO_3 + SiO_2$$

Since publication of Gibbs's classic work in 1967, for example, the number of papers dealing with Amazon River (and Orinoco) geochemistry alone has approached 100 (see Stallard and Edmond, 1983; Edmond *et al.*, 1995; Gaillardet *et al.*, 1999). Similarly, V. S. Subramanian and his colleagues and students have studied and summarized fluvial characteristics for most of the major rivers draining the Indian subcontinent, and there have been

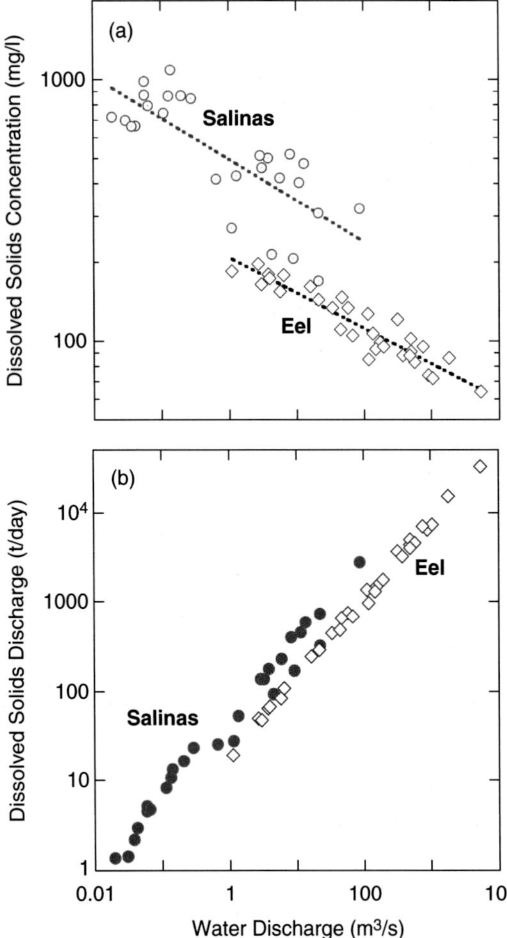

Figure 2.32. Concentrations (a) and discharges (b) of dissolved solids from two California rivers with relatively similar drainage basin areas but markedly different runoffs and annual dissolved-solid discharges, the Salinas River at the Spreckels gauging station (10 600 km^2 upstream basin area, 27 mm/yr runoff, mean annual dissolved solid discharge 0.17 Mt/yr) and the Eel River at Scotia (8000 km^2, 810 mm/yr, 1 Mt/yr). Because water discharge varies by 4 orders-of-magnitude, compared to 2- to 10-fold range in dissolved-solid concentrations, dissolved-solid loads primarily reflect water discharge. For any given discharge the Salinas River's dissolved-solid concentrations tend to be a factor of 2–5 higher that the Eel's, but the Eel's 30-fold greater water discharge equates to a 6-fold greater annual dissolved-solid discharge. Data from USGS NWIS.

similar comprehensive studies in northern Russia (Huh *et al.*, 1998; Gordeev, 2000; Rachold *et al.*, 2003), the Ganges–Brahmaputra (Galy and France-Lanord, 2001; Krishnaswami and Singh, 2005; Hren *et al.*, 2007), and many Chinese rivers (see Zhang *et al.*, 1992, 1994), as well as important global summaries by Michel Meybeck (Meybeck, 1979, 1988, 1994; Meybeck and Ragu, 1996). The US Geological Survey has systematically sampled

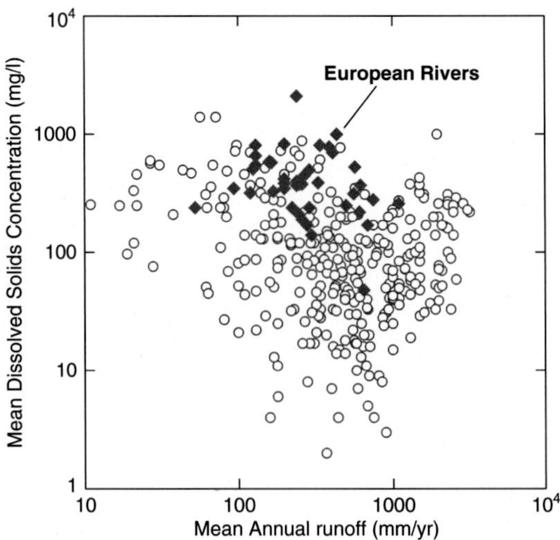

Figure 2.33. Mean concentration of dissolved solids vs. mean runoff for 300+ global rivers. Note the elevated concentrations of European rivers (red diamonds).

many US rivers for more than 40 years, and although the effort has slackened in recent years, the complete USGS database is available on the internet (http://waterdata.usgs.gov/nwis/qw). Similarly, geochemical data for many larger rivers can be accessed on the GEMS website (http://www.gemswater.org/). The most complete geochemical database, however, remains the one compiled for UNEP by Meybeck and Ragu (1996).

Before beginning our discussion, one important difference between suspended and dissolved solids should be noted. Whereas the concentration of suspended sediments in any river system tends to increase with river flow (Fig. 2.12b), dissolved-solid concentrations (*Qdc*) become diluted as flow increases (Fig. 2.32a), although dissolved discharge tends to increase with increased flow (Fig. 2.32b). As with suspended sediment discharge, a reliable dissolved solid rating curve requires measurements taken throughout the year, during both low and high flow. Calculating total dissolved solid (TDS) from the Salinas or Eel River based only on measurements taken during low or high flow, for instance, could result in concentrations or discharges far too high or too low (Fig. 2.32). In some arid rivers, declining dissolved-solid concentrations at lower runoffs may reflect the decreased role of interstitial water in arid climates (Dunne, 1978). Total dissolved solid concentrations for global rivers show no obvious trend with runoff (Fig. 2.33), as previously noted for HCO_3 and Cl by Holland (1978) and Kump *et al.* (2000). European rivers, however, tend to have elevated dissolved solid concentrations relative to runoff, a point discussed later in this chapter and in Chapter 4.

Table 2.9. *Highest and lowest fluvial dissolved-solid concentrations (in bold) in decreasing order; global mean is 100 mg/l. Note the number of high-concentration European rivers, in large part reflecting chemical pollution (see Chapter 4). Low-concentration rivers mostly drain shield-dominated rocks, primarily in high latitudes in Canada, Scandinavia, and Russia.*

River	Country	Area (× 10³ km²)	Elevation	Climate	Runoff (mm/yr)	Dissolved load (× 10³ t/yr)	Dissolved yield (t/km²/yr)	Dissolved conc. (mg/l)
Severn	England	6.8	Upland	Te-H-W	380	8.7	1300	**3400**
Weser	Germany	46	Upland	Te-SA-W	240	3.6	565	**2100**
Taizhe	China	25	Upland	Te-A-S	72	2.5	100	**1400**
Huanghe	China	750	High Mt	Te-A-S/Au	57	21	28	**1400**
Kodori	Georgia	2	Mountain	Te-W-S	1950	0.39	195	**1000**
Tevere	Italy	17	Mountain	Te-H-W	435	7.5	440	**1000**
Krishna	India	260	Mountain	Tr-A-S	260	22	85	**880**
Jucar	Spain	22	Upland	Te-A-W	200	1	45	**830**
Sakarya	Turkey	57	Mountain	Te-SA-W	98	2.9	51	**810**
Guadalquivir	Spain	56	Mountain	Te-SA-W	130	5.9	105	**810**
Rhine	Netherlands	220	Mountain	Te-H-C	340	60	270	**810**
Global								**100**
Dramselv	Norway	17	Mountain	SA-H-S	590	0.07	4	**7**
Attawapiskat	Canada	50	Lowland	SA-H-S	400	0.14	3	**7**
Quoich	Canada	29	Lowland	A-A-S	180	0.03	1	**6**
Fuchunjiang	China	54	Lowland	Te-H-S	685	0.18	3	**5**
Ellice	Canada	17	Upland	A-SA-S	160	0.01	0.6	**4**
Tauyo	Russia	25	Mountain	SA-H-S	440	0.05	2	**4**
Coppename	Suriname	20	Upland	Tr-W-S	750	0.06	3	**4**
Skiensvassdraget	Norway	10	Mountain	SA-W-S	900	0.03	3	**3**
Kola	Russia	38	Lowland	A-H-S	370	0.03	1	**2**

Environmental controls on dissolved-solid discharge to the coastal ocean.

The quantity and quality of dissolved solids in river water depend on the same environmental factors that dictate the discharge of suspended solids: climate, basin relief, basin geology, and drainage basin area. Highest average concentrations occur in European and Eurasian rivers, whereas lowest concentrations are found primarily in high-latitude rivers (Table 2.9). Which environmental factor plays a greater role, however, continues to be debated.

Citing Arrhenius's law – the dependence of reaction rates on temperature – and his own GEOCARB model, Robert Berner (see Berner, 1994; Berner and Berner, 1987) argued for primacy of climate in chemical weathering. John Edmond, in contrast, favored the dominant role of bedrock lithology (Edmond and Huh, 1997). Both schools of thought have had their advocates, some workers using elements of both the climatic and lithologic arguments. Bluth and Kump (1994), while favoring climatic control, pointed out the importance of physical breakdown of rock and removal of overlying soil in promoting rapid chemical breakdown. Dunne (1978) showed that the rate of chemical denudation in Kenyan river basins increases as a power function of runoff, and, equally important, that the percentage of dissolved silicate increases with runoff. Riebe *et al.* (2001a), in contrast, observed that chemical weathering of granites in the Sierra Nevada is closely dependent on the rate of physical erosion (and thereby inferring a tectonic control), but has little correlation with climate. White and Blum (1995) concluded that SiO_2 and Na fluxes are a function of temperature, precipitation and runoff, such that warm wet watersheds have very rapid weathering rates; K, Ca, and Mg delivery reflects ion exchange, nutrient recycling, and rock lithology. Anderson *et al.* (1997) showed that chemical weathering in glaciated regions can be as great as in warmer climates, presumably reflecting the abundance of ice-produced fine-grained rock flower as well as high runoff (Tranter, 2004), and Huh (2003) similarly failed to find any temperature dependence on silicate weathering. Kump *et al.* (2000) also pointed out the importance of runoff rather than temperature in determining the rate of chemical erosion rates.

Plants also play an important role in chemical weathering through nutrient exchange, acid production by respired CO_2, and the physical reworking of soils by plant roots (Bluth and Kump, 1994), roles that clearly increased during

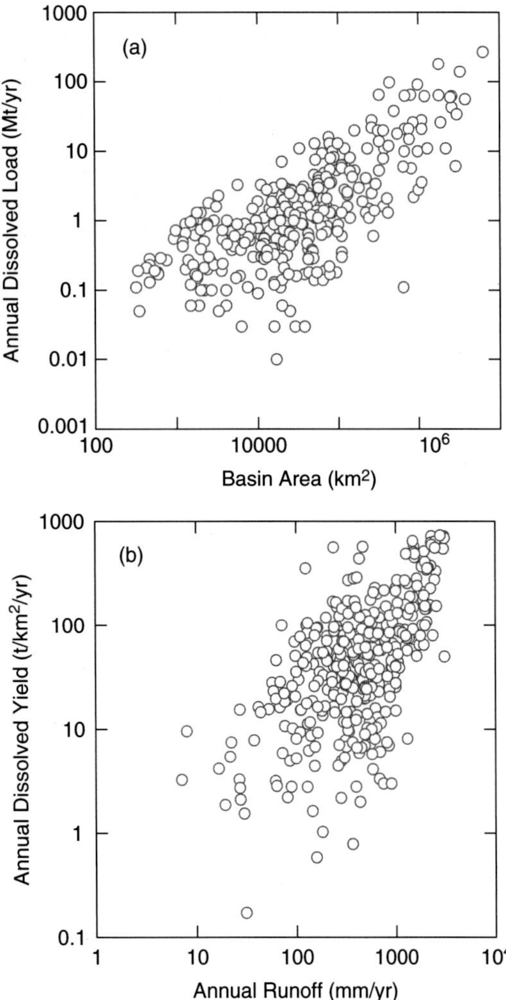

Figure 2.34. Variations of annual dissolved-solid discharge with basin area (a) and of annual dissolved-solid yield with runoff (b).

the Phanerozoic (Drever, 1994; Kelly *et al.*, 1998; Berner, 2004; Berner *et al.*, 2003; Balogh-Brunstad *et al.*, 2008, and references therein).

The importance of considering multiple environmental factors becomes obvious when we look at total dissolved load compared to basin area (Fig. 2.34a) and runoff (Fig. 2.34b). The annual dissolved loads from the 400 rivers listed in our appendices range from ~0.01 (Ellice River, Canada) to 270 Mt/yr (Amazon River) (Table 2.10). Although large rivers tend to have higher dissolved-solid discharge than small rivers, the scatter is considerable. Dissolved yields, which range from 1 to 600 t/km^2/yr (Table 2.11), tend to increase with increasing runoff (Fig. 2.34b) as well as with the level of human intervention (Chapter 4). The annual yield of dissolved solids from the Amazon is ~ 40 t/km^2 compared with >500 t/km^2/yr for some Taiwanese rivers.

By utilizing morphology, basin area, geology, and runoff, as was done with sediment loads and yields earlier in this chapter, we find an excellent relationship between basin area and both dissolved load and yield for rivers draining tropical wet young mountains (Fig 2.35). In this case the rivers range in size from the very small rivers/streams draining northern Papua New Guinea (Raymond, 1999) to the Amazon and Brahmaputra rivers, five orders of magnitude range of basin area.

Some rivers that drain tropical wet old mountains have dissolved loads similar to or, in some cases, greater than similar-size young mountain rivers; but others have considerably lower loads and yields (Fig. 2.36a, b). Although dissolved loads and yields vary greatly, most rivers draining arid mountains carry about an order of magnitude less dissolved solids than similar-size tropical wet young mountain rivers (Fig. 2.36c, d). In subarctic and arctic rivers we find a similar relation as we do for sediment loads and yields: semi-arid rivers draining the older mountains of the Eurasian and Canadian Arctic have an order of magnitude lower dissolved yields (see Millot *et al.*, 2003) than rivers draining the wet, young mountains of southeastern Alaska and Iceland, which have dissolved values closely approximately those of similar-sized tropical wet rivers (Fig. 2.36e, f).

Using a similar approach to that used in our discussion of suspended sediment, if we hold basin area, morphology, and temperature more or less constant, we note a log-linear relation between runoff and dissolved yield (Fig. 2.37). Despite the scatter in the plots, dissolved yields of mountainous and non-mountainous rivers are not too dissimilar, in contrast to suspended sediment yields. The wide scatter in dissolved yields, however, underscores the strong influence of rock type in chemical weathering, as discussed later in this chapter. Dissolved yields for many high-runoff temperate and arctic rivers appear to be as high or higher than those for tropical rivers, suggesting the importance of precipitation in chemical weathering. At high rates of precipitation and runoff, however, easily weatherable soils may become sufficiently depleted in dissolvable ions to decrease chemical weathering (Stewart *et al.*, 2001).

Reduced physical weathering (see below) or a thick sequence of slowly accumulating sediment (hence depletion of movable ions) also can result in reduced chemical weathering (Stallard, 1985; Anderson *et al.*, 2002). In the Hong River (Vietnam), chemical denudation (0.074 mm/yr) is less than half that of physical denudation (0.180 mm/yr) during the summer wet season, but greater (0.016 vs 00.011 mm/yr) during the dry winter, in large part due to the 16-fold reduction in physical erosion during the dry season (Moon *et al.*, 2007).

Table 2.10. *Highest (above) and lowest (below) average annual dissolved-solid loads (in bold) of the ~320 rivers in our database. Note that most rivers with high dissolved loads drain high mountains, all with large drainage basins and/or high runoff. In contrast, rivers with low dissolved loads drain a variety of watersheds, ranging from lowlands to high mountains, very low runoff to high runoff.*

River	Country	Area (× 10^3 km^2)	Elevation	Climate	Runoff (mm/yr)	Dissolved load (× 10^3 t/yr)	Dissolved yield (t/km^2/yr)	Dissolved conc. (mg/l)
Amazon	Brazil	6300	High Mt	Tr-W-S	1000	**270**	43	43
Changjiang	China	1800	High Mt	Te-H-S	500	**180**	100	200
Mississippi	USA	3300	High Mt	Te-SA-Sp	150	**140**	42	290
Irrawaddy	Burma	430	High Mt	Tr-W-S	1000	**98**	230	230
Ganges	Bangladesh	980	High Mt	STr-H-S	500	**91**	93	190
Danube	Romania	820	High Mt	Te-H-S	260	**80**	98	380
Mekong	Vietnam	800	High Mt	Tr-H-S	690	**65**	81	120
Salween	Burma	320	Mountain	Tr-H-S	660	**65**	200	310
MacKenzie	Canada	1800	High Mt	SA-SA-S	170	**64**	36	210
Brahmaputra	Bangladesh	670	High Mt	STr-W-S	940	**63**	94	100
Parana	Argentina	2600	Mountain	STr-SA-C	180	**62**	24	135
St. Lawrence	Canada	1200	Mountain	SA-H-S	280	**62**	52	180
Ellice	Canada	17	Upland	A-SA-S	160	**0.01**	0.6	4
Esk	England	0.31	Upland	Te-H-W		**0.02**	52	
Souss	Morroco	16	Mountain	STr-A-W	19	**0.03**	2	97
Skiensvassdraget	Norway	10	Mountain	SA-W-S	900	**0.03**	3	3
Santa Ana	USA	6.3	High Mt	STr-SA-W	110	**0.03**	5	44
Kola	Russia	38	Lowland	A-H-S	370	**0.03**	1	2
Quoich	Canada	29	Lowland	A-A-S	180	**0.03**	1	6
Tauyo	Russia	25	Mountain	SA-H-S	440	**0.05**	2	4
Chehalis	USA	3.3	Mountain	Te-W-W	1000	**0.05**	15	15
Exe	England	0.6	Lowland	Te-W-W	870	**0.05**	85	98

Global and regional dissolved-solid fluxes

Because drainage basin area has a lesser role in determining dissolved-solid yields and fluxes, calculating global fluxes for dissolved solids is somewhat easier than it is for suspended sediments. As with other fluvial data, however, our numbers are subject to constant revision as new data are gathered and old data are revised or rejected. Moreover, the present distribution of dissolved-solid measurements is strongly geographically biased, reported dissolved data for North American rivers exceeding all of those from Oceania, South America and Africa combined. As new data from unrepresented areas become available – or perhaps we should say *if* such data become available – estimated regional and global estimates almost surely will change.

Rivers annually discharge about 3.8 Bt of total dissolved solids (TDS) to the global ocean (Fig. 2.38), the greatest flux being delivered from Asian rivers. Comparing arrow widths in Fig. 2.38 with those in Figs. 2.9 and 2.30, however, one cannot help but be struck by the high concentrations and discharge from European rivers (see also Figs. 2.33, 2.39, and 2.40). Total dissolved solid discharge from North American rivers is also higher relative to the discharge of water or suspended sediment (compare Fig. 2.38 with Figs. 2.10 and 2.30).

North and Central America

Most North American rivers are characterized by low dissolved yields; southeast Alaska and parts of Central America (although based on few data) are the only areas where annual yields exceed 100 t/km^2 (Fig. 2.40a). Lowest annual yields (<10 t/km^2/yr) are found in southeast Texas rivers and in streams draining the Canadian arctic islands. Much of the rest of the North America has dissolved solid yields between 10 and 100 t/km^2/yr. With the exception of the MacKenzie River (35 t/km^2/yr), northern Canadian rivers discharging directly or indirectly (i.e. through James, Hudson and Ungava bays) have a collective annual dissolved yield of 16 t/km^2/yr, almost exactly same as the yields from Eurasian Arctic rivers (see below). Three areas in North America account

Table 2.11. *Highest and lowest dissolved-solid yields, in descending order, of global rivers (in bold); annual global mean is 38 t/km²/yr. With exception of the Weser River (polluted) all rivers drain small mountainous basins in western Pacific Ocean high-standing islands and have runoffs 1500–3200 mm/yr. Most low-yield rivers, in contrast, are found in arid to subarid regions, with runoffs <200 mm/yr.*

River	Country	Area (×10³ km²)	Elevation	Climate	Runoff (mm/yr)	Dissolved load (×10³ t/yr)	Annual dissolved yield (t/km²/yr)	Dissolved conc (mg/l)
Lanyang	Taiwan	0.98	High Mt	STr-W-S	2900	0.72	**735**	260
Progo	Indonesia	2.5	High Mt	Tr-W-S	3100	1.8	**720**	230
Hsiukuluan	Taiwan	1.8	Mountain	STr-W-S	2300	1.3	**720**	310
Kaoping	Taiwan	3.3	High Mt	STr-W-S	2600	2.3	**700**	290
Manavgat	Turkey	1.3	Mountain	STr-H-W	3200	0.9	**690**	220
Waiau	New Zealand	2	Mountain	Te-W-C	1500	1.3	**650**	430
Hualien	Taiwan	1.5	Mountain	STr-W-S	2500	0.96	**625**	250
Peinan	Taiwan	1.6	High Mt	STr-W-S	2300	1	**625**	270
Tungkang	Taiwan	0.47	Mountain	STr-W-S	2300	0.28	**600**	255
Homathko	Canada	5.7	Mountain	Te-W-S	1500	3.3	**580**	375
Weser	Germany	46	Upland	Te-SA-W	240	3.6	**565**	2100
Linpien	Taiwan	0.34	Mountain	STr-W-S	2500	0.19	**560**	220
Choshui	Taiwan	3.1	High Mt	STr-W-S	2000	1.6	**520**	260
Ooi	Japan	1.3	Mountain	STr-W-S	1500	0.66	**510**	330
Global							**38**	
Tauyo	Russia	25	Mountain	SA-H-S	440	0.05	**2**	4
Nadym	Russia	64	Lowland	A-H-S	280	0.14	**2**	8
Back	Canada	94	Lowland	A-SA-S	179	0.18	**2**	11
Comoe	Ivory Coast	78	Lowland	Tr-SA-S	100	0.17	**2**	21
Senegal	Senegal	270	Lowland	Tr-A-S	81	0.6	**2**	27
Nile	Egypt	2900	High Mt	STr-A-W	28	6.1	**2**	76
Souss	Morroco	16	Mountain	STr-A-W	19	0.03	**2**	97
Rio Grande	USA	870	High Mt	STr-A-SP	21	2.2	**2**	120
Kola	Russia	38	Lowland	A-H-S	370	0.03	**1**	2
Quoich	Canada	29	Lowland	A-A-S	180	0.03	**1**	6
Ellice	Canada	17	Upland	A-SA-S	160	0.01	**0.6**	4
Colorado	Mexico	640	High Mt	Tr-A-S	31	0.11	**0.2**	550

for most of fluvial delivery of dissolved solids: southeast Alaska (120 Mt/yr; small drainage areas, high yields), the Arctic (220 Mt/yr; large drainage areas, low yields), and the Mississippi basin (200 Mt/yr; large drainage area, moderately low yields).

South America

The lack of dissolved solid data prevents much discussion of dissolved-solid delivery from South American rivers. Most of northern South America is characterized by dissolved-solid yields between 30 and 100 t/km²/yr; rivers in this region, primarily the Amazon and Orinoco, discharge 390 Mt/yr, ~70% of the total export from South America (Fig 2.40b). Most southern rivers, in contrast, have yields between 10 t/km² and 30 t/km²/yr; collectively they discharge an estimated 85 Mt/yr (Fig. 2.40b). The lack of data for western South American rivers means that the values on Figs. 2.38, 2.39, and 2.40 are only educated (we would like to think) guesses based primarily on regional rainfall and runoff.

Europe

Old bedrock (Fig. 2.27), generally low elevations (Fig. 2.28), and moderately low runoff (Figs. 2.10 and 2.11) mean that European rivers should have a small dissolved-solid discharge. In fact, European rivers have some of the highest dissolved-solid yields in the world, averaging 80 t/km² annually; the Weser River (Germany) until recently had a mean annual yield of 565 t/km²/yr, and southern European rivers average >100 t/km²/yr) (Fig. 2.40c). These high yields can be explained both by the prevalence of carbonate and evaporate rocks throughout central Europe as well as anthropogenic influence (e.g. mining and industry; see Chapter 4). In contrast, Scandinavian

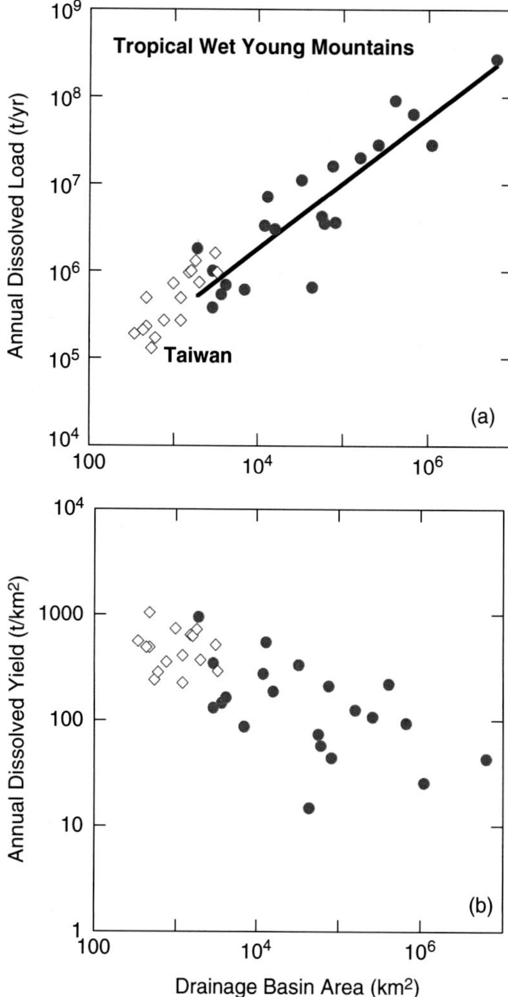

Figure 2.35. Variation of annual dissolved-solid loads (a) and yields (b) for tropical rivers draining wet young mountains. In contrast to suspended sediments (see Fig. 2.23), dissolved-solid loads and yields in Taiwanese rivers (blue diamonds) display no obvious differences from other tropical wet young mountain rivers.

rivers have annual dissolved yields <10 t/km^2/yr, reflecting low level of chemical weathering and relatively little human impact.

Africa

Based on data from only 22 rivers, the integrated dissolved-solid yield for Africa is estimated to be <15 t/km^2/yr, lower than for any other continent, reflecting the dominance of cratonic rocks and generally low rainfall throughout much of the continent. The five largest rivers (Congo, Nile, Niger, Zambesi, Orange) have a collective annual TDS yield of 7.7 t/km^2/yr, compared with an annual a global dissolved yield of ~35 t/km^2/yr. Interestingly, west African rivers have lower annual dissolved yields than rivers draining to the east (9 vs. 26 t/km^2/yr; Fig. 2.39), reflecting their shield-dominated watersheds (Fig. 2.28).

Southern Asia and Oceana

Rivers draining southern Asia and Oceania deliver roughly 35% of the dissolved solids reaching the ocean but, as with water and sediment discharge (Figs. 2.11 and 2.31), regional differences are considerable (Fig. 2.40d). Highest yields are found in rivers draining the high-standing islands of Oceania, yields in many Taiwan rivers exceeding 500 t/km^2/yr (Table 2.11). Owing to their large watershed areas, however, the major Himalayan rivers collectively discharge half again as much as the oceanic high-standing islands (Fig. 2.40d); six of the ten global rivers with the highest dissolved-solid discharges (Changjaing, Irrawaddy, Ganges, Mekong, Salween, Brahmaputra) are Himalayan in origin (Table 2.10). Lowest dissolved yields are seen in the Murray River basin and adjoining SE Australia, with an average annual TDS yield <10 t/km^2/yr.

Eurasian Arctic

As might be expected from the discussion earlier in this chapter, the dissolved yield for Eurasian (Russian) Arctic rivers is low – 18 t/km^2/yr – but it still exceeds the annual suspended-sediment yield (12 t/km^2/yr Fig. 2.30). Smaller rivers (e.g. Olenyok, Taz, Pyr, Mezen, Anabar) have higher annual dissolved-solid yields than larger rivers (Ob, Lena, Yenisei; 22 vs. 12 t/km^2/yr), which may reflect in part slightly greater runoff (245 mm/yr vs. 190 mm/yr) from these smaller rivers.

Lithologic control of dissolved-solid delivery

The climate vs. bedrock lithology debate about which controls chemical weathering – and thereby the dissolved yield and character of river water – has been addressed in a number of insightful studies (e.g. Bluth and Kemp, 1994; Gaillardet *et al.*, 1999; Dupré *et al.*, 2003). Axtmann and Stallard (1995), for example, noted a similarity in total cation yields in rivers draining temperate glaciers in the South Cascades and similar sized rivers in tropical Puerto Rico; but compositions are markedly different, the subglacial Cascade rivers, for instance, having higher K/Na and Ca/Na ratios. In the following paragraphs we argue that bedrock lithology is the controlling factor in the character and quantity of total dissolved delivery, but that climate – particularly precipitation – controls the rate of chemical weathering.

Dissolved silica and chemical weathering

The best place to begin this discussion is by looking at data compiled by Michel Meybeck (1994), in which he showed the clear role of geology in determining the concentrations

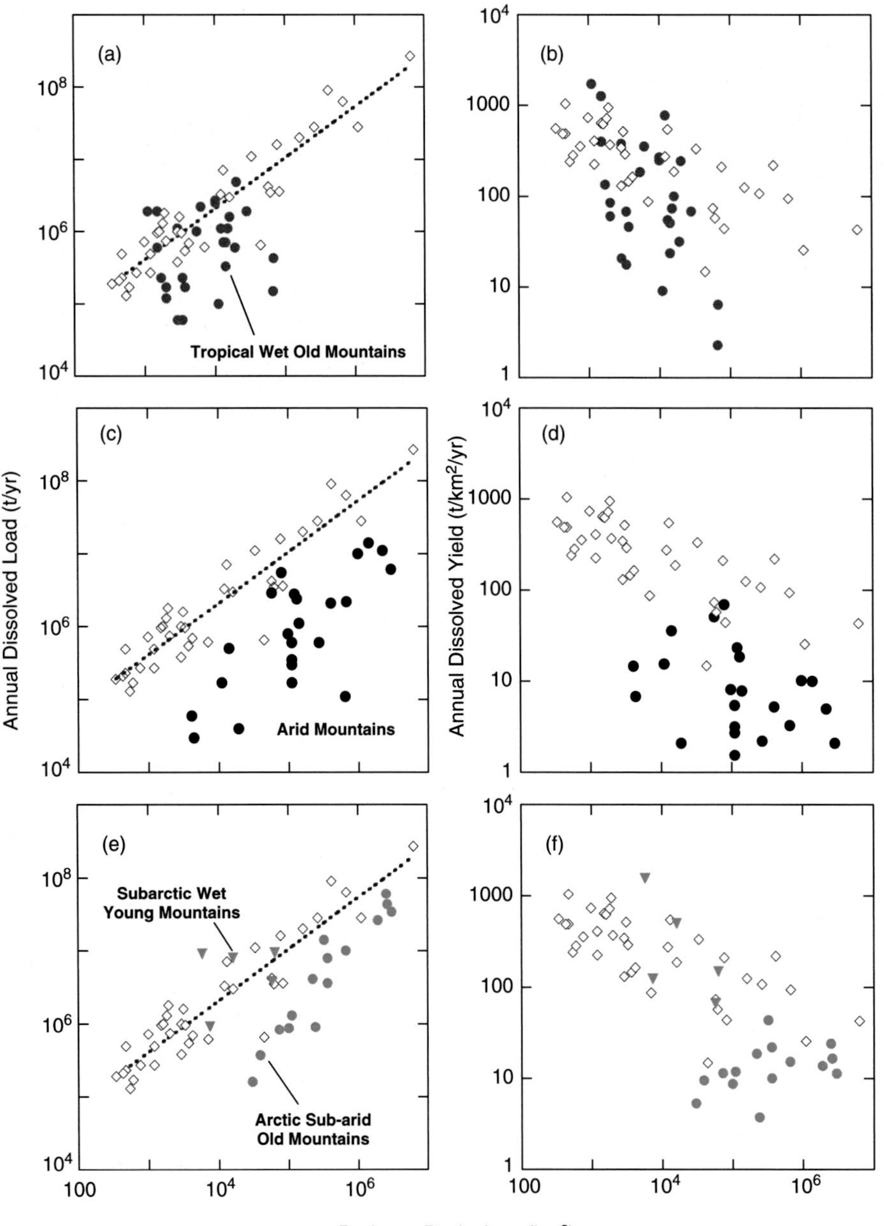

Figure 2.36. Variation of annual dissolved loads (left) and yields (right) with basin area for tropical rivers draining wet old mountains (red dots; a, b), rivers draining arid mountains (black dots; c, d), and rivers draining wet subarctic (green triangles) and semi-arid arctic mountains (green dots) (e, f) compared with dissolved loads for tropical wet young mountain rivers (blue diamonds and trend line; see Fig. 2.35).

in streams draining different lithologies. Streams draining predominantly granites and gneisses have an order of magnitude less concentration of total cations than streams draining carbonate rocks, and two orders of magnitude less than streams draining evaporitic rocks (Fig. 2.41).

Depending on rock type, bicarbonate concentrations in river waters can vary by an order of magnitude, sulfate by two orders of magnitude, and chloride by more than three orders of magnitude (Meybeck, 1994, his Table 4.1). The only major ion whose concentrations appear to remain relatively independent of lithology is dissolved silica: with the exception of streams draining volcanic rocks, river waters generally contain 100–150 mmol/l SiO_2 (Fig. 2.41), reflecting the comparative insolubility of silica in fresh water. Using Meybeck's data, streams draining gneisses and evaporates tend to have similar silica concentrations (Fig. 2.41), although rivers draining back arc and volcanic islands can have elevated silica levels (Hartmann *et al.*, 2007). Relative concentrations also can vary seasonally with runoff, more soluble ions (e.g. calcium and carbonate) showing greater seasonal change than less soluble ions (e.g. silica) (Tipper *et al.*, 2006).

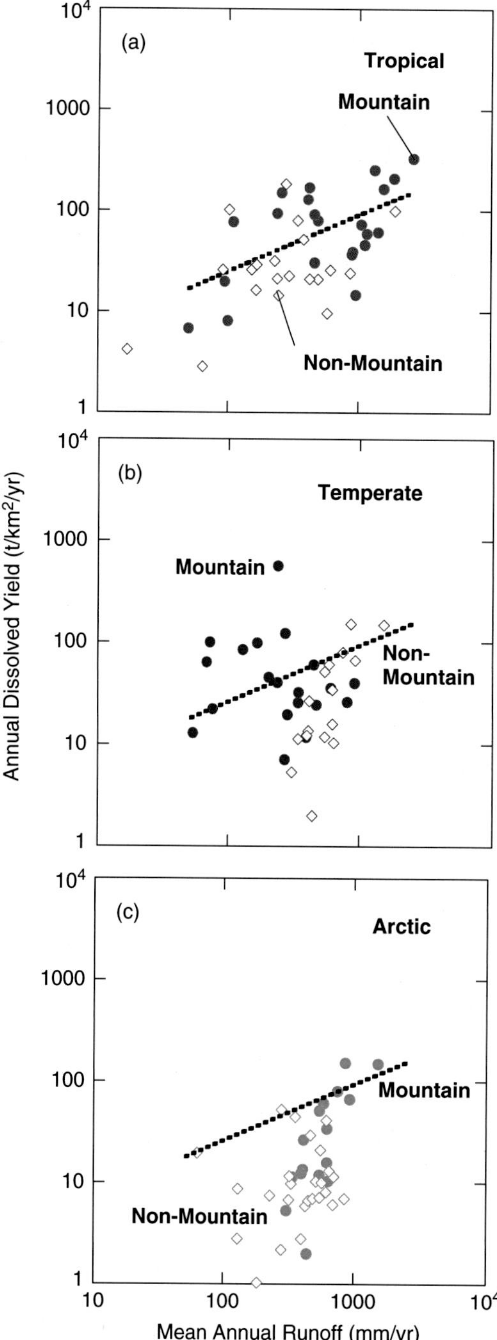

Figure 2.37. Variation of annual dissolved-solid yields and runoffs for (a) tropical, (b) temperate, and (c) arctic medium-size (20 000–100 000 km^2) mountain and non-mountain rivers. For comparison, dashed lines delineate yield-runoff trend for medium-size tropical mountain rivers.

If temperature were a prime controlling factor in chemical weathering, one should expect that rivers draining the Russian and Canadian arctic would display considerably lower concentrations of total dissolved solids than, say, tropical mountain rivers. Such is not the case (Fig. 2.42a), suggesting that some other factor – we would argue lithology – must play a critical role. The role of lithology, however, is not restricted to small rivers, even large watersheds can be dominated by single lithologies: the Pearl River (Zhujiang) drains mostly (80%) carbonate rocks, the Don and Yukon >80% shale, the Colorado, Niger, Orange, and Senegal >50% sand and sandstone, and the Amur, Godavari, Huanghe, and St. Lawrence >50% shield rocks (Amoitte-Suchet *et al.*, 2003).

Given its relatively constant composition in river water (Fig. 2.41), dissolved silica therefore appears to be the best measure of chemical weathering. Other than Icelandic rivers (which drain young volcanic soils) and arid rivers (which are transport-limited), silica concentrations show a far more logical climatic progression than other dissolved solids (compare Figs. 2.42a and b): silica concentrations in Arctic rivers are lower than in northern European rivers and on average an order of magnitude lower than in tropical rivers.

The correlation between mean annual dissolved silica discharge for tropical wet young mountainous rivers and drainage basin area (Fig. 2.43a) is the best ($R^2 = 0.98$) of any correlation derived in this chapter; yield shows a similarly close correlation with runoff (Fig. 2.43b). Dissolved-silica yields for tropical wet old mountain (and non-mountain) rivers and some subarctic wet young mountain rivers are similar to those for tropical wet young mountain rivers (Fig. 2.44a), whereas dissolved-silica yields from arid rivers and semi-arid Arctic rivers tend to be less by an order of magnitude (Figs. 2.44b and c, respectively).

The role of runoff (as a proxy of the precipitation responsible for both dissolution and dissolved solid removal) in determining dissolved silica flux is illustrated in Fig. 2.45; included in this plot are rivers draining mountains and non-mountains, young rocks and old rocks. Over the >3-order-of-magnitude variation in runoff in Fig. 2.45 is a 500-fold range in silica yield; the R^2 between runoff and dissolved silica flux for medium- sized tropical rivers is 0.97, essentially the same as seen for smaller rivers (see Fig. 2.43). Silica fluxes at low runoff are similar for all sizes of tropical watersheds, even for very small rivers; for example, 14–70 km^2 watersheds in the eastern Caribbean (McDowell *et al.*, 1995) and for seasonal runoff in Himalayan rivers (Tipper *et al.*, 2006). The increase of silica yield with increasing runoff, however, is more pronounced in smaller rivers; at a runoff of 1000 mm/yr, small (1–5 × 10^3 km^2) rivers have ~3- to 4-fold greater silica yield than large (>5 × 10^5 km^2) rivers. The clear implication from Figs. 2.43 and 2.45 is that although bedrock geology plays a major role in determining total dissolved yield, it plays a relatively minor role with respect to silicate weathering. Given the relatively constant dissolved silica concentrations in rivers draining

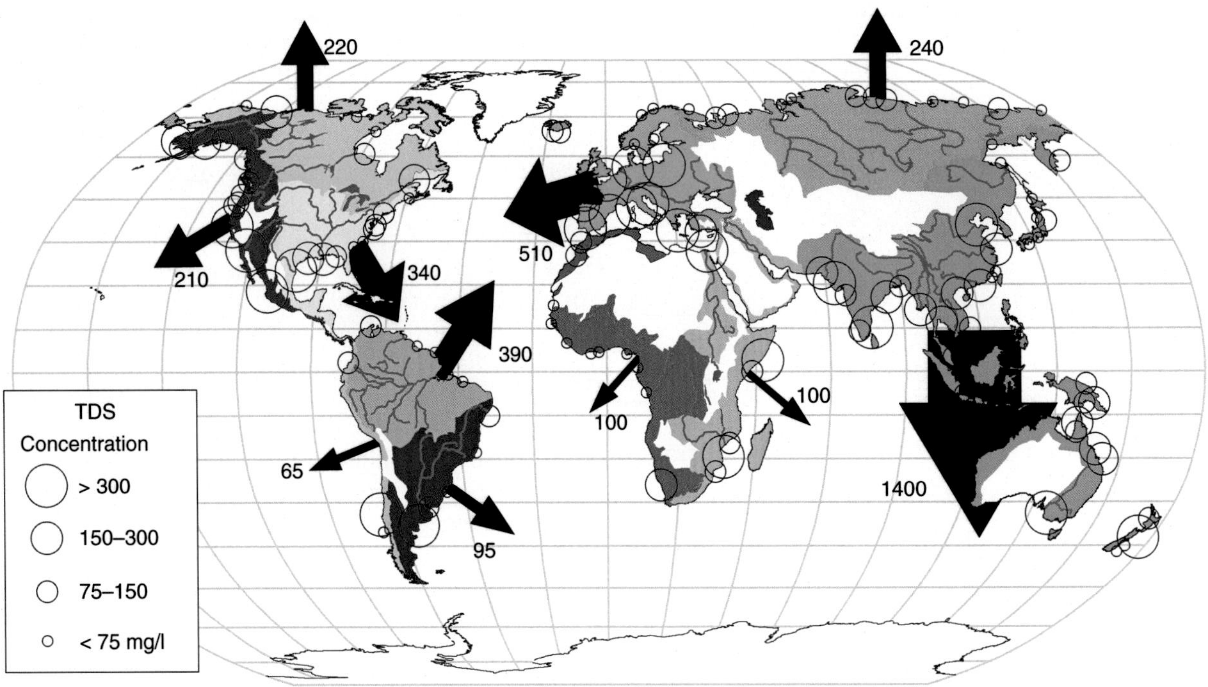

Figure 2.38. Fluvial delivery of total dissolved solids to the global coastal ocean. Circles represent average dissolved-solid concentrations for selected rivers. Most data from Meybeck and Ragu (1996). Because of the lack of data for western South America rivers, dissolved-solid and dissolved-silica discharges shown here and in subsequent figures are no more than educated guesses, based largely on climatic considerations.

different types of bedrock (Fig. 2.41), this should not be surprising.

Utilizing data from the large number of medium-sized rivers, in Fig. 2.46 we compare runoff and dissolved silica yield for tropical rivers as seen in Fig. 2.43 with those from temperate and subarctic/arctic rivers; again, basin morphology and bedrock geology are not considered. The increase in subarctic/arctic silica yields with runoff is more pronounced than for tropical/subtropical rivers: at runoffs <100 mm/yr, silica yield is about an order of magnitude less than in tropical rivers, but at runoffs >1000 mm/yr, dissolved-silica yield is similar to that in tropical rivers ($R^2 = 0.71$). In cold arid and semi-arid watersheds, such as in the Canadian and Russian Arctic, silicate weathering is particularly low, but in humid and wet watersheds, such as coastal Alaska, silicate weathering approaches that seen in more tropical areas. The rapid increase in silica yield with increased precipitation – i.e. chemical weathering – no doubt illustrates the effect of mechanical erosion, as fresh surfaces become exposed to chemical weathering (Gaillardet *et al.*, 1999; Huh, 2003; Dupré *et al.*, 2003, their Fig. 4; Anderson, 2005). The rate of tectonic uplift, which affects the rate of physical erosion, and the rate of precipitation appear to exert lithologic control on chemical weathering in the New Zealand Southern Alps as well as the Brahmaputra River by exposing more easily weathered minerals, (Jacobson *et al.*, 2003; Hren *et al.*, 2007).

The role of temperature in chemical weathering can be viewed in several ways (see Kump *et al.*, 2000, for a discussion). Taken together with the above discussion, Figure 2.46 suggests that at low runoff, cold temperatures can retard chemical weathering, but the temperature effect wanes as runoff increases, and it may play a minimal role when annual runoff approaches 1000 mm/yr. As pointed out by Huh (2003), this may simply reflect the increased exposure of freshly exposed rock surfaces to chemical weathering, but also may reflect increased weathering by plants (Berner *et al.*, 2003) in wetter environments. The increased role of precipitation in warmer environments differs from the climatic effect proposed by White (White and Blum, 1995; White, 2004), in which solute flux varies in a linear fashion with precipitation but exponentially with temperature.

In tropical rivers the delivery of total dissolved-solids can vary considerably with basin area (or runoff) whereas dissolved-silica delivery shows little scatter (Fig. 2.43). Having similar runoffs, 240–250 mm/yr, the tropical Cauweri (India), Chao Phyra (Thailand), and Sassandra

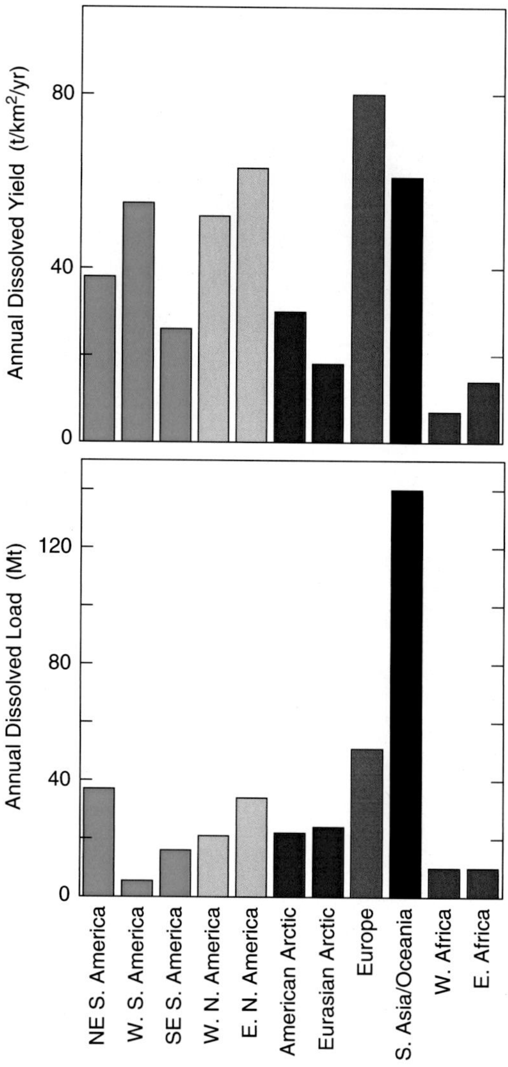

Figure 2.39. Fluvial dissolved yields (a) and discharges (b) to the global ocean from the various regions shown in Fig. 2.38.

(Ivory Coast) rivers have similar dissolved-silica yields, 4.0–5.0 $t/km^2/yr$, whereas their total dissolved yields range from 17 to 95 $t/km^2/yr$ (data in Meybeck and Ragu, 1996).

Because it is less dependent on bedrock character than is total dissolved-solid (TDS) flux, dissolved-silica flux (SiO_2) provides a measure of the climatic role in chemical weathering; at low runoff, temperature plays an important role, but at high runoff, chemical weathering appears to be driven primarily by snow and rain, owing to (we assume) the role that precipitation plays in physical erosion. Total dissolved-solid flux, in contrast, reflects bedrock character (as well as the effects of anthropogenic activity; see Chapter 4), and is an integrated measure of chemical weathering on the watershed and total solute delivery to the coastal ocean. Superimposed on this lithology–climate interaction are human activities, which are discussed further in Chapter 4.

These differences are perhaps best seen in Fig. 2.47, which compares TDS/SiO_2 concentrations in rivers draining European, Arctic and developing countries. Developing country rivers (in South America, Africa, Asia, Oceania) have an average TDS/SiO_2 trend of ~5:1. TDS concentrations in European rivers, in contrast, show no relationship to SiO_2; whereas the former range between 5 mg/l and 800 mg/l, the latter are generally <5 mg/l, underscoring the roles of lithology and (more importantly, human activities) on TDS fluxes. Arctic rivers fall somewhere in between, reflecting a lesser impact from human activity.

Physical and chemical weathering

In the preceding pages we alluded to the dependence of dissolved-solid flux on physical weathering (see Millot *et al.*, 2002 and references therein) by which fresh rock and grain surfaces are exposed to chemical weathering. In Fig. 2.48 we show the relation between TDS (Figure 2.48a) and dissolved SiO_2 (Figure 2.28b) yields to sediment yield for the 100 or so rivers for which we have complete data. The green diamonds in this figure represent data from relatively natural rivers that drain tropical wet mountains, African watersheds, and the Canadian and Eurasian Arctic. Many of these rivers with annual sediment yields >100 $t/km^2/yr$ have TSS/TDS and TSS/SiO_2 ratios of >10 and >1 $t/km^2/yr$, respectively, whereas many natural rivers with annual sediment yields <100 $t/km^2/yr$ have TSS/TDS and TSS/SiO_2 ratios <10 and <1. A recent model proposed by Gabet and Mudd (2009), which predicts decreased chemical weathering when sediment erosion exceeds 100 $t/km^2/yr$ owing to decreased fresh minerals exposed in a thin regolith, seems to run counter to the data shown in Fig. 2.48. Dissolved TSS/TDS ratios in most European rivers are <1 (Fig. 2.48a), whereas in terms of dissolved SiO_2, most European rivers are scattered around – or below – the 1:10 dissolved silica/sediment yield (Fig. 2.48b), suggesting that chemical weathering in European watersheds is much lower than dissolved solid yields would suggest. The total dissolved-solids and dissolved-silica yields for arid rivers, in contrast, are very low, most SiO_2 yields <1 $t/km^2/yr$, indicative of the degree to which low precipitation and runoff limit chemical weathering.

Global and regional dissolved silica flux

As silica fluxes and yields are measures of chemical weathering, it is not surprising that the high-rainfall tropical areas in Southern Asia (and Oceania) and northeast South America account for 60% of the chemical weathering in watersheds draining into the global ocean (Figs. 2.49

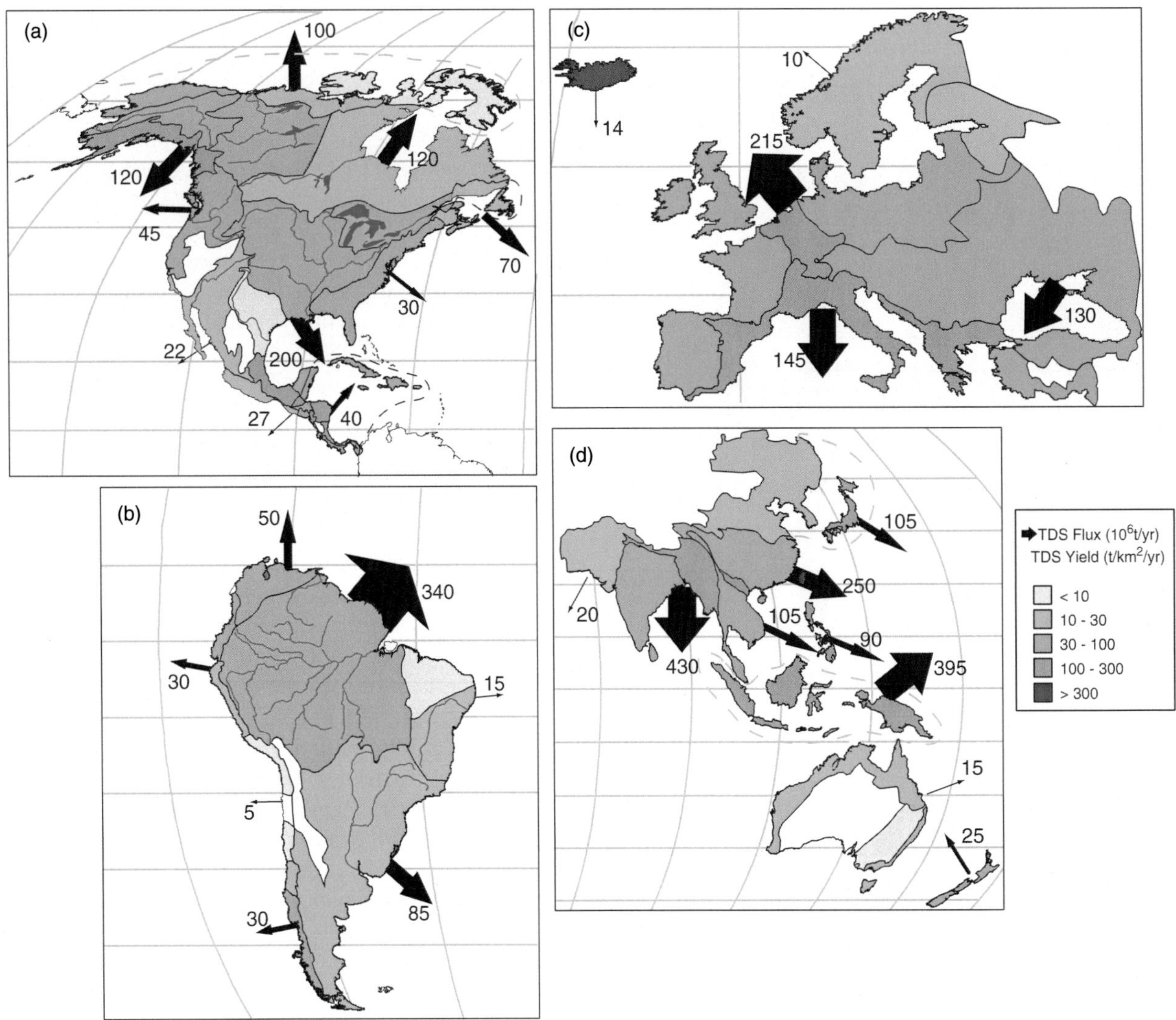

Figure 2.40. Regional yields and fluvial delivery of total dissolved solids from (a) North and Central American, (b) South American, (c) European, and (d) Austral-Asian watersheds to the coastal ocean. Yields for the various polygons are color-coded, from <10 to >300 t/km^2 year. Arrows in each panel are proportional to dissolved solid discharge in that panel. Where data are meager, we extrapolated yields and loads from adjacent polygons. Large areas with similar physiographies, geology and climates are bordered by broad black lines. Africa and the Eurasian Arctic are not shown because of the general uniformity of their physiography, geology, and climate (see Figs. 2.38 and 2.39).

and 2.50). Globally, total dissolved-solid to dissolved-silica ratios vary greatly. The global TDS/SiO_2 ratio is ~11 (3700 vs. 330 Mt/yr), which approximates the mean ratio in Asian and western North American, and east African rivers (11–12.5). TDS/SiO_2 ratios are much lower in South American (5.3–6.8) and western African (3.6) rivers, and much higher in Eurasian, eastern North American, and European rivers (24, 35, and 58, respectively) (Fig. 2.52).

North and Central America

Because of relatively low runoff (together with the old craton lithology of many watersheds), most North American arctic rivers are characterized by low levels of chemical weathering, as evidenced by annual dissolved-SiO_2 yields <1 t/km^2/yr. Similar low rates are seen in the low-precipitation watersheds in southeastern Texas, stretching across to southern California and into northwestern Mexico (Fig. 2.51a). In contrast, west coast US and Canadian rivers, from central California up to southeastern Alaska, have annual SiO_2 yields >3 t/km^2/yr, and in the Pacific northwest annual yields exceed 10 t/km^2/yr. Southern Mexico and northern Central American rivers also have annual SiO_2 yields >10 t/km^2/yr. Regionally TDS/SiO_2 ratios in North American vary considerably, although many rivers have ratios around

15–20 (e.g. the St. Lawrence and Susquehanna watersheds), reflecting the importance of limestone weathering as well as human activities.

South America

Lacking data but extrapolating from other watersheds with similar environmental characteristics, we anticipate that the highest dissolved-SiO_2 yields (>10 t/km²/yr) in South American rivers are from northwest rivers (Fig. 2.51b), where the mountains are high (and young), runoff exceeds 1000 mm/yr, and thus physical erosion is high (Figs. 2.30 and 2.31). Northeastern South American rivers (including the Amazon, Orinoco, and Magdalena watersheds) also exhibit high rates of chemical weathering, with annual dissolved-silica yields ranging between 3 and 10 t/km²/yr. Lowest annual yields (<0.3 t/km²/yr) are found in the very arid central west coast and low-elevation, arid central Brazilian rivers (Fig. 2.51b). South American generally rivers have TDS/SiO_2 ratios ranging from 5 to 6, about half the global average (10–11) (Fig. 2.52).

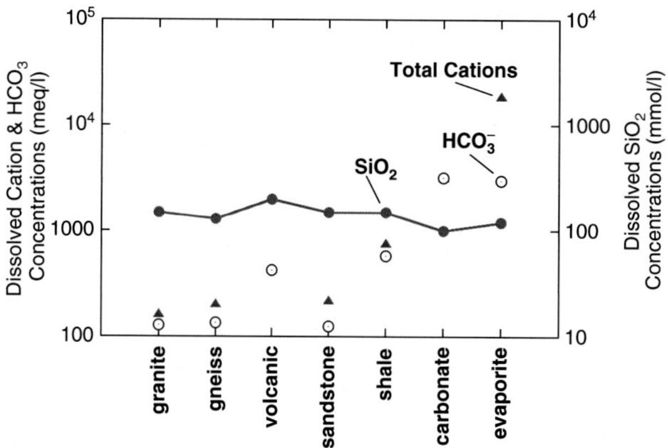

Figure 2.41. Variability of major dissolved constituents in the water of streams draining various rock types. Data from Meybeck (1994).

Europe

European rivers generally have low rates of chemical weathering, the annual SiO_2 yield for much of Europe being <3 t/km²/yr (Fig. 2.51c). Lowest weathering rates are seen in eastern Europe, particularly the southern Russian rivers (<0.3 t/km²/yr) (Fig. 2.51c). Heavy rainfall, a generally mild climate, easily denuded volcanic rocks, and geothermal activity (Kristmannsdottir *et al.*, 2002), in contrast, give Icelandic rivers the highest dissolved-silica yields (>10 t/km²/yr) in the North Atlantic region. Owing to low concentrations of dissolved solids, the TDS/SiO_2 ratio in Scandinavian rivers is ~10, about the global average, whereas northern, southern and eastern European rivers have TDS/SiO_2 average ratios of 70–120 (Fig. 2.52), the ratio of 600 for the Weser being the highest we have found.

Africa

Chemical weathering in west African rivers is more intense than in the east – collective annual SiO_2 yields of 2 vs. 1 t/km²/yr (Fig. 2.50), presumably reflecting the west's wetter climate (Fig. 2.2). This contrasts with lower dissolved-solid yields in western rivers (Fig. 2.39), presumably owing to their shield-dominated watersheds (Fig. 2.28). It is probably not surprising to see that the arid North African rivers have low TDS and dissolved-SiO_2 yields.

Southern Asia and Oceania

Low runoff in Pakistan, northeastern Asia, and Australia translates to levels of chemical weathering generally <0.3 t/km²/yr SiO_2 (Fig. 2.51d). Rivers draining India and much of Indochina and China, in contrast, have annual yields 3–10 t/km²/yr. Highest dissolved-silica yields are seen in areas with intense precipitation and rapidly eroding mountains (i.e. sediment yields >1000 t/km²/yr; see Fig. 2.31d) – that is, in Indonesian, New Guinean, Taiwan, Philippine, and Japanese watersheds. Dissolved silica in these rivers

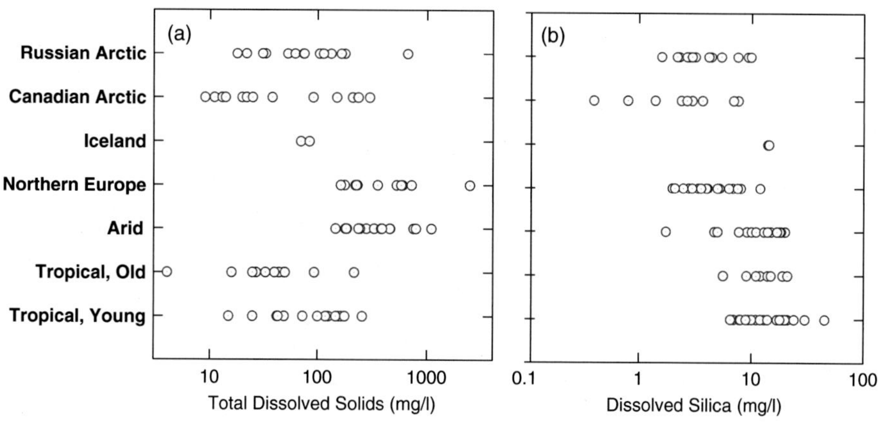

Figure 2.42. Concentrations of total dissolved solids (a) and silica (b) in rivers from different climates. Note the scatter and lack of climatic signal for total dissolved solids compared to a much tighter cluster of dissolved-silica concentrations. Dissolved silica data primarily from Meybeck and Ragu (1996).

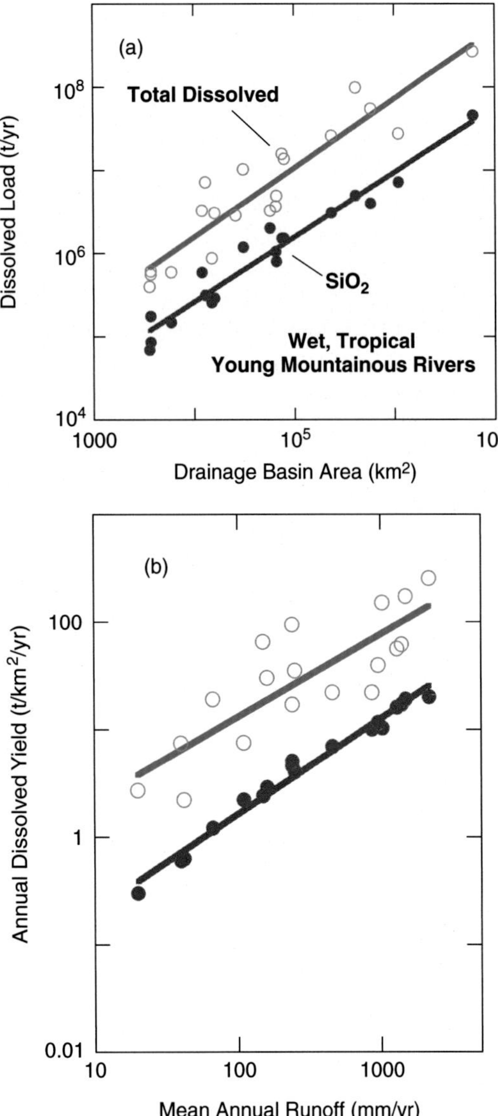

Figure 2.43. Dissolved-silica discharge relative to basin area (a) and yield relative to runoff (b) for rivers draining tropical wet young mountains. Dissolved-silica data from Meybeck and Ragu (1996). For comparison, total dissolved-solid discharge and yield are also shown.

averages >10 t/km^2/yr. The oceanic high-standing islands alone, we calculate, may account for as much as 20% of the global flux of fluvially derived dissolved silica. TDS/SiO_2 ratios in Himalayan rivers are about 20 (Fig. 2.52), compared with ~6 for rivers draining oceanic islands (compare Figs. 2.40d and 2.51d).

Eurasian Arctic

Eurasian (Russian) Arctic watersheds have low levels of chemical weathering, generally <3 t SiO_2/km^2/yr

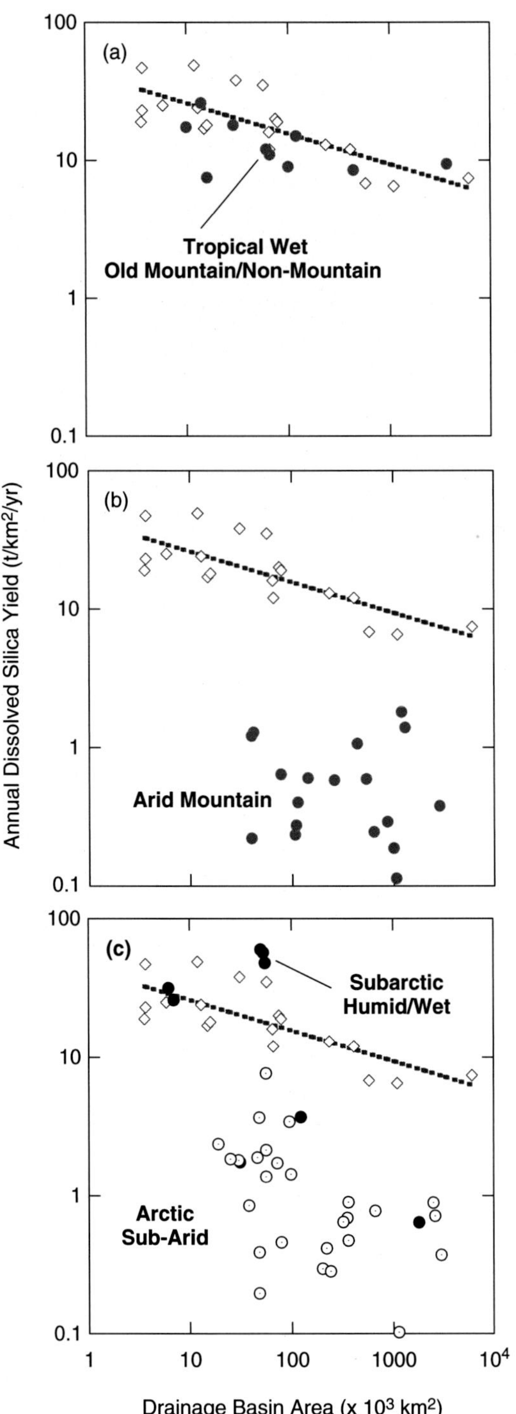

Figure 2.44. Dissolved-silica yields (vs. drainage basin area) for rivers draining tropical wet old mountains and uplands (a), tropical and subtropical arid mountains (b), and subarctic and arctic watersheds (c). For comparison rivers draining tropical wet young mountains are shown as open blue diamonds and their trend as a power function curve, as derived from data shown in Fig. 2.43. Dissolved-silica data primarily from Meybeck and Ragu (1996).

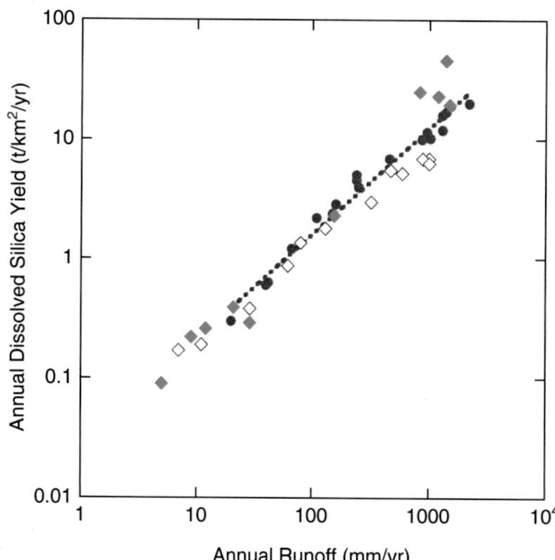

Figure 2.45. Variation of dissolved-silica yields versus runoff for small (1000–5000 km^2; green diamonds), medium-size (20 000–100 000 km^2; red dots) and large (>500 000 km^2; blue diamonds) tropical and subtropical rivers. Most values are derived from silica concentrations reported by Meybeck and Ragu (1996); some US river data obtained from the USGS NWIS website.

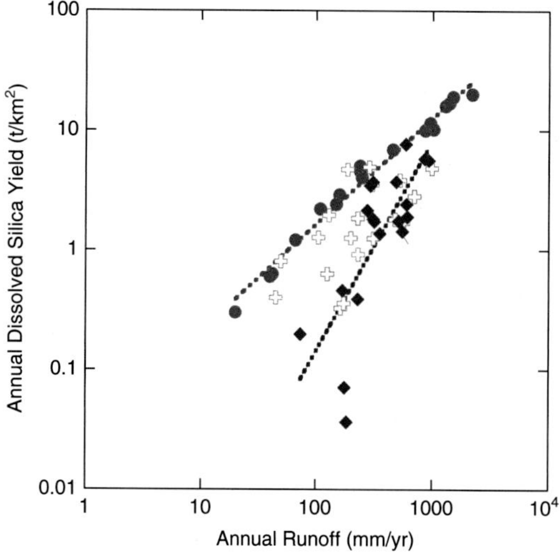

Figure 2.46. Dissolved-silica yields versus runoff for tropical/subtropical (red dots), temperate (green crosses) and subarctic/arctic (blue diamonds) medium-size (20 000–100 000 km^2) rivers. Red dots and the associated regression curve for tropical/subtropical rivers are the same as those shown for medium rivers in Fig. 2.45. Values derived from silica concentrations reported by Meybeck and Ragu (1996).

(Fig. 2.50), whereas many rivers have dissolved-solid concentrations >50 mg/l (Fig. 2.47). We assume that the difference in these two ratios largely reflects the low levels of chemical weathering and the dominance of limestone in many pan-Arctic watersheds as well as the local influence of human activities, particularly mining and manufacturing.

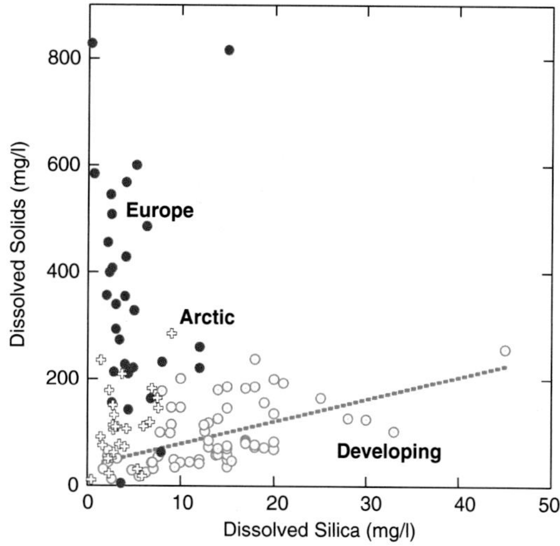

Figure 2.47. Dissolved-silica and total dissolved-solid concentrations in rivers draining pan-Arctic watersheds (blue crosses; assumed to be relatively undeveloped), European watersheds (red dots; assumed to be heavily industrialized), and watersheds in developing countries (green circles). Several European rivers with particularly high dissolved-solid concentrations (e.g. Weser ~2100 mg/l) have been omitted. Data from Meybeck and Ragu (1996).

Summing up

Because the preceding discussion in this chapter may appear in places a bit more serpentine than originally intended, we summarize here some of the key points regarding fluvial discharge to the global ocean.

Water

River discharge (or runoff) can be thought of as the difference between basin-wide precipitation and evapotranspiration, although along its course a river also can lose water through storage, consumption and/or interbasin transfer (see Chapter 4). Runoff for most rivers ranges between 100 mm/yr and 1000 mm/yr, although some arid rivers can have mean annual runoffs <25 mm/yr (e.g. the Murray–Darling in Australia, Orange in South Africa) and in some high-runoff rivers it can exceed 3000 mm/yr (e.g. New Zealand, Taiwanese, Indonesian, and New Guinean rivers). Given the global distribution of precipitation (Fig. 2.2), it is not surprising that rivers draining northwest South America and southeast Asia/Oceania account for nearly 2/3 of the freshwater discharged to the coastal ocean (Figs. 2.9, 2.10, and 2.11).

Delivery of suspended sediments and dissolved solids

The erosion and fluvial transport of particulate and dissolved solids reflect the cumulative effects of drainage basin size and morphology (in part a surrogate for tectonics),

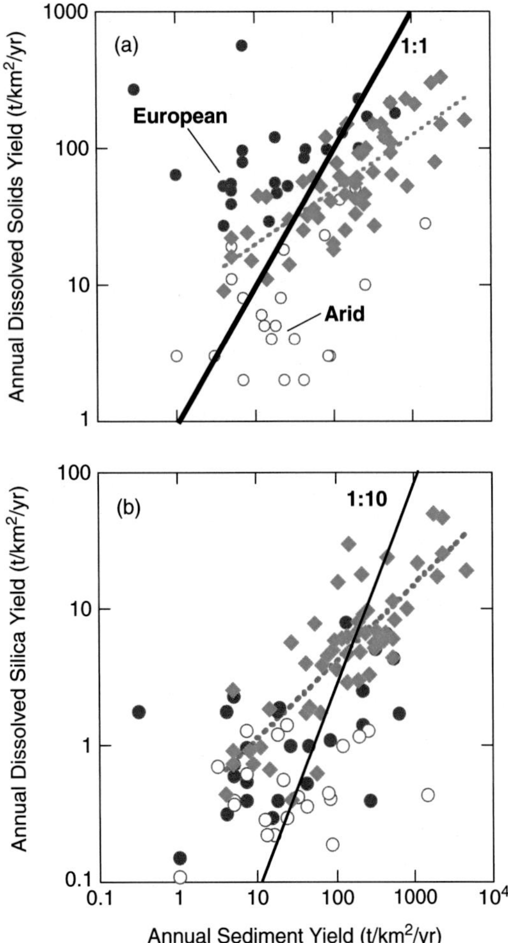

Figure 2.48. Comparison of (a) total dissolved-solid yields and (b) dissolved-silica yields vs. sediment yields of global rivers. Green diamonds represent yields from tropical and temperate mountain and upland rivers; red dots from European rivers; and red circles from arid rivers. Dashed lines show green diamond trends. Rivers left of diagonal 1:1 and 1:10 lines in (a) and (b), respectively, indicate rivers in which dissolved-solid transport and chemical weathering (b) have equal or greater importance than physical weathering.

lithology and climate, as well as the impact of human activities. Mountains represent the major sources of fluvial sediment, in large part because of their high susceptibility to earthquakes, steep slopes and resulting landslides, and orographically enhanced rainfall. Although arid rivers can have high concentrations of suspended sediment (see Table 2.5), they often are transport-limited except during rare flood events (see Chapter 3).

Because of high elevations, small drainage basin areas, heavy rainfall, generally erodible rocks (Peucker-Ehrenbrink and Miller, 2003), as well as long-time human impact (see Chapter 4), rivers draining SE Asia and the high-standing islands of Oceania account for as much as 60–70% of the present-day sediment discharge to the global ocean (Figs. 2.29–2.31); small rivers draining western North and South America also have high sediment yields.

The fate of sediments derived from these active-margin rivers stands in stark contrast to the sediments discharged by passive margin rivers. In the latter case much of the eroded sediment can be stored in floodplains and/or deltaic, estuarine or coastal environments; offshore escape occurs during sea-level regression or low-stands. In contrast, sediment discharged from active-margin rivers (e.g. from the west coasts of Sumatra, Java, North and South America) often bypasses the narrow shelves even during high stands of sea level, particularly during the episodic floods that are common to many small watersheds (see Chapter 3). Some of the sediment derived from active-margin rivers may be deposited in troughs and trenches, reincorporated into the subduction conveyor belt (von Huene and Scholl, 1991), and ultimately uplifted and reincorporated into coastal mountains (Lundberg and Dorsey, 1990).

As with suspended sediment, the quantity and character of dissolved-solid discharge depend greatly on the lithology of the source rock; in contrast, dissolved silica tends to reflect the rate of chemical weathering. Temperature plays a major role in chemical weathering in low-runoff environments, but its role appears to diminish with increasing runoff (Fig. 2.46). The correlation between sediment and dissolved yields (Fig. 2.48; see also Stallard, 1995a; Gaillardet *et al.*, 1997, 1999; Millot *et al.*, 2002; Lyons *et al.*, 2005) demonstrates the dependence of chemical weathering on the extent of physical weathering, which exposes new surfaces and uncovers soils to renewed weathering. Complete removal of soils, however, can remove microenvironments conducive to weathering, thereby lowering the rate of chemical weathering (Anderson *et al.*, 2002).

Although the global delivery of total dissolved solids (3.8 Bt/yr) is only 1/5 that of the global suspended solids, 38% of the 308 rivers for which we have both sediment and dissolved data are dissolved-dominated. In fact, 28% of the rivers in our database have dissolved yields >2-fold greater than sediment yields (Table 2.12). As mentioned previously, in part this reflects the bias of the available database. Dissolved-dominated rivers are particularly pervasive in Europe and Eurasia, rivers with TDS/TSS ratios >2 representing 65% and 47%, respectively, of all European and Eurasian rivers in our database (Table 2.12, Fig. 2.53); far fewer dissolved-dominant rivers are noted in Africa, South America, Oceania or Asia. If data were more globally distributed such that African and Oceanic rivers (in particular) were better represented, the percentage of dissolved-dominant rivers seen in Table 2.12 almost certainly would be lower.

Our biased database aside, several other reasons help explain the global distribution of dissolved-dominated rivers: (1) low suspended-sediment yields due to low rates of physical erosion (e.g. rivers draining Precambrian shield

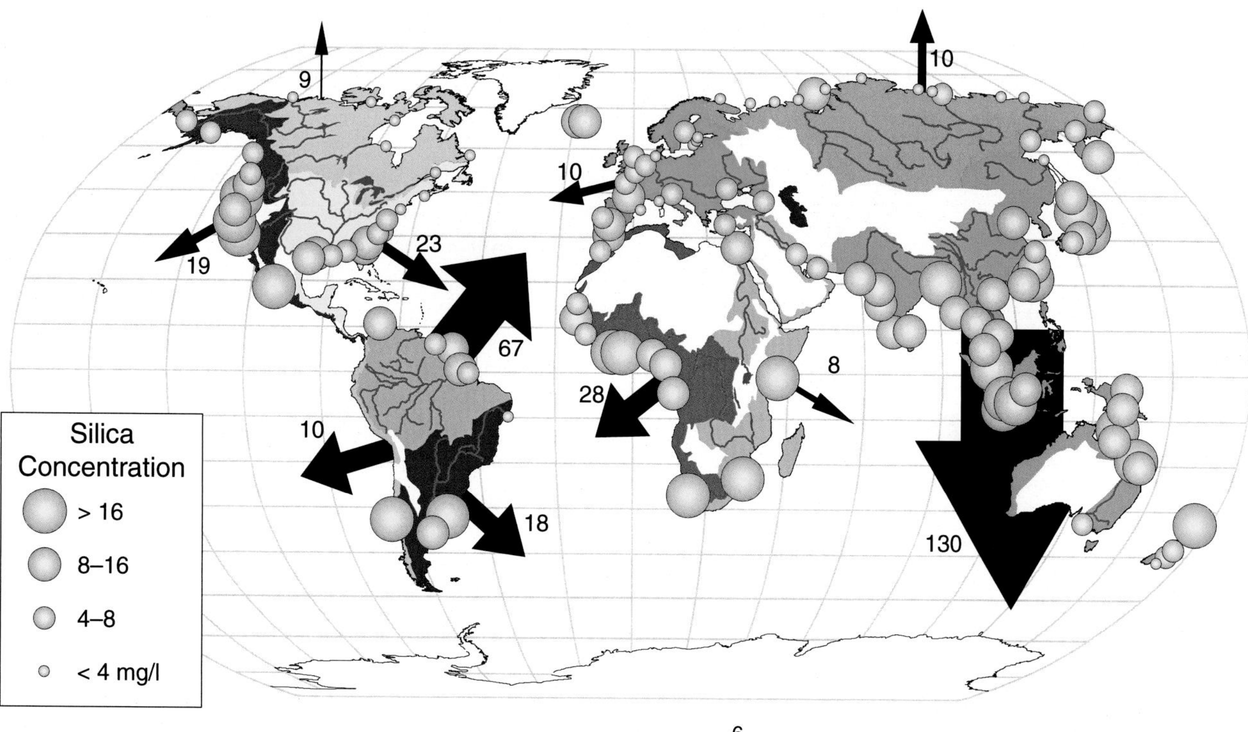

Figure 2.49. Fluvial discharge of dissolved silica to the global coastal ocean. Dots represent average concentrations for selected rivers. Most data are from Meybeck and Ragu (1996).

rocks in northern Canada, NE South America) or sediment trapping by dams (e.g. Ebro, Rhone, Indus) and/or natural lakes (e.g. St. Lawrence, Rhine); (2) higher rates of chemical weathering relative to physical weathering due to either the nature of the rocks (e.g. limestone or evaporites) or human activity. Although there are many rivers in Europe and some in Eurasia in which chemical pollution has greatly elevated dissolved yields, the mean global dissolved yield of the dissolved-dominated rivers is the same as it is for sediment-dominated rivers (36 t/km^2/yr). Rather, it is the difference in physical delivery (or lack thereof) of sediment that seems to be the key factor in determining whether a river is sediment- (average annual sediment yield of 190 t/km^2/yr) or dissolved-dominated (average annual sediment yield of 33 t/km^2/yr).

It comes as no surprise that within any type of river – e.g. wet, young mountain rivers, or arid mountain rivers – suspended-sediment discharge and dissolved-solid discharge tend to increase with drainage basin area; but dissolved discharge increases faster than sediment discharge (Fig. 2.54). The decreasing TSS/TDS ratio with increasing basin size suggests a subtle but important difference in the source and delivery of fluvial suspended sediment and dissolved solids. Most sediment delivered to the coastal ocean is ultimately derived from mountain erosion, particularly young mountains exposed to heavy precipitation. Some of the eroded sediment, however, may be stored somewhere between the sediment's source (generally in mountains) and the ocean, either temporarily or permanently downstream within the river channel or on the river's floodplains. The amount of sediment stored tends to be directly proportional to floodplain area and average basin gradient; hence the inverse relation between sediment yield and basin area (e.g. Fig. 2.15). In contrast, chemical weathering occurs not only in the exposed hinterland but also within stored sediments at lower elevations (Johnsson and Meade, 1990; Johnsson *et al.*, 1991; Meade, 2008, his Fig. 4.4). The combination of downstream storage of suspended sediments and the increase in dissolved solids thus explains the not-uncommon downstream decrease in the TSS/TDS ratio.

Three end members: wet, arid, and arctic

Even a casual glance at the figures throughout this chapter should convince a cynical reader of the global importance of mountainous rivers – particularly those draining young mountains with high precipitation and runoff (>750 mm/yr) – in the delivery of water, sediments and dissolved solids to the coastal ocean. Occupying an estimated 14% of the cumulative watersheds draining into the global ocean, wet young mountain rivers account for about 45% of the freshwater discharge (although much of the discharge may come from lower elevations within the

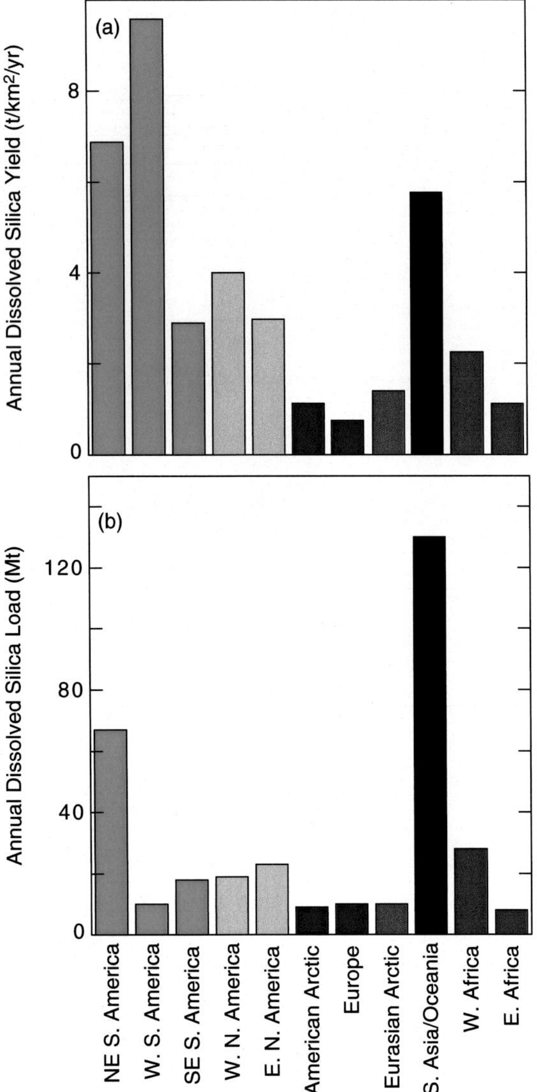

Figure 2.50. Dissolved-silica yields (a) and loads (b) discharged to the global ocean from the regions shown in Fig. 2.49.

drainage basin) and 62% of the sediment; they discharge a relatively low percentage of total dissolved solids (38%, reflecting watershed lithology and lack of anthropogenic pollution) but 60% of the dissolved silicate, underlining the importance of runoff and physical erosion in dictating chemical weathering (Fig. 2.55).

By contrast, watersheds draining arid mountains collectively occupy nearly as much watershed area (12% vs. 14%), but, given (by definition) their low runoff, they account for only about 1% of the global freshwater flux (Fig. 2.55). Low runoff suggests relatively low levels of mechanical and chemical weathering and transport, as seen in their moderate sediment discharge (~10% of the global total) and very low levels of dissolved-solid and dissolved-SiO_2 discharge (2% and 1%, respectively). Rivers draining semi-arid polar watersheds collectively occupy a slightly larger global area, about 18%, and account for about 13% of the freshwater discharge. But they discharge only ~2% and 6% of the global sediment and dissolved SiO_2 (Fig. 2.55), the result of low runoff and the low-lying old mountains that they drain. Together, these three end members occupy about 44% of the land area draining into the global ocean, and account for ~60%, 70%, 52%, and 70% (respectively) of the water, suspended sediment, dissolved solids, and dissolved silica discharged to the ocean.

This comparison, of course, is not meant to diminish the regional and even global importance of very large rivers (e.g. Congo, Mississippi, Yangtze, Danube) or the many smaller rivers that collectively contribute significant amounts of water or sediment. But it does put into perspective the importance of mountainous rivers. Without mountains – especially young mountains in regions of active tectonism (and thus active physical erosion and chemical weathering) – the discharge of sediment and dissolved solids, would be dramatically reduced.

Importance of small mountainous rivers

Throughout this chapter we have documented the key role that small mountainous rivers play in the discharge of suspended and dissolved solids to the global ocean, a point made in a number of earlier papers (e.g. Milliman and Syvitski, 1992; Milliman *et al.*, 1999; Syvitski and Milliman, 2007; Lyons *et al.*, 2005; Leithold *et al.*, 2006). As a first-order quantitative estimate as to the global role of small mountainous rivers, which we define as having basin areas smaller than 10 000 km^2, we estimate that these rivers drain a cumulative global area of 10 million km^2, about 10% of the global watershed draining into the ocean. Although regionally runoffs may be less than 100 mm/yr (e.g. southern California, northwest Africa), globally small mountainous rivers have a much greater runoff (640 mm/yr) than the other 90% of the global rivers (300 mm/yr; Table 2.13); collective discharge from small mountainous rivers is ~6.5 km^3 (18% of the global total). A much greater role, however, is their discharge of sediment, about 8.7 Bt, ~45% of the global total; mean global sediment yield is 870 t/km^2/yr compared with 115 t/km^2/yr for the remaining 90% of the world. The calculated discharge of dissolved solids by small mountainous rivers (1.1 Bt/yr) represents a dissolved yield of 110 t/km^2/yr compared with 30 t/km^2/yr for the other world rivers.

The reader should note, however, that because few small mountainous rivers have been monitored for sufficiently long periods – if monitored at all – these estimates may be subject to considerable revision. Our estimated dissolved-solid discharge for rivers draining western South America (Fig. 2.40), to take an extreme example, is based on few data but rather comparisons with similar

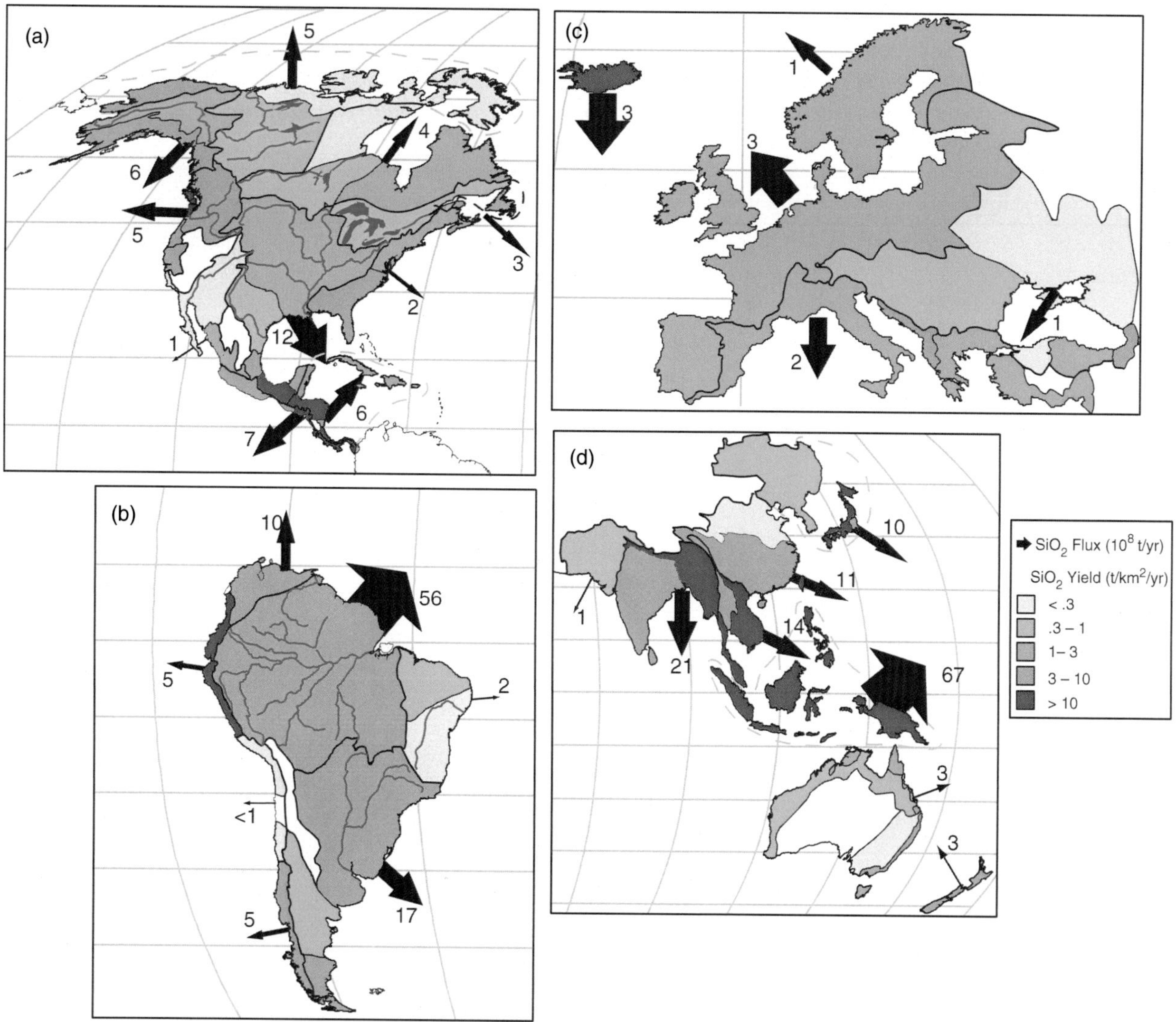

Figure 2.51. Calculated regional yields and fluvial delivery of dissolved silica from (a) North American, (b) South American, (c) European, and (d) Austral-Asian rivers to the coastal ocean. Annual yields for the various regions are color-coded, from <0.3 to >10 t/km²/yr. Arrows in each panel are proportional to dissolved solid discharge in that panel. Where data are meager, we extrapolated yields and loads from adjacent polygons. Large areas with similar physiography, geology, and climate are defined by black lines. Africa and the Eurasian Arctic are not shown because of the general uniformity of their physiography, geology, and climate (see Figs. 2.49 and 2.50).

rivers from around the world. Moreover, with only two exceptions, which are discussed in some detail in Chapter 3, the impact of episodic events on small mountainous rivers, is undocumented, even though events should result in a greater discharge than during non-events (see Fig. 3.47). The Choshui River in Taiwan, in fact, may be the only river for which dissolved-solid discharge has been measured during a heavy rainfall event (see Fig. 3.53). But even though the numbers in Table 2.13 can only be considered first-order estimates, they do emphasize the degree to which small mountainous rivers are so dominant, particularly in lands bordering the Pacific and easternmost Indian Ocean.

Measures of landscape denudation

Calculated sediment loads and yields often have long been used to define rates of denudation (Dole and Stabler, 1909; Gilluly, 1955; Judson and Ritter, 1964). Assuming a sediment/rock specific gravity of 2.0 (done here for simplicity), a sediment yield of 2000 t/km²/yr equates to a denudation rate of 1 mm/yr. As most of the annual yields discussed in this chapter fall between ~100 t/km² and 10 000 t/km²/yr, most basin-wide denudation rates range between 0.05 mm/yr and 5 mm/yr. As recognized by Schumm (1963), apparent denudation rates (i.e. yields) increase with decreasing basin area and increasing relief (as stressed throughout this chapter).

Table 2.12. *Rivers from our database for which the ratio of dissolved-solid yield to suspended-sediment yield exceeds 1 (TDS/TSS >1) (columns 3–4), and those for which the ratio exceeds 2 (columns 6–7). Numbers in parentheses (columms 4 and 7) represent rivers whose decreased sediment loads (mainly because of damming) caused the shift from sediment-dominated to dissolved-dominated.* **Bold** *rivers are more or less natural rivers; italic rivers are those in which dams have significantly reduced sediment loads.*

Continent	# Rivers in database	Rivers with TDS/TSS>1		Examples	River with TDS/TSS > 2		Examples
Africa	19	2	(0)	**Congo**	0	(0)	
S America	15	6	(0)	**Uruguay**	1	(0)	**Approuaque**
Asia	73	12	(4)	*Pearl*	5	(3)	*Haihe* *Indus* *Krishna*
Oceania	32	5	(2)	**Waitaki**	4	(1)	**Murray–Darling**
N/C America	74	25	(2)	*Colorado*	14	(2)	*Columbia* **St. Lawrence** **Susquehanna**
Eurasia	38	22	(6)	*Neva*	12	(6)	**Lena** **Ob** *Don*
Europe	57	42	(4)	*Rhone*	35	(2)	**Rhine** **Thames** *Ebro*
Global	**308**	**97**	**(18)**		**71**	**(14)**	

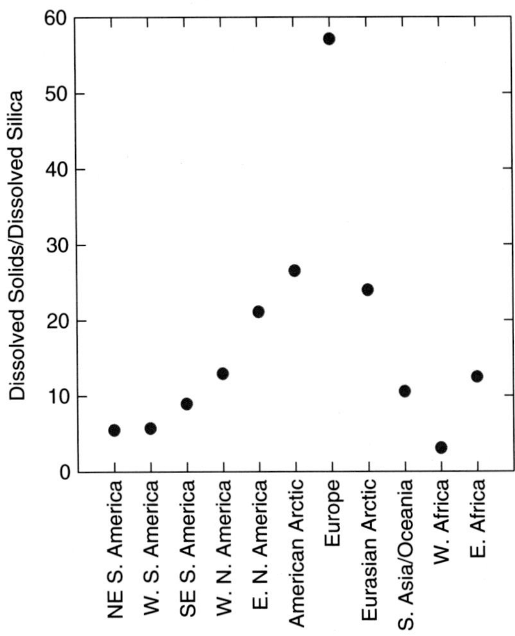

Figure 2.52. Ratio of total dissolved solids to dissolved silica for the 11 areas shown in Figs. 2.39, 2.40 and 2.49, 2.50.

Dole and Stabler (1909) were amongst the first to calculate denudation rates for the continental USA, finding a range between 0.014 mm/yr (western Gulf of Mexico) and 0.057 mm/yr (Colorado River basin). Using a more extensive and up-to-date database, Judson and Ritter (1964) reported denudation rates ranging from 0.04 mm/yr (North Atlantic rivers) to 0.17 mm/yr (Colorado), although Judson (1968) acknowledged that, "erosion rate is not uniform over the entire basin, but it is convenient…to assume that it is." Nevertheless, calculated denudation rates – based on solid and dissolved yields divided by an assumed mean soil and rock specific gravity of 2.0 – can serve a useful purpose. Africa and the Eurasian Arctic, for example, have the lowest rates of denudation: 0.017–0.020 mm/yr (Table 2.14), whereas Asia/Oceania has a mean denudation rate of 0.31 mm/yr. But regional denudation rates within Asia/Oceania vary greatly, from 0.025 mm/yr in Australia to 4.9–5.7 mm/yr in Taiwan. The highest-yield river in Taiwan, the Erhjen, has a present-day basin-wide denudation of 35 mm/yr, perhaps 60 mm/yr if one considers elevations higher than 200 m as the major sediment sources (Kao and Milliman, 2008). Similarly, although Europe has a continent-wide average denudation of 0.11 mm/yr, Albania's average rate of denudation (1.7 mm/yr) is 300 times that of Sweden's (0.006 mm/yr).

The average present-day global denudation rate (0.11 mm/yr) is 6 times greater than the long-term Phanerozoic denudation (0.017 mm/yr), but only 1/5 of that during the Pliocene (Table 2.14). More disconcerting, perhaps, is the observation that present-day denudation is about twice that of global soil production. In large part this discrepancy reflects anthropogenic soil loss (0.6 mm/yr) (Wilkinson and McElroy, 2007), a point discussed further in Chapter 4.

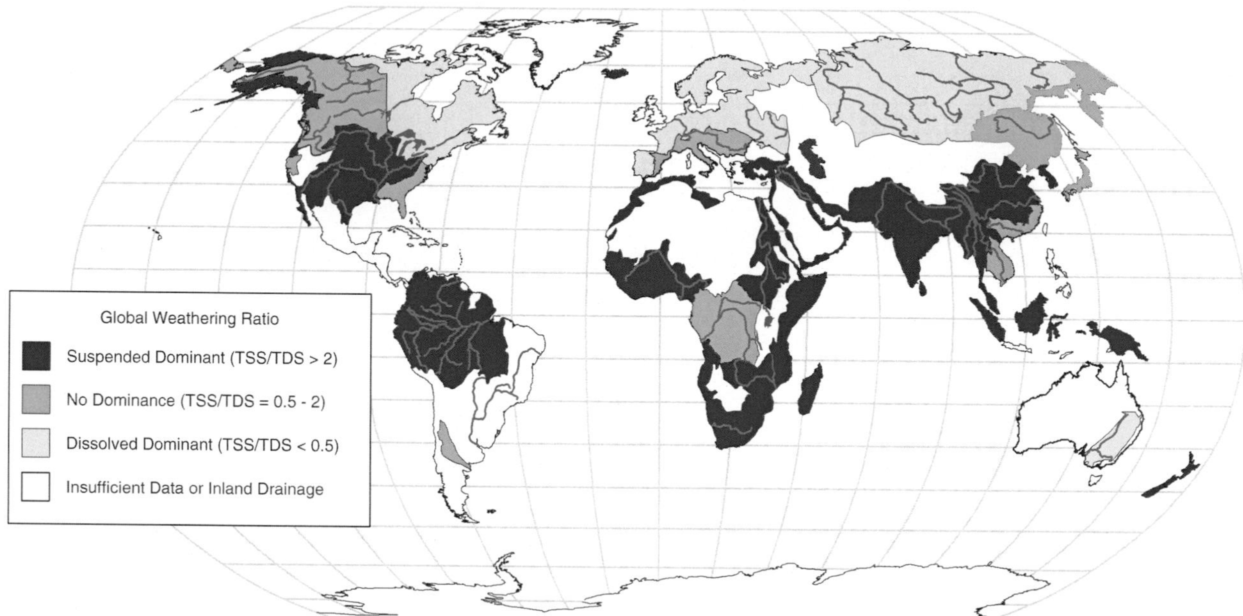

Figure 2.53. Global distribution of suspended- and dissolved-dominant rivers, defined as rivers with TSS/TDS >2 and <0.5, respectively.

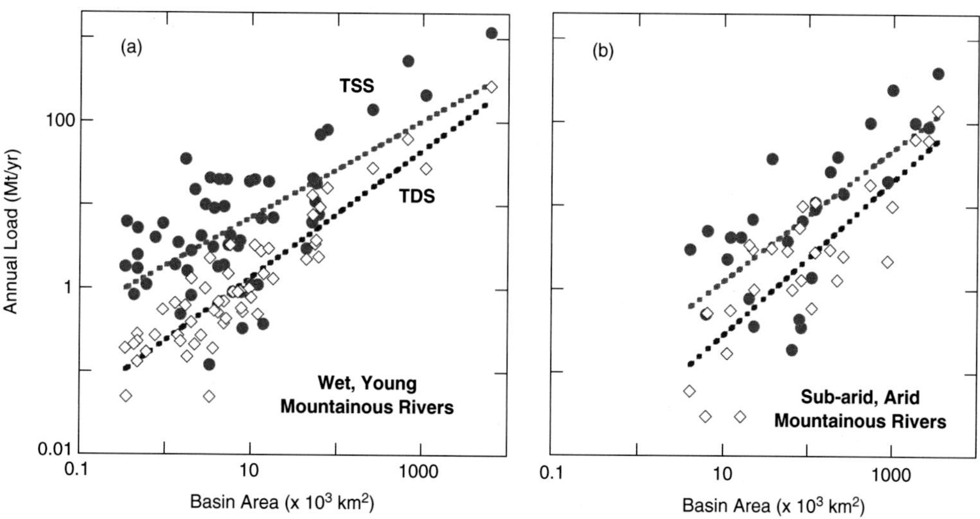

Figure 2.54. Suspended-sediment (red dots) and dissolved-solid (blue diamonds) delivery from high-runoff (runoff >750 mm/yr) young mountainous rivers (a) and arid (runoff <250 mm/yr) mountainous rivers (b) vs. drainage basin area. Note that dissolved-solid delivery increases with increasing drainage basin area somewhat faster than suspended-solid delivery.

As stated above, basin-wide denudation in many ways is a misleading term because most drainage basins (particularly large basins) display a wide variety of denudation rates, part reflecting variation in relief, climate, lithology, as well as sediment storage; sub-basins with active sediment storage can have negative denudation rates. The basin-wide denudation of 0.07 mm/yr for the Amazon watershed, for example, belies the fact that most erosion occurs in the Andes, <10% of the watershed, and much of the remaining 90% of the basin actually stores sediment (see Dunne *et al.*, 1998; Aalto *et al.*, 2006). Assuming that at least 1/2 of the sediment eroded from the Andes is stored downstream (Aalto *et al.*, 2006), total erosion in the Andes part of the Amazon basin would be ~2–3 Bt/yr, meaning a denudation rate of ~2 mm/yr. Sediment accumulation in 90% of the Amazon basin would then be ~1–2 × Bt/yr, equating to a mean basin-wide accumulation of ~0.2–0.3 mm/yr.

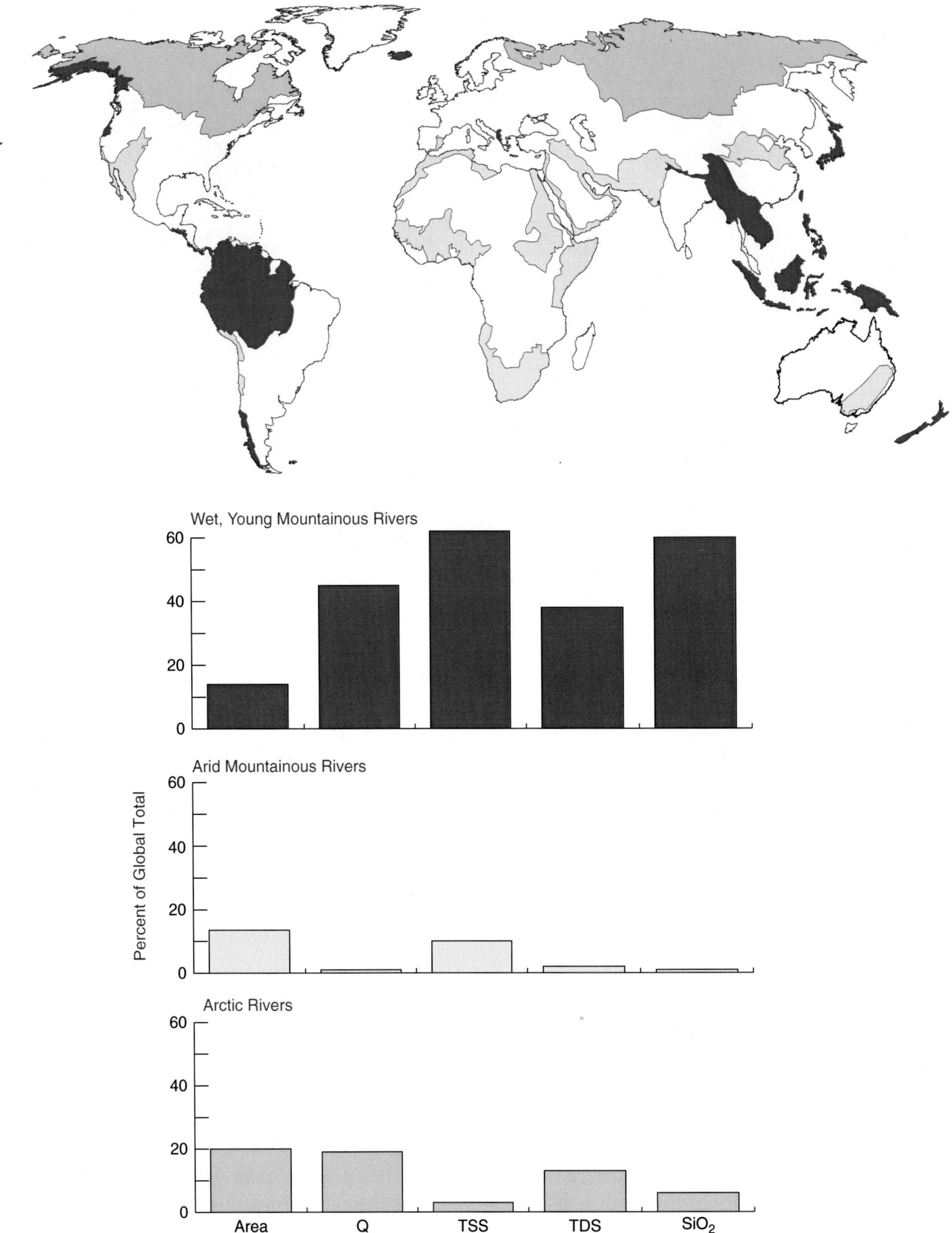

Figure 2.55. Global percentages of cumulative drainage basin area, discharge (Q), suspended-sediment (TSS), dissolved-solid (TDS), and dissolved-silica discharges from wet young mountainous (red), arid mountainous (yellow), and Arctic (blue) rivers; global distributions are shown on the map. The three types of rivers each account for 12–18% of the global watershed draining into the coastal ocean.

Table 2.13. *Cumulative basin areas, discharges/runoffs, sediment loads/yields, and dissolved loads/yields of small mountainous rivers (SMR, defined as having watersheds smaller than 10 000 km², compared to other global rivers. As a first approximation, we assume that small mountainous rivers collectively drain about 10 million km² of the ~105 million km² emptying into the global ocean (see Fig. 1.3). Data are derived largely from figures presented earlier in this chapter as well as data in the appendices.*

	Global area × 10^6 km²	Discharge (runoff) × 10^3 km³/yr (mm/yr)	Sed. load (yield) Mt/yr (t/km²/yr)	Diss. load (yield) Mt/yr (t/km²/yr)
SMRs	10	6.4 (640)	8.7 (870)	1.1 (110)
Other	95	29.6 (300)	10 (115)	2.7 (30)

Table 2.14. *Calculated regional and national/local (in italics) denudation rates compared to modern and long-term rates of global denudation (in bold). Denudation is calculated as solid plus dissolved yields divided by an assumed mean soil and rock specific gravity of 2.0; soil loss calculation assumes a mean specific gravity of 1.5. Owing to the relative dearth of data, dissolved-solid yields for Swedish, Albanian, Australian, and Taiwanese rivers are assumed to be 10, 200, 10 and 500 t/km²/yr, respectively.*

Regional and national basin-wide denudation (mm/yr)	
Africa	0.017
Eurasian Arctic	0.02
N. America	0.07
Europe	0.11
Sweden	0.006
Albania	1.7
S. America	0.1
Asia/Oceania	0.31
Australia	0.025
Taiwan	4.9–5.7[d]
Erhjen	35–60[d]
Global modern and long-term denudation (mm/yr)	
Phanerozoic continental denudation[a]	0.017
Global soil production[b]	0.06–0.08
Modern global fluvial average	0.11
Pliocene denudation[c]	0.53
Anthropogenic soil loss[a]	0.6

Data [a]from Montgomery and Brandon (2002), Wilkinson and McElroy (2007), Montgomery (2007); [b]Wakatsuki and Rasyidin (1992), Troeh *et al.* (1999); [c]Hay *et al.* (1988), Wilkinson and McElroy (2007); [d]sediment yield from total watershed area and >100 m in elevation, respectively, Kao and Milliman (2008).

Table 2.15. *Calculated mean suspended-sediment and dissolved-solid yields and denudation rates for the Mississippi River. Data from USGS NWIS.*

	Annual TSS yield (t/km²)	Annual TDS yield (t/km²)	Denudation (mm/yr)
Upper Mississippi	34	45	~0.03
Missouri	206	18	~0.10
Ohio (minus Tennessee)	19	98	~0.045
Tennessee	19	49	~0.03
Arkansas	145	35	~0.07
Red	285	35*	~0.13
Mississippi (at mouth)	124	44	~0.06

But denudation can shift dramatically in response to climate change: using U-series nuclides, Dosseto *et al.* (2006) have concluded that while chemical weathering in the lowland regions of the Amazon basin reflects present-day conditions, physical erosion in the Andes has experienced periodic perturbations in response to climatic shifts.

A more accurate calculation similarly can be made for the Mississippi drainage basin, which Judson and Ritter (1964) listed as having a basin-wide denudation rate of 0.05 mm/yr and we list as 0.06 mm/yr. The major sub-basins within the Mississippi, however, show a fairly wide variation of mean denudation rates, from ~0.03 for the upper Mississippi and Tennessee rivers to ~0.13 mm/yr for the Red River (Table 2.15). Within the individual rivers, calculated denudation tends to decline downstream.

As more data become available, particularly with sub-millimeter measurements of erosion (Hartshorn *et al.*, 2002) or short-lived isotopes such as ^{10}Be and ^{137}Cs (Walling and He, 1999) used to determine actual rates of sediment removal and accumulation, respectively, it may be possible to calculate more accurately local variations in denudation. In the meantime, using downstream discharge values to calculate basin-wide denudation may misrepresent (in many cases, wildly so) the actual removal and accumulation of sediment and dissolved solids.

3 Temporal variations

Small showers last long,
But sudden storms are short

W. Shakespeare

The past is over

G. W. Bush

Introduction

The numbers and broad arrows shown in Figures 2.9, 2.30, 2.38, and 2.49 portray mean annual river discharges of water, suspended sediment, dissolved solids, and dissolved silica, respectively, to the global ocean; they are based on the database presented in the appendix. Because these data for the most part represent long-term means, discharges for any given year, month or hour can vary greatly. The large discharge arrows emanating from southern Asia in Fig. 2.11d, for example, primarily reflect fluvial response to intense summer monsoon precipitation and Himalayan snowmelt. For seven months of the year much of southern Asia can be considered arid, and arrows on Fig. 2.11d would thus be correspondingly smaller. Over longer periods of time – years to decades – slight shifts in climatic drivers (e.g. El Niño-Southern Oscillation or the North Atlantic Oscillation) can lead to significant shifts in river flow (Knox, 1995; Liu and Wang, 1999). As we delve further into the past and as our links to present-day measurements and environmental conditions become less certain, the uncertainties of river flow increase. How, for example, did local, regional or global river discharge change during the last glacial maximum (LGM); or how did Asian river discharge respond to the Miocene uplift of what is now the Himalayas?

In this chapter we discuss temporal variations in river discharge. We feel on safest ground in discussing global changes over the past 50–60 years, which represents the limits of the database for many global rivers (see Chapter 1). We then proceed to longer-term change, most of our discussion focusing on the past 20 000 years. We conclude the chapter with a discussion of episodic events in terms of both erosion and sediment transport, which collectively can help determine a river's character.

Short-term cycles and their atmospheric and oceanic drivers

Where available, long-term river gauging can show interesting cycles and trends, sometimes marked, sometimes subtle. In Chapter 1, for example, we noted that the annual sediment flux from the Santa Clara River in southern California varies by orders of magnitude, from essentially zero in the early 1950s to 46 Mt in 1969 (see Fig. 1.6). Even in larger rivers annual water discharge can vary significantly. In the Columbia River, for example, discharge varies from 100 km^3/yr to 230 km^3/yr, and several longer-term 50-yr cycles also can be noted (Fig. 1.7). Global long-term trends, unfortunately, are difficult to gauge because, except for a handful of northern European rivers, few available measurements are available prior to the mid to late nineteenth century. As shown in Chapter 2, much of the global discharge flows from tropical and subtropical rivers, but with the exception the Nile and Yangtze, none of their discharge records pre-date the twentieth century. Many rivers, in fact, are represented by less than 50 years of discharge data (Figs. 1.4a, 1.5).

Probably the best place to begin this discussion is in the contiguous United States, where the extensive and accessible US Geological Survey database (USGS/NWIS) gives us access to countrywide discharge since 1950. Collectively the 55 rivers represented in Figures 3.1, 3.2, and 3.3 drain 4.7×10^6 km^2 of watershed, or about 60% the total area of the contiguous USA. Mean annual discharge for individual rivers ranges from 490 km^3/yr (Mississippi River) to <0.1 km^3/yr (San Luis Rey and Santa Ynez rivers). To put US rivers into a global context, mean annual cumulative discharge for the 55 rivers is 1100 km^3/yr, about 15% that of the Amazon River.

Although annual aggregate discharge varies greatly (e.g. 790 km^3 in 1981, 1350 km^3 in 1983), between 1951 and 2000 fluvial discharge trends increased everywhere except for the west coast north of San Francisco Bay (Figs. 3.1, 3.2, 3.3), which, ironically, has the highest runoff in the conterminous USA (see also Lins, 1997). Atlantic rivers experienced low discharges in the mid 1960s and high in the mid 1970s, with fluctuating but gradually increasing values, following a pattern not dissimilar from that displayed by the

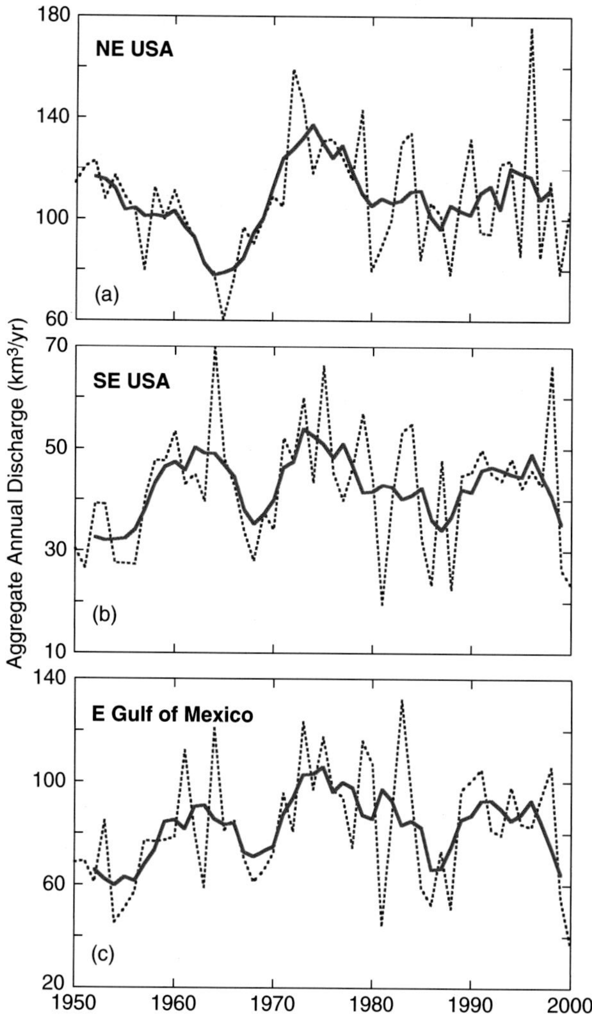

Figure 3.1. Interannual variation in aggregate annual discharge (blue dashed line) and 5-yr running average (solid red line), 1951–2000, for selected rivers draining the NE Atlantic seaboard (a) Androscoggin, Kennebec, Penobscot, Merrimack, Connecticut, Hudson, Delaware, Susquehanna, Potomac, Rappahannock, York, and James rivers; SE Atlantic (b) Roanoke, Pamlico, Cape Fear, Peedee, Edisto, Savannah, Ogeechee, Altamaha, and Satilla rivers; and the eastern Gulf of Mexico (c) Suwanee, Apalachicola, Choctawhatchee, Escambia, Mobile, Pascagoula, and Pearl rivers. The Santee River is excluded because of an incomplete 50-yr record.

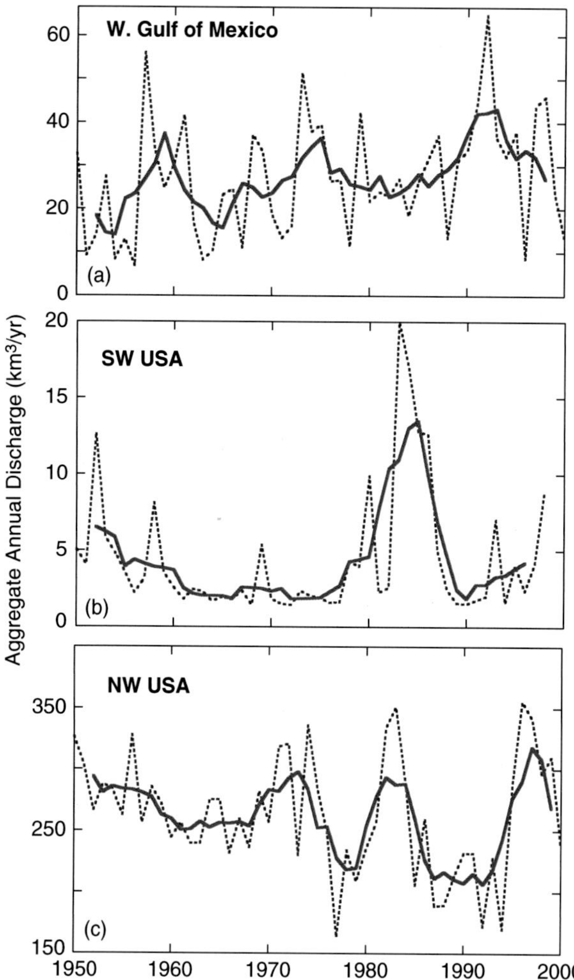

Figure 3.2. Interannual variation in aggregate annual discharge (blue dashed line) and 5-yr running average (solid red line), 1951–2000, for selected rivers draining the western Gulf of Mexico (a) Sabine, Neches, Trinity, Brazos, Colorado, Guadalupe/San Antonio, and Nueces, rivers; southern California (b) San Luis Rey, Santa Margarita, Santa Clara, Santa Ynez, Santa Maria, Salinas, and Pajaro rivers; and northern California and the Pacific Northwest (c) Russian, Eel, Mad, Klamath, Smith, Rouge, Umpqua, Nehalem, Columbia, Skagit, and Snohomish rivers. The Rio Grande, Colorado (Arizona), and Sacramento/San Joaquin rivers are excluded because dams and water diversion have greatly decreased – or eliminated – freshwater discharge from these rivers.

North Atlantic Oscillation (see below). Discharge from SE Atlantic and the eastern Gulf of Mexico (North Carolina to coastal Mississippi) rivers showed similar highs (early 1960s, mid 1970s, mid 1990s) and lows (mid 1950s, late 1960s, late 1980s) (Fig. 3.1), suggesting a ~20-yr return cycle. Mississippi flow followed a similar pattern, but over the 50-yr period its discharge increased by ~30% (Fig. 3.3; Milliman *et al.*, 2008). Rivers draining Texas (into the western Gulf of Mexico) and southern California have the lowest annual runoffs in the continental USA, but between 1970 and 2000 Colorado River (near Grand Canyon), upper Rio Grande (at Albuquerque, NM), and southern California river flow increased dramatically (Fig. 3.2).

What caused the interannual patterns portrayed in the three previous figures, such as the dramatic drop in NE river discharge in the late 1960s (Fig. 3.1), and why was

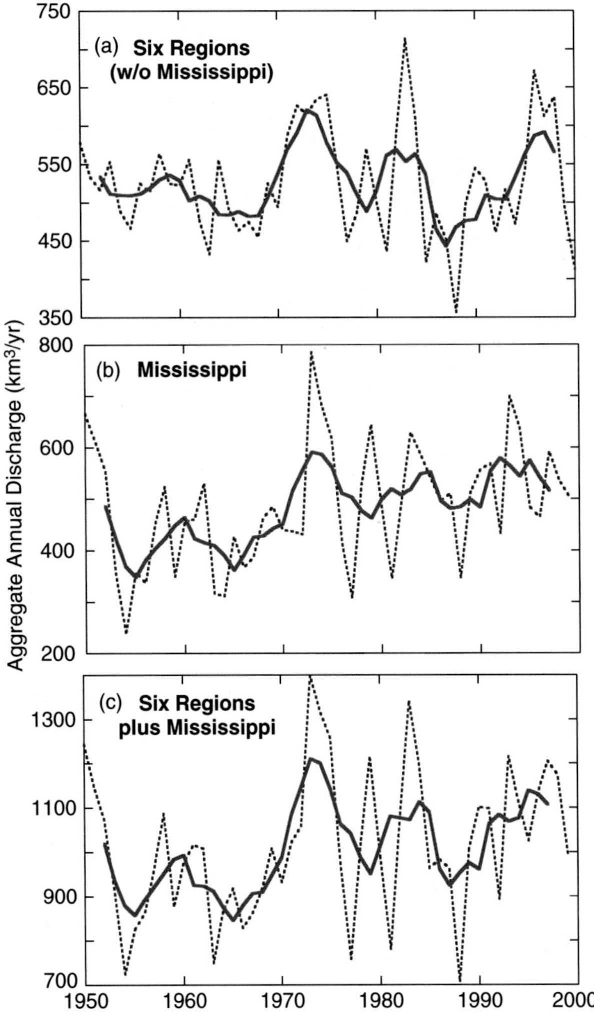

Figure 3.3. Interannual variation in aggregate annual discharge (blue dashed line) and 5-yr running average (solid red line), 1951–2000, for the six regions shown in Figs. 3.1 and 3.2 (a); annual discharge from the Mississippi River (as measured at Vicksburg, MS) (b); and collective discharge, including the Mississippi (c).

the cumulative discharge from SW rivers around eight times greater in 1983 than it had been two years previously (Fig. 3.2)? To address these and other global changes in fluvial discharge, we must look at the atmospheric and oceanic forcing factors that drive climatic cycles. Until a few decades ago, such a discussion would have been mercifully short, since few climatic drivers had been recognized, let alone named. Prior to 1990, El Niño was perhaps the only widely recognized phenomenon; even the term La Niña was known to relatively few researchers. Since then a number of climatic drivers have been identified, and their acronyms – e.g. PDO, NAO, AO, AMO – have been increasingly referenced in the scientific and popular literature. In the past few years there also has been an increased awareness of the various teleconnections within the Earth's climate system. Late Pleistocene aridity in northern Asia, for instance, may have been synchronous with iceberg events in the North Atlantic (Porter and An, 1995). In the following paragraphs we discuss some of the more prominent present-day climatic drivers, realizing that other drivers – as well as teleconnections between various drivers – almost certainly will be identified in the foreseeable future. For a succinct discussion of climatic drivers, the reader is referred to Trenberth *et al.* (2007).

El Niño-Southern Oscillation (ENSO) and the Pacific Decadal Oscillation (PDO)

The direction and magnitude of equatorial trade winds in the Indian and Pacific oceans are perhaps the greatest factors in controlling interannual variability of global weather (Dettinger and Diaz, 2000). The Pacific and Indian oceans are characterized by a warm pool of surface water that normally resides in and around the Indonesian Archipelago. The relaxation of easterly trade winds, often first noted in the northern Indian Ocean off southwestern Asia, can result in an eastward shift of the warm-water pool, by which surface-water temperatures in the tropical eastern Pacific can increase by 2–4° (Fig. 3.4). These El Niño events (the Spanish name for the Christ child, as these events were first noticed – off Peru – during the Christmas season) can lead to torrential rainfalls and high river runoffs throughout western North America and South America. But in other areas, notably the western Pacific region, much of central Asia, northern South America, and Africa, El Niño years bring decreased rainfall and lower river discharge. With time, as the trades intensify, the warm pool again shifts westward, resulting in what are commonly termed La Niña conditions. Rainfall patterns (and thus river runoff) shift in response, resulting in greater rain in Asia and Oceania and less rainfall throughout much of the eastern Pacific. The entire cycle is termed the El Niño-Southern Oscillation (ENSO).

The intensity of the ENSO signal can be quantified in several ways. One is the air pressure differential between Tahiti and Darwin, Australia. Higher pressure in the western Pacific (Darwin) means a negative Southern Oscillation Index (SOI), indicative of El Niño; higher pressure in Tahiti indicates La Niña conditions. ENSO also can be defined on the basis of wintertime (December–March) deviations – the season of strongest atmospheric circulation – from mean sea-surface temperature (SST) in the tropical eastern Pacific (5N–5S; 120–170° W) (Niño index 3.4). Using this definition, El Niño is defined as an SST +0.5° or more above mean SST, and La Niña is −0.5° C or lower than mean SST (Fig. 3.4). When SST anomalies range between −0.5 and +0.5, the pseudo-Spanish term Nada Niño is sometimes used. A complete ENSO cycle generally lasts between five

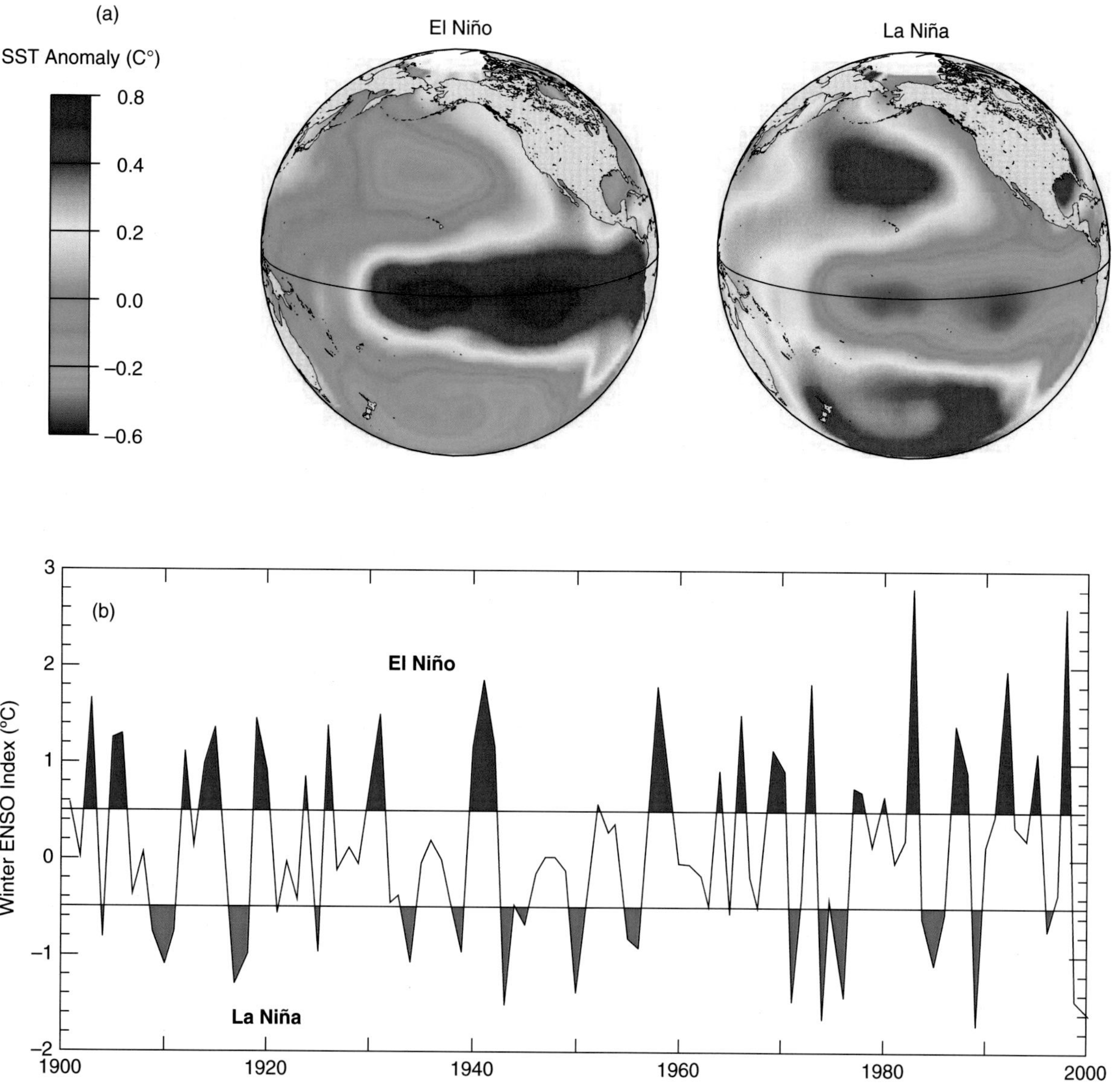

Figure 3.4. (a) Distribution of sea-surface temperature (SST) anomalies during El Niño and La Niña phases of ENSO; after N. Mantua (http://jisao.washington.edu/pdo/). (b) Wintertime (December-March) ENSO Index (Niño index 3.4), 1901–2000. During this 100-yr period there were approximately 20 El Niño and 15 La Niña events, most lasting one year. The number of events – as well as their strength and length – appears to have increased during the last half of the century.

and eight years (Fig. 3.4), and over the past 50 years there have been roughly equal numbers of El Niño, La Niña, and Nada Niño events.

ENSO is closely related to the longer-term Pacific Decadal Oscillation (PDO), which can be thought of as an extra-tropical low-frequency ENSO (Deser *et al.*, 2004) although it seems to be at least partly derived from the tropics (Newman *et al.*, 2003; Shakun and Shaman, 2009). The PDO, which is closely linked to the Inter-decadal Pacific Oscillation and the North Pacific Index (see discussion by Trenberth *et al.*, 2007), is marked by the east-west shift of the warm pool of North Pacific surface water (Fig. 3.5) caused by fluctuations in the strength of the Aleutian low-pressure system. Shakun and Shaman (2009) have suggested the presence of a southern-hemisphere equivalent to the PDO. PDO is measured as SST anomalies at latitudes >20° N; any influence of global warming on the anomaly is minimized by removing monthly mean global average SST anomalies. With the shift in warm-water and cold-water pools, the paths of the jet stream and ensuing storms are altered (Mantua *et al.*, 1997). During weakened low atmospheric circulation, the warm pool

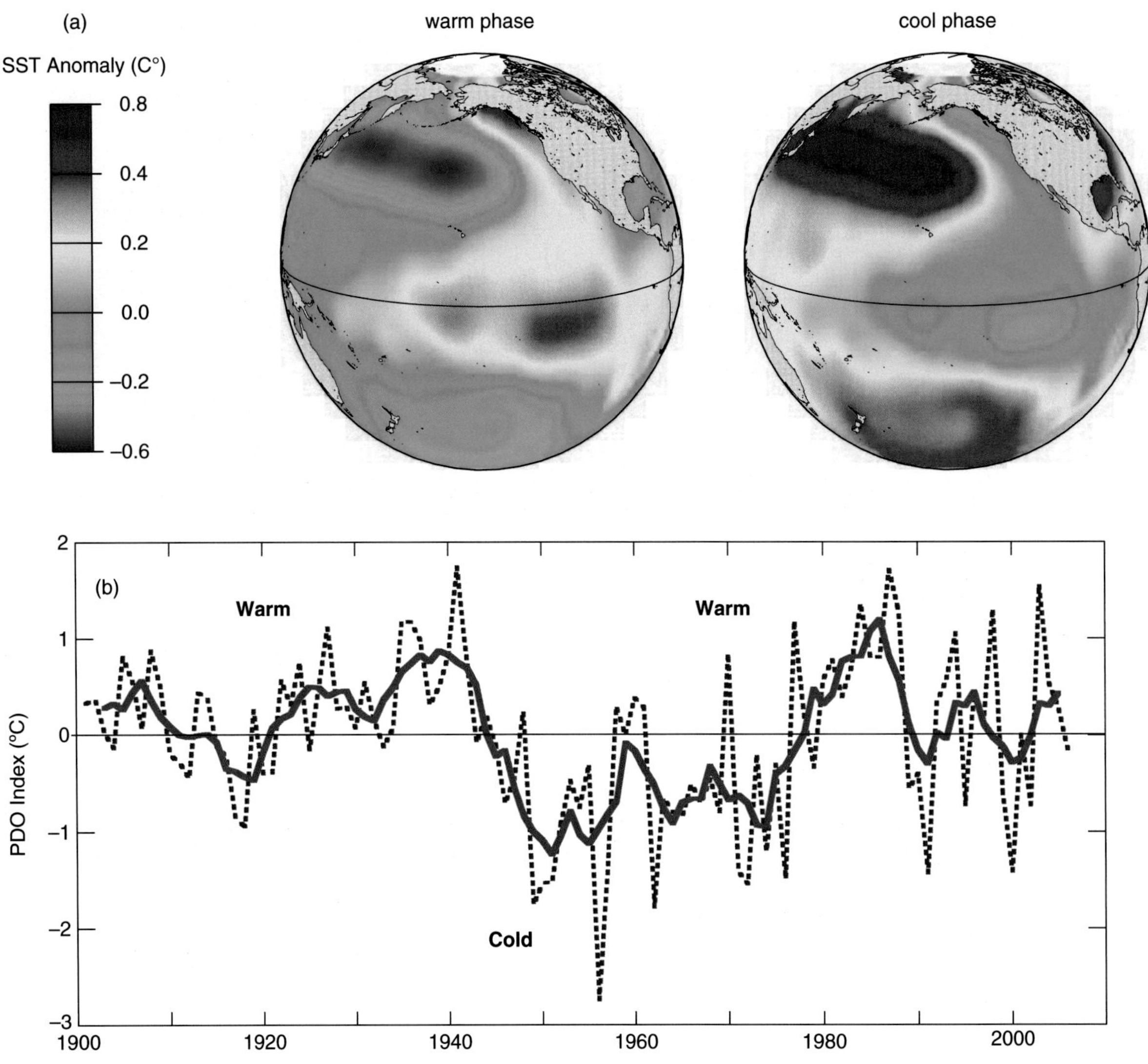

Figure 3.5. (a) Distribution of warm and cool phases of the Pacific Decadal Oscillation (PDO); after N. Mantua (http://jisao.washington.edu/pdo/). (b) Interannual variation in PDO index (dashed blue line), 1900–2006. Solid red line represents 5-yr running mean.

remains in the west and SST off western North America is cooler than average: the PDO cold phase. In response to increased strength of the Aleutian low-pressure system, the warm pool shifts eastward, resulting in a PDO warm phase (Hare, 1996; Mantua *et al.*, 1997; Mantua and Hare, 2002) (Fig. 3.5).

In contrast to ENSO, the PDO is inter-decadal in length, each phase lasting 20–40 years. Over the past 100 years there have been two cold PDO phases, 1890 to 1924 (although this is open to question) and 1947 to 1977, and two warm phases, 1925 to 1947 and one beginning in 1977 (Mantua *et al.*, 1997; Minobe, 1997) (Fig. 3.5). The connection between PDO and ENSO can be seen by the increase in El Niño anomalies after the shift to the warmer PDO in 1976 (see discussion by Alverson *et al.*, 2003, pp. 118–120). Based on the increased frequency and severity of droughts in the American southwest and throughout Mexico in recent years (Stahle *et al.*, 2009), we suspect that the latest PDO warm phase may have ended some time between 1999 and 2005. A more exact timing of the shift will become clearer when the database is longer.

Together, ENSO and PDO have great control on weather patterns throughout much of the tropical and subtropical world, not only precipitation but also evapotranspiration (Meza, 2005). Flow of US West Coast and some Gulf Coast rivers, for example, is largely controlled by ENSO and PDO (Latif and Barnett, 1994; Vega *et al.*, 1998). In the US southwest, river discharge is generally higher during El Niño years (Kahya and Dracup, 1994), but the ratio

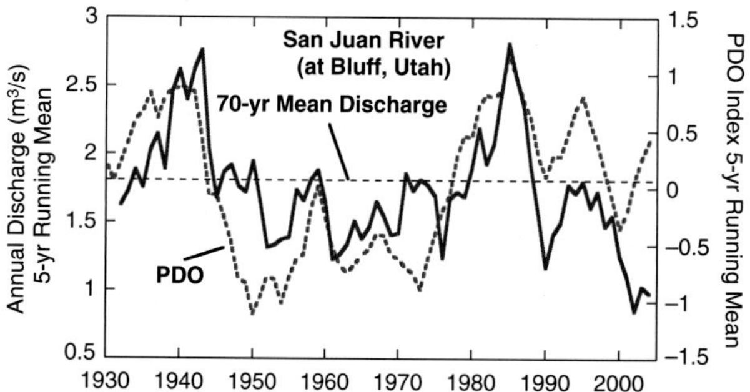

Figure 3.6. Five-year running means of San Juan River (SJ on Fig. 3.7 map) annual discharge (as gauged at Bluff, Utah), blue line, and the Pacific Decadal Oscillation (PDO) index, red dashed line. Data from USGS and US National Oceanic and Atmospheric Administration (NOAA).

of El Niño to non-El Niño floods decreases sharply north of 32° N (Andrews *et al.*, 2004) and the influence of La Niña increases, such that stream flow in the northwest tends to be lower during El Niño years (Cayan and Webb, 1992; Shabbar, 2006).

The influence of PDO, however, should not be minimized: El Niño-related discharge in southwest US rivers, for example, tends to be greater during warm PDO phases, as seen by the close correlation between PDO and the annual discharge of the San Juan River, a tributary to the Colorado River (Fig. 3.6) and southern California rivers (Figs. 3.7 and 3.8). Along the North American west coast, the influence of the warm PDO decreases northward, its influence – at least in terms of annual discharge – apparently ending at ~30°N. North of ~40° N La Niña periods tend to be wetter during cold PDO Phases (Tootle and Piechota, 2006; Figs. 3.7 and 3.8). Wintertime discharge from southeastern Alaskan rivers is greater during warm-PDO years, and greater in summer during cold-PDO years (Neal *et al.*, 2002), particularly in glaciated basins (Hodgkins, 2009). Given the N–S shift in the impact of ENSO and PDO, discharge in N–S flowing rivers, such as the Colorado, can reflect the influence of different drivers, northern tributaries experiencing increased flow during La Niña years and southern tributaries having higher flow during El Niño years, particularly during the PDO warm phase (Kim *et al.*, 2006; Cañon *et al.*, 2007; Ropelewski and Halpert, 1989; Kahya and Dracup, 1993, 1994). The correlation of river flow with El Niño and warm PDO continues eastward, throughout southeastern Texas and into southern Georgia and Florida (Kurtzman and Scanlon, 2007).

In northern South America precipitation and river runoff are greater during La Niña years, as seen in the Magdalena, Orinoco, and northern tributaries of the Amazon (Waylen, 1995, his Fig. 3.10; Krasovskaia *et al.*, 1999; Restrepo and Kjerfve, 2000b; Gutierrez and Dracup, 2001; Foley *et al.*, 2002; Aalto *et al.*, 2003). To the south we again find a positive correlation with the warm PDO and El Niño (Depetris *et al.*, 1996; Amarasekera *et al.*, 1997; Meza, 2005), although this trend shows some regional anomalies. As in North America, South American precipitation and river discharge appear to be greatest when ENSO and PDO are in phase (Kayano and Andreoli, 2007). Parana River (Argentina) discharge, for example, increased sharply in the late 1970s, the same time as the increased discharge from the San Diego River in California, and coincident with the advent of the warm PDO (compare Fig. 3.9 with Fig. 3.6). In Peru, on the other hand, the effect of El Niño floods decreases markedly south of about 8° S; the mean El Niño/La Niña flood ratio in the Piura River (5°13'S) is about 30, whereas farther south in the Pisco River (13°40'S), the El Niño/La Niña flood ratio is only 1.2 (Waylen and Caviedes, 1987, their Table 3), not dissimilar from what we see in the south–north transition in California (Fig. 3.8).

Because of its west-east migration as well as the fact that the ENSO Index is referenced to the eastern Pacific, ENSO effects in eastern South America and Africa can lag the ENSO signal by as much as a year, precipitation declining during El Niño years (Jury, 2003; Love *et al.*, 2008). Major droughts in southern Africa and southern Asia often occur during the second year of an ENSO warm event (Dilley and Heyman, 1995). As with the Colorado River, N–S-flowing rivers can display a mixed signal; the Blue Nile flow increases during La Niña whereas the White Nile shows little correlation with the ENSO index (Amarasekera *et al.*, 1997). Throughout much of Africa, however, North Atlantic oceanic and atmospheric drivers can be important, as pointed out in the following section.

While the effects of ENSO on Middle East precipitation are somewhat less certain, precipitation in western and eastern Turkey increases during El Niño years (Karabork *et al.*, 2007). Throughout much of tropical Asia and Oceania, rainfall and river flow increase (decline) often one year prior to La Niña (El Niño) conditions being noted in the eastern Pacific, perhaps as a result of suppressed convection (Xu *et al.*, 2004; Zhang *et al.*, 2007; Shaman and Tziperman, 2005) (Fig. 3.10). In Bangladesh, cyclone frequency and magnitude decline considerably during El Niño years (Choudhury, 1978).

Precipitation and river flow throughout southern Asia increase during La Niña years. The effect of ENSO, however, is not entirely restricted to the tropics; it can be felt as far north as the Yellow River (Huanghe) watershed in northern China (Wang *et al.*, 2006) and as far south as eastern Australia – in Queensland and New South Wales, although not in Victoria (Verdon *et al.*, 2004) (Fig. 3.10). ENSO influence on New Zealand rivers appears to be

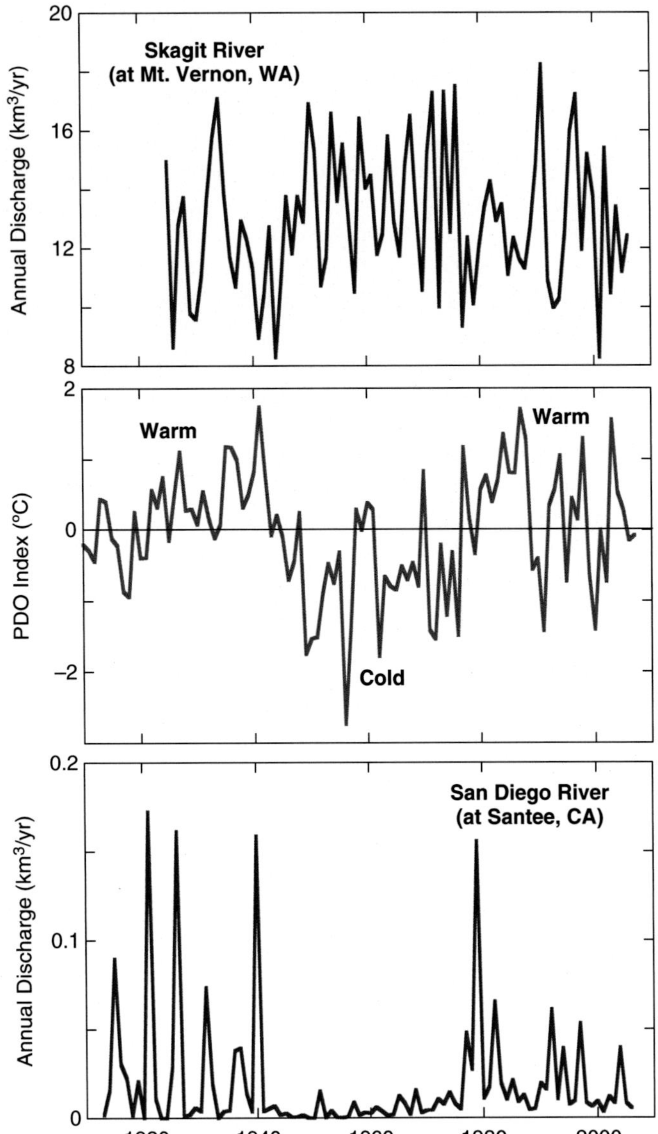

Figure 3.7. Annual discharge of the Skagit (S; northern Washington) and San Diego (SD; southern California) rivers relative to the annual PDO index. Note the marked difference in San Diego discharge during cold and warm PDO phases, as well as in response to El Niño and La Niña events. Interannual fluctuations of Skagit discharge appear more regular during the cold-phase PDO (1947–1977) than during warm phases. Data from USGS and NOAA.

highly variable in part because of the mountainous terrain (Mosley, 2000).

North Atlantic Oscillation (NAO) and Northern Annular Mode (NAM)

Another major driver of weather patterns that directly impacts river flow, albeit primarily in the Atlantic and Arctic oceans (Hurrell, 1995; Dettinger and Diaz, 2000), is the North Atlantic Oscillation (NAO). The NAO refers to the redistribution of atmospheric mass between the Arctic and subtropical Atlantic, which in turn reflects and affects significant changes in the heat and moisture transport between the Atlantic and peripheral continents, thereby influencing the number and intensity of storms (see Hurrell, 1995; Hurrell *et al.*, 2003). The Northern Annular Mode (NAM), also known as the Arctic Oscillation (Thompson and Wallace, 1998), reflects winter sea-level pressure over the entire northern hemisphere and therefore is highly correlated with the NAO (Fig. 3.11). Its role in determining wintertime circulation in the northern hemisphere seems uncertain (Deser, 2000; Trenberth *et al.*, 2007). Here we primarily refer to the NAO, recognizing its connectedness with the NAM.

The NAO index is commonly reported as the difference between wintertime (December–March) normalized sea-level pressure anomalies at Lisbon, Portugal, and those at Reykjavik, Iceland (Hurrell, 1995) (Fig. 3.11). Gibraltar, however, may be a more suitable southern reference station

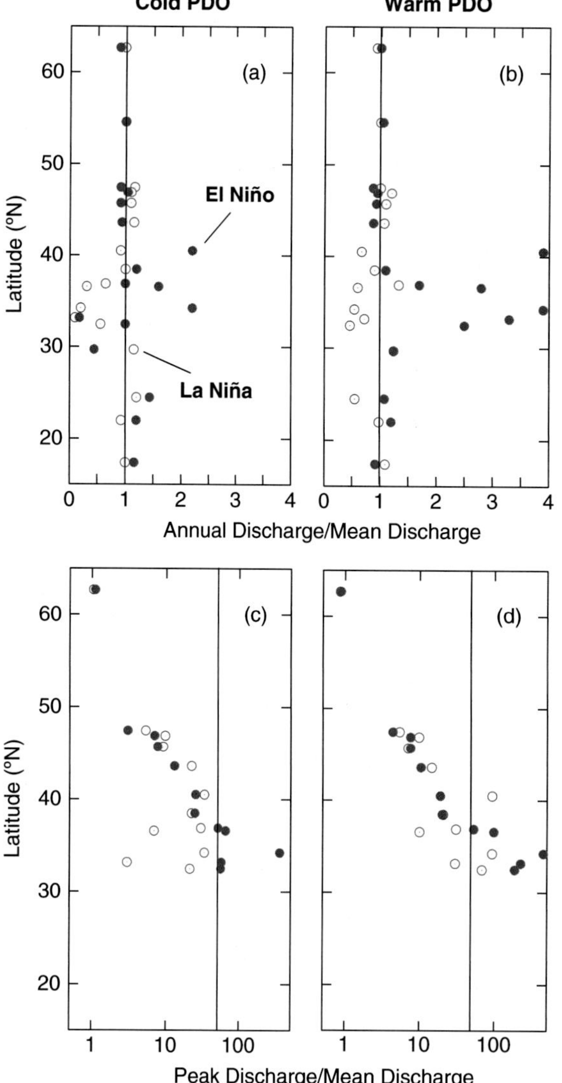

Figure 3.8. El Niño (red dots) and La Niña (blue circles) annual discharge of North American west-coast rivers relative to mean annual discharge (a, b), and peak discharge relative to mean annual discharge (c, d) during cold-phase and warm-phase PDO. Note the northward decrease of peak/mean discharges during both cold and warm PDO phases; to the south rivers have significantly higher annual and peak discharges during El Niño years, whereas to the north La Niña discharge tends to be somewhat higher. Cold PDO La Niña events occurred in 1951, 1955, 1956, 1965, 1968, 1971, 1974 and 1976; warm PDO La Niña events in 1984, 1985, 1986, 1996, 1999, and 2000. Cold PDO El Niño events occurred in 1952, 1958, 1959, 1964, 1966, 1969, 1970, and 1973; warm PDO El Niño events in 1977, 1978, 1980, 1983, 1987, 1988, 1992, 1995, and 1999. The remaining years experienced neutral ENSO signals (−0.5 to +0.5; see Fig. 3.4). Note that peak vs. mean discharge axes in (c) and (d) are logarithmic. Data from USGS and Global River Data Centre (GRDC).

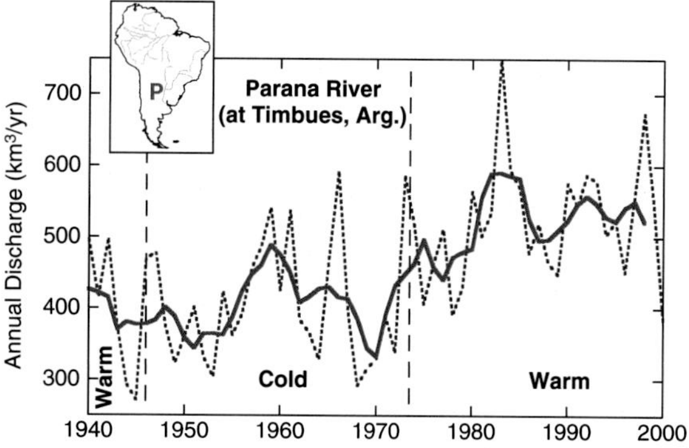

Figure 3.9. Annual discharge (blue dashed line) of the Parana River (P), 1940–2000, as measured at Timbues, Argentina; solid red line represent 5-yr running mean. Dashed vertical lines give the approximate dates of cold and warm PDO phases. Data from GRDC and P. Depetris (personal communication).

than Lisbon because of its longer historical record (Jones *et al.*, 1997). A positive NAO index translates to increased westerly circulation of moist air over North America, resulting in increased precipitation – often during winter storms – across eastern North America and northern Europe, and drier conditions in northeastern Canada and southern Europe (Fig. 3.11). In contrast, negative NAO indices coincide with fewer high-latitude storms and increased precipitation in southern Europe, extending into Spain (López-Moreno *et al.*, 2007), down into northern and central Africa as far south as the Congo River basin (Fig. 3.12), and as far east as western Asia. There does not appear to be a significant relation between NAO and ENSO, although there may be indirect links affecting North Atlantic sea-surface temperatures (Hurrell *et al.*, 2003, and references therein).

The 145-year Reykjavik–Lisbon sea-level pressure record shows four periods during which both NAO index amplitude and period had distinctly different characteristics. Between 1864 and 1900, the running 5-yr mean of the NAO index fluctuated between −1 and +0.6, and each cycle, although not well-defined, lasted 7–9 years (Fig. 3.11). From 1900 to 1940, the NAO index displayed two 20-yr periods during which amplitudes were significantly greater than before, −1.2 to +2.0. Between 1940 and 1965 the NAO indices ranged between −1.2 and +1.2, each cycle lasting 8–10 years. From 1965 to 2000 the NAO index was characterized by four 5–8-yr-long periods during which the index increased, from +1.4 in the mid 1970s to +3.2 in the early 1990s (Fig. 3.11).

The NAO influence on precipitation and runoff is seen in northeastern and mid-Atlantic USA rivers. Note, for instance, the close correlation between the depressed NAO index in the mid 1960s and the sharp decline in river runoff

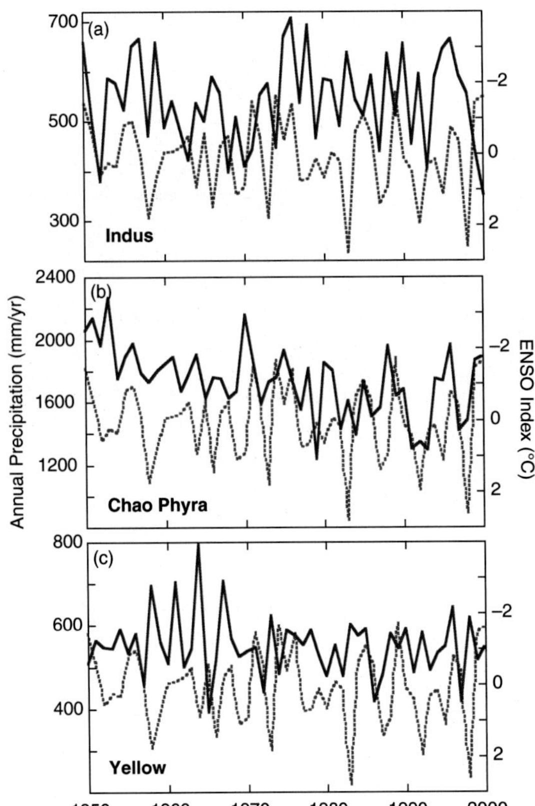

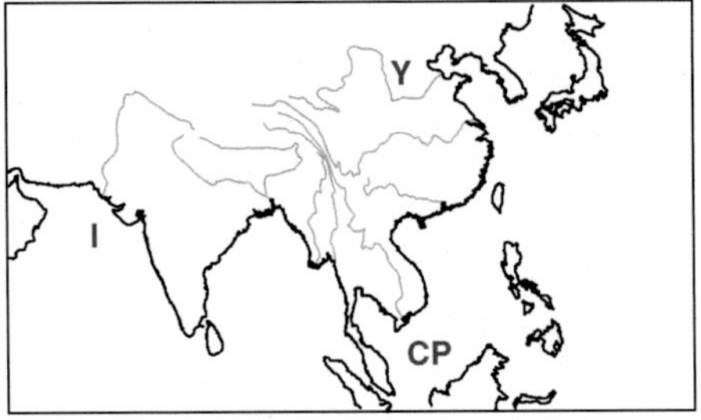

Figure 3.10. Annual basin-wide precipitation in the Indus (I), Chao Phyra (CP) and Yellow River (Y) watersheds (solid blue lines) vs. winter ENSO index (red dashed lines). Because extensive damming and irrigation have affected discharge in all three rivers, precipitation is a better measure of natural runoff and therefore of ENSO climatic impact. The ENSO index is reversed so that La Niña now reads positive. Annual precipitation data extrapolated for the three river basins by K.-H. Xu, using CRU (Climate Research Unit, University of East Anglia) archived data.

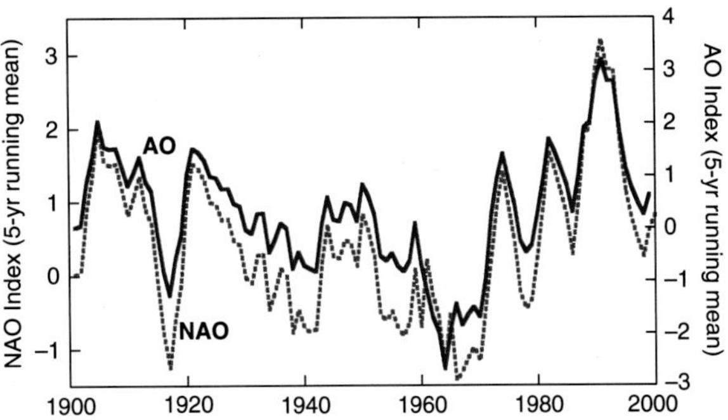

Figure 3.11. Wintertime (December–March) indices (5-yr running means) for the North Atlantic Oscillation (NAO; red line) and the Northern Annual Mode/Arctic Oscillation (NAM/AO; blue line), 1864–2000. The offset between these two curves mostly reflects different scales of their vertical axes; the only noticeable discrepancy between the two occurred in the 1960s. Data from J. Hurrell (www.cgd.ucar.edu/cas/jhurrell/indices.data.html#naopcdjfm).

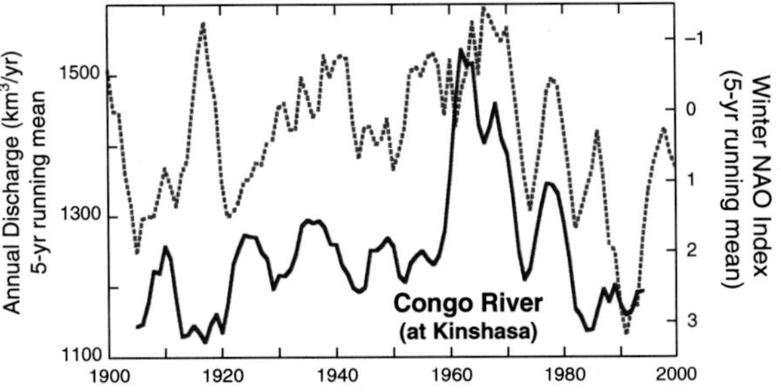

Figure 3.12. Annual Congo River (C on Fig. 3.18) discharge (5-yr running means; solid blue line) vs. 5-yr running means of the NAO index (dashed red line). For ease of comparison, the NAO index axis has been reversed. Note the negative correlation after ~1960 and positive correlation prior to ~1930.

throughout much of eastern North America (Fig. 3.13). Farther north, as seen in the Churchill River in Labrador, the positive correlation between NAO and annual discharge breaks down (Fig. 3.13), reflecting the NAO-induced circulation patterns.

As implied above, there is an inverse relation between the NAO index and river runoff in Europe (Fig. 3.14), which extends into Turkey (Kiely, 1999; Hanninen *et al.*, 2000; Lloret *et al.*, 2001; Quadrelli *et al.*, 2001), the Don River basin in southern Russia (Milliman *et al.*, 2010; Fig. 3.14), southward into northern Iran, westward into Pakistan and northwestern India (Syed *et al.*, 2006), and southward to the Congo River (Fig. 3.12). The northern limit of this correlation appears to be somewhat north of the Gota River

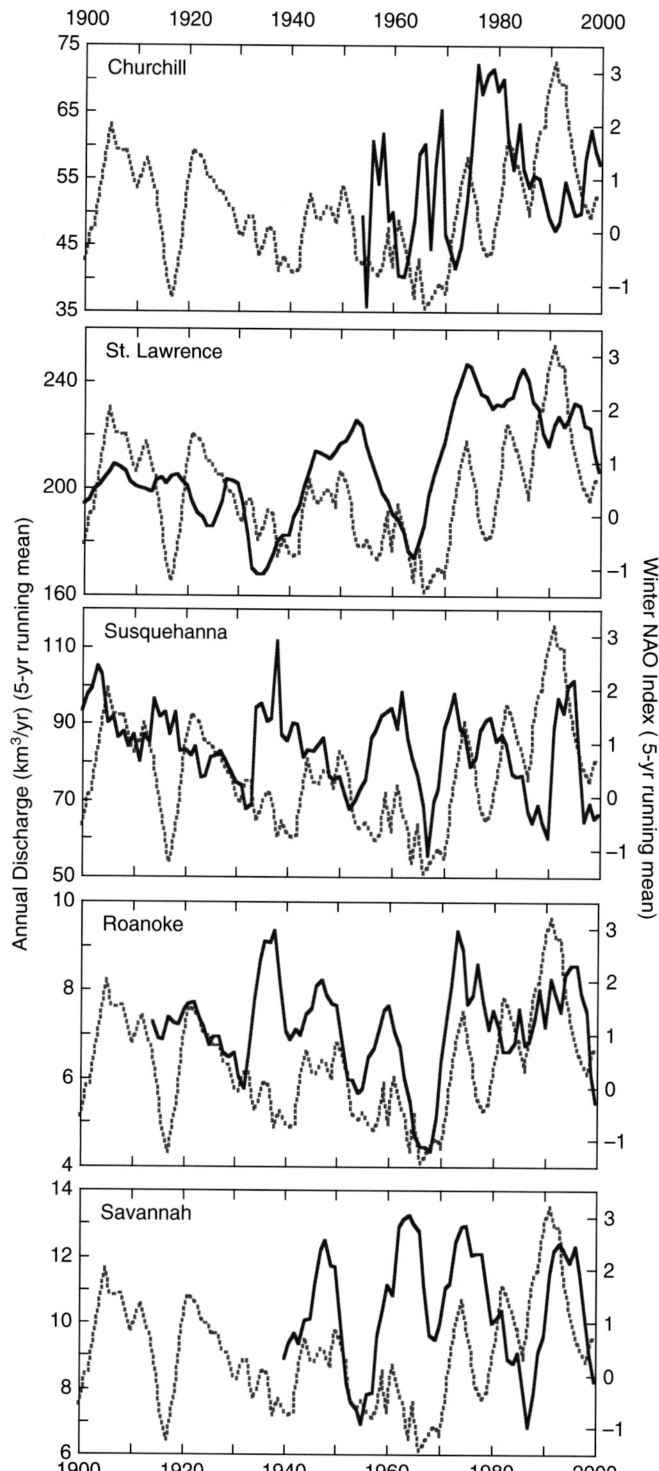

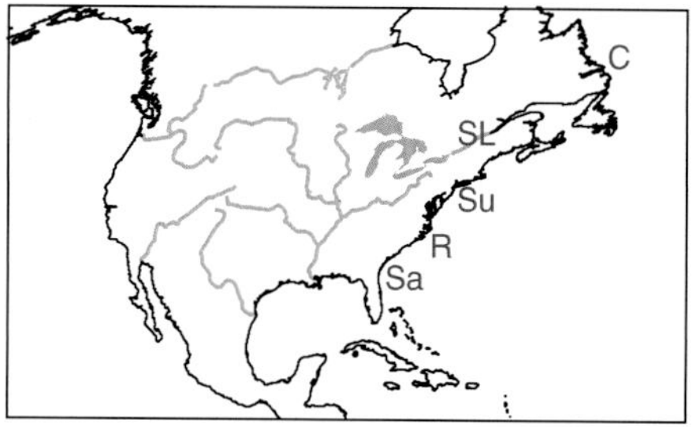

Figure 3.13. North to south annual discharges (5-yr running means; solid blue lines) of the Churchill (C), St. Lawrence (SL), Susquehanna (Su), Roanoke (R) and Savannah (Sa) rivers, 1900–2000, compared to the annual winter (December–March) NAO Index (5-yr running means; red dashed lines). In northeastern Canada, as shown by the Churchill's discharge record, there is little relation between NAO and river runoff. Within the St. Lawrence's watershed and farther south, however, the NAO appears to be the prime driver of long-term precipitation and runoff, as shown by the marked decrease in NAO and river runoff in the mid to late 1960s, high discharge in the mid 1970s, etc. Note that correlation between NAO and river discharge is less clear prior to ~1940 or after ~1985. Data from GRDC, NOAA and USGS.

(Figs. 3.14), as western Arctic rivers appear to be directly correlated with NAO (see Fig. 3.16).

A closer inspection of Figures 3.12–3.14, however, suggests that the influence of the NAO/AO on Atlantic watershed precipitation – and thereby river flow – is more complex than might appear at first glance. The close direct correlation between NAO index and river runoff in eastern North American rivers, for instance, began about

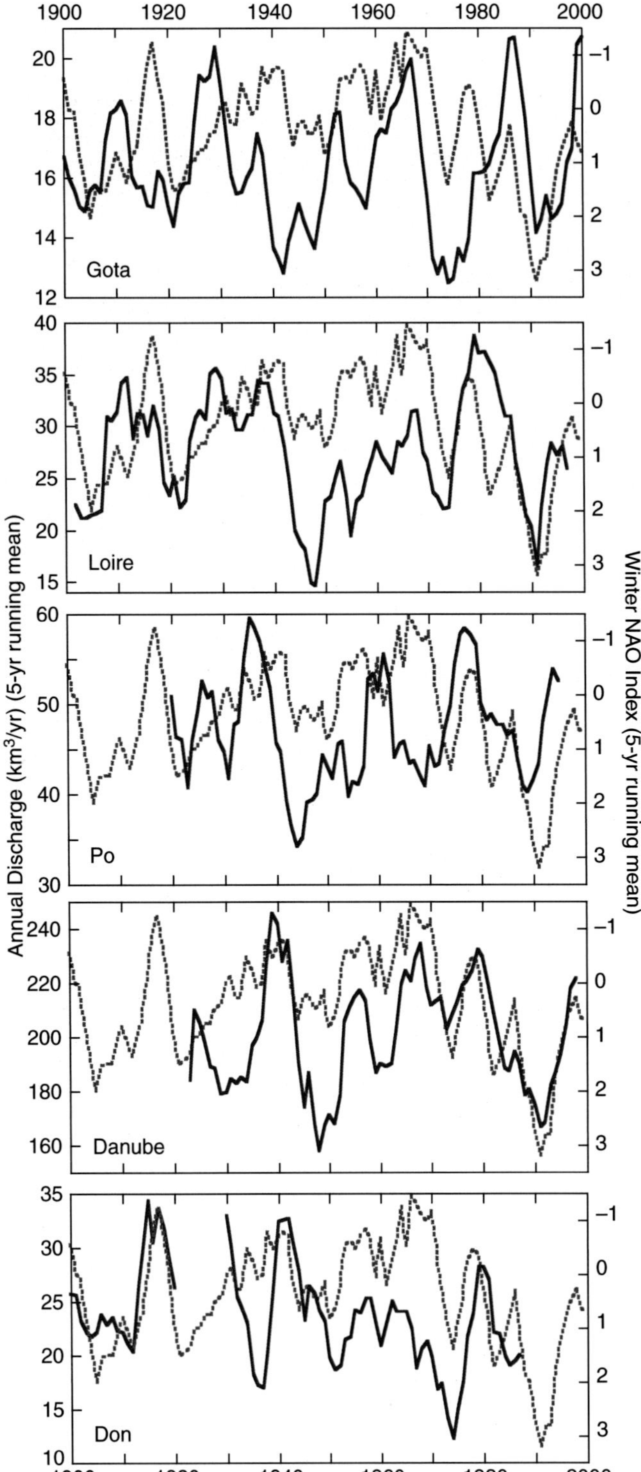

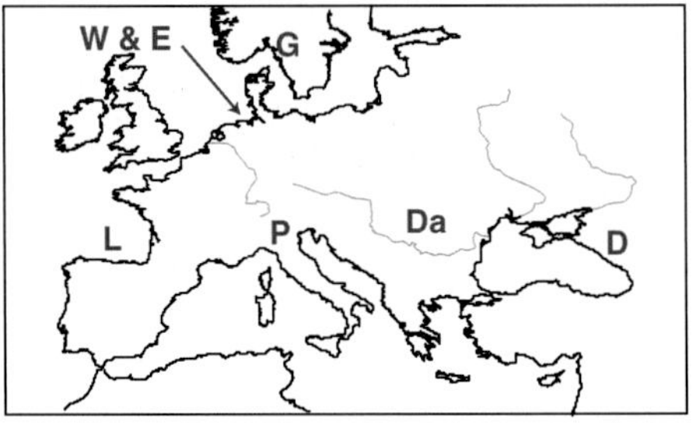

Figure 3.14. Annual discharge (5-yr running means; solid blue lines) for the Gøta (G), Loire (L), Po (P), Danube (Da) and Don (Do) rivers, 1900–2000, compared to the annual winter (December–March) NAO index (5-yr running means; dashed red lines). For comparison, NAO index is reversed so that negative NAO reads positive. Note the close inverse relation between the NAO index and river discharge, as evidenced, for example, by the low discharges during the mid 1970s and early 1990s. Data from GRDC and NOAA.

1940 and was not uniform. The low NAO index in the early 1940s, for instance, is reflected by decreased discharge of the St. Lawrence, but not in Susquehanna or Roanoke discharge; and the low NAO index around 1917 is not reflected in either the St. Lawrence or Susquehanna (Fig. 3.13).

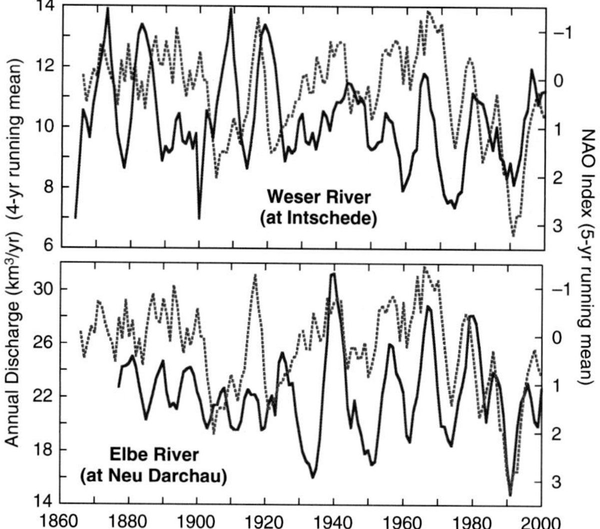

Figure 3.15. Time series of 5-year running means of the Weser River (at Interschede) and Elbe River (at Neu Darschau) ((W and E, respectively, in map in Fig. 3.14) discharge (solid blue lines) vs. NAO Index (dashed red lines; axis reversed for ease of comparison). Note a close inverse correlation of Weser discharge and NAO after about 1930, prior to which the two appear to have been directly correlated. In contrast, Elbe discharge at Neu Darchau, roughly 100 km to the north and east, has remained more or less inversely correlated with NAO since the 1870s. Data from NOAA and GRDC.

The situation is more complex in Europe, where there appears to have been several changes in the relation between NAO and river discharge. Prior to 1920, for instance, Weser River discharge (as gauged at Intschede, northwest Germany) was generally inversely related to the NAO index. But between 1875 and 1910 there was a direct relation between discharge and the NAO index (Fig. 3.15). It was only after about 1950 that Weser River discharge and the NAO index were inversely related. Similar trends are noted in the Gøta River to the north. In contrast, the Elbe River (at Neu Darchau, roughly 100 km north and east of Intschede) has displayed a generally constant relation with NAO since the mid 1870s (Fig. 3.15). At the southern extreme (as far as we can tell) of the NAO influence, the Congo River shows a direct relation with NAO prior to ~1935, in contrast with a strong inverse correlation between winter/spring NAO and river discharge after 1945 (Fig. 3.12; see also Todd and Washington, 2003). Although we can think of several possible reasons why the correlation between and river discharge and the NAO index would change with time, such as changes in the longer-period Atlantic Multidecadal Oscillation (see below), the most likely seems to be latitudinal shifts in the westerly circulation of moist air, thereby influencing the pathways and intensities of precipitation.

In the Arctic, the correlation between NAO and precipitation/runoff is less clear. Influence from the NAM/AO appears to affect discharge of the Amur River (Ogi and Tachibana, 2006). River discharge in Hudson Bay (Déry and Wood, 2004) and river discharge to the Eurasian Arctic (Peterson *et al.*, 2002) also have been correlated with the NAM/AO. But an east-to-west transit of the Arctic basin (Fig. 3.16) shows that the imprint of the NAO ranges from strongly positive (Lena) to modest (Severnaya Dvina) to negative (Kolyma). Superimposed on Arctic discharge trends is a long-term increase in river discharge (Peterson *et al.*, 2002).

Atlantic Multidecadal Oscillation (AMO)

The Atlantic Multidecadal Oscillation (AMO; Kerr, 2000) represents the cyclic warming and cooling of the North Atlantic surface waters, each cycle apparently lasting 65–75 years. A warm phase lasted for at least 30 years prior to 1900, followed by a cool phase until 1930. Between ~1930 and ~1965, there was another warm phase, then a cool phase until 1995, followed by the current warm phase (Fig. 3.17).

The AMO's control on precipitation and river runoff remains relatively undocumented, in part because it only recently has been recognized as a major climate-driver (Thompson and Wallace, 2001). Extensive droughts throughout the conterminous USA may be associated with the convergence of positive AMO and negative PDO (McCabe *et al.*, 2004). Sutton and Hodson (2005) also have observed a correlation between sea-surface temperature and the frequency of droughts. Kelly and Gore (2008) noted that river flow in Florida was greater during AMO warm periods than cold periods. But the AMO's influence may be much wider. Zhang and Delworth (2006), for instance, report a close relationship between AMO and rainfall in the Sahel of Africa and in India.

The two cold AMO periods in the twentieth century appear to correspond with periods of low Sahel precipitation, as seen in the Senegal and Niger drainage basins (Fig. 3.18), although these rivers are also influenced by ENSO. A somewhat clearer positive relation is seen in between AMO and precipitation in Orange River basin (Fig. 3.18) although there is also some influence from ENSO and the Southern Annular Mode (see next section). A negative relation between AMO and precipitation is seen in the Sao Francisco basin in Brazil, high annual precipitation tending to occur in years with low AMO indices (Fig. 3.18).

Southern Annular Mode (SAM)

Also known as the Antarctic Oscillation (Thompson and Wallace, 2000), the SAM index represents the difference

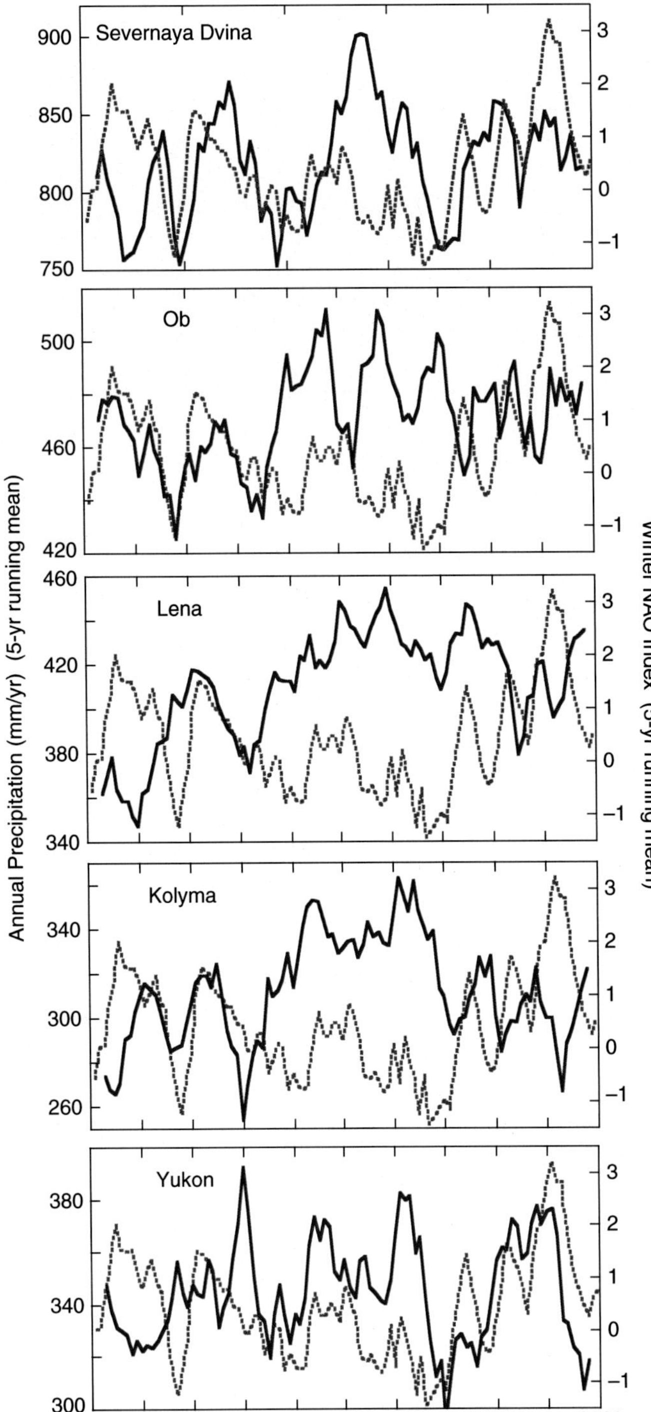

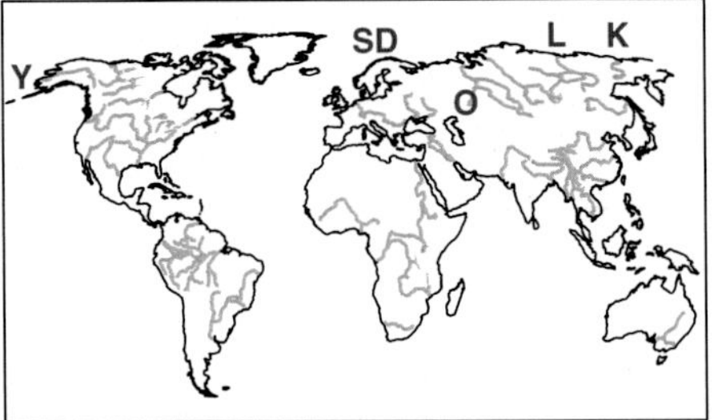

Figure 3.16. Temporal variation of North Atlantic Oscillation (NAO; dashed red lines) and basin-wide precipitation (solid blue lines) in the Severnaya Dvina (SD), Ob (O), Lena (L), Kolyma (K) and Yukon (Y) watersheds. Precipitation is used here because of gaps in reported discharge data. Precipitation data from CRU, University of East Anglia.

in geopotential sea-level height between Antarctica (65°S) and southern hemisphere mid-latitudes (40°S). As such, it provides a measure of the changes in the main belt of sub-polar westerly winds, which intensify during the positive phase of SAM (Gillett *et al.*, 2006; Trenberth *et al.*, 2007). The annual SAM index varies from −4 to +3.5. Prolonged periods of positive SAM occurred in the 1930s and early 1940s, and more recently in the late 1970s to late 1980s (Fig. 3.19).

There is an inverse relation between the winter (JJA) SAM index and precipitation/river discharge in south-eastern Australia and western New Zealand (Meneghini *et al.*, 2007; Vera and Sylvestri, 2009; Wallace, 2009), as well as in southernmost South America (although the trend is subtle; Fig. 3.20). In contrast, precipitation and river runoff seem to increase throughout much of Australia, southern Argentina, and southeastern Africa during periods of positive SAM (Fig. 3.20). In the mid-latitudes of Australia SAM may play as important a role as ENSO in driving climate variability (Watterson, 2009). The reader should be warned, however, that the relatively poor database for southern hemisphere rivers as well as the short period in which SAM has been recognized as a potentially important climate driver prevent us from painting a clearer picture of SAM and its role in influencing precipitation and river runoff.

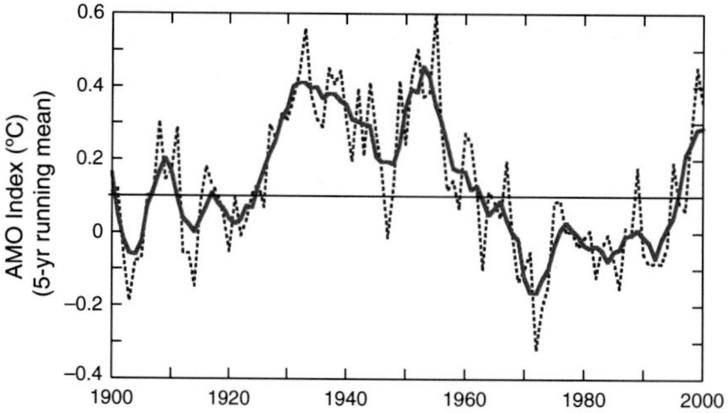

Figure 3.17. Atlantic Multidecadal Oscillation Index (AMOI), 1900–2000 (dashed blue line); solid red line represents 5-yr running mean. Data from the Kaplan Extended SST anomaly data (http://ingrid.ldgo.columbia.edu/sources/.kaplan/.extended/).

Summary

The oceanic and atmospheric factors controlling precipitation and river discharge are continually being re-evaluated and redefined as new data are brought to light and new temporal and spatial trends are identified. Similarities in temporal trends for PDO, AMO and SAM – e.g. low indices around 1920, late 1940s early 1970s, and highs in

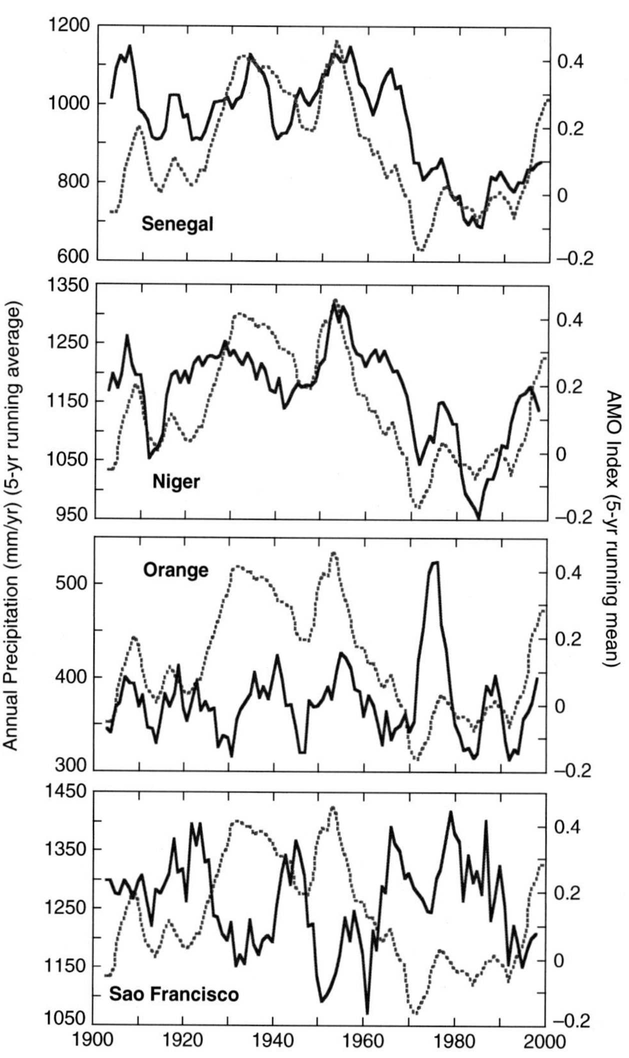

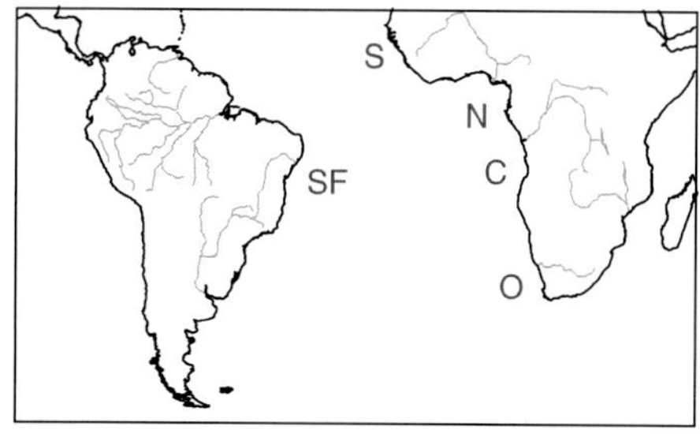

Figure 3.18. Temporal variation in the AMO (red dashed lines) and river basin precipitation in the Senegal (S), Niger (N), Orange (O) and Sao Francisco SF) river basins (solid blue lines), 1900–2000. Due to incomplete gauging data – together with damming and irrigation – we use precipitation rather than river discharge. Note the generally positive correlation (albeit with some temporal lag) between the AMO and the African river basin precipitation in contrast to a negative correlation in the Sao Francisco basin.

the 1930s and 1960 (compare Figs. 3.5, 3.17, and 3.19) – suggest a possible connection, although these similarities also could simply represent an artifact of highly filtered data (K. Trenberth, P. D. Jones, written communication).

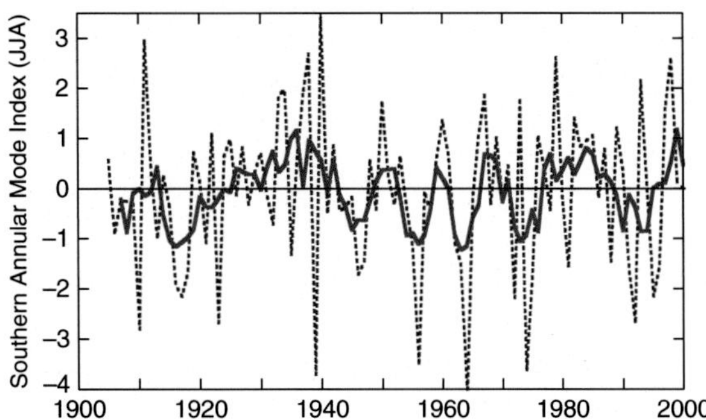

Figure 3.19. Annual wintertime (June, July, August) Southern Annular Mode (SAM) index (dashed blue line) and 5-yr running mean (solid red line), 1905–2005.

At present we must be content to say that two short-period (ENSO and NAO) and three longer-period (PDO, AMO and SAM) factors seem to drive much of the temporal and spatial variation seen in global precipitation and resulting river flow.

Correlating these drivers to short-term and long-term precipitation and discharge, we detect clear global distributions in which one or more climatic driver plays a major role in river discharge (Fig. 3.21). Because global precipitation is greatest in the tropics and subtropics (see Figs. 2.2, 2.3, 2.4, and 2.8) and is largely controlled by La Niña (Fig. 3.21), it is not surprising to find a close correlation between global precipitation and the ENSO signal, although often with a one-year lag (Fig. 3.22). Global precipitation during la Niña years tends to be 1–2% greater than during El Niño years. ENSO's impact also extends into higher latitudes, as seen by its teleconnections with North Atlantic precipitation and runoff (Emile-Gray *et al.*, 2007). Locally PDO appears to accentuate the ENSO effect, as seen in southwest US river runoff (Figs. 3.6–3.8).

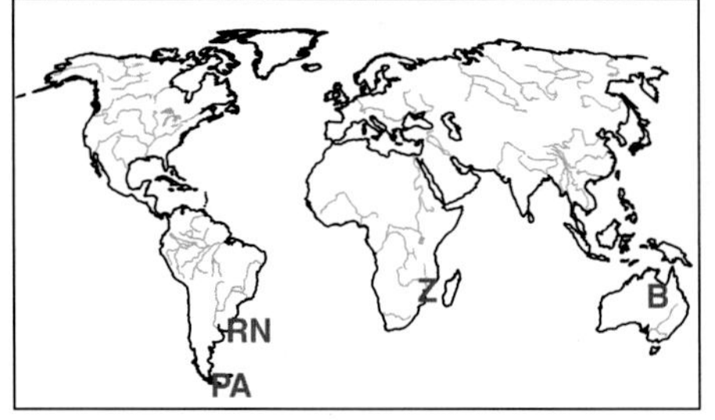

Figure 3.20. Temporal variation in annual river runoff and basin-wide precipitation vs. SAM index, 1900–2000, in eastern Australia (Burdekin River; B), southern Chile (Punta Arenas; PA), southern Argentina (Rio Negro; RN), and southeastern Africa (Zambesi River; Z). Discharge data from GRDC, Punta Arenas precipitation data from CRU, University of East Anglia.

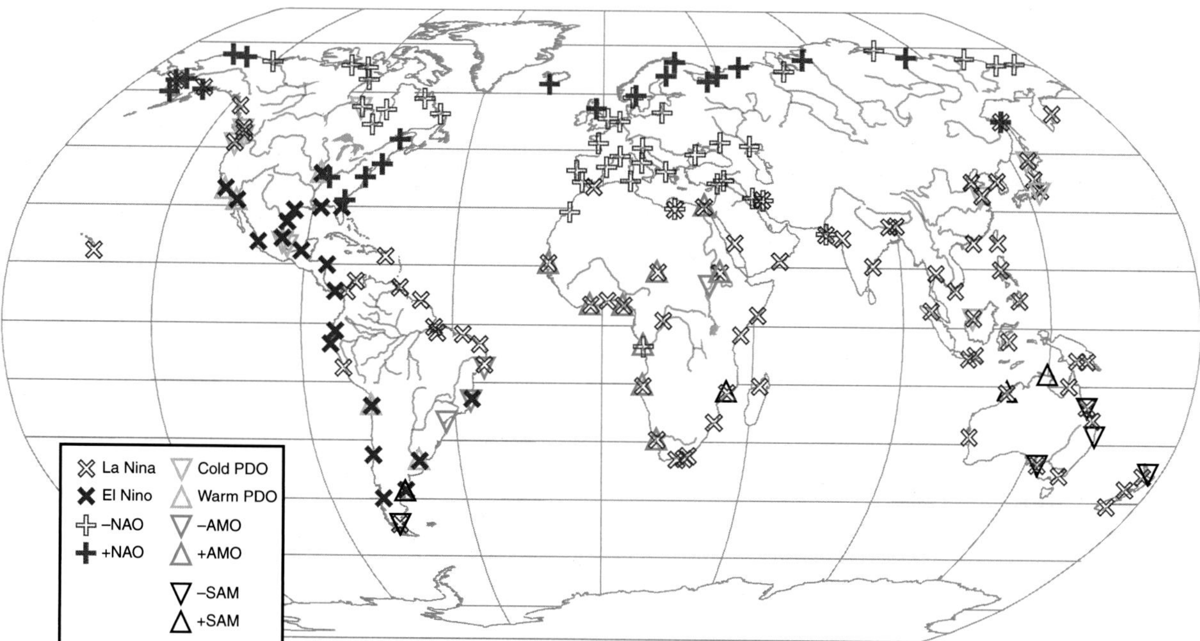

Figure 3.21. Global distribution of primary climatic signals that drive precipitation and river discharge. Maximum river discharges in rivers, delineated by solid crosses, occur primarily during years with El Niño or positive NAO indices. Similarly, rivers can respond to positive/warm PDO, AMO, and SAM indices (upward triangles) or negative/cool (downward triangles). Multiple symbols (e.g. El Niño and warm PDO) means that two drivers appear primarily responsible for temporal variations in river discharge. Most discharge data from GRDC and USGS, with additional data and correlations from and by Ropelewski and Halperb (1989), Kahya and Dracup (1994), Amarasekera *et al.* (1997), Thompson and Wallace (2000), Foley *et al.* (2002), Mantua and Hare (2002), Todd and Washington (2003), McCabe *et al.* (2004), Déry and Wood, (2004), Andrews *et al.* (2004), Verdon *et al.* (2004), Kelly (2004), Meza (2005), Wang *et al.* (2006), Ogi and Tachibana (2006), Tootle and Piechota (2006), Kim *et al.* (2006), Woodhouse *et al.* (2006), López-Moreno *et al.* (2007), Cañon *et al.* (2007), Kurtzman and Scanlon (2007), Xu *et al.* (2007), Karabork *et al.* (2007), Syed *et al.* (2006), among others.

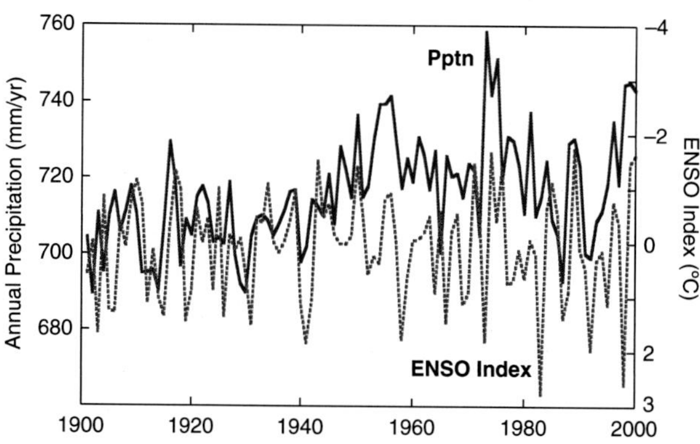

Figure 3.22. Temporal trends in global on-land precipitation (solid blue line) and winter ENSO index, 1901–2000 (dashed red line). Note that the ENSO index is reversed so that La Niña now reads positive. The slight temporal lag between global precipitation and ENSO index reflects the difference between an ENSO cycle first appearing off southern Asia and it being recorded as an SST shift in the eastern Pacific. Precipitation data from CRU, University of East Anglia.

The NAO/AO also plays a prominent role in northern hemispheric precipitation and runoff. While most easily recognized in rivers draining into the North Atlantic, it extends northwards into the Arctic and westwards to southwestern Asia and the northern Pacific Ocean. Least documented are the AMO, which extends throughout the tropical and extratropical Atlantic-draining watersheds, and the SAM, whose influence may extend into southern Africa (Figs. 3.20 and 3.21). How our understanding of these global trends will evolve depends on future research, which in part will depend on accessing and comprehending a better database.

Temporal variation in global discharge and runoff

Post-1950

A detailed analysis of temporal trends in discharge for 140 globally dispersed rivers shows that the overall global fluvial discharge for the second half of the twentieth century remained virtually unchanged (~0.5% ΔQ); discharge from 97 of these rivers, however, changed by 10% or more (Fig. 3.23) (Milliman *et al.*, 2008). For

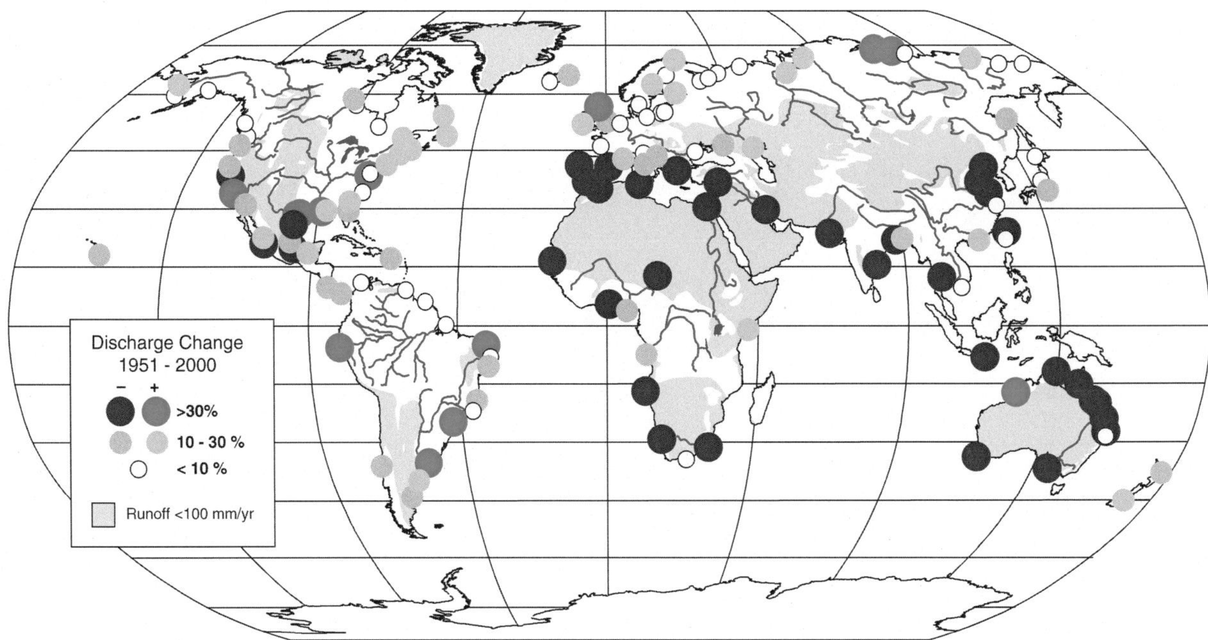

Figure 3.23. Discharge trends for 140 global rivers, 1951–2000; modified from Milliman *et al.* (2008), with additional data from Favier *et al.* (2009), Isik *et al.* (2008), and Mernild *et al.* (2008). Trends for many non-US rivers are based on as few as 40 years of data, but in all cases the data extend at least into the mid 1990s. Nearly all 50-yr changes greater than 10% are statistically significant (Milliman *et al.*, 2008). Note that many of the greatest changes are seen in low-runoff rivers, reflecting the impact that changes in precipitation or increased water withdrawal can have on these rivers (see Chapter 4). Many high-runoff rivers, in contrast, had changes less than 10%, the prime exceptions being the Mississippi and Parana, where PDO/ENSO-driven climate resulted in significant increases in basin-wide precipitation.

approximately 2/3 of the rivers in this database, discharge trends primarily reflect changes in basin-wide precipitation. For the other 1/3, however, change in river discharges reflects anthropogenic influence in the drainage basins, primarily the withdrawal of water for irrigation, but also the possibility of climatic change. Rivers draining lower mid-latitude arid and semi-arid regions in Africa, Asia, and Australia showed the greatest declines in runoff during the second half of the twentieth century, in response to decreased precipitation and increased water withdrawal. In contrast, many rivers with the greatest increases in discharge were located in the Americas, reflecting the shift to positive PDO in the 1970s (see Figs. 3.5–3.7, 3.9).

Twentieth century

The 1951–2000 record does not necessarily represent a sufficiently long period to encompass a complete PDO or AMO cycle. For a complete PDO cycle, for example, one would need a discharge record that extends back to the early 1940s (see Fig. 3.5). Unfortunately, as stated earlier, river discharge data for first half of the twentieth century are sparse and geographically biased (e.g. Fig. 1.4), making it difficult to extend global river discharge back much before 1950. Various attempts to model pre-1950 discharge (Labat *et al.*, 2004; Milly *et al.*, 2005), at best, have proved controversial (Legates *et al.*, 2005; Peel and McMahon, 2006; Milliman *et al.*, 2008). Assuming that global discharge accurately traces precipitation, however, we can use the latter, as assembled by the Climate Research Unit at the University of East Anglia, England, to delineate a longer-term record of global discharge.

During the twentieth century, we calculate global on-land precipitation to have increased by ~3%, significantly less than the 7.7% increase calculated by Gerten *et al.* (2008). Heaviest global precipitation occurred in the late 1970s, lowest in 1902, 1930, and the late 1980s and early 1990s. The five highest precipitation years in the twentieth century occurred within one year of a La Niña, whereas nine of the ten lowest precipitation years occurred during El Niño (Fig. 3.22).

The greatest changes in global precipitation (15%) occurred in the southern mid-latitudes, particularly in the Americas and Australasia (15–20%; Fig. 3.24), much of this reflecting decreased precipitation during the latter half of the century (compare Figs. 3.25a and b). The century-wide decrease in Euro-African precipitation, extending

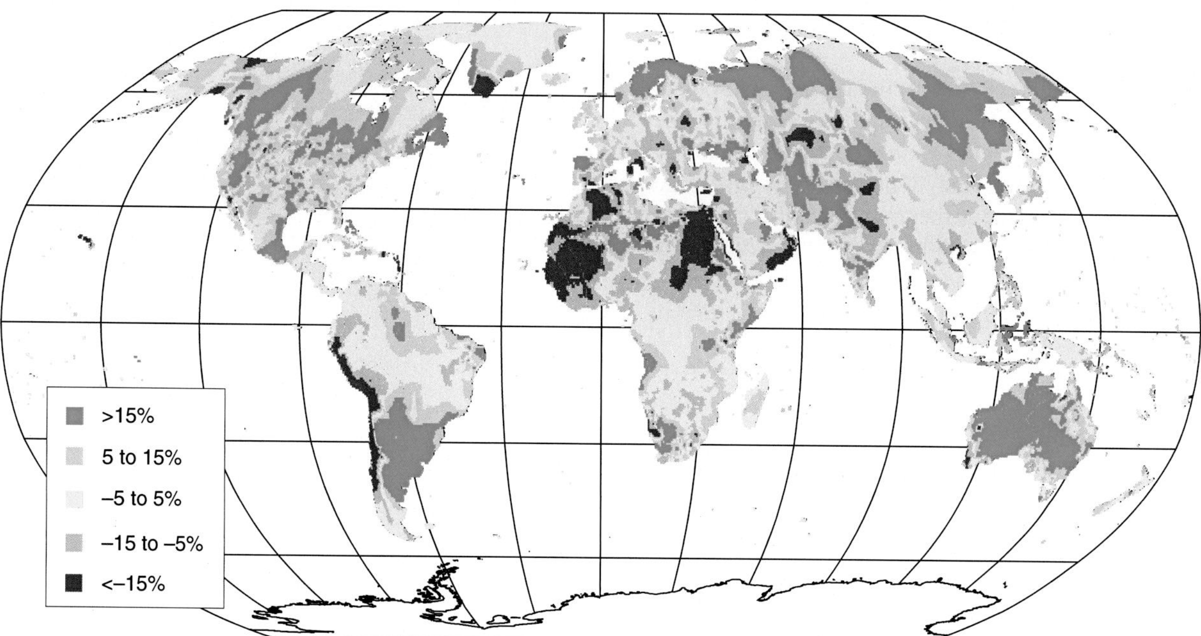

Figure 3.24. Twentieth-century global precipitation change. Figure courtesy of K. H. Xu; data from CRU, University of East Anglia.

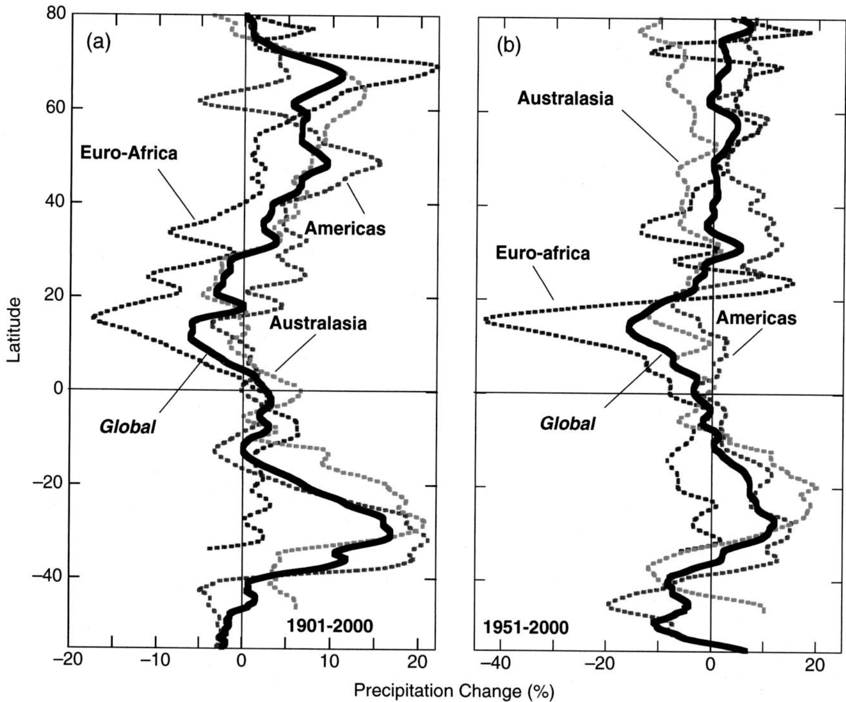

Figure 3.25. Latitudinal change in precipitation for the Americas (red), Euro-Africa (blue), Australasia (green) and global (black), 1901–2000 (a) and 1951–2000 (b). Data from CRU.

from the equator northward to 40 N, in large part reflects a post-1975 decline in precipitation. In contrast, precipitation increased significantly in the southern mid-latitudes in both South America and Australia, as well as in higher latitudes in Asia and Europe. Otherwise there were few regions in which precipitation changed by more than 5% (Figs. 3.24 and 3.25). See New *et al.* (2001) for a detailed analysis of the CRU database.

Pre-twentieth century changes in precipitation and runoff

Projecting back to earlier times, climatic and hydrologic databases become increasingly unreliable prior to the late nineteenth century, thus making their applicability in documenting changes and trends in river discharge increasingly tenuous. One is therefore forced to use other proxies – such as coral growth-bands (Isdale *et al.*, 1998), paleolimnology (e.g. lake levels, pollen), paleoceanography (Herbert *et al.*, 2001), ice cores (e.g. GISP cores in Greenland), splenothems (Bar-Matthews *et al.*, 2003), or tree rings (Anchukaitis *et al.*, 2006; Woodhouse *et al.*, 2006, and references therein) – to estimate precipitation patterns and thereby river runoff. Recent interest in global climate change has fortunately reaped many new paleoclimatic data, particularly with respect to climate during and following the last glacial maximum (LGM), as witnessed by international programs such as PAGES (Past Global Changes; http://www.pages.unibe.ch).

Of particular use in delineating precipitation trends has been the Palmer Drought Severity Index (PDSI), which measures the intensity, duration, and spatial extent of drought. PDSI values, which are derived from measurements of precipitation, air temperature, and local soil moisture, have been standardized to facilitate inter-regional comparisons. Delineating periods of heavy and light rainfall by relating the PDSI to tree rings has been particularly useful in the western USA (Fig. 3.26). Haston and Michaelsen (1994), for instance, found evidence in Santa Barbara (CA) of increased frequency of rainy years in the late 1500s and early 1600s, perhaps in response to the Little Ice Age. The severe AD 900–1300 drought in the North American southwest, in which as much as 60% was drought-stricken, appears to coincide with during the Medieval Warm Period (Cook *et al.*, 2004; see also Woodhouse *et al.*, 2006). Tree ring data, however, must be used with caution, for although they can delineate wet years, they do not necessarily reflect heavy rainfall events, which could help dictate both short-term and long-term sediment discharge.

In the next sections we reverse the direction of our discussion, beginning with the last glacial maximum (LGM), followed by a consideration of the climatic change during the Holocene. Finally, we reverse direction once again with a brief discussion of the pre-Pleistocene.

As a guide to late glacial, post-LGM and Holocene fluvial climate – as well as land erosion and sediment delivery – we refer the reader to Figure 3.27, which presents a schematic view of water discharge, land erosion, and sediment flux to the global ocean over the past 20 000 years. We hasten to point out that the plot is conceptual not quantitative, as indicated by the lack of numbers on the y-axis.

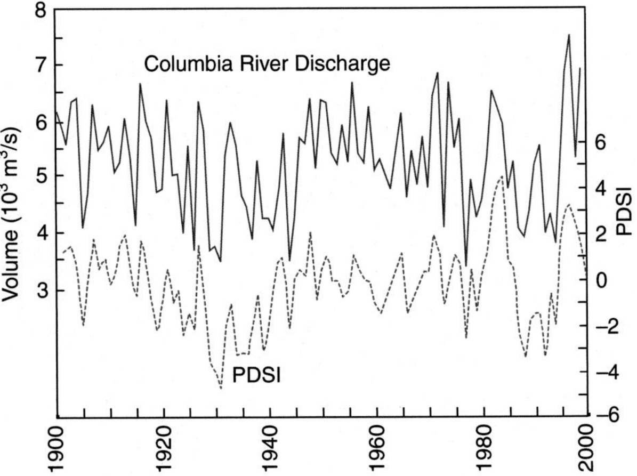

Figure 3.26. Comparison of the Palmer Drought Severity Index (PDSI) and Columbia River discharge, northwest USA. Adapted from Dai *et al.*, 2004.

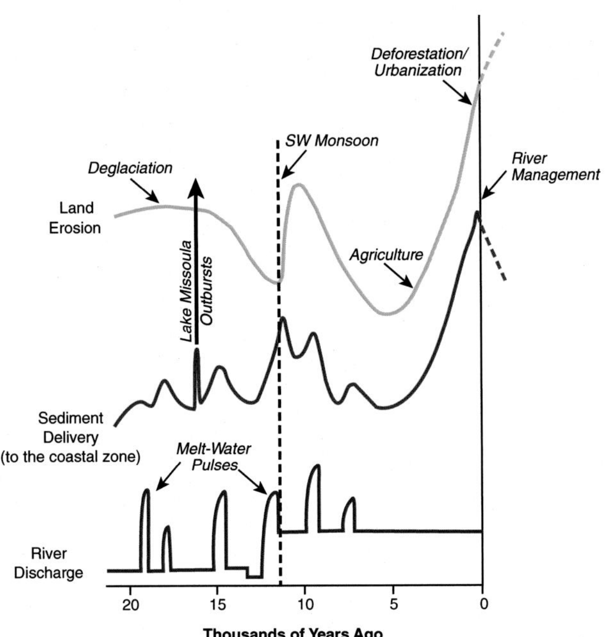

Figure 3.27. Temporal variations in post-LGM (Last Glacial Maximum) river discharge, land erosion, and sediment flux to the global ocean. Although conceptual, as evidenced by the lack of y-axis numbers, this post-LGM plot is not dissimilar from the Black Sea runoff history suggested by Degens and Ross (1972) and is also similar to long-term sediment fluxes from northern Europe summarized by Dearing and Jones (2003, their Fig. 2).

Last Glacial Maximum (LGM) and post-LGM

We can assume that river discharge to the global ocean was significantly reduced during the LGM. Many high-latitude drainage basins in the northern hemisphere were glaciated. Moreover, in response to the cooling of the Asian land mass and a corresponding increased intensity of Northeast

Monsoon, southern Asia was more arid (Williams, 1985; Sirocko *et al.*, 1991; von Rad *et al.*, 1999a, b). Loess and saline paleo-lake deposits in northern China (see review by Baker *et al.*, 1995), for instance, suggest that the Yellow River may have ceased flowing during the LGM (Xia *et al.*, 1993). Similarly, the southward shift of the Inter-tropical Convergence Zone (ITCZ) may have caused long-term droughts in northwest Africa (Itambi *et al.*, 2009). The LGM climate in South America is less clear, some workers citing evidence for decreased rainfall in the Amazon, Orinoco, and Magdalena headwaters, others finding little change in the LGM climate (see discussions by COHMAP Members, 1988; Markgraf, 1993; Thomas and Thorp, 1995; Colinvaux *et al.*, 1999; Baker *et al.*, 2001; Jaeschke *et al.*, 2007). Northwestern South America, whose runoff presently can exceed 4–6 m/yr (Fig. 2.5), received less precipitation in response to the southward shift of the ITCZ (Pahnke *et al.*, 2007). Taking into account all these observations, it is likely that river discharge to the global ocean during the LGM may have been at least 20–25% lower than present-day discharge.

Not everywhere, however, did fluvial discharge decline, especially as the LGM was terminating. Syvitski and Morehead (1999), for instance, calculated that the sediment load of the Eel River (California) may have been ~25% higher 18 ka BP than at present, largely in response to greater precipitation in northern California. A wetter climate in Mediterranean Sea watersheds may have resulted in increased sediment loads (Leeder *et al.*, 1998); Mediterranean river sediment loads increased again in the late Holocene in response to deforestation (see Chapter 4).

While it is difficult to quantify global river discharge during the LGM, judging from the 120 m drop in global sea level, there were at least 40 000 000 km^3 of water stored within the LGM ice sheets. Assuming that it took 10 000 years to store this quantity of terrestrial ice, this would equate to 4000 km^3/yr of decreased freshwater discharge, interestingly, the same amount as presently discharged into the Arctic Ocean. If less intense SW monsoons in southern Asia, which are chiefly responsible for 11 000 km^3 of freshwater discharged annually, resulted in 25–50% decline in freshwater discharge, then decreased discharge from Arctic and southern Asia rivers alone would have been responsible for a ~7000–10,000 km^3/yr decline in global discharge. LGM global river discharge therefore might have been less than 30 000 km^3/yr.

As glaciers began melting, about ~20 ka BP, much of the trapped water was released and post-LGM sea level rose about 120 m in the ensuing 13 000 years. The rise was not uniform but was punctuated by as many as six rapid rises (Fairbanks, 1989; Bard *et al.*, 1990; Liu and Milliman, 2004). During the two primary melting events (14 ka BP and 11.3 ka BP; Fairbanks, 1989; Bard *et al.*, 1990), each lasting about 100–250 years, sea level rose ~50 mm/yr (in contrast, present-day sea-level rise is 2–3 mm/yr). Assuming that the 20-m sea-level transgression during MWP 1A (representing a volume of about 6 000 000 km^3) occurred over about 200 years, this would equate to about 30 000 km^3/yr of increased discharge, roughly the same as we have estimated for annual LGM discharge. As most glaciers were located in eastern North America and northwestern Eurasia, much of this freshwater presumably was discharged directly or indirectly into the North Atlantic Ocean, into which presently <3000 km^3 of river water is discharged annually.

The rapid draining of glacial lakes following the LGM (Teller, 1987; Knox, 1995) also witnessed regional fluxes of freshwater to the global ocean. A prime example is the Lake Missoula flood(s) ~15 ka BP, probably the most extreme fluvial event(s) in the recent geologic past, which is discussed in greater detail later in this chapter. Emiliani *et al.* (1975) noted a marked freshening of the NW Gulf of Mexico at ~14.6 ka BP (Erlingsson, 2008), which they concluded to be related to the glacial melt from the Mississippi River upper watershed. Similarly, rapid evacuation of glacial Lake Aggasiz in northeastern North America saw peak discharges perhaps reaching 100 000 m^3/s (c.f. Knox, 1995), which may have led to the Younger Dryas cooling event (Broecker, 1989; Teller and Leverington, 2004). Thieler *et al.* (2007) have suggested that the Hudson shelf valley south of Long Island (NY) formed as a result of a catastrophic meltwater event around 13.35 ka BP that involved the draining of several glacial lakes; peak discharge may have approached 500 000 m^3/s, the entire event lasting for ~80 days (Donnelly *et al.*, 2005; Thieler *et al.*, 2007). For comparison, the neighboring Hudson River presently has a mean discharge of ~500 m^3/s.

Presumably land erosion and sediment fluxes (slightly lagging land erosion, we assume) increased during periods of high discharge. As the climate stabilized and formerly glaciated areas were revegetated, however, we assume that both erosion and (subsequently) sediment discharge declined, only to increase about 5 ka in response to human activities (Fig. 3.27).

Holocene

Beginning ~11 ka BP, during the early Holocene climatic optimum, there was significant strengthening of the southwest Asian Monsoon (Sirocko *et al.*, 1991), which resulted in increased precipitation throughout much of southern Asia (Steinke *et al.*, 2006; Colin *et al.*, 2010). For example, this period saw a significant increase in the sediment discharge from the Ganges-Brahmaputra and Indus River systems (von Rad *et al.*, 1999a; Goodbred and Kuehl, 2000; Goodbred, 2003; Rahaman *et al.*, 2009).

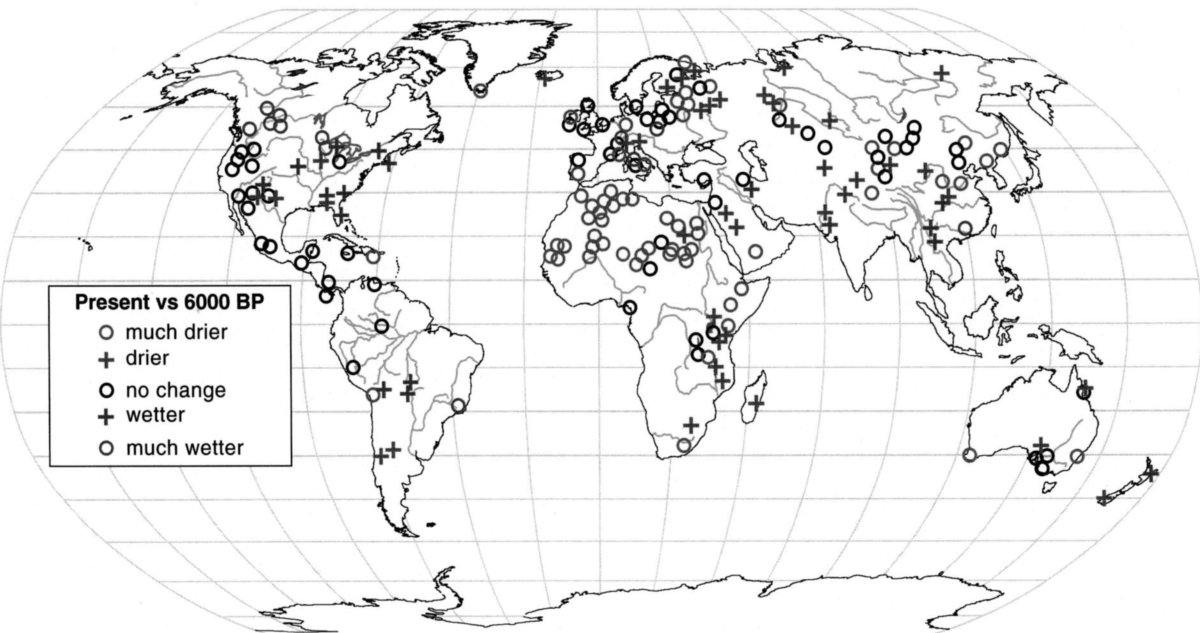

Figure 3.28. Precipitation change from mid-Holocene (6 ka BP) to modern (pre-Industrial Revolution) based on lake and paleolake data. After Braconnot *et al.* (2004).

Intensification of the SW Monsoon in southern Asia may have resulted in a 3000-5000 km^3/yr increase in fluvial discharge, but, unlike the meltwater pulses, this would represent a long-term increase, not a few-hundred-year spike. Other significant changes in Holocene climate occurred in lower latitudes, particularly the Middle East and the African tropics (COHMAP, 1988; Knox, 1995; Gasse, 2000; Claussen and Bolle, 2004; c.f. Steffen *et al.*, 2003, their Fig. 2.10). Because of a warmer climate, rainfall and runoff in Eurasia and North Africa during the late Pleistocene and earliest Holocene were sufficiently high that a number of rivers, which presently are inactive, discharged directly into the Atlantic or Mediterranean (see Fig. 2 in Hoelzmann *et al.*, 1998). Vörösmarty *et al.* (2000, their Table 5), for example, list four presently inactive North African basins whose combined drainage area exceeds 4 000 000 km^2.

This is not to say that flow was continuously high: fluctuations in lake levels in the Ziway–Shala Basin in Ethiopia, for example, indicate periodic episodes of rain and aridity throughout much of the Holocene (Street-Perrott and Perrott, 1990). Early Holocene precipitation in northern Sudan, which presently receives only about 25 mm of rainfall annually, was sufficiently high to move the southern edge of the eastern Sahara 500 km northward. As a result, the drainage basin area of the Nile River may have expanded by as much as 100 000 km^2 (Pachur and Kropelin, 1987), further augmented by contributions from the Melik and Kalabska wadis (Hoelzmann *et al.*, 1998). In east Africa, post-LGM climate seems to have remained cool and dry until ~14 ka BP, after which it was generally warm and moist until about 4–5 ka BP (Kiage and Liu, 2006). The water transferred by these now-extinct rivers may have caused greater stratification in the Mediterranean, perhaps helping create conditions conducive for sapropel formation (Rossignol-Strick *et al.*, 1982).

For the land area surrounding the Arabian Sea, Singh (1991) cited many data to indicate that a relatively wet early Holocene in western India was followed by a substantial decrease in precipitation ~8.0 ka BP (Staubwasser *et al.*, 2002). Paleohydrologic estimates from lake cores in the Thar Desert in NW India show that lake levels again rose about 6.3 ka BP, and then desiccated about 4.8 ka BP (Singh, 1991). Ironically, early civilization may have thrived along the Indus River because of – rather than in spite of – these arid conditions in NW India (Enzel *et al.*, 1999).

By 5.5–4.0 ka BP, the climate throughout much of northern Africa, the Middle East to central Asia had turned arid (Fig. 3.28), and most of the verdant Sahara had disappeared. This appears to have been caused by the southward shift in the ITCZ, perhaps related to changes in orbital forcing (Kutzbach and Guetter, 1986; Wanner *et al.*, 2008). This shift in the ITCZ affected the summer monsoonal rains, as well as other – as yet undefined – mechanism(s) that may involve some form of atmospheric–vegetation coupling (see Claussen and Bolle, 2004, various papers in Issar and Brown, 1998, and Issar, 2003, for detailed discussions). Russell *et al.* (2003) found a ~725-yr climatic cycle during the late Holocene in central Africa, which they suggest was linked to Indian and Pacific monsoons, inferring at least a regional if not global climatic teleconnection (Gasse, 2000).

Using radiocarbon-dates from fluvial units from the UK, Spain, and Poland, Macklin *et al.* (2006) identified 12 periods within the Holocene when flooding events have occurred in two or more of these areas, suggesting regional climate shifts. Prior to 5.0 ka BP, for instance, Macklin *et al.* suggest, flooding periods coincided with IRD (ice-rafted debris) events, indicative of climate warming, whereas the 8.2 ka BP cold event coincided with a period with relatively few floods. Two of these periods, 4.8 and 1.9 ka BP, appear to lie close to the 4.7 and 1.7 ka BP ages of flood deposits found in western India (Sridhar, 2007). Similarly, two droughts noted in East Africa by Kiage and Liu (2006), 8.3 and 5.2 ka BP, correspond closely with a period of significantly fewer floods in Spain (Macklin *et al.*, 2006). Whether this is coincidence or reflects teleconnections in climate shift is not clear.

A much firmer teleconnection is seen in what appears to be a global drought, what Helama *et al.* (2009) term a mega-drought, during the Medieval Climate (warm) Anomaly (~AD 950–1350), followed by a wetter period during the Little Ice Age (~AD 1600–1750). The Medieval drought can be delineated in tree rings in northern Europe (Helama *et al.*, 2009) and the western United States (Cook *et al.*, 2007; Seager *et al.*, 2007), in east Africa lake cores (Verschuren *et al.*, 2000; Russell and Johnson, 2007) and in paleoclimatological deep-sea proxies off Peru (Rein *et al.*, 2004, 2005).

There is increasing evidence to suggest that the influence of ENSO changed in intensity during the mid to late Holocene. According to laminated lake sediments from southern Ecuador (Moy *et al.*, 2002), ENSO events first appeared about 7.0 ka BP, and increased in frequency about 4.8 ka BP; with high frequencies occurring about 1500, 1200, and 900 BP. Data from other proxies and from other areas show similar trends in terms of a late Holocene intensification of ENSO, but the timing remains subject to considerable debate (Tudhope *et al.*, 2001; Moy *et al.*, 2002; Woodroffe *et al.*, 2003; Gagan *et al.*, 2004).

Pre-Pleistocene

It is not difficult to envision long-term climate change as a series of superimposed sinusoidal curves, each with distinctly different amplitudes and periods. Mountain building and subsequent erosion occur at frequencies of 10^5 to 10^7 years, superimposed on which are 10^4 to 10^6-yr-long rhythmic cycles driven by orbital processes, resulting in alternating glacial and interglacial periods (Zachos *et al.*, 2001). Superimposed on these are still shorter-term climatic cycles, ranging down to the ocean-driven cycles discussed in the first part of this chapter.

One might therefore question how well the data derived from present-day rivers might correspond to the longer record? Unfortunately, the jury is still out, in large part because our ability to quantify fluxes and sediment budgets worsens as we delve further into the geological past. In one of the only attempts to compare modern with pre-Pleistocene sediment fluxes, Métivier and Gaudemer (1999) showed a close correlation between present-day sediment discharges for some of the largest rivers in southern Asia and the mass accumulation in their sedimentary basins over the past 2 million years. Métivier and Gaudemer concluded that "...present-day average discharge at the outlet (of these very large rivers) has remained constant throughout the Quaternary...." Short-term sediment fluxes, however, have likely fluctuated considerably, from low delivery during the arid LGM to high delivery during the early Holocene to moderate delivery prior to human modification of river basins (see Fig. 3.27).

Events

Many geologists and hydrologists accept – even if they may not tout – a catastrophist's view of the world: normal, everyday processes do little more than connect the rare events or disasters that ultimately define the long-term record. Bettersaid: millenia of boredom interrupted by brief periods of sheer terror.

In fluvial systems, episodic events can occur over very short time intervals, often hours to weeks. The most obvious example is the annual flood during which disproportionate amounts of the river's water and sediment are discharged. In larger rivers the flood can last months, while in small rivers the event may be over in a few hours. Catastrophic events such as typhoons may play even greater roles in the discharge of both water and sediment. Supply events, such as landslides, fires, cyclonic storms and volcanic eruptions, often occur less frequently, perhaps only once every century or millennium, but their impacts sometimes can last for decades or longer. Such extreme events include "high-energy megafloods," in which exceptionally large amounts of water are discharged over short periods of time (Baker, 2002).

In the following paragraphs we discuss events that erode and supply sediment to the fluvial system (supply-related events) and those events that transport and deliver water and sediment downstream (transport-related events). An earthquake, for example, can initiate a landslide or debris flow, but its immediate effect on the river watershed may be transport-limited, requiring an intense rainstorm to erode and transport the sediment downstream. A particularly heavy rainfall can exert two controls on landslides: triggering the landslide itself, and then providing the water by which the landslide-derived sediment is

Figure 3.29. Active slump scars on sheep-grazed pastureland, Waipaoa River watershed, North Island of New Zealand. Close inspection shows evidence of previous slumps, some of which are again vegetated. Photograph courtesy of J. P. M. Syvitski.

transported down stream (Montgomery and Buffington, 1997; Hovius *et al.*, 2000).

Supply-related events

Mass wasting

Downslope mass movement plays a defining role in the erosion and transport of sediment within many river basins (Hovius and Leeder, 1998) and also helps moderate topographic relief (Stock and Dietrich, 2003). In the broadest sense, mass movements can range from slow-moving downgradient slumps to isolated rock falls to the catastrophic "falling away" of entire hillsides. Nolan *et al.* (1995) and Reid and Dunne (1996) both classified mass movements on the basis of whether movement is continuous and relatively slow (earthflow, slump, creep) or episodic and rapid (debris slide and debris avalanche). The latter type of downslope sediment movement usually involves the physical removal of rock – i.e. the production of new sediment – whereas the former also can remobilize sediment previously eroded and temporarily stored (Reid and Dunne, 1996). Swantson *et al.* (1995) found that soil creep in the Redwood Creek basin (California) ranged from 1–2.5 mm/yr, whereas block glide averaged between 2.9–16 mm/yr. Earthflows can be significantly more rapid, involving both translational and rotational movement (Harden, 1995); downslope rates can vary between 0.01–29 m/yr (Kelsey, 1978; Swanson *et al.*, 1979), the size and speed of mass movement depending in part on soil moisture (Nolan and Janda, 1995). In a 5-year study of two earthflows in Redwood Creek basin, California, Nolan and Janda (1995) reported that >90% of the sediment delivered to the stream was by mass movement, <10% by strictly fluvial processes. Large landslides account for most of the down-slope volume transfer, smaller landslides accounting for only a fraction of the cumulative down slope transfer (Kelsey *et al.*, 1995), their relative importance often dependent on topographic gradient (Vanacker *et al.*, 2007). Landslides also can over-stress river channels, leading to stream avulsion (Korup *et al.*, 2004). Several examples of mass wasting on the North Island of New Zealand are shown in Figs. 3.29 and 3.30.

Figure 3.30. Gully erosion in the Waiapu River watershed, North Island of New Zealand. Steep erodible terrain, periodically heavy rains, combined with deforestation can lead to highly unstable slopes and accelerated rates of erosion. Mean annual sediment yield of the Waiapu watershed is 21 000 t/km^2. Linear rows of trees in the lower right show an attempt to retard erosion. Photo courtesy of R. Gammish.

Mass wasting can occur in response to either seismic events (e.g. earthquakes) or intense rainfall (Allen and Hovius, 2000; Reid and Page, 2003) by which excess pore pressure along bedding surfaces can facilitate failure. The timing and magnitude of landslides depend on the amount of rainfall as well as local topography, soil thickness, and diffusivity of the soil (Iverson, 2000). In Puerto Rico, short-duration, high-intensity rainfall primarily results in shallow soil slips and debris flows, whereas long-duration, low-intensity rains produce larger, deeper debris avalanches and slumps (Larsen and Simon, 1993). Larsen and Simon (1993, their Fig. 4) suggested that tropical landslides require greater intensity, longer duration storms than do temperate landslides, although the reasons are not clear. Monthly rainfalls in excess of 150–200 mm in China seem necessary to initiate debris flows (Wang, 1992).

The erosional importance of mass movements is obvious. Hovius *et al.* (2000) concluded that most of the sediment transported by Taiwanese rivers is derived from hillslope mass wasting, and Keefer (1994) found that earthquake-induced landslides on Oahu, Hawaii, result in mean erosion rates of 0.55 mm/yr, 3.5 times greater than long-term slope erosion. High-magnitude, low-frequency landslides are primarily responsible for a regional denudation rate of 9 + 4 mm/yr in the western Southern Alps of New Zealand (Hovius *et al.*, 1997), although in the Motueka River peak sediment discharge reflects contributions from both gully erosion and landslides (Hicks and Basher, 2008). Landslides on the North Island, in contrast, seem responsible for only ~15% of the river's suspended sediment (Reid and Page, 2003); gully erosion is the main supplier of river sediment (see Fig. 3.30). However, Page *et al.* (1999) calculated that about half of the 26 Mt of sediment transported by the Waipaoa River (North Island, New Zealand) during Cyclone Bola was via landslides.

Vegetation (or the lack of it) plays an important role in the frequency and speed of mass movements. In northeastern New Zealand, measured earthflows are 1–3 orders-of-magnitude faster in grassed areas than in reforested areas, the result of decreased soil cohesion (absence of tree roots) and high water content due to decreased evaporative moisture loss (Marden *et al.*, 1992; Page *et al.*, 2000) (see Fig. 3.30). Much of the sediment transported by the Eel River during the infamous 1964 Christmas storm was directly related to landslides along the middle reaches of the river, particularly on sparsely vegetated steep slopes (Brown and Ritter, 1971), which in part may have been caused by intense logging activity (Best, 1995) that helped destabilize the hill slopes (Harden, 1995).

Mass movement, of course, does not infer immediate transfer of sediment to the river. In the Canadian Arctic, for instance, Lamoureux (2000) concluded that small-scale landslides provide sediment that is re-eroded and delivered to rivers during subsequent years. Similarly, Pearse and Watson (1986) and Matsuoka *et al.* (2008) found that much of the earthquake-induced landslide material on the northwestern part of the South Island of New Zealand and in northern Japan (respectively) is temporarily stored (much of it in landslide-dammed lakes), which ultimately serves as a sediment source for rivers in subsequent years. During and after the 1964 flood, the sediment-rating curve of the Eel River increased by a factor of 2–3 (Fig. 3.31; Brown and Ritter, 1971), suggesting that much of the sediment eroded by slumps and landslides remained stored within the basin and was released during subsequent years. In the Takiya River (Japan), post-flood sediment-rating curves increased an order-of-magnitude relative to pre-flood rating curves (Whitaker *et al.*, 2008). It took several years for the Eel River to return to pre-1964 conditions (Brown and

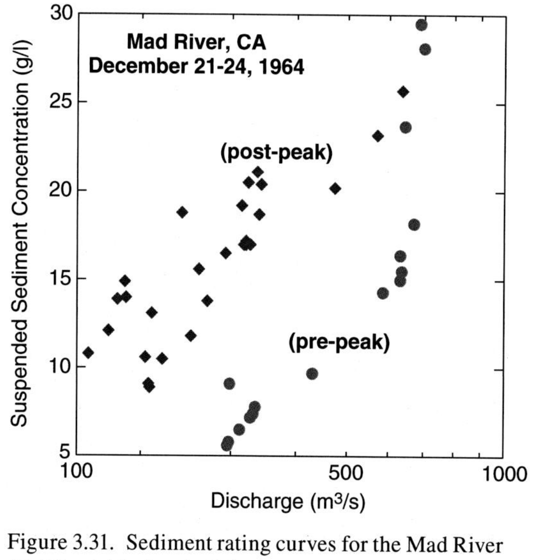

Figure 3.31. Sediment rating curves for the Mad River (California) before (red dots) and after peak (blue diamonds) discharge during the 100-yr flood in December 1964. Note that the *y*-axis is linear. Data from USGS, Waananen *et al.* (1970), and Brown (1973).

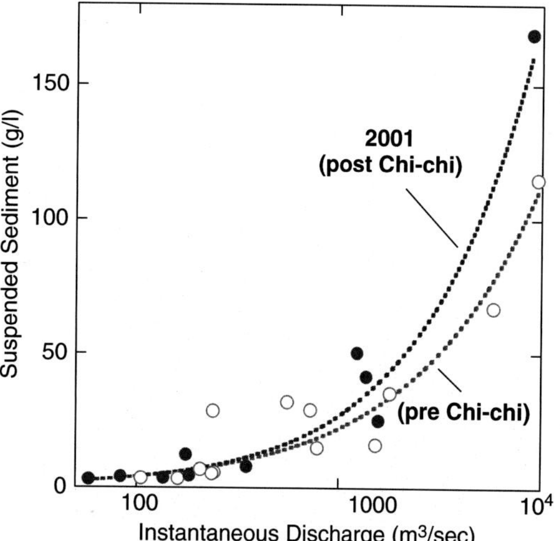

Figure 3.32. Sediment rating curves for the Choshui River before (1994) and during Typhoon Herb (1996; red circles), and during Typhoon Toraji (2001; blue dots), two years after the Chi-chi earthquake (blue dots). During and after the 2004 typhoon Mindulle, the river appears to have returned to pre-Chi-chi conditions.

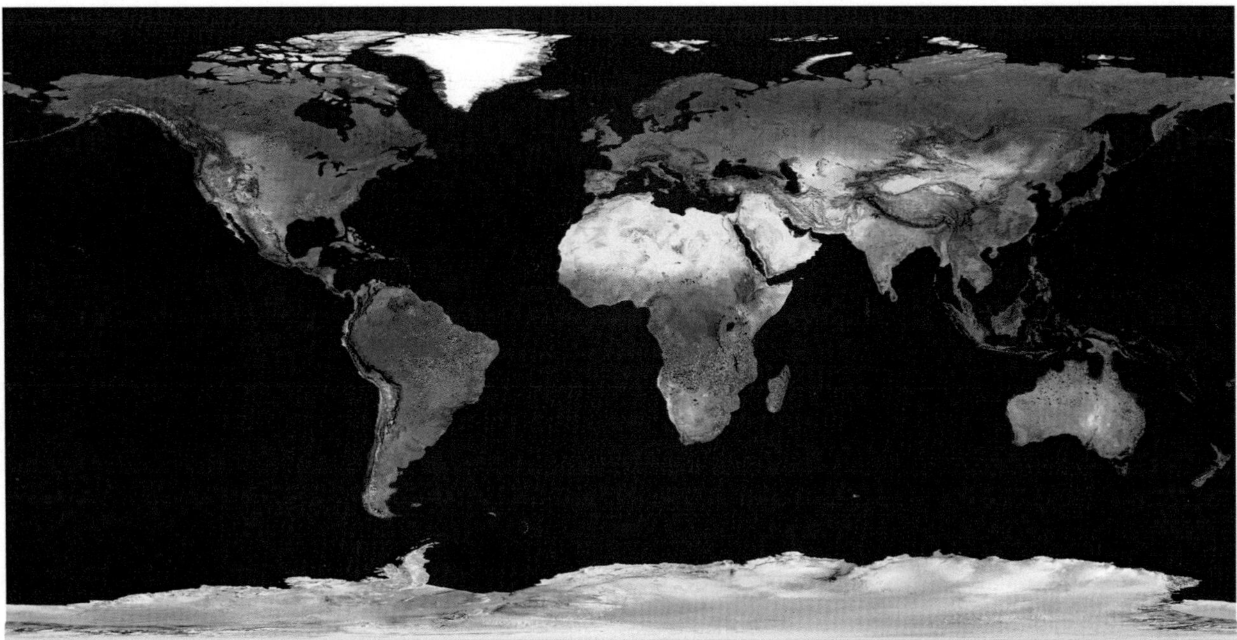

Figure 3.33. NASA mosaic of nighttime Earth, October 2002. Although fires (in red) are noted throughout the world, they seem especially concentrated in southern Africa, central Brazil, and eastern Europe.

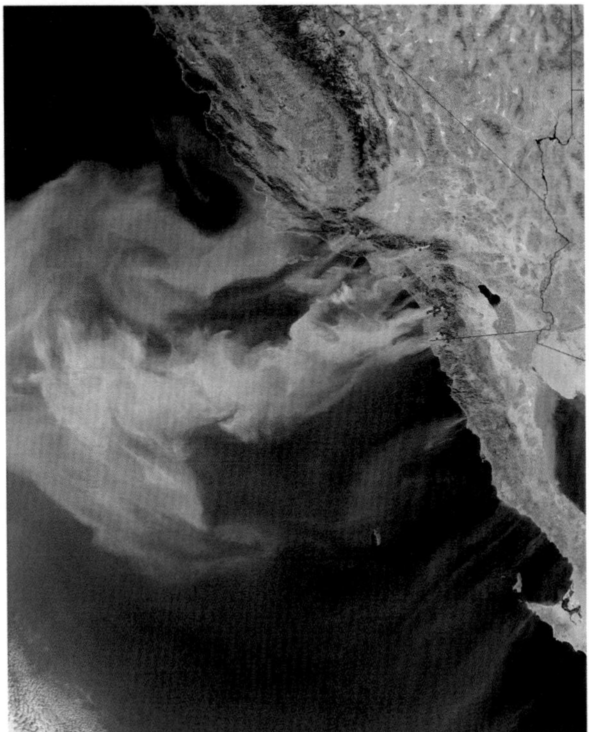

Figure 3.34. NASA-generated image from the NASA-MODIS platform, October 27, 2007, showing areas of active fires (red marks), the smoke being blown westward over the coastal ocean by Santa Ana winds. To the south one can detect what appear to be lower-elevation dust storms blowing westward from Baja California. Note the Colorado River, far right, which discharges – or, better said, once discharged – into the Gulf of California. Enhanced NASA image courtesy of J. A. Warrick, USGS.

Ritter, 1971; Sloan *et al.*, 2001); the neighboring Redwood Creek took even longer (Nolan and Janda, 1995; Nolan *et al.*, 1995). Hicks and Basher (2008) have suggested that the 2–3 years needed for the Motueka River (New Zealand) to recover to normal TSS concentrations was more related to the stabilization of riparian sediment storage than to the healing of erosional scars.

Dadson *et al.* (2004) calculated that the Chi-chi earthquake, a 7.6 earthquake that struck central western Taiwan in October 1999 and caused more than 20 000 landslides, increased sediment yields along the Choshui River by as much as four-fold, a watershed that already had been exposed to intense erosion during Super Typhoon Herb in 1996 (Milliman and Kao, 2005). Downstream, however, the Chi-chi earthquake seems to have increased Choshui sediment discharge by only ~50% (Fig. 3.32). Nevertheless, typhoons Toraji and Nori in 2001, the first major typhoons to strike the island after Chi-chi, collectively transported 250 Mt of sediment from the Choshui River (Dadson *et al.*, 2004; Milliman and Kao, 2005). This equates to basin-wide sediment erosion of ~80 000 t/km^2 in little more than the seven days that the two typhoons collectively lasted. It was only five years after the earthquake, in 2004, that the erosional effects inherited from the Chi-chi earthquake may have begun to moderate (Milliman *et al.*, 2007). The effects of particularly large landslides, however, may last for centuries: some of the 7-km^3 landslide that occurred 8.8 ka BP, for instance, still remains within the Fly River watershed (Blong, 1991; cf. Swanson *et al.*, 2008).

(a)

(b)

Figure 3.35. Up-stream views of City Creek, near Highland, CA, before (a) and after (b) a fire in autumn 2003. The fire, which burned most of the creek's 54 km^2 watershed, was followed in December by heavy rains that produced floods and debris flows emanating from the burned areas. Pictures courtesy of J. A. Warrick.

Fire

A NASA mosaic of nighttime Earth shows a number of interesting anthropogenic features, such as spatial concentrations of fishing ships and oil-well related gas fires in Russia and the Middle East. But perhaps the most surprising feature is the prominence of fire – shown on the NASA mosaic in red – throughout much of the world (Fig. 3.33). The northern limits of Sub-Sahara Africa are alight with red, as is much of eastern Europe, central Brazil, southern Africa, Asia, and Cuba. Most fires are probably human-initiated, but no matter the origin, their collective effect can significantly increase sediment erosion. A NASA image in Fig. 3.34 shows the regional impact of fire in southern California. The actively burning areas (delineated by red marks) are emitting smoke that easterly Santa Ana winds blow over the coastal ocean.

Fires have a number of deleterious effects on the watershed. They can destroy vegetation and root structures that retard erosion. Condensation of vaporized organic compounds can form hydrophobic layers that repel absorption of rainfall, hence increasing Horton overland flow and subsequent erosion (Mooney and Parsons, 1973; Savage, 1974; Dunne and Dietrich, 1980; Greene *et al.*, 1990; Inbar *et al.*, 1998). Onda *et al.* (2008) report that runoff from a California site increased substantially during the first heavy rains following a burn, as ash compressed, crusted and penetrated into potential flow paths. Fire can also increase the concentrations of dissolved solids and nutrients (Bayley *et al.*, 1992; Williams and Melack, 1997; Petrone *et al.*, 2007). Combined with destruction of biotic crusts that help retard erosion and decrease rainwater penetration, increased surface flow can accelerate soil erosion (Fig. 3.35), although the actual impact of a fire depends strongly upon the physical and climatic environment (see Moody and Martin, 2009 and references therein).

In the central Oregon Coast Range, on the other hand, Jackson and Roering (2009) found that discontinuities in the hydrophobic layer facilitated rainfall infiltration; fire also generated increased dry ravel transport that resulted in a significant lowering of the landscape.

Although moderate rainfall events can mobilize sediment (Florsheim *et al.*, 1991), the combination of heavy rain and fire-scorched landscape can result in massive erosion and debris flows, which in some areas can fill fluvial channels, thereby increasing the catastrophic effect of periodic floods (Keller *et al.*, 1997). Wilson (1999) found that high-intensity rains on experimental plots in Tasmania produced the greatest sediment yields from plots affected by wildfire. Marqués and Mora (1992) reported a two-order-of-magnitude increase in soil loss following a fire in forested land in Spain, and Pereira (1973) and Mooney and Parsons (1973) cited three-order-of-magnitude increases in soil erosion after fires in Australia and Arizona. Solute loss in the Pyrenees in Spain was less pronounced, perhaps reflecting the Pyrenees's relatively arid climate. (Not surprisingly, many of the controlled fire experiments have been performed in low-rainfall environments).

The duration of time that a burn event can have on a watershed varies greatly, from a few to many years. A controlled burn experiment in two experimental plots in the Spanish Pyrenees, for instance, resulted in significant increases in both water runoff and solid erosion; in one plot the increase lasted two years, in the other seven years (Cerdà and Lasanta, 2005).

One might assume that human activity has heightened the frequency and thereby the erosive impact of fires, but fragmentary evidence suggests that this may not always be

(a)

(b)

(c)

Figure 3.36. (a) Lahar flowing in the Sacobia Valley, August 1991, following a Mount Pinatubo (Philippines) eruption. Judging from the color of the flow, suspended sediment concentrations probably exceeded 1 kg/l. (b) Boulder transported and deposited by Mt. St. Helens lahar, May 1980; mass is estimated to be at least 50 000 t. (c) Comparing the observer with the mudline on a tree along the Toutle River, Washington, lahars formed during the May 1980 Mt. St. Helens eruption locally were at least 8–9 m thick. Photos (by T. J. Cadadevail and Lyn Topinka) courtesy of USGS.

the case. A sediment core from the Santa Barbara Basin in southern California indicates that within a 560-year period there were about 20 large (Santa Ana) fires, with a 20–30 yr recurrence, the intensity and frequency of which did not appear to change with the increased human activity in southern California (Mensing *et al.*, 1999).

Volcanism

In addition to producing earthquakes that can facilitate downslope mass movement, volcanic eruptions can produce huge amounts of fine-grained material, much of it transported as ash. The eruption of Mount Unzen (Kyushu, Japan) in 1990, for instance, deposited an average of 40 m of pyroclastic flow in nearby areas. Low rainfall resulted in much of the erosion occurring as debris flows, but heavy rains in 1993 caused as much as 40 m of channel erosion in the Mizunashi River (Mizuyama and Kobashi, 1996). Collapse of a caldera wall at Mount Bawakaraeng (Sulawesi) resulted in the release of $230 \times 10\ m^3$ of sediment within the Jeneberang River basin (727 km^2), one

(a)

(b)

Figure 3.37. The Upper Muddy River fan east of Mt. St. Helens, seen October 1980 (a), was covered by a broad lahar surface soon after the May 1980 eruption. Even five months after deposition, some visible signs of erosion are apparent. By October 1981 (b), the lahar surface was being actively eroded, forming a badland topography. Photos (by Lyn Topinka) courtesy of USGS.

immediate result being a 400-fold increase in river turbidity (Dhanio *et al.*, 2008).

One common result of a volcanic eruption is the formation of a lahar, an Indonesian word that describes a rapidly flowing mixture of volcanic debris and water. In contrast to debris flows, lahars generally have lower sediment concentrations, although concentrations of ~60 g/l have been noted (Lavigne and Suwa, 2004) (Fig. 3.36a). Moreover, lahar sediment comes not from a point source but rather is eroded over a wide area within the watershed. Because of their high suspended-sediment concentrations, lahars tend to exhibit non-Newtonian flow, although flow properties tend to be unsteady (Fig. 3.36b; see also Lavigne and Thouret, 2003). Velocities often reach several m/s, sometimes exceeding 10 m/s and often many meters deep (Fig. 3.36c). Other factors that can influence the character of a lahar flow are the slope and gradient of the basin, the volume and physical properties of the pyroclastic sediments, and vegetation cover (Lavigne and Thouret, 2003). Major *et al.* (2000) suggested the following model for watershed response to explosive volcanic eruptions: fresh volcanic debris erodes rapidly, resulting in badland topography (Fig. 3.37). Established channels then expand and erosion rates decrease as slope gradients diminish; subsequently channels become discharge-dependent so that changes in discharge can greatly affect bank stability.

One of the most-documented lahars resulted from the Mount St. Helens eruption in May 1980, when mudflows raised the level of the Cowlitz River (which flows into the Columbia River) by 7 m in less than 2 h. As a result, the capacity of the river channel to transport water was reduced by 90% (Lombard *et al.*, 1981). Peak velocities in the North Fork of the Toutle River, a branch of the Cowlitz River, exceeded 15 m/s, and sediment concentrations ranged from 75 g/l to 90 g/l (Janda *et al.*, 1981; Waitt *et al.*, 1983). Most of the erosion in the Toutle watershed was associated with sheet wash and rill erosion. Collins and Dunne (1986) predicted that ultimately mass wasting should be the major sediment-redistribution process at Mount St. Helens, but 20 years after the eruption fluvial erosion still predominated (Major, 2004). In the first five years after the eruption, the Toutle River discharged 260 Mt of sediment; sediment yield on the Toutle watershed during this period was ~40 000 t/km^2. Within this period about 90% of the Toutle's load reached the Cowlitz River downstream (data from Dinehart, 1997). (Lahars associated with eruptions from Mount Rainer within the past 6000 years have extended as far as the present city limits of Seattle and Tacoma, underscoring the potential hazard of future occurrences (Sisson *et al.*, 2001).)

Lahar sediment yield is assumed to decrease gradually 10–20 years after a volcanic eruption, in response to the erosion and removal of the most easily eroded volcanic debris. Although ~80% of the transported sediment in the 20 years following the 1980 Mount St. Helens eruption was transported in ~5% of the total lapsed time, most of it occurred when flow exceeded mean annual flow but less than 2-yr floods (Major, 2004). For a few days in August 1980, several months after the initial Mount St. Helens eruption, sediment concentrations in the North Fork of the Toutle River exceeded 500 g/l and were greater than 200 g/l as late as 1984. Concentrations decreased after 1982, and by 1988 concentrations rarely exceeded 10 g/l (Fig. 3.38), reflecting decreased erosion, perhaps more the result of increased infiltration of water and decreased erodibility of the tephra rather than increased vegetation (Collins and Dunne, 1986). By the mid-1990s, however, the weather became wetter (presaging the advent of a cold-phase PDO?) and sediment yield again increased

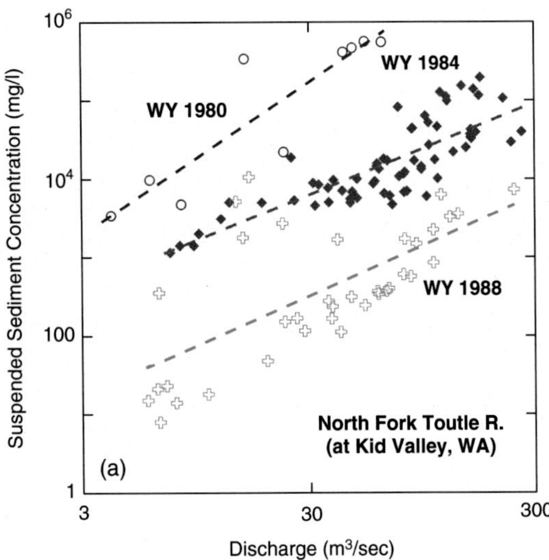

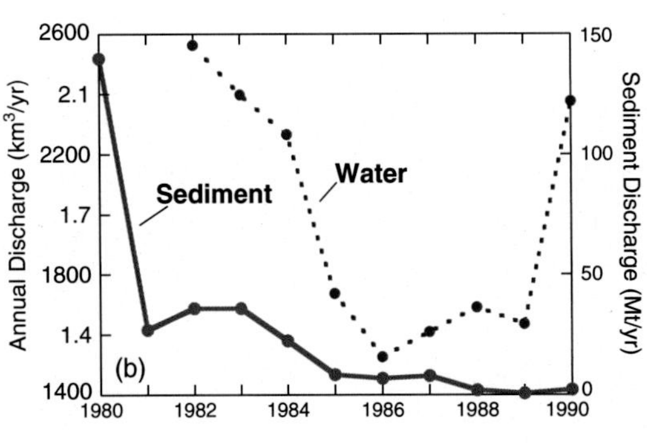

Figure 3.38. (a) Sediment rating curves for the North Fork of the Toutle River (at Kid Valley, WA; upstream drainage basin area 450 km^2) following the 1980 eruption of Mount St. Helens. Concentrations for several months after the eruption exceeded 550 g/l; eight years later measured concentrations were generally less than 5 g/l. (b) Water discharge (Q; blue dashed line) and suspended sediment concentrations (Qc; red solid line) in the North Fork of the Toutle River for the 10 years following the Mount St. Helens eruption. Data courtesy of R. L. Dinehart, USGS.

(Major *et al.*, 2000). The sediment yield of the South Fork of the Toutle River also decreased steadily after 1981 (Collins and Dunne, 1986), from about 50 000 t/km^2/yr in the early 1980s to less than 1000 t/km^2/yr in 1994. Yields in the 1990s, however, increased sharply in response to increased precipitation and flooding (Major *et al.*, 2000).

The eruption of Mount Pinatubo (Philippines) in June 1991 was an even more spectacular event than Mount St. Helens, disgorging 5–6 km^3 of pyroclastic flow and tephra material (see Fig. 3.36 left). The Pasig–Potrero drainage basin, for instance, filled with ~50 m, locally as much as 200 m, of "sediment" (Montgomery *et al.*, 1999; Hayes, 1999 and references therein). Not surprisingly, many large lahars were generated around Mount Pinatubo itself. For the 10 years before the 1991 eruption, the Pasig River (21 km^2 basin area) carried an annual load of about 0.02 Mt/yr (an annual yield of about 1000 t/km^2/yr); but over the remainder of 1991 alone, it transported nearly 100 Mt of sediment, an annual sediment yield of ~5 000 000 t/km^2/yr (Janda *et al.*, 1996; Umbal, 1997). In 1992 the river experienced 62 lahar events during which it transported an additional 50 Mt of sediment (Arboleda and Martinez, 1992). Umbal (1997) predicted that over the 10 years following the Pinatubo eruption, the Pasig would transport a total amount of sediment exceeding 1000 Mt, equivalent to an average annual denudation rate of 3 m/yr; changes in watershed configuration, in fact, may have rendered this prediction too low (Hayes, 1999). For the eight rivers draining the volcano, the combined sediment load for the first three years was about 3000 Mt (Janda *et al.*, 1996), a figure not too dissimilar from the estimated discharge of the Amazon River over the same time span, but from a river system smaller by a factor of about 10 000.

In response to the eruption of Santa Maria Volcano in 1902, the Samala River (drainage basin area of 1500 km^2) in southwestern Guatemala discharged a volume of sediment over the next 20 years that allowed its delta to prograde 7 km across the inner Pacific continental shelf. The delta's estimated volume growth – 4 km^3 in 20 years (Kuenzi *et al.*, 1979) – equates to an mean annual sediment load of 250 Mt/yr, or an annual sediment yield of ~160 000 t/km^2/yr. Assuming an exponential decrease in sediment load following a catastrophic event (Schumm and Rea, 1995), the loads and yields immediately after Santa Maria's eruption almost certainly were much higher.

The effect of a volcanic eruption can persist long after the eruption itself. Heavy rains around Mt. Vesuvius (Italy) in 1998, for example, triggered numerous landslides and lahars, killing more than 150 people (Pareschi *et al.*, 2000).

Increased concentrations of stratospheric aerosols derived from major volcanic eruptions also can have a significant effect on global precipitation – and thereby fluvial runoff. Trenberth *et al.* (2007) have shown that for the year following the Mt. Pinatubo eruption in 1991 there was a 5–8% decrease in precipitation and fluvial discharge to the global ocean. La Niña-impacted watersheds

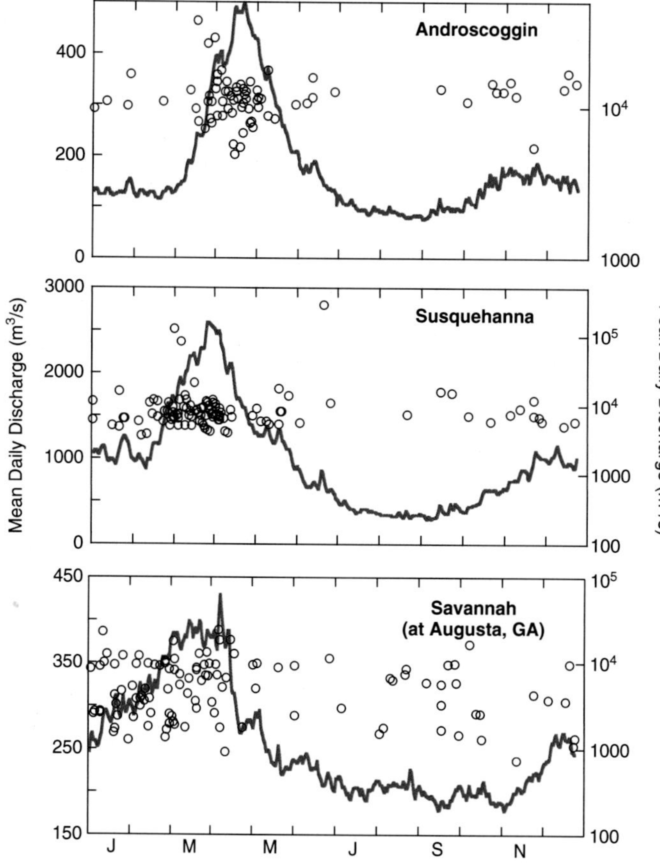

Figure 3.39. Long-term (60–100 years) mean daily discharge (red lines) and peak annual discharges (open circles) far three US east-coast rivers (Androscoggin (A), Susquehanna (Su), Savannah (Sa)) that have similar runoffs (410–570 mm/yr); locations shown on the insert map. Peak Savannah River discharges are distributed throughout the year, reflecting less winter snow accumulation (compared with the Androscoggin and Susquehanna) and greater impact from subtropical cyclonic events during summer and early autumn. Data from USGS/NWIS.

and rivers were especially affected by low-precipitation conditions.

Transport-related events

Annual floods

The magnitude of a flood – and its impact – ultimately depends on climate and on the geomorphic and geological configuration of the drainage basin. Flow in most low-latitude and many mid-latitude rivers reflects a direct response to precipitation; if it rains, a river's flow will increase. Seasonal flow in many higher-latitude and higher-elevation rivers also reflects the timing of snow and ice melt. US rivers provide excellent examples of annual flow patterns, not only because of the diversity of their climates and climatic drivers but also because of the ease of access to the USGS/NWIS database. In Figs. 3.39 and 3.40 we show the mean daily and annual peak discharges for three east-coast and three west-coast US rivers. The east-coast rivers (Androscoggin, Susquehanna and Savannah) have roughly similar runoffs (410–570 mm/yr) whereas the west-coast rivers (Santa Clara, Eel, and Skagit rivers) display runoffs varying from 44 mm/yr to 1750 mm/yr. The lag between precipitation and runoff with increasing latitude (*sensu* Dettinger and Diaz, 2000) is seen on the east coast: the Savannah River (Georgia) generally begins flooding in early March, although June–October tropical storms can generate periodically high discharge (Lecce, 2000). The Susquehanna (Pennsylvania) also floods in March; and the northernmost Androscoggin (Maine) in mid March to early April (Fig. 3.39).

In contrast, the three west-coast rivers drain dramatically different watersheds, the Santa Clara (California; 44 mm/yr) having a dry Mediterranean climate and the Skagit (Washington; 1750 mm/yr) a wet subalpine climate. The Santa Clara usually discharges the bulk of its water in early January to early March during winter storms, highest discharges occurring during El Niño and warm PDO periods (see Figs. 3.7 and 3.8); for the rest of the year discharge is minimal. Peak discharge on the Eel River (northern California) occurs in between early December and late March, whereas the Skagit River (whose discharge also is influenced by upstream dams) shows two distinct peaks (Fig. 3.40): late autumn, when precipitation events often occur as rain or snow that subsequently melts, and early summer, from the melting of the previous winter's snow pack.

Climate not only defines mean annual discharge but also interannual variations in discharge. The Rappahannock

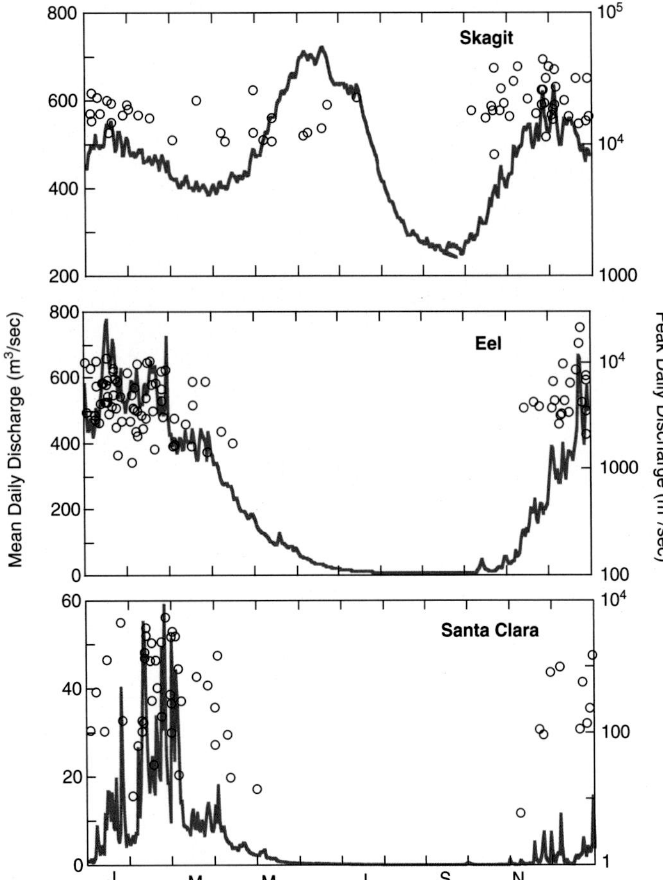

Figure 3.40. Long-term (60–100 years) mean daily discharge (red lines) and peak annual discharges (open circles) of three west-coast rivers (Skagit (S), Eel (E), Santa Clara (SC)) with markedly different runoffs (44–1750 mm/yr). Note the tight cluster of peak events for the Santa Clara and (particularly) the Eel (mostly in winter), compared with the Skagit, whose peak flows cluster in early summer and late autumn. Data from USGS/NWIS.

River (Virginia), for example, drains the same size watershed as Santa Clara River, 4100 km^2, but discharges an order-of-magnitude more water (2.1 km^3/yr vs. 0.18 km^3/yr; Fig. 3.41a). Because of its mild, humid climate, however, the Rappahannock shows relatively little interannual variation, whereas the Santa Clara's yearly discharge, which is strongly influenced by Pacific storms, can range from nearly zero (several years in the early 1950s) to 1 km^3/yr (1969) (Fig. 3.41a). Because it is more susceptible to periodic events, often related to PDO and ENSO cycles, the Santa Clara's peak daily discharge can be greater than the Rappahannock's: peak Santa Clara discharge (1969: 2600 m^3/s) was 40% greater than the Rappahannock's highest measured daily discharge (1972: 1600 m^3/s) (compare Figs 3.41b and 3.41c).

In very large rivers seasonal discharge patterns tend to be modulated by local events throughout the watershed. Mean daily discharge of the lower Mississippi at Vicksburg (just under 30 000 m^3/s), for example, peaks in late April, about one month after peak spring discharge of the Ohio River, the Mississippi's chief tributary. Springtime discharge in the upper Mississippi normally occurs in May, whereas the Missouri and Arkansas discharges generally remain relatively low throughout the year (Fig. 3.42a), owing in part to flow regulation by dams on their mainstreams and tributaries. But for any given year the timing and relative importance of the various tributaries can change dramatically. In 1993, which saw >100-yr floods throughout the upper Midwestern US (Parrett *et al.*, 1993), the lower Mississippi showed three distinct peaks, the first two primarily reflecting Ohio and upper Mississippi flow, the third peak, in early August, reflecting the record flow of the upper Mississippi and Missouri (Fig. 3.42b).

The impact of floods on a river system can be illustrated in a number of ways. The Francou index – the relationship between maximum observed flow and drainage basin area (Francou and Rodier, 1967) – facilitates in comparing the relative magnitude of floods in different river basins (Rodier and Roche, 1984; Herschy, 2003) (Fig. 3.43a). This index also gives some indication of the upper limits of meteorological flooding (O'Connor *et al.*, 2002): for watersheds larger than ~100 km^2 there appears to be a level above which no discharge has been measured, even though the more than 280 rivers shown in Fig. 3.43 collectively represent more than 6000 years of measured discharge data (Herschy, 2003). This hypothetical line seems to predict

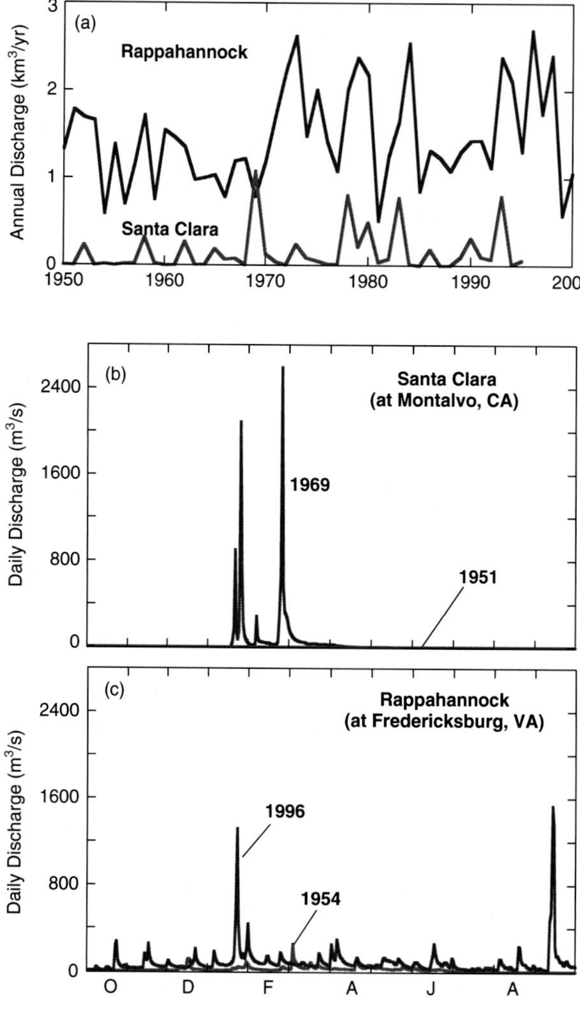

Figure 3.41. Mean annual discharges for the Santa Clara (red) and Rappahannock (blue) rivers, 1951–2000 (a), and daily discharges for the Santa Clara (b) and Rappahannock (c) during lowest (red; 1951, 1954) and highest (blue; 1969, 1996) discharge years. In most years the Rappahannock discharges one- to two- orders-of-magnitude more water than the Santa Clara, but occasionally the Santa Clara's peak daily discharge exceeds that of the Rappahannock. (Note that horizontal scales in (b) and (c) portray the water year (October to September) rather than calendar year.

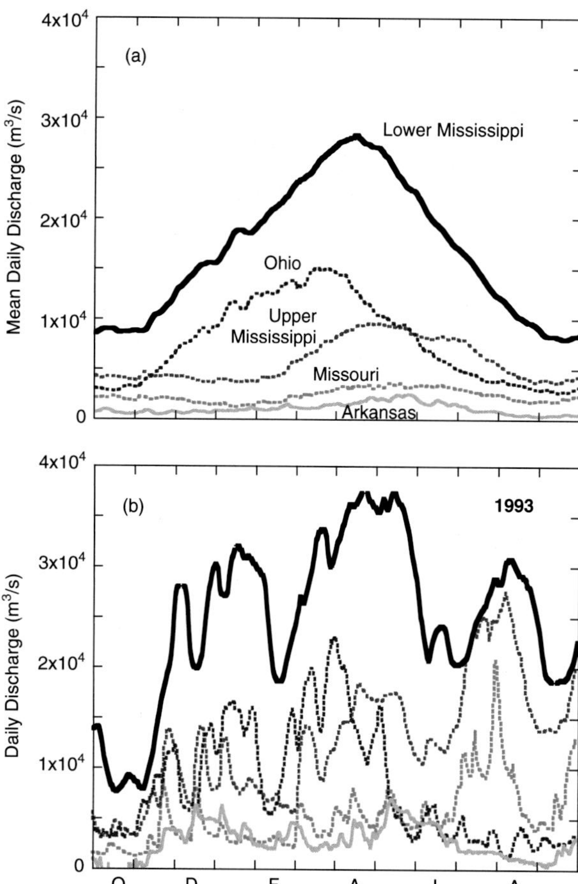

Figure 3.42. (a) Long-term (>50 yr) mean daily discharge of the lower Mississippi River (at Vicksburg, MS) and its tributaries: the Ohio/Tennessee River (at Metropolis, IL), the upper Mississippi (at Thebes, IL), the Missouri (at Hermann, MO), and the Arkansas (at Little Rock, AR). (b) Daily discharge of the same five rivers in 1993, which saw >100-yr floods throughout the upper midwestern USA. Data from USGS/NWIS.

the maximum discharge that a precipitation-initiated flood can achieve in any given drainage basin area; that is, the maximum impact that a meteorological event can have on river discharge (O'Connor *et al.*, 2002). According to this plot, maximum precipitation-derived discharge in a river with drainage basin of 10 000 km^2 should be no greater than ~37 000 m^3/s, and maximum discharge from a drainage basin of 1 000 000 km^2 should be no greater than ~207 000 m^3/s.

Viewed in terms of peak runoff (discharge per unit basin area), Fig. 3.43b shows a strongly inverse relationship with basin area. Peak runoff in the Amazon basin, measured at Obidos, is only ~0.2 mm/h, whereas maximum runoff in rivers with basins smaller than 100 km^2 can exceed 100 mm/hr. The inverse relation between peak runoff and basin area reflects the fact that the storms that produce heavy rainfall (e.g. thunderstorms, cyclonic storms) are often no more than 20–100 km in diameter (Leopold, 1994). For a river with a basin area less than 1000 km^2, such a storm can affect the entire watershed, passing over the basin within a matter of minutes to hours. A large typhoon, for example, can encompass nearly the entire island of Taiwan. A similar-size storm, in contrast, would envelope only a tiny fraction of the Amazon's 6 400 000 km^2 drainage basin.

White (1996) suggested that frequent thunderstorms may be responsible for supplying much of the sediment

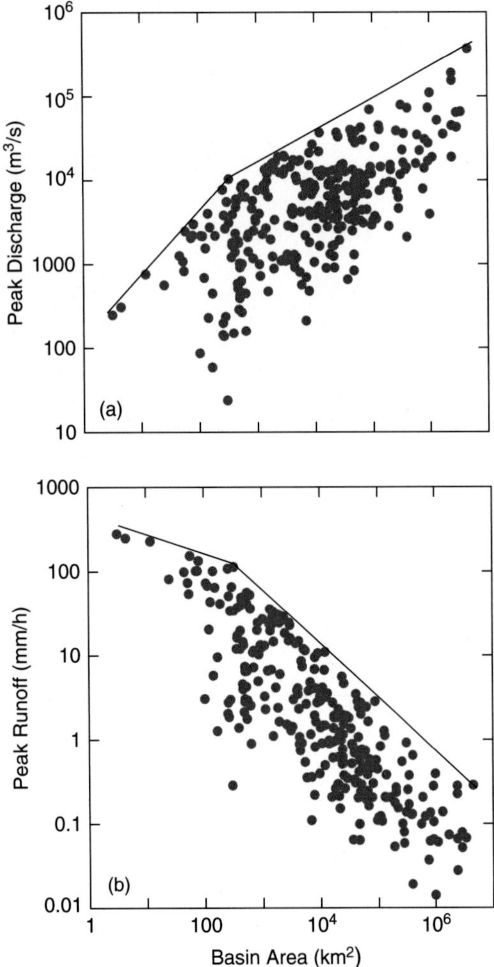

Figure 3.43. Maximum observed precipitation-derived discharge (a) and runoff (b) vs. drainage basin area for selected global rivers. Note the relatively log-linear relationships, above which no meteorologically-induced floods discharges have been measured. Data from Rodier and Roche (1984), Herschy (2003), and USGS/NWIS.

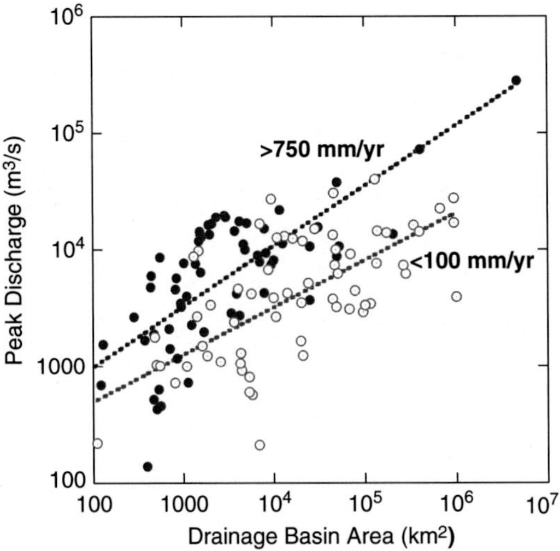

Figure 3.44. Peak measured discharges of high-runoff (>750 mm/yr; blue dots) and arid (runoff <100 mm/yr; red circles) rivers. Although arid rivers generally have lower peak discharges, occasionally their peak discharges can approach or surpass those of high-runoff rivers. Data from Rodier and Roche (1984), Herschy (2003), and USGS/NWIS.

to the channel system (presumably much of it by way of local landslides – c.f. Hovius *et al.*, 1997), whereas larger storms – such as typhoons – may be responsible for actual channel erosion and downstream transport. Not only do extreme storms erode and transport large amounts of sediment, but they also may play an important role in the delivery of dissolved solids and organic carbon (Goldsmith *et al.*, 2008). Hilton *et al.* (2008) estimate that since 1970 77–92% of the non-fossil POC transported in Taiwan occurred during typhoon-induced floods.

It might seem axiomatic to state that high-runoff rivers generally experience higher peak flow than low-runoff rivers (Fig. 3.44). But there are instances shown in Fig. 3.44 where low-runoff rivers have experienced higher peak discharges than high-runoff rivers. Similarly, the significance of peak flow increases in smaller rivers and with decreasing runoff (Fig. 3.45a). Peak estimated discharge of the Amazon, for instance, is about 1½ times its mean discharge – although an increase in Amazon discharge (at Obidos) from 175 000 m^3/s to ~280 000 m^3/s (R. H. Meade, personal communication) is indeed impressive. Peak discharge in a small humid or high-runoff river, in contrast, can be 10-fold to 100-fold greater than its mean discharge. In small arid rivers, where overland flow can greatly exceed subsurface groundwater flow (see Becker *et al.*, 2004), the peak-to-mean ratio may exceed 1000 (Fig. 3.45a). If we look at rivers with drainage basins between 1000 km^2 and 5000 km^2 in area, the peak-to-mean ratio increases from 3–100 for rivers with annual runoffs >1000 mm/yr to 900–10 000 for rivers with mean runoffs <10 mm/yr (Fig. 3.45b). As such, infrequent floods often account for much – if not most – of an arid river's discharge (see McMahon, 1979).

Figure 3.46 provides a graphic example of the differences cited above. Runoff of five US mid-Atlantic rivers is more or less constant (~400–600 mm/yr), but the small Patuxent River (348 km^2) achieves half of its 50-year cumulative discharge in about half the time required for the much larger Ohio River (500 000 km^2) (Fig. 3.46a). By contrast, the five California rivers in Fig. 3.46b have roughly similar watershed areas (1200–3700 km^2) but their annual runoffs vary from 9 mm/yr (Santa Maria) to 2000 mm/yr (Smith). Nearly half of the Santa Maria River's aggregate discharge over a 47-year record can be accounted for by only 180 days of peak river flow, ~1% of the cumulative time, whereas half of the aggregate discharge for the Smith

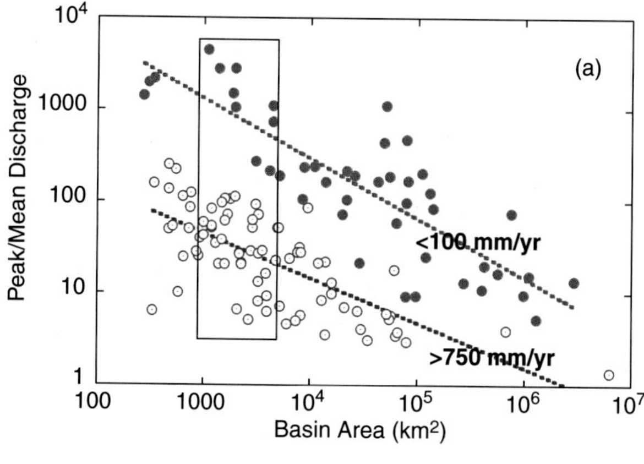

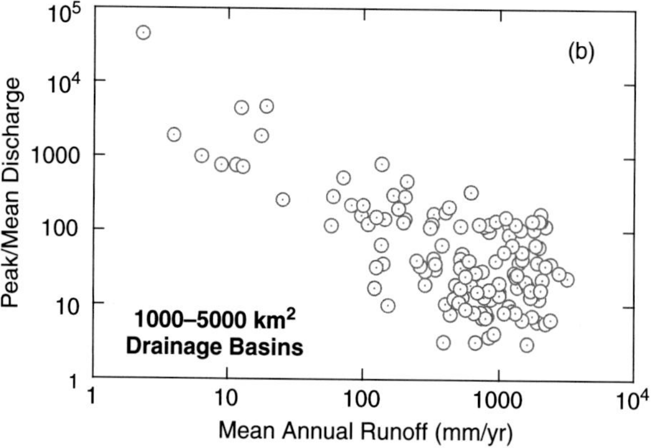

Figure 3.45. (a) Maximum recorded discharge relative mean discharge for high-runoff (mean runoff >750 mm/yr) and low-runoff (<100 mm/yr) rivers. While some of the scatter in maximum/mean ratios is due to the difference in length of gauging records, much of it reflects the range in runoff for both the high- and low-runoff rivers. In (b) we show the peak-to-mean ratios for rivers vs. mean annual runoff for rivers with drainage basins 1000 to 5000 km^2 in area, as delineated by box in (a). Data from Herschy (2003) and USGS/NWIS..

River required about 1700 days, nearly 10% of the cumulative time (Fig. 3.46b).

Although the database is scant, sediment discharge during periodic events tends to be greater from small watersheds than from large watersheds (see also Meybeck, 2003; Gonzalez-Hidalgo *et al.*, 2009). Peak calculated daily sediment discharge of the Mississippi River (13 300 000 km^2 basin area) is 4.5 Mt (January 15, 1950), whereas the Santa Clara River (4100 km^2) discharged 18 Mt on February 25, 1969, and the Eel River (8000 km^2) 52 Mt on December 22, 1964 (Fig. 3.47). In fact, the 22 largest daily sediment loads for US rivers have measured in three California rivers – the Eel, Santa Clara, and Salinas (Farnsworth and Milliman, 2003). Gonzalez-Hidalgo *et al.* (2009) have found that in California the three recorded largest events have contributed nearly 40% of the cumulative sediment load. This is not to say that periodic events necessarily play key roles in the erosion and transport of sediment in all river basins. In the wet Waipaoa River (North Island of New Zealand), lower-magnitude events appear to have a greater cumulative effect on sediment transport than the less frequent, higher-magnitude events, although the roles of deforestation and changing landuse in the removal of stored sediment in this and other watersheds are not completely understood (Trustrum *et al.*, 1999).

To a large extent the impact of peak events on small drainage basins reflects the erosive power of flash floods (often resulting in landslides and other downslope mass movements), in part the inability of small basins to absorb local precipitation events (Gonzalez-Hidalgo *et al.*, 2009). In two large watersheds in the Queen Charlotte Islands (British Columbia) four storms between 1891 and 1978 were responsible for 85% of the mass wasting (Hogan and Schwab, 1991; c.f. Chatwin and Smith, 1992). In the Redwood Creek basin (730 km^2), a major rain event in December 1964 produced a sediment three-day load of ~130 Mt, correlating to a peak sediment yield of ~5000 t/km^2/day. The river's channel widened by as much as 100 m and aggraded by as much as 4.5 m, increasing the amount of stored sediment 1.5-fold (Nolan and Marron, 1995; Madej, 1995).

Peak rainfall events in small watersheds can cause tributary damming and subsequent breaching, resulting in flood bores that can greatly magnify the impact of the rainfall itself; during these peak discharges, sediment concentrations can reach or exceed 100 g/l (Schick and Lekach, 1987). In contrast, broad flood peaks in very large rivers not only reflect the decreased impact of peak events (Figs. 3.43b and 3.47) but also the storage of sediment during falling stages of river flow and its remobilization during rising flow (R. H. Meade, personal communication).

As shown in Figures 3.41, 3.44–3.47, peak events tend to have a particularly great impact on small rivers draining arid regions. Flood years in arid regions can result in orders-of-magnitude greater mass-movement denudation than during an average year (Starkel, 1972; Gupta, 1988). The erosive power of intense rainfall in arid regions reflects the often bare soil surface, which intensifies surface flow as well as being more erodible, and because evaporation is minimal during such short-term rainfalls (Schick, 1988).

The effect of peak discharge events in arid regions is illustrated in photographs of the Ventura River (California, basin area 480 km^2, mean runoff 110 mm/yr) during summertime low flow (~5 m^3/s, Fig. 3.48a) and during an El Niño flood (1000 m^3/s; Fig. 3.48b), both photographs taken in 1998. Suspended sediment concentrations during low flow were <1 g/l, whereas they exceeded 35 g/l during the flood; daily sediment discharge in early June 1998 was ~200 t/day, on February 23, 3 000 000 t/day (J. A. Warrick, personal communication).

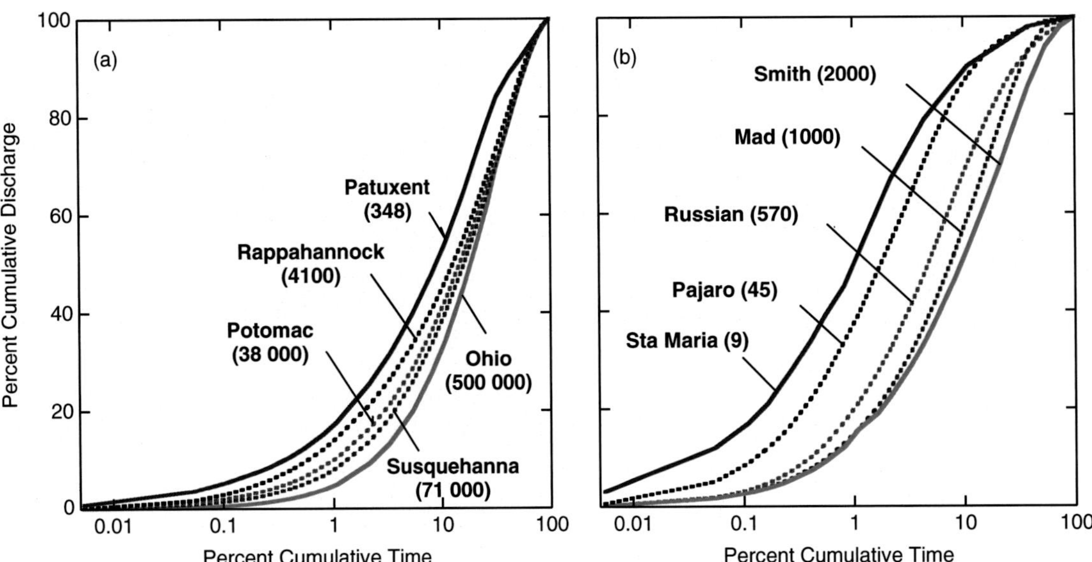

Figure 3.46. Time required in the 50-yr period, 1951 and 2000, for Mid-Atlantic east coast rivers with similar runoffs (a) and similar-sized small California rivers with varying runoffs but similar basin areas (b) to achieve levels of cumulative discharge. Numbers in parentheses in (a) indicate basin areas (km^2), and in (b) mean annual runoffs (mm/yr). Data from USGS/NWIS.

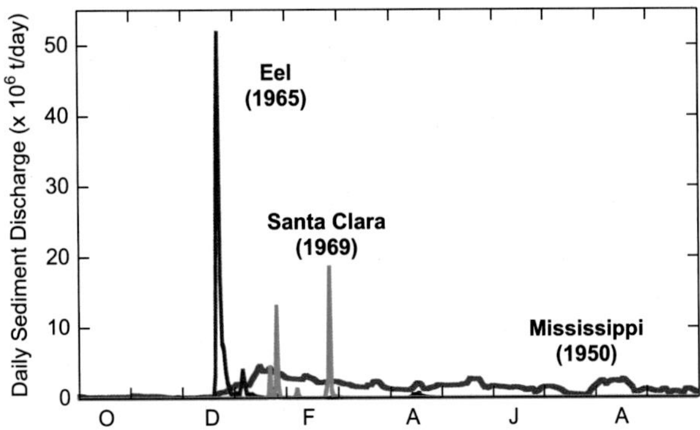

Figure 3.47. Daily discharge for peak discharge years for the Santa Clara (green; water year 1969), Eel (blue; 1965) and Mississippi (red; 1950) rivers. (It should be noted that the database for the "untamed" Mississippi effectively ends in 1953 with closure of dams on the Missouri River; see Chapter 4.) Peak daily loads represent 40% (18 of 46 Mt), 35% (52 of 150 Mt) and 0.9% (4.5 of 500 Mt) of the Santa Clara's, Eel's, and Mississippi's record annual loads, respectively. As discussed below, most of the Santa Clara's sediment was discharged at hyperpycnal concentrations (>40g/l). During peak discharge the Eel may have reached hyperpycnal levels, but the Mississippi sediment concentrations never exceeded 1 g/l. Data from USGS/NWIS.

The effect of climate is also demonstrated in four California rivers whose runoffs ranged from 52 mm/yr to 3600 mm/yr for the periods during which their sediment loads were monitored (Fig. 3.49). In the arid San Luis Rey River (52 mm/yr runoff), a single event (Feb. 26, 1969) accounted for 40% of the 11-yr cumulative sediment load. In contrast, the highest discharge event for Mad River (runoff 3600 mm/yr) (Dec. 22, 1964) accounted for only 7% of the cumulative load (Fig. 3.49). The San Luis Rey River discharged 50% of its 11-yr cumulative load in only 10 days, but this represented only 0.5 Mt, whereas during its 10 peak-discharge days the Mad River discharged 20 Mt.

The extent to which a flood can affect an arid river is seen in the ephemeral Wadi El Arish, Sinai, a river whose flow is so low that it probably has never been gauged or monitored. Nevertheless a 1975 flood discharged enough sediment to form a 500 m × 300 m delta extending into the Mediterranean Sea (Schick, 1988). Detailed clay mineral analyses, in fact, indicate that the El Arish and adjacent small coastal wadis account for a considerable portion of the shelf sediment deposited off Israel (Sandler and Herut, 2000). Stow and Chang (1987) similarly estimated that 2/3 of the sand delivery to the southern Californian coastal zone from the small (896 km^2) San Dieguito River is contributed during 25-, 50- and 100-yr floods, all of which (although their paper was written before ENSO became a household word) presumably are related to El Niño events. In the badlands of semi-arid Spain, most of the sediment is eroded in response to high-intensity rainfall events, but these events are sufficiently rare that the overall denudation rate remains low (Canton *et al.*, 2001).

Figure 3.48. Lower Ventura River, California (looking downriver) in (a) early June 1998 (flow ~5 m^3/s) and (b) February 23, 1998 (flow 950 m^3/s). Photographs courtesy of J. A. Warrick.

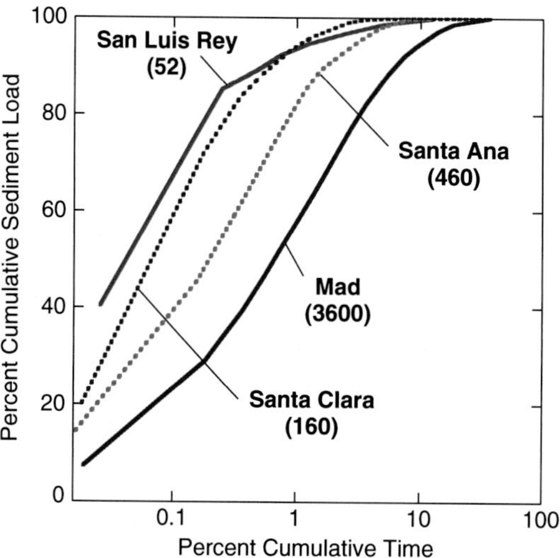

Figure 3.49. Cumulative sediment discharge vs. cumulative time for four California rivers: San Luis Rey (water years 1968–78, 1984), Santa Ana (1968–88), Santa Clara (1968–81, 1984, 1985), and Mad (1960–74). Basin areas range from 1400 (Mad) to 6300 (Santa Ana) km^2. Each river experienced at least one major discharge event during the monitoring. Note that Santa Clara, and Santa Ana River runoffs (bold numbers in parentheses) during measured periods were different than their long-term means (see appendix). Data from USGS/NWIS.

Monitored peak discharge events

Monitoring peak precipitation and discharge events is not an easy task either in terms of measurement procedures or personal safety. As a result, few peak events have been documented. Below we discuss two monitored events – the winter 1969 storms in Southern California (arid climate) and the 2004 Typhoon Mindulle in Taiwan (high-runoff climate).

Santa Clara River, January–February 1969

The storms that struck southern California in January and February 1969 provide an excellent example of the effect that episodic rains can have on a small, arid watershed. Perhaps for the first time, major storms were sampled hourly through what now seem like almost super-human efforts by the USGS (Waananen, 1969).

The Santa Clara River is classified an arid river (44 mm/yr), although occasionally annual precipitation is sufficiently intense that the watershed can be considered – at least temporarily – subarid. The arid nature of the river and its watershed is noted by the fact that between June 12, 1968, and January 18, 1969, there was no measured discharge at the river mouth (Montalvo gauging station). Beginning on January 18, early rains, with intensities approaching 25 mm/h, were mostly absorbed into the dry ground. These heavy rains therefore had little effect on Santa Clara discharge. A second wave of rains reached the coastal mountains on January 24, when 61 mm of rain fell, followed the following day by 178 mm. Flow at the river mouth, which lagged the peak rainfall at Ojai (near the headwaters of the Santa Clara) by about 5–6 h, on the January 25 approached 4000 m^3/s and sediment concentrations exceeded 160 g/l; sediment discharge for the day was 13 Mt.

Because the ground was now saturated with water, a somewhat less intense storm a month later (229 mm between February 23 and February 25) resulted in equally high flow. Total sediment discharge during this second storm approached 25 Mt, and suspended sediment concentrations reached or exceeded 50 g/l for 15 consecutive hours; peak concentrations approached 150 g/l (Fig. 3.50; see Warrick and Milliman, 2003). Total sediment discharge over the 32-day period between January 25 and February 26 was 45 Mt, by far the greatest load ever recorded in the Santa Clara.

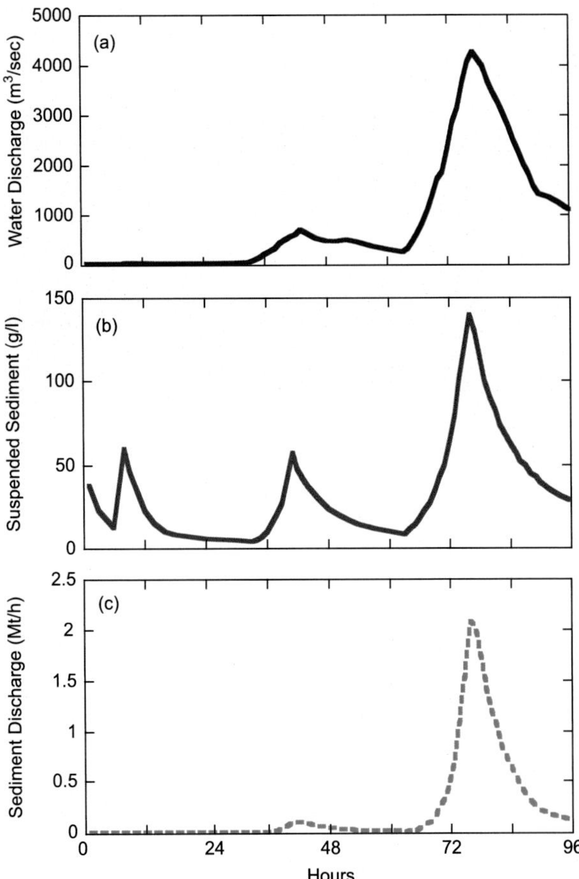

Figure 3.50. February 24–26, 1969, discharge (a), suspended sediment concentrations (b), and calculated sediment loads (c), Santa Clara River at Montalvo, California. Suspended sediment concentrations >40 g/l greater are considered to be at hyperpycnal levels. Data from Waananen (1969).

To put these two events in perspective, single-hour sediment discharges during both storms reached 2–3 Mt/h (Fig. 3.50c), and during the peak 20 h of each event, the Santa Clara discharged about as much sediment as it had during the previous 20 years. Moreover, most of the sediment was discharged at hyperpycnal concentrations (>40 g/l; Fig. 3.50b; see Warrick and Milliman, 2003). A considerable portion of the sediment discharged from the Santa Clara and neighboring rivers was deposited on the adjacent shelf and in the offshore Santa Barbara and Santa Monica basins as a prominent turbidite layer (Drake *et al.*, 1972; Gorsline, 1996).

Choshui River, July 2004

Philippine and Taiwanese rivers have some of the highest sediment yields in the world (see Chapter 2), and their greatest discharges usually occur during major typhoons (White, 1992; Milliman and Kao, 2005; Kao and Milliman, 2008). Lying in "Typhoon Alley," Taiwan is particularly susceptible to typhoons. Milliman and Kao (2005) calculated that between 1979 and 1998, 11 Taiwanese rivers collectively experienced at least 100 hyperpycnal events, most of them typhoon-related, many enhanced by the erosion of landslide deposits (Hovius *et al.*, 2000; Dadson *et al.*, 2004, 2005; Galewsky *et al.*, 2006). Because these estimates were based on infrequent observations and because many events last only a few hours, the actual number of hyperpycnal events experienced by Taiwanese rivers is certainly much greater. Based on sparse data, Milliman and Kao (2005) estimated that during Super Typhoon Herb (August 1996), the Choshui River (watershed area 3300 km^2) discharged an estimated 125 Mt of sediment, peak concentrations exceeding 150 g/l. Typhoon Toraji in July 2001, the first major typhoon after the M-7.6 Chi-chi earthquake two years earlier, transported more than 200 Mt of sediment, with peak concentrations probably exceeding 200 g/l (Dadson *et al.*, 2004, 2005). Given the abruptness with which precipitation and resulting discharge can shift and vary, however, these calculations can be considered only rough estimates.

Between July 2–5, 2004, Taiwan was struck by Typhoon Mindulle, which dropped more than 1 m of rain. Rainfall intensities exceeded 20 mm/h during the first phase of the storm, and then, after a one-day lull, they exceeded 60 mm/h on (July 4; Fig. 3.51a). As with the Santa Clara River in 1969, the first phase of Typhoon Mindulle resulted in peak Choshui discharges (8 km from the river mouth) decidedly lower than those seen during the second phase (3500 vs. 6500 m^3/s; Fig. 3.51b), as the ground initially absorbed much of the rain. Sediment concentrations during the first phase of Mindulle increased to 190–200 g/l (Figs. 3.51c, 3.52), of which >40 g/l were sand, perhaps reflecting erosion of the 1999 Chi-chi earthquake landslide debris (Milliman *et al.*, 2007; see above). Sediment concentrations during the second phase of the storm were <100 g/l even though peak discharge was greater than during the first phase. Total sediment discharge during the 3-day period of Typhoon Mindulle was 72 Mt, about 35–40% of it sand (Milliman *et al.*, 2007). Peak hourly loads exceeded 1.5 Mt/h during both the first and second phases of the typhoon (Fig. 3.51d).

During Typhoon Mindulle and for 80 days following the typhoon, the Choshui was also monitored for suspended sediment, particulate organic carbon (POC), and dissolved solids by S. J. Kao and his colleagues from Academia Sinica, Taiwan (see Goldsmith *et al.*, 2008). During these 80 days, the Choshui transported 90 Mt of suspended sediment and 0.25 Mt of dissolved solids. Some 85% of the cumulative 80-day suspended load was transported during the four days of Mindulle (57 Mt) and three days of Typhoon Aere (22 Mt); in contrast, the two typhoons

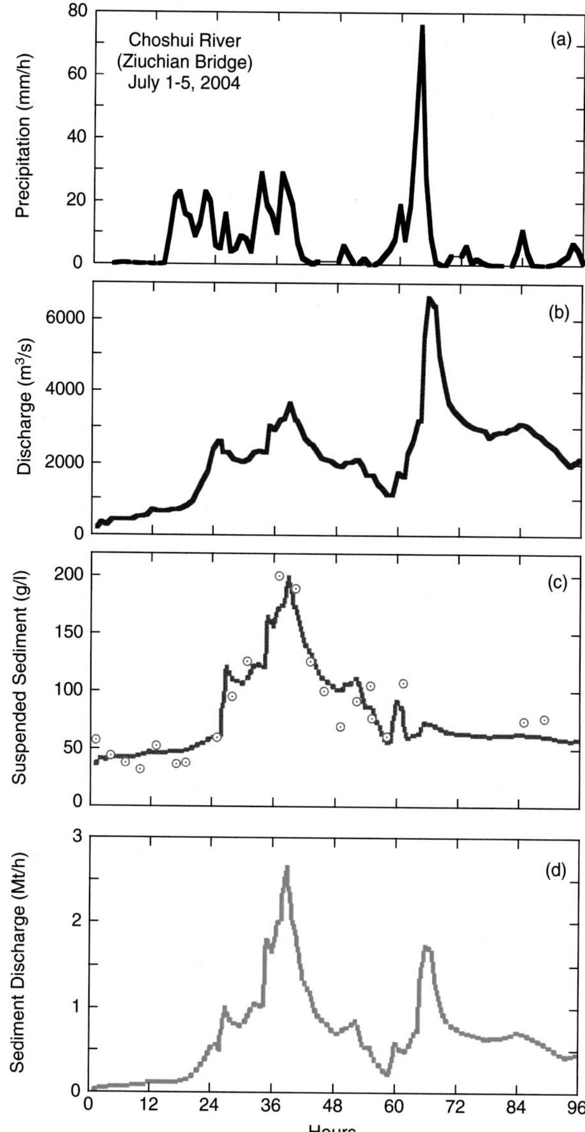

Figure 3.51. Precipitation (a), discharge (b), suspended sediment concentrations (c), and calculated sediment loads (d) in the Choshui River (Taiwan) during Typhoon Mindulle, July 1–5, 2004. Open circles in (c) represent direct measurements made by S.J. Kao and colleagues at Jangyang Bridge; the red curve in (c) is based on a stratified sediment-rating curve (see Kao *et al.*, 2005; Milliman *et al.*, 2007).

Figure 3.52. Choshui River (Taiwan) looking downstream to the Shuili Bridge, 40 km from the river's mouth and 32 km upstream from Ziiuchian Bridge, during Typhoon Mindulle, July 4, 2004. Measured discharge and suspended sediment concentrations exceeded 6000 m^3/s and 50 g/l, respectively. Photograph by Jil Chian, courtesy of S. J. Kao.

transported only 40% (0.14 Mt) of the 80-day dissolved load (Fig. 3.53c). During low flows before Mindulle, dissolved-solid concentrations were ~15% of suspended solids (0.22 g/l vs. 1.5 g/l at 100 m^3/s) whereas during the latter stages of Mindulle they were about 400-fold less (0.15 g/l vs. 60 g/l) at a discharge of 4000 m^3/s (compare Figs. 3.53a and b). Viewed in another way, dissolved and suspended discharges at 100 m^3/s were 100 and 700 t/h, respectively, whereas at 4000 m^3/s they were 1800 and 720 000 t/h (Fig. 3.53c). The less dissolved-solid transport during high-flow events clearly reflects the the decrease in dissolved-solid concentration with increased flow (Fig. 3.53b) due to dilution (Fig. 2.32) and less contact time for chemical weathering reactions to approach equilibrium before the water is purged (Burns *et al.*, 2003). In contrast to suspended solids, therefore, any increase in dissolved-solid discharge during floods results from increased flow. Thus, whereas the 79 Mt discharge during the two typhoons was twice the annual average (40 Mt), the 0.14 Mt discharge of dissolved solids represented only about 10% of the annual mean (1.6 according to Li, 1976).

Although the typhoon-generated floods did not greatly affect the discharge of dissolved solids, their composition changed significantly. Just prior to Mindulle Ca/Na molar ratios in the Choshui ranged between 1 and 2, but then increased to 3.5–5 during the initial stages of Mindulle, not dissimilar from groundwater values measured near the river (S. J. Kao, unpublished data). Throughout the remainder of Mindulle, Ca/Na ratios declined to ~1. They then again increased to 4–5 for four days after Mindulle. Most dissolved Na (hence low Ca/Na ratios) appears to have been derived from surface-weathered layered silicates, not from the salts in typhoon rains, as evidenced by low Cl concentrations (Milliman *et al.*, in preparation). By July 12, one week after Mindulle, Ca/Na molar ratios declined to 1–2, but then increased to 4–5 during Aere, indicating a return of groundwater discharge, being pushed out, we assume, by typhoon-induced flow. Three weeks after Aere Ca/Na molar ratios again had declined to 1–2 (Milliman *et al.*, in preparation).

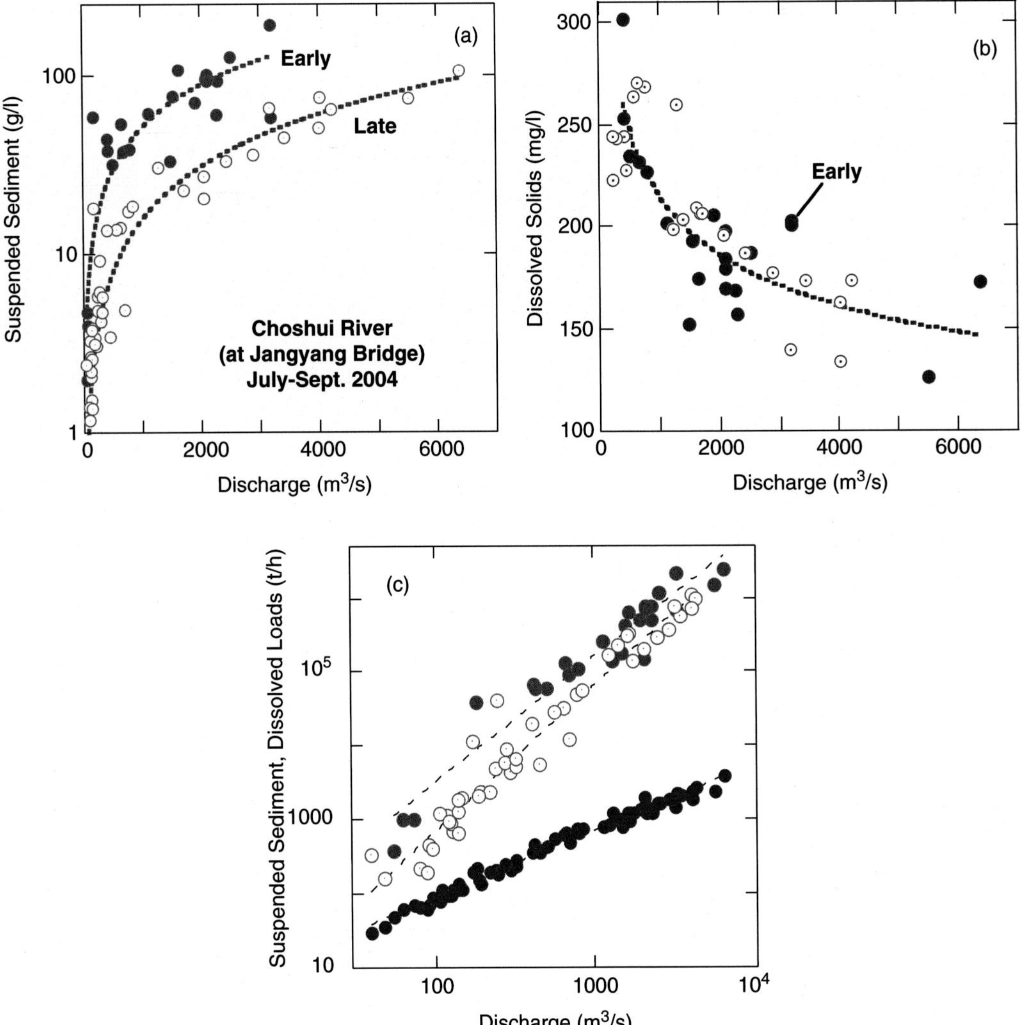

Figure 3.53. Suspended-sediment (a) and dissolved-solid (b) rating curves for the Choshui River (Taiwan) at Jangun Bridge, 50 km upstream from the river mouth, July 1–September 18, 2004, based primarily on data collected during and immediately after typhoons Mindulle (July 2–5) and Aere (August 24–26). Suspended sediment concentrations increased as a power function of river flow, although decreasing sharply after the early stages of Mindulle (early stages – solid dots, later stages – open circles). In contrast, dissolved concentrations decreased with discharge and showed no major shifts with time. (c) With increased discharge, calculated suspended load increased relative to dissolved load, reflecting the different trends of rating curves shown in (a) and (b). Data courtesy of S. J. Kao, Academia Sinica, Taiwan.

Typhoon Mindulle also provided the opportunity to monitor the response of seafloor morphology and sediment seaward of the Choshui River mouth, in the eastern Taiwan Strait, to the typhoon. Repeated monthly sediment coring surveys showed that the previously sandy coastal sediments were covered by as much as 1–2 m of mud immediately after Mindulle. But within one month after the typhoon the sediment deposited by Mindulle appears to have been resuspended and transported northward by the north-flowing Taiwan Warm Current. By the following spring there was no visual evidence of Mindulle-deposited sediment off the Choshui River mouth (Milliman *et al.*, 2007).

Hyperpycnal flows

Hyperpycnal defines suspended-matter concentrations that are sufficiently dense to allow fluvial water to sink beneath coastal marine waters at the river mouth, thereby minimizing estuarine mixing; 40 g/l is generally considered the lower limit of hyperpycnal concentrations (Mulder and Syvitski, 1995; Imram and Syvitski, 2000). Instantaneous concentrations as high as 1500 g/l have been noted in the loess plateau of China (Walling, 1999), 950 g/l during tropical cyclone Kina in Fiji (Kostaschuk *et al.*, 2003), and greater than 700 g/l in several south-western Taiwanese rivers (Taiwan Water Resources Agency, WRA, 1992).

Mulder and Syvitski (1995) identified nine global "dirty rivers," which they defined as having average annual sediment concentrations greater than 10 g/l, the inference being that hyperpycnal levels were attained periodically. The data in our appendices bring the number up to 21; those with mean concentrations >15 g/l are listed in Table 2.5. Mulder and Syvitski (1995) also listed a number of rivers that may produce hyperpycnal plumes at least once every 100 years in response to "maximum possible" flooding events. But because many of these rivers are poorly monitored – if monitored at all – this estimate almost certainly is minimal. Mulder *et al.* (1998) have estimated that during peak flows, the Var River (southern France), which has a mean sediment concentration of 7.7 g/l, can discharge sediment/water concentrations in excess of 200 g/l. During the 1969 winter storms in Southern California, measured sediment concentrations in the San Luis Rey, Santa Margarita, Callegas Creek, Santa Clara and Santa Maria ranged from 13 (Santa Margarita) to 160 g/l (Santa Clara). Because such events may last only a few hours, it seems likely that most – if not all – rivers and creeks along this 300-km stretch of coast discharged sediment concentrations that reached or exceeded 40 g/l.

The impact of hyperpycnal events on the adjacent ocean floor is unclear, since such events tend to be short-lived, particularly in smaller drainage basins. If flow is sufficiently dense and fast and if the shelf gradient is sufficiently steep, the sediment might bypass the littoral cell, perhaps eroding as it transits to deeper water (Mulder and Syvitski, 1995; Mulder *et al.*, 2003) and perhaps, as claimed for the central Sea of Japan (Nakajima, 2006), to be hundreds of kilometers from their source. Sediment discharged during the 1969 southern California events, for instance, appears to have reached offshore basins within a matter of days to weeks (Drake *et al.*, 1972; Gorsline, 1996), but whether the sediment bypassed the continental shelf immediately or was temporarily stored on the shelf is not clear (Warrick and Milliman, 2003; Warrick and Farnsworth, 2009). Using satellite imagery, Mertes and Warrick (2001) found that surface plumes off southern Californian rivers during the 1998 El Niño winter could account for only about 1–2% of the total river sediment output, the inference being that much of the remaining 98–99% escaped as hyperpycnal flow that sank beneath the ocean surface or was rapidly deposited. Given the fact that hyperpycnal events often are caused by heavy rains, there are few aerial photographs that can illustrate the discharge of a hyperpycnal flow into the ocean.

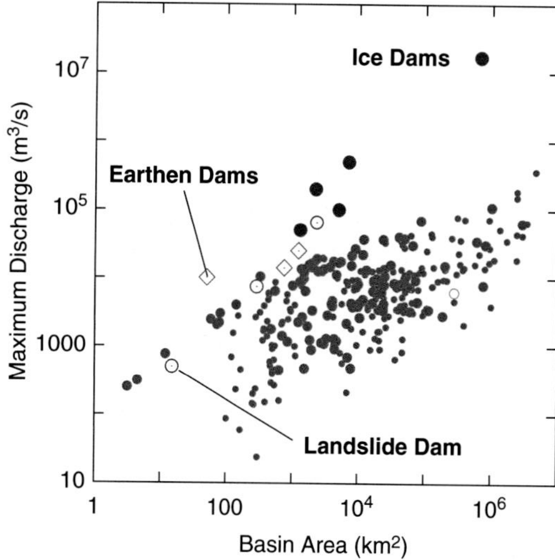

Figure 3.54. Maximum discharges measured or estimated from natural (ice dams: blue circles and landslide dams: blue dots) and man-made earthen-dam (open blue diamonds) failures. For comparison, we also show the trend of maximum discharge vs. drainage basin area caused by precipitation events (red dots; see Fig. 3.43). Ice-dam-related discharges are estimated from geological evidence (Baker *et al.*, 1995). While maximum discharge relative to small earthen dam failure appears to be no greater than that initiated by precipitation-generated floods, landslide and (particularly) ice-dam failures appear capable of generating discharges 1- to 2-orders- of- magnitude greater than precipitation-generated events. Data from Pitlick (1993), O'Connor *et al.* (2002), and Tomasson (2002).

Figure 3.55. Small ephemeral lake behind a landslide dam created on the Te Arai catchment, North Island, New Zealand, about two weeks after the passage of Cyclone Bola, March 1988; the landslide was 10 ha in area. The lake is shown a few days before it was breached by a trench dug by a local farmer (M. Page, personal communication). The white areas in this photograph provide graphic evidence of the phenomenal number of landslides generated by Bola, a 100-yr event. Photograph courtesy of N. Trustrum, New Zealand Institute of Geological Science Ltd.

Table 3.1. *Sediment flux from various types of episodic events.*

Event	Date	River	Basin area (km)	Sed. flux (Mt)	Event length	Sed. yield (t/km^2 per event)
Meteorological floods						
Storm Agnes	6/1972	Susquehanna	72 000	10	3 days	140
El Niño Rain	2/ 1969	Santa Clara	4100	19	2 days	4600
Rain	12/1964	Eel	8000	120	3 days	15 000
Typhoon Herb	8/ 1996	Choshui	3000	84	2 days	25 000
Volcanic eruptions						
Mt. St. Helens	5/1980	N. Toutle	735	55	2 yrs	75 000
Santa Maria	1902	Samala	1500	5000	22 yrs	3 000 000
Pinatubo	1991	Pasig	21	100	1 yr	4 000 000
Mt. Unzen	1990	Mizunashi Ck.	12	8	3 yrs	700 000
Dam failure						
Earthen Dam	1982	Roaring	~15	0.026	5 yrs	1650
Jokulhlaup	1918	Myrdalsjokul	~2000	800		4 000 000
Glacial Dam	15 (kaBP)	Columbia	250 000	10^6	10 days	~4–8 000 000

Data from USGS, Wannanen (1969), Milliman and Kao (2005), Dinehart (1997), Hayes (1999), Suwa and Yamakoshi (1999), Kuenzi *et al.* (1979), Pitlick (1993).

Recognizing flood events in marine sediments is not easy, since the thickness of a flood-derived layer may be far thinner than the depth of sediment mixing (20–100 cm in shallow water depths). As such, it can be mixed quickly into underlying sediments by the benthos. The best quantitative estimates of the frequency and magnitude of paleoevents often come from anoxic basins close to the probable sediment source. For instance, Schimmelmann *et al.* (1998) suggested that a 1–2 cm-thick gray silt layer in the Santa Barbara Basin (off southern California) was deposited at or near the beginning of the Little Ice Age during a major flood on the Santa Clara River, perhaps involving the draining of a large lake in the Mojave Desert.

Natural dams and their failure

Most floods owe their high discharges to intense precipitation, melting of snow pack, or a combination of the two. Much greater flow, however, can be gained when a natural or artificial lake is breached. The result can be one- to two-orders-of-magnitude greater discharge than that stemming from a storm or snow-melt (Fig. 3.54). Natural events that can dam a river include beaver dams, log or ice-jams, and earthquake- or rainfall-generated landslides (see Fig. 3.55).

High discharge coming from a breached dam may last only a few hours, but the effect in terms of erosion, transport and discharge can much greater than that transported during a normal precipitation-related flood. The July 1982 failure of a man-made earthen dam in Rocky Mountain National Park (Colorado), for example, resulted in a peak discharge in Roaring River about 30 times the 500-yr flood (Jarrett and Costa, 1986). The alluvial fan deposited by the flood represented the equivalent of 28 000 years of bed load (Pitlick, 1993). Subsequent erosion along the Roaring River resulted in an annual bed load at least 1000 times pre-flood levels, but within five years the loads had decreased by >95%, suggesting a rather rapid return to previous conditions (Schumm and Rea, 1995). Table 3.1 includes estimated sediment discharge from several dam failures. It should be noted, however, that some natural dams can last ~1000 years and some may never experience catastrophic breaching (Korup *et al.*, 2006).

Beaver dams

The North American beaver, *Castor canadensis*, as might be implied by its presence on the MIT logo, is commonly viewed as the prime example of a natural engineer. Beavers construct their dams from wood and mud, thereby entrapping bodies of water in which they live. The size of the beaver-dam "reservoir" depends in part on stream order and (thereby) the flow of the stream (Naiman *et al.*, 1988). By modulating stream flow and trapping sediment, beaver dams can dramatically affect stream morphology. Either decay or sudden floods can result in the breaching of beaver dams, sending trapped pond waters downstream. How much sediment actually escapes ponds is open to speculation, but recent studies suggest that most of the sediment remains more or less in place (Butler and Malanson, 2005). More important may be the release of reduced nutrients (e.g. NH_4) that are flushed downstream with the escaping pond water.

Assuming a sediment volume of 200–500 m^3 per beaver pond (Butler and Malanson (2005), pre-Columbian beaver ponds in North America, estimated to have been between 60 and 400 million (Naiman *et al.*, 1988; Butler, 1995), would have retained ~1.2–20 Bt of sediment. Decimation of the beaver population in the eighteenth and nineteenth centuries reduced the number of ponds considerably, perhaps by an order of magnitude relative to pre-Columbian populations.

Debris dams, logjams and icejams

Debris, defined as tree trunks, limbs, and twigs as well as rocks and boulders, can alter stream flow and hydraulics as well as stream-bed morphology (Mutz, 2003; Montgomery and Piégay, 2003; Lancaster and Grant, 2006). Logjams, for example, are a major factor in determining the distribution of bedrock vs. alluvial channel morphology on the Willapa River (WA) (Massong and Montgomery, 2000).

Icejams normally occur in high-latitude rivers susceptible to mid-winter thaws. Melting ice cover breaks down into smaller blocks that are transported downstream until they jam behind an obstacle (e.g. ice, rock or bridge) that hinders further downstream transport. Water builds up behind this temporary dam until it either tops or breaks through the jam, resulting in a sudden release of water, often at much greater velocities than a rain-initiated flood (Beltaos and Burrell, 2002). Such icejam floods can erode banks and channels, and produce gullies in overtopped floodplains (Gay *et al.*, 1998), resulting in elevated suspended concentrations and loads (Prowse, 1993). If climate warming continues in the foreseeable future, icejams and their related flood hazards could increase substantially (Beltaos, 1999), at least in the short term.

Landslide and volcanic dams

Landslides can block a river valley, thereby forming a dam. Although most landslides are generated by heavy rainfall or earthquakes (see Fig. 3.55), some can be formed by pyroclastic flows associated with volcanic episodes. A list of landslide dams compiled by Ermini and Casagli (2003; their Tables II–IV list more than 30 in Italy alone) shows dam heights as great as 500 m and reservoir volumes as great as 2 km^3. Depending on the lifespan of the dam, upstream floods are not uncommon. More dangerous is dam failure by which outbreaks of water can cascade downstream (Costa and Schuster, 1988). Depending on construction and composition of the dam, water can escape either by piping or by overtopping, the latter mode of escape being more common (Macias *et al.*, 2004).

The breaching of a 70-m-high landslide dam on the Dadu River (southwest China) in 1786 killed about an 100 000 people; estimated peak discharge may have been as great as 250 000 m^3/s (Dai *et al.*, 2005). A debris avalanche produced by the Mount St. Helens eruption in 1980 blocked the North Fork of the Toutle River, forming a temporary lake as deep as 15 m and with a volume of 4 000 000 m^3 (Costa and Schuster, 1988). Subsequent escape of water and pumice from the lake created a lahar with peak speeds of 15–20 m/s that was carried at least 35 km down valley (Waitt *et al.*, 1983).

Sudden breaks in such dams have resulted in a number of massive flow events in the Markham River valley on the northeastern coast of Papua New Guinea, and similar events have been noted in New Zealand (Pearse and Watson, 1986). The eruption of El Chichon volcano (Mexico) in March 28–April 4, 1982, led to the formation of a pyroclastic dam behind which 26 000 000 m^3 of water accumulated (Macias *et al.*, 2004). When the dam broke in late May of that year, peak discharges exceeded 10 000 m^3/s, during which at least 17 Mt of sediment were transported down the Magdalena River (a tributary of the Grijalva River), much of it at hyperpycnal concentrations (Macias *et al.*, 2004).

Glacial surge

Proglacial streams can discharge seasonally or interannually considerable quantities of sediment. Vatne *et al.* (1995), for instance, measured suspended sediment concentrations in proglacial streams in Svalbard, an area that is considered to have low sediment yield, that exceeded 4 g/l. During a glacial surge in 1963–64, the Jokulsa a Bru in Iceland transported 25 Mt (basin-wide denudation rate of 13.7 mm), and in 1966, the sediment load was 6.6 Mt (Bjornsson, 1979). A major ice-surge and subsequent retreat of the Bering Glacier between 1967 and 1993 resulted in sediment accumulation that locally exceeded 10 m (Molnia and Carlson, 1995).

Jökulhlaups

Jökulhlaup is the Icelandic word to describe a glacial outburst flood. All jökulhlaups involve the rapid melting of ice and abrupt release of trapped water: the melting of an ice-dammed lake, the draining of meltwater formed during a subglacial volcanic eruption (forming lahar-like flows seen, for instance, during the Mount St. Helens eruption in 1980), or the draining of a subglacial lake in a geothermal region (Bjornsson, 1992). Although within the broader definition of the term a jökulhlaup does not necessarily require the sudden release of geothermal heat, most present-day jökulhlaups are generated by volcanic activity. Not surprisingly, most jökulhlaup-like events have been documented in Iceland; comprehensive reviews have been given by Bjornsson (1992) Tomassan (1991), and Russell *et al.* (2006).

During volcanic events in an ice-bound area, the amount of water production (and its ultimate escape) can be impressively high. The Gjalp subglacial eruption at Iceland's Vatnahjökull in October 13, 1996, resulted in the melting of 3 km^3 of ice, in one location melting through 500 m of ice in 30 h (Gudmundsson *et al.*, 1997). Jökulhlaup-induced flows have been recorded up to 13 m/s, sediment loads greater than 70 t/s, and suspended sediment concentrations greater than 35 g/l (Tomasson *et al.*, 1980). The sediment transported during a jökulhlaup tends to be dominated by silt, sand and fine gravel (Maizels, 1989).

During a 1934 jökulhlaup, total sediment transport to the Skeidarasandur was on the order of 150 Mt (Bjornsson, 1992). Jonsson *et al.* (1998) calculated that the November 1996 jökulhlaup on the Skeiarrsandur discharged 3.4 km^3 of water (peak discharge 52 000 m^3/s), resulting in a sediment delivery to the downstream sandur of about 180 Mt; average sediment thickness of the Skeidararsandur flood plain was 4 m (Smith *et al.*, 2002; Maria *et al.*, 2000). During peak discharge, suspended sediment concentrations exceeded 100 g /l (Snorrason *et al.*, 2002). Snorrason *et al.* (2002) estimated that 180 Mt of sediment were discharged in 47 h. Given a drainage basin area of 1500 km^2 (Oddur Sigurdsson, 2001, written communication), this single event had a sediment yield of more than 100 000 t/km^2 in just two days. An October 1918 jökulhlaup from the Myrdalsjökull was even greater, with a peak discharge of 200 000 m^3/s and flow depths of 70 m. The cumulative volume of sediment transported was estimated by Larsen and Asbjornsson (1995) to have been at least 0.6 km^3, about the volume of the Great Wall of China.

How much jöklhlaup-derived sediment is actually transported beyond the alluvial plains (called Sandurs in Iceland) and reaches the sea is not known, but the 1934 jökulhlaup produced a delta that extended 3 km into the sea (more than 8 m thick near the coastline) (Bjornsson, 1995). The 1996 jökulhlaup added 7 km^2 of new coastline (Sigurdsson *et al.*, 1998), and data from several gravity cores suggest that sediment was transported at least 10 km from and to a thickness of 10 cm (Maria *et al.*, 2000). Sigurdssen (cited by Baker, 2002) documented a large turbidite-rich submarine fan complex 500 km south of Iceland that apparently has been formed by exceptionally large jökulhlaups.

Lawler (1994; Lawler and Wright, 1996) has estimated that the sediment load of the Jökulsa a Solheimasandi glacial river in Iceland has decreased by half in recent years. Although the causes are not clear, Lawler has speculated that flow is becoming more seasonal owing to warming summertime temperatures as well as the possible depletion of proglacial sediment supply.

Although jökulhlaup-derived deposits have been recognized in other glacial and sub-glacial areas, understandably only a few actual measurements have been made during actual jöklulhlaup events. The 1986 failure of the Hubbard Glacier ice dam across the entrance of Russell Fjord in Yakutat Bay in SE Alaska released a 25-m-high head of water, discharging 6.5 km^3 of water into Disenchantment Bay, cutting a 7.5 km-long channel, and depositing a 1–2-m-thick sediment drape across the bay (Cowan *et al.*, 1996). A series of jökulhlaup events emanating from neoglacial Lake Alsek (western Canada) and entering the Gulf of Alaska via the Alsek River (southeastern Alaska) have deposited a thick – in places 80 m thick – sequence of Holocene sediments in the Alsek Sea Valley (Milliman *et al.*, 1996).

The geological record provides a number of spectacular examples of jökulhlaups, particularly those occurring in North America at the end of the last glacial maximum, 15–20 ka BP. The periodic draining of Glacial Lake Agassiz, which occupied about 900 000 km^2 in central Canada between the Rocky Mountains and the present Great Lakes, discharged large volumes of water to the western Atlantic (Teller and Thorliefson, 1987). Early discharge events were directed southward along the Mississippi River valley and into the Gulf of Mexico (Emiliani *et al.*, 1975; Kehew and Lord, 1987). Erosion and transport of sediment deposited along the Mississippi floodplains more than 12 kyr ago may represent part of the present-day Mississippi sediment supply (R.H. Meade, personal communication). It has been speculated that a subsequent eastward draining of Canadian glacial lakes may have freshened the North Atlantic and thus prevented formation of the North Atlantic Deepwater, leading to the Younger Dryas cold event 11.5 ka BP (Broecker, 1989).

Megaflood and glacial-lake outburst deposits also have been noted in such diverse locales as the Altai Mountains in south-central Siberia (Carling *et al.*, 2002), along the southern flank of the Himalayas (Kuhle, 2002), and on Mars (Baker *et al.*, 1992; Baker, 2002). Interestingly, the timing of the last of the Altai outburst floods, 15.8 kaBP, is coincident with a freshwater spike in the Kara Sea, and also occurring about the same time as the Lake Missoula floods (discussed in the following paragraphs), but several hundred years before deglaciation of the Ulagan Plateau (Reuther *et al.*, 2006; Vogt, 1997).

Unquestionably the most impressive jökulhlaup – and perhaps the most significant fluvial event in the past 25 000 years – occurred as a series of floods that drained glacial Lake Missoula and discharged through the Columbia River onto the Oregon continental margin (Bretz, 1930, 1969; Baker and Bunker, 1985; Craig, 1987; Atwater, 1987) about 15 000 calendar years ago (Fig. 3.56). Based on the probable dimensions of Glacial Lake Missoula (area ~7700 km^2, maximum lake depth 610 m;

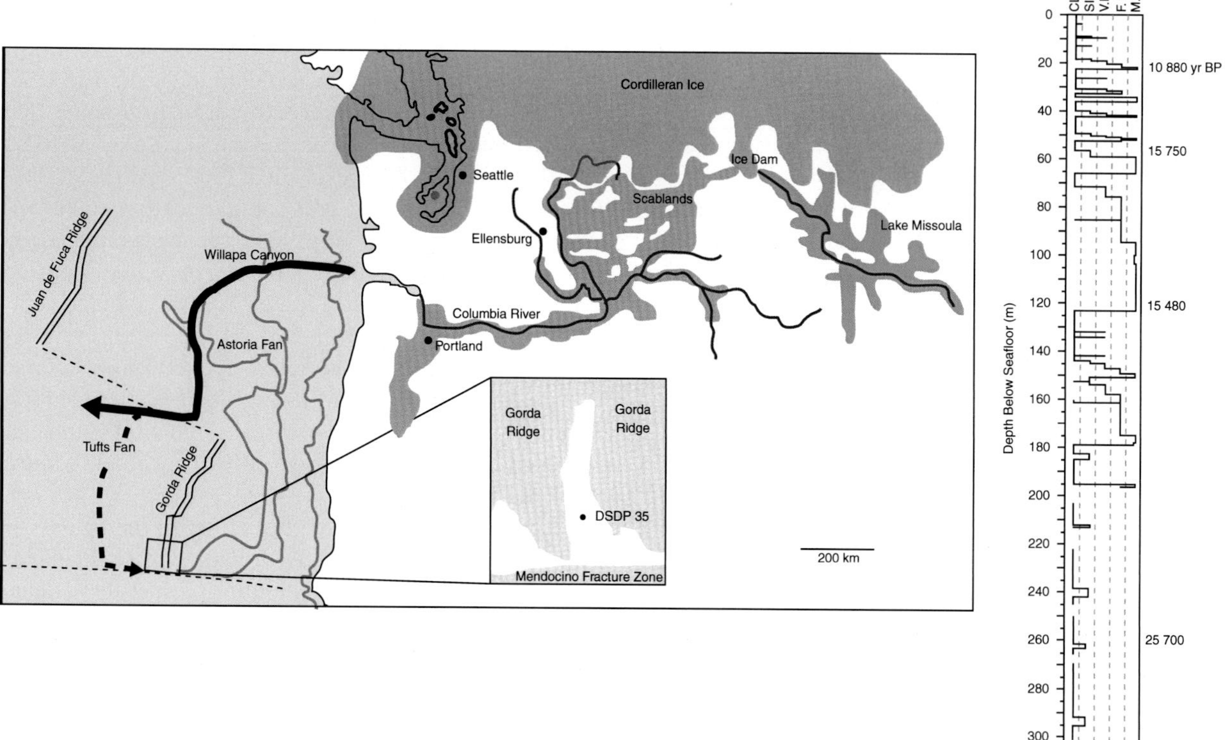

Figure 3.56. Path of the Lake Missoula floods. After being discharged into the Pacific Ocean at the present mouth of the Columbia River, sediment was transported down across the Astoria Fan as a massive turbidity current, some of which overtopped the Cascadia Channel between the Juan de Fuca and Gorda ridges and flowed south along the western margin of the Gorda Ridge and then eastward along the Mendocino Fracture Zone. Arrowed line shows probable path of flow. An ODP core from Escanaba Trough (lower right) penetrated a series of sand layers, the thickest of which, 57 to 121 m below the seafloor, has been dated at 15.5 ka BP ^{14}C years (18 ka BP calendar years) (after Zuffa *et al.*, 2000).

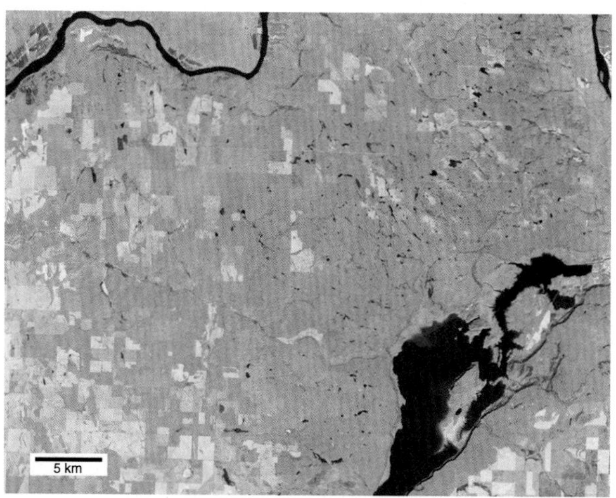

Figure 3.57. Channeled scablands, eastern Washington. Light geometric patterns indicate patches of irrigated cropland, contrasting with the darker terrain which has limited soil, presumably eroded during outbreak events. River in upper left is the Columbia River; lake in lower right is part of Banks Lake.

volume ~2000–2500 km^3, half that of present-day Lake Michigan), peak discharge from the Columbia River is estimated to have been somewhere between 1 and 13 × 10^6 m^3/s (Clarke *et al.*, 1984; Benito and O'Connor, 2003), 5–60 times greater than mean Amazon River discharge; peak velocities almost certainly exceeded 10 m/s. This means that for a few hours (days?) 15 000 years ago, 90% or more of the freshwater entering the global ocean was discharged from a point source – the Columbia River – in western North America. This "megaflood" was not a single event: 20 or more outbreak events may have occurred over a 2000–2500-yr period (Craig, 1987; Benito and O'Connor, 2003; Clague *et al.*, 2003). Lopes and Mix (2009) also have identified a number of freshwater spikes in sediment cores taken seaward of the Klamath River (northern California) that suggest a series of earlier megafloods, 20 000 years ago, perhaps related to the sudden release of water from another glacial lake or to a growth phase of the Cordilleran Ice Sheet.

Much of the sediment transported during the Missoula flood appears to have been eroded from eastern Washington, an area now termed the channeled "scablands" in recognition of its scoured and barren morphology (Fig. 3.57). Estimating the magnitude of sediment discharged during these mega-events is subject to considerable artistic license and perhaps is best done as a back-of-the-envelope calculation; but long cores recovered by the Ocean Drilling Program in Escanaba Trough, just north of the Mendecino Fracture Zone, 800 km probable travel-distance from the Columbia's mouth, penetrated a series of prominent sand-rich turbidite layers, one of which consisted of 57 m of fine- to medium-grained sand (Fig. 3.56). Carbon-14 dates from the top and bottom of this thick turbidite layer gave ages of 15.7 ka and 15.5 ka (calendar ages ~18 ka BP), respectively (Brunner *et al.*, 1999; Zuffa *et al.*, 2000; Normark and Reid, 2003; Lopes and Mix, 2009), suggesting instantaneous deposition of this 64-m-thick sequence. The 64-m layer apparently was deposited during the first major outbreak of Lake Missoula; overlying sand layers, "only" 1–5 m thick, may represent smaller events, which seems reasonable if one assumes that most of the available sediment along the outburst path had been eroded during the first outbreak (Atwater, 1987). The most probable delivery route to the Escanaba Trough was by the over-topping of the 170-m deep Cascadia Channel (Griggs *et al.*, 1970; Normark *et al.*, 1997), southward flow along the western edge of the Gorda Rise, and then eastward along the Mendocino Fracture Zone (Zuffa *et al.*, 2000; W. Normark, oral communication) (Fig. 3.56).

Total sediment volume of the 18-ka-BP turbidite layer in the Escanaba Trough is 90 km^3 (Zuffa *et al.*, 2000). Since most of the transported sediment presumably remained within the Cascadia Channel and ultimately was discharged westward onto the Tufts Fan (Fig. 3.56), it does not seem beyond reason to assume that total sediment discharged during the first Lake Missoula Jokulhlaup could have exceeded 1000 km^3 (i.e. 1200–1500 Bt) (W. Normark, oral communication). The volume of sediment discharge also can be estimated by assuming that the 2000–2500 km^3 of water scouring eastern Washington and Oregon had an average sediment concentration of 100–200 g/l. The total event-related load would thus have been ~200–500 Bt, a volume of 160–400 km^3. Assuming an average 1 m of eroded soil during the Lake Missoula outbreaks, the total volume of sediment removed from the 40 000 km^2 scablands alone would have been ~40 km^3.

No matter which estimate lies closer to the truth, it is difficult to conceive of so vast a volume of sediment, as it represents ~2–100 times the total present-day annual river sediment discharge to the global ocean. Such a thorough scouring of the middle and lower reaches of its watershed, on the other hand, might help explain why a river as large as the Columbia (670 000 km^2) has such a small pre-dam sediment discharge: 15 Mt/yr prior to dam construction. Most large global rivers having such low annual sediment yields – 22 $t/km^2/yr$ – drain Arctic or arid climates or drain predominantly lower-elevation Pre-Mesozoic terrain.

4 Human activities and their impacts

Any water that escapes to the sea is wasted

Josef Stalin

Whiskey is for drinking; water is for fighting over

Mark Twain

Introduction

By their very nature, rivers have played key roles in human history by helping to determine, feed, extend, and confine the sites of civilization. Renourishment by fluvial waters, their soils and nutrients were necessary for sustainable agriculture. Rivers could both unite and separate, fostering communication and transportation but also serving as boundaries between cities, states, and countries. It is not surprising that the earliest Middle-eastern and Asian civilizations began along the Indus, Tigris–Euphrates and Nile rivers. Using geomorphic and historical evidence, Schumm (2005) argues that much of the political stability in early Egypt can be owed to the lower Nile having a straight, nonmeandering course, whereas channel avulsion along the Euphrates may help explain the upheavals and conflicts experienced by the Sumerians. Prior to railroads and automobiles 150 years ago, most major population centers were located on or near rivers. Paris, London, Berlin, Moscow, New York, Shanghai, Bangkok, and Calcutta are only some of the more obvious examples.

With the exception of a few sparsely inhabited areas, such as the high Arctic, few drainage basins have been spared some impact from human activity, a conclusion also reached by Wohl (2006) in her analysis of the human impact on mountainous rivers. One interpretation of the two characters that portray the Chinese word for "river" (jiang) (江) is that they originally represented the characters for "water" and "work", graphic evidence of the importance that the early Chinese placed on their rivers.

Crutzen and Stoermer (2000) have termed the present Cenozoic epoch as the Anthropocene, acknowledging the extent to which human activities have played an ever-increasing role in defining the Earth's environment. Identifying the starting date of the Anthropocene, of course, depends on the context to which it is referred. Crutzen and Stoermer suggest AD 1784, the year when James Watt invented the steam engine, thus accelerating the burning of fossil fuels. In terms of accelerating river use and misuse, one might view 1950 as more relevant starting date, coincident with the exponential increased use of fertilizers and river management (Meybeck, 2002), although perhaps human impact is better considered geographically as well as temporally. Deleterious land-use practices, for example, have had a far longer and more extensive impact in Asia, although evidence of human impact in Europe can be seen (perhaps best in lake cores) as early as 6000 BP (Dearing and Jones, 2003). Human impact in North and South American watersheds accelerated after 1492, but may well have been more widespread in pre-Columbian times than commonly believed (Williams, 2003); irrigation systems in Peru, for example, date back more than 3500 years (Baade and Hesse, 2008). Paradoxically, pollution (particularly chemical pollution) and dam building generally began affecting many new world rivers before those in Asia.

The importance of rivers and water in contemporary society is obvious – and seems to be increasing. Rarely does a week go by when the news media do not run a story on water, the lack of it, the cost of it, or the too-much-of-it. The variety and depth of human interactions (and problems) regarding rivers (and water in general) can be gleaned from a sampling of news headlines, principally from the *New York Times* (*NYT*), over the past several years:

- Time to Move the Mississippi, Experts Say (*NYT*, Sept. 19, 2006)
- A Troubled River Mirrors China's Path to Modernity. "The polluted Yellow River is being sucked dry by factories, growing cities and farming – with still more growth planned." (*NYT*, Nov. 19, 2006)
- Governments Deal New Blow to Drought-stricken California Farmers. "...authorities released projections on Friday showing that little or no water would be available from federal sources this year for agricultural use." (*NYT*, Feb. 20, 2009)
- Chilean Town Withers in Free Market for Water. "... They say mining companies have polluted their water and bought up water rights." (*NYT*, Mar. 15, 2009)
- Two headlines on the Business Day section of *NYT*, Sept. 30, 2009:

Criticism is Mounting Over Flood Premiums
Solar Power in a Water War: Alternative Energy Projects Stumble on a Need for Water

- Plan Outlines Removal of Four Dams on Klamath River (*NYT*, Oct. 1, 2009).
- China's Three Gorges Dam Unlikely to Hit Capacity: Amid drought in country's central region, officials scrap water-level target, fueling complaints of environmental harm (*Wall Street Journal*, Nov. 18, 2009)
- In Bolivia, Water and Ice Tell of Climate Change (NYT, Dec. 14, 2009)
- A Watershed for India and Pakistan: Tensions rise as radicals traditionally focused on Kashmir independence turn their attention to river rights...Pakistan's water situation is reaching crisis proportions... "Water flows or blood." (*Washington Post*, May 28, 2010)
- Vital River is Withering and Iraq has no Answers (*NYT*, June 13, 2010).
- Pakistan Flood Sets Back Years of Gains on Infrastructure (*NYT*, Aug. 27, 2010).

As human population grows, especially in developing countries and most especially in regions with low rainfall and little water availability, the need for a sustainable and sanitary water supply becomes ever more acute. In the twentieth century, when the global population increased more than three-fold, both the number of dams and the consumption of artificial fertilizer increased by more than two orders of magnitude. Vörösmarty *et al.* (2000) estimated that globally 1.8 billion people in 1985 lived in a high degree of "water stress," and they speculated that the combination of climate change and increased population could increase this number to 3.8 billion by the year 2025. The combined pressures of population and watershed changes cannot help but impact rivers and their water. In their analysis of sediment and water discharge patterns of 145 rivers, Walling and Fang (2003), for instance, found that about half of the rivers showed evidence of significant recent change, principally declining sediment discharge in response to dam construction and reservoir retention, but some rivers experienced increased sediment discharge in response to changing land use.

As present and future water supply (and its quality) generates increasing world-wide concern, the amount of relevant literature has literally exploded. The interested reader cannot help but note the plethora of recent scientific and popular literature that has focused on the overall effect of human impact on the environment, and within the context of this book, water (see, for example, Turner *et al.*, 1990; Bonell *et al.*, 1993; Hillel, 1994; McCully, 1996; Zebidi, 1998; Gleick, 1993; Gleick, 2000a, b; Herschy and Fairbridge, 1998; de Villiers, 2000; Subramanian, 2001; Grove and Rackham, 2001; Postel and Richter, 2003, to name only a few). The publication explosion is, if anything, more dramatic in the scientific literature. International Association of Hydrological Sciences special publications (better known as IAHS red books) annually issue a number of topic-specific tomes. A LUCC report (Nunes and Augé, 1999) and a comprehensive review by Goudie (2000) present relevant discussions on land-use and its impact on the natural environments; their bibliographies can be especially useful, although now ten years old. Newson's 1997 *Land, Water and Development* discusses similar problems, but more from the standpoint of the watershed. Of particular interest to North American readers is Jeffrey Mount's 1995 *California Streams and Rivers: The Conflict Between Fluvial Process and Land Use*, which discusses both fluvial processes and problems as they relate to present-day California.

Our discussion in this chapter centers around the effects of land-use and river management on the fluvial system, particularly how they affect the quantity and quality of the downstream flow of water, suspended sediments, and dissolved solids. How has water discharge or its seasonal pattern changed during the Anthropocene? How has changing land-use affected the erodibility and transport of both suspended and dissolved solids? What has been the effect of point-source (e.g. mining, sewage) or disseminated (e.g. farming) pollutant inputs? How have dams and irrigation affected the discharge of water as well as suspended and dissolved constituents?

The Ebro River (southern Spain) can serve as an example of the anthropogenic impacts on a river and its drainage basin. Deforestation and land clearing, initially by Romans in the first and second centuries AD, increased during the tenth and eleventh centuries in response to accelerated agricultural activities, particularly by the Moors. Further land clearing in the fifteenth and sixteenth centuries and increasingly large herds of domestic animals led to accelerated erosion, peaking (according to Guillén and Palanques, 1997; their Fig. 3) around 1600. Over the past century dam construction along the Ebro has resulted in a ~50% decrease in water discharge, and the river's present sediment discharge (about 0.1 Mt/yr) is less than 1% of what it was in the late nineteenth century (20–25 Mt/yr) (Guillén and Palanques, 1997) (Fig. 4.1a). Moreover, seasonal patterns of discharge have changed drastically after dam construction in the early 1960s. Average peak spring discharge is now <15% of its historic values (Fig. 4.1b). These recent changes have been augmented by farmland abandonment, leading to a reduction in soil erosion (Garcia-Ruiz, 2010).

Although the Ebro is particularly well documented (Nelson, 1990; Palanques *et al.*, 1990; Mariño, 1992; Guillén and Palanques, 1997), its history is not dissimilar from that of many other European, Asian and North

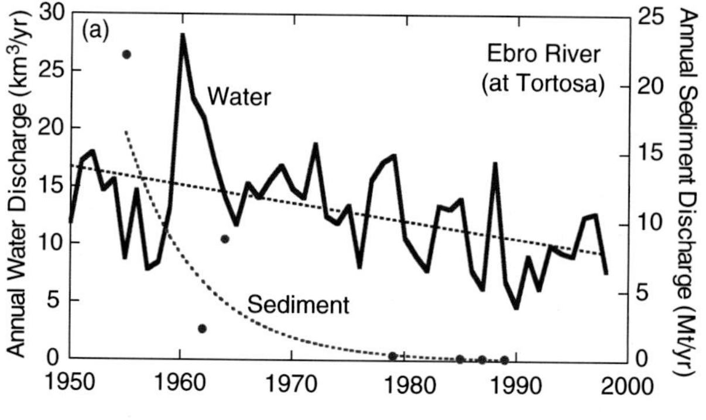

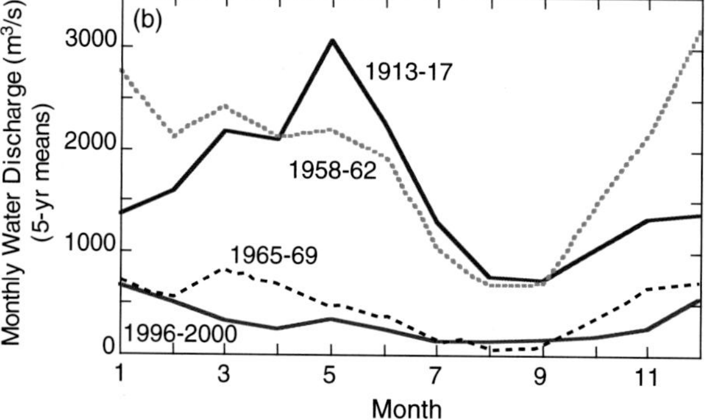

Figure 4.1. (a) Water (blue line) and sediment (red dots) discharges from the Ebro River, 1951–2000, as measured at Tortosa, Spain. (b) Changes in the Ebro's monthly water discharge, 1913–2000. Data from Ebro River Basin Authority (http://195.55.247.237/saihebro/) and Guillén and Palanques (1997).

American rivers and their waterhelds. At the end of this chapter, we discuss in greater detail four basins: the Nile, Mississippi, Yellow and Yangtze rivers. But we first begin with a discussion of anthropogenic activities and impacts that can affect the quantity and quality of river flow.

Watershed land-use

Since the middle part of the twentieth century, perhaps the most important agent in determining the delivery of water and sediment to much of the coastal ocean has been man (Clark and Wilcock, 2000, and references therein). The regional and global impacts of land use and land-use change have led to an ever-growing number of non-governmental focus groups, such as IGBP (discussed in Chapter 1), SCOPE (Scientific Committee on Problems of the Environment), LUCC (Land-use and Land-cover Change; Nunes and Augé, 1999), and GWSP (Global Water System Project; Vörösmarty *et al.*, 2004). While many of these studies have directly or indirectly addressed the effects of global climate change, non-greenhouse impacts from more local human activities have proved – at least to us – particularly illuminating and far more threatening. A UNEP-funded study in the early 1990s, for instance, concluded that land-use, augmented by population growth, especially in North Africa, will probably have a greater effect on the coastal regions in the Mediterranean basin than the combined impacts of projected climate change and sea-level rise (Milliman *et al.*, 1992). Nearly 20 years later these speculations appear to have been borne out.

Land-use and its impacts on rivers and their drainage basins – particularly deforestation, farming, mining and modern construction – can accelerate both watershed runoff and erosion. In contrast, river "management," such as canals, irrigation ditches, and dams, can divert water and/or trap sediment, delaying or curtailing its downstream escape. As Des Walling (1997, 2006) has pointed out, after many centuries of increased sediment delivery due to poor land-use practices, dam construction over the past 50 years is bringing global fluxes back to pre-agriculture levels. Said another way, while erosion may continue increase – both regionally and globally – more sediment is being stored on land and less is escaping to the coastal zone (Trimble, 1999; see Fig. 3.27).

Early human impact on the terrestrial environment was primarily through the planned burning of land to facilitate hunting (Pyne, 1991, Meyer and Turner, 1994) and farming. Extensive deforestation also may have facilitated in local climatic changes (Fu and Yuan, 2001). In the present-day Amazon basin, for instance, the removal of forest cover can lead to increased temperatures, decreased rainfall, and therefore decreased fluvial runoff (Shukla *et al.*, 1990).

Land-use can affect the fate of both the flowing water and the rate of sediment erosion. Lacking vegetation, freshly tilled soil is more erodable than untilled soil (Fig. 4.2). Some general examples are shown in Fig. 4.2, and Goudie (2000) cites several other examples that find two- to three-order-of-magnitude increases in anthropogenically enhanced erosion. For any non-believers, the following paragraph gives only a few examples, presented in somewhat random order.

Following widespread farming of the northern China loess plateau, about 2500 years ago, the Yellow River's sediment discharge increased three- to ten- fold (Milliman *et al.*, 1987; Saito *et al.*, 2001; see below). Sediment yields in the deforested tropical highlands of Sri Lanka appear to be one- to two- orders-of-magnitude faster than natural rates of rock erosion and soil production (Hewawasam *et al.*, 2003), and conversion from tropical forest to cocoa and palm oil plantations in Malaysia led to increased water runoffs of 157% and 470%, respectively (Abdul Rahmin, 1988). Rates of erosion in Iceland increased by an order of magnitude following settlement in the late ninth century, owing in large part to the deforestation and subsequent agriculture (Bjørnsson, 1995); similar rates were noted after Post-European settlement in the Waipaoa River watershed, on the North Island of New Zealand (Kettner *et al.*,

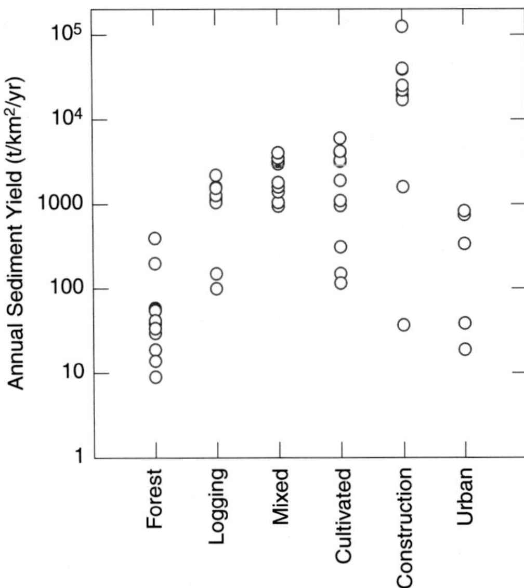

Figure 4.2. Sediment yields resulting from various types of land-use in Malaysia, Sabah, Java, and construction and urbanization in USA and Japan. Data from Douglas (1996) and compiled by Goudie (2000).

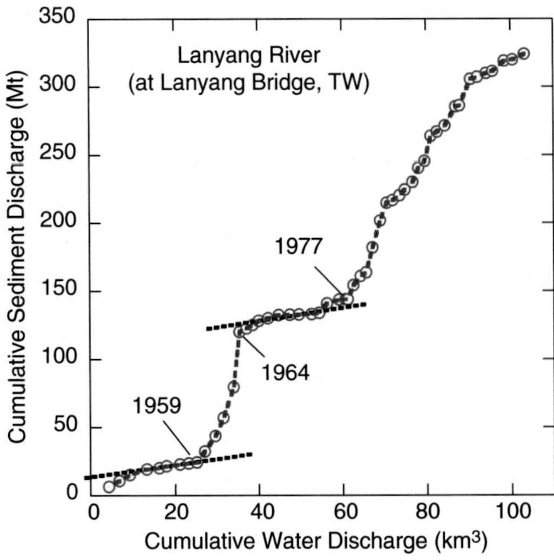

Figure 4.3. Cumulative water and sediment discharges from the Lanyang River, NE Taiwan, 1950–2000. The order-of-magnitude increase in sediment discharge between 1960 and 1964 (22 Mt/yr) most likely reflects increased runoff and erosion due to highway construction. From 1964–77, sediment discharge (2.6 Mt/yr) was similar to that for the 10 years prior to 1960–64 road construction (2.7 Mt/yr). The steady increase in sediment discharge after 1977 reflects further highway construction, urbanization and increased agricultural activity (S. J. Kao, personal communication).

2007). Even though Australian sediment delivery ranks amongst the lowest in the world, Wasson *et al.* (1996, their Table 2) estimated that yields for various anthropogenically impacted watersheds have increased 2- to 375-fold. Bork *et al.* (1998) showed that extensive deforestation in medieval Germany increased runoff to the extent that a 1000-yr rainfall event in 1342 resulted in river discharges as much as 200 times greater than modern measurements. These catastrophic floods, together with the bubonic plague, which struck soon after the floods, wiped out a significant portion of the population such that farmland was abandoned and ultimately reforested, thereby leading to an eventual decrease in runoff and erosion.

In a thought-provoking paper, Hooke (2000) calculated that over the past 5000 years humans have moved (both intentionally and unintentionally) a volume of sediment equivalent to a 4000-m-high mountain range, 40 km wide and 100 km long. Global per-capita human displacement of earth has increased by about an order of magnitude since the Egyptian dynasties (earth-removal rate in the United States being five times higher than the worldwide rate), as the global population has increased 6000-fold (Hooke, 2000). According to Hooke's Fig. 4, humans presently move more than 100 Bt of earth annually, some five-fold greater than the fluvial delivery of sediment to the oceans. Holdgate *et al.* (1982; c.f. Goudie, 2000) put a much higher value on the anthropogenic movement of rock and soil: 3000 Bt/yr, which to us intuitively seems more likely. Most of this sediment, however, probably does not travel far from where it was eroded before it is deposited.

This does not mean that all rivers have been seriously affected by human-accelerated erosion. For example, it is doubtful that the Amazon, Congo, Orinoco, or Magdalena rivers have seen much change in their discharge patterns. Even the Mississippi River's sediment load may not be much greater than it was before the arrival of European trappers and settlers (Meade and Moody, 2008, 2010). Taken in total, McLennan's (1993) and Syvitski *et al.*'s (2005) estimates of a 12–13-Bt/yr pre-human sediment discharge – compared with 19–20 Bt/yr prior to widespread dam construction during the past 50 years – do not seem unreasonable. Walling (2008), by contrast, has assumed that sediment flux increased by 160% due to human activities, but that much of this sediment has been trapped in reservoirs behind dams.

The extent to which changes in land-use can affect the character of a river is well illustrated in the character of the rivers draining Taiwan, which have some of the highest sediment yields in the world (Table 2.11). Prior to 1960, the Lanyang River (basin area 980 km^2), in northeastern Taiwan had an annual sediment discharge of ~2.7 Mt/yr, a sediment yield of ~3000t/km^2/yr. Following highway construction in 1963, sediment discharge increased by nearly an order of magnitude, averaging 22 Mt/yr between 1960 and 1963 (Fig. 4.3). Average sediment yield during this four-year period was 23 000 t/km^2/yr. In 1977 sediment discharge increased once again and for the following 25 years averaged more than 7 Mt/yr, reflecting a combination of further highway construction, urbanization, and increased agricultural activities on steeper slopes within the drainage basin (Kao and Liu, 2002; Fuller *et al.*, 2003; Kao and Milliman, 2008). While the Lanyang may be an extreme example, its detailed monitoring gives some quantitative measure of the extent to which a drainage basin (particularly a small river draining high, easily erodable mountains) can respond to human activity.

Deforestation and logging

Often the first human change to a natural landscape is the removal of trees in order to harvest wood (building materials, wood pulp, energy) and to clear land for farming, roads, and construction. Erosion rates on logged land can be as much as several orders of magnitude greater than on unlogged land, the increase depending on the terrain, soil, rainfall intensity, and degree of logging; cultivated land often experiences even greater rates erosion (see Fig. 4.2). In coastal British Columbia, for instance, Brardinoni *et al.* (2003) found that the mean sediment yield in logged areas was about 30-fold greater than in forested areas, although locally yields increased by almost three orders of magnitude.

The removal of trees translates into less water being lost via evapotranspiration; therefore more precipitation can directly reach the soil. In a controlled experiment, Miyata *et al.* (2009) found that forest-floor cover in Japanese cypress plantation forests prevents 95% of the soil erosion by raindrops. The impact from logging, however, also depends on the types of trees removed (e.g. deciduous vs. evergreens; native vs. non-native; Mohammad and Ada, 2010) and the timing of precipitation events (Jones, 2000). Snow, for instance, is likely to accumulate in greater thickness and melt more rapidly in a treeless setting (Bowling and Lettenmaier, 2001). The relation between logging and storm-related peak flows, however, is difficult to quantify (Rothacher, 1973; Harr *et al.*, 1975; Lewis *et al.*, 2001; Moore and Wondzell, 2005, and references therein), in no small part because monitoring studies are rarely sufficiently long to sample extreme meteorological events (Dunne, 2001).

Decreased leaf litter and thereby less capacity to absorb moisture mean that logged areas are more likely to deliver increased surface runoff during heavy precipitation events (Penman, 1963; Rothacher, 1970). Increased soil moisture and piezometric levels (Keppler *et al.*, 1994) and fewer root structures (thus decreased sediment binding) lead to increased erosion. Decreased slope stability can initiate downslope mass movements and shallow landslides (Sidle and Wu, 2001, and references therein). Reducing vegetation cover by 1/3 (from 55% to 20%) in Zimbabwe, for instance, led to a five-fold increase in soil erosion (Livingstone, 1991).

Landslides in the Waipaoa River Basin (North Island of New Zealand) increased by an order of magnitude following logging of native forest and its conversion into pasture land (Trustrum *et al.*, 1999; Page *et al.*, 2000; Reid and Page, 2003) (see Fig. 4.6). Landslide-induced erosion in the Waiapu watershed, north of the Waipaoa, has been even more severe, its annual sediment yield of 21 000 t/km^2/yr ranking as one of the highest in the world (Figure 3.31; see Table 2.7). Although it has greater runoff than the Waipaoa, the relatively pristine Motu River watershed, less than 50 km north of the Waipaoa, has a yield 1/3 that of the Waipaoa basin (Hicks *et al.*, 2000).

Increased sediment delivery from a river system also can result in increased channel aggradation, which in turn can lead to increased overbank flooding. In the Skokomish River, western Washington, for example, increased land-use (primarily logging and road construction) led to increased river channel elevation – and thus flooding – despite no perceived change in peak discharge (Stover and Montgomery, 2001).

The process of logging itself can have a great impact on the landscape, even when post-fire forests are logged (Silins *et al.*, 2008). The tires of tree-removal and road-building equipment can compress and effectively destroy microbiotic layers that tend to retard erosion (Bresson and Valentin, 1994; Eldridge, 1998; Uchida *et al.*, 2000). In Tasmania, areas that had been mechanically disturbed by logging machinery showed increased sediment yield, even greater than areas that had been recently burned; Wilson (1999) attributed this to the disturbance of the biotic crust.

Logging roads and their drainage ditches also can alter runoff patterns and thereby both subsurface and surface flow, increasing peak flows (Bowling and Lettenmaier, 2001, and references therein) if not total discharge volume (Jones, 2000). Diverted flow by plugged culverts accounted for 68% of the road related erosion along Garrett Creek, a tributary to Redwood Creek (Best, 1995). Landslides in the proximity of logging roads can increase in frequency by a factor of 3–300 (Wolfe and Williams, 1986; Sidle *et al.*, 1985). Although there was no observed increase in landslides, Lane and Sheridan (2002) noted a 3.5-fold increase in sediment load downstream of a forest road stream crossing in Central Highlands of Victoria, Australia. The impact of road construction, however, varies greatly between watersheds (Brown and Krygier, 1971; Wright *et al.*, 1990; Montgomery, 1994). Larsen and Parks (1997) concluded that most impact was centered within an 85-m swath on either side of the road in Puerto Rico, landslides occurring outside that swath being less frequent by a factor of five. Luce and Black (1999) found that sediment production in logging roads is proportional to the length of the road and the square of the slope, as well as being dependent on soil texture (coarser sediments being less erosive) and vegetative cover.

The type of construction, usage, and drainage of the road also can influence sediment yield (Thompson *et al.*, 2008). Older roads and drainage ditches have much lower sediment yields than freshly cleared areas (Luce and Black, 2001, and references therein). Heavily used roads in the Pacific Northwest, for example, can have two orders of magnitude greater sediment yield than abandoned roads (Reid and Dunne, 1984). On the other hand, discontinued road maintenance can lead to an increased failure of drainage structures and thus increased sediment erosion (Chatwin and Smith, 1992).

While deforestation clearly has accelerated over the past 100–500 years, the impact of local, artisan deforestation was obvious long before the intrusion of western civilization. In a swamp in the western highlands of Papua New Guinea, for instance, Hughes *et al.* (1991) cited an 8- to 14-fold increase in sediment accumulation rates 9000 years ago, which they related to deforestation and the subsequent advent of agriculture. Replacement of subsistence farming with organized plantations in recent years has increased the sedimentation rate by another 16 times. Altogether, according to Hughes *et al.* (1991), modern erosion is >200 times greater than it was 9000 years ago. Similarly, Page and Trustrum (1997) and Kettner *et al.* (2007) found that erosion in the Waipaoa watershed increased two- to three-fold following Maori settlement ~600 years ago.

Williams (1992) estimated that throughout human history deforestation has resulted in the clearing of about 7 000 000–8 000 000 km^2 of land, equating to about 15% removal of the global forest. Ramankutty and Foley's (1999) numbers are higher; they calculate that since 1700 there has been a nearly 9 000 000 km^2 net loss of global forest and woodland (nearly 17% loss), with a corresponding 14 000 000 km^2 (340%) increase in cropland area. But the timing of deforestation and agriculture has varied greatly throughout the world. Wide-scale farming in parts of southern Asia, initiated more than 3000 years ago, may have increased sediment delivery rates of some watersheds by as much as an order of magnitude (Milliman *et al.*, 1987). In Europe, extensive land-use was first practised in Mediterranean watersheds (McNeill, 1992; Wainwright and Thornes, 2004; Hooke, 2006) and as much as 1000 years later in the north. Sestini (1992), for instance, estimated that the sediment discharge from the Po River began accelerating about 2000 years ago in response to Roman mining, subsequently increasing again in response to other land use. Throughout much of the New World, in contrast, organized land-use – at least on a large scale – did not begin in earnest until the arrival of Europeans in the seventeenth century.

Changes in forest/woodland and cropland area over the past 200–300 years have been most dramatic in the New World and southern Asia, with about half of it occurring in the past 100 years. In recent years global deforestation appears to have slowed, perhaps even reversed in parts of North America and Europe, whereas it continues to increase in Central and South America (Fig. 4.4) and southern Asia (although apparently not in China, where forest biomass has increased due to changes in land-use policies; Fang *et al.*, 2001), where natural rates of erosion are already high (see Chapter 2). Over this period there has been a corresponding increase in cropland area, increasing between 1700 and 1980 from 860 000 km^2 to 4 000 000 km^2 in southern and southeastern Asia, and from 100 000 km^2 to 3 400 000 km^2 in the New World (Richards, 1990). Using Richards's estimates, tropical rain forests decreased by ~25% in area during the twentieth century.

Three contrasting examples of land cover can be seen in the eastern and western USA and southeast Asia (Fig. 4.5). In the eastern USA between 1800 and 1900 the increase in cropland was commensurate with the loss of forested land (Fig. 4.5a). In response to nineteenth-century farming, suspended sediment yields of the Susquehanna River may have increased 3- to 4.5-fold (Reed *et al.*, 1997), but in the past century erosion in much of the eastern USA has decreased (Wolman, 1967) as the east has been urbanized and the amount of cropland has decreased. In the western USA, deforestation during the later half of the nineteenth and early twentieth centuries was slight, as most of the cropland expansion occurred on prairie land. But for the past 100 years both forest and cropland areas have remained relatively constant (Fig. 4.5b). In SE Asia (by which Ramankutty and Foley, 1999, mean Thailand, Cambodia, Vietnam, Malaysia, the Philippines, and Indonesia) forest and woodland losses have accelerated in the last 100 years, in consort with increased farming (Fig. 4.5c). Between 1943 and 1993, for example, the percentage of forest cover in the northern mountains of Vietnam decreased from 95% to 17% because of land-clearing and war; 60–65% of this land is now barren (Thanh *et al.*, 2004). Legal and illegal logging in the Philippines, which peaked in the 1960s and 1970s, has left only 13% of the country's forests remaining, greatly increasing the severity of landslides and flash floods caused by typhoons. Typhoon Winnie, which struck the northeastern Philippines in December 2004, for instance, resulted in a presumed death toll of nearly 1000 (Carlos H. Conde, *New York Times*, December 3 and 4, 2004). While the highest rates of deforestation occur in the tropics (coincident, unfortunately, with high rates of rainfall and therefore erosion; FAO, 1999), whether or not erosion is actually increasing is debatable owing to of the lack of adequate monitoring (R. Defries, in Steffen *et al.*, 2003, p. 99; Foley *et al.*, 2005). Monitoring of a series of reservoir

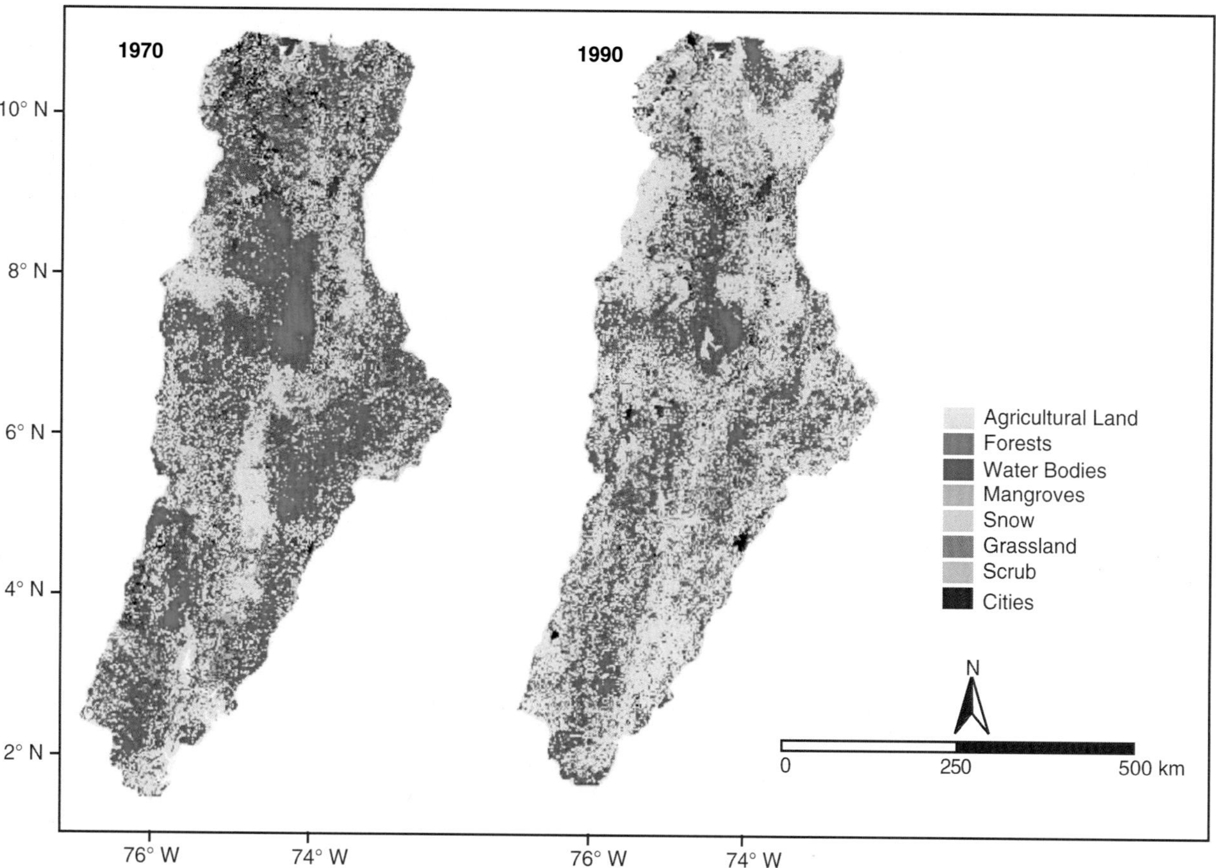

Figure 4.4. Change in land cover in the Magdalena watershed, Colombia, between 1970 (left) and 1990 (right). Note the decrease in native vegetation (forest and grassland) and the increase in cultivated land. Maps courtesy of J. D. Restrepo.

catchments in southeast Asia, however, has indicated a 2.5 to 6% annual increase in annual sediment yield (Abernethy, cited in Walling, 1999).

The effects of logging also can be felt for some time after the logging activity has ceased. Intense rainfalls in the steep Pacific Northwest mountains can create shallow landslides up to 10 years after logging (Montgomery *et al.*, 2000). In the Redwood Creek basin in northern California, much of downslope mass movement recorded during and after the December 1964 storm (see Chapter 3) reflected active tree harvesting during the preceding 15 years, as evidenced by the fact that similar or larger storms in the late nineteenth century had far less impact than the 1964 storm (Harden, 1995). The effect in some areas, however, can be mitigated in a relatively short time through reforestation. For instance, the planting of the upper reaches of the Waipaoa watershed (Fig. 4.6) reduced erosion – mostly downslope movement – by an order of magnitude (Marden *et al.*, 1992). C. Chen *et al.* (2004) noted decreased runoff in some Taiwanese rivers following reforestation, but they admit that this also may have been caused by increased air temperatures and thus increased evapotranspiration. Marden and Rowan (1993) concluded that landslide response to Cyclone Bola (North Island of New Zealand) decreased after about eight years following reforestation (c.f. Reid and Page, 2003; Marden *et al.*, 2008). Recovery times from logging activities in the western foothills of Taiwan (Chang and Slaymaker, 2002) and in coastal British Columbia (Brardinoni *et al.*, 2003) took substantially longer, about 20 years in each case.

Agriculture

By "agriculture," we mean both the farming of crops and the grazing by domestic animals. Since many cultivated areas were originally forest, farming may only exacerbate the environmental change brought about by logging. Agricultural crops such as corn or cotton can alter the ability of soil to absorb rainfall, and, given its rough surface and the absence of plant roots and other organic binding matter, tilled soil is much more susceptible to erosion than is natural pastureland or forest. Dead plants and roots can help retard erosion, perhaps by as much as 100-fold (Pimentel *et al.*, 1995 and references therein). The practice of removing excess plant material from fields (for use as fodder or fuel), a common practice throughout much of the developing

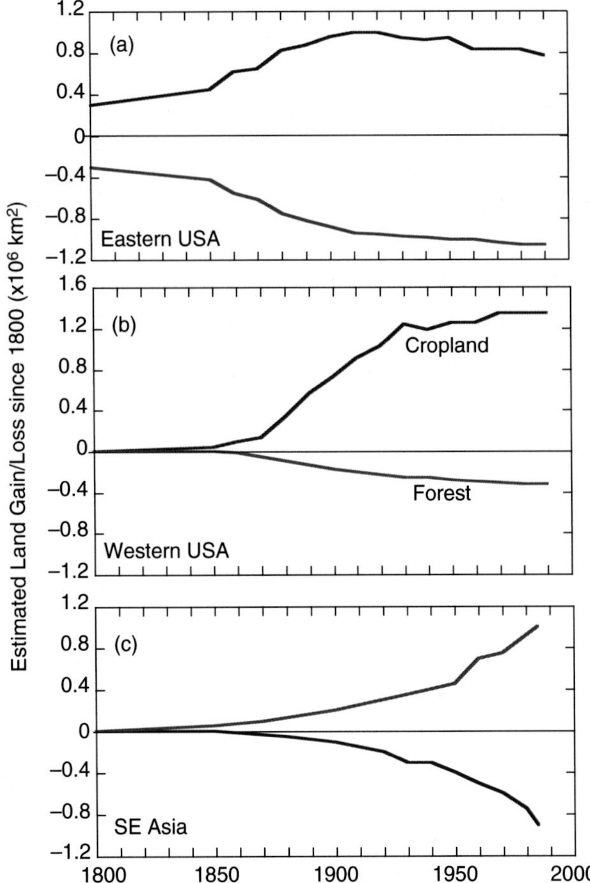

Figure 4.5. Land-use change as measured in areas occupied by forestland (red) and cropland (blue) from 1800–1900. (a) Eastern USA, (b) western USA, and (c) SE Asia, 1800–1990. Data from Ramankutty and Foley (1999).

Figure 4.6. Gully erosion in the upper Waipaoa River (North Island, New Zealand) watershed, largely due to deforestation and sheep farming in the late nineteenth and early twentieth centuries. Tree planting, as seen by the regular tree patterns, has helped to somewhat reduce erosion. Photo courtesy of J. P. M. Syvitski. Another example of human-induced erosion can be seen in Fig. 3.29.

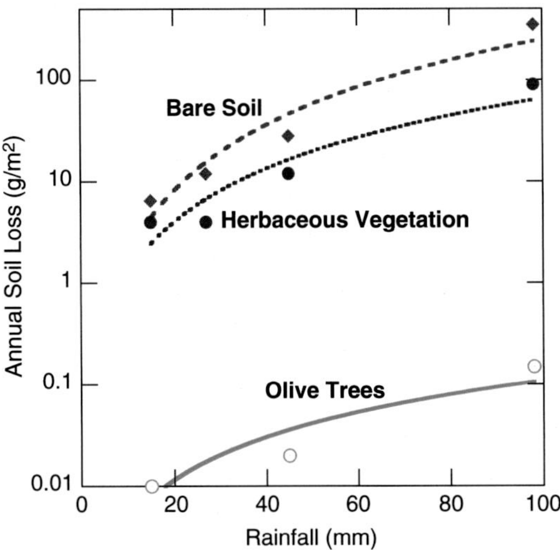

Figure 4.7. Soil loss in various experimental plots in response to different intensities of rainfall in the Mediterranean region. Olive trees represent more or less natural conditions, whereas vines, herbaceous vegetation, and bare soil represent various degrees of human intervention. Erosion rates in the vineyards in part are elevated because of the steep terrain (in which erosion rates are higher) and because rows often run perpendicular rather than parallel to topographic contours. Data from Grove and Rackham (2001; after Kosmas *et al.*, 1996). Tabulations of human-facilitated erosion rates around the Mediterranean Basin can be found in Wainwright and Thornes (2004, pp. 181–183).

world, can therefore lead to increase soil erosion. Moreover, the removal of organic matter from soils can decrease the infiltration of rainwater, thus increasing surface water runoff and thereby erosion. In the Mediterranean basin, bare soil can experience two- to three- orders-of-magnitude greater soil loss than will terrain planted with olive trees (Fig. 4.7). Although oenophiles may want to suppress the notion, vineyards also have high rates of soil loss, in part because they often are located on steep hillsides. On the plus side, viticulture uses much less water than many other types of crops. In regions where the cultivated land is under irrigation, some to much of the eroded soil may be deposited in irrigation ditches, thereby hindering the effective movement of water within the irrigation network.

Sometimes the combination of bad weather and poor land-use practices can result in particularly erosive events. A case in point is the prolonged drought in the early 1930s (coincident with a cold PDO, by the way), combined with the poor farming practices throughout the US west, which resulted in the infamous "Dust Bowl" from which large amounts of fertile soil were literally blown away. The extreme rainfall events and subsequent accelerated erosion in fourteenth-century Germany, mentioned earlier in this chapter, occurred when a large percentage of the land was being farmed or lay fallow (Bork *et al.*, 1998; Lang *et al.*, 2003, their Fig. 7).

Anthropogenically enhanced erosion can also occur on flood plains, which are normally considered to be areas of net deposition (Asselman and Middelkoop, 1995). Visser *et al.* (2002), for instance, report that sediment loss on sugar-cane cultivated land in the Herbert River (Australia) flood plain was 390 t/m^2 between December 1999 and May 2000, more than twice the average yield of the entire basin. As mentioned previously, pasture-based agriculture, particularly sheep grazing, in New Zealand resulted in a rapid "gullization" of the New Zealand landscape and accelerated erosion (Marden *et al.*, 2008; Litchfield *et al.*, 2008 and references therein). Kettner *et al.* (2008, their Fig. 4) calculate a five- to six-fold increase in Waipaoa basin sediment yield in response to deforestation and sheep-based agriculture.

Landuse on the Russian plains offers a similar bottom line, but with a much different history. The amount of sown land reached a maximum by 1887, but then decreased in the first half of the nineteenth century in response to war and revolution. Collective farming, which resulted in less ownership responsibility for the land itself, led to increased erosion, particularly after collectivization in 1928 (Sidorchuk and Golosov, 2003). Total erosion in the past 300 years, according to Sidorchuk and Golosov, has been 100 km^3 of soil, equating to an average loss of 8000 t/km^2; but most of the eroded soils have probably remained with their respective watersheds and not transported farther.

Herds of domestic animals, particularly sheep, are effective grazers, munching away all but the plant stubble, thereby increasing the erosive effect of rain on the often bare soil. In a controlled experiment in arid western India, peak runoff in heavily grazed land was 10 times greater than ungrazed land, and sediment yield was about 40 times greater (Sharma, 1997). In New Zealand, sheep populations as recently as the early 1990s numbered about 80 million, underscoring their impact on soil erosion. Extensive grazing also can decrease plant productivity by affecting the plants' ability to absorb moisture, thus increasing the potential for surface runoff. Moreover, as Trimble and Mendel (1995) have pointed out, heavy hoofed animals, such as cows and water buffalo, can compact the soil, not unlike logging machinery, decreasing rainfall penetration and increasing surface runoff.

(It is not only domestic animals that facilitate erosion. Butler (2006) points out the deleterious effects that natural animals had on the pre-European US western plains. In addition to beavers and their dams, discussed in Chapter 3, buffalo wallows numbered more than 100 million, each wallow resulting in the displacement ~23 m^3 of sediment (Butler, 2006). Prairie dogs may have been even more effective sediment movers: Butler estimates they may have displaced 5–65 tons of sediment per hectare, or 500 to 6500 t/km^2. Eradication of these animals and introduction of domestic animals therefore may not have had as deleterious effect on erosion rates – at least in the prairie-dog infested and buffalo-rich high plains – as one might imagine.)

The effect of enhanced erosion from agricultural activities can be seen from the fact that, on average, US croplands annually lose 1700 t/km^2 of soil (pastureland loses 600 t/km^2/yr) (Barrow, 1991; Pimentel *et al.*, 1995), whereas the average rate of soil production is 100 t/km^2/yr (Troeh and Thompson, 1993). Iowa cropland, for instance, is being lost at the rate of 3000 t/km^2/yr, and about 1/2 of the productive topsoil has been lost in the last 160 years (Klee, 1991; Pimentel *et al.*, 1995). To put these figures into perspective, the global sediment yield of fluvial sediment discharge to the coastal zone is ~190 t/km^2/yr. According to Pimentel *et al.* (1995), about 90% of US cropland is losing soil faster than it can be produced; they estimate that the 1 600 000 km^2 of US cropland loses 4 Bt of soil annually (volumetrically equivalent to 6–7 Great Walls) and 130 km^3 of water. In the Illinois River watershed, which drains into the upper Mississippi, one result of the extensive agriculture-induced erosion has been the silting in of bottomland lakes, the 30-km-long Lake Peoria alone losing more 70% of its water storage capacity during the twentieth century (Demissie, 1996). The rate of soil erosion, as might be expected, is even greater in Asia, averaging 4000 t/km^2/yr (Barrow, 1991). The amount of anthropogenically enhanced erosion by the deforestation and farming of the loess plateau in northern China is particularly illuminating: Assuming that the Yellow River's sediment load was elevated 5 to 10 times relative to its pre-human load, total erosion over the past 2000 years has presumably as much as 2–3 trillion tons!

In the conterminous United States and throughout much of Europe, total land area being actively farmed has declined in recent years. Using Waisanen and Bliss's (2002) detailed county-by-county analysis, total US land area farmed in 1940 was twice that in 1880, but as of 1997 total farmed area had decreased 20% relative to 1940. In some Rocky Mountain states the farmed area has remained more or less constant (e.g. Utah, Nevada) or actually increased (Montana). But in the Northeast it has fallen to less than half the 1940 totals, and even the mid-western states have seen a 15% decline in farmland (Waisanen and Bliss, 2002, their Table 2).

A number of recent studies have illustrated the positive effects that improved land-conservation measures can have in stemming soil loss. Within a 20-year span, for example, changing soil conservation practices in a small catchment on the Russian plains resulted in a 2.5- to 2.8-fold decrease in soil loss (Golosov *et al.*, 2008). Similarly, excluding cattle from rangeland in two small catchments within the Burdekin River watershed (eastern Australia) resulted in a 50% decrease in sediment yield, even though runoff remained unchanged for the five years following cattle exclusion (Hawdon *et al*., 2008). Changing landuse in southern Europe, such as abandonment of agricultural land, has led locally to reduced sediment fluxes (e.g. Garcia-Ruiz, 2010), aided by wide-spread construction of dams and irrigation systems that have trapped much of the eroded sediment (see Hooke, 2006).

One example of the see-saw balance between agriculture vs. conservation is seen in the Nahal Hoga, an ephemeral stream in Israel that drains from the Judean Mountains into the eastern Mediterranean. Prior to 1948, the landscape was mostly bare owing to lack of water management and (particularly) overgrazing. For the next 17 years, grazing was limited in favor of water-conserving cultivation, one result being the reappearance of native vegetation. Then, as grazing once again increased, water conservation decreased and, even though some soil conservation practices were continued, sediment yields again increased. In response to increased soil loss, beginning in the early 1980s, grazing areas were once again reduced, vegetation increased, soil conservation practices were extended, and sediment and water yields again declined (Rozin and Schick, 1996).

Urbanization

Almost by definition, the urban landscape is unnatural, as summarized recently by Chin (2006). Diurnal and annual temperature ranges in the urban setting, for example, tend to be considerably greater than in the natural or rural environment. Paved surfaces can reduce the infiltration of water, one result being an increased tendency for short-term accentuated flood-related runoff. Urban rivers, with the greater influence of impervious surfaces (i.e. roads and parking lots) on their runoff, tend to react to peak events in a non-linear manner, thereby increasing the possibility of channel alteration and increased erosion (Galster *et al*., 2006). Overflowing storm sewers are not uncommon in many urban settings during and following heavy rains, and in some situations overflow can be extreme (Savini and Krammerer, 1961), thus enhancing the possibility of more intense erosion than would be felt in rural or natural settings (Fig. 4.2). Intensified runoff and increased down-stream erosion resulting from unbanization have helped facilitated increased local erosion in the Chesapeake Bay watershed (Wolman and Schick, 1967; Brush, 2001), resulting locally in a two- to five-fold increase in sedimentation rates in the Bay (Colman and Bratton, 2003).

The inability of urban surfaces to absorb natural and anthropogenic waste means that much of it is transferred to adjacent waterways. A rather graphic example of the problem was provided by Ponting (1991; cited in Goudie, 2000), who calculated that domestic pets deposit 0.35 Mt of solid waste (equivalent to the annual suspended load of the Severn) and 1 billion liters of urine on British streets.

Sediment yields in urban environments, however, are small compared to the high rates achieved during the construction phases of urban development, often two- to four- orders-of-magnitude greater than rates in forest and grassland areas. Construction-related sediment yields exceeding 10 000t/km^2/yr are not uncommon (Douglas, 1996; Fig. 4.2). The dramatic increase in sediment discharge from the Hoshe basin in central Taiwan primarily, for example, reflected an increase number of landslides owing to road construction (Chang and Slaymaker, 2002). Although an uneven sampling scheme led to some uncertainties regarding the accuracy of comparison, Ismail (1996; his Fig. 6) showed that sediment production in an area undergoing urbanization in Sg. Relau catchment (Malaysia, 11.5 km^2 drainage area) was about three times greater than areas affected by hill agriculture.

Whereas urban construction can lead to accelerated erosion – and thus sediment accumulation – once construction projects are completed, erosion can decline rapidly, although the many diverse data gathered by Chin (2006, her Fig. 11) suggest that it may take as much as one to two decades before there is a significant decrease in sediment production. During this waning period, sediment emplaced during the construction phase can be eroded, leading to channel widening.

Mining

Mining can have many effects – mostly negative – on a river, particularly the increased suspended-sediment and dissolved-solid concentrations and discharge (Balmurugun, 1991; Douglas, 1996). An obvious impact is the dumping of mining debris either directly into the river or on steep slopes, where much of it ultimately moves downslope and into the river. The amount of mining- and industry-related material that may enter a river can be impressive. One potash mine in the Alsace, for instance, dumped an average of 15 000 t of spoil daily over the span of more than 50 years (Meybeck, 1998). Modern earth-moving machines have the capacity of removing efficiently huge quantities of sediment. The 2 Bt of debris and metals removed by world's largest copper mine, Hull-Rust-Mahoning Mine in Hibbing, Minnesotta, is dwarfed by the estimated 30 Bt excavated by the Syncrude mine in the Athabaska oil sands in northern Canada (J.P.M. Syvitski, personal communication).

Hydraulic mining of gold in California from the late 1850s to the mid 1880s resulted in the production of more than 1 billion m^3 of rock debris (Gilbert, 1917; Fig. 4.8a). Aggradation of tailings in the headwaters of one of the tributaries to the Bear River exceeded 40 m in just nine years (James, 1994; Fig. 4.8b). The result was an order-of-magnitude increase in the sediment load of the Sacramento River. Agricultural activities added additional sediment to

Figure 4.8. (a) Effects of hydraulic mining at North Bloomfield, California, 1909, showing the degree of erosion caused by the mining itself as well as subsequent erosion of exposed surfaces. Mining operations essentially had ceased 20 years before this picture was taken. (b) Mine tailings on bank of Deer Creek near Nevada City, California, 1908. Note hat, presumably that of G. K. Gilbert, in the middle of picture for scale. Photos (by G. K. Gilbert) courtesy of USGS.

the river system (Meade *et al.*, 1990; McPhee, 2004). After cessation of wide-spread hydraulic mining in the 1880s, sediment yield quickly fell roughly in half; even 100 years later, however, the sediment yield remains twice as high as it was before 1850 (Wright and Schollhamer, 2004).

In the eastern USA, relaxed environmental regulations in the late 1990s and the first decade of this century saw a sharp increase in "mountaintop removal" of rock, locally as much as 150 m in West Virginia, to reach coal seams. Assuming a triangular-shaped cross section, 2.5 km^2 of mountaintop would yield about 200 million m^3 (~400 Mt) of rock and sediment, most of it dumped into adjacent river valleys. Because of the removal of surface soils, gold mining in the Kolyma River basin (Russia) has led to a significant increase in the river's sediment load since 1966 (Bobrovitskaya, 1996; Walling, 2006).

In addition to increased sediment discharge, mining also can release noxious substances into fluvial waters. A large open-pit copper–gold mine in the upper reaches of the Ok Tedi River (tributary to the Fly River in the western Papua New Guinea highlands) locally elevated the Ok Tedi's river channel by 10 m, but also resulted in an order-of-magnitude increase in copper concentrations in Fly River alluvial sediments (Hettler *et al.*, 1997). Other impacts of mining activities are discussed in the following section on chemical pollution.

Mining also can serve as a precursor to other human activities that may have even greater impact on the fluvial environment. For example, tin mining on the Malaysian peninsula resulted in increased basin erosion, but the subsequent planting of rubber plantations in previously mined areas appears to have resulted in even greater erosion. More recent clear-cutting has led to frequent floods; in addition, the altered hydrological regime has led to inadequate water supply during dry seasons for downstream rice farms (Brookfield *et al.*, 1992).

Considering the fact that urban areas are often located on or near rivers, a logical source of sediment for road and building construction can be adjacent river beds. The Arno in western Italy, for example, carries a well-sorted sand that can be used for construction purposes. Despite a decreased sediment load because of river damming, the streambed has been mined extensively, lowering of the river bed in the lower reaches of the river by as much as 2–4 m over the past century (Billi and Rinaldi, 1997). Sand mining combined with decreasing sediment discharge has resulted in the lowering of China's Pearl River's channel bed by 10 m in some locations (Lu *et al.*, 2007). Another aspect of gravel and sand mining can be the removal of river armor, which increases potential sediment movement and thus channel deepening (see Winkley, 1994; Schumm, 2007).

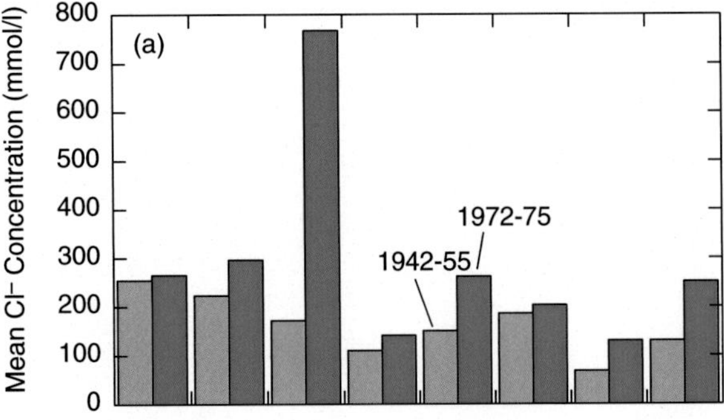

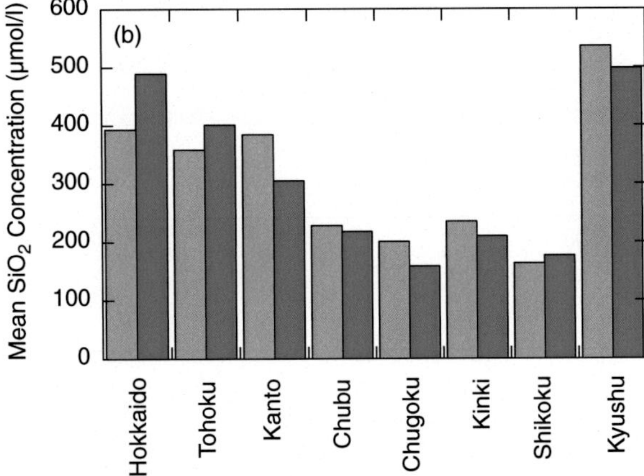

Figure 4.9. Comparison of (a) dissolved chloride (Cl^-) and (b) silica (SiO_2) concentrations in rivers from eight regions in Japan (Hokkaido, Tohoku, Kanto, Chubu, Chugoku, Kinki, Shikoku, Kyushu) for the periods 1942–1955 (orange) and 1972–1975 (blue). Note that dissolved silica (b) concentrations changed relatively little during this period, whereas chloride levels (a) increased in all eight regions. In Kanto, which includes heavily industrialized Tokyo and Yokohama, it increased more than four-fold. Data courtesy of T. Kimoto and A. Harashima.

Chemical pollution

To the general public, the most noticeable impact of human activities on watersheds is evidence by the declining water quality in many rivers. Elevated nutrient concentrations and resulting eutrophication, elevated concentrations of heavy metals, the presence of pesticides and other anthropogenic compounds, and high concentrations of coliform bacteria are some of the more obvious examples. Some pollutants (e.g. chemical fertilizers and pesticides) come from agricultural activities within the watershed, other wastes come from mining and industrial activity, and some come from the direct discharge of waste (human, agricultural, and industrial) into the river. Many point sources, such as storm sewers and effluent from sewage-treatment (or, worse, raw sewage), are located in or near urban centers, and often are concentrated in the lower portions of the watershed (e.g. Cairo on the Nile, Hamburg on the Elbe, New York City on the Hudson, Philadelphia on the Delaware, Shanghai on the Yangtze, etc.). Other pollutants, such as those related to agricultural activities, are considered non-point sources, often entering the river in the more rural upper reaches of the watershed.

As stated in Chapter 1, we have here studiously avoided discussion of nutrients, organic carbon and trace elements, both because of the voluminous available literature and because of our relative ignorance on these subjects. However, such topics cannot be ignored when discussing chemical pollution of rivers. The following discussion should be considered only a sampling of the problem.

Understanding the effect of human activities on dissolved solids within a watershed often necessitates a basin-wide approach. One of the best-documented basin-wide studies is the Piren–Seine program (Meybeck *et al.*, 1989; Meybeck, 2002). Meybeck (2002) has pointed out, for instance, that for every m^3 of river flow during periods of low discharge, the waste of 250 000 Parisians enters the Seine. Other basin-wide studies include the Mississippi (Meade, 1995; Goolsby and Pereira, 1996), the Rhine and Elbe (Vink *et al.*, 1999; de Wit, 1999), and the Murray–Darling River in Australia (MDBMC, 1999).

One of the most obvious signals of anthropogenic activity in rivers is the increased presence of chloride. In addition to reflecting watershed lithology, aridity, and influx of marine aerosols, Cl in river water can come from mining activities, industrial waste, and road salt. Just as the concentrations and flux of silica can provide a quantitative estimate of chemical weathering (see Chapter 2), the ratio of dissolved Cl to SiO_2 can provide a rough estimate of the level of chemical pollution within a river. A comparison of dissolved Cl and SiO_2 concentrations in Japanese rivers following World War II with those in the mid 1970s, by which time industrial output had risen markedly, provides a particularly good example (Fig. 4.9). Dissolved silica concentrations in most regions remained more or less constant during this period, whereas Cl levels in all eight regions increased; in three watersheds Cl concentrations more than doubled.

Table 4.1. *Comparison of historic and recent dissolved solid concentrations in three rivers whose watersheds have experienced varying degrees of human activity. Historic data are summarized in Livingstone (1963) and recent data are from Meybeck and Ragu (1996), USGS NASQAN and Weser River Basin Commission (http://www.fgg-weser.de/en/weser_water_quality_monitoring_program.html)*

	Amazon		Mississippi			Weser			
	1903	1988–1995	1905–1906	1979–1993	1998–2005	1893	1990	1995	2005
Ca^{2+}	5.4	5.4	34	39	35	52	120	76	79
Mg^{2+}	0.5	0.9	8.9	12	11	8.7	195	34	36
Na^{+}	1.6	1.8	14	14	18	29	1538	167	125
K^{+}	1.8	0.8		3.1	3	5.4	76	20	21
Cl^{-}	2.6	2.1	10	19	20	49	2860	314	252
SO_4^{2-}		4.5	25	41	45	64	459	119	122
HCO_3^{-}	18	21	116	98	104	124	159	149	?
Totals	43	43	223	276	236	342	7397	2874	<2800

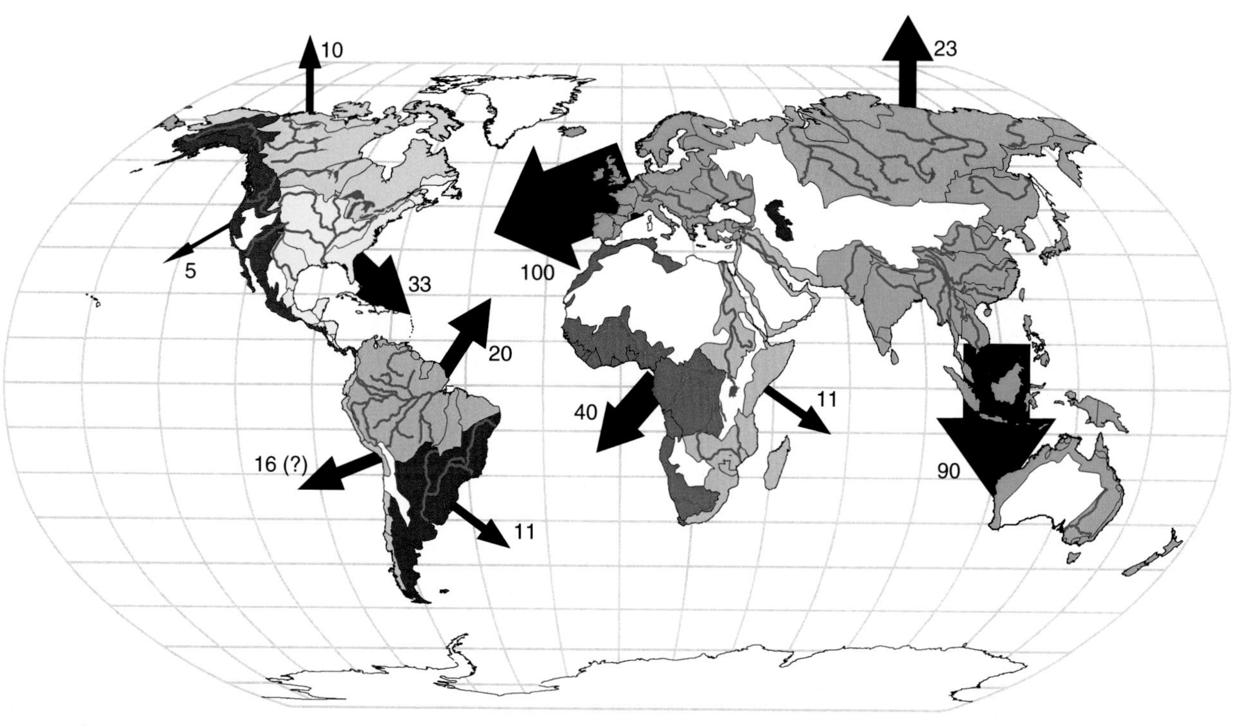

Figure 4.10. Fluvial discharges of dissolved chloride to the global coastal ocean. As shown in other global maps in Chapter 2, widths of the arrows are proportional to the annual discharge (in Mt/yr). Note that estimate for western South American river flux is based on few data.

By the late twentieth century dissolved-solid concentrations in many European rivers reached improbable levels. In the late 1990s, for example, the Weser River, in northern Germany, had chloride concentrations exceeding 1000 mg/l (Table 4.1), meaning that this relatively small river actually discharged more chloride than the Amazon River. As shown in Figs. 2.10 and 2.29, western Europe rivers discharge relatively little water or sediment to the global ocean, and the collective dissolved silica delivery from European rivers represents only ~3% of the global total (Fig. 2.49). By contrast, total dissolved solids discharged from European rivers account for nearly 15% of the global total (Fig. 2.39), and nearly 28% of the dissolved Cl (Fig. 4.10). Although concentrations of both dissolved Cl and SiO_2 reflect both lithology and climate, the ratio of the two gives a good indication of the relative levels of human pollution in the watershed.

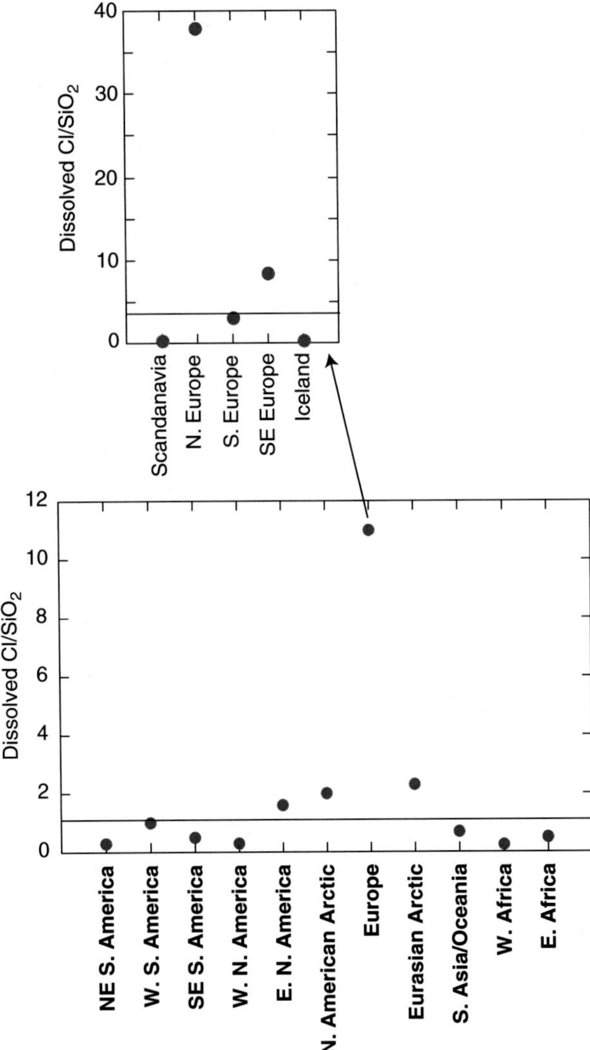

Figure 4.11. Cl^-/SiO_2 ratios from various global watersheds. Data from Meybeck and Ragu (1996).

Northern European rivers have average Cl/SiO_2 ratios >35, whereas more pristine rivers draining NE South America and western North America have mean Cl/SiO_2 ratios of ~0.3 (Fig. 4.11), a 100-fold difference in Cl flux.

The impact of human activites on watersheds also can be delineated by comparing historic and recent fluvial data. For rivers whose drainage basins have not been subjected to large-scale human activity, historic and recent dissolved concentrations appear similar. In the Amazon River, for example, total dissolved solid and specific ion concentrations have been nearly constant for the past 90 years (Table 4.1). At the other end of the spectrum is the Weser River, whose total dissolved solid concentrations increased seven-fold between 1893 and the early 1970s, and Na and Cl levels increased by more than 50-fold in response to potash mining. A basin-wide cleanup initiated in the early 1990s led to a remarkable 65% decrease in dissolved solids, but an even greater decrease in dissolved Na and Cl^-(90%) (Table 4.1). Dissolved Cl^- and Na decreased another 20% by 2005, but overall dissolved-solid concentrations declined only slightly. Despite this marked improvement in water quality, 2005 levels were still 7–15 times greater than they were in 1893. A somewhat middle ground is seen in many other rivers. Dissolved concentrations in the Mississippi, for example, have increased by "only" ~20% during the past century; some ions (e.g. Ca^-, Na^+) have remained more or less constant, but others (particularly Cl^-) nearly doubled (Table 4.1).

Nutrient levels (specifically, nitrogen and phosphorous) have increased dramatically in many rivers over the past 50 years, largely in response to increased application of fertilizers to agricultural lands and phosphate in detergents. Point-source inputs also include the discharge of raw sewage or treated wastewater. As a result, fluvial nitrate concentrations range over more than two orders of magnitude – from 8.6 mg/l (Trent River, in England) to the pristine 0.019 mg/l (Moise River, Canada) (Meybeck and Ragu, 1996; Meybeck, 1998). Two obvious examples of this anthropogenically induced nutrient enrichment in large watersheds are the Mississippi and Nile rivers, both of which are discussed later in this chapter.

Recent modifications in mining activities across western Europe have resulted in decreased discharge of some contaminants, but other concentrations have increased. Etchanchu and Probst (1988), for instance, reported that between 1971 and 1984, the Garonne River (SW France and one of France's least polluted large rivers) had "only" a 14–19% increase in Cl^-, SO_4^{2-}, and K^+, but a 78% increase in NO_3^-, which they attributed almost entirely to the increased use of fertilizers.

One less obvious example of river pollution – one can almost consider it inverse pollution – is the draw-down of silica and other nutrients by freshwater phytoplankton within man-made bodies of water, such as reservoirs behind dams. The impact on the river's estuary and coastal waters can be significant. Silica drawdown behind the Iron Gates dam on the Danube River, for example, may have led to a major change in the ecosystem in the northwestern Black Sea, silicate-producing diatoms giving way to flagellates (Humborg *et al.*, 1997), and there is increasing evidence of falling silica levels in other dam-impacted rivers (Humborg *et al.*, 2000; Conley *et al.*, 2000; Li *et al.*, 2007). In the Mississippi River, silica concentrations fell by nearly 50% between 1950 and 1970, presumably in response to the increased damming of upstream tributaries, whereas nitrate-nitrogen concentrations (reflecting increased fertilizer application) more than doubled (Turner and Rabalais, 1991). The degree to which this change in the nutrient character of fluvial waters might affect hypoxia or anoxia in coastal waters is a matter of debate, as discussed in following sections of this chapter.

Unfortunately, the general lack of long-term data precludes any global synthesis or prediction regarding chemical pollution (Meybeck, 1998). The Global Environmental Monitoring System (GEMS) had hoped to solve this impasse by collecting analytical data for 82 global rivers (Fraser *et al.*, 1995), but this assumed that all participating countries would share similar data, which clearly has not been the case (see Chapter 1). Even if the comprehensive data were collected, it may be years before meaningful trends could be extrapolated on a regional or global basis.

Recent changes in land-use and fluvial response

Throughout many parts of the economically developed world there has been a concerted effort in recent years to reduce land erosion and water pollution. Reduction in agricultural land-use (in part owing to urbanization) and additional control of both erosion and water flow (i.e. dams), for example, have resulted in significant decreases in suspended-sediment concentrations in most rivers draining central Japan (Siakeu *et al.*, 2004). Allowing agricultural land to revert back into a naturally vegetated landscape not only can decrease the erosion per se, but also increase water retention, thereby decreasing peak discharge events (Piégay *et al.*, 2004) as well as possibly reducing non-event discharge. The dramatic decrease in Yellow River discharge since the 1970s, for instance, can be explained in part by the widespread reforestation of the loess plateau (Wang *et al.*, 2007; Z. S. Yang and J. P. Liu, personal communication). In reforested areas in the Yellow River basin sediment loads in gullies have been reduced by as much as 92% (Li, 1992).

Trimble and Crosson (2000) cite evidence for declining soil loss (some of it dramatic) in parts of the USA over the past six decades, in large part owing to improved land-use and soil-conservation practices. The fact that not all rivers have responded to changes in land-use may be explained by the buffering capacity (largely in flood plains) of river basins, which for years had stored much of the increased sediment load. Clark and Wilcock (2000), for instance, showed that between 1830 and 1950, in response to deforestation and increasing cultivation of deforested terrain, runoff in northeastern Puerto Rican rivers increased by 50% and sediment supply to the river channels increased by more than an order of magnitude. The shift away from agriculture since then has decreased sediment erosion (although not affecting runoff), so that streambeds have degraded as sediment derived from the previous 120 years has been reworked and transported downstream. Thus improved land practices may not be felt in terms of decreased sediment discharge until the stored sediment is removed (Trimble, 1977; Walling, 1997, 1999).

Decreased sediment discharge is particularly clear in European rivers, where land-use has evolved significantly over the past century. Coastal accretion along the Italian Adriatic coast, for example, began in response to land utilization during the Bronze Age and accelerated during the Renaissance. Over the past 100 years, however, much of the coast has eroded at an increasing rate as sediment discharge has deceased owing to of improved agricultural techniques, reforestation of former farms, and the mining of river beds for sand and gravel (Coltorti, 1997; Simeoni and Bondesan, 1997). The recovery of a former agricultural land, of course, depends in part on morphology of the basin. Begueria (2006), for instance, found that whereas reforested farmland in the Pyrenees experienced less shallow landslide activity, former arable fields on valley slopes showed no change in landslide activity, even on revegetated plots.

Reforestation and revegetation, while able to decrease erosion, may not be a universal blessing. For example, pine tree (*Pinus radiata*) reforestation at several sites in Chile has resulted in less water perculation into the soil and greater evapotranspiration loss of water (Huber *et al.*, 2008) and a decrease in summer runoff (Little *et al.*, 2009).

Frangipane and Paris (1994) report that sediment discharge from the Ombrone River (central western Italy) decreased from 6–7 Mt/yr in the mid nineteenth century to about 1 Mt/yr at present. Agriculture in the Ombrone watershed actually increased during this 150-yr span, but grazing decreased markedly (from 37% of the basin to 7%), and forest cover correspondingly increased from 29% to 40%. Farther north, sediment discharge from the Arno River, which flows through Florence, increased from about 2 Mt/yr prior to the sixteenth century to 7.5 Mt/yr between 1500 and 1800, but has since fallen to about 2.7 Mt/yr over the past 50 years (Billi and Rinaldi, 1997).

Owing to its long record, the Rhine River offers an excellent example of the varied chemical response of a river to changing land-use (Fig. 4.12). A comparison of data collected in 1854 and the latter part of the nineteenth century indicates that, until the early part of the twentieth century, the Rhine was relatively pristine in terms of dissolved constituents. Potash mining in the Alsace, together with other mining and industrial activities throughout the river basin, raised Cl levels (and to a lesser extent SO_4, Na, K, and Ca; van der Weijden and Middelburg, 1989, their Table 7) in the 1920s. A comparison of nitrate concentrations between 1854 (measured at Mainz by J. W. Gunning; Livingstone, 1963, his Table 36) and those measured in the 1930s, however, suggests little effect increase in dissolved-nutrient concentrations over the 80-yr period. Between 1930 and the end of World War II, however, dissolved ammonia (indicative of domestic waste disposal) nearly tripled and nitrate increased, but less rapidly (Fig. 4.12). Between the 1950s and 1980s, in large part because of economic post-World

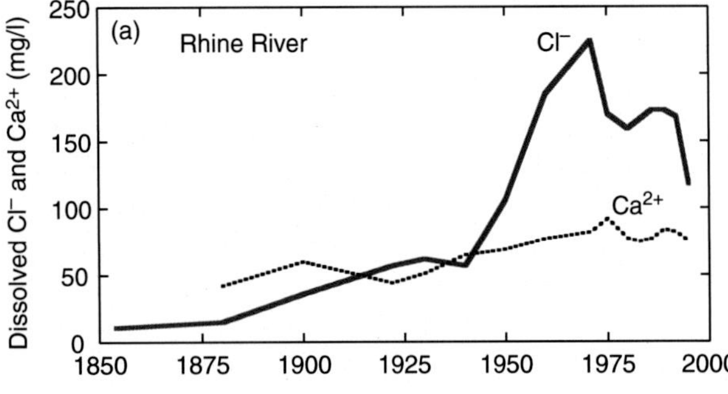

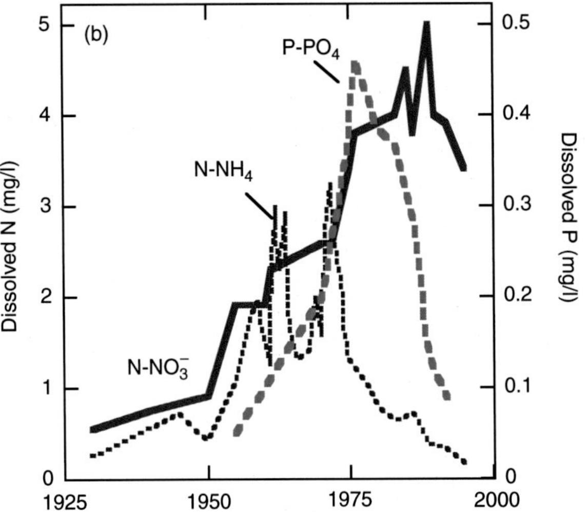

Figure 4.12. Changing concentrations of dissolved constituents in Rhine River water as measured at Arnhem (NW Germany) and Lobith (eastern Netherlands), 1850–2000. Data from Livingstone (1963), van der Weijden and Middelburg (1989), and GEMS.

War II recovery, all dissolved constituents in Rhine water increased, nitrate concentrations (because of accelerated use of artificial fertilizers) increasing ten-fold. After 1980, significant decreases in pollutant concentrations were noted. Decreased phosphate levels in large part reflected increased use of phosphate-free detergents. Ammonia levels also decreased, as most domestic waste was subject to increased treatment; by the early 1990s ammonia concentrations were lower than they had been in the 1930s (Fig. 4.12). Only recently was there a hint at decreased nitrate concentrations in the Rhine, but whether this reflects a long-term trend is not clear. Another hopeful indicator of improved water quality in the Rhine is the marked decrease in Cl concentrations, indicative of the improved mining and industrial practices seen most graphically in the Weser River (see Table 4.1).

Both solid and dissolved loads from European and North American rivers should continue to decline in coming years in response to more conservation-friendly land-use practices. In contrast, suspended and dissolved yields across much of southern Asia and Oceania may increase in response to increased deforestation and agriculture, as well as growing industrial capacities throughout the region. Unfortunately, without long-term monitoring of these rivers – and access to the data – we can only speculate about present and future trends.

Shared rivers

Because of the active and passive roles that rivers and their watersheds play in agriculture, water supply, transportation and defense, many rivers inevitably flow through many states or countries. Many US rivers delineate state boundaries, from New England (New Hampshire–Vermont) to the Pacific northwest (Oregon–Washington). As vexing as interstate disputes can be, however, they are minor compared to international disputes regarding land and water use. The Brahmaputra river discharges to the Bay of Bengal through Bangladesh, but 93% of the river's watershed lies outside Bangladesh – in India, Nepal and (to a lesser extent) China. How India, Nepal, and China utilize the river's water can therefore have major – perhaps deleterious – impacts on their downstream neighbor.

Similarly, the impact of Swiss economic and social activities on the Rhine can be felt downstream in France and Germany, and how Germans and French regulate the Rhine's flow can have a great effect on the Dutch. Building dikes and levees to minimize the amount of overbank flooding in Germany, for instance, probably has been a main contributor to flooding in the Netherlands.

In a not-too-proud moment in the United States' history, the damming of the Colorado River and the subsequent parceling out of upstream water to various western states in the 1930s all but wiped out agriculture on the Colorado River delta, once one of the most fertile areas in Mexico. Allocation of Colorado River discharge within the USA, as formulated in the Colorado River Compact, was based on river discharge in the 1920s, which unfortunately coincided with high warm-PDO precipitation and runoff (Young, 1995). Over the past 500 years, in fact, the Colorado River basin has tended more towards drought than flood (Woodhouse *et al.*, 2006).

Probably the region most affected by the problems of shared rivers is the Middle East, where the disagreements about present and future use of the Tigris/Euphrates, Jordan and Nile rivers have affected and will continue to affect diplomatic relations between Turkey, Syria, Iraq, Jordan, Israel, Palestine, Egypt, Sudan, and Ethiopia (Kliot, 1994). It is doubtful, for instance, that US policy-makers thought through the implications that the possibility of a greater Kurdistan might have on Turkish water resources before they decided to invade Iraq in 1993. Hillel (1994) has presented a most readable history and political analysis of Middle East water-use problems, arguing that water access and water rights almost certainly will play an ever-increasing role in Middle East politics and conflicts. More academically directed discussions have been presented by Biswas (1994), Shady *et al.* (1996), and Scheumann and Shiffler (1998).

River control

Perhaps no aspect of the environmental debate generates so many points of view and stirs so many emotions as the control of river flow. The level and tone of discussion of such academic matters as fluvial fluxes or episodic events pale in comparison to the passion raised by such controversial topics as irrigation and dams. Literally overnight, obscure animals and plants become poster children for endangered species, and any economic rationale for construction can be lost in less-than-logical reasoning. At the same time, the world has become more and more aware of the global imbalance of water supply and the "looming crisis" (to use the words of a UNESCO report; Zebidi, 1998) regarding the future of water use. And the discussion only broadens and intensifies. To cite David Brooks's (2000, p. 157) *Says Law*, "The more people are saying, the more there is to be said." Such book titles as *Rivers for Life* (Postel and Richter, 2003); *Water: The Fate of Our Most Precious Resource* (de Villiers, 2000) and *Silenced Rivers* (McCully, 1996), as well as such web pages as, "An act of economic and environmental nonsense" (International Rivers Network – http://irn.org/programs/) give only a hint of the emotional level of the debate.

Because of the obvious dependence of humans on rivers and their waters, there are many reasons to modify or divert natural river flow – channel "improvement," irrigation, mechanical (historically) and hydro-electric power (which is assumed to be cleaner and less harmful to the environment than either fossil-fuel or nuclear power generation), flood control, improved navigation, and recreation. And there are many ways to control flow: levees, channel dredging, dams, etc.

A few random tidbits illustrate the importance of human control on river discharge and its quality. Total water withdrawal during the twentieth century in the USA alone increased roughly nine-fold to about 550 km^3/yr. Dynesius and Nilsson (1994) calculated that about 3/4 of the cumulative discharge of world's northern rivers is affected by river constructions. At present ~10% of the global fluvial runoff is withdrawn annually (Gleick, 2000a), and present-day global storage capacity of fluvial water is about 16–18% of annual fluvial discharge (Vörösmarty and Sahagian, 2000). Regionally water storage ranges from as much as 30% in North America to as little as 9% in Europe (Table 4.2), and in recent years Africa and Asia/Oceania have seen significant increases in large-reservoir storage capacity. Within a single region, of course, the storage of water ranges considerably. In North America, for instance, reservoir storage ranges from less than 5% of annual runoff throughout much of northern Canada and Alaska to more than 100% in the Rocky Mountains and Southwest. (Storage greater than runoff infers that groundwater has been extracted and added to the stored river water.)

Table 4.2. *Global distribution and cumulative storage in large (>0.5 km^3) reservoirs. Data from Vörösmarty* et al. *(1997b).*

	Number	Total maximum storage capacity (km^3)
North America	175	1184
South America	96	806
Europe	88	430
Asia	201	1480
Oceania	19	54
Africa	43	915
Totals	**622**	**4869**

Levees, channelization, and dredging

One of the most common forms of river control is the emplacement of levees or dikes to prevent overbank flooding as well as to promote the self-scouring of the river channel, which in turn can facilitate river navigation. By containing floodwaters, levees allow organic- and nutrient-rich floodplains to be farmed. But levees are a two-edged sword: by preventing the lateral escape of mud-laden floodwaters, levees also deprive floodplains from receiving periodic renourishment as well as the sediment that would, under natural conditions, compensate for floodplain subsidence (see Mississippi River discussion below). Drainage ditches within the leveed floodplains permit reclamation of wetlands and irrigation of farm fields, but they also can facilitate the oxidation of organic-rich sediments and underlying peat deposits, thus accelerating lowland subsidence. Deverel and Rojstaczer (1996) estimated that 75% of the subsidence in the Sacramento–San Joaquin Delta (central California), which locally has subsided more than 5 m over the past 150 years, has stemmed from peat oxidation, the remaining 25% from dewatering of underlying sediments. As a result, much of the increasingly urbanized Sacramento–San Joaquin Delta is presently below river level, and locally below sea level. Failure of the 1700-km-long levee system on the delta could lead to flooding not unlike that experienced by New Orleans during Hurricane Katrina (Mount and Twiss, 2005; J. F. Mount, oral communication).

Historically, channel straightening was a relatively inexpensive means by which the accelerated flow could clean a channel of excess sediment, as well as lower the water table and thereby drain adjacent wetlands for agricultural use. The engineering of 16 channel cutoffs along the lower Mississippi between 1930 and 1950, for instance, resulted in a 243-km reduction in channel length between Memphis (TN) and Baton Rouge (LA) (Winkley, 1994; Goudie, 2000). Channel modification, however, can have unforeseen consequences. The straightening of the Mississippi, for instance, helped lead to the capture of the Atchafalaya River system, provoking a long-standing problem discussed later in this chapter and eloquently related by McPhee (1989). The use of levees and breakwaters along the lower reaches of the Mississippi River has helped maximize river flow and thus self-flush the river channels; but they also have prevented the lateral escape of flood-transported sediment onto the Mississippi delta, a major reason for the rapid loss of Louisiana wetlands (see below).

Inter-basin transfer

Most large countries contain both regions of high and low precipitation, the high-runoff US Pacific Northwest and arid southwest being an obvious example. Annual precipitation in northern India and southern China can range from 2000 mm/yr to 10 000 mm/yr, whereas rainfall in southeastern India and northern China can be less than 200 mm/yr. One obvious way to minimize the disparity between water-excess and water-poor regions is to transfer water from the former to the latter. The southward diversion of freshwater from Owens Lake by William Mulholland was an early example of interbasin transfer, and it subsequently provided the water necessary for Los Angeles' development in the early twentieth century. See Reisner (1993), for a detailed – if somewhat biased – history; better still, see Roman Polansi's classic movie *Chinatown*. Pumping Colorado River water westward into southern California, in many ways was simply a logical continuation of the Owens Lake transfer. In recent years, however, increased need of water in the rapidly populating and developing southwest has necessitated southern California to look farther a field for its water – some plans have even called for water diversion from the Columbia River, Canada, and southeastern Alaska (Reisner, 1993).

Large-scale water diversions also have been proposed to transfer water to water-scarce southern India, northeastern Brazil, and northern China, where a portion of Yangtze River flow would be channeled northward into northern China (p. 160). Partial diversion in 1976 of the Ganges River in India resulted as much as 60% of the seasonal flow being channeled into the Hoogly River. Among the impacts of this diversion were reduction of surface water resources in Bangladesh, alteration of aquatic breeding grounds, and perhaps most importantly, increased dependence on arsenic-rich ground water, thus putting at risk the health of the 75 million down-river Bangladeshis and Indians (Adel, 2001). Not to be deterred, India has planned a complex series of interbasin transfers in which water would be borrowed from one river, only to be repaid with water from another. If all the projects are ever completed, the map of India would be begin to resemble a bowl of spaghetti (Fig. 4.13), with profound ramifications; see Gong (1991), Subramanian (2004) and Minza *et al.* (2008) for discussions.

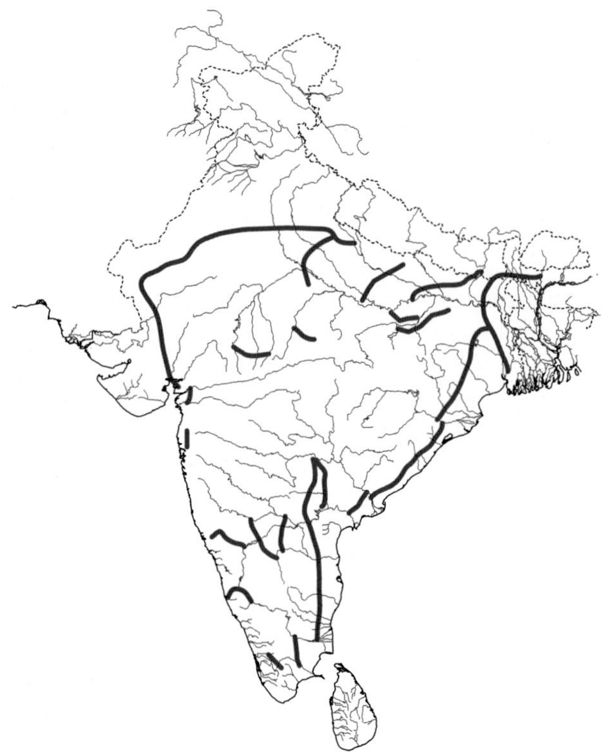

Figure 4.13. Proposed interbasin transfers of river water in India (red lines). After World Wildlife Fund (2007).

Irrigation

For millennia irrigation has provided dry agricultural land with water, making lush green Edens out of "inhabitable" desserts (see Fig. 4.14). Gleick (2000a) has calculated that, by the end of the twentieth century, 2 700 000 km^2 of land were irrigated throughout the world. The impacts of irrigation and its corresponding reduction of freshwater discharge have been particularly severe in arid rivers and their watersheds. The Barwon–Darling River, a major tributary to the River Murray in SE Australia, for instance, experienced a 42% decrease in annual runoff over a 60-year period (Thoms and Sheldon, 2000), similar to that seen in the Sakaraya River discussed below. In more arid rivers such as the Indus, Colorado, and Nile, discharge reduction has approached 100% (see below). Not surprisingly, Asian countries have been particularly active in irrigating their low-precipitation lands. In their digital analysis of global irrigation areas, Siebert *et al.* (2005) calculated that Asia has 1 880 000 km² of irrigated land area, China and India accounting for 540 000 km² and 570 000 km², respectively.

Diverting river flow for irrigation, unfortunately, can impart deleterious effects on both the irrigated land and the river draining it. In particularly arid climates – where irrigation is especially necessary (see Figs. 2.8 4.14 and 4.16) – extensive irrigation can accelerate water loss through both evaporation from the irrigation ditches and evapotranspiration from the vegetation. Even though the irrigation channels may ultimately again merge with the river downstream, freshwater can be appreciably diminished in both volume and quality. Milliman *et al.* (2008) documented a number of global rivers in which, over the past 50 years, runoff had decreased at a much more rapid pace than any change in basin-wide precipitation (Figure 4.15), indicating a net consumption of water within the watershed. Nearly all of these "deficit" rivers were extensively dammed and irrigated, as indicated by flow regulation and irrigation indices (Nilsson *et al.*, 2005).

Another result of irrigation can be the water-logging of soil, resulting in the build-up of highly saline soils and eventual desertification. The demise of the Sumerian civilization, for example, may have resulted from the excessive irrigation and subsequent salinization of the arid Fertile Crescent (see Hillel, 1994, for a popular account). Minimizing such salinization of irrigated lands remains a major challenge for irrigation engineers.

By changing the flow regime, irrigation channels also can serve as effective sediment traps, filling drainage ditches

Figure 4.14. The striking contrast between the Algodones Dunes and the farmed fields of Morelos, Mexico, results from waters fed by irrigation channels stemming from the American Canal (lower left and lower right), which also carries Colorado River water to California's Imperial Valley. Yuma, Arizona, is just out of the picture below. Photo courtesy of J. P. M. Syvitski.

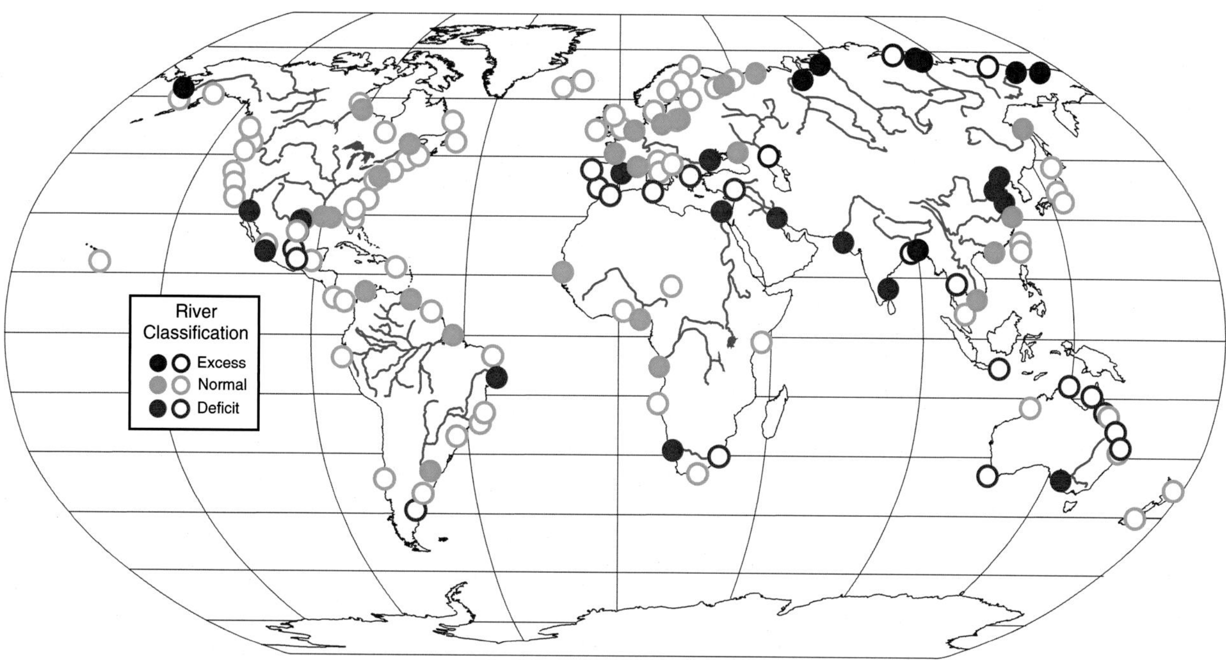

Figure 4.15. Global distribution of normal, deficit and excess watersheds, 1951–2000; after Milliman *et al.* (2008). The 50-yr discharges for "normal" rivers reflect primarily precipitation trends, whereas discharge trends in "deficit" rivers reflect the consumption of water resulting from damming, irrigation and interbasin water transfers (see Figs. 4.13, 4.14, and 4.16). The Don River (southern Russia) is classified as a deficit river even though its runoff was reduced in the later 1940s; decrease in Sacramento River (CA, USA) flow reflects downstream transfer to the San Joaquin River. High-latitude and high-altitude "excess" rivers have experienced increased discharge despite generally declining precipitation, perhaps the result of changed seasonality of discharge, decreased storage, decreased evapotranspiration, or some combination of the three.

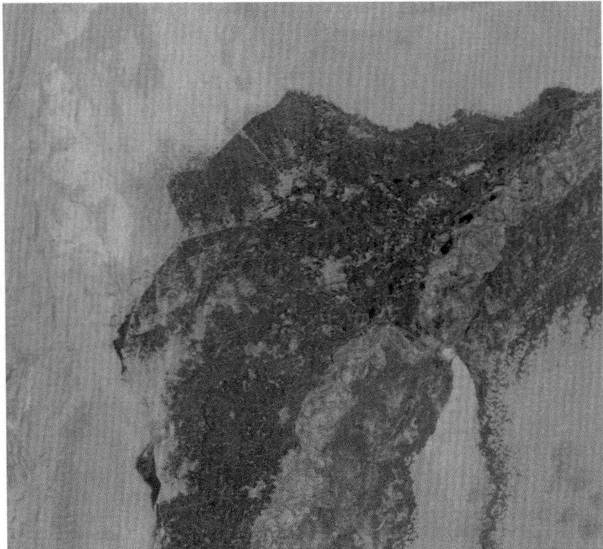

Figure 4.16. Agricultural field irrigation water from extensive ditches in the lower Indus River watershed. Note the contrast between desert terrain and agricultural lands. Image courtesy of NASA.

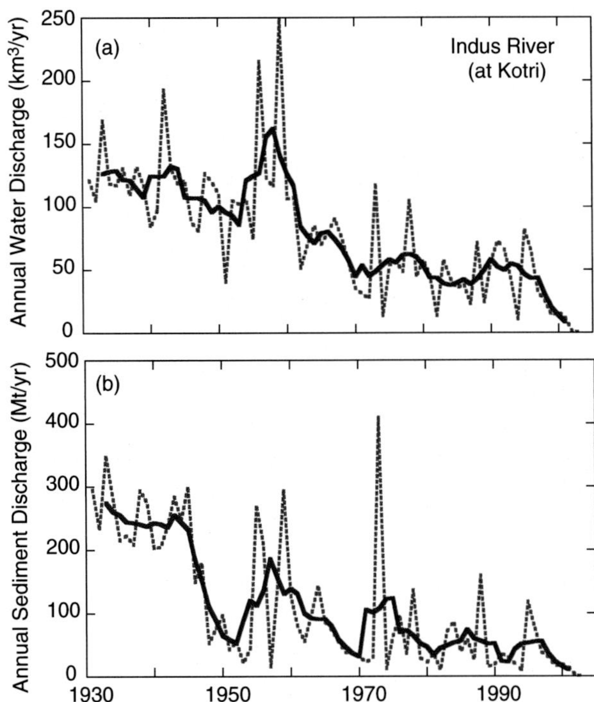

Figure 4.17. Annual water (a) and sediment (b) discharges from the Indus River (as monitored at Kotri, near the river mouth), 1931–2002. Annual values are shown as a dashed red lines; solid blue lines represents 5-yr running means. The marked decrease in sediment discharge in the late 1940s coincided with initiation of irrigation barrages along the river (Milliman *et al.*, 1984). Water discharge remained unchanged until the early 1960s, when construction of several large dams in the upper reaches of the river trapped and rerouted some of the flow. By the year 2000, both water and sediment discharges were less than 5% of their pre-1947 levels. Note, however, the continued interannual variability in both water and sediment discharge. Data courtesy of Asif Inam.

and thereby decreasing sediment discharge of the river. The construction of barrages and irrigation channels along the Indus River in the late 1940s (Fig. 4.16), for example, had little effect on freshwater discharge to the Arabian Sea, but collectively these constructions reduced sediment delivery by the river by about 75% (Fig. 4.17b). Construction of the Mangla and Tarbela dams in the early 1960s did reduce the Indus's freshwater discharge (Fig. 4.17a), but it had relatively little effect on the river's sediment discharge, which had been diminished 15 years before. By the late 1990s very little sediment or water was reaching the Indus Delta (Fig. 4.17), and erosion of the delta had begun (Kravtsova *et al.*, 2009). One particularly vivid example of the deleterious effects of river damming can be seen in the Yellow River, discussed later in this chapter. The reader should also keep in mind that the trapping ability of irrigation ditches can last far longer than the civilizations that built them: ancient irrigated fields, more than 3000 years old, for instance, still trap sediment in the Rio Grande de Nazca in southern Peru (Baade and Hesse, 2008).

Dams and their reservoirs

Dams and their reservoirs are a particularly effective means of regulating river flow, providing energy, and storing water (see discussion by Magilligan and Nislow, 2005, and references therein). Magilligan *et al.* (2003), for instance, found an average 60% reduction in 2-yr floods in many US dammed rivers.

Although dams have marked the human-altered landscape for thousands of years, they proliferated greatly over the past 100 years with the advent of the machines and building materials required to construct such large structures. Dam construction in the first few decades of the twentieth century was centered in Europe and, to a lesser extent in North America. During the middle part of the century, North America provided the main building sites. Dam construction accelerated globally in the 1950s (Fig. 4.18), following the global economic downturn in the 1930s and World War II, in part spurred by improved economic times and the end of colonial occupation throughout much of Africa and Asia. By 1970 more than half of the present-day large US dams had been built, and by 1990 more than 95%. By the year 2020, in fact, more than 85% of all large dams will be more than 50 years old (Hossain *et al.*, 2009; see their Fig. 1). In contrast, only about 60% of existing large African dams had been built by 1980, and about 20% have been in operation only since 1995 (Fig. 4.18).

During the last quarter of the twentieth century, Asia accounted for most dam construction (Fig. 4.18), particularly

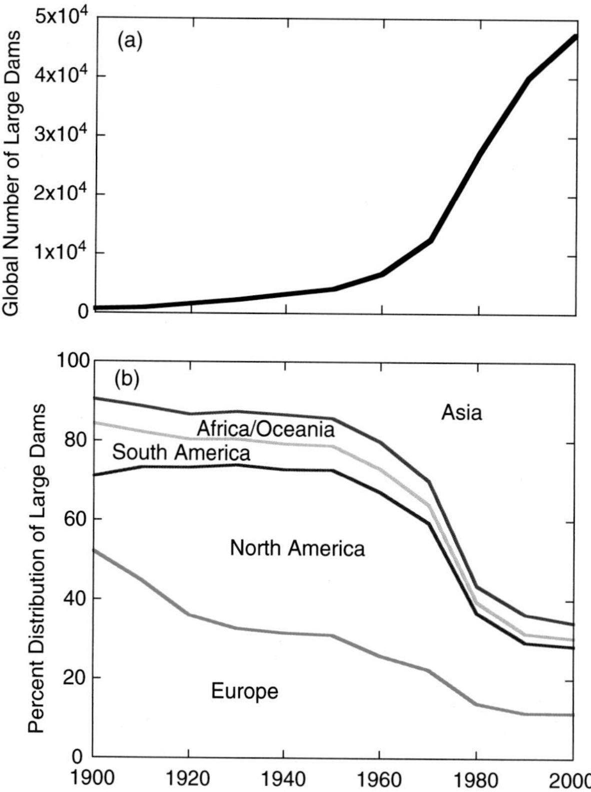

Figure 4.18. Number of large dams (>15 m in elevation) built throughout the twentieth century (a), and their global distribution (b).

dams in China and India. Data from the International Commission on Large Dams (ICOLD) show that as of 1999 Canada had no new large (>15 m in height) dams under construction, whereas China and India had 330 and 650, respectively. Of the 43 countries responding to an ICOLD questionnaire in 1999 (Australia, most of South America, and countries within the former Soviet Union were generally absent from this database), a total of 47 425 large dams were listed, of which China had 26,094 – 55% of the total, a remarkable number considering that China had only three large dams in 1949. Of the more than 10 000 dams in the world higher than 30 m, China claims 45%.

Reckoning the number of dams in the US depends on exactly what classification is used. The single best source for data is probably the compilation by ICOLD, which is restricted to large dams higher than 15 m. The National Inventory of Dams (NID: http://crunch.tec.army.mil/nid/webpages/nid.cfm), by contrast, lists dams that have significant hazard potential, or dams with low hazard potential but that are either higher than 1.8 m with 62 000 m^3 of storage or 8 m high with more than 18 500 m^3 of water storage (Graf, 1999). Using these criteria, the NID has registered nearly 77 000 dams in the United States (Stallard, 1998, cites 68 000) of which 81% are of earthen construction. The Mississippi drainage basin alone contains more than 39 000 dams that match the NID guidelines, but the actual number seems far higher (see below). Perhaps equally important in terms of their cumulative effect of trapping water and sediment may be the numerous smaller dams and impounded ponds – millions estimated in the USA alone (Renwick *et al.*, 2005) – many of them built by farmers, private land owners, as well as by state agencies. Not surprisingly, the timing of dam construction across the United States reflects its settlement. In 1850 dams were mostly constrained to the eastern seaboard. Western dam construction began in the latter part of the nineteenth century but reached its maximum in the mid twentieth century (Fig. 4.19). In total there are about 75 000 US dams that are capable of storing the equivalent of one or more years' worth of runoff (Graf, 1999). The highest concentration of US dams remains in the northeast, many of which were constructed in the nineteenth century as mill dams (Graf, 1999).

Increased environmental concerns, questionable economics, and the progressive filling of the most likely sites have all contributed to the decline in dam building in North America and Europe. In fact, in recent years there seems to have been as much debate in North America about the removal of older dams as to the construction of newer dams. Some dams have simply outlasted their purpose and expected lifetime – old mill dams in New England and the middle Atlantic states being obvious examples – and some reservoirs have been partially or completely filled, thus negating their potential to produce power or store water. The costs of removing dams – both in economic and environmental terms – however, are considerable. How does one effectively remove the dam (which often was built in constricted parts of the river, thus limiting access), and where does one dump the sediment trapped behind the dam? Depending on how the 2–5 000 000 m^3 of sediment trapped behind Matilija Dam (Ventura River, California) – a dam built for 4 million dollars in 1947 – is removed, the dam-removal project could cost 150 million dollars or more. The need for water control in the USA, however, continues to increase, and the largest dam in California, the 180-m high Shasta Dam may be raised another 2–60 m in response to increased water needs, both agricultural and for human consumption (*New York Times*, July 8, 2001).

Impacts

The immediate implications of river control projects are cost and displacement (often forced) of native inhabitants. Tom Zeller (*New York Times*, November 19, 2000), for instance, reported that 10 proposed dams (none in North America or Europe) had an estimated construction cost of 54 billion US dollars and would displace more than 1.7 million people. Dam construction and site development for reservoirs also can bury historic and archaeological sites

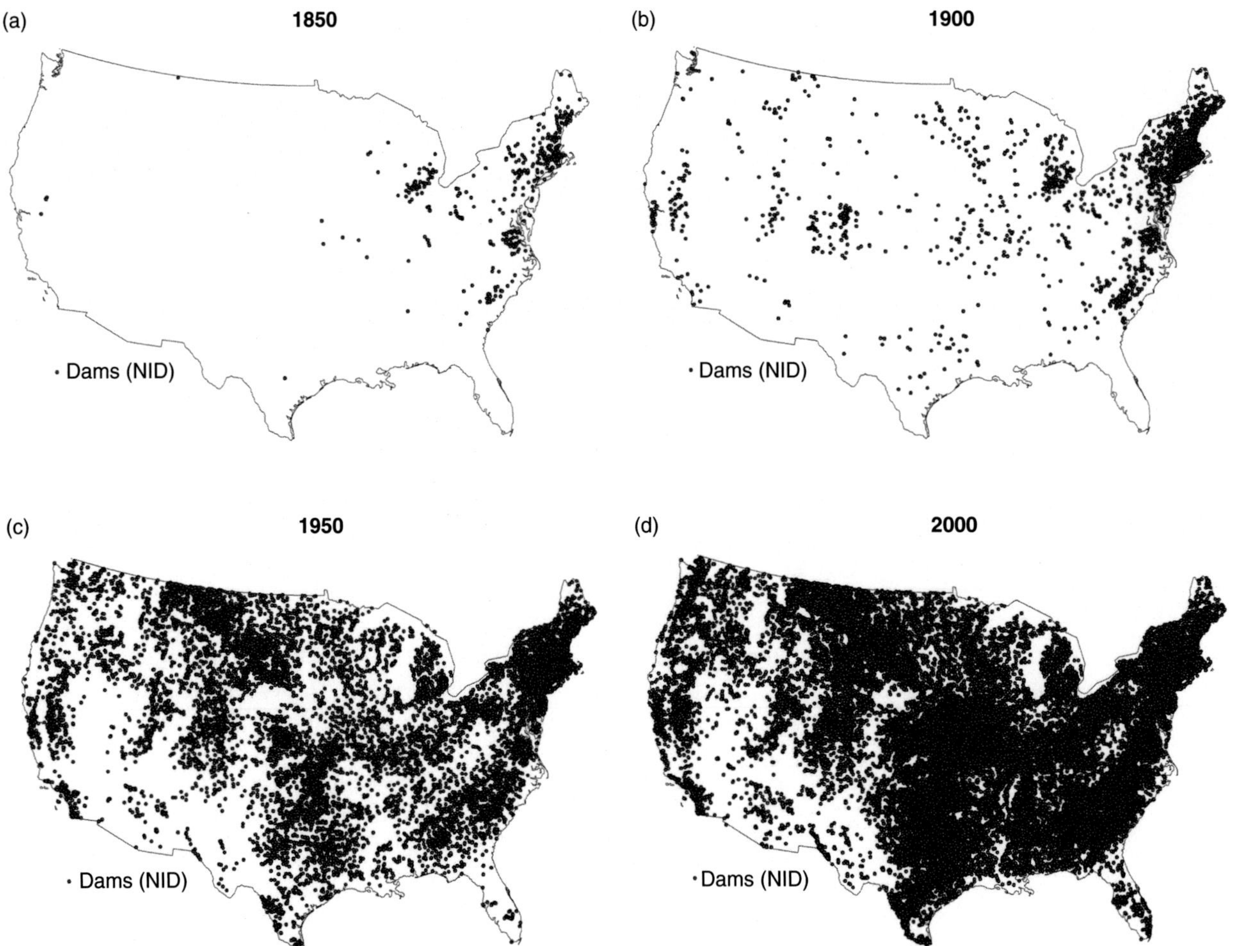

Figure 4.19. Dam construction across the United States from 1850 to 2000. Figure courtesy of J. P. M. Syvitski.

(such as in the flooding of Lake Nasser on the Nile) and can lead directly or indirectly to large-scale deforestation, thereby increasing (at least temporarily) erosion. Moreover, the decay of the inundated terrestrial vegetation can markedly decrease the oxygen content of the reservoir water, thus affecting water quality, at least in the short term. Some to much of the sediment may be trapped in the reservoir, thereby decreasing reservoir capacity.

The social, economic, and perhaps political impact of dam construction can be judged from a September 30, 2001 (p. A7) article in the *New York Times*. Several proposed dams in China's Yunnan Province, on the upper reaches of the Mekong River, would "lift" the 43 million people in Yunnan out of poverty through hydro-electric power generation and flood control as well as by opening shipping channels to northern Laos and Cambodia. By eliminating seasonal flooding these dams might facilitate continued "development" of the Mekong watershed (Wang and Lu, 2008). But these dams also could greatly impact regional fisheries by preventing some fish species from migrating upstream and also diminish seasonal farming in downstream countries. "Chinese officials call these concerns exaggerated, though they admit that some environmental damage is inevitable. Still, they say, the dams are necessary to power Yunnan's industrialization and improve living conditions" (*New York Times*, Sept. 30, 2001). Here is the debate over dams in a nutshell: economic need vs. environmental (and, thus, ironically, often long-term economic) concerns.

Whether or not dams are beneficial over the long term is a difficult and highly charged question. However, even the most fervent dam supporters acknowledge that dams and other human water-diversion projects have invariably altered not only the amount of fresh water discharge reaching the ocean but also water quality and the rate at which the water is discharged. Vörösmarty *et al.* (1997a) estimated that the 633 largest reservoirs (storage capacities greater than 0.5 km^3) have a cumulative storage capacity of nearly 5000 km^3, with a discharge-weighted residence time of 0.21 years. Regionally, the residence times vary greatly, in North America ranging from a few weeks to more than

3 years. Reservoirs in Morocco, Algeria, and Tunisia have total capacities of 10 km^3, 4.3 km^3, and 2.3 km^3, respectively (Lahlou, 1996), which approximates or surpasses the average river discharge from each country (14 km^3/yr, 4.6 km^3/yr, and 1 km^3/yr; see Appendix). Vörösmarty *et al.* (1997a) calculated that their 633 large reservoirs collectively intercept about 40% of the global fluvial discharge. Sediment retention within the reservoirs, Vörösmarty *et al.* (1997a; 2003) speculate, may result in the interception of 25% or more of the global flux.

Increased dam construction has resulted in increased storage of reservoir water, leading to an aging of river water (Vörösmarty *et al.*, 1997b). Ignoring inter-basin transfers of water, water discharge, and retention in much of the USA and Canada follow rather closely meteorological runoff: in sparsely populated regions with runoff >250 mm/yr, present-day flow tends to approximate natural outflow, whereas people in arid areas tend to consume much of their natural water supply. The Colorado Basin presents an extreme example, discharging only about 9% of its natural flow to the Gulf of California; the rest is consumed or exported (to southern California). In fact, according to data presented by Hirsch *et al.* (1990, their Table 7), ground water extraction hides the fact that the Colorado Basin is running an annual water deficit. In fact, groundwater is being extracted faster than it is being recharged. We discuss this further in the last few pages of this chapter.

Although river managers often argue that river impoundment has limited environmental impact and that river runoff to coastal environments is essentially wasteful, Rozengurt and Haydock (1993) have concluded a river probably cannot afford to have more than 25–30% of its water diverted for other uses without having a severe effect on the watershed and coastal zone. In his analysis of 137 very large American dams, Graf (2006) noted not only a marked decrease in maximum flow but also a 79% decrease in downstream floodplain areas as well as impacts on smaller, less diverse riparian ecosystems due to the dammed rivers' reduced geomorphic complexities. Downstream climate may even be affected: construction of the Alcántara dam on the transnational Tagus River in 1969, for instance, has been suggested to have led to increase in downstream droughts (López-Moreno *et al.*, 2009). Even flood control must be carefully planned and implemented. Balamurugun *et al.* (1999) concluded that regulation of the Ångeman River in Sweden has led to greater summer floods when the reservoir is full prior to the flood.

To some degree large dams may be capable of altering local precipitation patterns. In part this can be owed to greater moisture content due to evaporation of reservoir waters. But more importantly, dams often result in changed land-use and land-cover by providing, for example, the waters necessary for irrigation, which in turn can lead to increased farming and urbanization. Such dam-induced changes may be particularly important in facilitating extreme precipitation events (see Hossain *et al.*, 2009).

Even if the annual discharge of a dammed river is not affected, trapping water during high flow and releasing it during low flow can change the seasonality of discharge (see examples in Figs. 4.20 and 4.31). In doing so, the

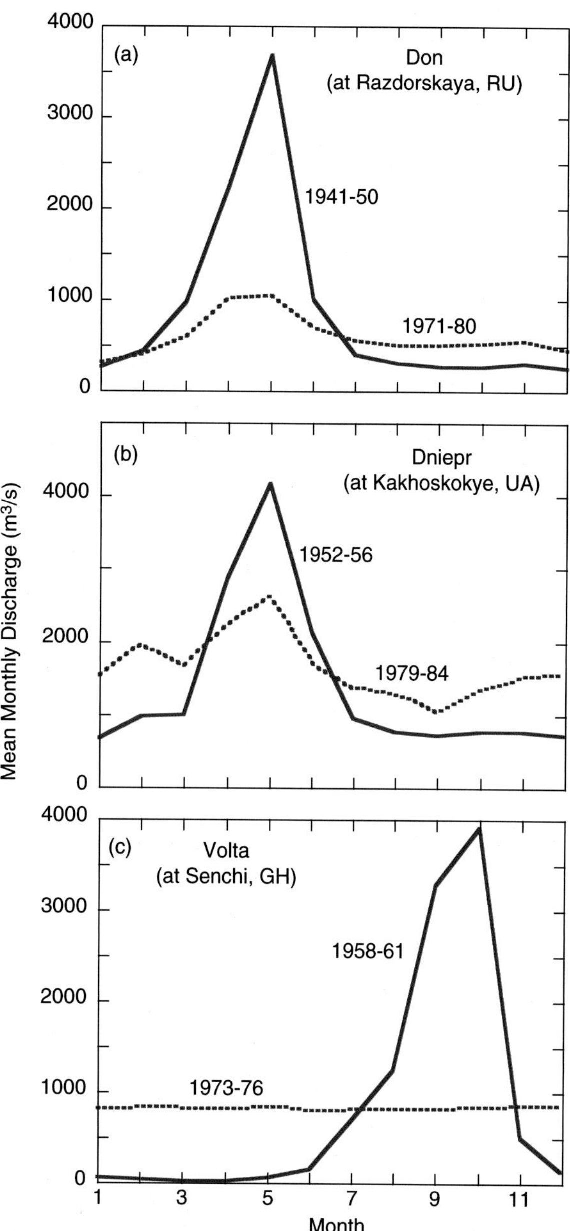

Figure 4.20. Average monthly discharges of the (a) Don, (b) Dniepr, and (c) Volta rivers before (red) and after (blue) dam construction. The Don experienced a net decrease in annual discharge, whereas the Volta and Dniepr both experienced mild declines. All three rivers, however, experienced significant changes in seasonal discharge. Data from GRDC and UNESCO.

downstream riparian zone is more likely to be disconnected from the river; sediment transport and channel maintenance, both of which rely on bank-full discharge, can be particularly affected (Wolman and Miller, 1960). Prior to construction of Ghana's Volta High Dam in the mid 1960s, the Volta River was a highly seasonal river, with more than 90% of its annual discharge during the four-month wet season. Annual discharge changed little after dam construction, but the seasonality of discharge was lost (Fig. 4.20). Although farmers probably were quite happy to have a continual supply of water, the change in seasonality presumably affected coastal productivity, estuarine flushing, and (assuming decreased over-bank escape of fluvial sediment) also increased the effects of local and regional subsidence.

Box 4.1 A tale of two rivers

In the early 1960s the Sakaraya River in northwest Turkey, which drains an area of 56 000 km^2, discharged 6–10 km^3 of water and ~10 Mt of sediment annually to the Black Sea (Fig. 4.21a). These numbers may have been somewhat lower than historic means, as the Sariyar hydroelectric dam, built in 1950–1956, almost certainly siphoned off some of the river's sediment and perhaps decreased its water discharge. Following construction of other dams in the late 1960s and early 1970s, both runoff and sediment load steadily declined. As of 1999 water discharge had fallen nearly 50% from its earlier levels, in

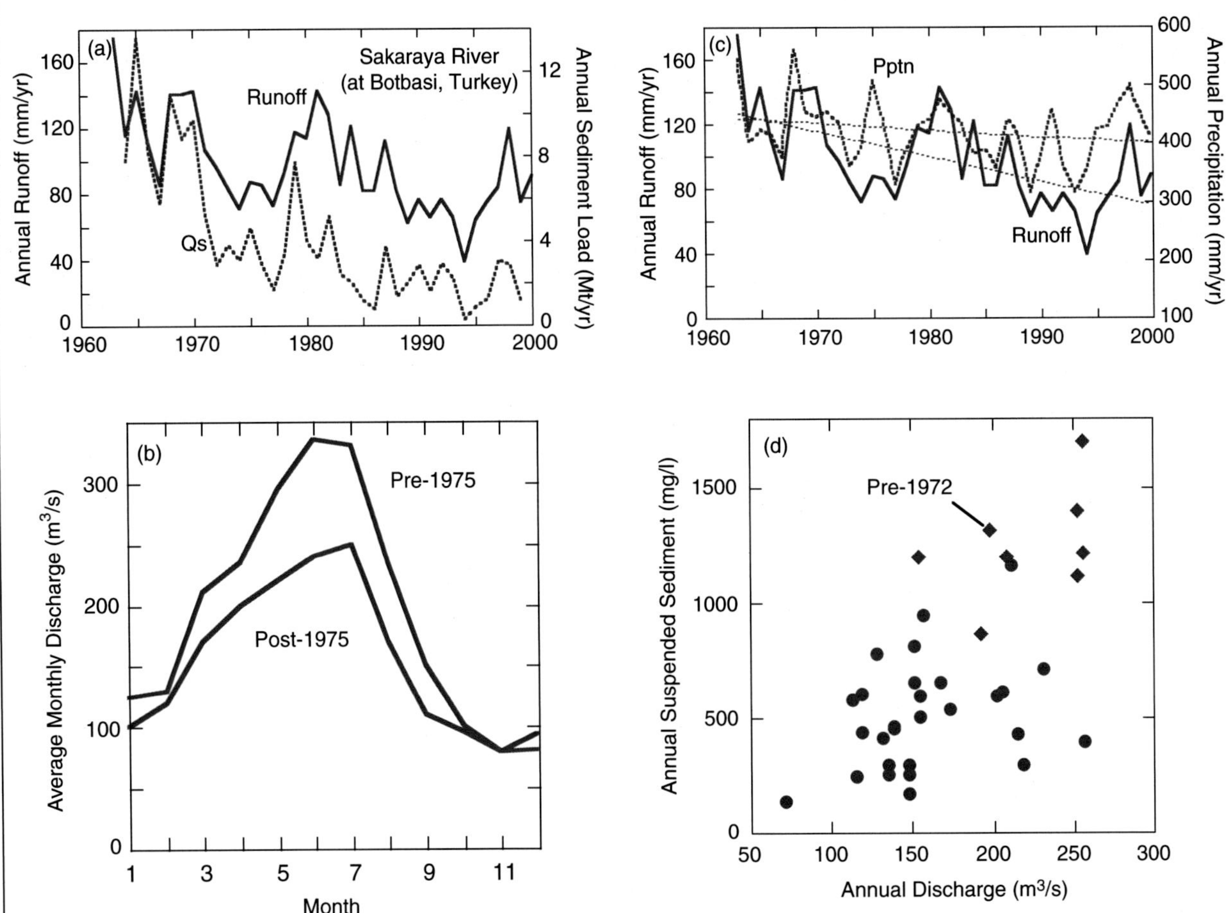

Figure 4.21. (a) Sakaraya River (gauged at Botbasi, Turkey) water runoff (blue) and sediment discharge (red), 1963–1999. (b) Average monthly Sakaraya discharge before (blue) and after 1975 (red). (c) Comparison of annual runoff (blue) and basin-wide precipitation (red). (d) Annual average suspended sediment concentration vs. discharge; note the prominent drop of both parameters after 1972 (red dots) in response to dam construction and increased irrigation. Data from Isik *et al.* (2008).

contrast to a <10% decline in regional precipitation (Fig. 4.21c). Decreased water discharge is partly due to evaporation from the dammed lakes (combined surface area >50 km^2) but also by increased use of water for irrigation (Isik *et al.*, 2008). As a result, greatest decrease in discharge has occurred during the growing season, spring to early summer (Fig. 4.21b). The dam-effect is particularly stark in terms of sediment load and average suspended sediment concentration, which decreased by ~90% and 75%, respectively (Fig. 4.21a, d). Decreased downstream sediment concentrations have led to increased scouring, which, together with increased sand withdrawal for construction, has resulted in channel widening and deepening by as much as 7 m (Isik *et al.*, 2008).

In contrast to the Sakaraya River basin, whose semi-arid climate has required intensive irrigation, precipitation in the Pearl River watershed (450 000 km^2) in southern China is sufficiently great that irrigation is less necessary. Many of the 9000 dams built along the Pearl and its tributaries over the past 55+ years were designed primarily for flood control and hydroelectric energy. Between 1955 and the mid 1960s, annual sediment discharge of Pearl River more or less followed that of water discharge, averaging ~70 Mt/yr. In the ensuing 30 years, owing primarily to increased land-use, sediment loads increased at a higher rate than river flow (Fig. 4.22b). Average sediment discharge during this 30-yr period was ~85 Mt/yr (Fig. 4.22a). After the mid 1990s, improved land conservation, construction of the Yantan Dam in 1992 (see discussion by Dai *et al.*, 2008), increased irrigation, as well as decreased precipitation, led to decreases in both water and sediment discharge. By 2004 and 2005 annual water discharge had dropped ~1/3 relative to the 1960s, and sediment load now averaged ~35 Mt/yr (Fig. 4.22). Dai *et al.* (2008) estimate that the Pearl River dams trap presently 400–600 Mt of sediment annually, more than an order-of-magnitude greater than the present-day river sediment discharge.

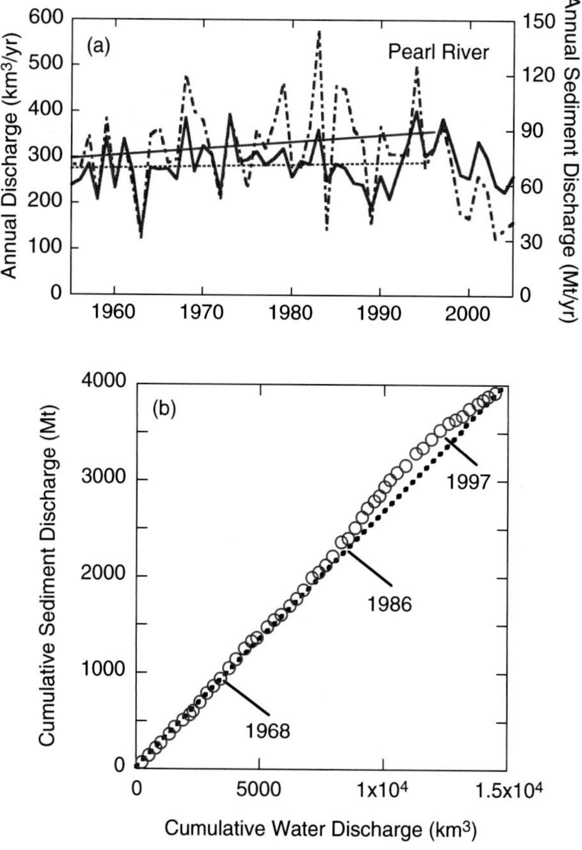

Figure 4.22. Annual (a) and cumulative (b) water (blue) and sediment (red) discharges from the Pearl River, 1955–2005. Note that annual sediment discharge closely followed water discharge until about 1965, after which, in response to increased land-use, it increased to ~85 Mt/yr (1965–1995). By 2004 and 2005, in response to increased sediment trapping behind large dams (Dai *et al.*, 2008), annual sediment load had declined to 35 Mt/yr. Water discharge after 1997 was ~33% of its levels in the early 1960s. Data from Dai *et al.* (2008).

In his detailed analysis of the global time-series database, Des Walling (2006; Walling and Fang, 2003) has observed that about half of 145 major rivers investigated have shown a significant change in sediment load over the past century. Most rivers experienced decreased sediment delivery. Examples of decreased fluvial sediment fluxes owing to dam construction (as well as irrigation and changing land-use) are so often cited and so well known that we need to list only a few representative examples. Stallard (1998) estimated that 43 000 reservoirs on the US National Inventory of Dams trap 1.2 Bt of sediment annually. When the millions of smaller impoundments are also considered, an estimate of 2.4–4 Bt/yr (Smith *et al.*, 2001; Renwick *et al.*, 2005) seems reasonable. Decreased sediment discharges from the Yellow and Nile rivers are discussed in detail in the following section, but numerous other rivers – such as the Rio Grande and Colorado River (USA/Mexico) and the Indus River (see p. 134) – show decreases that are just as dramatic. The Hoa Bin Dam on the Da River, large tributary to the Song Hong (Red River) in Vietnam, traps 83% of the sediment load.

Table 4.3. *Examples of decreased sediment loads of global rivers resulting from dam construction and irrigation. Changes in Mediterranean river sediment discharge are shown in Table 4.4. Data from appendix.*

River	Country	Basin area ($\times 10^3$ km^2)	Previous load (Mt/yr)	Present load (Mt/yr)	% Decrease
Colorado	Mexico	640	120	0.1	100
Yisil Irmak	Turkey	65	19	19	99
Kizil Irmak	Turkey	79	17	0.44	97
Krishna	India	250	64	4	95
Liaohe	China	220	39	2.7	95
Rio Grande	USA	870	20	0.6	95
Grijalva	Mexico	50	24	1.3	95
Indus	Pakistan	980	250	<20	>90
Huanghe	China	750	1000	100	90
Chao Phraya	Thailand	160	30	3	90
Volta	Ghana	400	19	1.6	90
Limpopo	Mozambique	410	33	6	80
Orange	South Africa	1000	89	17	80
Song Hong	Vietnam	490	80	25	70
Yenisei	Russia	2600	13	4.1	70
Zhujiang	China	490	80	25	70
Mississippi	USA	3300	470	145	70
Changjiang	China	1800	470	180	60
Narmada	India	99	70	30	60
Danube	Romania	820	67	42	35
Totals		**15 473**	**2974**	**607**	**80**

Partly as a result, about half of the coastal sites adjacent to the Red River's mouth are eroding at 5–10 m/yr (Thanh *et al.*, 2004). As dam construction continues throughout Africa and southern Asia – the latter area responsible for 60–70% of the global sediment flux – the amount of sediment and water trapped by dams will continue to increase.

The amount of fluvial sediment trapped behind dams and deposited in and along irrigation ditches is difficult to quantify, and even for those rivers for which we have "before" and "after" discharges, the numbers are hazy. Syvitski *et al.* (2005) calculated that retention within reservoirs has reduced the global river-sediment flux by 3.6 Bt/yr. However, the 34 rivers shown in Tables 4.3 (global) and 4.4 (Mediterranean Sea), which collectively drain 19 million km^2 of land area, show a 2.5 Bt/yr decline in sediment discharge over the past 50 years, a 75% reduction. The four major rivers in China (Huanghe, Changjiang, Zhujiang, and Liaohe) alone account for nearly 1.3 Bt/yr reduction in sediment discharge, an 80% reduction. The 78 rivers in our database for which we list post-dam values show a total reduction of sediment discharge of nearly 3.5 Bt/yr. Since these rivers account for less than half the total pre-dam sediment discharge calculated in Chapter 2, and many rivers remain undocumented in terms of pre- and/or post-dam values, we assume that global decreased sediment discharge attributable to reservoir retention may exceed 5 Bt/yr.

Box 4.2 Mediterranean Sea and its rivers

Few regions are as diverse in climate and landscape, or have been impacted by so many human activities, as the watershed surrounding the Mediterranean Sea. Climate ranges from cold moist alpine to hot arid desert, precipitation from <100 to >3500 mm/yr, and river runoff from <10 (Libya) to >1000 mm/yr (Greece, Albania) (Fig. 4.23). Records of human occupation date back into the Paleolithic, and the impact of human activities and their environmental impacts are found as early as the Early Bronze Age (see McNeill, 1992, and Wainwright and Thornes, 2004, for thorough discussions). By the Greco-Roman period, much of Mediterranean landscape had been deforested and

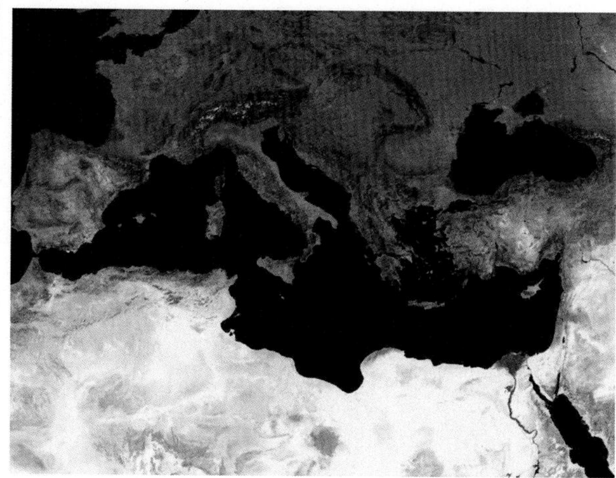

Figure 4.23. Mediterranean Basin, illustrating the stark difference between humid to moderate- to high-runoff regions in the north and arid regions in the south. Note the green, ribbon-like lower Nile River in the southeast. NASA image.

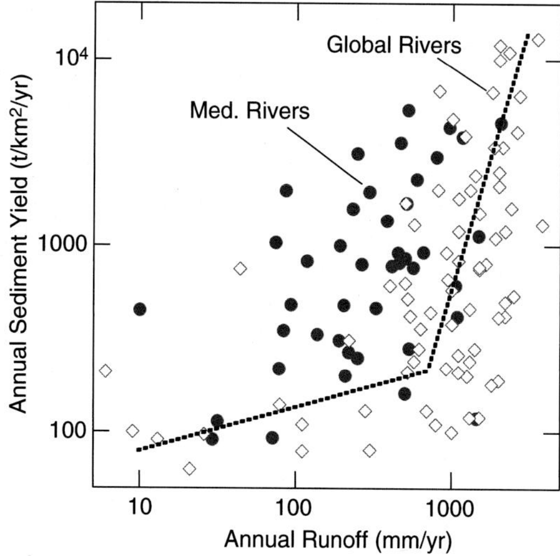

Figure 4.24. Sediment yield vs. annual runoff for small young mountainous rivers (drainage basins 1000–10 000 km²) in the Mediterranean basin (red dots) compared to sediment yields from similar-sized rivers in western North America, Japan, Taiwan, Indonesia, Philippines, and New Zealand (open diamonds).

farmed as well as (locally) mined (reviewed by Wainwright and Thornes, 2004). Anthropogenic impacts, particularly in the northern Mediterranean watersheds, appear to have lessened somewhat during the Middle Ages, only to accelerate again in the seventeenth century.

One rather obvious impact of this long history can be seen in the high sediment yields of small rivers that drain young mountains compared to small mountainous rivers from other parts of the world (Woodward, 1995; Hooke, 2006, and references therein). Mediterranean rivers with runoffs between 200 mm/yr and 300 mm/yr (e.g., Arno, Goksu, Reno, Simeto, Ombrone, and Mazufran), for example, had pre-dam sediment yields of 270, 250, 790, 1900, 3100, and 1600 t/km²/yr, respectively, whereas rivers in other parts of the world with similar runoffs (Wairua and Awatare rivers in New Zealand, Diego River in Mexico) have or had (pre-dam) yields of 310, 130, and 80 t/km²/yr (Fig. 4.24). Woodward (1995) estimated that about 75% of the sediment yields in the upper watersheds of Mediterranean rivers (which are generally the most susceptible to high rates of erosion) result from human activities, perhaps the highest of any global climatic zone (see also Dedkov and Mozzherin, 1992). But the anthropogenic influence on erosion is uneven throughout the Mediterranean. Soil loss in the Pindus (Greece), Lucanian Appenines (Italy) watersheds is much greater than in the western Taurus Mountains (Turkey). Erosion is particularly high in the Rif Mountains (Morocco), where soil loss exceeds soil formation by an order of magnitude (McNeill, 1992).

In recent years the damming of many Mediterranean rivers as well as climate change has resulted in dramatic declines in the discharge of both water and sediment. Most obvious is the Nile River following construction of the Aswan Dam in the mid 1960s (see p. 143), but the discharge from many other rivers also declined during the second half of the twentieth century. Of the 20 long-term discharge trends that we have for Mediterranean rivers, only two (Segura and Rhone) show a statistically significant increase in discharge between 1951 and 2000, whereas 14 show decreased discharges of 30% or more (Fig. 4.25a). A 2003 UNEP report estimates that, since the beginning of the twentieth century, freshwater discharge to the Mediterranean has dropped by half. Declining precipitation explains decreased discharge of some rivers (e.g. Var and Llobregat), but in many rivers the decrease seems to be related primarily to dams and irrigation (see Milliman *et al.*, 2008) (Fig. 4.25b). A similar explanation applies to several Black Sea rivers, although, again, the available database is incomplete (Milliman *et al.*, 2010).

Decreased water discharge, combined with changing land-use practices, has resulted in reduced erosion rates in many watersheds, which in turn has led to dramatic declines in sediment discharge (Liquete *et al.*, 2009). The 14 rivers (exclusive of the Nile) in Table 4.4, which drain a total area of 445 000 km², have experienced a collective 75% drop in their sediment discharge; decreases range from 13% (Ceyhan River) to >95% (Ebro and Asi rivers) (UNEP/MAP, 2003). Added to this is the 98% decrease of sediment discharge from the Nile River in response to the Aswan Dam (Table 4.4). The impacts of these decreased sediment loads include deepening river channels (aided by river-sand mining), eroding shoreline, and (in the case of the Nile, Ebro, and Rhone rivers) eroding deltas.

Table 4.4. *Pre- and post-dam changes in annual water discharge (Q) and sediment discharge (Q_s) in Mediterranean rivers. Data from appendix.*

River	Basin area (km^2)	Country	Pre-dam Q (km^3/yr)	Post-dam Q (km^3/yr)	Pre-dam Q_s (Mt/yr)	Post-dam Q_s (Mt/yr)	% Q_s loss
Ebro	87 000	Spain	50	17	18	0.15	99
Rhone	96 000	France	54		59	6.2	89
Ombrone	3200	Italy	0.8		10	1.9	81
Tiber	17 000	Italy	7.4		1.3	0.3	77
Tronto	1200	Italy	0.3		1.2	0.6	50
Pescara	3300	Italy	1.7	0.9	2.2	1.2	45
Po	74 000	Italy	46		15	10	33
Semani	5300	Albania	5.6		30	16	47
Drini	13 000	Albania	12		16	2.1	87
Vijose	6700	Albania	6.4		29	8.3	71
Asi	23 000	Turkey	2.7		19	0.36	98
Ceyhan	21 000	Turkey	7		5.5	4.8	13
Cheliff	44 000	Algeria	1.3		8	4	50
Moulouya	51 000	Morocco	1.3	0.2	13	1	92
Totals (w/o Nile)	**262 700**				**227**	**57**	**75**
Nile	2 900 000	Egypt	80	<<30	120	2	98
Totals (w/ Nile)	**3 162 700**				**347**	**59**	**83**

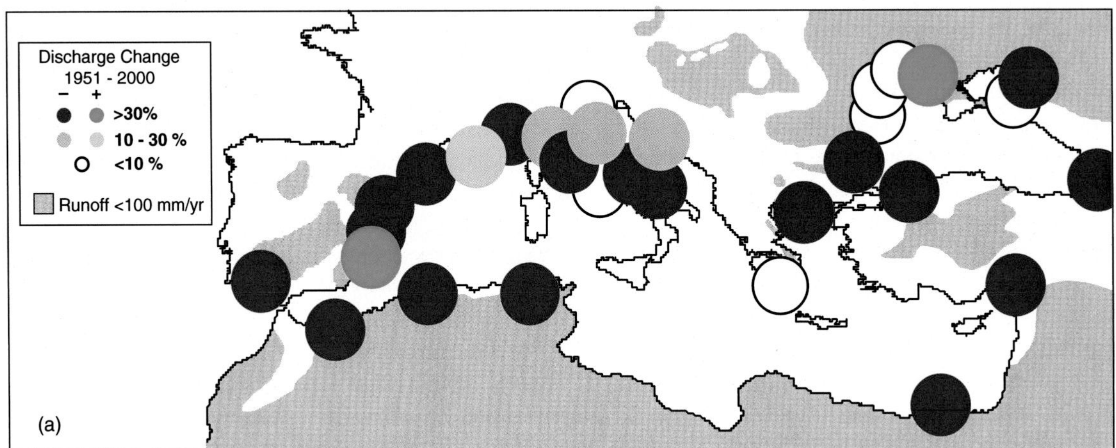

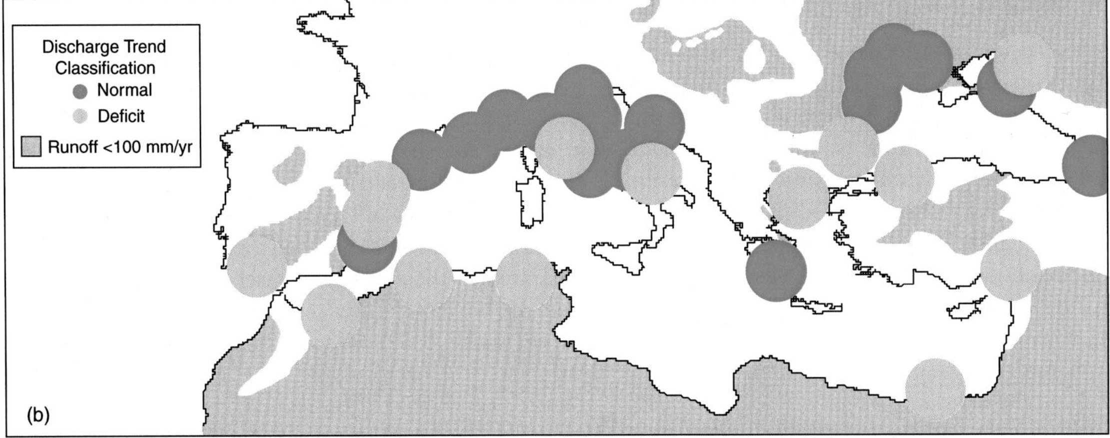

Figure 4.25. Change in water discharge (a) and distribution of normal and deficit rivers (b, as defined by Milliman *et al.*, 2008) based on a comparison of 50-yr discharge and precipitation trends for rivers draining into the Mediterranean and Black seas, 1951–2000. For those rivers not represented by a complete set of discharge data, the trend was extrapolated for the full 50 years (see Milliman *et al.*, 2008).

There is also a public health aspect to dams, reservoirs, and irrigation. Year-round water – both standing and flowing – favors the breeding of disease-spreading vectors. Prime among these are malaria and rift valley fever, which are spread by mosquitoes, and schistosome-bearing mollusks responsible for *Bilharzia*. Blackflies, who spread river blindness (*Onchocerciasis*), breed in rapidly flowing waters, such as spillways associated with dams. The causes, distribution, and potential solutions to these debilitating and often fatal diseases, which are particularly rampant in Africa and southern Asia, have been discussed thoroughly by Jobin (1999).

Finally, although there has been considerable discussion about this subject, it seems possible that sudden changes in reservoir water levels can produce sufficient stress on neighboring faults to initiate an earthquake (Talwani, 1997). Chen (2009) has suggested and Ge *et al.* (2009) have calculated that static stress and increased pore pressure in the Zipingpu Reservoir on the Minjiang (Min River) in Sichuan Province may have initiated the 2008 Wenchuan earthquake, in which almost 90 000 people were killed.

The reader should not assume that we are necessarily anti-dam zealots. Far from it: dams, when intelligently conceived, designed and constructed, provide a multitude of useful and ecologically practical functions to the societies that they serve. On the other hand, many dams have had at least short-term negative impacts on their river systems, and many much longer negative impacts. With care and planning many of these impacts can be minimized, but this requires that potential downsides are not glossed over in cost-benefit or impact analyses. Case studies discussed in the following section help display both the benefits and dangers of river control and utilization.

Case histories of four river systems

In the following pages we discuss four river systems in which human and natural changes have had large impacts. The rivers we discuss – Nile, Yellow and Yangtze, and Mississippi – are all large, but the human impacts on these rivers vary considerably, as do the long-term prognoses for these rivers and their drainage basins.

Nile River

Natural setting

The Nile River, with its source waters in Lake Victoria (White Nile) and the mountains of Ethiopia (Blue Nile), serves as the prime source of water for many East African countries, even though its natural annual runoff over the 2 900 000 km^2 watershed averages only ~30 mm/yr. The Blue Nile, which joins the White Nile at Khartoum, Sudan, has a mean annual runoff of about 150 mm/yr, even though the lower half of its watershed receives little precipitation. The Atbara River enters the Nile about 300 km north of Khartoum; it contributes about 13% of the Nile's water, but accounts for 25% of the sediment discharge (Foucart and Stanley, 1989; Woodward *et al.*, 2007).

Discharge from the White Nile tends to be more or less uniform throughout the year, ranging from 550–600 m^3/s (April–May) to 1150–1250 m^3/s (Sept–Dec) (Fig. 4.25). In contrast, Blue Nile and Atbara discharges are highly seasonal, with average February–April discharges less than 100 m^3/s and August–September discharges exceeding 5000 m^3/s (Fig. 4.26). Between 1917 and 1922 and again beginning in 1963, the White Nile experienced relatively high discharge (Fig. 4.27), reflecting the influence of the AMO (see Fig. 3.17). In contrast, Blue Nile annual discharge is highly variable, reflecting the influence of ENSO on basin precipitation (see Fig. 3.21).

Even a casual glance at a satellite image of Egypt (Fig. 4.28) underlines that the fact that the Nile provides the only green in an otherwise brown, arid landscape, extending northward to the verdant Nile Delta. Herodotus's observation that Egypt was "the gift of the Nile," still rings true. During the late nineteenth century, the Nile annually discharged 100–120 km^3 of water past Aswan but, because of changing precipitation patterns throughout eastern Africa, perhaps augmented by construction of the Aswan Low Dam, freshwater discharge decreased to about 85 km^3/yr after 1900 (Fig. 4.29). The delta and coastal lands adjacent to the delta prograded slowly but continually into the eastern Mediterranean Sea (Fig 4.30a). During the high-discharge years in the nineteenth century, the sediment load of the Nile may have averaged ~200 Mt/yr (El-Sayed, 1993). How much sediment actually reached the coastal zone prior to dam and barrage construction is a matter of conjecture, presumably much of it being deposited as overbank deposits during annual floods, but during this period the Rosetta Promontory prograded ~30–40 m/yr (Sestini, 1991; Nixon, 2003 and references therein).

Aswan High Dam and its impacts

In order to increase the amount of arable land in Egypt and to decrease the negative impacts of floods, a Low Dam was constructed at Aswan, in southern Egypt, in the early part of the twentieth century, about the same time that precipitation throughout northeastern Africa was declining. By 1964 the river's measured sediment load at Aswan had decreased from 200 Mt/yr to 120–160 Mt/yr (El-Sayed, 1993; Fanos, 1996 and references therein). Increased barrage construction along the lower course of the river also sequestered increasing amounts of fluvial sediment, so that by the early part of the last century, the

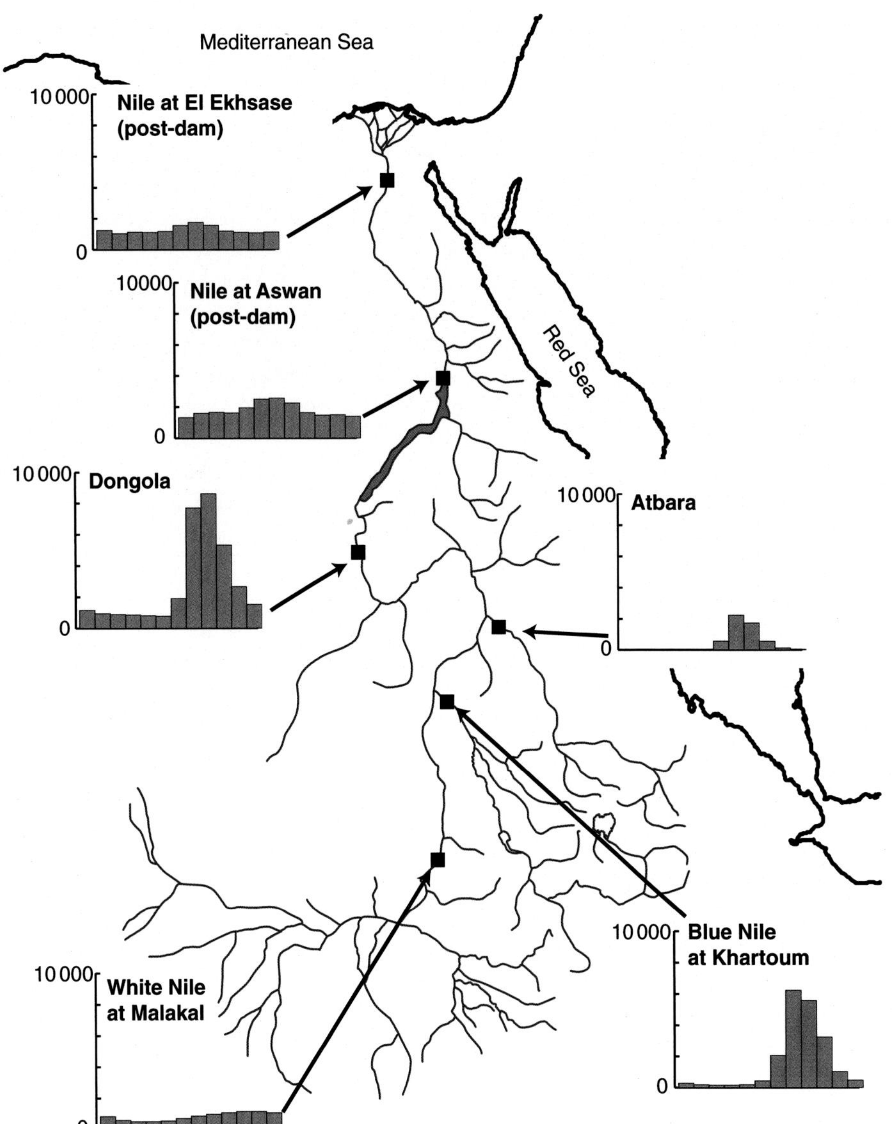

Figure 4.26. Map of the middle and lower Nile, showing monthly runoff (in m^3/s) for the White and Blue Nile as well as at several downstream stations. Note the greater (although more seasonal) discharge from the Blue Nile, also the pronounced change in monthly discharge downstream of the Aswan dam and the decreased discharge between Aswan and the edge of the delta. Discharge data from Global River Discharge Database (www.rivdis.sr.unh.edu) and Sutcliffe and Parks (1999).

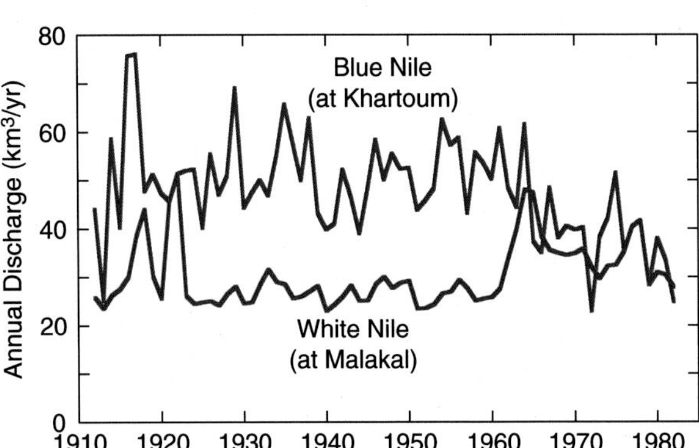

Figure 4.27. Annual discharges from the White Nile (red; at Malakal, Sudan) and Blue Nile (blue; at Khartoum, Sudan), 1912–1982. Data from Global River Discharge Database (www.rivdis.sr.unh.edu).

Rosetta and Damieta promontories had begun to erode. By 1964, the northern limb on the Rosetta Promontory had retreated nearly 1 km (Fig. 4.30b), with an average erosion of 14 m/yr and 20 m/yr on the western and eastern sides of the promontory, respectively (Torab and Azab, 2007).

The amount of water stored behind the Low Dam, however, never exceeded about 5 km^3, only about 6% of the total annual Nile discharge. This lack of water storage, combined with an ever increasing need for more arable land to feed Egypt's rapidly expanding population and the increased need for hydro-electric power, and fueled by nationalistic exhortations, provided more than enough rationale to initiate construction of the Aswan High Dam in the early 1960s. Behind the 111-m high and 3830-m wide dam, the 500-km-long lakes Nubia (in Sudan) and Nasser (in Egypt) have a

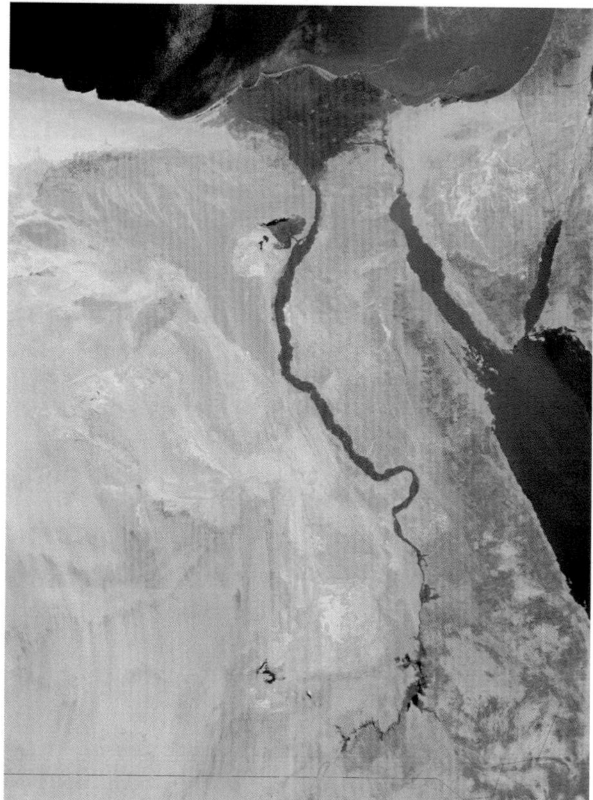

Figure 4.28. NASA Satellite image of the lower Nile, from northern Sudan to the Mediterranean Sea.

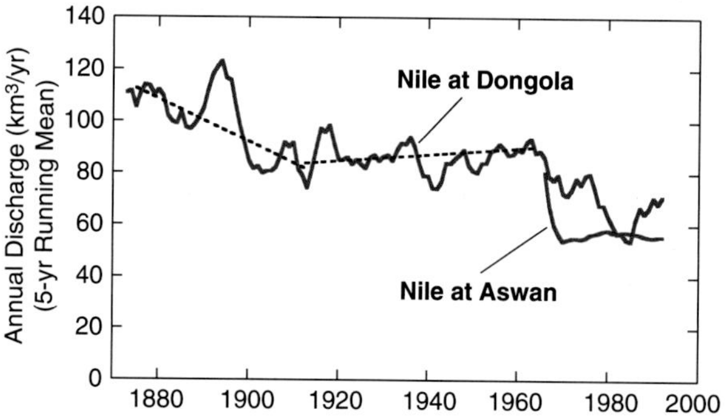

Figure 4.29. Five-year running-mean of Nile River discharges at Dongola (blue; Sudan) and Aswan (red; Egypt), 1873–1994. Note the dramatic decline in Aswan discharge following closing of the High Dam in 1964.

total storage capacity of 162 km^3, nearly two years worth of pre-High-Dam Nile discharge.

After completion of the High Dam in 1968, discharge below Aswan was set at 55.5 km^3/yr based on the 1900–1960 average discharge of 74 km^3/yr (Fig. 4.29), Egypt's agreed share of the Nile flow, the Sudan claiming the other 18.5 km^3/yr. About 8–10 km^3 of water are lost annually through evaporation (data from El-Sayed, 1993; Sutcliffe and Parks, 1999), and there is additional loss via groundwater seepage and flow diversion for irrigation. Despite these losses, controlled release from Lake Nasser has meant that flow at the Aswan gauging during the first half of the calendar year, when Nile discharge was normally low, as much as doubled; in contrast, post-High-Dam summer discharge fell by more than half (Fig. 4.31). Over the past 40 years, much of the Nile flow downstream of Aswan has been utilized for irrigation, and little water, suspended sediments or dissolved solids presently reach the Mediterranean coast.

The positive effects of the Aswan High Dam on Egypt cannot be minimized. Parceling out the river's discharge evenly throughout the year (Fig. 4.31), combined with increased irrigation, has resulted in a doubling of Egypt's arable land, although this figure is somewhat misleading since much of the existing agricultural land was compromised by waterlogging and salinization, as well as urbanization and the "mining" of land by the brick industry (El-Sayed, 1993). More importantly and in itself perhaps the best supportive argument for the High Dam, Egypt may not have survived the 1980s drought as successfully as it did without access to the water stored behind the High Dam. During the peak drought years, the volume of Lake Nasser dropped by nearly 80 km^3, and by 1986 the reservoir contained less than one year's worth of water (Fig. 4.32). Without this "surplus" water from Lake Nasser, the drought might have resulted in a 25% loss in Egyptian agriculture, which could have been even greater if the Sudan had used its full allocation of water (El-Sayed, 1993). As the level of Lake Nasser fell (Fig. 4.32), of course, hydroelectric power generated by the High Dam decreased accordingly, and in 1987 the High Dam accounted for only 18% of Egypt's total electric power consumption (El-Sayed, 1993).

Unfortunately, the negative impacts from the Aswan High Dam have been so widely – and often wildly – reported that the High Dam is often pictured as the poster-child of why not or how not to build a large dam. The literature concerning the High Dam is so considerable (Wahby and Bishara, 1981; Sestini, 1991; El-Sayed, 1993; Howell and Allan, 1994; Sutcliffe and Parks, 1999; Collins, 2002) that by 1997 D. J. Stanley and his colleagues had compiled a 65-page bibliography of relevant references, most written after 1970.

The most immediate impact of the High Dam was the forced migration of some 150 000 Nubians and the subsequent drowning of many valuable archaeological sites. Ironically this led to an increased awareness of the fragility of some of the more famous ruins, resulting in their removal to higher grounds. While downstream flow of water has become more evenly distributed throughout the entire year, there has been a precipitous decline in the amount of water reaching the delta and coastal

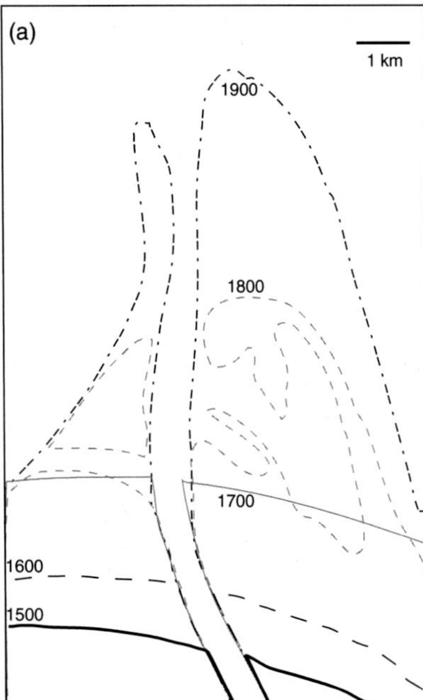

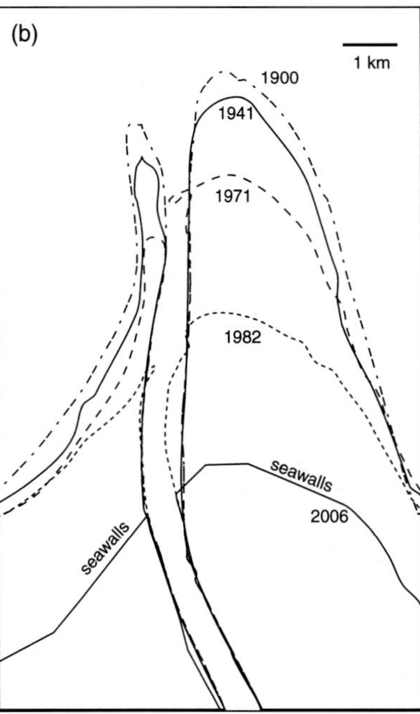

Figure 4.30. (a) Progradation (1500–1900) and (b) retreat (1900–2006) of the Rosetta Promontory. Between 1800 and 1900, the shoreline prograded about 4 km, after which it initially retreated at about 30–40 m/yr. Following construction of the Aswan dam (1964), erosion accelerated to ~100 m/yr. Shoreline retreat has since been (temporarily) controlled by extensive seawall construction. After Sestini (1991) and Fanos (1996).

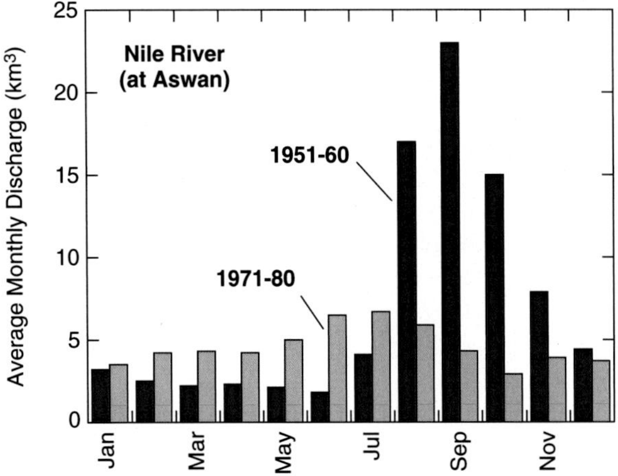

Figure 4.31. Mean monthly Nile discharge at the Aswan gauging station before (blue; 1951–1960) and after (orange; 1971–1980) construction of the Aswan High Dam.

waters. According to Stanley (1996), only about 1/2 of the Nile River water released downstream of the High Dam reaches Cairo, of which 1/3 is lost by infiltration and evaporation, and 1/3 moves through a 10 000-km network of canals and drains within the delta, meaning that only ~15% of the Aswan flow reaches Damietta Branch of the river (Wahby and Bishara, 1981; El-Sayed, 1993; and references therein). A quasi-synoptic monitoring of the river downstream of Aswan (Awadallah and Soltan, 1995) confirms these numbers (Fig. 4.33). Even though considerable sediment accumulates in deltaic canals and drains, increased discharge of pollutants (Stanley, 1996) has led to a downstream increase in suspended-sediment and dissolved-solid concentrations (Fig. 4.33). Moreover, the abatement of annual floods has meant that salts cannot be washed away from the arable soils, thus increasing soil salinity throughout much of the lower Nile and delta. Only a few years after closure of the High Dam, for instance, chloride$^-$ concentrations in Nile waters increased more than three-fold (Kempe, 1988).

Decreased water discharge had particularly severe impact on biological productivity of the continental shelf waters off the Nile, the so-called "Nile Bloom." Between 1962 and 1983 total Egyptian fishery catch decreased by about 66%, shrimp and prawn landings dropping nearly 85% (Dowidar, 1988). Over the ensuing years, however, total fishery catches gradually increased and now are actually higher than pre-dam levels, although shrimp and prawn landings have still not recovered (Nixon, 2003, his Fig. 2). The increased productivity presumably has resulted from the increased application of artificial fertilizers to agricultural fields (prior to 1964 most fertilizers were natural, coming mostly via annual flooding), but also discharge of untreated human sewage directly into the river from a rapidly expanding Cairo megalopolis (Nixon, 2003).

The continuous presence of fresh water in the irrigation canals also greatly increased the proliferation of snails that carry the schistosome parasite associated with *Bilharzia* (Schistosomiasis). Although *Bilharzia* had been a persistent disease in the Nile valley, historically the schistosome parasite snails transmitted a relatively mild urinary disease. But intensified irrigation has resulted in the so-called Nile

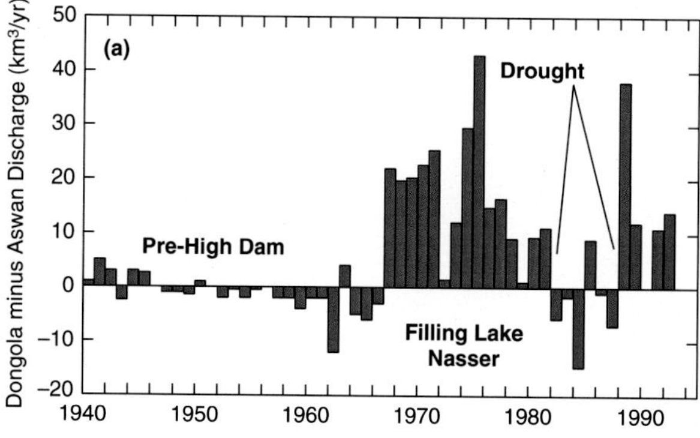

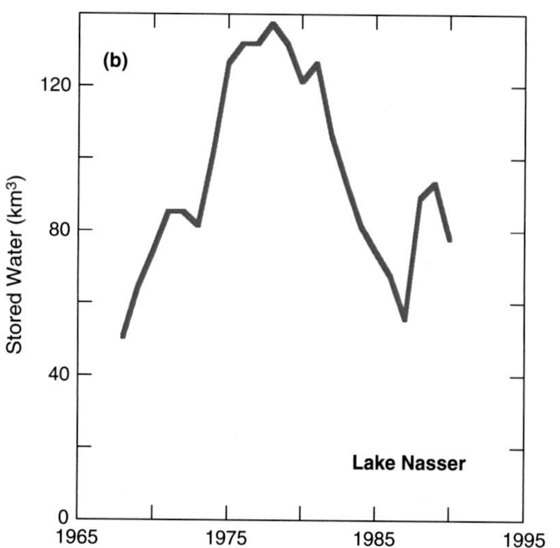

Figure 4.32. (a) Difference in discharge between Dongola, Sudan, and Aswan, Egypt, upstream and downstream of the High Dam. Water "lost" between Dongola and Aswan in the 1970s and early 1980s was primarily related to the filling of Lake Nasser. (b) Note the decrease in lake volume during the 1980s in response to a prolonged drought, as also evidenced by the five-year span when discharge at Aswan was generally greater than it was upstream.

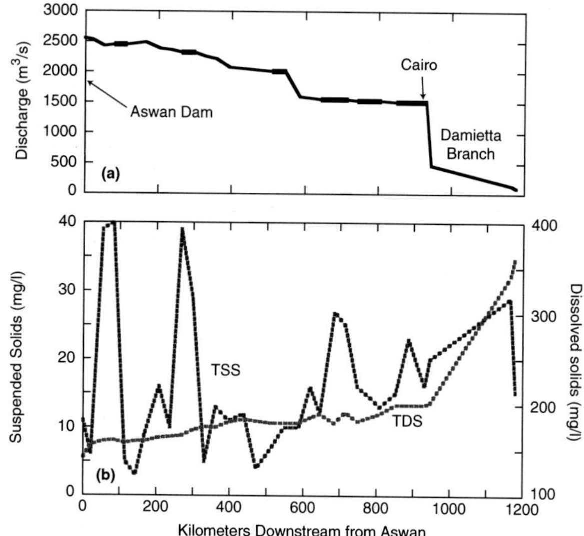

Figure 4.33. Spatial change in (a) water discharge, and (b) suspended-solid (blue) and dissolved-solid (red) concentrations from Aswan into the Damietta Branch of the Nile River. Data from Awadallah and Soltan (1995).

shift (Jobin, 1999), in which the populations of a more virulent form of intestinal parasite literally exploded. According to Jobin (1999, his Table 13.2), more than 1/3 of the human population in the Nile delta has *Bilharzia*, including about 1/2 of the farmers.

Finally, we come to the problem of decreased sediment loads and their role in shoreline erosion, a much-discussed topic over the past 30 years. Much (if not most) of the Nile's sediment load, presently estimated to be about 120 Mt/yr, is trapped in the upper reaches of Lake Nasser/Nubia; by 1990 a 240 km^2 delta had prograded into the lake (El-Sayed, 1993). One result of this upstream loss of suspended sediment has been increased downstream erosion, locally resulting in the undercutting of bridges and other river structures. Moreover, decreased turbidity of river waters has affected plankton productivity and thereby water quality (White, 1988). Extensive erosion has occurred off the Rosetta and Damietta promontories (Smith and Abdel-Kader, 1988; Sestini, 1991; Fanos, 1996), the Rosetta promontory retreating by more than 3 km since 1970 (Fig. 4.30). In contrast to the 14–20 m/yr of regression experienced by the Rosetta Promontory between 1900 and 1964, post-High-Dam regression (1964–2006) has averaged about 100 m/yr. Similarly, erosion of the Damietta Promontory increased from 40 to 100 m/yr following Aswan High Dam construction (Torab and Azab, 2007). Considering the overall stability of the delta, it is still not clear if erosion yet exceeds accretion (Fanos, 1996, his Figs. 14, 15). In time, however, it almost surely will.

Bottom line

Other problems regarding the Nile River still await Egypt. For example, there is considerable popular support within Egypt to utilize all of the Nile's water so that little or none ultimately escapes to the Mediterranean Sea, mimicking Stalin's dictum cited at the beginning of this chapter. For instance, there are currently plans for the diversion of 10% of the Nile to Toshka, an uninhabited desert region southwest of Aswan (Bohannon, 2010).

No matter how elevated the nutrient levels are within the Nile River, they can provide little boost to offshore productivity if no river water reaches the Mediterranean. Coastal erosion is almost certain to increase, although construction

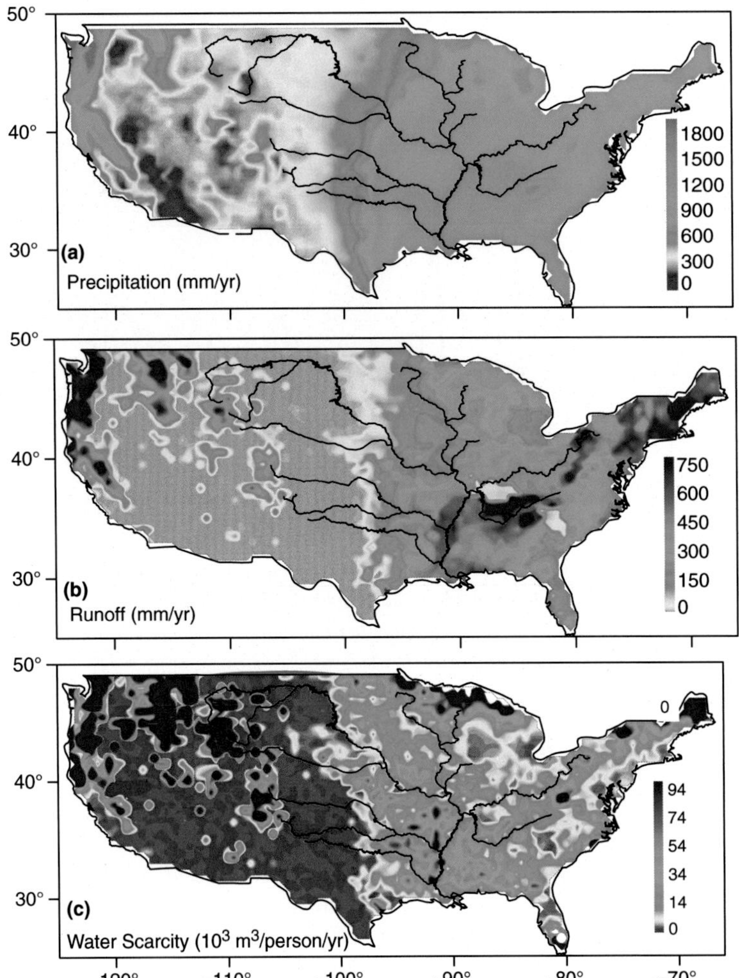

Figure 4.34. Precipitation (a), runoff (b), and water availability (defined as renewable water supply divided by population) (c) in the conterminous USA. Figure courtesy of K. H. Xu.

of shoreline protection structures has reduced, at least temporarily, promontory erosion (Frihy *et al.*, 2003; Torab and Azab, 2007). Moreover, if the removal of underground water accelerates the rate of natural subsidence (about 3–5 mm/yr; Stanley and Warne, 1993), coastal retreat could increase substantially in the next 50 years, particularly if one takes into account any future sea-level rise due to global warming (Milliman *et al.*, 1989, and references therein).

Perhaps even more ominous, and maybe with even a shorter fuse, is the ever-increasing population in Egypt as well as the prospect of increased water utilization by nations upstream of Egypt. Based on a 100-yr agreement between the Sudan and Egypt that was revised in 1959, the two countries claimed rights to the river's water based on a "no-harm" rule (Dellapenna, 1996). As the populations of these two countries increase, however, partitioning of Nile water almost certainly will become more contentious. A September 25, 2010 headline in the *New York Times* encapsulates the problem: "Egypt and Thirsty Neigbors at odds over the Nile: Upstream countries want more water." This situation is bound to be exacerbated if (more probably *when*) Ethiopia ... begins damming its Blue Nile, which presently is mostly undammed (Hillel, 1994; Howell and Allan, 1994). A May 16, 2010, story from Associated Press tells the story of four Nile countries (Tanzania, Rwanda, Uganda as well as Ethiopia) signing an agreement over upriver water use. Other Nile countries, such as Kenya and the Democratic Republic of the Congo, said they would support the agreement. Neither Egypt nor Sudan sent representatives to the meeting.

The Mississippi River

A diverse drainage basin

The Mississippi River, with a watershed of 3.3×10^6 km^2, drains about 42% of the contiguous United States, its headwaters originating in such distant states as Virginia, New York, Minnesota, Montana and Oklahoma, as well as southern Alberta and Saskatchewan (Figs. 4.34, and 4.35). Given its wide geographic range, the diversity of its tributary rivers is not surprising. Flowing from the north and east, precipitation in the generally low-lying Upper Mississippi, Ohio,

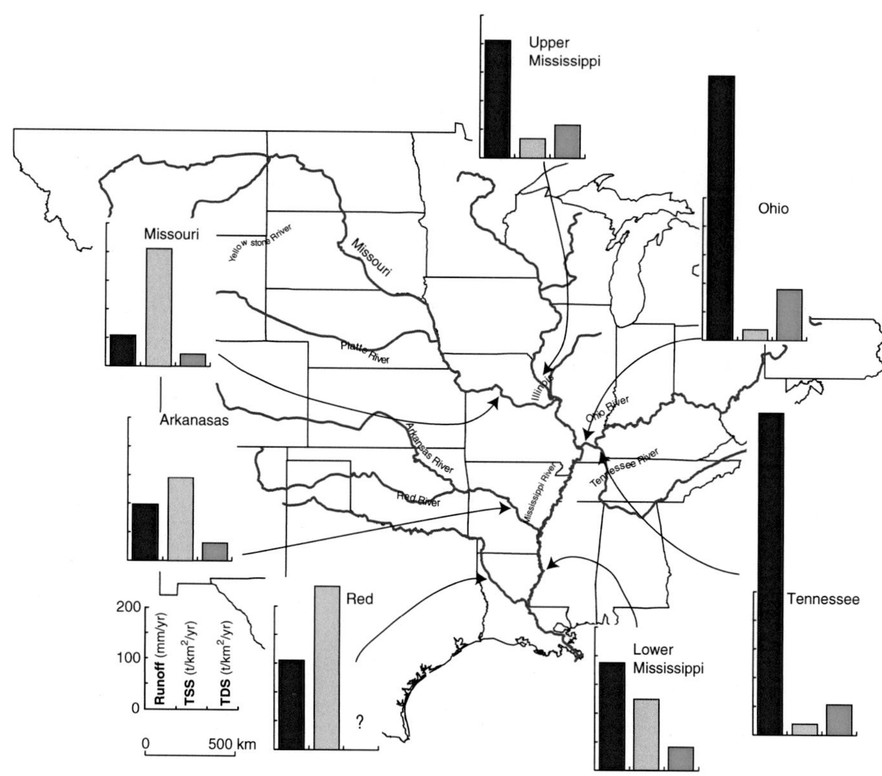

Figure 4.35. Mean annual runoffs and sediment and dissolved-solid yields in the mainstem and tributaries of Mississippi River watershed. Note the markedly different fluvial regimes in the east (high runoff, low sediment yields) and the west (low runoff, high sediment yields). For example, basin-wide runoff of the Ohio River is an order of magnitude greater than that for the Missouri River, whereas basin-wide sediment yield for the Ohio River is nearly an order-of- magnitude less than that for the Missouri. Data from USGS/NWIS.

and Tennessee watersheds generally exceeds 1000–1200 mm/yr (Fig. 4.34); before European settlers arrived, most of the eastern watersheds were dominated by forest (Fig. 9.5 of Knox, 2007). Despite their high runoffs (450–500 mm/y), the Ohio and Tennessee have annual sediment yields <25 t/km^2/yr (Fig. 4.34). The high runoff combined with an industrialized east and agriculturally based Midwest, however, explain the relatively high dissolved SiO_2 and total dissolved yields for the Upper Mississippi (1.5, 50), Ohio (1.9, 90), and Tennessee (2.0, 53 t/km^2/yr).

In contrast to the eastern watersheds, the eastward-flowing Missouri, Arkansass and Red rivers drain arid to semi-arid climates. Precipitation in the headwaters of many tributaries is less than 300 mm/yr, runoffs ranging from 50 mm/yr to 150 mm/yr (Figs. 4.34, and 4.35). Unlike the Mississippi's eastern tributaries, whose interannual flow is dictated largely by the North Atlantic Oscillation, the western rivers are influenced primarily by the Pacific Decadal Oscillation and ENSO (Fig. 3.21). Because western rivers drain younger and more erodable rocks, basin-wide sediment yields approach or exceed 200 t/km^2/yr (Fig. 4.35). The low runoff from western rivers, as well as the general lack of mining and industrial pollution, however, translates to lower levels of chemical weathering (dissolved SiO_2 yields in the Missouri and Arkansas watersheds being 0.5 and 0.4 t/km^2/yr, respectively) than seen in the eastern watersheds, even though dissolved concentrations can be significantly higher (e.g. 132 mg/l SO_4 in the Missouri; Table 4.5).

The lower Mississippi near Baton Rouge, LA, reflects the combination of the two diverse sources, resulting in moderately high runoff and sediment and dissolved yields (Fig. 4.35; Table 4.5). Because its discharge usually occurs in late March to early April, the Ohio/Tennessee dominates early seasonal flow of the Mississippi. The more northerly Upper Mississippi provides much of the flow in May and June. On average, the Missouri and Arkansas contribute relatively little water to the lower Mississippi (see Fig. 3.42a). On any particular year, however, the picture can change dramatically. Record floods on the Upper Mississippi and lower Missouri, for instance, contributed significantly to the Lower Mississippi's discharge throughout the summer of 1993, the Ohio and Tennessee playing relatively minor roles (see Fig. 3.42b; Parrett *et al.*, 1993)

When Europeans began settling the Mississippi Valley in the early nineteenth century, the Mississippi River was still adjusting to sediment-laden runoff events generated after the last glaciation. Much of the pre-European-settlement landscape of the main branch of the Mississippi and its western branches was dominated by grasslands (Knox, 2007). The result was a broadly meandering river with a wide variety of hydrologic and geomorphic regimes, reflecting the redistribution of post-glacial sediments (Fisk, 1947; Knox, 2007). The river was also responding to the

Table 4.5. *Dissolved constituents in the Mississippi River and its major tributaries. Data from USGS, NASQAN.*

Dissolved constituent	Upper Mississippi	Ohio	Tennessee	Missouri	Arkansas	Lower Mississippi
Ca (mg/l)	49	35	19	57	36	35
Mg (mg/l)	20	10	3.6	18	10	11
Na (mg/l)	16	17	1.4	45	45	18
K (mg/l)	3	2.5	1.6	5.7	3.5	3
Cl (mg/l)	26	21	7	18	63	20
So_4 (mg/l)	35	59	11	132	55	45
SiO_2 (mg/l)	8	4.6	3.5	10	4.2	6.4
SiO_2 yield (t/km^2/yr)	1.5	1.9	2	0.5	0.4	1.2
Total TDS (mg/l)	**290**	**210**	**94**	**400**	**320**	**236**
Total TDS yield (t/km^2/yr)	**50**	**90**	**53**	**21**	**31**	**44**

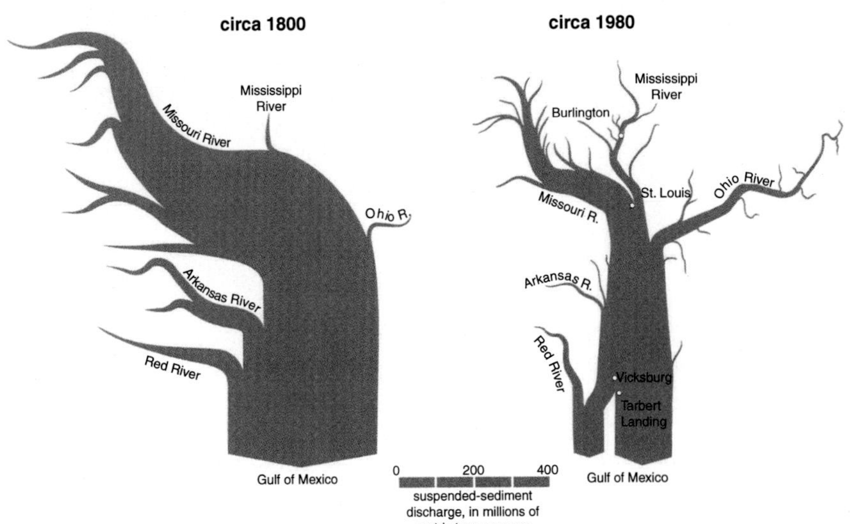

Figure 4.36. Estimated sediment loads of the Mississippi River, 1800 (left) and 1980 (right). Note the increased load of the Upper Mississippi and Ohio due to mining and agriculture, and decreased loads of the Missouri and Arkansas due primarily to damming. From Meade and Moody (2008).

New Madrid Earthquake (1811–1812), the impact of which is perhaps still being felt (Winkley, in Wolman *et al.*, 1990; Schumm *et al.*, 2002). Finally, sediment excavation and redistribution of the prairie by buffalo and prairie dogs (see Chapter 3) as well as the impacts of fires resulted in relatively high rates of soil erosion throughout the western watersheds. R. H. Meade (Meade and Moody, 2008, 2010, and references therein) has estimated that the Mississippi's sediment discharge in the early ninenteenth century may have been near the 400 Mt/yr measured in the mid twentieth century (Fig. 4.36).

Human impact: erosion and sediment delivery

Compared with the Nile, Yangtze, and Yellow rivers, large-scale human activities have occurred relatively recently throughout the Mississippi watershed. But over the past 200 years human activities have intensified to the extent that the northern and eastern parts of the watershed can be considered as strongly impacted by humans, particularly through agriculture and mining. The extent to which mining and industry have affected the Ohio River's high dissolved-solid yield is difficult to quantify, but judging from Fig. 4.35 and Table 4.5, they have not been inconsiderable.

As the Mississippi basin's rich valleys and flood plains began to be farmed in the early nineteenth century, and as its waterways were plied by larger riverboats, there was an ever-increasing need for channel dredging, construction of major cutoffs (16 between 1929 and 1942 alone; Winkley, 1994), new levees, dikes, and revetments. As in other river systems, human activity has facilitated erosion but also has decreased sediment delivery to the lower reaches of the Mississippi. Revetment emplacement, for example, has reduced bank erosion and caving upstream from Red River Landing by more than 90% (Kesel, 2003). Where distributary channels have been altered or closed, the ability of the Mississippi's channels and flood plains to mitigate river flooding has been diminished. As channels have shoaled and flood stages have grown higher, levees have

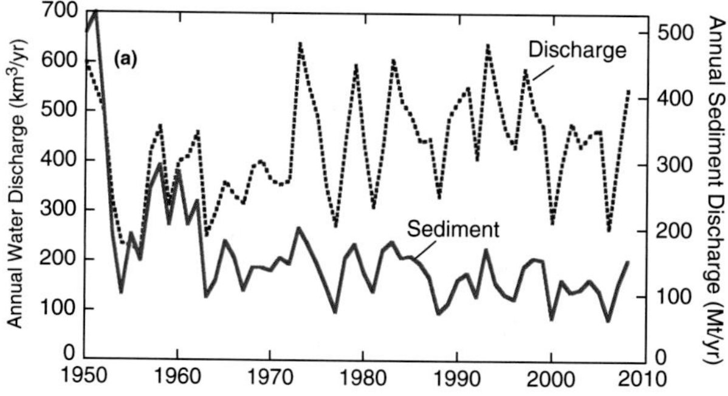

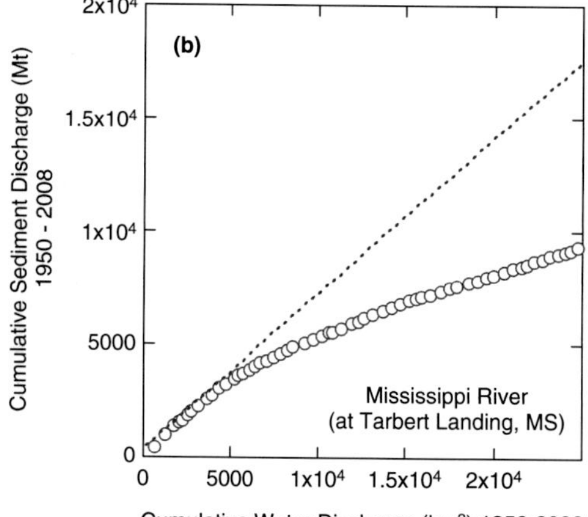

Figure 4.37. (a) Annual water (blue) and sediment (red) discharges of the Mississippi River (at Tarbert Landing, MS), 1950–2008. Note that water discharge increased during the second half of the twentieth century, whereas annual sediment discharge fell abruptly after dam closures on the Missouri River in 1953, and has continued to decline. This change also can be seen in a cumulative plot (b), which indicates a decline in sediment relative to water discharge. The gradualness of the trend reflects the slow reaction of the Lower Mississippi River to the abrupt closures of the dams, 2700 km upriver. Data from the US Army Corps of Engineers (http://www.mvn.usace.army.mil/cgi-bin/wcdata1.pl).

been necessarily heightened, but channelized flow means that floods have more regularly topped the river's levee system. As a result, during the twentieth century the Lower Mississippi slowly changed from an actively meandering river to one in which the flood plain played an increasingly diminished role as either a sediment sink or a sediment source (Kesel, 2003; Meade and Moody, 2010).

Another major impact on the Mississippi River has been the construction of dams along its tributaries. Rivers in South Dakota, Nebraska, Kansas, Oklahoma, Arkansas, Missouri, and Iowa, all of which ultimately drain into the Mississippi or its tributaries, have a total of 23 000 dams, 1220 of which are higher than 15 m. If one includes those parts of other states that lie within the Mississippi drainage basin, the total number of dams undoubtedly surpasses 40 000.

At about the same time as dam construction in the Mississippi basin was accelerating in the 1940s, land conservation measures began reducing land erosion throughout much of the watershed. Although decreased upland erosion does not necessarily transfer to less sediment delivery because of continued erosion of stored sediment (see Trimble, 1977, 1983), decreased erosion combined with the construction of large dams and levees has greatly affected the Mississippi's sediment discharge. Particularly critical were the completion of the Fort Randall and Gavins Point dams on the South Dakota portion of the Missouri River in 1953 and 1955 (Meade and Parker, 1985; Meade and Moody, 2010). By the late 1960s sediment loads of the Missouri and Arkansas rivers had fallen by more than 75% (Keown *et al.*, 1986; Wells, 1996, his Fig. 15), and the annual load in the lower Mississippi (combined with the Atchafalaya River) was less than 1/3 that it had been in the early 1950s (Figs. 4.36 and 4.37). Freshwater discharge, on the other hand, increased during the same period (Fig. 4.37a). From 1981 through 2007, sediment discharge continued to decline in a stepwise fashion, the load in 2007 being about 120 Mt/yr (Horowitz, 2010), about 30% of pre-1950 levels. The 1993 flood accentuated the decline by effectively removing much of the still accessible erodable sediment (Horowitz, 2010). A plot of cumulative sediment vs. water discharge (Fig. 4.37b) suggests that if the river had not been dammed and engineered, total sediment discharge between 1950 and 2008 would have been about 2.2 Bt, whereas the river only discharged about 40% of that amount past Tarbert Landing in lower Mississippi.

A comparison of the annual discharge of both water and sediment in 1950 and 1975 (Fig. 4.38a, b) shows the clear difference between the pre- and post-1953 Mississippi. While total water discharge in the two years was roughly comparable, the pre-dam river was much more "episodic" in its discharge of both water and sediment, with several clear hysterisis loops (Fig. 4.38c). What is more, the sediment load in 1975 was only ~30% that of 25 years earlier, in large part explained by the marked decrease in suspended sediment concentrations.

Human impact: contaminant fluxes

About 2/3 of the total USA cropland lies within the Mississippi drainage basin, and historically many of America's heavy industries have been located within the Ohio River watershed. The intense level of human activity over the past 150 years therefore has played a key role in defining the character of the present-day Mississippi. A report on the various contaminants within the Mississippi's various basins and tributaries is found in USGS Circular 1133 (1995) (http://pubs.usgs.gov/circ1133).

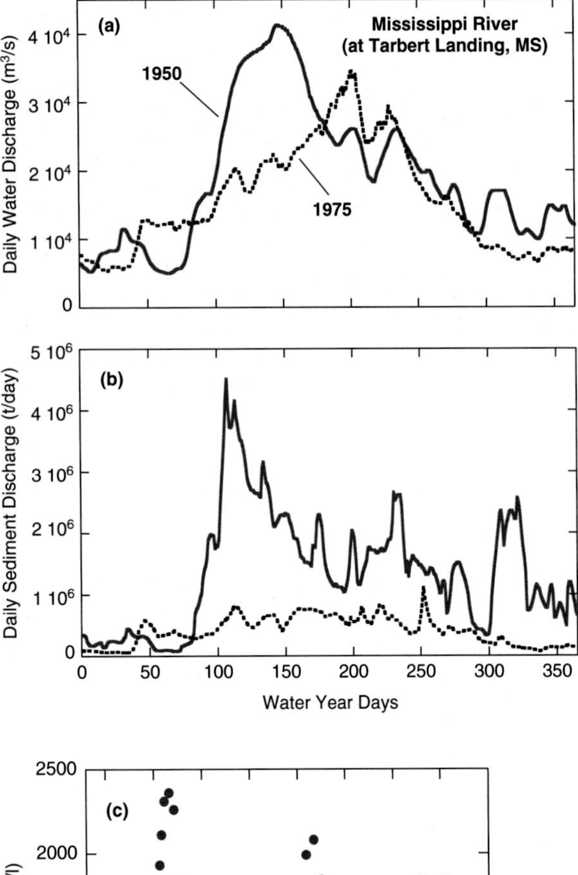

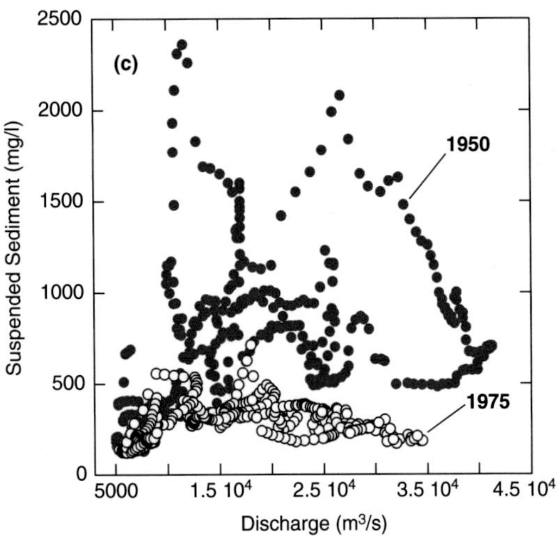

Figure 4.38. Daily water (a) and sediment (b) discharges from the Mississippi River at Tarbert Landing, MS, prior to major dam construction on the Missouri River (red; water year 1950) and 30 years after dam closure (black; water year 1975). Water year extends from 1 October to 30 September the next year. (c) Derived rating curves for the Mississippi River 1950 and 1975. Note that peak sediment concentrations in 1975 never exceeded about 300 mg/l, nearly seven-fold lower than highest concentrations in 1950. Data from the US Geological Survey (www.water.usgs.gov).

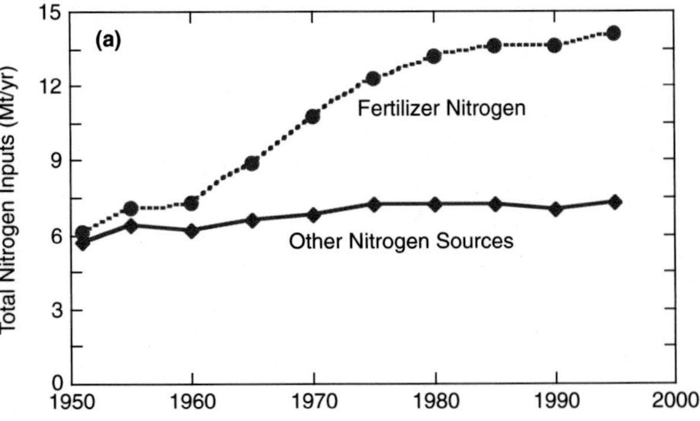

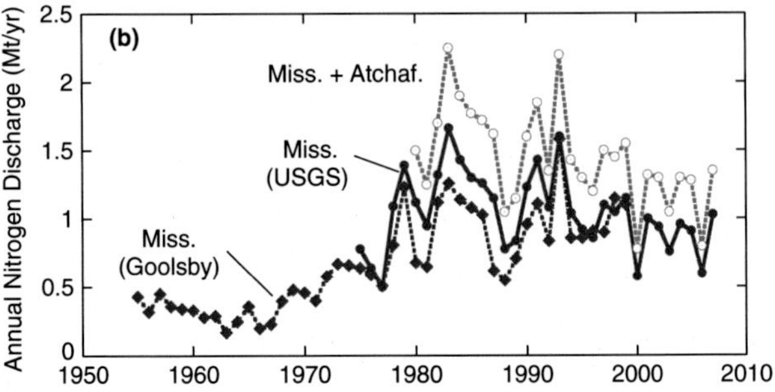

Figure 4.39. (a) Nitrogen inputs to the Mississippi watershed, 1951–1997. Data from Goolsby *et al.* (1999). (b) Nitrogen discharges of the Lower Mississippi (measured at St. Francis, LA) and calculated by Goolsby *et al.* (1999) and Aulenbach *et al.* (2007) as well as from the entire the Mississippi–Atchafalaya system (http://toxics.usgs.gov/hypoxia/mississippi/flux_ests/delivery/).

One of the most obvious impacts on the river's water quality has been the elevated concentrations of nutrients. For the first 100 years of ever-increasing agriculture throughout in the watershed – particularly in the Ohio and Upper Mississippi basins – agricultural production often could be increased simply by farming more land. Renourishment of the soil relied heavily on legumes, pasture and application of animal manure. During the past 80 years, and especially since 1950, however, rising farm production has required the increased application of artificial fertilizers (Fig, 4.39a; see also Goolsby *et al.*, 1999; Goolsby and Battaglin, 2001). In the early 1950s less than 6 Mt of fertilizer nitrogen were applied annually throughout the Mississippi basin; by 1990 this number had increased to more than 12 Mt/yr. In parts of central Indiana, Illinois, and Iowa applications exceed 10 t/km^2/yr (Antweiler *et al.*, 1995).

An obvious outcome of the increased application of artificial fertilizers was the ever-increasing nitrogen levels in Mississippi waters, delivery in the last years of the twentieth century exceeding 2 Mt/yr, about six-fold higher than in the 1950s (Fig. 4.39b). As artificial fertilizers have been the only nitrogen source whose use increased significantly during that period (Fig. 4.39a), it seems safe to assume that the increased nitrogen levels were primarily the result of increased fertilizer application.

One impact of the increased nitrogen flux has been a marked increase in the extent and severity of hypoxia in the shelf waters off Louisiana and east Texas (Turner and Rabalais, 1991; Goolsby *et al.*, 1999; Rabalais and Turner, 2001; Rabalais *et al.*, 1998, 2007 and references therein). Although other mechanisms have been suggested (Bianchi *et al.*, 2008), increased nitrogen discharge from the Mississippi seems to be a major culprit (Boesch *et al.*, 2009), perhaps combined with greater water clarity, the result of reduced suspended sediment concentrations. Analogous to the Danube River and adjacent Black Sea coastal waters (Humborg *et al.*, 1997), decreased silicate concentrations in Mississippi waters may have helped promote offshore dinoflagellate production.

The Mississippi basin also accounts for most of the pesticides applied to US croplands, about 60% of which are herbicides. Annual herbicide applications listed by Goolsby and Pereira (1996, their Fig. 39) total ~115 000 t/yr, and Goolsby and Pereira (1996) have estimated that between 1987 and 1989, 21 000 t/yr of Atrazine alone were spread over the Mississippi drainage basin. The amounts of pesticides that are dissolved and transferred to the river, however, vary with the time of application and the amount of subsequent rainfall. Between April 1991 and March 1992, for instance, only 366 t of the herbicide Atrazine were exported out the lower reaches of the river (Goolsby and Pereira, 1996), although periodically levels approached or exceeded maximum contaminant levels (Goolsby and Pereira, 1996, their Fig. 41). During the great flood of 1993, elevated levels of Mississippi-derived atrazine were found as far away as Cape Hatteras, North Carolina, having been transported by the Loop Current out of the Gulf of Mexico and then northward by the Gulf Stream (Goolsby, 1994).

The historic emphasis on heavy industry along much of the Ohio River has resulted in locally high levels of heavy metals. Most heavy metals are associated with the silt and colloid fractions of the river water and not dissolved (Garbarino *et al.*, 1995). As a result, relatively few heavy metals are discharged out of the Mississippi River. Other examples of contaminants within the Mississippi River system can be found in USGS Circular 1133 (Meade, 1995).

Near-term future of the Mississippi watershed and its delta

One result of the upstream manipulations of the Mississippi watershed has been a corresponding loss of coastal wetlands – more than 5000 km^2 since 1900 in coastal Louisiana alone (Templet and Meyer-Arendt, 1988; Wells *et al.*, 1984). Wetland loss is partly the result of natural subsidence, which varies from a few millimeters to several tens of millimeters per year (Wells and Coleman, 1987; Dixon *et al.*, 2006), together with the decreased river discharge of suspended sediment, the channelization of river flow, and the extensive levee system, which collectively have lessened the lateral escape of sediment to the low-lying floodplains and delta. Moreover, offshore jetties at the mouths of the major distributaries increase offshore escape of sediment by funneling sediment farther offshore and out of the littoral zone. According to Templet and Meyer-Arendt (1988), even a 9% increase in the lateral escape of the river's present-day annual suspended load would help offset the regional subsidence (see also Blum and Roberts, 2009).

Unless there is a fundamental change in the management of the lower river basin, one probably can expect to see continued or perhaps accelerated (depending on the rate of ground water or petroleum removal) subsidence and resulting wetland loss in future years. As Wells (1996) has pointed out, the influence of man has not changed the underlying processes; it has only changed the rates at which they occur. Recently, Kim *et al.* (2009) calculated that if the levees downstream of New Orleans were cut, between 700 and 2700 km^2 of land (depending on a number of variables assumed in the models) could be added to the delta by the year 2100.

The underlying roots and answers to management problems on the Mississippi basin, however, are largely the same as those faced for many rivers: how can the negative effects of up-river management be modified so that they have minimal impact on the lower reaches of the river and the adjoining coastal zone? The ability of humans to change the flux of sediment, nutrients or contaminants to the river ultimately depends on land conservation, more effective renourishment of the farmland, better manufacturing practices, and more environmentally sensitive river engineering, including a strategic breaching of levees so that floodplains can once again be connected with the rivers that formed them (Opperman *et al.*, 2009). Whether or not farmers in Iowa, miners in West Virginia, and people living in flood-prone areas can change their way of life will undoubtedly determine the future health of the entire watershed and particularly the well-being of those people living downstream.

Yin and Yang: Yellow and Yangtze Rivers

Much like the east–west contrast in precipitation in the United States, water resources in China are non-uniform. Rainfall in southern China locally exceeds 2000 mm/yr, whereas it averages less than 500 mm/yr throughout most of the north, and locally <200 mm/yr (Fig. 4.40). Precipitation in China is also highly seasonal owing to its monsoonal climate. The wet monsoon migrates seasonally from south to north: maximum rainfall in southern China occurring in May and June, and in July and August in the north. Runoff shows the same south–north disparity: >1000 mm/yr in the south, <100 mm/yr in the north, particularly the northwest.

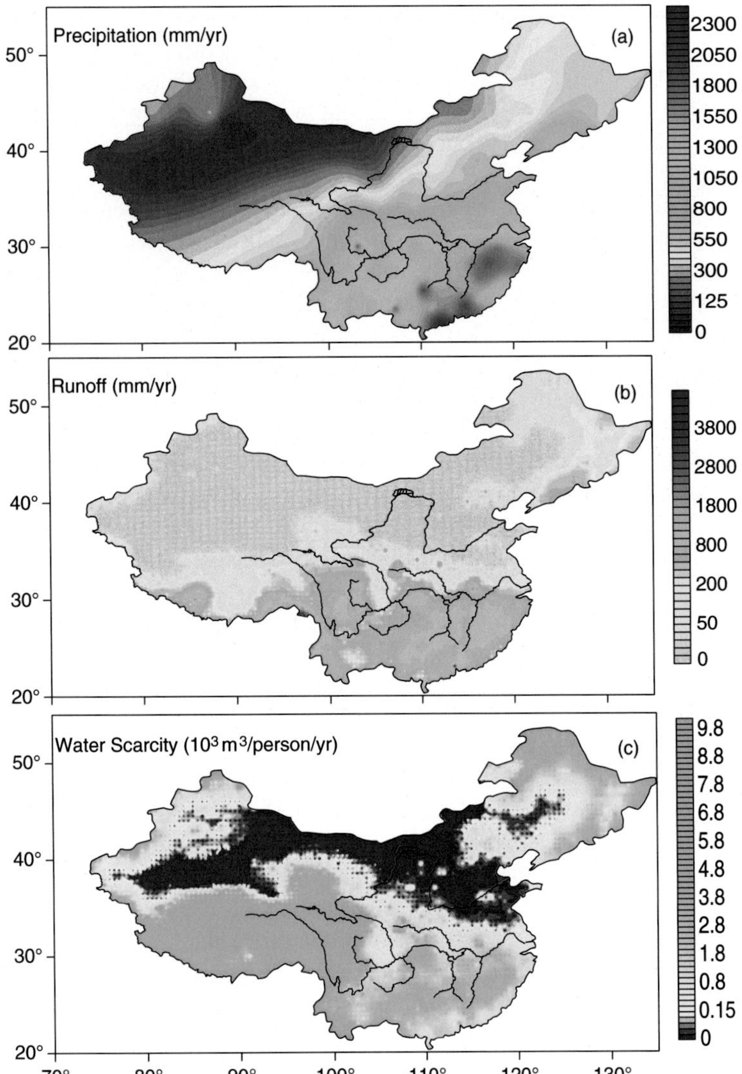

Figure 4.40. Distribution of annual precipitation (a), runoff (b), and water availability (renewable water supply divided by population (c) in China. Maps courtesy of K. H. Xu.

Although southern China is densely populated, the Pearl and Yangtze watersheds have annual water availabilities greater than 2000 m^3 per capita, compared with 350–750 m^3 per capita in the dry Haihe and Yellow watersheds (Postel *et al.*, 1996; Fuggle and Smith, 2000; Varis and Vakkilaninen, 2001). Water availability (Fig. 4.40 c) has become more probematic in recent years as the Chinese population has become increasingly urbanized and China's expanding industries have demanded a greater share of the water supply. Some rivers in northern China have devolved into what Xu (2004) has termed, "anthropogenic seasonal rivers."

The two most documented rivers in China are the Yangtze River, which effectively divides the wet south from the dry north much in the way that the Mississippi divides the wet eastern from the dry western USA, and the Yellow River. The two rivers thus are very different, the former being water-rich, the latter being water-poor. As such, human use and impact on the two basins have differed considerably, but the futures of the two basins ultimately may well be closely intertwined.

Water-rich Yangtze River (Changjiang)

Much has been written in recent years about the Yangtze River, particularly its response to the construction of nearly 50 000 dams throughout its watershed since 1950, nearly 30 – existing or proposed – standing higher than 100 m.[1] The closing of the world's largest dam – the Three Gorges Dam in 2004 – and its impact on the Yangtze and its estuary have received particular attention from both the scientific and popular press (Z. S. Yang *et al.*, 2006; S. L. Yang *et al.*, 2005, 2006; Xu *et al.*, 2006, 2007; Li *et al.*,

[1] One of the first dams in the Yangtze watershed was built on the Min River in 250 BC at Dujiangyan (Winchester, 2008).

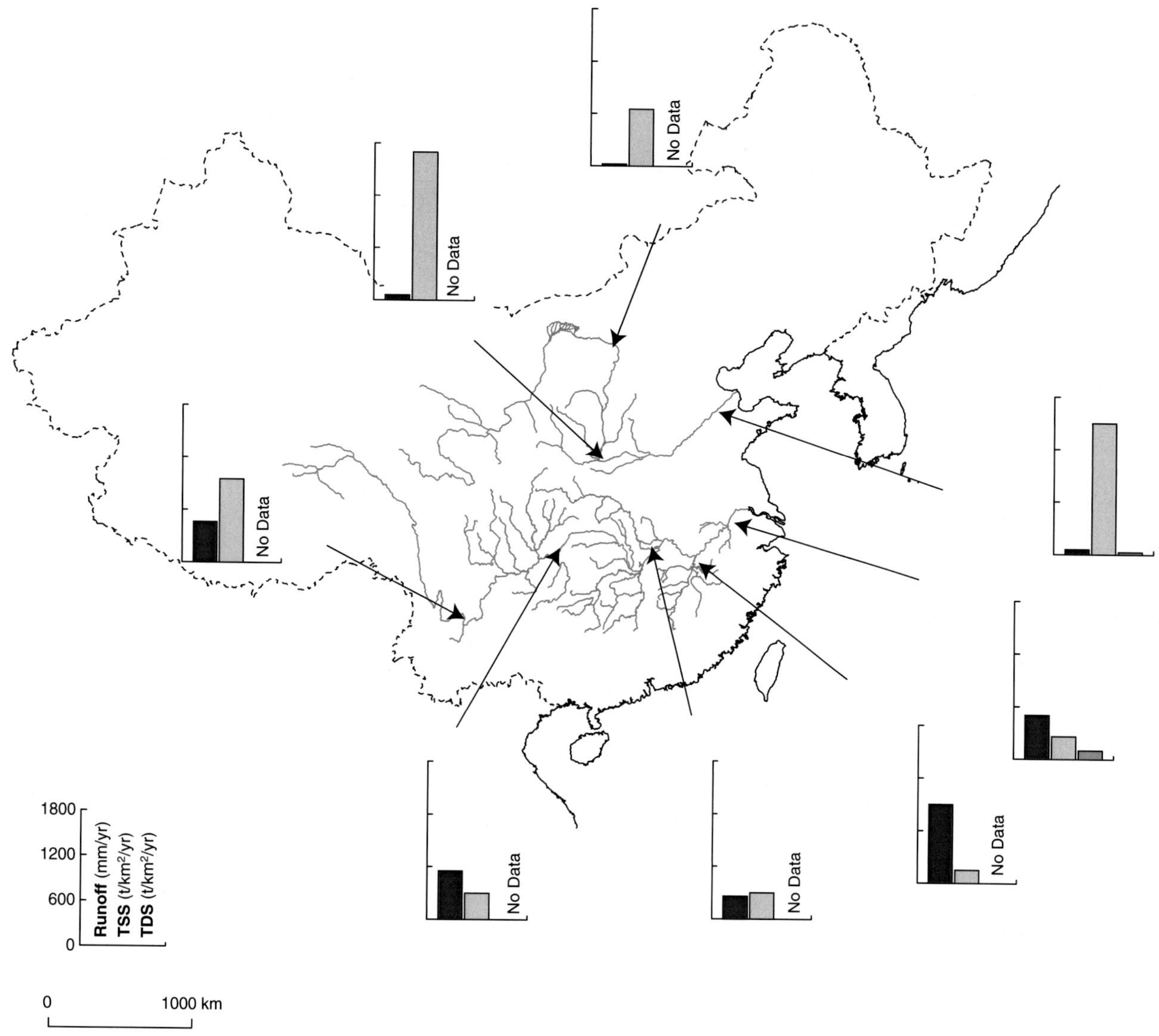

Figure 4.41. Mean annual runoffs and sediment and dissolved-solid yields in the mainstems and tributaries of the Yangtze and Yellow River watersheds.

2007) and for good reason: the Yangtze (Fig. 4.41) is Asia's largest river in terms of basin area (1 800 000 km^2), length (6300 km) and water discharge (900 km^3/yr), and third (behind the Ganges–Brahmaputra and Yellow) in sediment load (~470 Mt/yr prior to the early 1980s; Table 2.6) (Xu *et al.*, 2007). The Yangtze's southern tributaries have higher runoff than the northern tributaries, but lower sediment yields (Fig. 4.41). The timing of the wet season also changes within the Yangtze watershed, from April–June in the south and east to July–September in the north and west. At Datong, the seaward-most gauging station along the river, about 70% of the annual water discharge and 85% of the annual sediment discharge occur between May and October (X. Chen *et al.*, 2001; Xu *et al.*, 2007).

Over the past 50 years annual Yangtze runoff at Datong has generally fluctuated around 500 mm/yr, with no significant long-term change (Fig. 4.42a). Floods, however, seem to have increased in frequency in response to increased deforestation, increased wetland and lake reclamation (Xu and Milliman, 2009), and levee construction, while discharge during dry months has decreased (X. Chen *et al.*, 2001). By contrast, annual sediment discharge has declined substantially over the past 50 years, decreasing from ~500 Mt/yr through much of the 1950s and 1960s, to less than 400 Mt/yr in the mid and late 1990s, to less than 150 Mt/yr after construction of the Three Gorges Dam (Fig. 4.42a). The decreased sediment load in part reflects a decline in rainfall in the northern parts of the watershed,

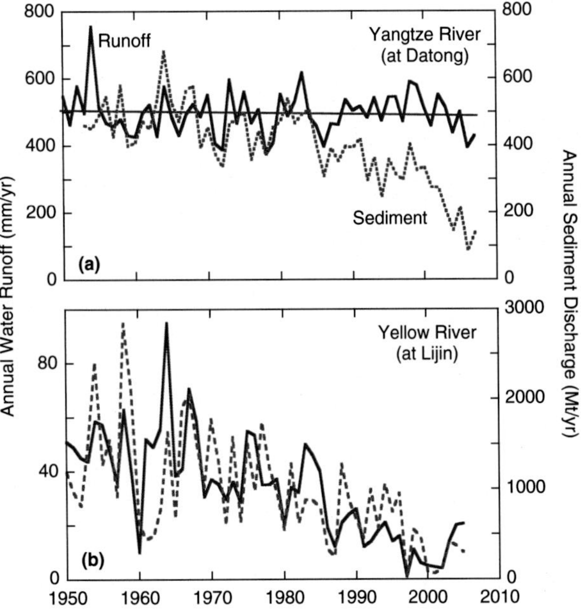

Figure 4.42. Annual water (blue lines) and sediment (red dashed lines) discharges of the (a) Yangtze River (Changjiang) at Datong and (b) Yellow River (Huanghe) at Lijin. Data courtesy of S. L. Yang, K. H. Xu, and Z. S. Yang; see also Wang *et al.* (2007).

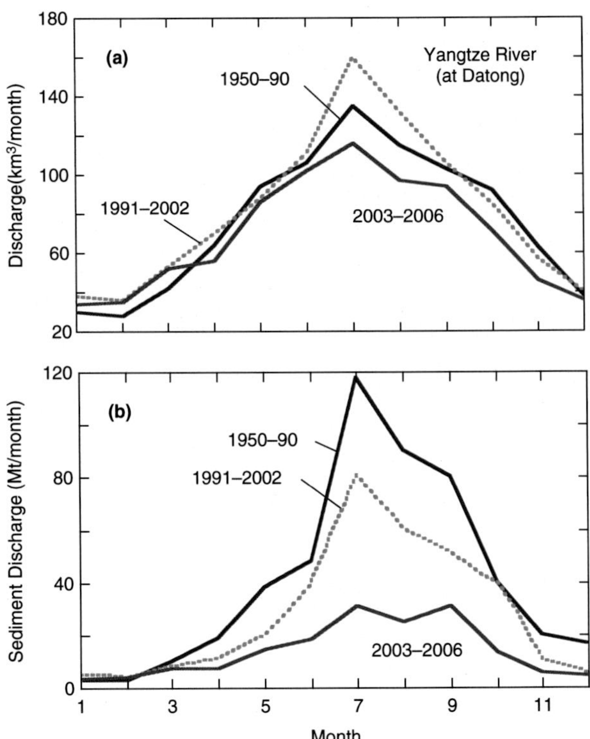

Figure 4.43. Monthly discharges of water (a) and sediment (b) from the Yangtze River at the Datong gauging station, 1950–2005. Water discharge has remained more or less constant over the 55-year period, whereas sediment discharge has declined by about 2/3; note also the seasonal shift in sediment discharge from mid summer to late summer and early autumn.

where sediment yield historically has been greatest. More important has been the effect of increased land conservation and the ever increasing number of dams (>50 000 as of the writing of this book) of which >12 000 stand higher than 15 m (Lu and Higgitt, 1998; S. L. Yang *et al.*, 2002; Z. S. Yang *et al.*, 2006; Xu *et al.*, 2007). While land-use changes and dam construction have not significantly affected water discharge on either an annual or monthly basis, annual sediment discharge has been significantly affected (Figs. 4.42a and 4.43b). Greatest declines are seen during high-discharge months (July and August), sediment discharge during low discharge months (December–March) showing little change (Fig. 4.43b). The relatively constant annual water discharge, together with the declining sediment discharge, indicates a decreasing concentration of suspended sediment, from nearly 0.6 g/l in the 1950s to about 0.2 g/l in recent years (Fig. 4.44a). The "lost" sediment presumably reflects deposition behind dams and changing land-use and improved conservation.

Historically, a series of lakes (the largest being Dongting Lake) along the Changjiang's midsection helped to absorb major flood pulses and considerable sediment discharge (Xu and Milliman, 2009). Increasing land reclamation and siltation, however, have substantially reduced lake areas, which, together with construction of new levees along the river and lake banks, have eliminated many escape routes for floodwaters. Rather than floodwaters being absorbed or spreading laterally, they have risen vertically, thereby displacing local inhabitants. As the river and lakes are increasingly "improved" through the building of new dikes (for example), the potential for dangerous floods presumably will continue to increase (Xu and Milliman, 2009; K. H. Xu, personal communication).

No discussion of the Yangtze can finish without mention of the Three Gorges Dam (TGD), 189 m high and forming a 600-km-long reservoir with a holding capacity of nearly 40 km³, which, in the process of its construction, displaced 1.2 million people and submerged 19 cities and 326 towns (Zeller, *New York Times*, November 16, 2000). Although the long-term impact by this gargantuan dam on the Yangtze watershed remains to be seen, since its closure in 2003 sediment discharge has declined to <150 Mt/yr and peak sediment discharge has shifted to late summer and early autumn (Fig. 4.43b; Xu and Milliman, 2009). Most noticeable has been the dramatic drop in sediment discharge at the Yichang gauging station, 70 km downstream from the TGD. In the 20 years prior to 2003, average sediment discharge at Yichang, about 70 km downstream from the TGD, gradually declined from >500 to <400 Mt/yr, about 90 Mt/yr greater than at Datong

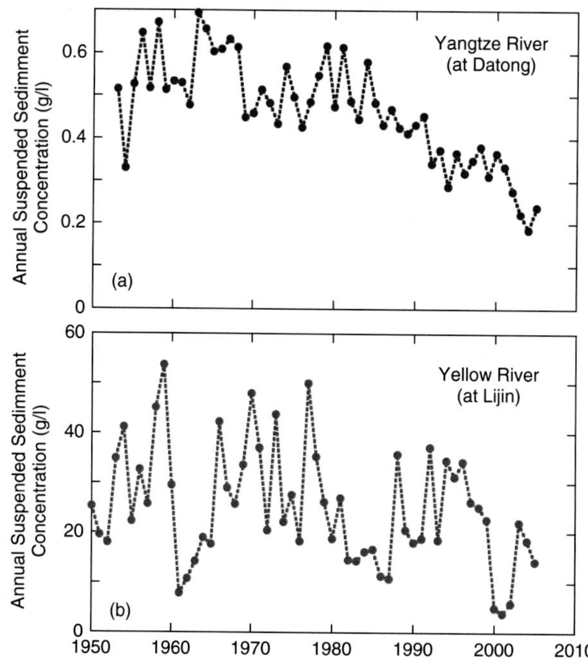

Figure 4.44. Annual suspended sediment concentrations in the Yangtze River at Datong (a) and the Yellow River at Lijin (b), 1950–2005. Depressed Yellow River concentrations in the early 1960s presumably reflect the closing of the Sanmenxia Dam.

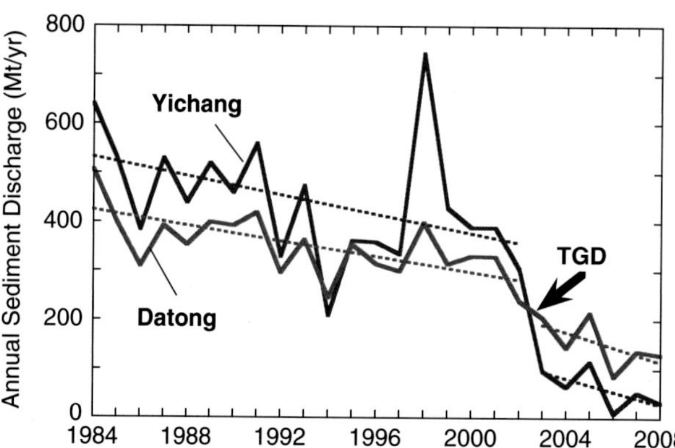

Figure 4.45. Annual sediment discharges at Yichang (blue; 70 km downstream from the Three Gorges Dam) and Datong (red; ~600 km landward of the Yangtze estuary), 1983–2008. The gradual change in sediment discharge at both stations after 1983–2002 was the result of the building of many upstream dams as well as increased land conservation. The precipitous decline in sediment discharge, particularly at Yichang, followed the closing of the Three Gorges Dam (TGD) in 2003.

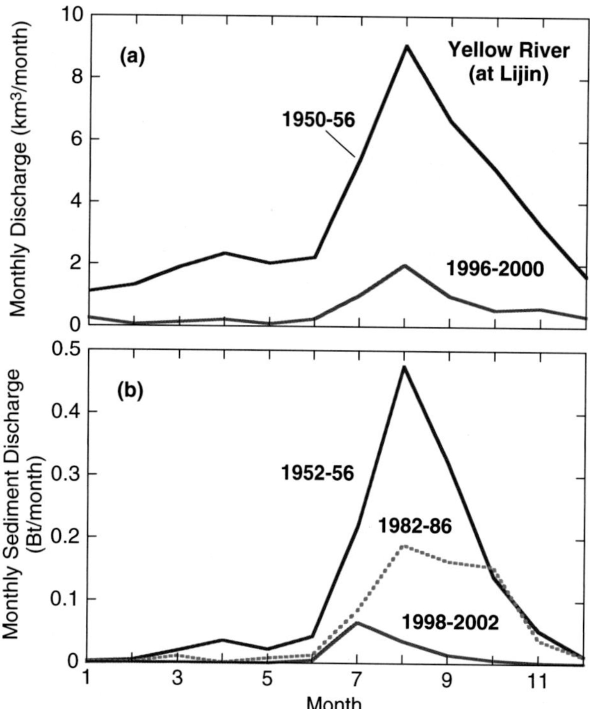

Figure 4.46. Monthly discharges of water (a) and sediment (b) from the Yellow River at the Lijin gauging station, 1950–2002. Cumulative water discharge for the first six months of the year decreased substantially, accounting for only about 15% of the annual discharge.

(Fig. 4.45). This downstream "loss" reflected flood plain and river channel accretion and lake sedimentation. With the closing of the TGD, sediment discharge past Yichang dropped precipitously to <100 Mt/yr, and since 2005 has averaged 35 Mt/yr. Sediment discharge at Datong, while also declining, has been about 100 Mt/yr greater than at Yichang (Fig. 4.44), indicative of channel and bank erosion downstream of the TGD (Xu *et al.*, 2006, 2007; Z. Yang *et al.*, 2007; Xu and Milliman, 2009, S.L. Yang, *et al.*, 2011). Mean particle size in the suspended sediment at Yichang decreased by roughly a factor of 3 after construction of the TGD, whereas 300 km downstream at Hankou, it increased 2-fold, indicative of sediment deposition behind the TGD and the corresponding downstream erosion (S. L. Yang *et al.*, 2011).

Another almost immediate result from the closing of the TGD has been a noticeable erosion of the Yangtze's submarine delta (S. L. Yang, *et al.*, 2011), and it seems likely that accelerated shoreline erosion may soon follow. C. Chen (2000) has suggested that reduced fresh-water discharge may also restrict estuarine circulation within the East China and Yellow seas, thereby decreasing upwelling and biological productivity in what has been one of the most heavily fished coastal areas in the global ocean.

Water-poor Yellow River (Huanghe)

Probably nowhere in China is the water problem more acute than in the Yellow River basin. Draining much of arid northern China (Figs. 4.40 and 4.41), the Yellow River had an annual runoff of 55–60 mm/yr during much of the

Figure 4.47. Aerial photograph of the loess plateau in the Yellow River watershed, showing the extent to which the river and its mid-stream tributaries have eroded the terrain. Courtesy of Li Rui, Institute of Soil and Water Conservation, China.

past century, but over the past 30 years it has declined to <20 mm/yr (Fig. 4.42b), marking it as perhaps the driest major river in the world. Nevertheless, the Yellow River remains a major supplier of water to this parched area and a major supplier of sediment (until recently) to the river's delta at the western edge of the Gulf of Bohai. Rainfall and thus river runoff are highly seasonal, occurring primarily in summer and early autumn, which often is marked by a sharp transition from drought to flood (Fig. 4.46). Because of its capacity to alleviate arid conditions but also its propensity to generate catastrophic floods, the river sometimes has been termed "source of joy, source of sorrow."

The Yellow River's watershed area is about 15% that of the Amazon, and its historic water discharge was several orders-of-magnitude smaller (57 km^3/yr vs. 6500 km^3/yr). But together with the Amazon and Ganges–Brahmaputra, the Yellow was the ranking world river in terms of sediment discharge: 1.1 Bt/yr (Qian and Dai, 1980; Milliman and Meade, 1983; Milliman and Syvitski, 1992). Particularly unique characteristics of this river are its high suspended-sediment concentrations (~ 25 g/l; see Fig. 4.44b) and a high sediment yield (1400 t/km/yr), the result of the middle course of the river passing through an easily erodible loess-dominated landscape. The rapid erosion of the loess plateau (Fig. 4.47) was greatly facilitated by human intervention. Prior to the deforestation and farming of the loess plateau about 2500 years ago, the river, then known as the Dahe (Great River), carried relatively little sediment. The increase of sediment load following the erosion of the loess plateau seems to have been 3- to 10-fold (Milliman *et al.*, 1987; Saito *et al.*, 2001; Xu, 2003), and by the Tang dynasty in the seventh century AD, deforestation, farming and the subsequent erosion on the loess plateau had transformed the relatively clear Dahe into the muddy Huanghe (Yellow River) (Ren and Zhu, 1994). A few examples indicate the degree of erosion in the Yellow River watershed. At Ansai, sediment yield in forested terrain with an average gradient of 22–28° is 37.5 t/km^2/yr, whereas in adjoining cropland it is 4400 t/km^2/yr (Ren and Zhu, 1994). Ren and Zhu also mentioned an area near Inner Mongolia where, between 1957 and 1988, 0.18% of the land surface was lost annually. One fluvial response to the increasingly mud-filled river was the rapid sedimentation within the lower stretches of the river, leading to increasing channel migration and annual flooding.

Because of the desire for water and the need to mitigate the annual floods, about 2500 years ago the Chinese began constructing a complex dike system along the lower reaches of the river. On average about 0.5–1 Bt of sediment were deposited annually along the lower reaches of the river, mostly as channel deposits, thereby elevating the river bed, which in turn necessitated the continual building of ever higher dikes (Ren, 1995). Xu (1994) estimated a ~30-fold increase in sediment accumulation on the lower parts of the flood plain since mid Holocene, more than 95% of it occurring in the past 1000 years. In some places the elevated river now sits 10 m or more above the flat surrounding flood plain, one result being that no tributaries flow into the lower Yellow River; rather water escapes from the river into the groundwater of the surrounding countryside. Gong and Xu (1987) estimate the lateral filtration rate as 320 000 m^3/yr/km of river length. Given the high rates of evaporation throughout northern China, groundwater salinity has increased, leading to local desertification (Gong and Xu, 1987).

Engineered channelization has not altogether stopped the river's meandering. Rather, the elevated riverbed has simply assured that any breaching of the dike could be catastrophic (Ren, 1992). A major flood in 1854 led to major breaks in the dikes, and the river shifted its mouth from the south to the north side of the Shandong Peninsula, a lateral distance of more than 300 km. In the process, tens of thousands of lives were lost. Now the river again flowed into the Gulf of Bohai, as it had 700 years previously (Ren, 1992). Progradation of the new northern delta has been rapid; between 1976 and 1992 the Yellow River main distributary channel to the Gulf of Bohai prograded 38 km, adding approximately 450 km^2 of area to the delta (Ren, 1995).

The need to control the high erosion and sediment discharge from the Yellow River drainage basin led to the initiation of three conservation and reclamation efforts in the 1980s. Reforestation on the loess plateau decreased runoff by 32%, declining further as the trees matured (Huang *et al.*, 2003). The reforested land is also less susceptible to peak flows during heavy rainfall events. Interestingly, gullies, which historically have been the sites of accelerated erosion, have proved to be particularly effective sites for tree and grass recolonization, and locally sediment yields have been reduced by more than an order of magnitude (Li, 1992). Runoff and erosion have also been reduced because of decreasing precipitation, basin-wide precipitation between 1950 and 2005 declining by about 12% (Wang *et al.*, 2007).

The past 60 years have also seen the increased construction of dams and reservoirs to store water for irrigation and to mitigate disastrous floods. As of the early 1990s, more than 2600 reservoirs of various sizes were located throughout the Yellow River watershed (Long *et al.*, 1994). The larger reservoirs also have retained a significant amount of the river's sediment, the Liujiaxia and Qingtongxia, in the upper Yellow River, alone retaining 70 Mt/yr (Ren and Zhu, 1994). At Samenxia, just down stream from the first major dam on the Yellow River, sediment discharge decreased from 1.78 Bt/yr between 1950 and 1960, to 1.36 Bt/yr in the 1970s, to 0.78 Bt/yr in the 1980s (Long *et al.*, 1994).

In addition, a considerable quantity of river water (and thus sediment) has been removed from the lower reaches of the river in response to increased agricultural and industrial utilization. Between 1949 and 1991, the amount of irrigated land within the Yellow River basin increased eight-fold, to more than 60 000 km^3, thereby diverting an additional 30 km^3 of water yearly from river flow (Ren, 1995). How much more water has been removed illegally is uncertain. One further result of the increased irrigation has been the increased salinity in the Yellow River as soil-salinized irrigation waters flow back into the river (J. Chen *et al.*, 2003).

Collectively, decreased precipitation, increased land conservation, and construction of dams and irrigation systems have resulted in a dramatic decrease in both water and sediment discharge from the Yellow River. The 1.1 Bt/yr sediment discharge calculated by Qian and Dai (1980) – and still cited by many workers – was based on measurements made between 1950 and 1975, when river flow was generally high (Fig. 4.41b). In 1963 alone, the river discharged more than 2.5 Bt of sediment past the Lijijn gauging. Since then, both water and sediment discharge have decreased steadily (Z. S. Yang *et al.*, 1998; Wang *et al.*, 2007). Between 1997 and 2002 annual water discharge averaged only 5 km^3/yr, due in part to a prolonged drought; it subsequently increased to about 18 km^3/yr between 2003 and 2005. Perhaps more impressively, between 2000 and 2002 average annual sediment discharge was less than 100 Mt/yr (Fig. 4.43b), in large part because of sediment retention in the Xiaolangdi reservoir (Wang *et al.*, 2007).

Low annual precipitation and low flow in the mid 1990s may have reflected in part the advent of a warm PDO cycle, perhaps superimposed by a 200-year precipitation cycle (Wu, 1992), but Wang *et al.* (2007) have calculated that decreased precipitation can explain no more than 30% of the decrease in water or sediment discharge. Human activities, particularly land conservation and dam construction explain most of the remaining 70%, according to Wang *et al.* (2007).

What is less easy to explain are the impacts on both water and sediment discharge by the increased removal of water for agriculture and industry (Z. S. Yang *et al.*, 1998). If sediment were preferentially removed via reservoir sedimentation, one would expect a corresponding decrease in average suspended sediment concentration, such as seen with the Yangtze River. In fact, suspended sediment concentrations in the Yellow River remained generally above 20 g/l, although with considerable interannual variation, between 1950 and 2000. It was only after 2000, perhaps the result of sediment trapping behind the Liujiaxia and Xiaolangdi reservoirs, that average annual suspended sediment concentrations fell below 20 g/l (Fig. 4.44b). This suggests that, at least until the present century, water and sediment were being removed in more or less equal amounts, which in turn suggests that much of the water and sediment removal occurred because of their removal – legal and illegal – for irrigation (Z. S. Yang, personal communication). This suspicion is strengthened by the fact that greatest declines in sediment loads occurred during the mid and late summer (Fig. 4.46b), when water is most needed for agriculture.

No matter what the cause of the decreased sediment discharge, delta progradation has decreased considerably (Mikhailova, 1998) and in fact, in recent years there has been active shoreline erosion as well as along many sections of the delta front (Wang *et al.*, 2007). At the same time, shrimp fishing, once a thriving industry, is now all but dead (Z. S. Yang, personal communication), the likely result of decreased nutrient output from the river.

What to do?

Sediment discharge from all of China's major rivers has decreased in recent years, a 70% reduction (1.3 Bt/yr) in the eight largest rivers according to Dai *et al.* (2009). Water discharge from northern rivers has also decreased significantly (Yang and Tian, 2009). Construction of more dams, such as the Longtan and Datingxia dams on the Pearl River, will decrease sediment discharge even further (Dai *et al.*, 2008). The most significant impacts, as detailed in this chapter, have occurred in the Yangtze and Yellow rivers. Human manipulation has decreased the Yangtze's sediment discharge by about 70% even though water discharge has remained relatively constant. After the closing of the Three Gorges Dam in 2004, the river channel along the lower course of the Yangtze began eroding (Xu *et al.*, 2007) and the submarine delta has shifted from one of accretion to one of erosion (S. L. Yang, personal communication). In the Yellow River there has been a dramatic decrease in both

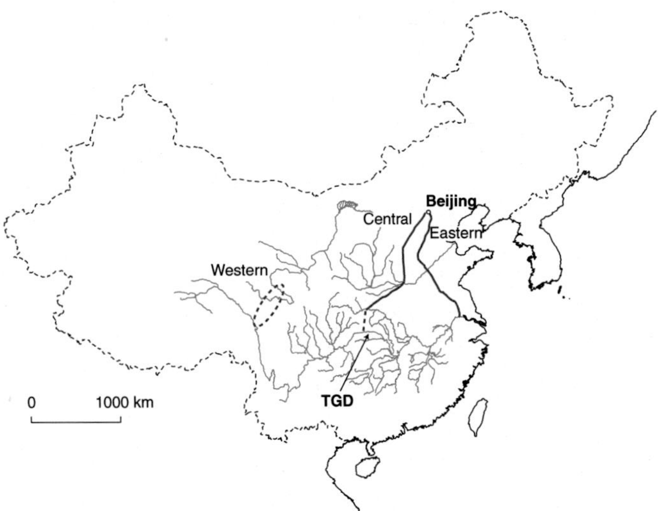

Figure 4.48. Proposed routes of three canals for the South–North Transfer Water Project. Actual timing of these projects, even the feasibility and precise location of the West Line, is still uncertain. (TGD represents the Three Gorges Dam).

water and sediment delivery since the early 1980s. It would not be an overstretch to declare that the Yellow is presently moribund – if not already dead – with respect to the discharge of both water and sediment to the coastal ocean.

Certainly a major important social and economic problem for China is the declining supply of freshwater to the Yellow River watershed. One answer to counter balance this is groundwater extraction, but recent groundwater utilization already has lowered the water table throughout the Yellow River and Haihe basins. The Yellow River delta in recent years has subsided as much as 20 cm/yr (Qi and Luo, 2007), and farther north, in Tianjin, local subsidence has exceeded 1.5 m (Han, 2003).

If groundwater is not able to provide sufficient water to the north, another option is construction of a series of south–north canals, the so-called South–North Transfer Water Project (see Z. Chen *et al.*, 2001, their Fig. 1), by which three separate diversions (Fig. 4.48) would ultimately divert as much as 10% of the annual Yangtze water discharge northward to the Yellow and Haihe rivers as well as Bejing. A 10% reduction in Yangtze discharge, however, could have considerable – even if still undefined – impact on the Yangtze estuary and the East China Sea.

Inevitably the Chinese face a most difficult task. The character of their two largest rivers, the Yellow and Yangtze, reflects a long history of agriculture, water utilization, and river management. But present and future changes in both watersheds may ultimately dwarf the historic patterns. Changing land-use practices almost certainly will lessen soil erosion and thereby decrease sediment discharge, but increased plant cover could lead to greater water retention and at the same time increase evapotranspiration; both would decrease stream and river runoff. Moreover, future damming and diversion of the Yangtze can only further muddle what already is a complicated situation. With the need to adequately feed and water a population of more than 1.3 billion (as of 2010) and with a vibrant economy, China's options appear limited. Almost certainly any solution will require policy and management decisions that will have major environmental, economic and social consequences. The next few decades should prove interesting.

Summing up and projecting to the future

Water use and the present "water crisis"

Humans presently withdraw about 3500 km^3 of freshwater annually for their use, most of which comes from rivers, slightly less than 10% of the annual global river discharge. In the Middle East, North Africa, and the US Southwest, to mention only the most obvious examples, the amount of renewable surface water is not sufficient to meet human needs. In some of these water-deficient areas, groundwater extraction can serve as an important source of freshwater, although some of these aquifers may have been last recharged thousands of years ago. In addition, poor land-use practices over the past three millennia – deforestation and agriculture, and more recently, urbanization – have accelerated land erosion as well as heightened the impact of episodic events on watersheds. In the past century, on the other hand, river control, most notably dams and irrigation, has decreased the discharge of both fluvial water and sediment. Some rivers now discharge less sediment to the coastal ocean than they did prior to the Anthropocene. The Colorado, Ebro, Nile, Indus, and Yellow, to mention only five rivers discussed earlier in this chapter, discharge much less water and sediment than they did prior to the advent of mid twentieth-century river "management."

Although many low-runoff, mid-latitude rivers have diminished low runoff (see Fig. 4.15), their collective decreased discharge has not significantly affected the

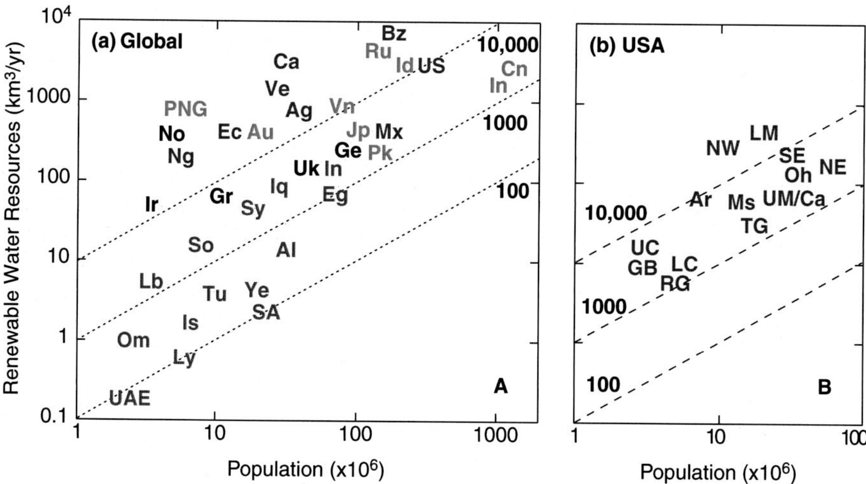

Figure 4.49. (a) Comparison of annually renewable water resources in Middle Eastern and North African countries (In=Iran, Gg=Egypt, Iq=Iraq, Sy=Syria, Al=Algeria, So=Somalia, Lb=Lebanon, Tu=Tunisia, Ye=Yemen, SA=Saudi Arabia, Is=Israel, Om=Oman, Ly=Libya, UAE=United Arab Emerites) vs. their populations in 2005. For global comparison, we show data for several European (black: Ru=Russia, Ge=Germany, Uk=Ukraine, No=Norway, Gr=Greece, Ir=Ireland), North and South American (Bz=Brazil, USA=United States of America, Ca=Canada, Ve=Venezuela, Ag=Argentina, Mx=Mexico, Ec=Ecuador, Ng=Nicaragua), and Austral-Asian (Cn=China, Id=India, Vn=Vietnam, Jp=Japan, Pk=Pakistan, Au=Australia, PNG=Papua New Guinea) countries. Diagonal dashed lines represent annual renewable water resources per person (m^3 per person per year). Note that several North African and Middle East countries have renewable resources about or less than 100 m^3 per person per year. (b) The United States has national average renewable water resources of 10 000 m^3 per person per year, but individual watersheds range from 1000 m^3 per person per year to greater than 20 000 m^3 per person per year (LM=Lower Mississippi, NW=Northwest, SE=Southeast, NE=Northeast, Oh=Ohio, UM/CA=Upper Mississippi and California, Ar=Arkansas, Ms=Missouri, TG=Texas Gulf Coast, UC=Upper Colorado, LC=Lower Colorado, GB=Great Basin, RG=Rio Grande). Most data from Gleick *et al.* (2006).

global water budget. The more important impact in low-lying coastal areas, many of them heavily populated (e.g. Italian Adriatic, coastal Egypt, Mississippi delta, Yangtze and Yellow river deltas), has been land loss owing to coastal erosion and uncompensated land subsidence (see Milliman and Haq, 1996; Syvitski *et al.*, 2009, and references therein), exacerbated where groundwater has been extracted to augment a diminished surface-water supply.

The minimal amount of freshwater required for daily human domestic use ranges from 20 to 50 liters per person – or about 6–15 m^3 per person per year. According to Gleick *et al.* (2006, their Data Table 2), the actual domestic use varies from <10 m^3 per person per year in countries like Angola, Haiti, and Cambodia to >200 m^3 per person per year in countries like Canada, Estonia, and New Zealand. Including non-domestic water uses such as agriculture and industry, annual water withdrawal ranges from <100 (many African countries) to >1000 m^3 per person per year (e.g. USA, Iraq, Guyana) (see Gleick's Data Table 2 on p. 230–236 of his 2006 book).

Countries that use more than 90% of their water for agriculture (mostly for irrigation) and whose renewable water resources are less than 1000 m^3 per person per year (e.g. Algeria, Kenya, Libya, Tunisia, Bahrain, Israel, Syria) may already be facing the problem of insufficient fresh water or they may well face this problem in the near future. Many of these water-deficient countries are located in the low runoff environs of the North Africa, the Middle East, and western Asia (see Fig. 2.8). Libya (0.6 km^3/yr renewable water, 5.8 million population) and Saudi Arabia (2.4 km^3/yr, 25 million people) already have renewable water supplies of about 100 m^3 per person/yr (Fig. 4.49a). Libya presently removes about 700% of its renewable water supply annually by extracting groundwater that was deposited during the mid-Holocene climatic optimum. How they will replace this non-renewable resource once it is diminished is not clear. Saudi Arabia, in contrast, has built major desalinization plants, something that may not be too difficult for an energy-rich country but is less feasible for water-starved, energy-poor countries such as Lebanon or Yemen.

The plot in Fig. 4.49a, however, does not take into account regional variations in water availability within each country, as graphically seen in Figs. 4.34 and 4.40. In the United States nationwide annual water renewal averages 10 000 m^3 per person per year, stemming in large part from the high discharges from the lower Mississippi and Columbia rivers and the relatively sparse populations within these two watersheds. The lower Colorado and Rio Grande watersheds, in contrast, have annual water supplies barely above 1000 m^3 per person per year (Fig. 4.49b), and irrigation in both basins accounts for 80–90% of the annual usage.

One option to the "water crisis," as Gleick (1993) has termed it, is to increase the supply of water. But how? A quick and obvious – although often wrong – answer is the extraction of groundwater. If the extracted groundwater can be renewed, then it can be viewed as another renewable resource. But when groundwater is removed at a faster rate than it can be renewed, the water table can drop and the aquifer and aquitards can be compacted, which in turn can lead to accelerated land subsidence. Groundwater pumping in Bangkok, for instance, has exceeded 0.7 km^3/yr, about two to three times the annual recharge. As a result, subsidence in some parts of Bangkok presently exceeds 100 mm/yr, and since 1978 areas in eastern Bangkok have subsided as much as 8 m (Phien-wej *et al.*, 2006). Such rapid subsidence has resulted in coastal inundation, destruction of mangrove forests and salt intrusion into aquifers (Nutalaya *et al.*, 1996; Phien-wej *et al.*, 2006). While Bangkok is perhaps the most notorious example of the perils from the over-extraction of groundwater, it is by no means the only one. Uncontrolled groundwater pumping in Shanghai during the late 1950s led to subsidence rates localy greater than 100 mm/yr (Ren and Milliman, 1996). This high rate of subsidence was subsequently reduced, and locally reversed, but by the mid 1990s groundwater removal in the southern Yangtze delta exceeded 1 km^3/yr, and local subsidence reached as much as 16 mm/yr (Zhang *et al.*, 2007), about five to six times the rate of global sea-level rise. Groundwater removal in Jakarta and Semarang (northcentral Java) has resulted in subsidence rates as great as 170 mm/yr (Habidin, 2005).

Worse yet is when the extracted groundwater is "fossil"; that is, water that was deposited in the geological past, such as much of the ground water contained within the Ogallala aquifer in the US Great Plains and the Nubian Sandstone in Libya. Tiwari *et al.* (2009) estimate that in northern India between 2002 and 2008, groundwater loss was 54 km^2/yr; at least some – perhaps most – of this groundwater being fossil. Extraction of fossil groundwater can be considered akin to the mining of a non-renewable resource, at least in terms of the human timescale. One result of fossil groundwater removal across the Great Plains, has been a 30-m lowering of the water table, thereby requiring more energy to extract it from ever-deepening aquifer.

If increasing the source(s) is not a likely solution to the water problem – and in many cases it is not – the other obvious option is to economize on the use of water. Because the bulk of withdrawn water in most countries is used for agriculture – 69% globally, according to 1987 data cited by Gleick (1993) – more efficient irrigation can provide an obvious – and immediate – solution to the problem. Depending on climate, soil type and crop type, drip-irrigation, to mention only one example, can require as little as one-tenth the amount of usable water as that needed for more traditional methods. Because drip-irrigation water can be delivered near the plants roots, there is less evaporation and up to 50% greater water retention in the soil. Drip-irrigation is used on nearly all of the agricultural land in Israel, compared with 20% in the USA and less than 2% in India.

Water conservation, no matter what form it takes, may require significant changes in diet and life style. For water-scarce areas such as southern India, this may necessitate a shift from rice to a more water-conservative staple, such as wheat; in fact, this already seems to be happening. Increased planting of less water-intensive crops, such potatoes in lieu of rice, has been suggested in China (*Washington Post*, May 31, 2010), but turning away from the centuries-old rice-based culture is not is not easy to envision. Meat production is another water-intensive agricultural activity that may need to rethought in some parts of the water-scarce world. Since neither of us plays golf, we hasten to point out that the watering of golf courses is a particularly blatant waste of water in water-deficient regions such as southern California and Saudi Arabia.

The future

Water supply and water use face even bleaker future prospects when one factors in population growth (Fig. 4.50a) and climate change. Many countries – again, mostly in North Africa, the Middle East, and western Asia – that remove more than 10% (the global average) of their annually renewable water also have population growth rates that exceed the global average (1.2% per year; global trends represented by dashed horizontal and vertical lines, respectively, in Fig. 4.50). Extreme examples are Yemen, Libya, and Saudi Arabia, which withdraw 120–720% of their renewable water annually and have population growth rates of 3.3%, 2.2%, and 1.95% per year, respectively. Without a more concerted effort to reduce population pressures, water deficits almost surely will be exacerbated to the foreseeable future.

In recent years, the United States, which presently withdrawals and uses about 15% of its renewable water annually, has experienced an annual population growth rate of about 0.9%. But in the lower Colorado River basin, which presently withdraws more than 100% of its annually renewable water (annual water consumption is actually slightly greater than water renewal, not an encouraging sign), has the highest annual population growth rate in the United States: 3.5%. Both its withdrawal and population growth are near those of Yemen (Fig. 4.50b). The adjacent upper Colorado River basin presently removes about 50% of its renewable water annually, and with an annual population growth rate of 2.5% it also may face a water deficit within the next 30 years. Assuming no significant change in water allocations in the southwest USA, Barnett and Pierce (2008) have predicted that Lake Mead, the largest reservoir on the Colorado River, might be dry by the year 2021. Rajagopalan

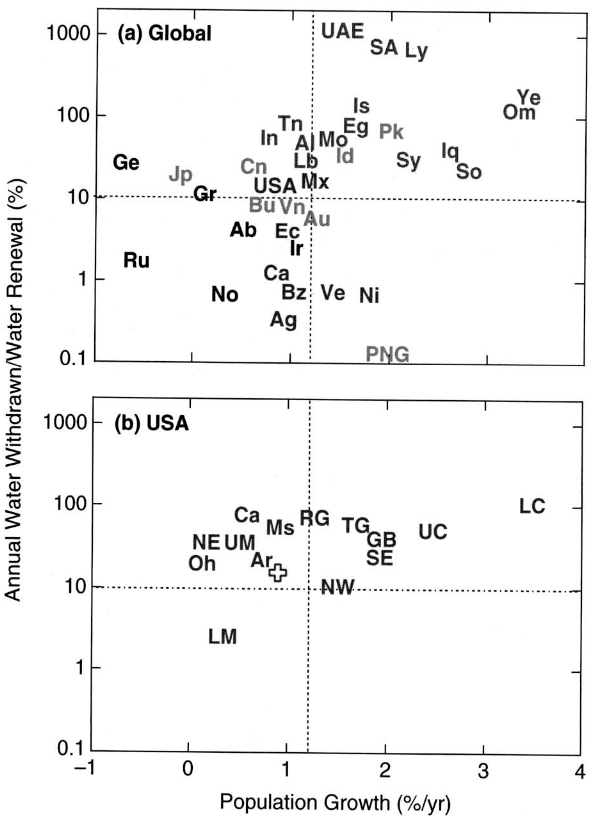

Figure 4.50. (a) Annual water withdrawal/renewable water in the same Middle Eastern, North African and other countries shown in Fig. 4.49 compared with their annual population growth rates. Global water withdrawal to renewable water is 10%, global population rate is 1.2%, as represented by dashed lines. Many Middle East and North African countries, as well as Pakistan and India, have extraction ratios approaching or exceeding 100%, the United Arab Emirates exceeding 1000%. (b) The USA (represented by blue open cross) annually withdraws about 15% of its renewable water supply and has a national growth rate of 0.9%. But several regions within the USA have much higher extraction ratios, the Lower Colorado (primarily Arizona and southern Nevada) extracting more than 100% of its renewable water (the difference being supplied by groundwater extraction); this region also has the highest growth rate in the USA, making it more or less equivalent to Oman and Yemen in (a). Same symbols as is Fig. 4.49. Most data are for the year 2005 as cited in Gleick *et al.* (2006).

et al. (2009) have questioned Bennett's and Pierce's conclusion, but if the PDO in fact did recently shift to a cold phase, as suggested in Chapter 3, the ensuing long-term drought could bring Barnett and Pierce's scenario to reality.

Superimposed on the local impacts of water use and availability is the looming spectre of global climate change. Predicted changes in precipitation, evapotranspiration as well as the frequency of catastrophic events – both floods and droughts – could considerably alter the flow patterns of world rivers (Arnell, 1996; Hirabayashi *et al.*, 2008; Goudie, 2006 and references therein), a subject thoroughly addressed in the Fourth IPCC report (Solomon, 2007). How climate change might affect precipitation and thus river discharge – locally, regionally and globally – however, is subject to considerable debate, depending in large part on the models used and the climatologic constraints placed on these models. Using a coupled ocean–atmosphere–land model, Manabe *et al.* (2004), for instance, predicted a 20–40% increase in Arctic river discharge for at least the next several centuries, substantial increases in tropical mountainous rivers such as the Amazon and Brahmaputra rivers, but reduced precipitation and runoff in many arid and semi-arid regions. Sun *et al.* (2007) reached similar conclusions; i.e. the water-rich may get richer and the water-poor poorer. The shift towards greater precipitation and runoff, together with higher temperatures, throughout the conterminous USA in the latter part of the twentieth century has been suggested as possibly reflecting the early stages of climate change (Mauget, 2003). On the other hand, although Revelle and Waggoner (1983; c.f. Goudie, 2000) predicted a 10% decrease in precipitation and a 76% decrease in water supply in the Rio Grande watershed with a 2° warming in the Rio Grande watershed, over the past 100 years precipitation has increased slightly, perhaps more in response to PDO and ENSO cycles (see Chapter 3) than to climate change; the decline in river discharge has resulted from damming and irrigation (Milliman *et al.*, 2008).

Coupled climate-system models also suggest more extreme precipitation events in the future. Some have argued that catastrophic storms could increase (Emanuel, 1987, 2005); others have pointed out that to date there has been little change in either storm activity or their intensity (Goldenberg *et al.*, 2001; Walsh, 2004; see discussion in Goudie, 2006).

Other impacts from climate change may not be immediately obvious but in the long run might have as great or greater impact than any change in precipitation. Climate warming, for instance, could lead to decreases in snowfall retention, resulting in earlier spring runoff in many higher-latitude rivers (Gleick *et al.*, 2000). A 2° rise in temperature in Lebanon, for example, could result in significantly less snow retention (the melt of which is a major source of spring runoff), which could therefore extend water-shortage periods by 15–30 days, heightening already serious water management problems (Hrieche *et al.*, 2007). Earlier spring melt over the past 50 years, in fact, has been observed throughout the western USA, 60% of which seems to be human-induced (Barnett *et al.*, 2008; Adam *et al.*, 2009). In the upper Mississippi River basin, earlier runoff reflects an earlier, warmer, moister spring, earlier summer dryness, and a warmer, moister autumn and early winter (Baldwin and Lall, 1999), something that also has been noted throughout much of the conterminous USA (Lins and Slack, 1999, and references therein).

Increased atmospheric CO_2 could also result in decreased plant respiration, resulting in decreased evapotranspiration, and thus, ultimately, increased river flow. The melting of permafrost and glaciers, to cite another example, could lead to increased short-term discharge from high-latitude and high-altitude rivers, but over the longer-term it could lead to groundwater recharge and thereby decreased surface runoff (Goudie, 2000). The increased discharge seen throughout much of the Arctic (Peterson *et al.*, 2002; McClelland *et al.*, 2006), however, does not seem related to permafrost melting (McClelland *et al.*, 2004), but rather might reflect decreased evapotranspiration, perhaps the result of increased atmospheric haze (Milliman *et al.*, 2008). Rapid retreat of Arctic and mountain glaciers also could lead to at least a temporary increase in water and sediment delivery (Meigs *et al.*, 2006), but over the long-term, discharge, particularly summer and autumn discharge, could decline considerably.

To complicate future predictions even further, what would happen if river damming led to increased groundwater extraction, thereby accelerating subsidence of coastal lowlands? Coupled with changing precipitation patterns, the future of river-dominated low-lying countries such as Egypt and Bangladesh may be one of increased land-loss from eustatic sea-level rise, subsidence and coastal erosion, thus accelerating existing social and economic problems in these heavily populated countries (Milliman *et al.*, 1989; Milliman and Haq, 1996; Syvitski *et al.*, 2009).

Within this bleak backdrop, however, one can glean a somewhat more optimistic picture, as suggested from the seminal studies of Gleick (1993, 1999) and Shiklomanov and Rodda (2003): water problems related to climate change in many parts of the world may only exacerbate the negative impacts by local and regional human activities. As such, local solutions could modulate many of the long-range problems. Changing land-use practices and how water is used, coupled with reduced population growth, could obviously help solve present and future "water crises." Allowing rivers to retain (or recover) their natural settings in terms of water and sediment discharge could do much to mitigate the negative impacts of climate change, whether these impacts prove to be as catastrophic as many predict or as benign as many of us hope. Increasingly, governmental agencies and planners will need to work in concert with earth scientists and hydrologists if they expect to address and solve problems that have no easy answers.

Appendices. Global river data base

Introduction

This appendix contains our global database for which we have divided the globe into seven broad regions, North America (along with Central America), South America, Africa, Europe, Eurasia, Asia (including, for ease of graphical presentation, Russia), and Oceania (Appendix Figure A.1). The broad regions are subdivided into even smaller areas (44 in total), each represented by a series of three maps and, in tabular form, the characteristics of rivers within each region; all 1534 rivers that empty into the coastal ocean.

Some of the 44 areas encompass a single country (e.g. Mexico, Australia); others encompass a number of smaller countries (e.g. Nicaragua, Costa Rica, and Panama; Ivory Coast, Ghana, Togo, and Benin). North and Central America are represented by eight areas (Canada; USA conterminous; USA Alaska; Hawaii and Fiji; Mexico; Belize, Guatemala, El Salvador, and Honduras; Nicaragua, Costa Rica, and Panama; Cuba, Jamaica, Hispaniola, and Puerto Rico), South America by five (Colombia and Venezuela; Ecuador and Peru; Chile and Argentina; Brazil; French Guiana, Suriname, and Guyana), Africa by eight (Egypt, Libya, and Sudan; Tunisia, Algeria, Morocco, and Western Sahara; Senegal, Gambia, Guinea Bissau, Guinea, Sierra Leone, and Liberia; Ivory Coast, Ghana, Togo, and Benin; Nigeria, Cameroon, Equatorial Guinea, Gabon, and Republic of the Congo; Democratic Republic of Congo and Angola; Namibia, South Africa, Mozambique, and Madagascar; Tanzania, Kenya, and Somalia), Europe by ten (Iceland; Scandinavia; Estonia, Latvia, Lithuania, and Poland; Germany, Belgium, and The Netherlands; United Kingdom; France; Portugal and Spain; Italy; Albania, Croatia, and Greece; Bulgaria, Romania, and Ukraine), Eurasia by three (Georgia and Turkey; Israel and Lebanon; Saudi Arabia, Yemen, Iraq, and Iran), Asia by six (Russia; Pakistan, India, Sri Lanka, and Bangladesh; Malaysia, Burma, Thailand, and Vietnam; China; Taiwan; Japan and Korea), and Oceania by four (Philippines; Indonesia and Papua New Guinea; Australia; New Zealand). The total number of countries represented in our database is 109.

For each of these 44 areas we present three maps, the first locating the rivers in the data base and their general drainage patterns. Regional morphology is also shown, based on the five categories used in the database and our river classification: coastal plain (<100 m), lowland (100–500 m), upland (500–1000 m), mountain (1000–3000 m), and high mountain (>3000 m) (see Chapter 2). A second map presents a simplified regional geology of the area (derived from Larsen and Pittman, 1985), showing the distribution of five types of rocks: Cenozoic Sedimentary/Metamorphic, Cenozoic Igneous, Mesozoic Sedimentary/Metamorphic, Mesozoic Igneous, and Pre-Mesozoic. The third map shows regional surface runoff (derived from Korzoun *et al.*, 1977) together with monthly flow histograms of some representative rivers within the area, mostly derived from RivDis (http://www.rivdis.sr.unh.edu).

Tables and data categories

For each area we present in tabular form the critical characteristics for the rivers shown in the preceding maps. Some of these entries can be found in other databases: e.g. basin area, mean annual water, sediment, and dissolve solid discharge. But we also show less commonly reported characteristics, such as basin morphology, climate, and geology of the headwaters, etc., as collectively they also help define the fluvial regime. Because of the limited space in the book itself, there are categories that we can only present in the digital database (www.cambridge.org/milliman); in the following sentences and paragraphs these are identified by italics.

River name

Many rivers are known by more than one name, the international name often being different from the local name. Most non-Chinese refer to the Yangtze and Yellow rivers, rather than the Changjiang and Huanghe. Similar problems can be found for the Rhine/Rhein and Danube/Donau. Indonesian rivers can be denoted by their Dutch or native spellings (e.g. Tjitarum vs. Citarum). Although many refer to the Ganges (Ganga to Indians) and Brahmaputra as separate rivers (or as a hyphenated name), Bangladeshis refer to the combined river system as the Padma. Rivers that drain more than one country or rivers that form the boundary between two countries also can have more than one name. Occasionally a river will have its name changed for political reasons, only to have it changed back again: the Congo River was the Zaire for

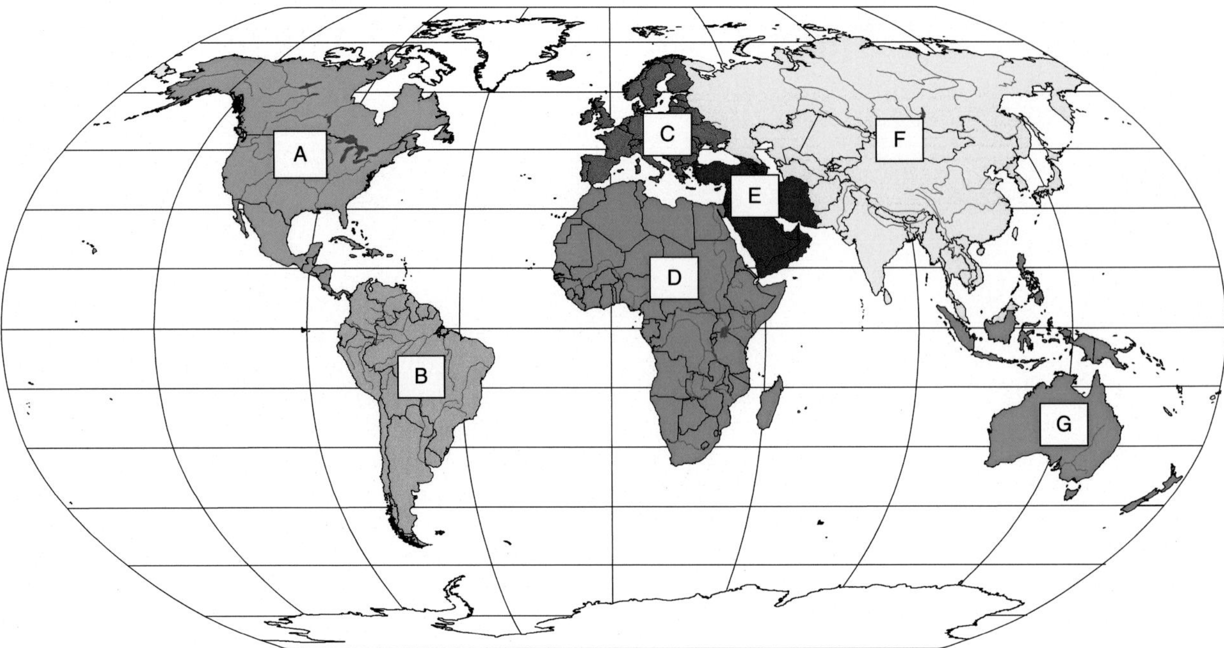

Figure A.1. Location map of the seven broadly defined regions of the database.

several decades, before reverting back to the Congo in the 1990s.

We have tried to maintain consistency in our reference to rivers, generally using the common name used in English-speaking countries. However, we admit to have succumbed to some personal biases, such as referring to the Changjiang and Huanghe rather than Yangtze and Yellow rivers – 1 billion Chinese cannot be wrong. *We include commonly used alternative names in the electronic version of this database;* Meybeck and Ragu (1996) also have listed many variations of river names in their Table II.

Country

By country, we refer to where the river exits to the coastal ocean (i.e. river's mouth) rather than the country of origin (i.e. headwaters), even though the river may mostly flow through another country. We therefore list the countries for the Rhine, Colorado, and the Ganges–Brahmaputra as The Netherlands, Mexico, and Bangladesh, respectively, rather than Germany, USA, and Nepal/India.

Ocean

In our original GLORI database (Milliman *et al.*, 1995) we listed the ocean into which the river emptied: Atlantic, Pacific, Indian, or Arctic. But as the list of rivers has increased, we have become more specific as to geographic location. Rather than listing the Arctic as the ocean into which a river drains, for example, we now list Beaufort Sea, Chukchi Sea, Kara Sea, White Sea, etc.

Basin area

The area of a river's drainage basin is assumed to refer to the entire area upstream of the river mouth. Although the total drainage basin areas of many large rivers are well-documented (although not consistent between various authorities – see Chapter 1), cited basin areas for other rivers may refer to the area upstream of a particular hydrologic station, which is smaller – sometimes much smaller – than the total drainage basin. Because we have taken some basin areas from the GRDC metadata list, the basin areas listed in our tables may be too small; the relative error may be greater in smaller drainage basins. Given the inexactitude of our numbers and in keeping with other numerical data listed in the database, we have rounded areas to the first two digits; a basin area of 127 500 km^2, for example, becomes 130 000 km^2.

River length

We consider river length to represent the distance from the farthest headwater to the river's mouth. As with basin area, however, lengths of some rivers are considered to end at the seaward-most gauging station. Others cited river lengths may begin at the confluence of certain tributaries, others at the mouth of a downstream lake. As such, rivers are often longer than their reported lengths would suggest.

Maximum elevation

As seen in Chapter 2, topography is an important controlling factor in determining the sediment load of a river (Ahnert,

1970; Pinet and Souriau, 1988; Milliman and Syvitski, 1992), in part because elevation can serve as a surrogate for a variety of physical factors that control erosion: tectonic activity, erosion by ice and glaciers, orographic precipitation, landslides, etc.

Some compendia (e.g. *Taiwan Hydrologic Yearbook*) list the maximum elevation of a river basin, but others refer to mean elevation or topographic relief of sub-basins. For many rivers we were forced to estimate maximum elevations from topographic maps. In our database we only report the first two digits of elevation – 1241 m, for instance, rounds to 1200 m.

Elevation categories (in digital database)

Based on maximum elevations, we divide river basins into five morphological classes.

***Coastal plain** (<100 m)*
***Lowland** (100–500 m)*
***Upland** (500–1000 m)*
***Mountain** (1000–3000 m)*
***High mountain** (>3000 m)*

Table A.1. *Categories and abbreviations used in climate classification.*

Temperature	
	Tropical (**Tr**) – summer and winter >20°
	Subtropical (**STr**) – summer >20°; winter >10°
	Temperate (**Te**) – summer >10°; winter >0°
	Sub Arctic (**SAr**) – summer >10°; winter <0°
	Arctic (**Ar**) – summer >0°; winter <<0°
Runoff	
	Arid (**A**) – Runoff <100 mm/yr
	Subarid (**SA**) – Runoff 100–250 mm/yr
	Humid (**H**) – Runoff 250–750 mm/yr
	Wet (**W**) – Runoff >750 mm/yr
Season of maximum runoff	
	Runoff more or less constant throughout the year (**C**)
	Maximum runoff generally in Winter (**W**)
	Maximum runoff generally in Spring (**Sp**)
	Maximum runoff generally in Summer (**S**)
	Maximum runoff generally in Autumn (**Au**)
	Desert (generally no flow; rainfall extremely episodic) (**D**)

Climate

There are many ways to classify climatic regimes, but to be useful, any classification should include watershed temperature and the magnitude and timing of flow. Meybeck *et al.* (1989), for example, include the following climates: glacial (water from glacier melt, normally in early summer); nival (snow melt, high flow in late spring, early summer); pluvial (peak flow in late autumn to early winter); dry tropical (high flow during summer rainy season, but less regular flow away from the equator); monsoon (contrasting dry and wet seasons); equatorial (regular, year-round runoff); desert (irregular, non-perennial rivers). The Köppen–Geiger (1954) system, a comprehensive classification used by many workers to categorize global climate, is based on air temperature (tropical, arid, warm temperate, snow, and polar, as well as the number of months during which temperatures exceed or are less than specific values) and the season of least rainfall. Haines *et al.* (1988) attempted to classify global rivers on the basis of seasonal variation in river flow, but the Köppen–Geiger classification, which was subsequently modified by Beckinsale (1969) and further simplified by Burt (1992), remains a favorite for many geographers and climatologists.

Unfortunately, the Köppen–Beckinsale–Burt climate classification has several flaws that ultimately persuaded us not to use it in our database. First, the symbols in this classification are sufficiently complex that one cannot easily distinguish various zones without referring to a classification key: quick, what is the difference between an AM and an AC/AW climate? Second, classifying a fluvial climate based on the timing of low flow seems counter-intuitive in terms of classifying rivers, since peak discharge more often defines the character of the river (and the estuary into which it enters). Accordingly we present a climatic classification that gives the three climatologic categories that we think best define the fluvial regime: mean air temperature, pre-diversion runoff (which effectively integrates precipitation and evapotranspiration over the entire watershed), and the season of maximum discharge. The categories used in this classification are shown in Appendix Table A.1.

In some rivers, however, average climate can be misleading. The Santa Clara River (between Los Angeles and Santa Barbara, California), for instance, has a subtropical climate and an average runoff less than 50 mm/yr, most of which occurs between January and early March. We therefore classify this river's climate as Subtropical–Arid–Winter (STr-A-W). In contrast, the Tijuana River, bordering Baja California and California, has been known to remain dry for many years; accordingly, we classify the Tijuana as STr-A-D (the D standing for desert). But during the cold PDO between the mid 1940s to late 1960s, the Santa Clara's precipitation and therefore river discharge were episodic, such that the Santa Clara could have been classified as STr-A-D. With the return of the warm PDO and re-intensification of El Niño events in the early 1970s, rainfall increased and the Santa Clara discharge in some years was greater than 250 mm/yr (STr-H-W). Fortunately, these climatic shifts are most apparent in smaller rivers, particularly subarid basins.

Another problem is that basin-wide climate may not take into account local climatic variations. This is particularly a problem in larger rivers, where different tributaries may have entirely different climates (e.g. the subarid Missouri and Arkansas in contrast to the humid Ohio, all of which drain into the Mississippi River; see Chapter 4). Moreover, mountainous parts of a river basin may have entirely different temperature and precipitation regimes than the lower portions of that basin.

Despite these uncertainties, the modified classification used in the database appears to delineate the long-term mean climatic factors that help control a basin's runoff.

Discharge (Q)

We list mean annual discharge measured at the seaward-most hydrologic station. These values are expressed in terms of km^3/yr (multiplying these discharge values by 32 gives an approximate mean flow in terms of m^3/s). Because the seaward-most station can be some distance landward of the river mouth, discharge can be under-estimated. The Amazon River, for example, includes 1700×10^3 km^2 of watershed (>25% of the area) seaward from Obidos, the river's downstream-most gauging station. Assuming an average runoff of 1 m/yr, a reasonable estimate considering basin-wide precipitation (see Fig. 2.2), the missing area downstream of Obidos may contribute 1700 km^3/yr of water to the Amazon's discharge, which is somewhat greater than the mean annual flow of the Congo! On the other hand, watersheds whose lower reaches are arid may lose considerable amounts of water through evapotranspiration, groundwater recharge or water removal for irrigation (e.g. Nile, Indus); as such they discharge less water than gauging station measurements would suggest.

Discharge for many rivers has changed appreciably in recent years because of river diversion, dam/reservoir construction and irrigation. Where data are available we list pre-diversion discharge in parentheses, as these pre-diversion data allow us to better define natural runoff (thus aiding our climate classification scheme mentioned above) as well as to delineate natural environmental factors that control discharge and load. Post-diversion discharge indicates how much fresh water, on average, is presently discharged to the coastal ocean.

Runoff (in digital database)

Runoff is calculated as the discharge divided by drainage basin area. Pre-diversion discharge, if available, is used to calculate runoff, since this helps define watershed climate.

Geology

The quantity and quality of both suspended and dissolved solids within a river system depend on the nature of the soils and rocks of the watershed, which is a function of both lithology (erodibility of mudstone > limestone > extrusive igneous > sandstone > intrusive igneous > metamorphic rocks) and age (younger rocks generally considered to be more erodible than older rocks). Assuming that much/most of the suspended and dissolved solids are derived from the higher elevations within the watershed, we have characterized the general surface geology of the headwaters using Larsen's and Pittman's (1985) bedrock geology of the world.

Cenozoic Sedimentary and Metamorphic (**Cen S/M**)
Cenozoic Igneous (**Cen Ig**)
Mesozoic Sedimentary and Metamorphic (**Mes S/M**)
Mesozoic Igneous (**Mes Ig**)
Pre-Mesozoic (**Pre-Mes**)

Sediment load (TSS)

We list mean sediment load (Q_s; 10^6 t/yr (Mt/yr)) transported annually by the river. Hidden in these values, of course, are many caveats, such as the validity of reported loads and the significance of event-driven discharge, etc., many of which are discussed in some detail in Chapters 1 and 3. Moreover, some/much of the sediment reported at an upstream station may not actually reach the ocean, but rather is deposited along the downstream flood plain. Nevertheless, the >750 sediment loads in our database serve as an important resource in terms of delineating environmental factors controlling sediment loads in various types of rivers discussed in Chapter 2.

Nearly all reported values for sediment discharge represent suspended loads, not total sediment load, which would include bed load. In most rivers, however, bed load is assumed to represent no more than 10% of the suspended load (Milliman and Meade, 1983), meaning that for first-order purposes we can assume that suspended load approximates total load. This assumption breaks down in small mountainous rivers, where steep terrain and flashy flow events can lead to much greater bed-load transport. As with other numbers, we report only the first two digits of the load.

Sediment yield (in digital database)

Sediment yield is the mean annual sediment load normalized to basin area, expressed as t/km²/yr, analogous to the relationship of runoff to discharge. Using sediment yield is helpful in relating loads of big and small rivers, but also in terms of visualizing the relative denudation of a river basin; see Chapter 2 for further discussion.

Sediment concentration (in digital database)

We calculate mean suspended sediment concentration (Q_{cs}), reported as mg/l, by dividing mean sediment load by mean discharge.

Dissolved-solid load (TDS)

As with sediment load, dissolved load (Q_d) values are expressed in Mt/yr. Approximately 80% of the numbers reported here come from the Meybeck and Ragu (1996) database.

Dissolved yield (in digital database)

We calculate this parameter by dividing dissolved load by basin area, the same as we have for sediment yield – Mt/km²/yr.

Dissolved concentration (in digital database)

We calculate dissolved concentration (Q_{cd}) in an analogous way to Q_{cs}, reported in mg/l.

References

We list the references from which we derived most or all of the values for each river. In some cases more recent references may have corrected earlier numbers.

Appendix A North and Central America

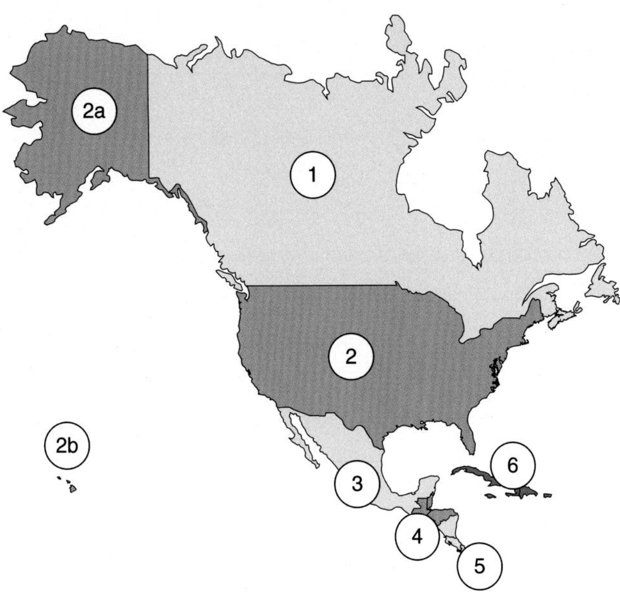

Figure A1

(1) Canada
(2) United States (contiguous)
(2a) United States (Alaska)
(2b) Hawaii plus Fiji
(3) Mexico
(4) Belize, Guatemala, El Salvador, Honduras
(5) Nicaragua, Costa Rica, Panama
(6) Cuba, Jamaica, Haiti, Dominican Republic, Puerto Rico

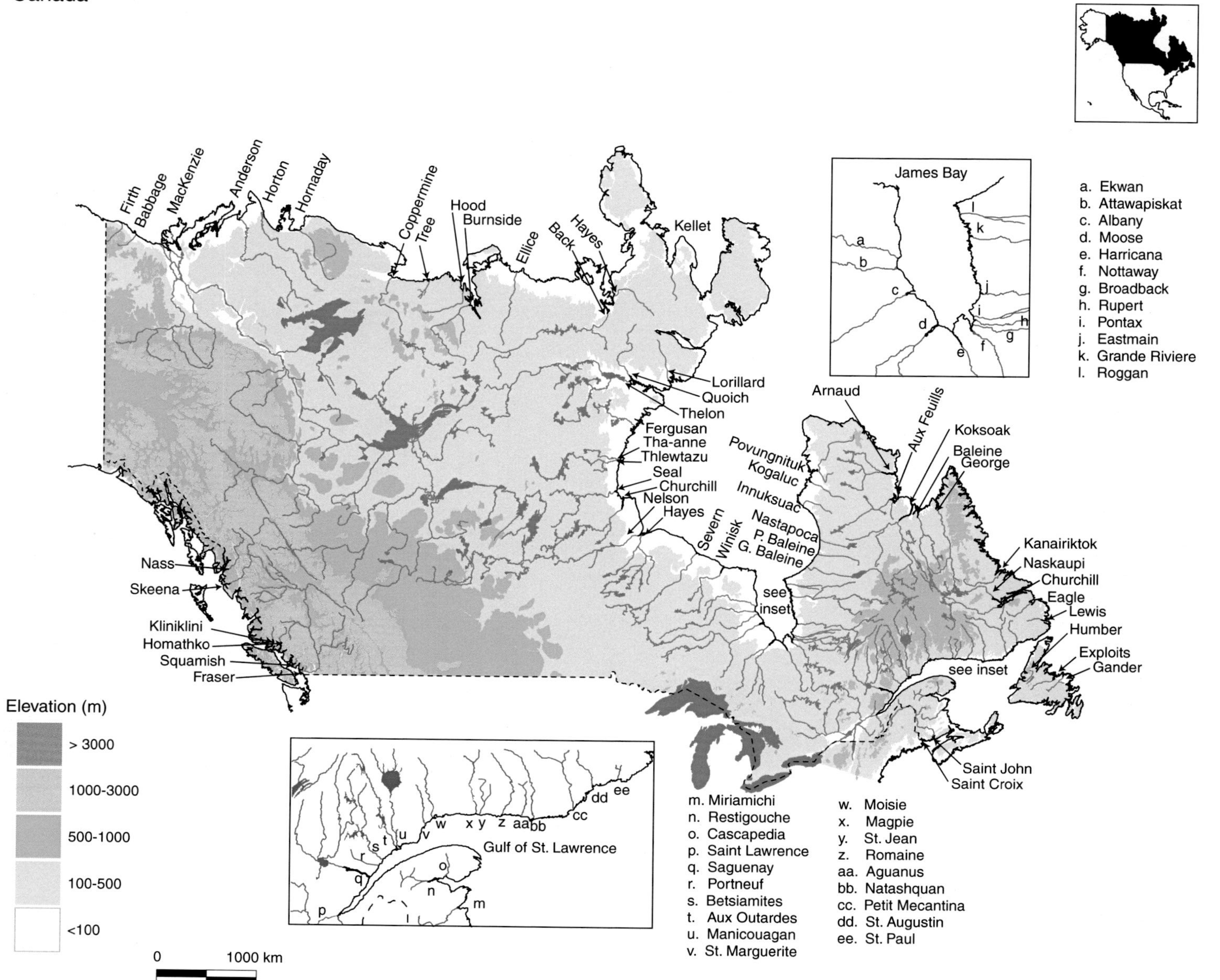
Canada
Firth
Babbage
MacKenzie
Anderson
Horton
Hornaday
Coppermine
Tree
Hood
Burnside
Ellice
Back
Hayes
Kellet
Lorillard
Quoich
Thelon
Fergusan
Tha-anne
Thlewtazu
Seal
Churchill
Nelson
Hayes
Severn
Winisk
Povungnituk
Kogaluc
Innuksuac
Nastapoca
P. Baleine
G. Baleine
see inset
Arnaud
Aux Feuills
Koksoak
Baleine
George
Kanairiktok
Naskaupi
Churchill
Eagle
Lewis
Humber
Exploits
Gander
see inset
Saint John
Saint Croix
Nass
Skeena
Kliniklini
Homathko
Squamish
Fraser
James Bay
a. Ekwan
b. Attawapiskat
c. Albany
d. Moose
e. Harricana
f. Nottaway
g. Broadback
h. Rupert
i. Pontax
j. Eastmain
k. Grande Riviere
l. Roggan
Gulf of St. Lawrence
m. Miriamichi
n. Restigouche
o. Cascapedia
p. Saint Lawrence
q. Saguenay
r. Portneuf
s. Betsiamites
t. Aux Outardes
u. Manicouagan
v. St. Marguerite
w. Moisie
x. Magpie
y. St. Jean
z. Romaine
aa. Aguanus
bb. Natashquan
cc. Petit Mecantina
dd. St. Augustin
ee. St. Paul
Elevation (m)
> 3000
1000-3000
500-1000
100-500
<100
0
1000 km

Figure A2

Canada

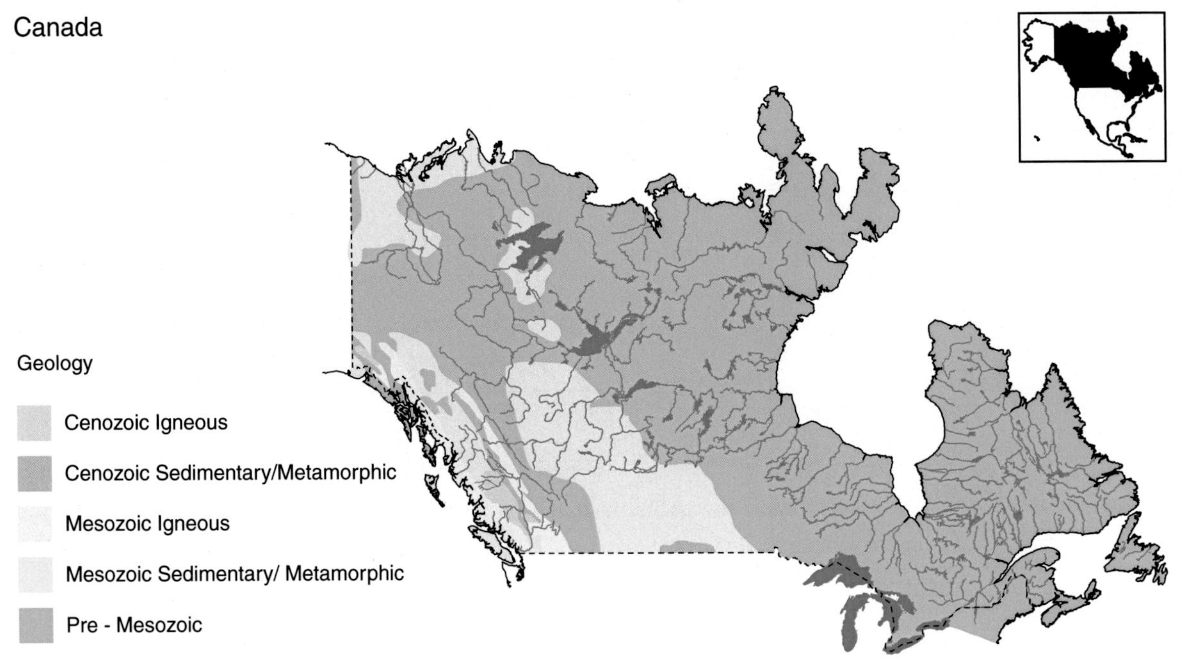

Surface Runoff (mm/yr)

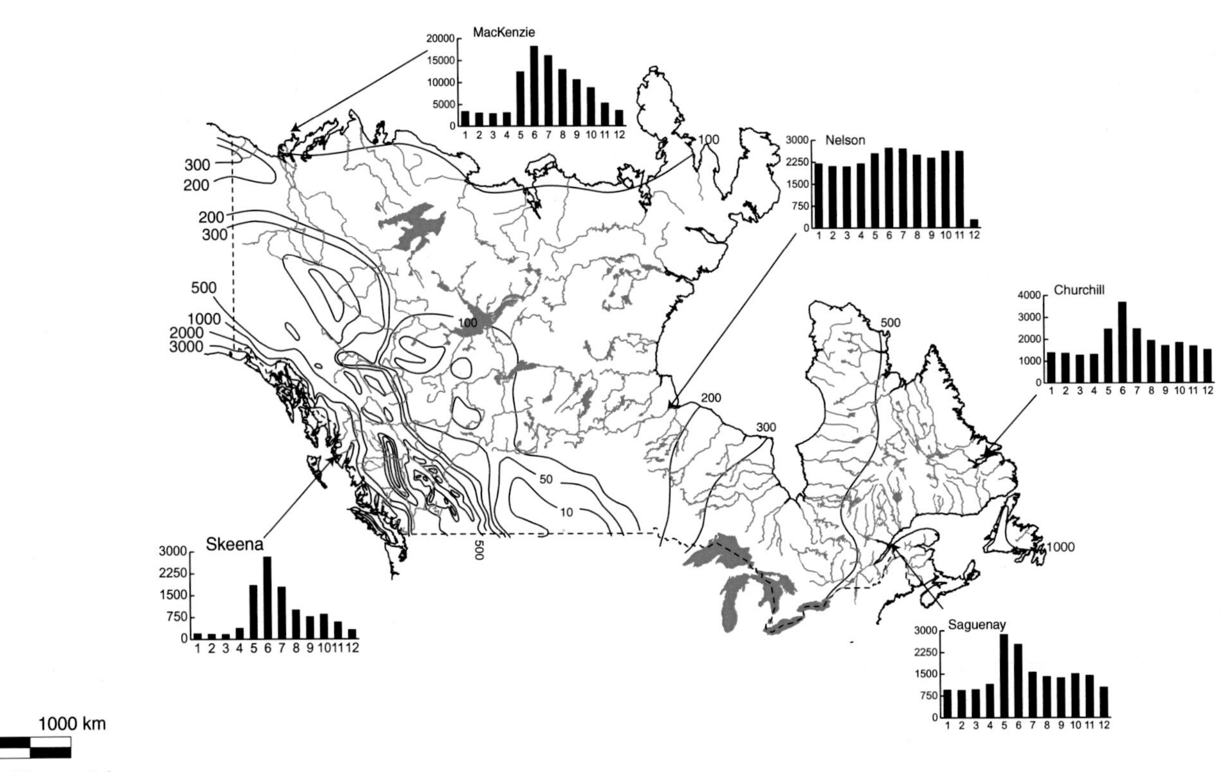

Figure A3

Table A1.

River name	Ocean	Area (10^3 km^2)	Length (km)	Max_elev (m)	Climate	Geology	Q (km^3/yr)	TSS (Mt/yr)	TDS (Mt/yr)	Refs.
Canada										
Aguanus	G. St. Lawrence	5.3	250	>500	SA-H-S	PreMes				
Albany	James Bay	140	980	>200	SA-H-S	PreMes	44		5.2	21, 36
Anderson	Beaufort Sea	56	750	<200	A-A-S	PreMes	3.5		1.1	20, 25, 27
Arnaud	Ungava Bay	49	380	>200	SA-SA-S	PreMes	21		0.29	21, 36
Attawapiskat	James Bay	50	760	>200	SA-H-S	PreMes	20	0.2	0.14	21, 27, 34, 36
Aux Feuilles	Ungava Bay	42	480	>200	A-SA-S	PreMes	19		0.28	10, 21
Aux Outardes	G. St. Lawrence	19	480	1300	SA-H-S	PreMes	11		0.11	3, 14, 18, 21
Babbage	Beaufort Sea	5	150	>1000	A-A-S	Mes S/M	1	3.5		11
Back	Arctic	94	970	>200	A-SA-S	PreMes	16		0.18	21
Baleine	Ungava Bay	32	430	>200	A-H-S	PreMes	21			
Betsiamites	G. St. Lawrence	19	440	>200	SA-H-S	PreMes	11			37
Broadback	James Bay	21	450	>200	SA-H-S	PreMes	12		0.21	10, 21
Burnside	Arctic	18	330	>200	A-SA-S	PreMes	4			12
Cascapedia	G. St. Lawrence	11	120	>200	SA-H-S	PreMes				
Churchill	Hudson Bay	290	1800	>200	SA-A-S	PreMes-Mes	26 (40)		3.4	26
Churchill	Labrador Sea	92	1000	550	A-H-S	PreMes	61			7, 13, 15, 26
Coppermine	Arctic	51	850	>200	A-A-S	PreMes	2.6 (11)		0.08	9, 10, 21, 27
Eagle	Labrador Sea	11	230	>500	SA-H-S	PreMes	7.5			5
Eastmain	James Bay	47	680	>200	SA-H-S	PreMes	29		0.38	26
Ekwan	James Bay	10	480	<200	SA-H-S	PreMes	3.5			27
Ellice	Arctic	17	290	>200	A-SA-S	PreMes	2.7		0.01	12, 21
Exploits	Atlantic (NW)	14	240	490	SA-H-S	PreMes				7
Fergusan	Hudson Bay	12		>200	A-SA-S	PreMes				20
Firth	Arctic	5.7	160	>200	A-SA-S	PreMes	1.1			14
Fraser	Pacific (NE)	230	1400	4000	Tr-H-S	PreMes - Mes	120	20	11	4, 9, 21, 23, 28
Gander	Atlantic (NW)	5	180	>200	SA-H-S	PreMes	3.8			12, 27
George	Ungava Bay	41	550	>200	A-H-S	PreMes	29		0.47	3, 21, 26, 36
Grand Baleine	James Bay	43	720	200	SA-H-S	PreMes	21		0.3	3, 9, 13, 21
Grande Riviere	James Bay	98	890	>200	SA-H-S	PreMes	66		0.7	20, 21
Harricana	James Bay	29	550	>200	SA-H-S	PreMes	18		1.2	21, 27, 36
Hayes	Hudson Bay	110	480	>200	A-SA-S	PreMes	18			21, 26, 27, 36
Hayes	Arctic	18		>200	SA-SA-S	PreMes	4			20
Homathko	Pacific (NE)	5.7	140	>1000	Te-W-S	Mes S/M	8.8	4.3	3.3	21, 32
Hood	Arctic	10	400	>100	A-SA-S	PreMes	2.6			20

Table A1. (*Continued*)

River name	Ocean	Area (10^3 km^2)	Length (km)	Max_ elev(m)	Climate	Geology	Q (km^3/yr)	TSS (Mt/yr)	TDS (Mt/yr)	Refs.
Hornadby	Beaufort Sea	15	300	>200	A-SA-S	PreMes	3			20
Horton	Beaufort Sea	33	440	>200	Ar-A-S	PreMes				1
Humber	Atlantic (NW)	7.9	120	700	SA-H-S	PreMes				7
Innuksuac	Hudson Bay	11	380	>200	A-H-S	PreMes	3			9, 20, 26
Kanairiktonk	Labrador Sea	13	26	<200	SA-H-S	PreMes	8.3			14
Kellet	Beaufort Sea	9.4		280	Ar-SA-S	PreMes				1
Kliniklini	Pacific (NE)	6.5	150	>1000	Te-SA-S	Mes I	10	5	0.52	21, 32
Kogaluc	Hudson Bay	11	190	<200	A-H-S	PreMes	5			26
Koksoak	Ungava Bay	130	870	>200	A-H-S	PreMes	54 (76)		1.6	21, 26, 27
Lewis	Atlantic (NW)	2	54	960	SA-H-S	PreMes	1.1	0.01		21
Lorillard	Hudson Bay	11		>100	A-H-S	PreMes				20
MacKenzie	Beaufort Sea	1800	4200	3600	SA-SA-S	Mes S/M	310	100	64	6, 21, 22, 34
Magpie	G. St. Lawrence	7.6	240	>500	SA-W-S	PreMes	5.7			14
Manicouagan	G. St. Lawrence	46	560	>500	SA-H-S	PreMes	32		0.28	13, 21
Miramichi	G. St. Lawrence	15	250	760	SA-H-S	PreMes	10			7, 26
Moisie	G. St. Lawrence	20	500	530	SA-W-S	PreMes	14		0.28	7, 21
Moose	James Bay	110	550	320	SA-H-S	PreMes	43	0.4	5.8	21, 24, 34
Naskaupi	Labrador Sea	18	220	>200	SA-H-S	PreMes	11			10
Nass	Pacific (NE)	21	380	2400	Te-W-S	Mes S/M	30			27, 29, 36
Nastapoca	Hudson Bay	13	360	<200	SA-H-S	PreMes	8			9, 20, 26
Natashquan	G. St. Lawrence	16	410	>500	SA-W-S	PreMes	13		0.12	10, 21, 27
Nelson	Hudson Bay	1100	2700	3400	SA-A-S	PreMes	89		21	1, 31
Nottaway	James Bay	66	780	>200	SA-H-S	PreMes	37	1	1.4	17, 21, 27
Petit Baleine	Hudson Bay	16	380	>200	SAr-H-S	PreMes	3.2			1, 26
Petit Mecatina	G. St. Lawrence	20	550	500	SA-W-S	PreMes	17		0.14	9, 10, 21, 36
Pontax	James Bay	6.1	300	>100	SA-H-S	PreMes				20
Portneuf	G. St. Lawrence	2		>500	SA-H-S	PreMes	2			14
Povungnituk	Hudson Bay	28	380	>200	SA-H-S	PreMes	12			9, 20, 26
Quoich	Hudson Bay	29		>200	A-A-S	PreMes	5.3		0.03	9
Restigouche	G. St. Lawrence	27	210	200	SA-H-S	PreMes				27
Roggan	James Bay	9.6		>100	SA-H-S	PreMes				20
Romaine	G. St. Lawrence	14	500	>500	SA-H-S	PreMes	11	0.16		1, 19, 27
Rupert	James Bay	43	610	>500	SA-H-S	PreMes	28		0.56	21, 26, 27

Table A1. (*Continued*)

Saguenay	G. St. Lawrence	88	700	1100	SA-H-S	PreMes	55	0.4	1.4	21, 34
St. Augustin	G. St. Lawrence	10	230	<450	SA-H-S	PreMes	6.2			9
St. Croix	G. of Maine	4.5		200	SA-H-S	PreMes	3			
St. Jean	G. St. Lawrence	5.6		>500	SA-H-S	PreMes	3.4	0.24		21
St. John	G. St. Lawrence	55	670	760	SA-H-S	PreMes	35			7, 8
St. Lawrence	G. St. Lawrence	1200	4000	1900	SA-H-S	PreMes	340	4.6	62	4, 11, 21, 26, 30, 35
St. Marguerite	G. St. Lawrence	6.2	340	>500	Te-H-S	PreMes				
St. Paul	G. St. Lawrence	6.3	180	<450	SA-H-S	PreMes				
Seal	Hudson Bay	48		>200	SA-SA-S	PreMes	11		0.36	10, 21, 26
Severn	Hudson Bay	100	980	>500	SA-SA-S	PreMes	15 (23)	0.22		21
Skeena	Pacific (NE)	55	580	2400	Te-W-S	Mes S/M	55	11	3.4	2, 21, 29
Squamish	Pacific (NE)	3.3	110	2400	Te-H-S	Mes S/M	7.8	1.8		16, 33
Tha-anne	Hudson Bay	29		>200	SA-H-S	PreMes	6.3			1, 20
Thelon	Hudson Bay	240	900	>200	A-SA-S	PreMes	44		1.1	9, 21, 24, 27
Thlewtazu	Hudson Bay	27		>200	SA-SA-S	PreMes	7			1, 14
Tree	Arctic	5.8		500	A-SA-S	PreMes				
Winisk	Hudson Bay	67	475	670	SA-H-S	PreMes	22			21, 23

References:

1. Arctic RIMS Website, rims.unh.edu/data.shtml; 2. Binda *et al.*, 1986; 3. Canadian National Committee, 1972; 4. Center for Natural Resources Energy and Transport (UN), 1978; 5. Clair *et al.*, 1994; 6. Culp *et al.*, 2005; 7. Cunjak and Newbury, 2005; 8. Dai and Trenberth, 2002; 9. Environment Canada, 1984; 10. Esser and Kohlmaier, 1991; 11. Forbes, 1981; 12. GEMS website, www.gemstat.org; 13. Global River Discharge Database, http://www.rivdis.sr.unh.edu/; 14. GRDC website, www.gewex.org/grdc; 15. Gresswell and Huxley, 1966; 16. Hickin, 1989; 17. Kranck and Ruffman, 1981; 18. Lajoie *et al.*, 2007; 19. Long *et al.*, 1982; 20. McClelland *et al.*, 2006; 21. Meybeck and Ragu, 1996; 22. Milliman and Meade, 1983; 23. Milliman, 1980; 24. Milner *et al.*, 2005; 25. Mulholland and Watts, 1982; 26. Natural Resources Canada, www.geobase.ca/geobase/en/data/nhn; 27.Rand McNally, 1980; 28. Reynoldson *et al.*, 2005; 29. Richardson and Milner, 2005; 30. Rondeau *et al.*, 2000; 31. Rosenberg *et al.*, 2005; 32. Syvitski and Farrow, 1983; 33. Syvitski and Saito, 2007; 34. J. P. M. Syvitski, personal communication; 35. Thorp *et al.*, 2005; 36. UNESCO (WORRI), 1978; 37. Walsh and Vigneault, 1986

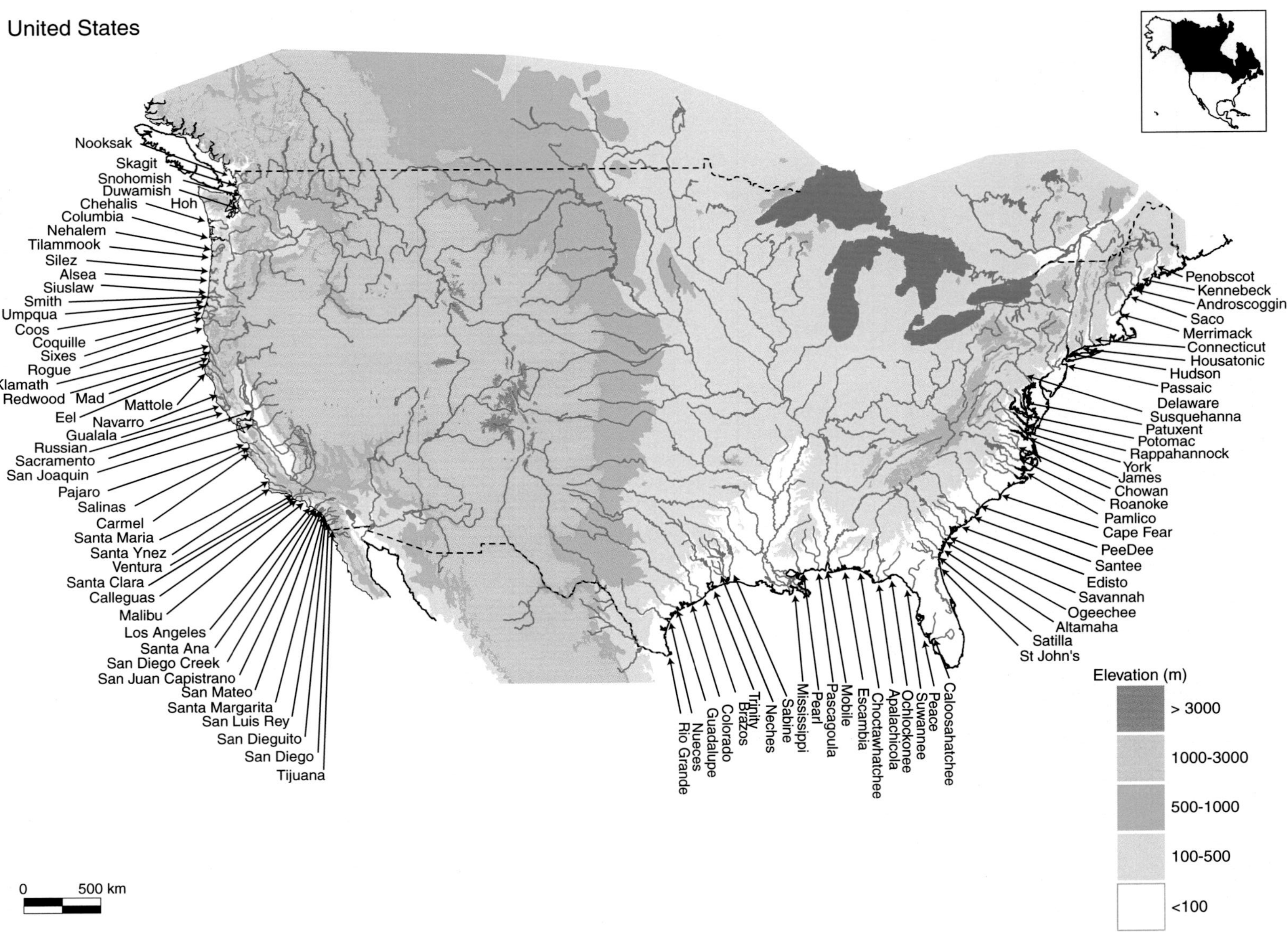
United States
Nooksak
Skagit
Snohomish
Duwamish
Chehalis
Hoh
Columbia
Nehalem
Tilammook
Silez
Alsea
Siuslaw
Smith
Umpqua
Coos
Coquille
Sixes
Rogue
Klamath
Redwood
Mad
Mattole
Eel
Navarro
Gualala
Russian
Sacramento
San Joaquin
Pajaro
Salinas
Carmel
Santa Maria
Santa Ynez
Ventura
Santa Clara
Calleguas
Malibu
Los Angeles
Santa Ana
San Diego Creek
San Juan Capistrano
San Mateo
Santa Margarita
San Luis Rey
San Dieguito
San Diego
Tijuana
Rio Grande
Nueces
Guadalupe
Colorado
Trinity
Brazos
Neches
Sabine
Mississippi
Pearl
Pascagoula
Mobile
Escambia
Choctawhatchee
Apalachicola
Ochlockonee
Suwannee
Peace
Caloosahatchee
Penobscot
Kennebeck
Androscoggin
Saco
Merrimack
Connecticut
Housatonic
Hudson
Passaic
Delaware
Susquehanna
Patuxent
Potomac
Rappahannock
York
James
Chowan
Roanoke
Pamlico
Cape Fear
PeeDee
Santee
Edisto
Savannah
Ogeechee
Altamaha
Satilla
St John's
Elevation (m)
> 3000
1000-3000
500-1000
100-500
<100
0
500 km

Figure A4

United States

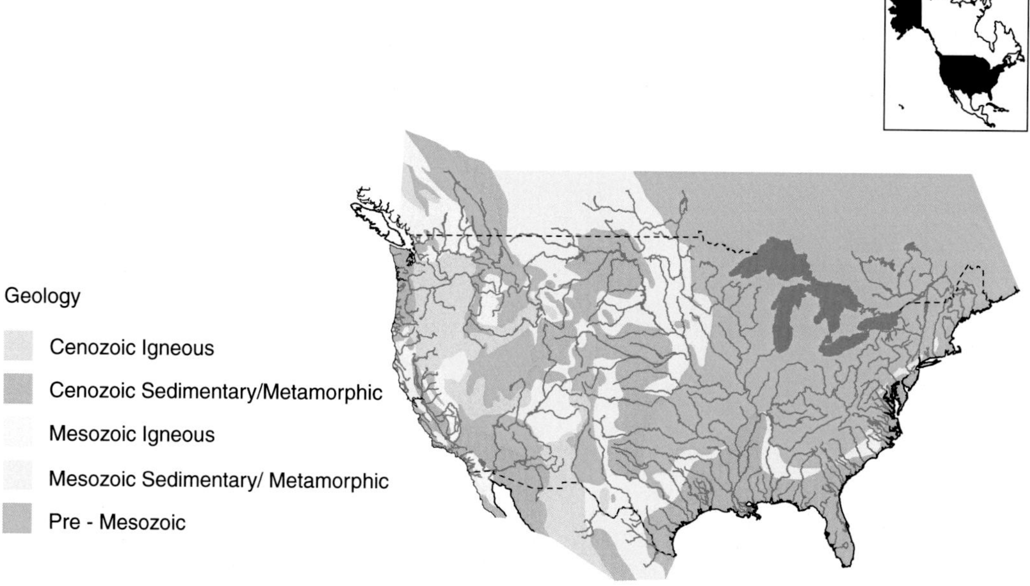

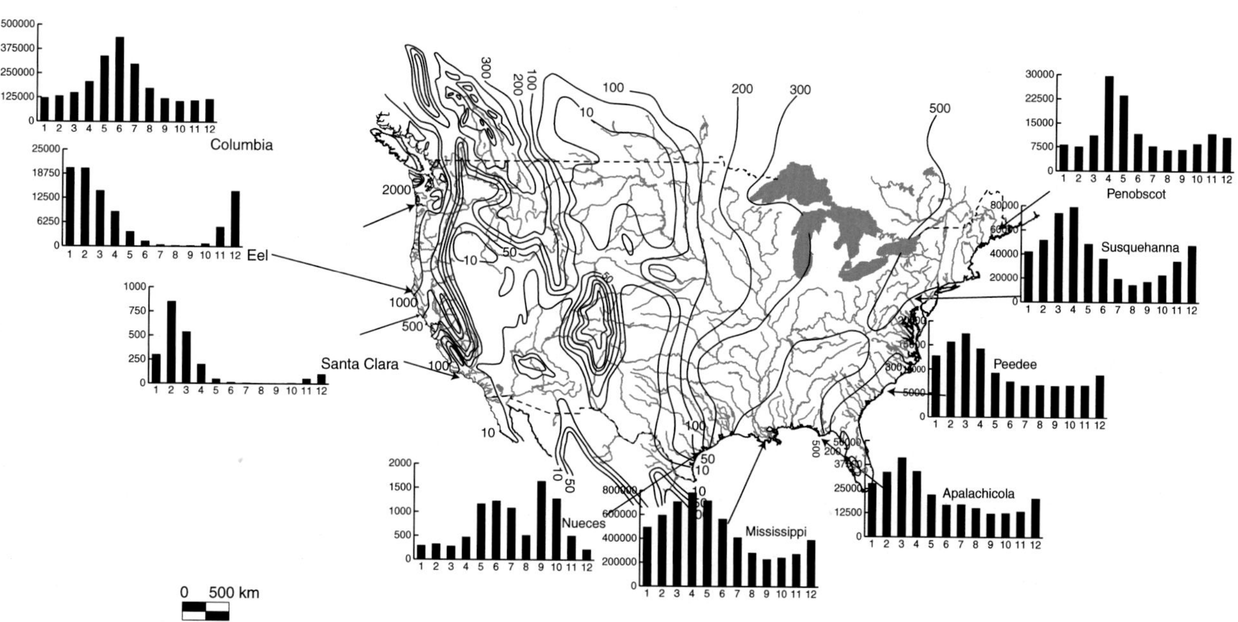

Figure A5

Table A2.

River name	Ocean	Area (10^3 km^2)	Length (km)	Max_elev (m)	Climate	Geology	Q (km^3/yr)	TSS (Mt/yr)	TDS (Mt/yr)	Refs.
United States of America										
Alsea	Pacific (NE)	0.86	65	>1000	Te-W-W	Cen S/M	1.3	0.16		37
Altamaha	Atlantic (NW)	35	220	370	Te-H-Sp	PreMes	12	1 (2.5)	0.9	19, 20, 24, 31, 33, 38
Androscoggin	Atlantic (NW)	8.8	280	1200	SAr-H-S	PreMes	5	0.17	0.3	1, 14, 28
Apalachicola	G. of Mexico	52	840	1100	Te-H-Sp	PreMes	22	0.17	1.1	15, 19, 24, 28, 33, 36, 38
Brazos	G. of Mexico	120	2000	1200	STr-A-Sp	Mes S/M	6	9.2	2.8	6, 10, 19, 23, 24, 35
Calleguas	Pacific (NE)	0.84		1200	STr-SA-W	Cen S/M	0.12	1		2
Caloosahatchee	G. of Mexico	3.6	120	<100	Tr-H-S	Cen S/M	1.1			28
Cape Fear	Atlantic (NW)	24	320	300	Te-H-Sp	PreMes	6.5	0.29	0.17	1, 19, 24, 25, 28, 31, 35, 38
Carmel	Pacific (NE)	0.63	55	>1000	Te-SA-W	Cen S/M	0.1	0.6		35
Chehalis	Pacific (NE)	3.3	180	>1000	Te-W-W	Cen	3.3	0.12	0.05	1, 19, 24, 35, 38
Choctawhatchee	G. of Mexico	12	270	180	STr-H-Sp	Cen S/M	6.3		0.28	20, 28, 36
Chowan	Atlantic (NW)	13	200	<100	Te-H-Sp	Cen S/M	8.8		0.3	20, 38
Colorado	G. of Mexico	110	1600	1200	STr-A-S	Cen-Mes-PreMes	2.4	1.4	0.6	5, 6, 19, 24, 30, 35
Columbia	Pacific (NE)	670	2000	4400	Te-H-W	Cen-Mes-PreMes	240	9.7 (15)	21	7, 23, 32, 35
Connecticut	Atlantic (NW)	28	650	1900	Te-H-S	PreMes	14 (17)	0.25	1	1, 14, 20, 24, 33
Coos	Pacific (NE)	1.6	100	>1000	Te-W-W	Cen	2.4	0.19		37
Coquille	Pacific (NE)	2.7	160	>1000	Te-W-W	Cen S/M-Mes S/M	2	0.29		37
Delaware	Atlantic (NW)	29	700	570	Te-H-Sp	PreMes	10	1	0.94	14, 19, 20, 24, 35
Duwamish	Puget Sound	1.2	120	1000	Te-W-W	Cen		0.24		9
Edisto	Atlantic (NW)	7.5	140	<200	Te-H-Sp	PreMes	2.4	0.02	0.11	1, 24, 28, 35
Eel	Pacific (NE)	9.5	320	2300	Te-W-W	Cen-Mes	7.7	19	1	4, 19, 24, 35
Escambia	G. of Mexico	11	370	180	STr-H-Sp	Cen S/M	5.9			36
Guadalupe	G. of Mexico	26	740	700	STr-A-W	Cen-Mes	2.1	0.3	0.6	1, 6, 19, 24, 28, 38
Gualala	Pacific (NE)	0.9		>1000	Te-H-W	Cen-Mes	0.4	0.27		37
Hoh	Pacific (NE)	0.65	90	>1000	Te-W-W	Cen S/M	2.2	0.72		35
Housatonic	Atlantic (NW)	4.2	200	250	Te-H-Sp	Cen	2.3	0.5	0.46	1, 3, 24, 35
Hudson	Atlantic (NW)	34	490	1600	Te-H-Sp	PreMes	15	0.2	1.3	1, 19, 23, 24, 28, 35, 38
James	Chesapeake Bay	21	540	1200	Te-H-Sp	PreMes	6.2	0.8	0.75	3, 16, 19, 24, 31, 35
Kennebeck	Atlantic (NW)	16	240	1200	Te-H-S	PreMes	7.9	0.65	0.4	1, 14, 20, 25, 28
Klamath	Pacific (NE)	41	460	2900	Te-H-Sp	Cen I	15	10	1.6	4, 17, 19, 24, 24, 28
Los Angeles	Pacific (NE)	2.1	80	2300	STr-SA-W	Cen S/M	0.24 (1.6)	0.24		29
Mad	Pacific (NE)	1.2	140	>1000	Te-W-W	Mes S/M	1.3	2.2		37

Malibu	Pacific (NE)	0.27		930	STr-A-W	Cen S/M	0.021			
Matolle	Pacific (NE)	1		>1000	Te-W-W	Cen	1.1	1.2		
Merrimack	Atlantic (NW)	12	220	1600	Te-H-S	PreMes	6.5	0.2	0.46	14, 19, 24, 33, 35
Mississippi	G. of Mexico	3300	5900	3700	Te-SA-Sp	Cen S/M-Mes S/M	490	210 (400)	140	7, 21, 22, 24
Mobile	G. of Mexico	110	1300	1300	STr-H-Sp	PreMes-Mes	60	4.5	6.3	19, 33, 35, 36
Navarro	Pacific (NE)	0.8	64	>1000	Te-H-W	Cen-Mes	0.42	0.24		37
Neches	G. of Mexico	25	650	150	STr-H-S	Cen S/M	5.6	0.54	0.67	6, 19, 20, 24, 38
Nehalem	Pacific (NE)	2.2	180	>1000	Te-W-W	Cen-Mes	2.3	0.22		
Nooksak	Puget Sound	2	120	>1000	Te-W-W	Cen S	3.3	1.6		9
Nueces	G. of Mexico	43	510	730	STr-A-SP	Cen S/M	0.72	0.71	0.18	5, 6, 19, 24, 28, 38
Ochlockonee	G. of Mexico	5.3	240	<100	STr-H-Sp	Cen S/M	1.4			28
Ogeechee	Atlantic (NW)	6.9	400	200	STr-H-S	Mes S/M	2.1	0.06	0.48	19, 31, 35, 38
Pajaro	Pacific (NE)	3.1		1700	Te-A-W	Cen S/M	0.08	0.3		8, 13
Pamlico	Atlantic (NW)	12	480	700	Te-H-Sp	PreMes	4.5	0.21	0.3	1, 5, 19, 20, 28, 30
Pascagoula	G. of Mexico	25	140	200	STr-H-Sp	Cen S/M	8.5	1.4	0.44	1, 20, 24, 28, 34, 36
Passaic	Atlantic (NW)	1.9	130	<100	Te-H-Sp	PreMes	1	0.03		18, 24, 28, 35
Patuxent	Chesapeake Bay	2.4	180	<100	Te-H-Sp	Cen S/M	1.1	0.04	0.85	1, 35
Peace	G. of Mexico	6	170	<100	Te-H-Sp	Cen S/M	1.5			
Pearl	G. of Mexico	23	780	210	STr-H-Sp	Mes S/M	10	1.2	0.32	10, 19, 24, 35, 36, 38
Peedee	Atlantic (NW)	42	570	1100	Te-H-Sp	PreMes	6.9	0.43 (0.86)	0.39	5, 19, 24, 27, 31, 35
Penobscot	Atlantic (NW)	17	410	1600	Te-H-S	PreMes	10	0.78	0.55	1, 3, 14, 19, 20, 22, 28
Potomac	Chesapeake Bay	38	660	1500	Te-H-Sp	PreMes	10	1.4	0.17	14, 19, 25, 33, 35
Rappahannock	Chesapeake Bay	4.1	300	>500	Te-H-Sp	PreMes	2.1	0.46	0.1	1, 3, 16, 21, 35
Redwood	Pacific (NE)	0.73	90	1600	Te-W-W	Mes S/M	0.88	0.94		16, 26, 35, 37
Rio Grande	G. of Mexico	870	3000	4200	STr-A-Sp	Mes	0.7 (18)	0.66 (20)	0.57 (2.2)	6, 19, 20, 33, 35
Roanoke	Atlantic (NW)	25	660	920	Te-H-Sp	PreMes	7.5	1 (2)	0.49	19, 20, 24, 31, 33
Rogue	Pacific (NE)	14	340	2900	Te-H-Sp	Cen	9	2.3	0.4	4, 19, 28, 35, 37, 38
Russian	Pacific (NE)	3.7	160	1300	Te-H-W	Cen	2.1	0.9	0.73	1, 4, 24, 35, 37, 38
Sabine	G. of Mexico	28	930	200	STr-SA-W	Mes S/M	6.7	0.75	0.6	6, 19, 24, 33, 35
Saco	Atlantic (NW)	4.1	210	>500	Te-H-S	PreMes	2.2			
Sacramento	**San Fran.Bay**	72	640	4300	Te-SA-W	Cen-Mes	20	2.3	1.6	4, 19, 24, 35, 38
Saint John's	Atlantic (NW)	8.7	460	<100	STr-H-Sp	Cen S/M	2.9	0.27	1.5	1, 24, 28, 38
Salinas	Monterey Bay	11	280	1900	STr-A-W	Cen S/M-Mes S/M	0.3	2.3	0.17	1, 11
San Diego	Pacific (NE)	1.1	70	2100	STr-A-W	Cen S/M	0.014	0.025 (0.1)		2, 29, 35, 38
San Diego Creek	Pacific (NE)	0.31	50	>500	STr-A-W	Cen S/M	0.029			

Table A2. (*Continued*)

San Dieguito	Pacific (NE)	0.9	90	1700	STr-A-W	Cen S/M	0.002 (0.008)	0.015 (0.17)		2, 29, 35
San Joaquin	San Fran.Bay	83	560	4400	STr-A-W	Cen-Mes	4 (18)	0.35	1.3	4, 11, 19, 20, 24, 33
San Juan Capistrano	Pacific (NE)	0.3		1900	STr-A-W	Cen S/M	0.01	0.089		13, 35
San Luis Rey	Pacific (NE)	1.4	80	2100	STr-A-W	Cen S/M	0.03 (0.11)	0.19		2, 13, 35
San Mateo	Pacific (NE)	0.34		>500	STr-A-W	Cen S/M	0.005			
Santa Ana	Pacific (NE)	6.3	150	3500	STr-SA-W	Cen S/M	0.68	0.16 (0.5)	0.03	1, 2, 4, 13, 24, 29, 35, 38
Santa Clara	Pacific (NE)	4.1	130	2000	STr-A-W	Cen S/M	0.12 (0.18)	3	0.06	1, 2, 13, 24, 35, 38
Santa Margarita	Pacific (NE)	1.9	80	2200	STr-A-W	Cen S/M	0.025 (0.04)	0.12		2, 13
Santa Maria	Pacific (NE)	4.5		2500	STr-A-W	Cen S/M-Mes S/M	0.026	0.93		
Santa Ynez	Pacific (NE)	2	120	2200	STr-A-W	Cen S/M-Mes S/M	0.1	2 (3)		2, 12
Santee	Atlantic (NW)	36	230	1800	Te-H-Sp	PreMes	13 (17)	0.86	0.88	20, 31, 33, 35
Satilla	Atlantic (NW)	7.2	320	110	Te-H-Sp	Cen S/M	2.1			28, 31
Savannah	Atlantic (NW)	27	500	1700	Te-H-Sp	PreMes	11	1 (2.8)	0.5	20, 24, 30, 31, 33
Siletz	Pacific (NE)	0.52	110	>1000	Te-W-W	Cen	1.3	0.06		
Siuslaw	Pacific (NE)	1.5	180	>1000	Te-W-W	Cen S/M	1.6	0.094		37
Sixes	Pacific (NE)	0.3	50	>500	Te-W-W	Cen S/M	0.5	0.39		35
Skagit	Puget Sound	8	260	2800	Te-W-C	Mes	14	0.33	0.52	19, 24, 35, 38
Smith	Pacific (NE)	1.6	80	>1000	Te-W-S	PreMes	3.2	0.3	0.23	1, 35, 37, 38
Snohomish	Puget Sound	3.9		>1000	Te-W-Sp	Mes	8	0.23		35
Susquehanna	Chesapeake Bay	71	710	960	Te-H-Sp	PreMes	32	1.8	4.3	14, 19, 20, 24, 33
Suwannee	G. of Mexico	25	400	140	STr-H-Sp	Cen S/M	6 (9.5)		1.3	20, 24, 28, 34, 36
Tijuana	Pacific (NE)	4.5	190	1100	STr-A-D	Mes I	0.04	0.47		2, 13, 35
Tillamook	Pacific (NE)	0.61		>1000	Te-W-W	Cen	0.76	0.08		
Trinity	G. of Mexico	46	960	360	STr-SA-Sp	PreMes-Cen	7	1.3	1.2	1, 6, 21, 28
Umpqua	Pacific (NE)	12	180	2800	Te-H-W	Cen	6.8	1.4	0.55	1, 4, 19, 20, 24, 37, 38
Ventura	Pacific (NE)	0.48	50	2000	STr-SA-W	Cen S/M	0.053	0.38 (0.8)		2, 13, 35
York	Chesapeake Bay	6.8	180	360	Te-H-Sp	Cen S/M	2.1	0.05		16, 31, 38

References:
1. Alexander *et al.*, 1996; 2. Brownlie and Taylor, 1981; 3. Bue, 1970; 4. Carter and Resh, 2005; 5. Curtis *et al.*, 1973; 6. Dahm *et al.*, 2005; 7. Degens *et al.*, 1991; 8. Eittreim *et al.*, 2002; 9. Embrey and Frans, 2003; 10. Hudson and Mossa, 1997; 11. IAHS/UNESCO, 1974; 12. Inman and Jenkins, 1999; 13. Inman *et al.*, 1998; 14. Jackson *et al.*, 2005; 15. Janda and Nolan, 1979; 16. Johnson and Belval, 1998; 17. Judson and Ritter, 1964; 18. Langland *et al.*, 1995; 19. Leifeste, 1974; 20. Meade and Parker, 1985; 21. Meade *et al.*, 1990; 22. Meybeck, 1994; 23. Milliman and Meade, 1983; 24. Moody *et al.*, 1986; 25. NOAA, 1985; 26. Nolan *et al.*, 1987; 27. Patchineelam *et al.*, 1999; 28. Rand McNally, 1980; 29. Sherman *et al.*, 2002; 30. Simmons, 1988; 31. Smock *et al.*, 2005; 32. Stanford *et al.*, 2005; 33. UNESCO (WORRI), 1978; 34. UNESCO, 1971; 35. USGS National Water Information System Website, waterdata.usgs.gov/nwis; 36. Ward *et al.*, 2005; 37. Wheatcroft and Summerfield, 2005; 38. Wilson and Iseri, 1969

Alaska

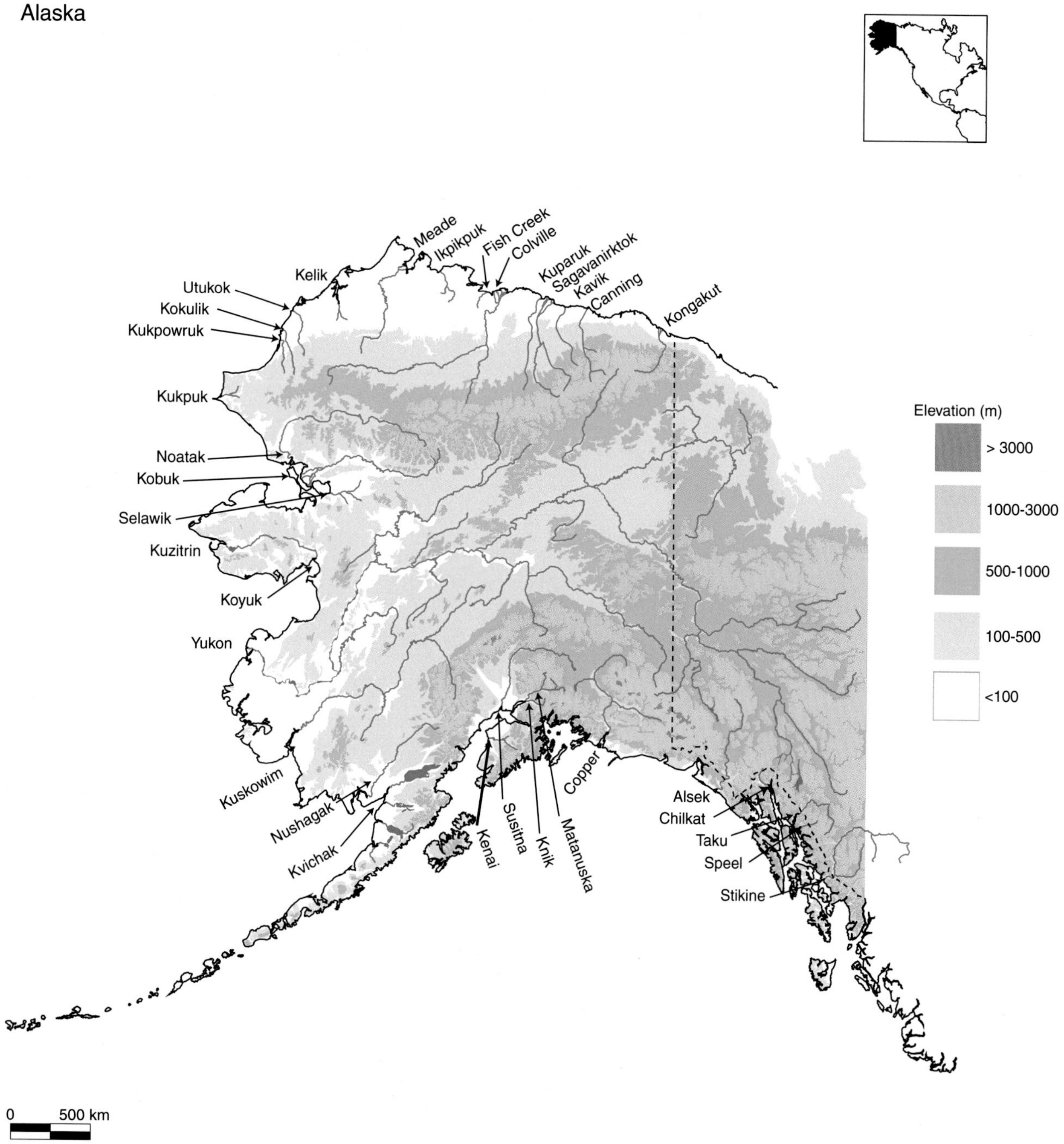

Figure A6

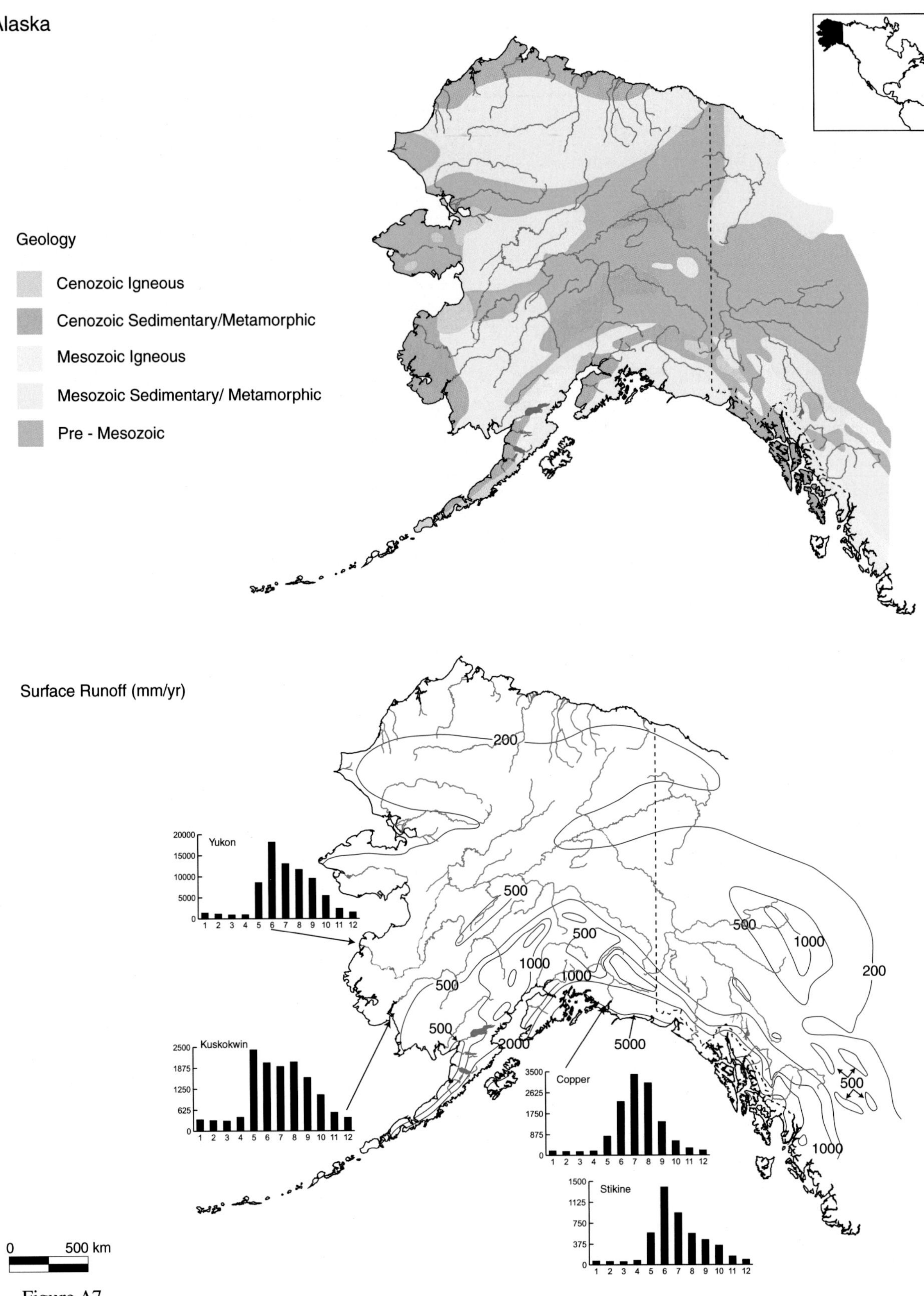
Alaska
Geology
Cenozoic Igneous
Cenozoic Sedimentary/Metamorphic
Mesozoic Igneous
Mesozoic Sedimentary/ Metamorphic
Pre - Mesozoic
Surface Runoff (mm/yr)
200
500
1000
2000
5000
Yukon
Kuskokwin
Copper
Stikine
0 500 km

Figure A7

Table A3.

River name	Ocean	Area (10^3 km^2)	Length (km)	Max_elev (m)	Climate	Geology	Q (km^3/yr)	TSS (Mt/yr)	TDS (Mt/yr)	Refs.
United States of America										
Alaska										
Alsek	G. of Alaska	28	210	>3000	SAr-W-S	Mes S/M	27			7, 17, 18, 19
Canning	Beaufort Sea	5.5	190	1000	Ar-H-S	Mes S/M				
Chilkat	Pacific (NE)	1.9	80	>1000	Te-W-W	Mes S/M	7.2	2.4		23, 24
Colville	Arctic	60	600	>1000	Ar-H-S	Mes S/M	25	6	1.6	3, 14, 15, 21, 24
Copper	G. of Alaska	63	460	4800	SAr-W-S	Cen S/M	96	70	9.4	1, 13, 17, 22, 23
Fish Creek	Arctic	6	130	<100	Ar-H-S	Cen S/M				
Ikpikpuk	Beaufort Sea	10	190	500	Ar-SA-S	Cen S/M				
Kavik	Arctic	5.5	150	<800	Ar-SA-S	Cen S/M				
Kelik	Arctic	6.7	110	900	Ar-SA-S	Cen S/M				
Kenai	G. of Alaska	5.2	130	>1500	Ar-H-S	Cen S/M				19
Knik	G. of Alaska	2.9	40	>1000	SAr-W-S	Mes S/M	6	6.8		23, 24
Kobuk	Chukchi Sea	31	480	2600	SAr-H-S	PreMes	17		1.6	1, 7, 17, 18
Kokulik	Chukchi Sea	5.5	200	1200	Ar-H-S	Mes S/M				
Kongakut	Beaufort Sea	5	150	2000	Ar-H-S	Mes S/M				
Koyuk	Bering Sea	31	580	>500	Ar-SA-S	Mes S/M				
Kukpowruk	Chukchi Sea	8.1	160	11	Ar-SA-S	Mes S/M	1.2			11, 16, 21, 23
Kukpuk	Arctic	5	150	900	Ar-SA-S	PreMes				
Kuparuk	Beaufort Sea	8.1	250	1500	Ar-SA-S	Mes S/M				16
Kuskokwim	Bering Sea	120	1100	>3500	SAr-W-S	PreMes	60	8.2	6.4	1, 7, 15, 17, 19, 22
Kuzitrin	Bering Sea	4.5	150	>1000	Ar-SA-S	PreMes				
Kvichak	Pacific (NE)	25	100	>1000	Te-H-Sp	Mes	18			7, 17
Matanuska	G. of Alaska	5.3	120	>1000	SAr-W-S	Cen S/M	10	6		
Meade	Beaufort Sea	10	240	>500	Ar-H-S	Mes S/M	1.2			2, 18
Noatak	Chukchi Sea	33	640	2600	Ar-H-S	Mes S/M	11			7, 18
Nushagak	Pacific (NE)	35	450	600	Te-W-SP	PreMes	33		1.1	1, 7, 17, 18, 19
Sagavanirktok	Beaufort Sea	15	290	2400	Ar-SA-S	Mes S/M	4	0.3		16, 20, 21, 23
Selawik	Chukchi Sea	14	260	1300	Ar-H-S	Mes S/M				
Speel	Pacific (NE)	0.58		>1000	Te-W-S	Mes S/M	3.1	2.4		23, 24
Stikine	G. of Alaska	57	540	>2900	SAr-W-S	Mes S/M	53	18	3.8	6, 8, 14, 19, 23
Susitna	G. of Alaska	52	510	6100	SAr-W-S	Cen S/M	45	21	7.6	9, 14, 15, 17, 19, 22, 23
Taku	G. of Alaska	30	250	>2300	SAr-W-S	Mes S/M	19			2, 19, 23
Utukok	Chukchi Sea	7.5	270	700	Ar-H-S	Mes S/M				
Yukon	Bering Sea	850	3700	6000	Ar-SA-S	PreMes	210	54	26	4, 5, 9, 10, 12, 17, 22

References:

1. Alexander *et al.*, 1996; 2. Arctic RIMS website, rims.unh.edu/data.shtml; 3. Arnborg *et al.*, 1967; 4. Bailey, 2005; 5. Brabets *et al.*, 2000; 6. Center for Natural Resources Energy and Transport (UN), 1978; 7. Dynesius and Nilsson, 1994; 8. GEMS website, www.gemstat.org; 9. Harrison, 2000; 10. Leifeste, 1974; 11. McNamara *et al.*, 1998; 12. Meade and Parker, 1985 13. Meade *et al.*, 1990; 14. Meybeck and Ragu, 1996; 15. Milliman and Meade, 1983; 16. Milner *et al.*, 2005; 17. Moody *et al.*, 1986; 18. Rand McNally, 1980; 19. Richardson and Milner, 2005; 20. Robinson and Johnson, 1997; 21. Trefry *et al.*, 2003; 22. UNESCO (WORRI), 1978; 23. USGS, 1994; 24. Wilson and Iseri, 1969

Due to the small size of these islands, global scale geology and runoff data are not available.

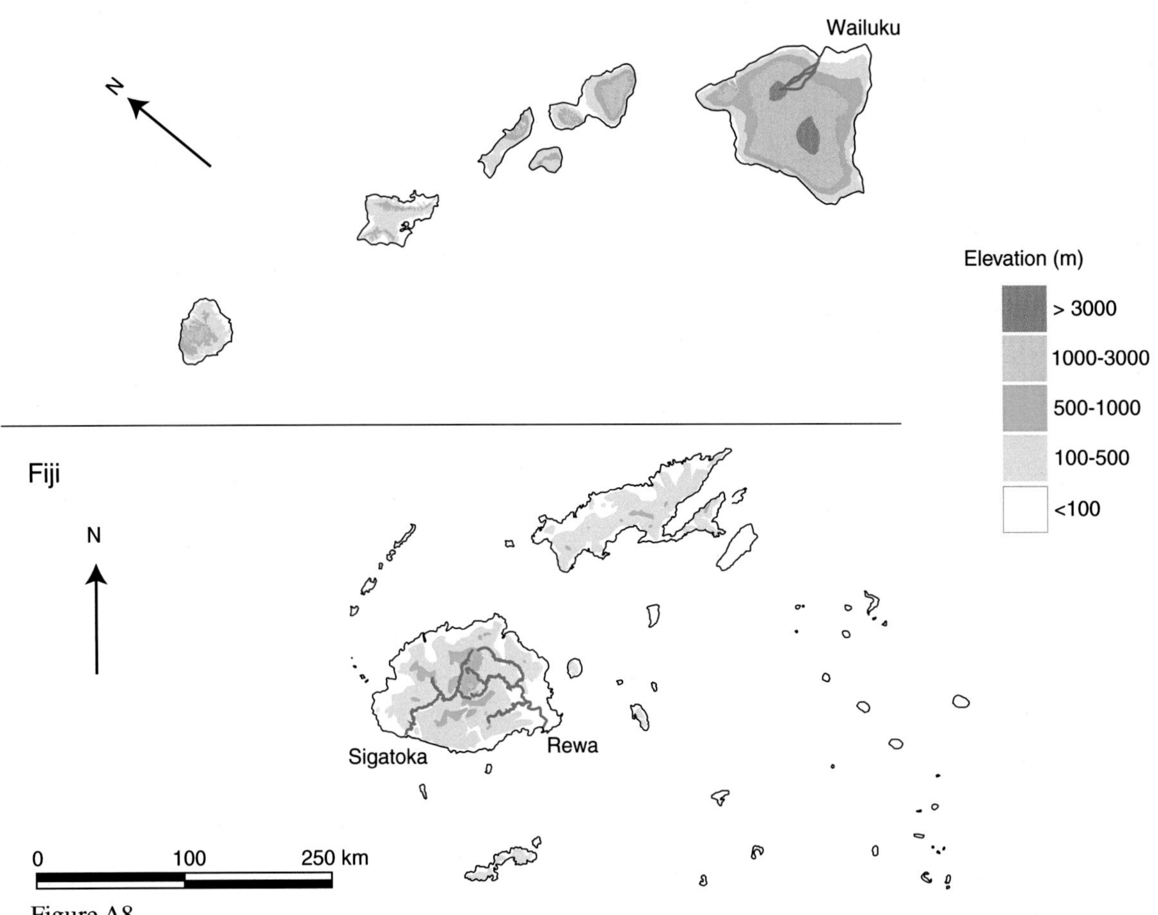

Figure A8

Table A4.

River name	Ocean	Area (10^3 km^2)	Length (km)	Max_elev (m)	Climate	Geology	Q (km^3/yr)	TSS (Mt/yr)	TDS (Mt/yr)	Refs.
USA										
Hawaii										
Wailuku	Pacific (Central)	0.65	42	3300	Tr-H-C	Cen I	0.33	0.02		
Fiji										
Vgiti Levu										
Rewa	Pacific (S)	2.9	140	1300	Tr-W-C	Cen	6	10	1	1
Sinngatoka	Pacific (S)	2.5	120	1100	Tr-W-C	Cen				

Reference:
1. IOC/WESTPAC, 1991

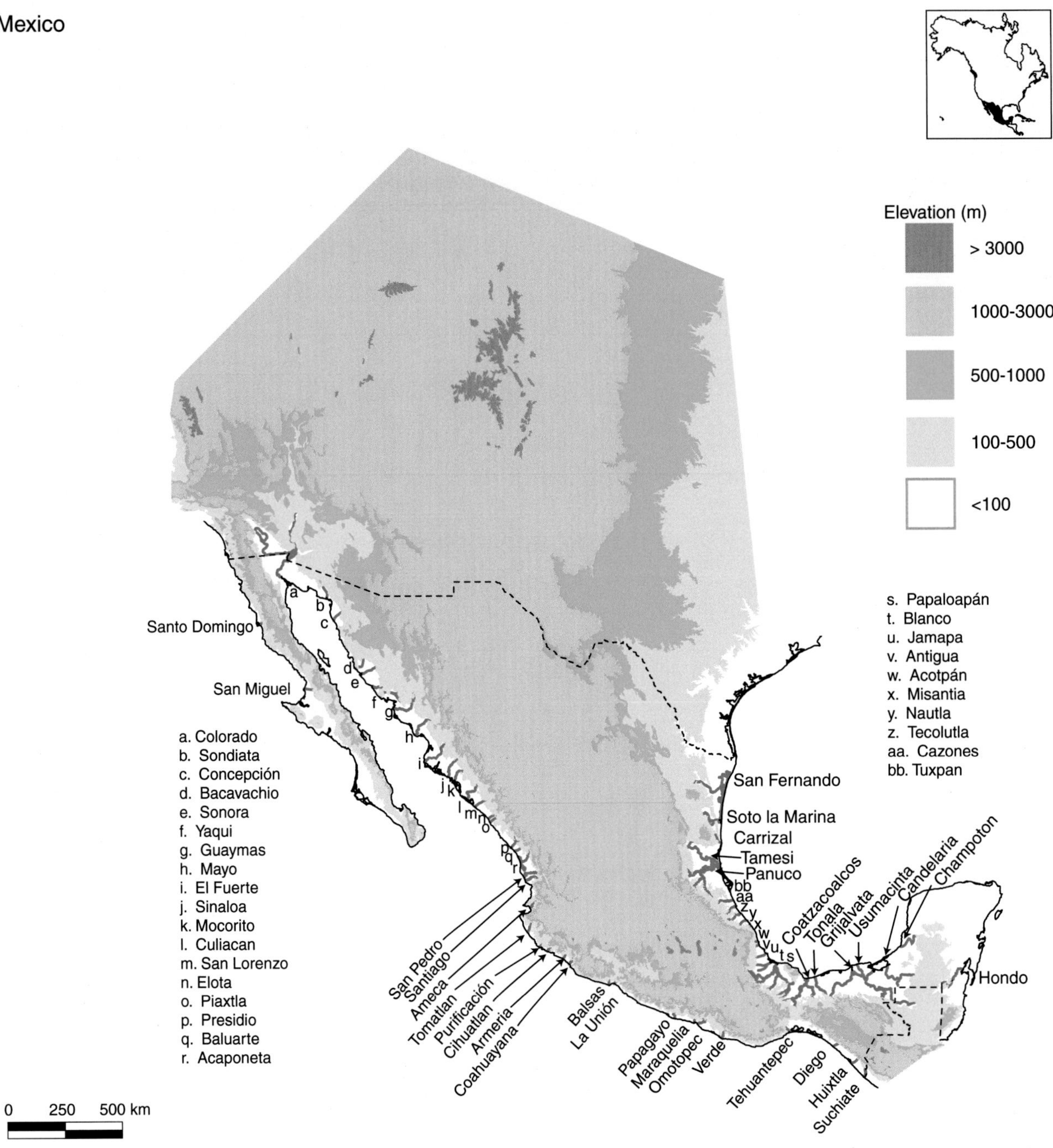
Mexico
Elevation (m)
> 3000
1000-3000
500-1000
100-500
<100
Santo Domingo
San Miguel
a. Colorado
b. Sondiata
c. Concepción
d. Bacavachio
e. Sonora
f. Yaqui
g. Guaymas
h. Mayo
i. El Fuerte
j. Sinaloa
k. Mocorito
l. Culiacan
m. San Lorenzo
n. Elota
o. Piaxtla
p. Presidio
q. Baluarte
r. Acaponeta
s. Papaloapán
t. Blanco
u. Jamapa
v. Antigua
w. Acotpán
x. Misantia
y. Nautla
z. Tecolutla
aa. Cazones
bb. Tuxpan
San Fernando
Soto la Marina
Carrizal
Tamesi
Panuco
Coatzacoalcos
Tonala
Grijalvata
Usumacinta
Candelaria
Champoton
Hondo
San Pedro
Santiago
Ameca
Tomatlan
Purificación
Cihuatlan
Armeria
Coahuayana
Balsas
La Unión
Papagayo
Maraquelia
Omotopec
Verde
Tehuantepec
Diego
Huixtla
Suchiate
0 250 500 km

Figure A9

Mexico

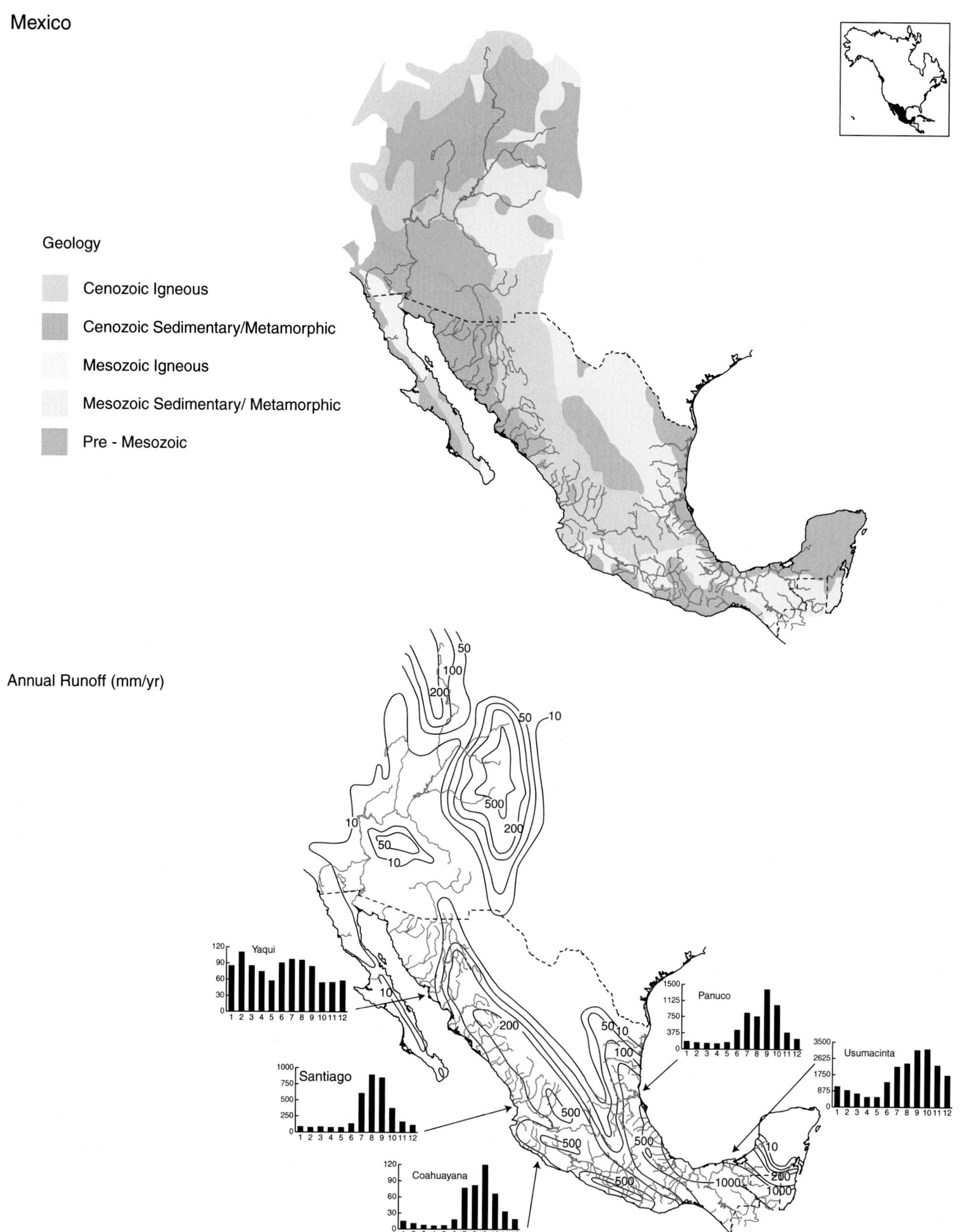

Figure A10

Table A5.

River name	Ocean	Area (10^3 km^2)	Length (km)	Max_elev (m)	Climate	Geology	Q (km^3/yr)	TSS (Mt/yr)	TDS (Mt/yr)	Refs.
Mexico										
Acaponeta	Pacific (NE)	5.1	230	2700	Tr-H-S	Cen I	1.4			1, 3
Actopán	G. of Mexico	2.5	90	> 1500	Tr-SA-S	Cen S/M	0.5	2.2		13
Ameca	Pacific (NE)	12	200	2600	Tr-SA-S	Cen I	2	4.2	0.56	1, 5, 10, 13
Antigua	G. of Mexico	3.3	115	>3000	Tr-W-S	Cen S/M	2.8	0.7		3, 5, 13
Armeria	Pacific (NE)	10	240	4300	Tr-SA-S	Cen I	1.2			13
Bacavachio	Pacific (NE)	6.8	150	2100	Tr-A-S	Cen S/M	0.04			5
Balsas	Pacific (NE)	120	770	4400	Tr-SA-S	Cen I	13	11	11	1, 4, 5, 10, 13
Baluarte	Pacific (NE)	5.4	130	2700	Tr-H-S	Cen I	1.9 (6.7)			5, 6
Blanco	G. of Mexico	3.8	175	4200	Tr-H-S	Mes S/M	1.8			5
Candelaria	G. of Mexico	14	250	370	Tr-H-S	Cen S/M	2 (3.6)			1, 5, 12, 13
Carrizal	G. of California	4.2	65	>500	Tr-SA-S	Mes I	0.84			5
Cazones	G. of Mexico	3.8	135	1800	Tr-H-S	Cen S/M	2.1			5, 13
Champoton	G. of Mexico	6.1	160	>200	Tr-SA-S	Cen S/M	0.68			5
Cihuatlan	G. of California	3.7	75	2900	Tr-SA-S	Cen I	0.89			5
Coahuayana	Pacific (NE)	7.5	200	4300	Tr-H-S	Mes S/M	1.9 (3)	4.6		3, 13
Coatzacoalcos	G. of Mexico	21	280	2000	Tr-W-S	Mes S/M	22			5, 12
Colorado	G. of California	640	3200	4100	Tr-A-S	Mes S/M	0.2 (20)	0.1 (120)	0.11	2
Concepción	G. of California	25		>200	Tr-A-S	Cen S/M	0.4			4
Culiacan	G. of California	18	490	> 2600	Tr-SA-S	PreMes	2.9	1.6		1, 5
Diego	Pacific (NE)	1	27	1400	Tr-H-S	Cen I	0.3	0.08		3
El Fuerte	G. of California	36	560	1300	Tr-SA-S	PreMes	5.2			1, 5, 7, 12, 13
Elota	Pacific (NE)	5	160	>1000	Tr-H-S	Cen I				12
Grijalva	G. of Mexico	50	320	3800	Tr-H-S	Cen I	23 (16)	1.3 (24)	4.6	1, 6, 7, 9, 10, 13, 15
Guaymas	G. of California	5.2		1700	Tr-A-S	Cen	0.04			5
Hondo	Atlantic (NW)	10	210	>100	Tr-SA-S	Cen S/M	1.3			4, 12, 13, 14
Huixtla	Pacific (NE)		46	>2500	Tr-A-S	Cen I	0.2	0.2		
Jamapa	G. of Mexico	3.3	125	4200	Tr-H-S	Cen S/M	1.9	0.4		1, 5
La Unión	Pacific (NE)	1.4	54	1000	Tr-SA-S	PreMes		0.06		3
Marequelia	Pacific (NE)	1.7	68	>1000	Tr-SA-S	PreMes		0.3		3
Mayo	G. of California	14	350	>2500	Tr-A-S	Cen I		0.94		3, 5, 12
Misantia	G. of Mexico	0.68	56	>2500	Tr-H-S	Cen S/M	0.43	0.3		3, 5
Mocorito	Pacific (NE)	2.8	100	1000	Tr-A-S	Cen S/M	0.13	0.48		3, 5
Nautla	G. of Mexico	3.4	120	3700	Tr-SA-S	Mes S/M	2.3			1, 11
Omotepec	Pacific (NE)	13	180	3300	Tr-H-S	PreMes	4.5			5
Panuco	G. of Mexico	85	500	2400	Tr-H-S	Mes S/M	19	6.6	10	1, 4, 5, 7, 8, 16
Papagayo	Pacific (NE)	8.2	160	2600	Tr-H-S	PreMes	3.7 (5.6)			1, 5
Papaloapán	G. of Mexico	46	350	>2000	Tr-W-S	Mes S/M	44	6.9		1, 4, 5, 13
Piaxtla	Pacific (NE)	11	220	2900	Tr-H-S	Cen I	1.3	4		1, 5, 13

Table A5. (*Continued*)

Presido	Pacific (NE)	5.2	175	2900	Tr-H-S	Cen I	1.8			5, 13
Purificación	Pacific (NE)	3	90	2000	Tr-H-S	Cen I	0.72			5
San Fernando	G. of Mexico	18	400	2600	Tr-A-S	Cen S/M	0.8	3.6		1, 5, 13
San Lorenzo	G. of California	10	315	3000	Tr-SA-S	Cen I	1.9 (5.2)	2.4		1, 5, 6, 13
San Miguel	Pacific (NE)	2	65	>1000	Tr-A-W	Cen I				4
San Pedro	Pacific (NE)	29	500	3300	Tr-A-S	Cen I	3.5			1, 5, 13
Santiago	Pacific (NE)	130	960	3600	Tr-A-S	Cen I	11			1, 5, 16
Santo Domingo	Pacific (NE)	2.1	85	>2500	Tr-A-S	Mes S/M	0.1			13
Sinaloa	Pacific (NE)	13	400	1800	Tr-SA-S	PreMes	2.2 (5.6)			5, 6, 13
Sondiata	G. of California	8		>1000	Tr-A-D	Cen-Mes				
Sonora	G. of California	29	375	2600	Tr-A-S	Cen S/M	0.17			5, 13
Soto la Marina	G. of Mexico	23	420	2500	Tr-A-S	Cen S/M	2.1			1, 5, 13
Suchiate	Pacific (NE)	1.8	72	>1000	Tr-W-S	Cen I	2.6	0.5		4, 13
Tamesi	G. of Mexico	18	400	1200	Tr-SA-S	Cen S/M	2.3			5, 12
Tecolutla	G. of Mexico	8.1	370	2900	Tr-W-S	Cen S/M	6.9			1, 5
Tehuantepec	Pacific (NE)	11	220	> 2000	Tr-SA-S	PreMes	1.2	4.6		13
Tomatlan	Pacific (NE)	2.8	80	2000	Tr-H-S	Cen I	0.64			5
Tonala	G. of Mexico	6	130	>100	Tr-W-S	Cen S/M	5.9			5
Tuxpan	G. of Mexico	5.9	160	1200	Tr-W-S	Cen S/M	2.6			5, 13
Usumacinta	G. of Mexico	51	430	4000	Tr-W-S	Mes S/M	67	6.2	13	4, 7, 10, 16
Verde	Pacific (NE)	19	340	3000	Tr-H-S	PreMes	6.1	8.1		1, 5, 13
Yaqui	G. of California	79	740	3200	Tr-H-S	Cen I	3.6			9

References:
1. Comission Nacional del Agua, 1990; 2. Curtis *et al.*, 1973; 3. de la Lanza Espino, personal communication; 4. GEMS website, www.gemstat.org; 5. Global River Discharge Database, http://www.rivdis.sr.unh.edu/; 6. GRDC website, www.gewex.org/grdc; 7. Hudson *et al.*, 2005; 8. Hudson, 2003; 9. IAHS/UNESCO, 1974; 10. Meybeck and Ragu, 1996; 11. NOAA, 1985; 12. Rand McNally, 1980; 13. Sistemas de Informacion, Geografica S. A. de C. V., 2007; 14. Terrain Analysis Center, 1995a; 15. UNESCO (WORRI), 1978; 16. van der Leeden, 1975

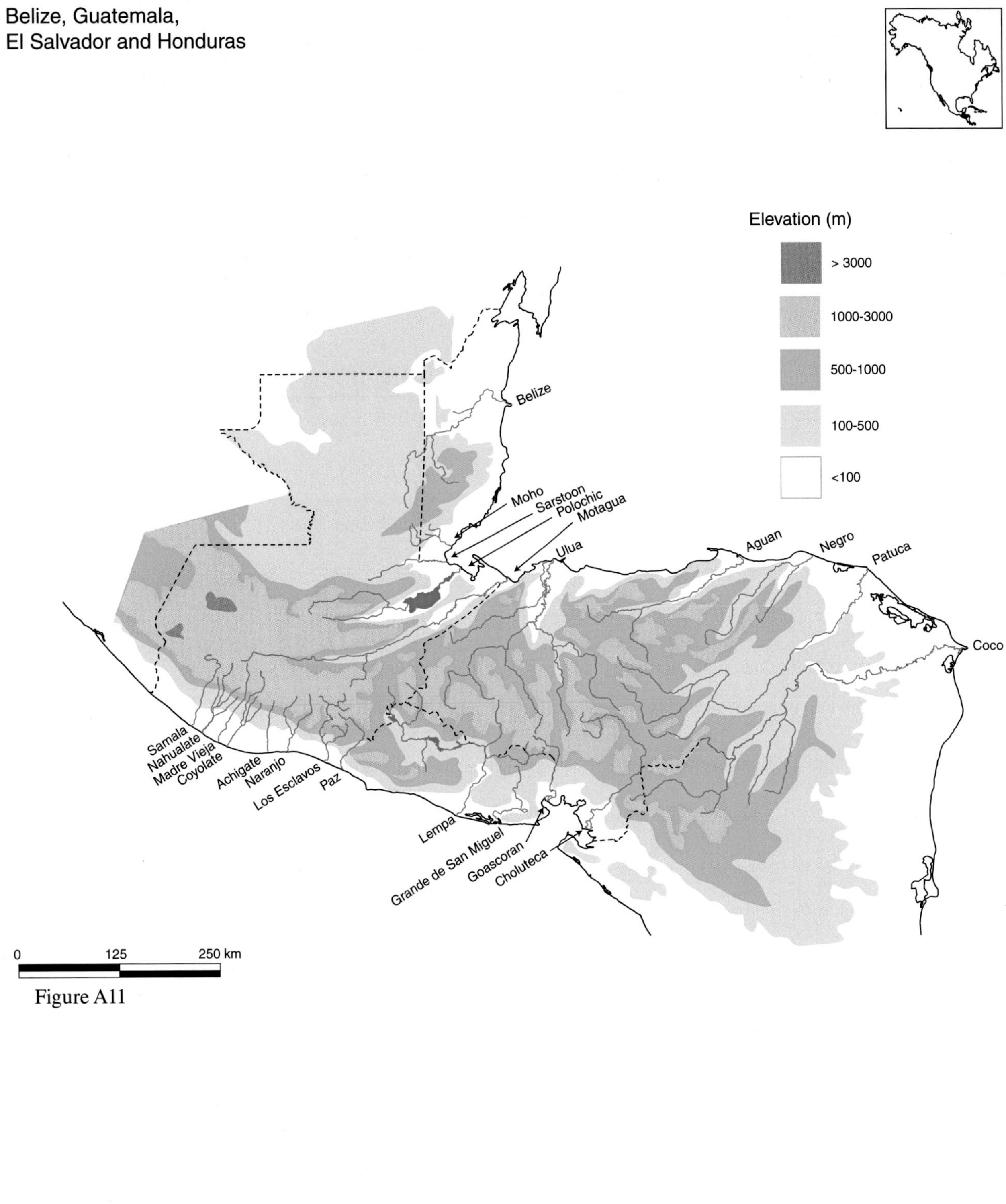

Belize, Guatemala,
El Salvador and Honduras

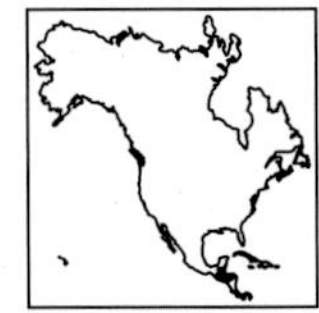

Elevation (m)

> 3000
1000-3000
500-1000
100-500
<100

Belize
Moho
Sarstoon
Polochic
Motagua
Ulua
Aguan
Negro
Patuca
Coco
Samala
Nahualate
Madre Vieja
Coyolate
Achigate
Naranjo
Los Esclavos
Paz
Lempa
Grande de San Miguel
Goascoran
Choluteca

0 125 250 km

Figure A11

Belize, Guatemala, El Salvador and Honduras

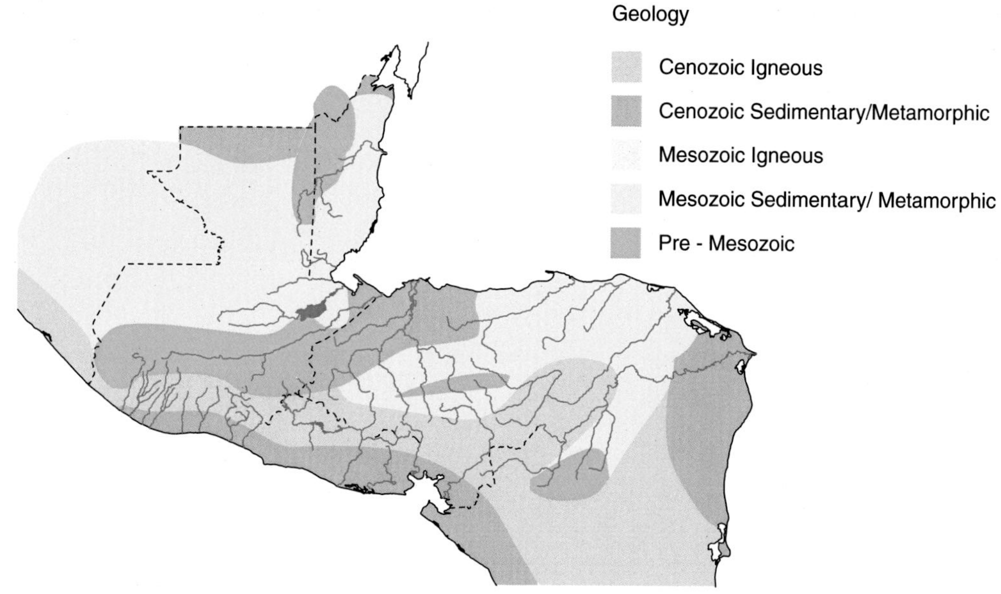

Surface Runoff (mm/yr)

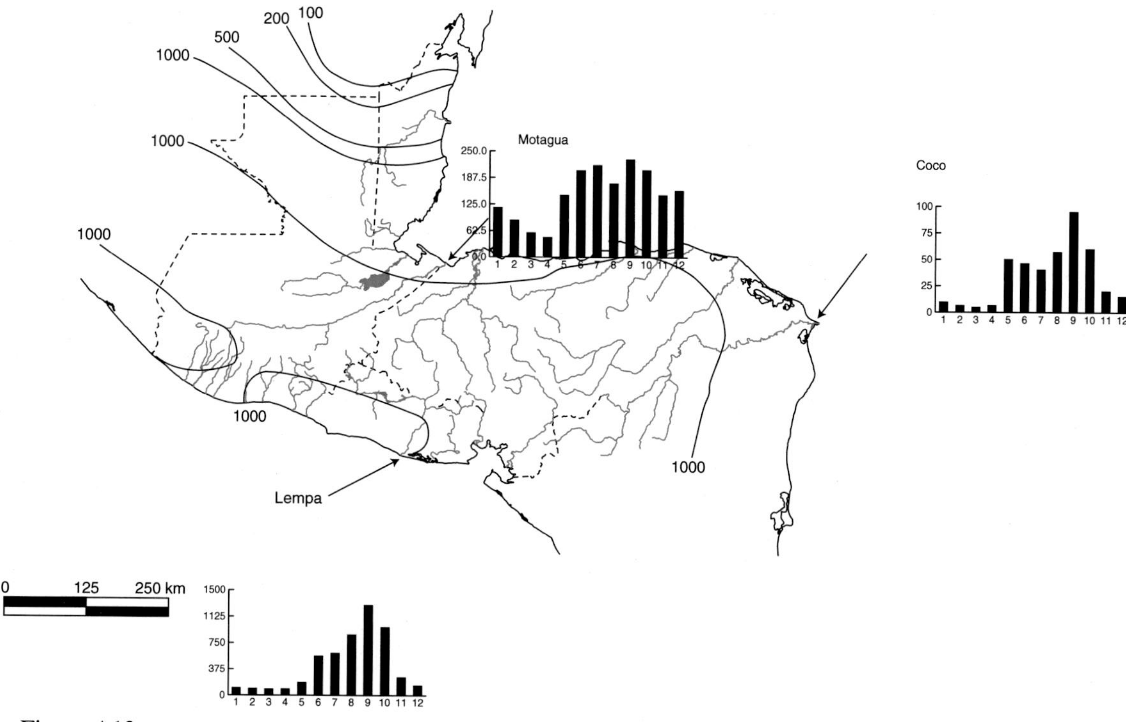

Figure A12

Table A6.

River name	Ocean	Area (10^3 km^2)	Length (km)	Max_elev (m)	Climate	Geology	Q (km^3/yr)	TSS (Mt/yr)	TDS (Mt/yr)	Refs.
Belize										
Belize	Caribbean	9.1	290	>200	Tr-H-S/Au	Mes S/M	2.5			7, 10
Moho	Caribbean	1.2		600	Tr-H-S/Au	Mes S/M	1.1			11
Sarstoon	Caribbean	2.2	140	>500	Tr-W-S/Au	Mes S/M				7
Guatemala										
Achigate	Pacific (E)	1.3	120	2100	Tr-H-Au	Cen S/M				6, 7
Coyolate	Pacific (E)	1.6	150	2100	Tr-H-Au	Cen S/M	0.8			6, 7, 11
Los Esclavos	Pacific (E)	3.2	120	>1000	Tr-H-Au	Cen S/M	0.33			3, 10
Madre Vieja	Pacific (E)	1	120	2200	Tr-H-Au	Cen S/M	0.3			5, 6, 7
Motagua	Caribbean	15	400	2000	Tr-H-Au	PreMes	6	7.5		6, 10, 13
Nahualate	Pacific (E)	1.8	160	3000	Tr-H-Au	Cen S/M				6, 7, 10
Naranjo	Pacific (E)	1.3	100	>1000	Tr-H-Au	Cen S/M				
Polochic	Caribbean	8.7	230	<100	Tr-H-Au	Mes S/M				6
Samala	Pacific (E)	1.5	140	2700	Tr-H-Au	Cen S/M	0.3			6, 8, 11
El Salvador										
Goascoran	Pacific (E)	2.5	120	>500	Tr-H-Au	Cen I	1.1		0.3	1, 2, 10, 11
Grande de San Miguel	Pacific (E)	2.4	40	>500	Tr-W-Au	Cen I	5.5		0.9	3, 11, 14
Lempa	Pacific (E)	18	320	>1000	Tr-W-Au	Cen I	14	7	1.3	2, 10, 11, 14
Paz	Pacific (E)	2.1	44	1500	Tr-H-Au	Cen S/M	1			6, 14
Honduras										
Aguan	Caribbean		240	>500	Tr-H-S/Au	Mes S/M				10
Choluteca	Pacific (E)	6.3	280	>1000	Tr-SA-Au	Cen I	1.1			3, 10
Coco	Caribbean	27	480	>500	Tr-W-S/Au	Cen I	36	6.5 (7.4)		3, 9, 10
Negro	Caribbean	2.8	240	>500	Tr-H-S/Au	Mes S/M				1, 10
Patuca	Caribbean	26	320	>500	Tr-W-S/Au	Cen I	20			4, 10, 12
Ulua	Caribbean	27	320	>500	Tr-H-S/Au	Cen I	17			4, 10, 12

References:
1. Center for Natural Resources Energy and Transport (UN), 1978; 2. CEPAL, 1971 (cf. van der Leeden, 1975); 3. CEPAL, 1972 (cf. van der Leeden, 1975); 4. CEPAL, 1973 (cf. van der Leeden, 1975); 5. ECLAC, 1990; 6. Global River Discharge Database, http://www.rivdis.sr.unh.edu/; 7. Instituto Geografico Nacional; 8. Kuenzi *et al.*, 1979; 9. Murray *et al.*, 1982; 10. Rand McNally, 1980; 11. Terrain Analysis Center, 1995a,b,c,d; 12. UNESCO (WORRI), 1978; 13. van der Leeden, 1975; 14. Vörösmarty *et al.*, 1996b

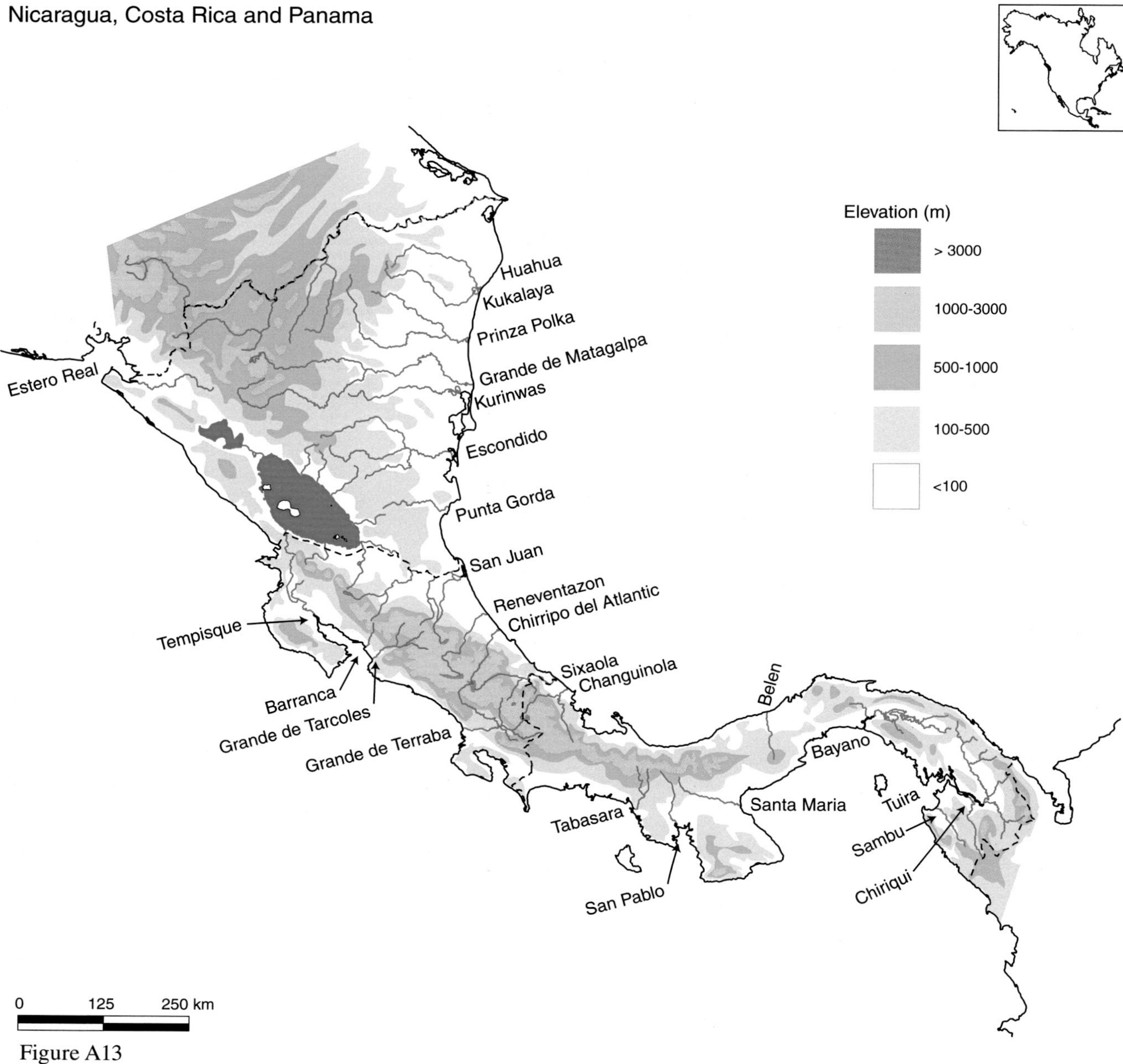
Nicaragua, Costa Rica and Panama
Elevation (m)
> 3000
1000-3000
500-1000
100-500
<100
Huahua
Kukalaya
Prinza Polka
Grande de Matagalpa
Kurinwas
Escondido
Punta Gorda
San Juan
Reneventazon
Chirripo del Atlantic
Sixaola
Changuinola
Belen
Bayano
Santa Maria
Tuira
Sambu
Chiriqui
San Pablo
Tabasara
Grande de Terraba
Grande de Tarcoles
Barranca
Tempisque
Estero Real
0
125
250 km

Figure A13

Nicaragua, Costa Rica and Panama

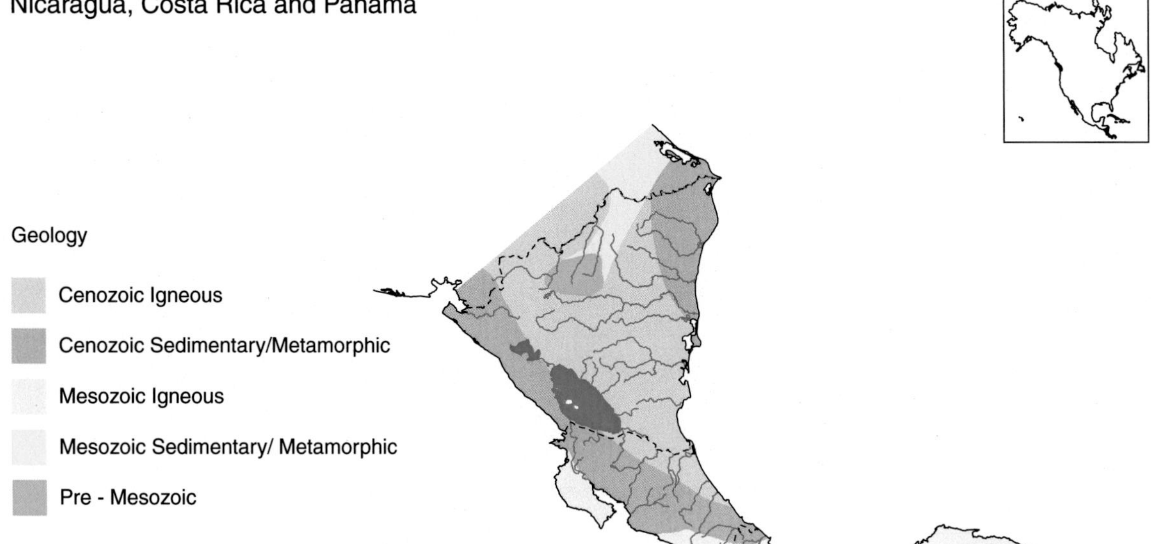

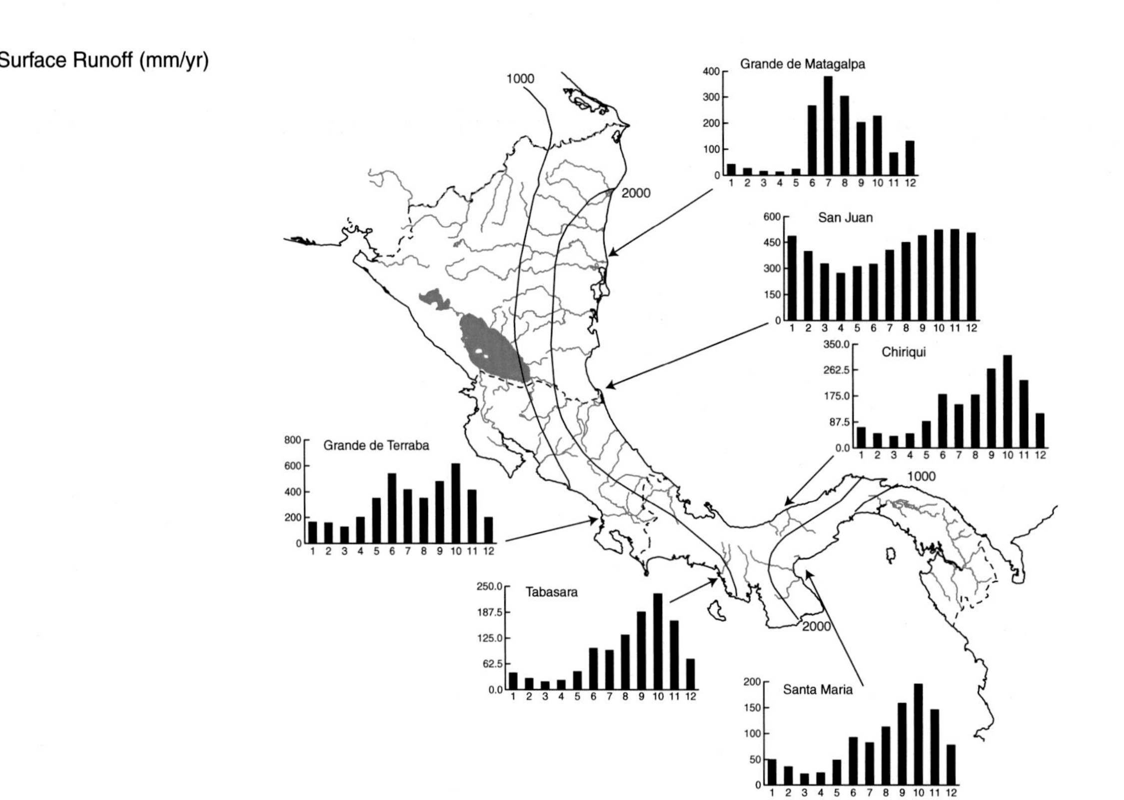

0 125 250 km

Figure A14

Table A7.

River name	Ocean	Area (10^3 km^2)	Length (km)	Max_elev (m)	Climate	Geology	Q (km^3/yr)	TSS (Mt/yr)	TDS (Mt/yr)	Refs.
Nicaragua										
Escondido	Caribbean	12	160	>500	Tr-W-S	Cen I	26	4.7 (5.4)		11, 12, 14
Estero Real	Pacific (E)	3	96	>500	Tr-W-Au	Cen I				12
Grande de Matagalpa	Caribbean	20	435	>500	Tr-W-S	Cen I	29	5.25 (6)		11, 14
Huahua	Caribbean	2.9	160	>200	Tr-W-S	Cen S/M				12
Kukalaya	Caribbean	2	123	>100	Tr-W-S	Cen S/M				
Kurinwas	Caribbean	3.7	143	>200	Tr-W-S	Cen I				
Prinza Polka	Caribbean	11	190	>200	Tr-W-S	Cen I	21	3.6 (4.2)		1, 11, 12
Punta Gorda	Caribbean			>500	Tr-W-C	Cen I				
San Juan	Caribbean	39	180	<200	Tr-H-C	Cen I	18	4.9		1, 5, 10
Costa Rica										
Barranca	Pacific (E)	0.2		2100	Tr-W-Au	Cen I	0.3			15
Chirripo del Atlanti	Caribbean	0.8		3800	Tr-W-Au	Cen S/M	1.9			7
Grande de Tarcoles	Pacific (E)	2.2		3400	Tr-W-Au	Cen S/M	5		0.5	4, 6, 8, 13, 15
Grande de Terraba	Pacific (E)	4.8	160	3800	Tr-W-Au	Cen S/M	10	1.9	0.7	3, 13
Reventazon	Caribbean	1.3	140	3300	Tr-W-Au	Cen I	3.4		0.2	6, 8, 13
Sixaola	Caribbean	2.7		>3000	Tr-W-Au	Mes I	7.4		0.1	13
Tempisque	Pacific (E)	1	129	>500	Tr-W-Au	Cen S/M	1.5			12, 15
Panama										
Bayano	Panama Bay	3.9		>100	Tr-W-Au	Cen S/M	5.7			2
Belen	Caribbean			>1000	Tr-W-Au	Mes I				
Changuinola	Caribbean	3		3600	Tr-W-Au	Mes I	6.4			1, 2
Chiriqui	Pacific (E)	1.4		>1000	Tr-W-Au	Mes I	4.5			2, 16
Sambu	Panama Bay			>200	Tr-W-Au	Mes I				
San Pablo	Pacific (E)	0.7	80	>1000	Tr-W-Au	Mes I	1.5			7, 12
Santa Maria	Panama Bay	1.5		1500	Tr-W-Au	Mes I	2.5			2, 6
Tabasara	Pacific (E)	1.1		>1000	Tr-W-Au	Mes I	3			2, 16
Tuira	Panama Bay	2.8	200	>500	Tr-W-Au	Cen S/M	3.2			

References:
1. Center for Natural Resources Energy and Transport (UN), 1978; 2. CEPAL, 1972 (cf. van der Leeden, 1975); 3. CEPAL, 1978 (cf. van der Leeden, 1975); 4. Fuller *et al.*, 1990; 5. Gleick, 1993; 6. Global River Discharge Database, http://www.rivdis.sr.unh.edu/; 7. GRDC website, www.gewex.org/grdc; 8. IAHS/UNESCO, 1974; 9. Krishnaswamy *et al.*, 2001; 10. Meybeck and Ragu, 1996; 11. Murray *et al.*, 1982; 12. Rand McNally, 1980; 13. Terrain Analysis Center, 1995b; 14. van der Leeden, 1975; 15. Voorhis *et al.*, 1983; 16. Vörösmarty *et al.*, 1996b

Cuba, Jamaica, Haiti,
Dominican Republic and Puerto Rico

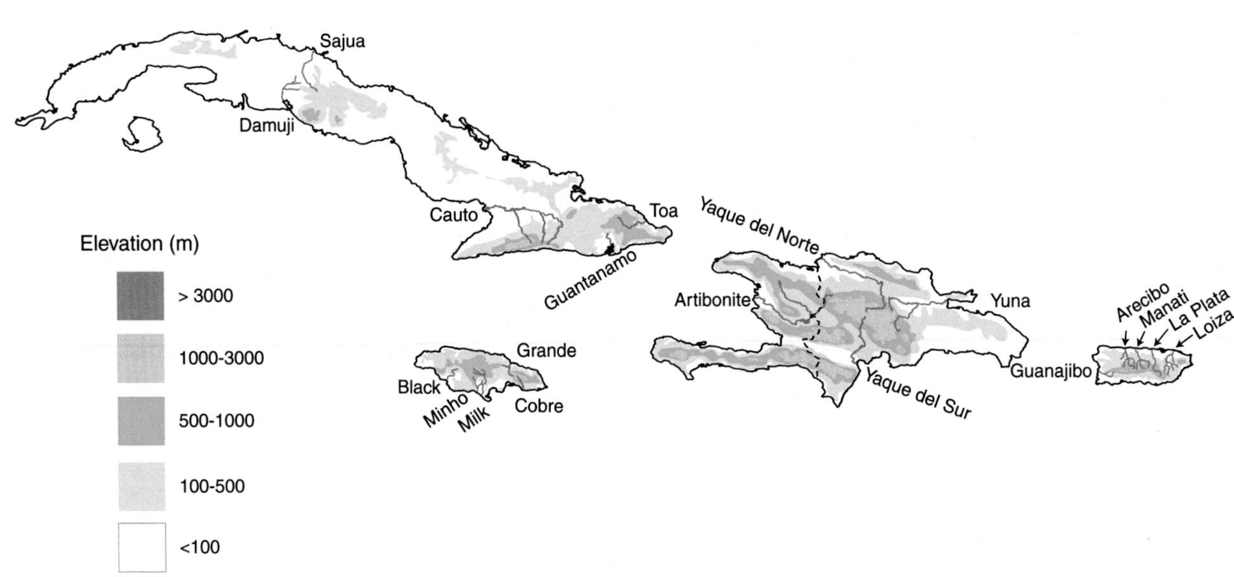

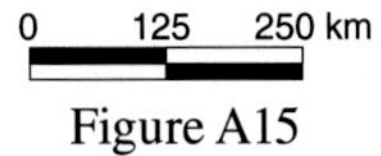

Figure A15

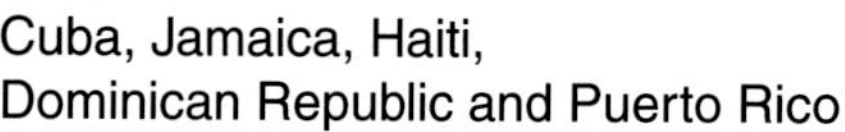

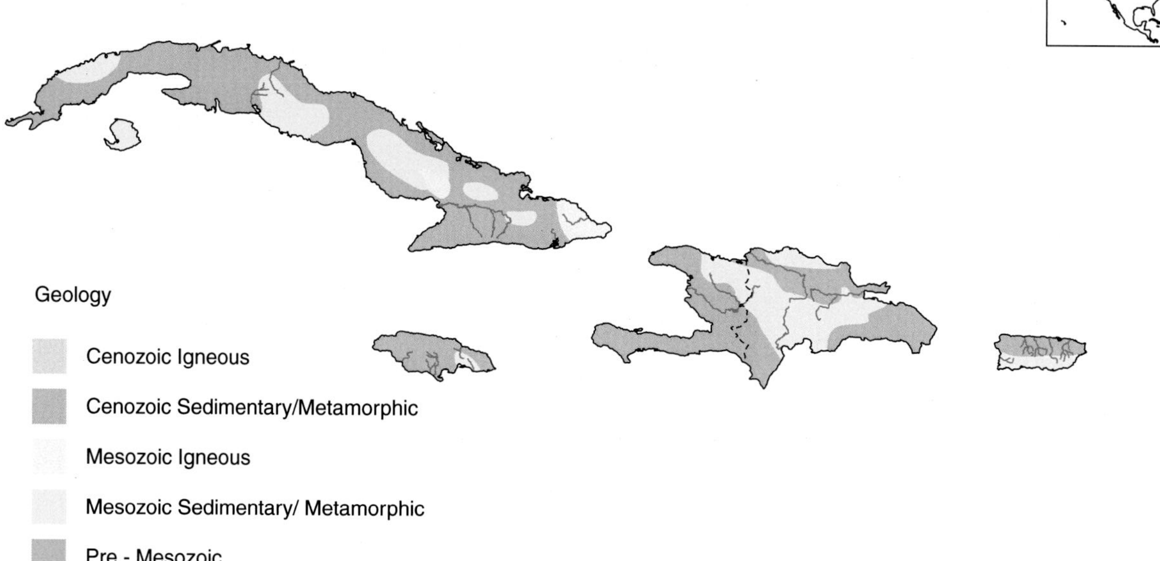

Surface Runoff (mm/yr)

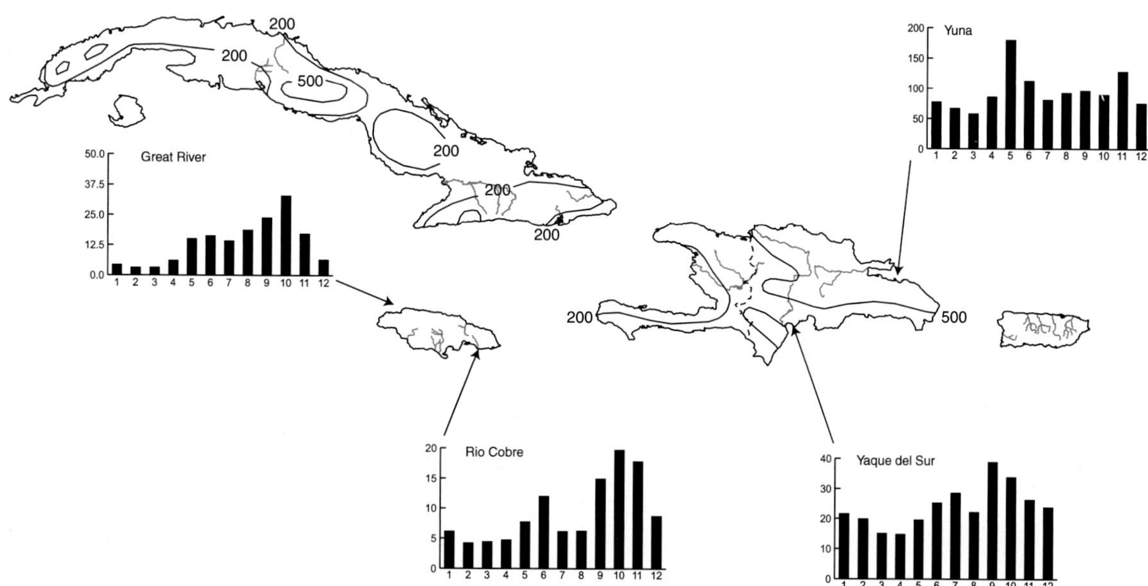

Figure A16

Table A8.

River name	Ocean	Area (10^3 km^2)	Length (km)	Max_elev (m)	Climate	Geology	Q (km^3/yr)	TSS (Mt/yr)	TDS (Mt/yr)	Refs.
Cuba										
Cauto	Caribbean	8.6	240	>500	Tr-H-S	Cen S/M				5
Damuji	Atlantic (NW)	0.85	56	<100	Tr-H-S	Cen S/M	0.3			6
Guantanamo	Caribbean	1.3	80	>500	Tr-H-S	Cen S/M	0.6			3
Sajua	Caribbean		120	>100	Tr-H-S	Cen S/M				5
Toa	Atlantic (NW)	0.75	61	>500	Tr-W-S	Cen S/M	1			6
Jamaica										
Black	Caribbean	0.6	53	>100	Te-W-S	Cen S/M				
Cobre	Caribbean	0.5	50	>500	Tr-H-S	Mes S/M	0.3			6
Grande	Caribbean	0.2	33	>1000	Tr-W-S	Cen S/M	0.8			6
Milk	Caribbean	0.7		>500	Tr-W-S	Cen S/M				
Minho	Caribbean	0.5	93	>500	Tr-H-S	Cen S/M				
Haiti										
Artibonite	Caribbean	7.9	240	>1000	Tr-H-S	Mes S/M	3.2			1, 4, 5
Dominican Republic										
Yaque del Norte	Atlantic (NW)	7	250	2600	Tr-H-S	Mes S/M	2			2
Yaque del Sur	Caribbean	5	180	2700	Tr-SA-C	Mes S/M	0.7			2
Yuna	Atlantic (NW)	5.1	100	<200	Tr-H-S	Mes S/M	2.9			2
Puerto Rico										
Arecibo	Atlantic (NW)	0.52	64	1300	Tr-W-S	Mes S/M	0.1			5
Guanajibo	Caribbean	0.31		>500	Tr-H-S	Mes S/M	0.21	0.045		3
La Plata	Atlantic (NW)	0.54	100	>1000	Tr-H-S	Mes S/M	0.28			3
Loiza	Atlantic (NW)	0.54	100	>500	Tr-H-S	Mes S/M	0.3	0.01		3
Manati	Atlantic (NW)	0.51		>1000	Tr-H-S	Mes S/M	0.38			3

References:
1. Center for Natural Resources Energy and Transport (UN), 1978; 2. Global River Discharge Database, http://www.rivdis.sr.unh.edu/; 3. GRDC website, www.gewex.org/grdc; 4. OAS, 1972 (cf. van der Leeden, 1975); 5. Rand McNally, 1980; 6. Vörösmarty *et al.*, 1996b

Appendix B South America

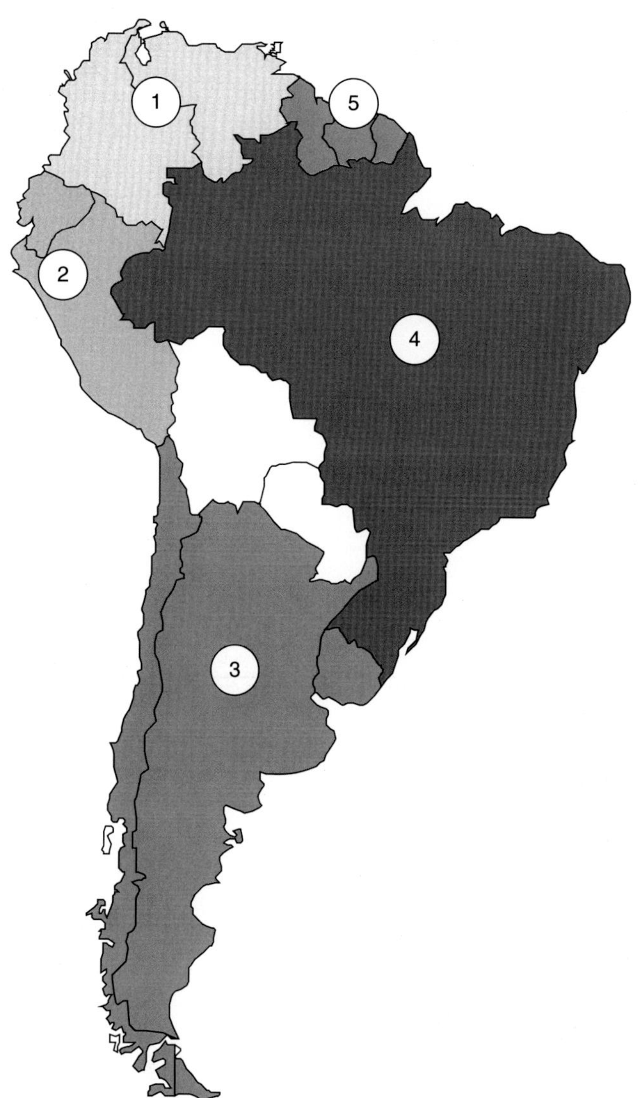

Figure B1

(1) Venezuela and Colombia
(2) Ecuador and Peru
(3) Chile, Argentina, Uruguay
(4) Brazil
(5) French Guiana, Suriname, Guyana

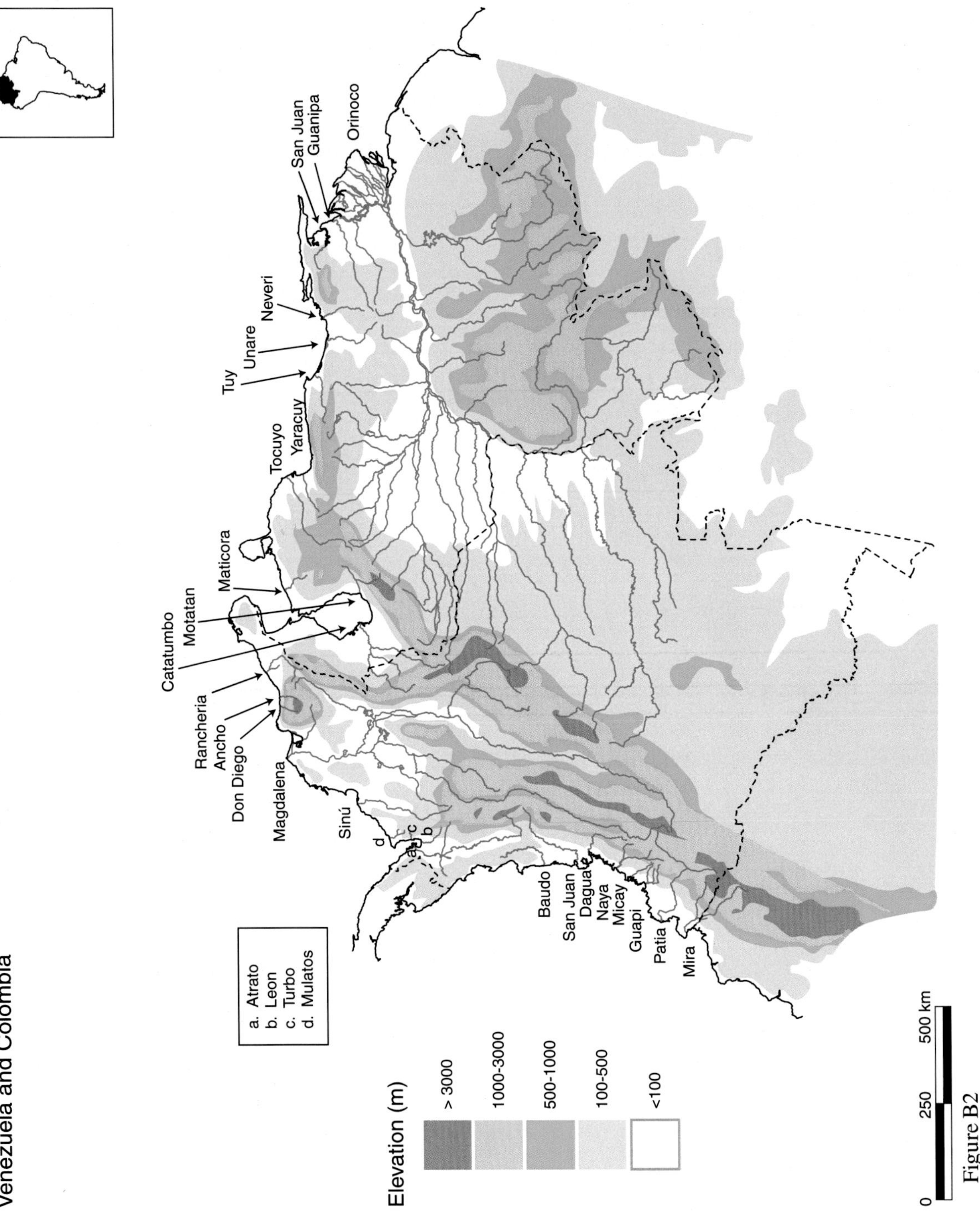

Venezuela and Colombia
Elevation (m)
> 3000
1000-3000
500-1000
100-500
<100
a. Atrato
b. Leon
c. Turbo
d. Mulatos
Sinú
Magdalena
Don Diego
Ancho
Rancheria
Catatumbo
Motatan
Maticora
Tocuyo
Yaracuy
Tuy
Unare
Neveri
San Juan
Guanipa
Orinoco
Baudo
San Juan
Dagua
Naya
Micay
Guapi
Patia
Mira
0
250
500 km

Figure B2

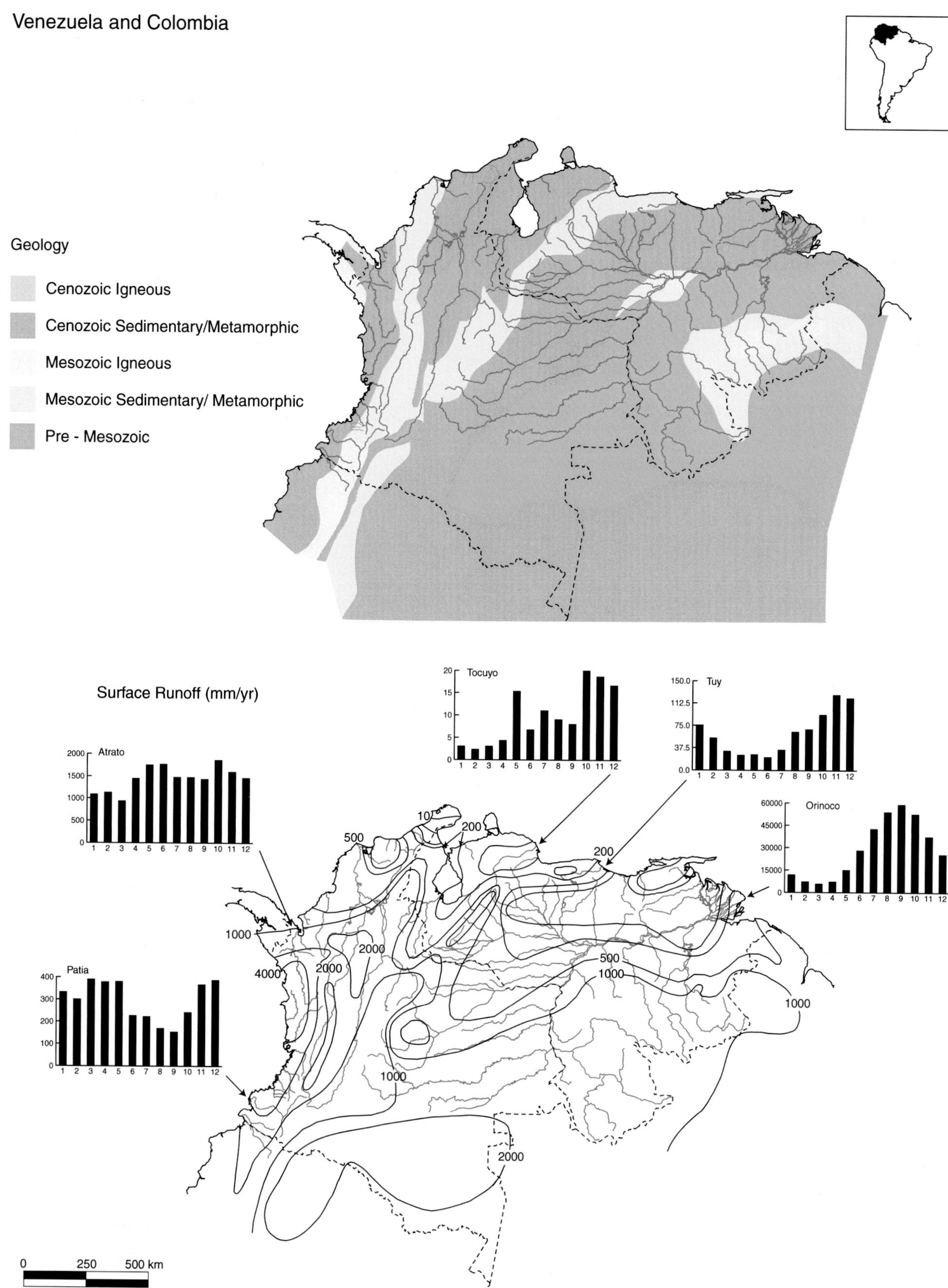
Venezuela and Colombia
Geology
Cenozoic Igneous
Cenozoic Sedimentary/Metamorphic
Mesozoic Igneous
Mesozoic Sedimentary/ Metamorphic
Pre - Mesozoic
Surface Runoff (mm/yr)
Atrato
Tocuyo
Tuy
Orinoco
Patia
0 250 500 km

Figure B3

Table B1.

River name	Ocean	Area (10^3 km^2)	Length (km)	Max_elev (m)	Climate	Geology	Q (km^3/yr)	TSS (Mt/yr)	TDS (Mt/yr)	Refs.
Venezuela										
Catatumbo	Caribbean	35	340	>500	Tr-H-W	Cen S/M	4			1, 2, 11
Guanipa	Atlantic (W)	10	270	>500	Tr-SA-S	Cen S/M				2
Maticora	Caribbean	3	140	>500	Tr-A-W	Cen S/M		5		2, 7
Motatan	Caribbean	4	130	>2000	Tr-H-W	Cen S/M	1			17
Neveri	Caribbean	1		>500	Tr-H-S	Cen S/M	1	0.3		2, 3, 7
Orinoco	Atlantic (W)	1100	2800	6000	Tr-W-S	Cen S/M	1100	210 (150)	28	6, 9, 10, 16, 17
San Juan	Atlantic (W)	8	180	>500	Tr-SA-S	Cen S/M-PreMes	1			2
Tocuyo	Caribbean	4	320	>500	Tr-SA-S	Cen S/M	0.34			5, 11, 17
Tuy	Caribbean	7	200	>1000	Tr-H-C	Mes S/M	2			8, 18
Unare	Caribbean	17	210	>200	Tr-SA-S	Cen S/M	2			2
Yaracuy	Caribbean	1	100	>500	Tr-SA-S	Mes S/M				17
Colombia										
Ancho	Caribbean	0.54		>500	Tr-W-W	PreMes	0.47	0.03		14
Atrato	Caribbean	36	710	>2000	Tr-W-W	Mes S/M	76	11		4, 10, 11, 12, 13, 18
Baudo	Pacific (E)	5.4	195	1800	Tr-W-Au	Cen S/M-Mes S/M	24			14
Dagua	Pacific (E)	1.7	100	>2000	Tr-W-Au	Mes S/M	4			11
Don Diego	Caribbean	0.52	48	>3000	Tr-W-S	PreMes	1.1	0.023		14
Guapi	Pacific (E)	2.9	110	>500	Tr-W-W	Cen S/M	11			14
Leon	Caribbean	1.2	46	>500	Tr-W-W	Cen S/M	2.6	1.4		14
Magdalena	Caribbean	260	1600	3300	Tr-W-C	Mes S/M	230	140	28	8, 10
Micay	Pacific (E)	4.4	60	>1000	Tr-W-C	Cen S/M-Mes S/M	19			2
Mira	Pacific (E)	9.5	270	>3000	Tr-W-C	Mes I	27	9.7		2, 13, 18
Mulatos	Caribbean	1	110	>500	Tr-H-S	Cen S/M	0.33	0.21		14
Naya	Pacific (E)	2	82	>2000	Tr-W-W	Cen S/M-Mes S/M	13			14
Patia	Pacific (E)	24	420	>2000	Tr-W-C	Mes I	40	21		4, 13, 18
Rancheria	Caribbean	2.2	170	>1500	Tr-SA-S	Cen S/M	0.39	0.1		14
San Juan	Pacific (E)	16	350	>2000	Tr-W-C	Mes I	82	16		2, 10, 11
Sinú	Caribbean	15	300	4000	Tr-W-W	Cen S/M-Mes S/M	12	4.2		2, 11, 13, 15
Turbo	Caribbean	0.16	20	>500	Tr-W-S	Cen S/M	0.12	0.073		14

References:
1. Center for Natural Resources Energy and Transport (UN), 1978; 2. ECLAC, 1990; 3. FAO Global River Sediment Yield Database, http://www.fao.org/nr/water/aquastat/sediment/; 4. Global River Discharge Database, http://www.rivdis.sr.unh.edu/; 5. GRDC website, www.gewex.org/grdc; 6. Hernandez, D. P., personal communication; 7. IAHS/UNESCO, 1974; 8. Jaffe *et al.*, 1995; 9. Meade, 1994; 10. Meybeck and Ragu, 1996; 11. Rand McNally, 1980; 12. Restrepo and Kjerfve, 2000b; 13. Restrepo and Löpez, 2008; 14. J. D. Restrepo, personal communication; 15. Serrano Suarez, 2004; 16. UNESCO (WORRI), 1978; 17. UNESCO, 1971; 18. Vörösmarty *et al.*, 1996c

Ecuador and Peru

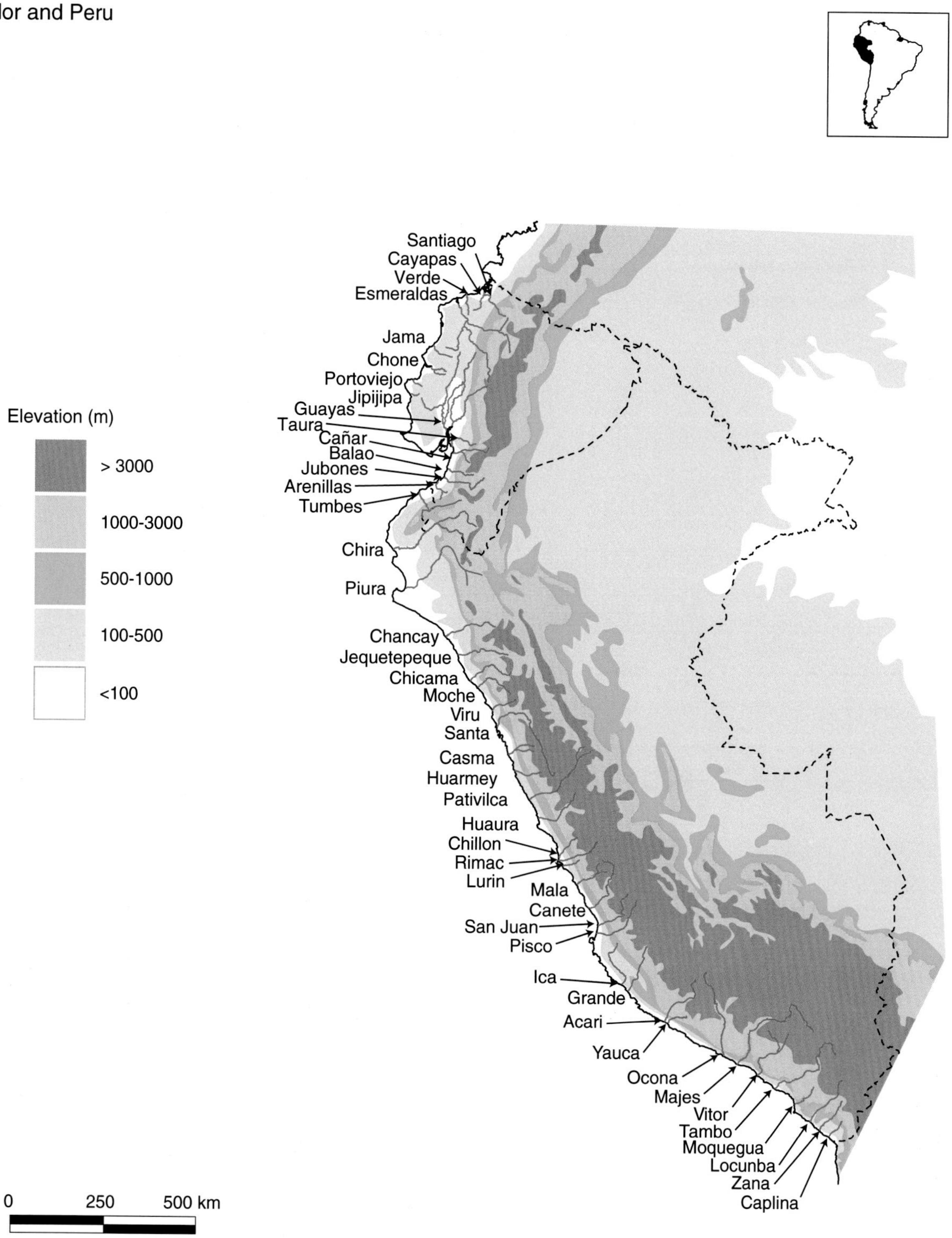

Santiago
Cayapas
Verde
Esmeraldas
Jama
Chone
Portoviejo
Jipijipa
Guayas
Taura
Cañar
Balao
Jubones
Arenillas
Tumbes
Chira
Piura
Chancay
Jequetepeque
Chicama
Moche
Viru
Santa
Casma
Huarmey
Pativilca
Huaura
Chillon
Rimac
Lurin
Mala
Canete
San Juan
Pisco
Ica
Grande
Acari
Yauca
Ocona
Majes
Vitor
Tambo
Moquegua
Locunba
Zana
Caplina

Elevation (m)

> 3000
1000-3000
500-1000
100-500
<100

0 250 500 km

Figure B4

Ecuador and Peru

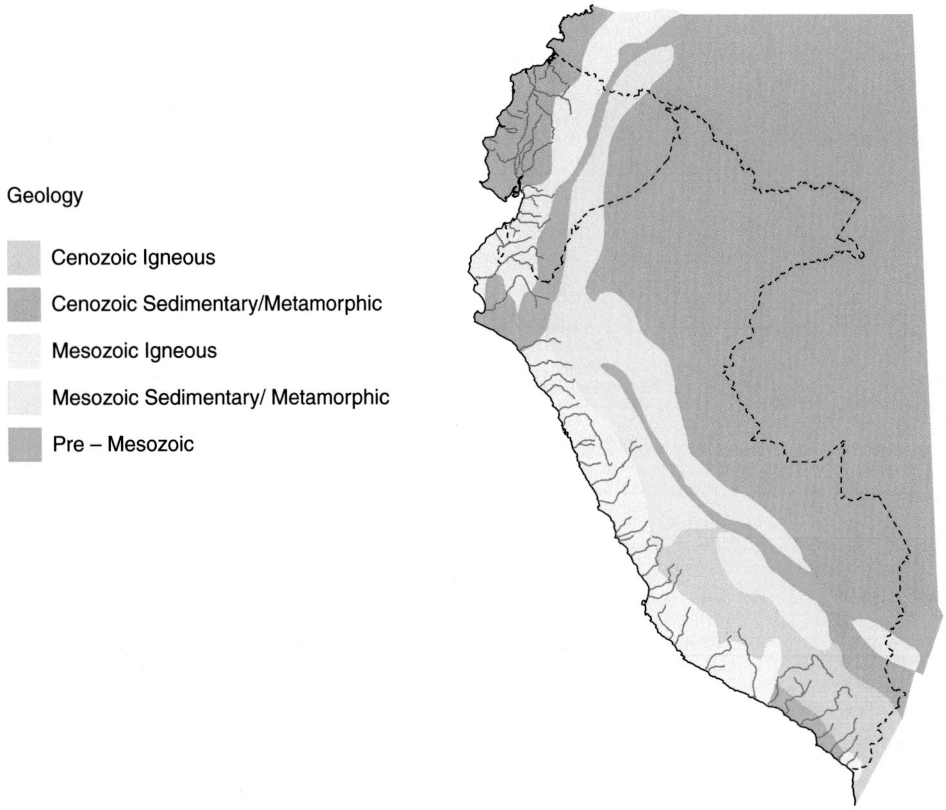

Geology

Cenozoic Igneous

Cenozoic Sedimentary/Metamorphic

Mesozoic Igneous

Mesozoic Sedimentary/ Metamorphic

Pre – Mesozoic

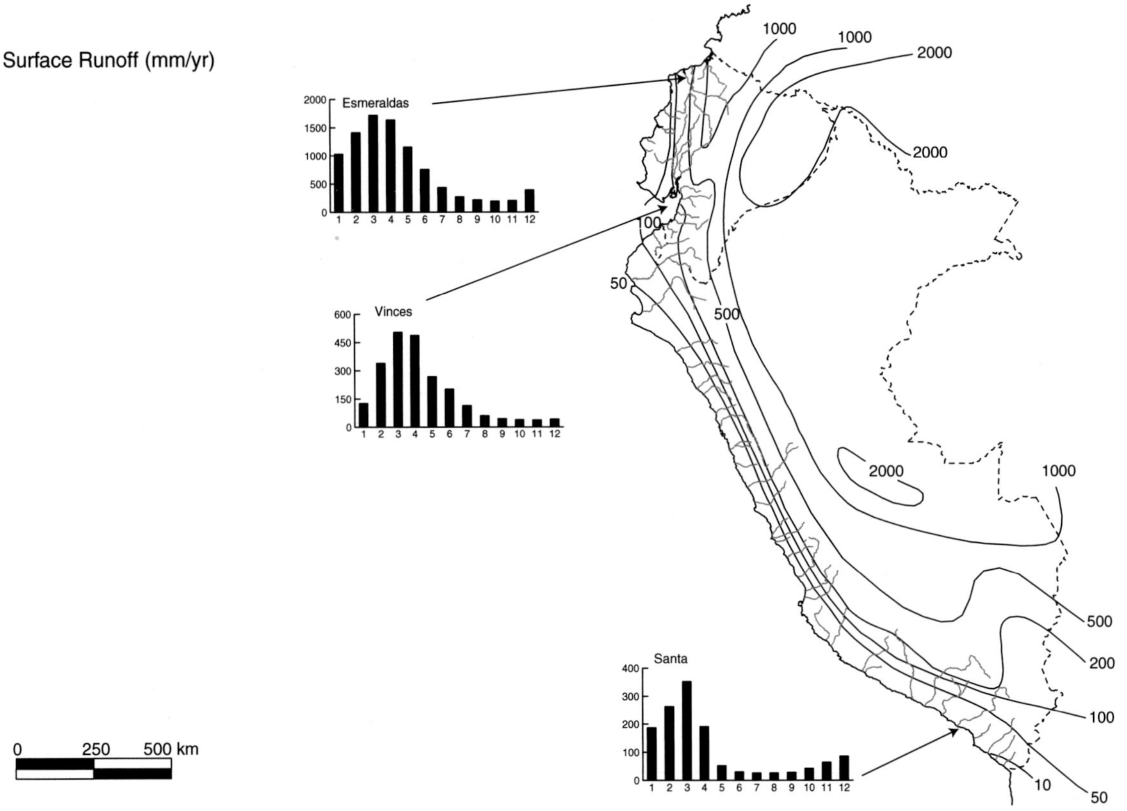

Surface Runoff (mm/yr)

Esmeraldas

Vinces

Santa

1000
1000
2000
2000
100
50
500
2000
1000
500
200
100
10
50

0 250 500 km

Figure B5

Table B2.

River name	Ocean	Area (10^3 km^2)	Length (km)	Max_elev (m)	Climate	Geology	Q (km^3/yr)	TSS (Mt/yr)	TDS (Mt/yr)	Refs.
Ecuador										
Arenillas	Pacific (E)	2.7		>500	Tr-SA-S	Mes I	0.6			7
Balao	Pacific (E)	3.4		>3000	Tr-H-S	Mes I	1.9			7
Cañar	Pacific (E)	2.4		>3000	Tr-H-S	Mes I	1.6		0.2	7
Cayapas	Pacific (E)	6		>100	Tr-W-S	Cen S/M	12			7
Chone	Pacific (E)	2.5	75	>300	Tr-H-S	Cen S/M	0.8			4, 7, 8
Esmeraldas	Pacific (E)	21	300	5300	Tr-W-S	Mes I	26		1.9	4, 7, 8
Guayas	Pacific (E)	32	390	>3000	Tr-W-S	Mes I	37			4, 6, 7
Jama	Pacific (E)	2.1	75	>200	Tr-A-S	Cen S/M	0.15			4, 7
Jipijapa	Pacific (E)	2.6		>200	Tr-A-S	Cen S/M	0.2			7
Jubones	Pacific (E)	4.1	160	3500	Tr-H-S	Mes I	1.8			4, 6, 7
Portoviejo	Pacific (E)	2.2		>200	Tr-SA-S	Cen S/M	0.4			7
Santiago	Pacific (E)	2.6	120	1500	Tr-W-S	Cen S/M				
Taura	Pacific (E)	2.3		>500	Tr-H-S	Mes S/M	0.9			7
Verde	Pacific (E)	2.3		>100	Tr-H-S	Cen S/M	1.3			7
Peru										
Acari	Pacific (E)	4.1	330	>4000	STr-A-S	Cen I	0.41			3, 6
Canete	Pacific (E)	6.7	190	>4000	STr-H-S	Mes	1.7			3, 6
Caplina	Pacific (E)	2.2	110	6400	STr-A-S	Cen I	0.04			3
Casma	Pacific (E)	2.9	100	4500	STr-A-S	Mes I	0.18			3
Chancay	Pacific (E)	5.2	100	>2000	STr-SA-S	PreMes	0.89			3
Chicama	Pacific (E)	4.8	130	4800	STr-SA-S	Mes	0.95			3
Chillon	Pacific (E)	2	80	>4000	STr-SA-S	Mes	0.29			3
Chira	Pacific (E)	20	160	>3000	STr-H-S	Mes	5	20		1, 3, 6
Grande	Pacific (E)	13	190	>4000	STr-A-S	Cen I	0.55			3, 6
Huarmey	Pacific (E)	2.1	90	>4000	STr-A-S	Mes I	0.12			3
Huaura	Pacific (E)	5.5	140	4700	STr-SA-S	Mes	0.88			3
Ica	Pacific (E)	7.4	160	>3000	STr-A-S	Mes	0.33			3, 6
Jequetepeque	Pacific (E)	4.2	140	>1000	STr-SA-S	Mes	0.9			3
Locunba	Pacific (E)	2.4	130	>4000	STr-SA-S	Cen I				
Lurin	Pacific (E)	2.5	90	>4500	STr-A-S	Mes	0.12			3

Table B2. (*Continued*)

River name	Ocean	Area (10^3 km^2)	Length (km)	Max_elev (m)	Climate	Geology	Q (km^3/yr)	TSS (Mt/yr)	TDS (Mt/yr)	Refs.
Majes	Pacific (E)	17	400	5000	STr-SA-S	Cen I	2.9			3, 5, 6
Mala	Pacific (E)	2.1	110	>4000	STr-H-S	Mes I	0.57			1
Moche	Pacific (E)	2.1	85	>3000	STr-SA-S	Mes	0.3			3
Moquegua	Pacific (E)	3.4	120	>4000	STr-A-S	Cen I	0.06			3
Ocona	Pacific (E)	15	240	5500	STr-A-S	Cen I				
Pativilca	Pacific (E)	4.7	160	3800	STr-H-S	Mes S/M	1.4			3
Pisco	Pacific (E)	4.4	240	>4000	STr-SA-S	Mes I	0.83			3, 6
Piura	Pacific (E)	13	240	>500	STr-A-S	Mes I	0.83			3, 6
Rimac	Pacific (E)	3.5	130	>3000	STr-H-S	Mes I	0.91			3, 6
San Juan	Pacific (E)	3.9	130	>4000	STr-A-S	Mes I	0.2			3
Santa	Pacific (E)	12	320	>3000	STr-H-S	Mes	4.7			3, 6
Tambo	Pacific (E)	9.5	250	>4000	STr-A-S	Cen I				
Tumbes	Pacific (E)	4.6	180	>1000	STr-W-S	Mes I	3.9			2, 3
Viru	Pacific (E)	2	75	3000	STr-A-S	Mes I	0.12			3
Vitor	Pacific (E)	16	260	>4000	STr-SA-S	Cen I				6
Yauca	Pacific (E)	4.5	80	>4000	STr-A-S	Cen I	0.27			3
Zana	Pacific (E)	2	150	>4000	STr-SA-S	Cen I	0.26			3

References:
1. Burz, 1977; 2. Center for Natural Resources Energy and Transport (UN), 1978; 3. CEPAL, 1968 (cf. van der Leeden, 1975); 4. Manchero, 1973 (cf. van der Leeden, 1975); 5. Meybeck and Ragu, 1996; 6. Rand McNally, 1980; 7. US Army Corps of Engineers, 1998; 8. UNESCO (WORRI), 1978

Chile, Argentina and Uruguay

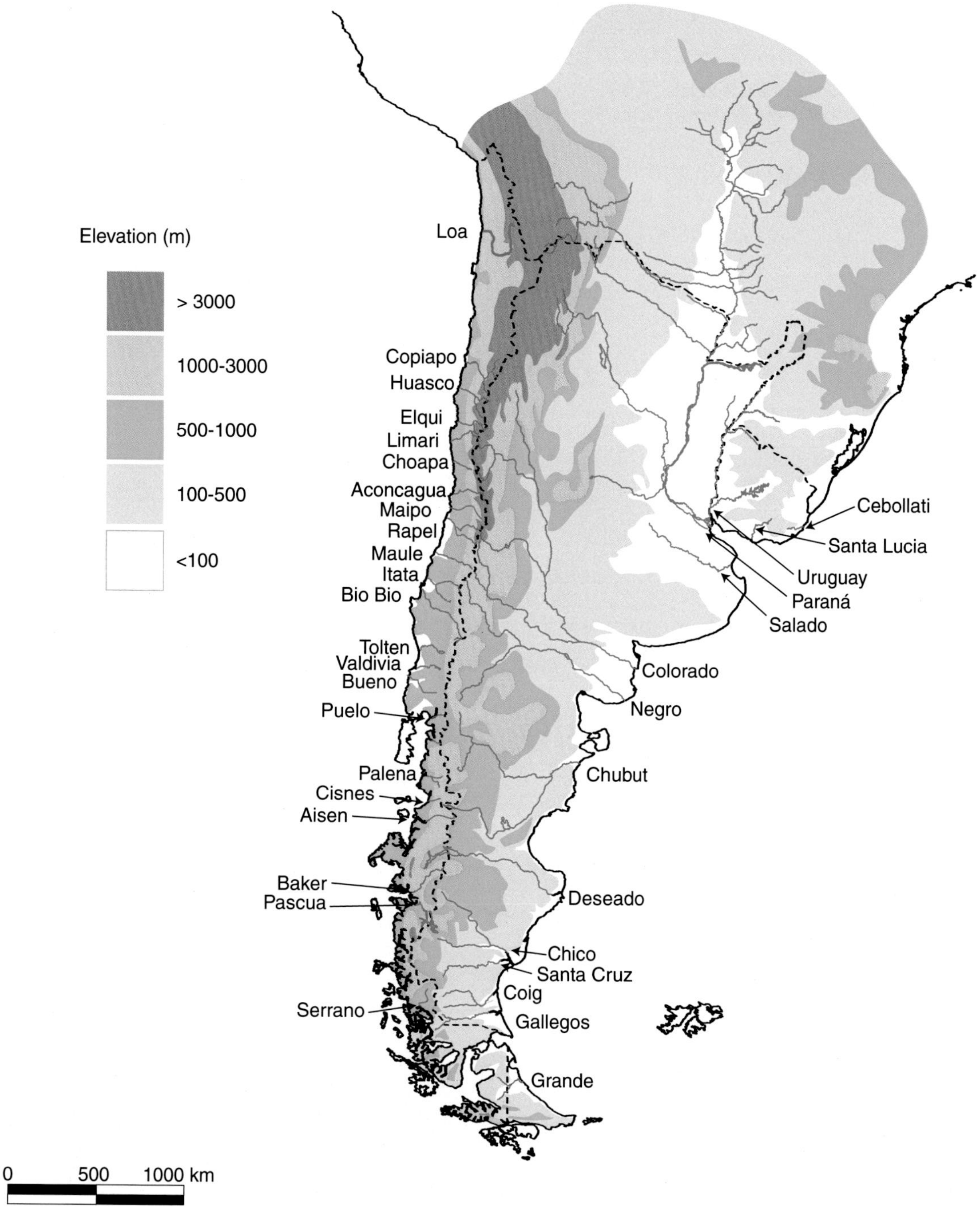

Elevation (m)

> 3000

1000-3000

500-1000

100-500

<100

Loa
Copiapo
Huasco
Elqui
Limari
Choapa
Aconcagua
Maipo
Rapel
Maule
Itata
Bio Bio
Tolten
Valdivia
Bueno
Puelo
Palena
Cisnes
Aisen
Baker
Pascua
Serrano
Cebollati
Santa Lucia
Uruguay
Paraná
Salado
Colorado
Negro
Chubut
Deseado
Chico
Santa Cruz
Coig
Gallegos
Grande

0 500 1000 km

Figure B6

Chile, Argentina and Uruguay

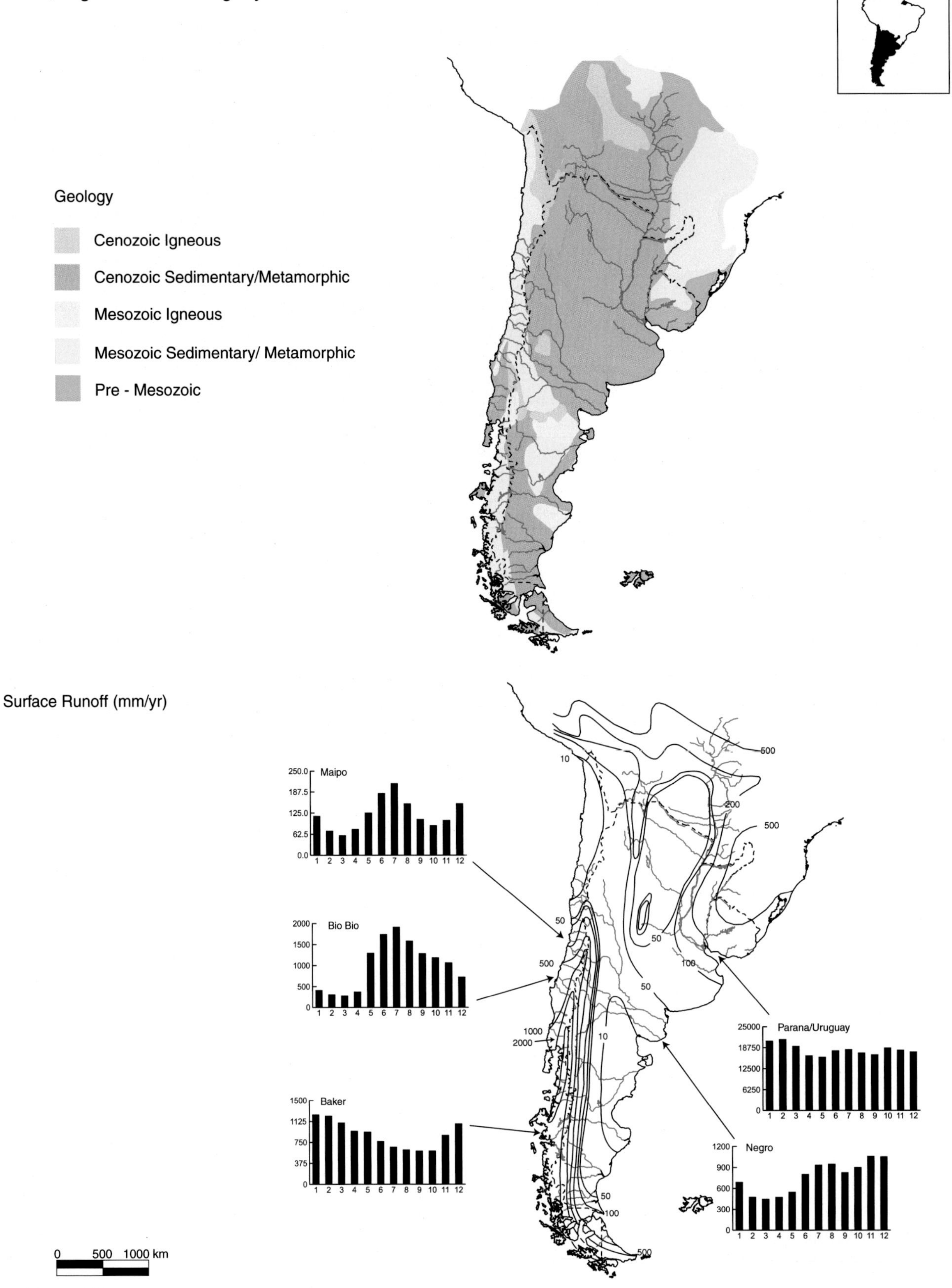

Figure B7

Table B3.

River name	Ocean	Area(10^3 km^2)	Length(km)	Max_elev(m)	Climate	Geology	Q(km^3/yr)	TSS(Mt/yr)	TDS(Mt/yr)	Refs.
Chile										
Aconcagua	Pacific (SE)	2.1	190	4300	Te-A-W	Mes I	1	0.5	0.5	7, 9, 18
Aisen	Pacific (SE)	4	110		Te-H-C	Mes I	1.4			7
Baker	Pacific (SE)	26	310	>1000	Te-W-C	PreMes	31			1, 7
Bio Bio	Pacific (SE)	24	380	>2000	Te-W-W	Cen I	33		1.8	7, 10, 14, 19
Bueno	Pacific (SE)	15	120	>200	Te-W-W	PreMes	11			7, 16
Choapa	Pacific (SE)	3.6	150	4000	Te-A-W	Mes I	0.34			7
Cisnes	Pacific (SE)	5.2	140	>1000	Te-W-C	Mes I	7.9			7
Copiapo	Pacific (SE)	5.1	170	>3000	Te-A-W	Mes I	0.07			7
Elqui	Pacific (SE)	9.6	210	>3000	Te-A-W	Mes I	0.25			16
Huasco	Pacific (SE)	7.2	150	5900	Te-A-W	Mes I	0.15			7
Itata	Pacific (SE)	11	180	>1000	Te-H-W	Mes I	5.9			14, 18
Limari	Pacific (SE)	11	110	4900	Te-A-W	Mes I	0.3			7, 10, 11
Loa	Pacific (SE)	33	440	4300	Str-A-W	Mes I				
Maipo	Pacific (SE)	15	230	5600	Te-H-W	Mes I	3.8		2.5	7, 10, 14
Maule	Pacific (SE)	22	240	2200	Te-H-C	Mes I	13			7, 19
Palena	Pacific (SE)	13	290	500	Te-H-C	Mes I				1, 18
Pascua	Pacific (SE)	15		3000	Te-H-W	Mes I	18			
Puelo	Pacific (SE)	8.6	150	>1000	Te-A-W	Cen S/M	21			3, 16
Rapel	Pacific (SE)	13	210	4900	SA-H-S	Mes I	5.4 (25)			7, 10, 14
Serrano	Pacific (SE)	9.1		>3000	Te-W-C	Mes S/M				
Tolten	Pacific (SE)	3	120	>1000	Te-W-C	Cen S/M	10			7
Valdivia	Pacific (SE)	11	150	>500	Te-W-W	Cen S/M	14			7, 19
Argentina										
Chico	Atlantic (SW)	17	410	>1000	Te-A-Sp	Cen S/M	0.9			
Chubut	Atlantic (SW)	40	850	>1000	Te-A-Sp	Cen S/M	1.3	0.6	0.05	6, 12, 14, 15, 16, 20
Coig	Atlantic (SW)	15	250	>500	Te-SA-Sp	Mes S/M	0.15	0.1	0.02	6
Colorado	Atlantic (SW)	22	1000	>3000	Te-A-Sp	Cen I	4.1	6.9	3	4, 6, 12, 15, 20
Deseado	Atlantic (SW)	14	610	>1000	Te-A-Sp	Cen S/M	0.15	0.5	0.05	6, 18
Gallegos	Atlantic (SW)	5.1	320	>500	Te-SA-Sp	Cen S/M	1	0.1	0.06	1, 6
Grande	Atlantic (SW)	8.4	100	>200	Te-A-Sp	Cen S/M	0.7			10
Negro	Atlantic (SW)	95	1000	>1000	Te-H-Sp	Cen I	27	13	5.4	6, 12,13, 14, 20
Paraná	Atlantic (SW)	2600	4800	>1000	STr-SA-C	Mes I	530	90	62	5, 8, 10, 14, 15, 17, 19, 20
Salado	Atlantic (SW)	40	640	>100	Te-A-Sp	Cen S/M	0.6			10, 11, 18
Santa Cruz	Atlantic (SW)	24	450	>500	Te-W-Sp	Mes I	21	0.7	0.05	6, 10, 20
Uruguay	Atlantic (SW)	370	1500	>1000	STr-H-S	Mes	140	10	11	15
Uruguay										
Cebollati	Atlantic (SW)	18	210	>100	Te-H-W	PreMes	3.2			18, 20
Santa Lucia	Atlantic (SW)	9	200	>100	Te-A-W	PreMes	0.5			2

References:
1. Center for Natural Resources Energy and Transport (UN), 1978; 2. CEPAL, 1972 (cf. van der Leeden, 1975); 3. Dai and Trenberth, 2002; 4. Depetris and Irion, 1996; 5. Depetris and Lenardon, 1982; 6. Depetris *et al.*, 2005; 7. Donoso, 1967 (cf. van der Leeden, 1975); 8. Drago and Amsler, 1988; 9. FAO Global River Sediment Yield Database, http://www.fao.org/nr/water/aquastat/sediment/; 10. GEMS website, www.gemstat.org; 11. Global River Discharge Database, http://www.rivdis.sr.unh.edu/; 12. Holeman, 1968; 13. IAHS/UNESCO, 1974; 14. Meybeck and Ragu, 1996; 15. Pasquini and Depetris, 2007; 16. Peucker-Ehrenbrink, 2009; 17. Probst, 1992; 18. Rand McNally, 1980; 19. UNESCO (WORRI), 1978; 20. UNESCO, 1971; 21. (http://www.fao.org/landwater/aglw/sediment/default.asp)

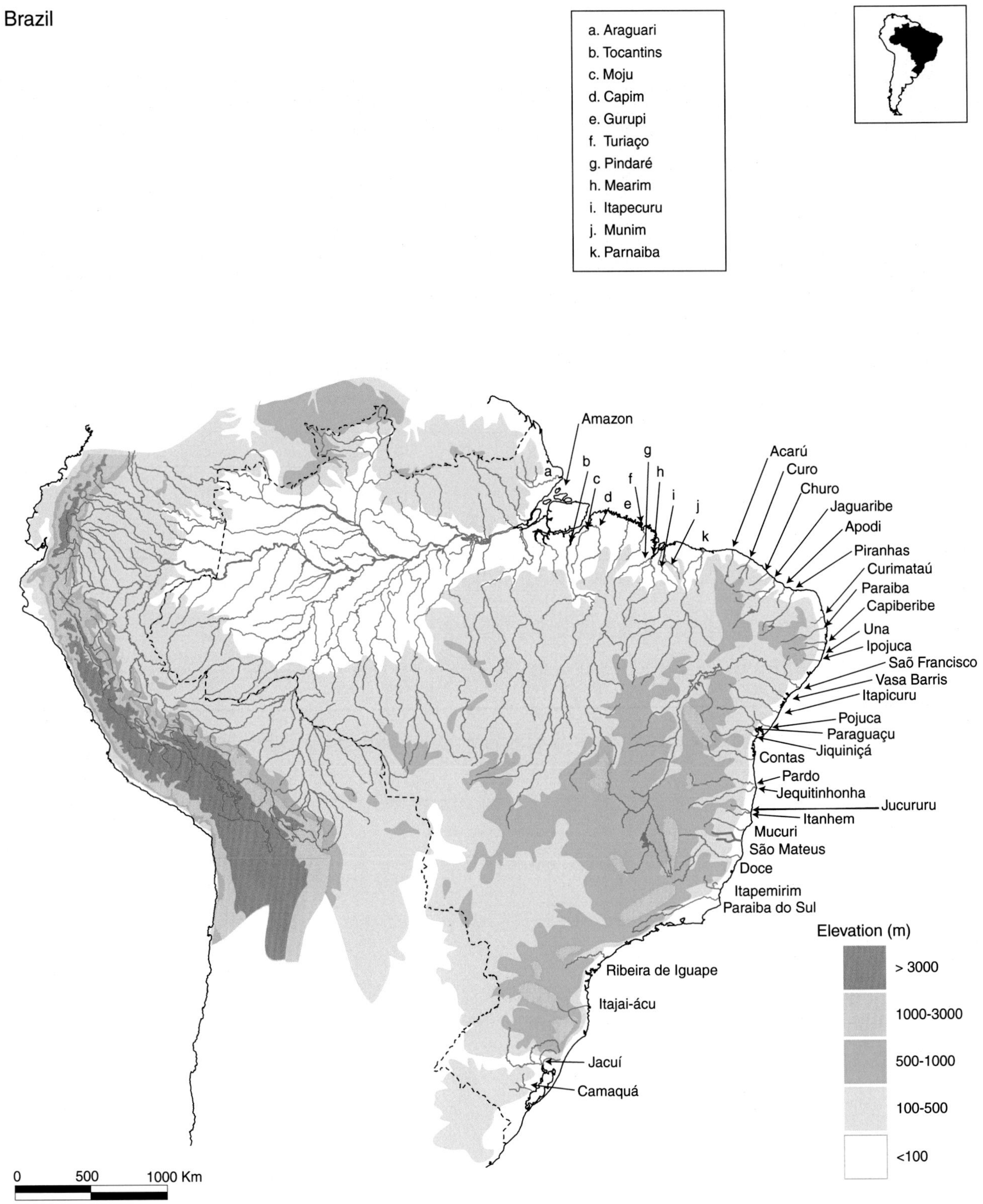

Brazil

a. Araguari
b. Tocantins
c. Moju
d. Capim
e. Gurupi
f. Turiaço
g. Pindaré
h. Mearim
i. Itapecuru
j. Munim
k. Parnaiba

Amazon
a b c d e f g h i j k
Acarú
Curo
Churo
Jaguaribe
Apodi
Piranhas
Curimataú
Paraiba
Capiberibe
Una
Ipojuca
Saõ Francisco
Vasa Barris
Itapicuru
Pojuca
Paraguaçu
Jiquiniçá
Contas
Pardo
Jequitinhonha
Jucururu
Itanhem
Mucuri
São Mateus
Doce
Itapemirim
Paraiba do Sul
Ribeira de Iguape
Itajai-ácu
Jacuí
Camaquá

Elevation (m)
> 3000
1000-3000
500-1000
100-500
<100

0 500 1000 Km

Figure B8

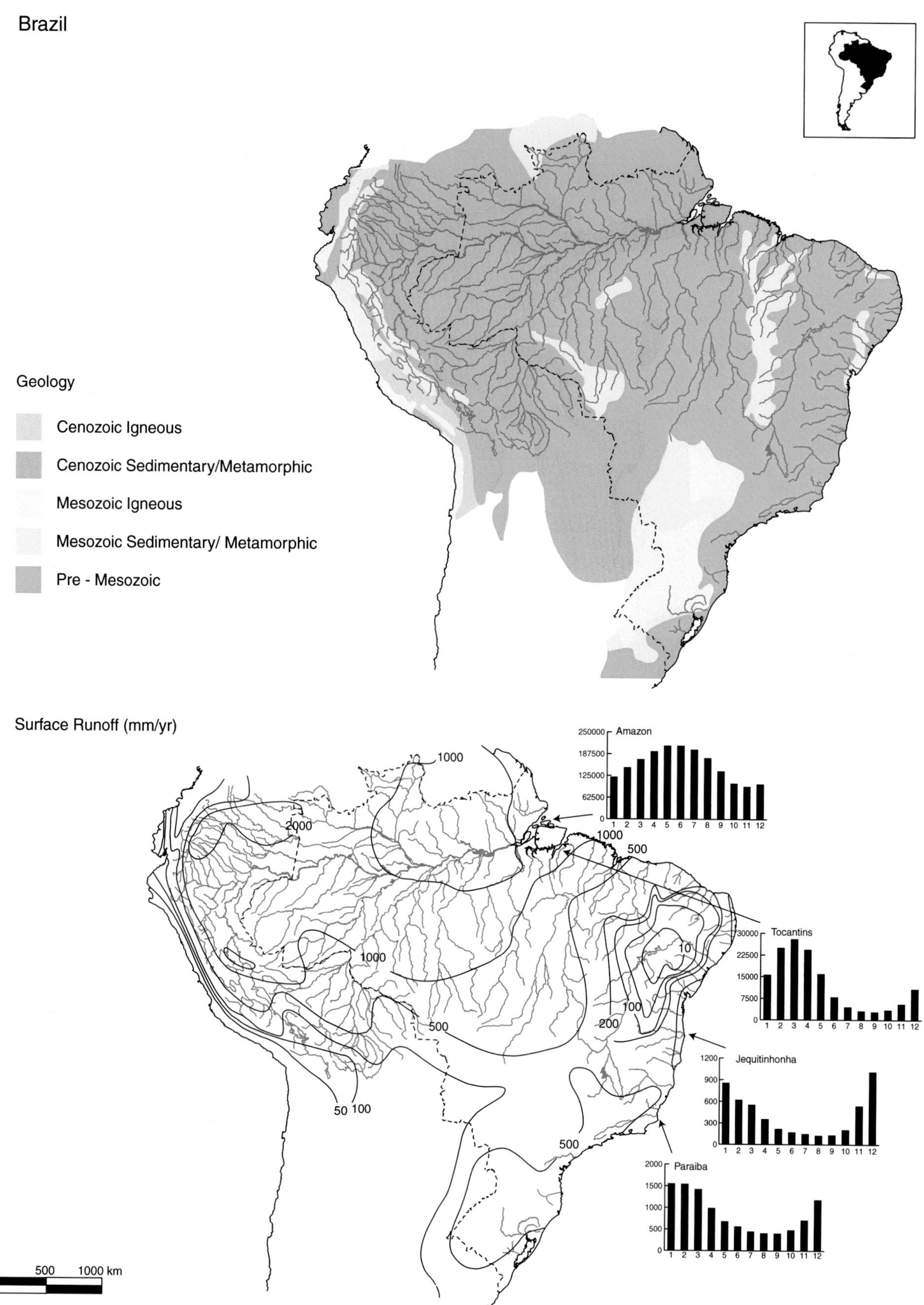
Brazil
Geology
Cenozoic Igneous
Cenozoic Sedimentary/Metamorphic
Mesozoic Igneous
Mesozoic Sedimentary/ Metamorphic
Pre - Mesozoic
Surface Runoff (mm/yr)
1000
2000
1000
500
1000
10
100
200
500
50 100
500
Amazon
Tocantins
Jequitinhonha
Paraiba
0 500 1000 km

Figure B9

Table B4.

River name	Ocean	Area (10^3 km^2)	Length (km)	Max_elev (m)	Climate	Geology	Q (km^3/yr)	TSS (Mt/yr)	TDS (Mt/yr)	Refs.
Brazil										
Acaraú	Atlantic (W)	32	240	400	Tr-H-S	PreMes	5			6
Amazon	Atlantic (W)	6300	6400	5500	Tr-W-S	Mes I	6300	1200	270	2, 11, 12
Apodi	Atlantic (W)	12	180	1000	Tr-SA-S	PreMes	1			1
Araguari	Atlantic (W)	43	400	300	Tr-W-S	PreMes	30	0.5		3, 12, 15
Camaquá	Atlantic (SW)	16	320	>200	STr-H-W	Cen S/M				
Capiberibe	Atlantic (W)	7.2	200	1000	Tr-H-S	PreMes	0.6			16
Capim	Atlantic (W)	38	660	200	Tr-W-S	PreMes-Mes S/M	25			14
Churo	Atlantic (W)	5		<500	Tr-SA-S	PreMes	0.5			
Contas	Atlantic (W)	56	490	>500	Tr-SA-S	PreMes	1.5	0.1		1, 3, 18, 19
Curimataú	Atlantic (W)	5	180	700	Tr-SA-S	PreMes	0.15			5
Curu	Atlantic (W)	7.5		>500	Tr-A-S	PreMes	0.1			
Doce	Atlantic (W)	76	850	>1000	Tr-H-S	PreMes	45	10		1, 15, 18, 20
Gurupi	Atlantic (W)	40	480	>200	Tr-H-S	Mes S/M	14	10		8, 15, 21
Ipojuca	Atlantic (W)	6.7	220	1000	Tr-H-S	PreMes				
Itajai-Ácu	Atlantic (SW)	15	300	1000	STr-H-Sp	PreMes	7	0.76		8, 17
Itanhem	Atlantic (W)	50	200	>500	Tr-H-S	PreMes	14			1, 18, 19
Itapecuru	Atlantic (W)	76	640	>200	Tr-SA-S	Mes S/M	9			3
Itapemirim	Atlantic (SW)	5.2		>500	Tr-H-S	PreMes	2.4			
Itapicuru	Atlantic (W)	43	400	>500	Tr-SA-S	PreMes	8	0.01		5, 15, 18
Jacuí	Atlantic (SW)	71	450	>500	STr-H-W	Mes I	41		0.69	5, 12, 15
Jaguaribe	Atlantic (W)	81	460	>500	Tr-A-S	PreMes	<1 (4)	0.06		1, 9, 10
Jequitinhonha	Atlantic (W)	68	800	1000	Tr-H-S	PreMes	23	5		8, 15, 16, 18, 20
Jiquiniça	Atlantic (W)	6		<500	Tr-H-S	PreMes	0.35			18, 19
Jucururu	Atlantic (W)	4.2	180	500	Tr-H-S	PreMes				
Mearim	Atlantic (W)	100	570	>200	Tr-H-S	PreMes	17	0.7		1, 12, 20, 21
Moju	Atlantic (W)	25	300	100	Tr-W-S	PreMes-Mes S/M				
Mucuri	Atlantic (W)	15	320	>500	Tr-H-S	PreMes	2.9	2.5		6, 18, 19
Munim	Atlantic (W)	28	230	150	Tr-H-S	Cen S/M	8.8			1, 5, 6
Paraguaçu	Atlantic (W)	60	480	>1000	Tr-SA-S	PreMes	4	0.2		3, 18, 19
Paraiba	Atlantic (W)	20	300	>500	Tr-SA-S	PreMes	4			

Table B4. (*Continued*)

River name	Ocean	Area (10^3 km^2)	Length (km)	Max_elev (m)	Climate	Geology	Q (km^3/yr)	TSS (Mt/yr)	TDS (Mt/yr)	Refs.
Paraiba do Sul	Atlantic (SW)	57	1100	1800	Tr-H-S	PreMes	28	4	1.2	8, 12, 18, 19
Pardo	Atlantic (W)	47	640	>1000	Tr-A-S	PreMes	3.8			15, 18
Parnaiba	Atlantic (W)	340	1400	?500	Tr-A-S	PreMes	40	3		1, 5, 21
Pindaré	Atlantic (W)	34	460	800	Tr-H-S	PreMes	6.4			1, 6
Piranhas	Atlantic (W)	43	400	>500	Tr-A-S	PreMes	2.7			1
Pojuca	Atlantic (W)	5		>100	Tr-SA-S	PreMes	0.9			
Ribeira do Iguape	Atlantic (SW)	31	320	>1000	STr-H-C	PreMes	27			1, 15
Saõ Francisco	Atlantic (W)	640	2700	1400	Tr-SA-S	PreMes	70 (80)	0.8 (15)	6	4, 5, 7, 12, 13, 18
Saõ Mateus	Atlantic (W)	12	190	>500	Tr-H-S	PreMes				18
Tocantins	Atlantic (W)	760	2700	1100	Tr-H-S	PreMes	370	75	15	3, 5, 12, 20
Turiaço	Atlantic (W)	13	350	300	Tr-H-S	Cen S/M				
Una	Atlantic (W)	6		>200	Tr-H-S	PreMes	1.4			
Vasa Barris	Atlantic (W)	16	380	500	Tr-A-S	PreMes	3.5			1, 6, 18

References:
1. Brazil Agencia Nacional de Aguas HidroWeb, http://hidroweb.ana.gov.br/; 2. Carvalho and Cunha, 1998; 3. Dai and Trenberth, 2002; 4. Depetris and Paolini, 1991; 5. GEMS website, www.gemstat.org; 6. Global River Discharge Database, http://www.rivdis.sr.unh.edu/; 7. Harrison, 2000; 8. B. J. Kjerfve, personal communication; 9. B. Knoppers, personal communication; 10. Lacerda and Marins, 2002; 11. Meade *et al.*, 1985; 12. Meybeck and Ragu, 1996; 13. Milliman, 1975; 14. Peucker-Ehrenbrink, 2009; 15. Rand McNally, 1980; 16. Rodier and Roche, 1984; 17. Schettini, 2002; 18. Souza and Knoppers, 2003; 19. Souza *et al.*, 2003; 20. UNESCO (WORRI), 1978; 21. Jennerjahn *et al.*, 2010

French Guiana, Suriname and Guyana

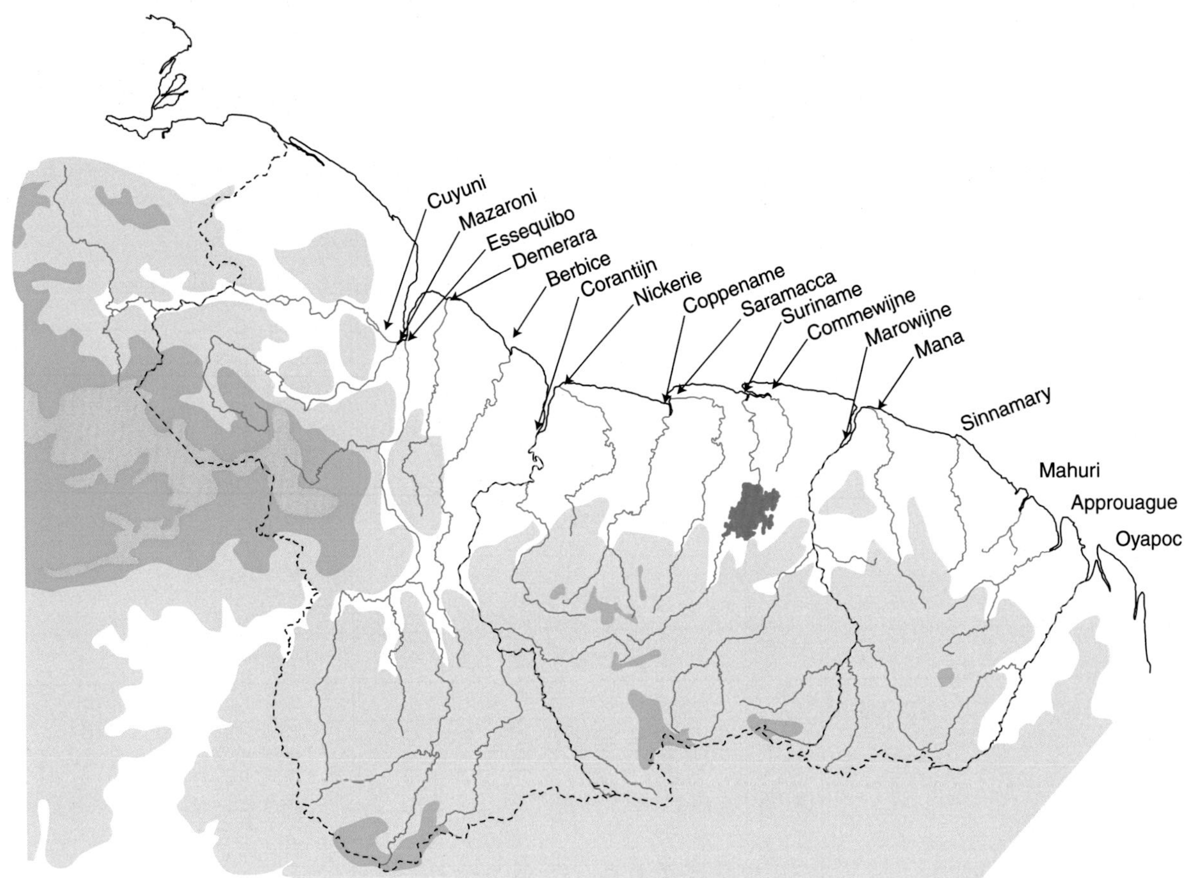

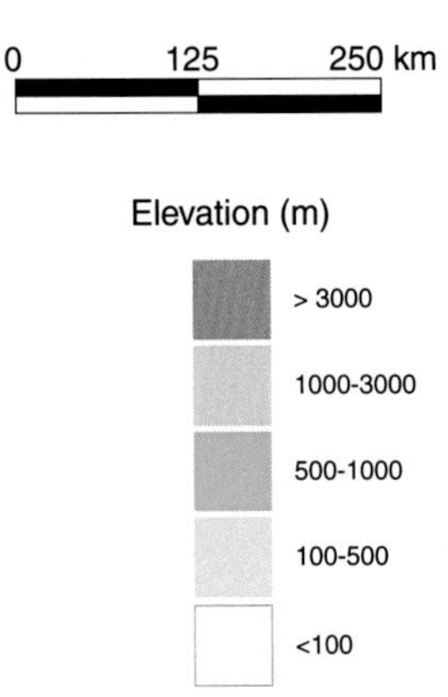

Figure B10

French Guiana, Suriname and Guyana

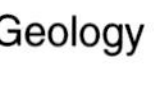

Geology

Cenozoic Igneous

Cenozoic Sedimentary/Metamorphic

Mesozoic Igneous

Mesozoic Sedimentary/ Metamorphic

Pre - Mesozoic

Surface Runoff (mm/yr)

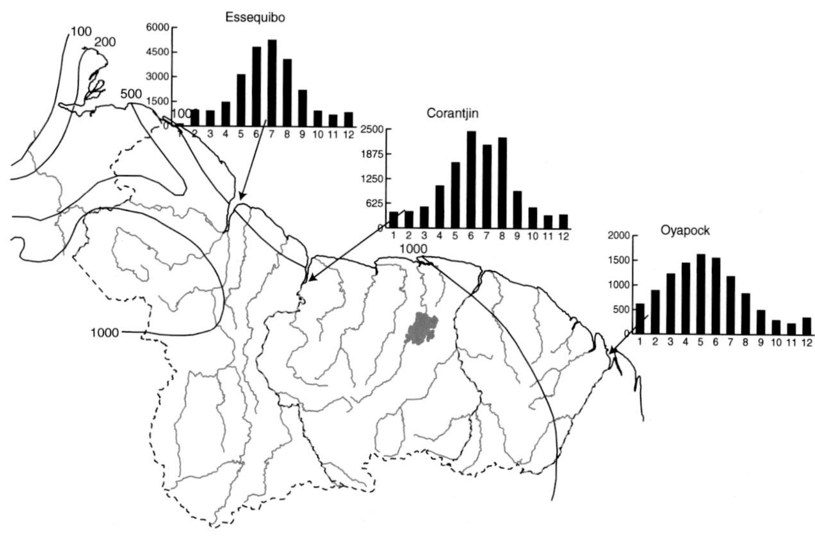

0 125 250 km

Figure B11

Table B5.

River name	Ocean	Area (10^3 km^2)	Length (km)	Max_elev (m)	Climate	Geology	Q (km^3/yr)	TSS (Mt/yr)	TDS (Mt/yr)	Refs.
French Guiana										
Approuague	Atlantic (W)	11	280	>100	Tr-W-S	PreMes	12	0.2	0.43	4, 7
Mahuri	Atlantic (W)	3.7	720	<100	Tr-W-S	PreMes	7.2		0.24	4, 7
Mana	Atlantic (W)	12	320	<100	Tr-W-S	PreMes	12	0.1		4, 6, 7
Marowijne	Atlantic (W)	66	720	>500	Tr-W-S	PreMes	57	1.4	1.6	4, 6, 7
Oyapoc	Atlantic (W)	30	420	>100	Tr-W-S	PreMes	28	0.5		1, 4, 6, 7, 10
Sinnamary	Atlantic (W)	6.5	250	<100	Tr-W-S	PreMes	9.1			4
Suriname										
Commewijne	Atlantic (W)	6.7	160	<100	Tr-H-S	PreMes	3.8			7, 9
Coppename	Atlantic (W)	20	410	>500	Tr-W-S	PreMes	15	0.4	0.06	4, 5, 7, 9
Nickerie	Atlantic (W)	9.7	320	<100	Tr-H-S	PreMes	6.3	0.2		4, 7, 9
Saramacca	Atlantic (W)	16	400	>200	Tr-H-S	PreMes	7.6	0.2		4, 5, 7, 9
Suriname	Atlantic (W)	16	480	>100	Tr-W-S	PreMes	14	0.3	0.33	4, 5, 9
Guyana										
Berbice	Atlantic (W)	11	560	<100	Tr-W-S	PreMes	11	0.2		4, 7, 9
Corantijn	Atlantic (W)	72	880	>200	Tr-H-S	PreMes	47	1.1		1, 4
Cuyuni	Atlantic (W)	54	560	>500	Tr-H-S	Cen S/M	33		1.4	2, 4, 7
Demerara	Atlantic (W)	1.6	330	380	Tr-W-S	Cen S/M	2.2			7, 9
Essequibo	Atlantic (W)	67	970	1400	Tr-W-S	PreMes	70	4.5	4.9	2, 3, 4, 8, 10
Mazaroni	Atlantic (W)	14	560	1000	Tr-W-S	PreMes	23		0.71	4, 7

References:
1. Center for Natural Resources Energy and Transport (UN), 1978; 2. Dai and Trenberth, 2002; 3. GEMS website, www.gemstat.org; 4. Meybeck and Ragu, 1996; 5. NEDECO, 1968; 6. ORSTOM, 1969 (cf. van der Leeden, 1975); 7. Rand McNally, 1980; 8. UNESCO, 1971; 9. van der Leeden, 1975; 10. Vörösmarty *et al.*, 1996c

Appendix C   Europe

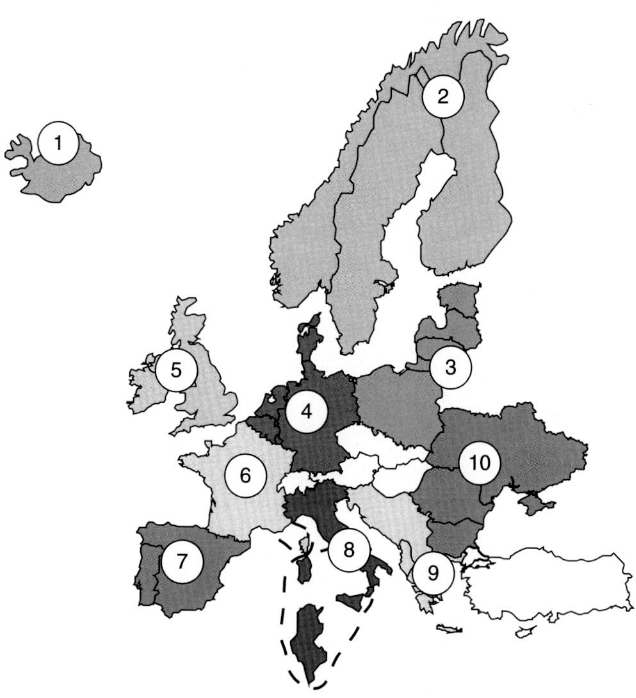

Figure C1

(1) Iceland
(2) Norway, Sweden, Finland
(3) Estonia, Latvia, Lithuania, Poland
(4) Germany, Denmark, The Netherlands, Belgium
(5) Scotland, England, Ireland
(6) France
(7) Spain and Portugal
(8) Italy
(9) Albania, Croatia, Greece
(10) Bulgaria, Romania, Ukraine

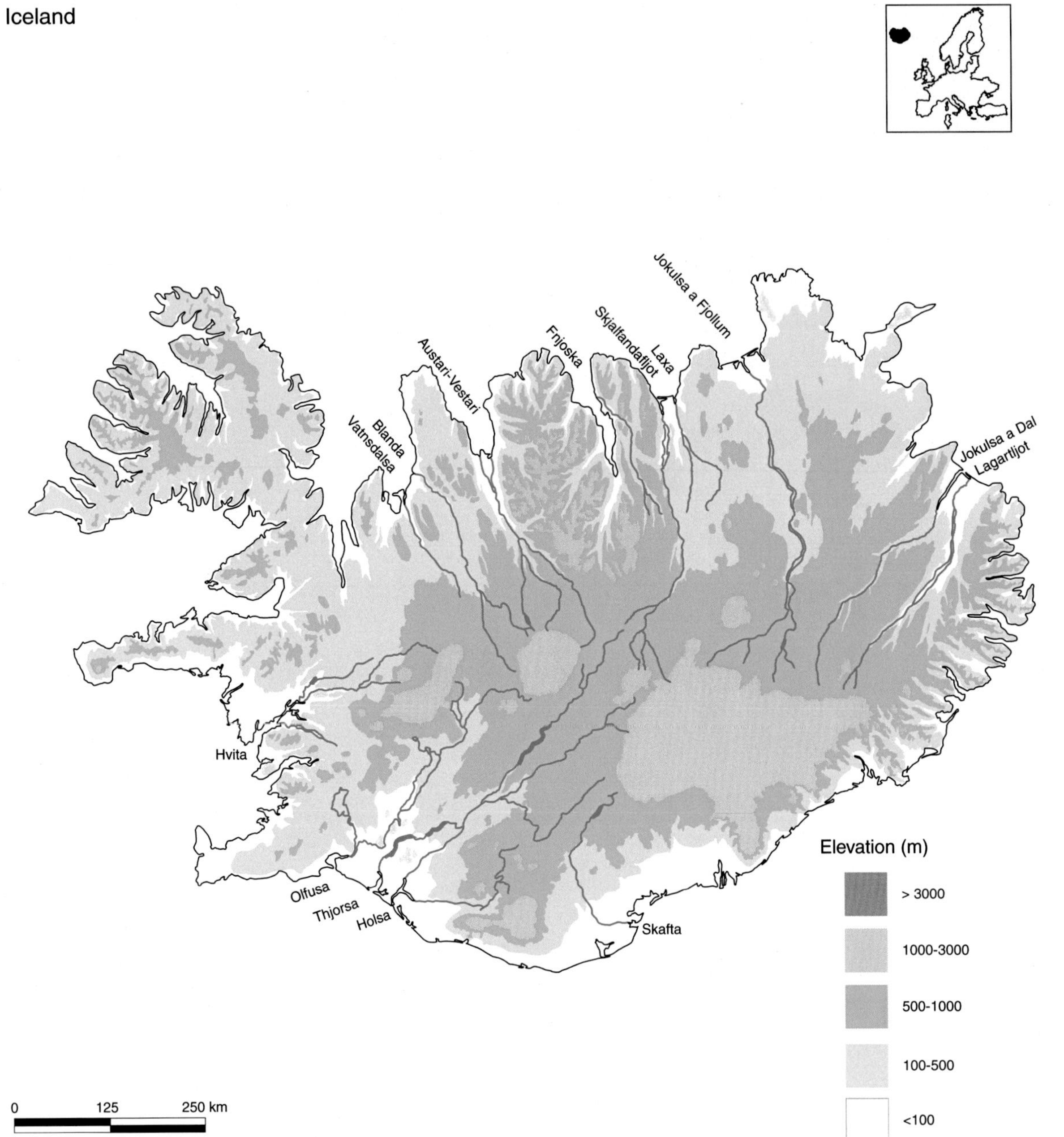

Iceland

Jokulsa a Fjollum
Laxa
Skjalfandafljot
Fnjoska
Austari-Vestari
Blanda
Vatnsdalsa
Jokulsa a Dal
Lagarfljot
Hvita
Olfusa
Thjorsa
Holsa
Skafta

Elevation (m)
> 3000
1000-3000
500-1000
100-500
<100

0 125 250 km

Figure C2

Iceland

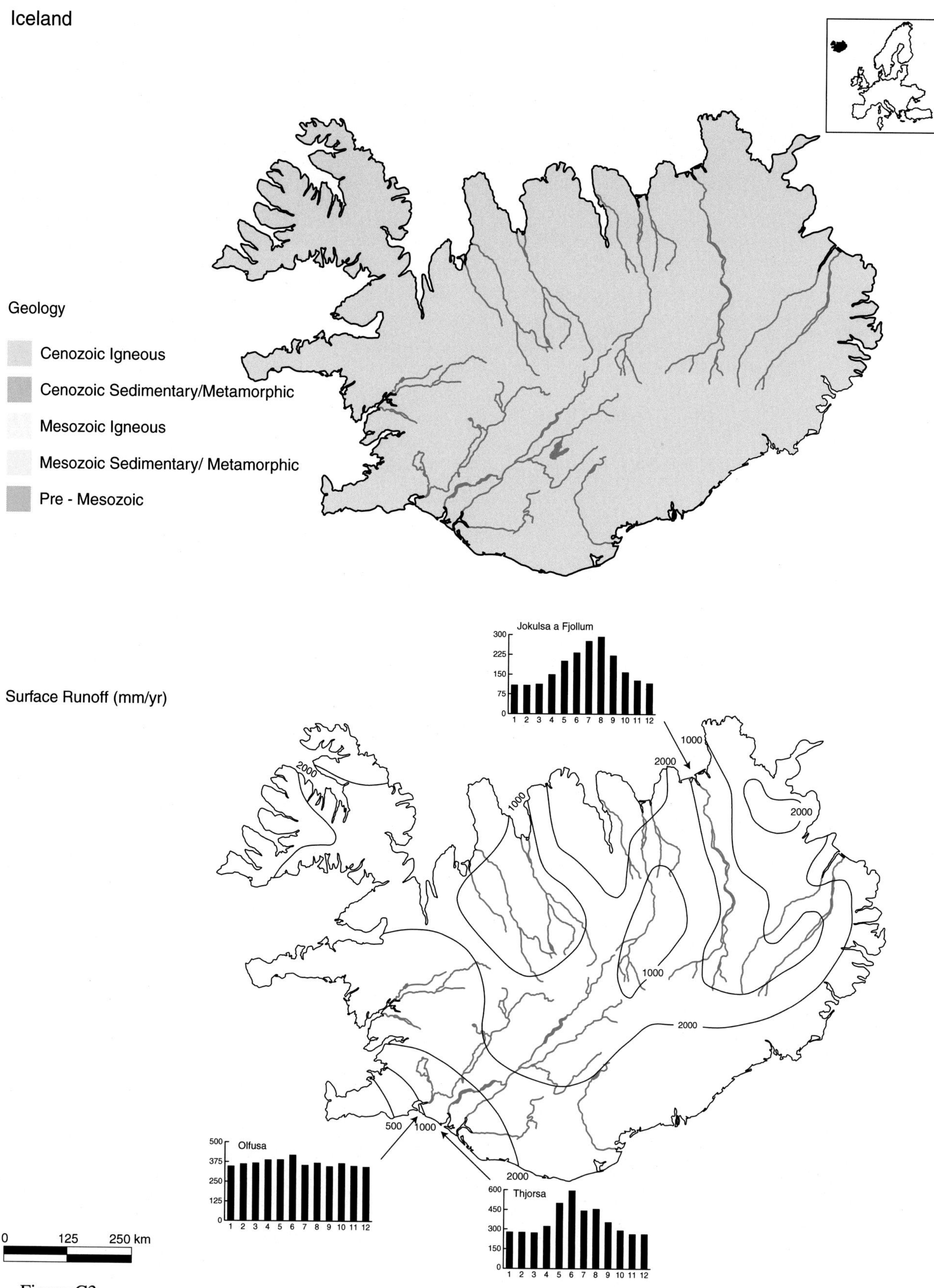

Figure C3

Table C1.

River name	Ocean	Area (10^3 km^2)	Length (km)	Max_elev (m)	Climate	Geology	Q (km^3/yr)	TSS (Mt/yr)	TDS (Mt/yr)	Refs.
Iceland										
Austari-Vestari	Arctic	1.9	110	1800	SA-W-S	Cen I	1.8	0.4		4
Blanda	Arctic	1.7	110	1800	SA-W-S	Cen I	1.4	0.5		4
Fnjoska	Arctic	1.1	100	1100	SA-W-S	Cen I	1.2			4
Holsa	Atlantic (N)	0.57	60	1500	SA-W-S	Cen I	1.6			4
Hvita	Atlantic (N)	2.2	130	1400	SA-W-S	Cen I	3.3	0.25		2, 4, 5
Jokulsa a Dal	Arctic	3.3	130	1500	SA-W-S	Cen I	4.5	9		4
Jokulsa a Fjollum	Arctic	7.1	200	2000	SA-W-S	Cen I	6	8		4
Lagartljot	Arctic	2.9	130	1800	SA-W-S	Cen I	3.7	0.07		4
Laxa	Arctic	2.4	75	1300	SA-W-S	Cen I	1.3			1, 2, 4
Olfusa	Atlantic (N)	6.1	160	1800	SA-W-S	Cen I	13	0.9	0.91	3, 4, 6
Skafta	Atlantic (N)	1.5	100	1800	SA-W-S	Cen I	3.6	4		4
Skjalfandafljot	Arctic	3.3	160	2000	SA-W-S	Cen I	2.6	0.25		4
Thjorsa	Atlantic (N)	7.4	230	2000	SA-W-S	Cen I	11	0.8 (3.7)	0.92	3, 4, 6
Vatnsdalsa	Arctic	1.5	70	>500	SA-W-S	Cen I	0.27			2

References:
1. Brittain *et al.*, 2009; 2. Malmstrom, 1958; 3. Meybeck and Ragu, 1996; 4. Orkustofnun Hydrological Service, 1997; 5. Rand McNally, 1980; 6. UNESCO (WORRI), 1978

Finland, Norway and Sweden

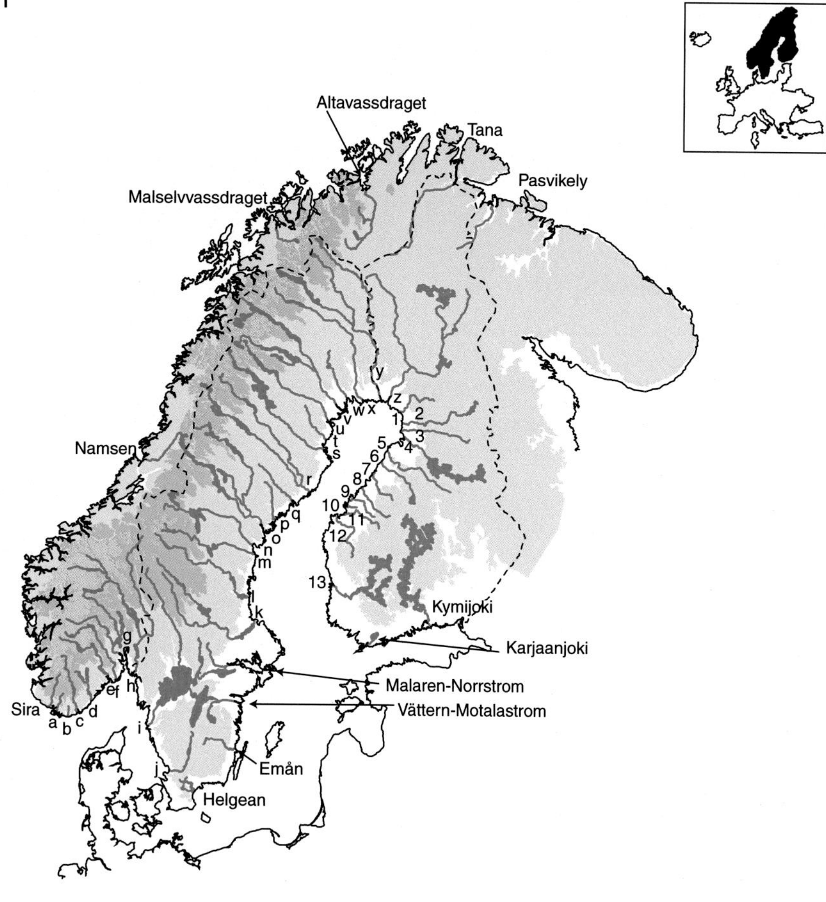
Altavassdraget
Tana
Pasvikely
Malselvvassdraget
Namsen
Sira
a b c d
e f g h i j k l m n o p q r s t u v w x y z
1 2 3 4 5 6 7 8 9 10 11 12 13
Kymijoki
Karjaanjoki
Malaren-Norrstrom
Vättern-Motalastrom
Emån
Helgean

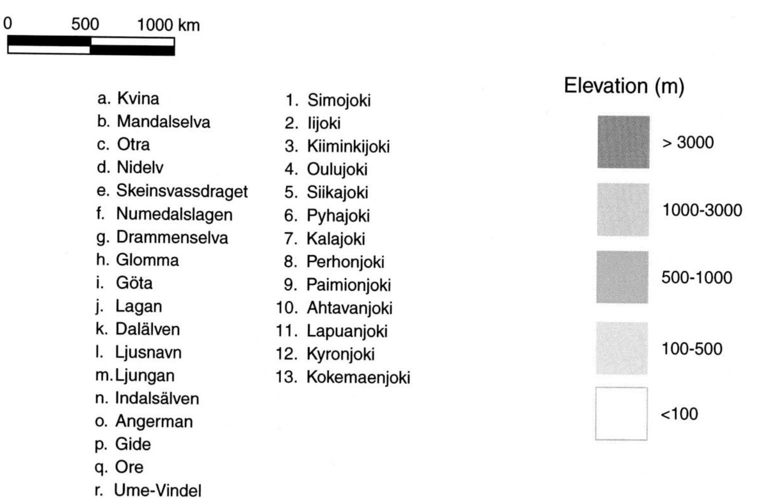
0 500 1000 km
a. Kvina
b. Mandalselva
c. Otra
d. Nidelv
e. Skeinsvassdraget
f. Numedalslagen
g. Drammenselva
h. Glomma
i. Göta
j. Lagan
k. Dalälven
l. Ljusnavn
m. Ljungan
n. Indalsälven
o. Angerman
p. Gide
q. Ore
r. Ume-Vindel
s. Skellefte
t. Byske
u. Pite
v. Lule
w. Rane
x. Kalixälven
y. Torne
z. Kemijoki
1. Simojoki
2. Iijoki
3. Kiiminkijoki
4. Oulujoki
5. Siikajoki
6. Pyhajoki
7. Kalajoki
8. Perhonjoki
9. Paimionjoki
10. Ahtavanjoki
11. Lapuanjoki
12. Kyronjoki
13. Kokemaenjoki
Elevation (m)
> 3000
1000-3000
500-1000
100-500
<100

Figure C4

Norway, Sweden and Finland

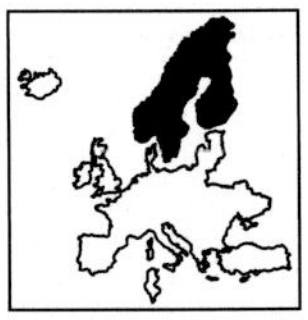

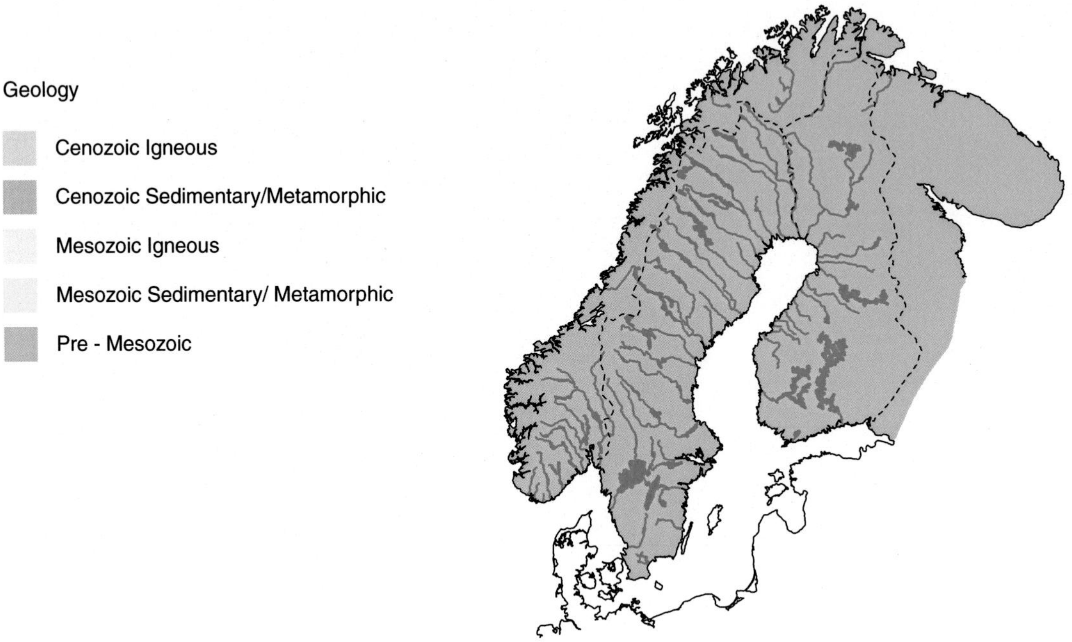

Geology

- Cenozoic Igneous
- Cenozoic Sedimentary/Metamorphic
- Mesozoic Igneous
- Mesozoic Sedimentary/ Metamorphic
- Pre - Mesozoic

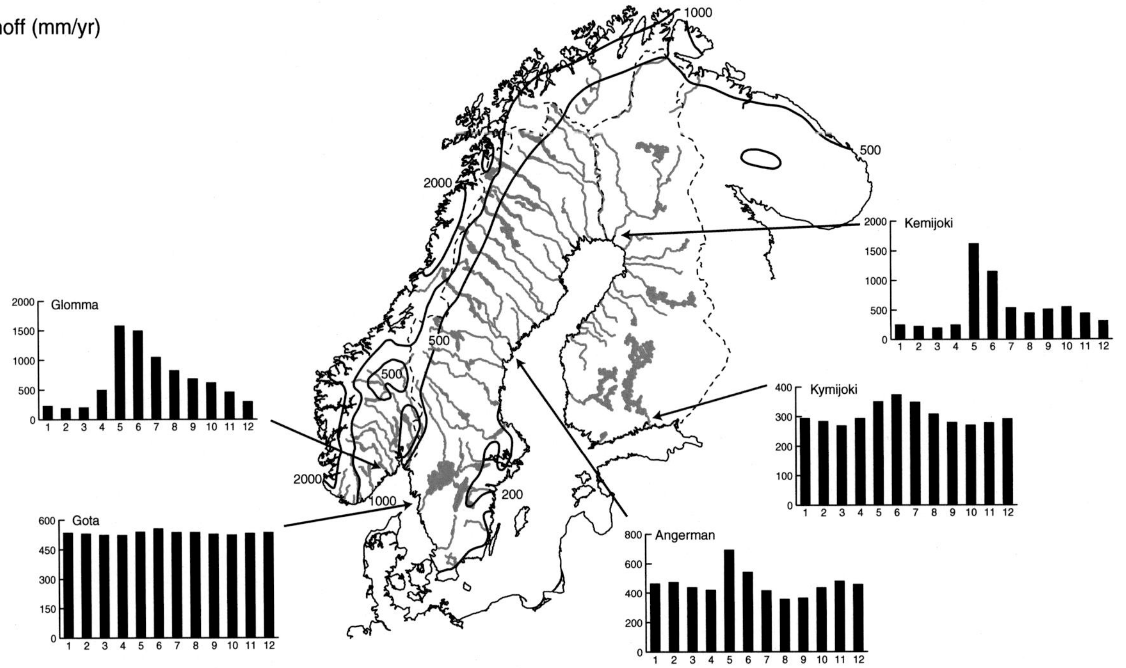

Surface Runoff (mm/yr)

Glomma

Gota

Kemijoki

Kymijoki

Angerman

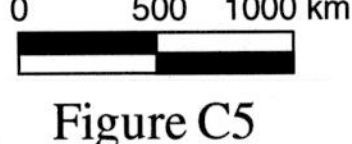

Figure C5

Table C2.

River name	Ocean	Area (10^3 km^2)	Length (km)	Max_elev (m)	Climate	Geology	Q (km^3/yr)	TSS (Mt/yr)	TDS (Mt/yr)	Refs.
Norway										
Altavassdraget	Arctic	7.4	190	>500	SAr-H-S	PreMes	2.6			4, 18, 21
Drammenselva	North Sea	17	300	1900	SAr-H-S	PreMes	10	1.8	0.07	16, 21
Glomma	North Sea	42	610	2500	SAr-H-S	PreMes	23	15	0.5	9, 16, 21
Kvina	North Sea	0.8	70	>1000	SAr-W-S	PreMes	2.7			4
Malselvvassdraget	Arctic	6	160	1700	Ar-W-S	PreMes	5.7			4, 21
Mandalselva	North Sea	1.7	120	>500	SAr-W-S	PreMes	1.5	0.001		1
Namsen	Norweigen Sea	6.3	190	>500	Ar-W-S	PreMes	8.5			4, 18, 21
Nidelv	North Sea	3.8		1500	SAr-W-S	PreMes				
Numedalslagen	North Sea	5.7	310	>1000	SAr-H-S	PreMes	3.7			4, 18, 21
Otra	North Sea	3.6	200	>1000	SA-W-S	PreMes				
Pasvikely	Barents Sea	21	145	400	SAr-H-S	PreMes	5.4	0.05		6, 16
Sira	North Sea	2.6	100	>1000	SAr-W-S	PreMes	3.7			4
Skiensvassdraget	North Sea	10		>1000	SAr-W-S	PreMes	9	1.5	0.03	4, 16, 19, 21
Tana	Barents Sea	16	320	530	SAr-H-S	PreMes	5.9		0.12	4, 18, 21
Sweden										
Angerman	G. of Bothnia	32	440	>500	SAr-H-S	PreMes	15	0.06	0.95	2, 16, 20, 22
Byske	G. of Bothnia	3.6	210	>200	SAr-H-S	PreMes	1.3			13,22
Dalälven	G. of Bothnia	29	520	>500	SAr-H-S	PreMes	10 (15)	0.03	0.3	2, 16, 22
Emån	Baltic Sea	4.5	220	>200	SAr-SA-S	PreMes	0.88			22
Gide	G. of Bothnia	3.4	240	>200	SAr-H-S	PreMes	1.1			13, 22
Göta alv	North Sea	50	720	1700	SAr-H-S	PreMes	17	0.13		6, 16
Helgean	Baltic Sea	4.8	180	>100	SAr-H-S	PreMes	1.5			22
Indalsälven	G. of Bothnia	25	420	>500	SAr-H-S	PreMes	14	0.13		6, 13, 16, 22
Kalixälven	G. of Bothnia	18	430	>500	SAr-H-S	PreMes	8.9	0.04		2, 4, 22
Lagan	North Sea	6.4	270	>200	SAr-H-S	PreMes	2.3			4, 22
Ljungan	G. of Bothnia	13	360	>500	SAr-H-S	PreMes	4.4	0.01		2, 13, 22
Ljusnan	G. of Bothnia	20	430	1600	SAr-H-S	PreMes	7.3			13, 22
Lule	G. of Bothnia	25	450	>1000	SAr-H-S	PreMes	16	0.04	0.26	2, 16, 20, 22
Malaren-Norrstrom	Baltic Sea	23		>200	SAr-H-S	PreMes	5.2			4, 22
Ore	G. of Bothnia	3	220	>500	SAr-H-S	PreMes	1	0.03		2, 22
Pite	G. of Bothnia	11	400	>1000	SAr-H-S	PreMes	5.4	0.04		2, 6, 13, 22
Rane	G. of Bothnia	4.1	210	690	SAr-H-S	PreMes	1.3	0.002		2, 4, 22
Skellefte	G. of Bothnia	12	400	>100	SAr-H-S	PreMes	4.9	0.009	0.08	2, 13, 16, 22

Table C2. (*Continued*)

River name	Ocean	Area (10^3 km^2)	Length (km)	Max_elev (m)	Climate	Geology	Q (km^3/yr)	TSS (Mt/yr)	TDS (Mt/yr)	Refs.
Torne	G. of Bothnia	40	570	2000	SAr-H-S	PreMes	12	0.1	0.37	2, 4, 16, 17, 22
Ume-Vindel	G. of Bothnia	26	460	>1000	SAr-H-S	PreMes	13			10
Vättern-Motalastrom	Baltic Sea	15	100	>100	SAr-H-S	PreMes	2.8			13, 22
Finland										
Ahtavanjoki	G. of Bothnia	2		>100	SA-H-S	PreMes	0.56	0.01		8, 17
Iijoki	G. of Bothnia	14	250	>200	SA-H-S	PreMes	5	0.014		8, 17
Kalajoki	G. of Bothnia	4.2	140	>100	SA-H-S	PreMes	1.3	0.04		8, 11, 17
Karjaanjoki	G. of Finland	2		>100	SA-H-S	PreMes	0.64	0.002		9, 11, 14, 16
Kemijoki	G. of Bothnia	51	550	300	SA-H-S	PreMes	17	0.07	0.5	6, 8, 9, 17
Kiiminkijoki	G. of Bothnia	3.8	180	>100	SA-H-S	PreMes	1.4	0.01		5, 8, 11, 15, 17
Kokemänjoki	G. of Bothnia	27	350	<100	SA-H-S	PreMes	9.3	0.06		6, 8, 11, 12, 17
Kymijoki	G. of Finland	37	600	<100	SA-H-S	PreMes	12	0.05	0.43	9
Kyronjoki	G. of Bothnia	4.9	160	190	SA-H-S	PreMes	1.5	0.05		7, 8, 11, 17
Lapuanjoki	G. of Bothnia	4.1	150	>100	SA-H-S	PreMes	1.3	0.03		7, 8, 11, 17
Oulujoki	G. of Bothnia	25	300	>100	SA-H-S	PreMes	8	0.024	0.17	7, 8, 17
Paimionjoki	Baltic Sea	1.1	100	80	SA-H-S	PreMes	0.33	0.09		8, 17
Perhonjoki	G. of Bothnia	2.7	130	>100	SA-H-S	PreMes	0.77	0.01		7, 8, 11, 17
Pyhajoki	G. of Bothnia	3.7	160	>100	SA-H-S	PreMes	1	0.16		11, 15, 17
Siikajoki	G. of Bothnia	4.4	150	>100	SA-H-S	PreMes	1.4	0.002		8, 11, 12, 14, 17
Simojoki	G. of Bothnia	3.2	180	>100	SA-H-S	PreMes	1.1	0.007		3, 7, 11, 16, 17

References:
1. Bogen, 1996; 2. Burman, 1983; 3. Centre for Natural Resources, Energy and Transport (UN), 1978; 4. Dynesius and Nilsson, 1994; 5. Ekholm, 1992; 6. Eurosion, 2004; 7. Finland Environmental Administration, Hydrological Yearbook, 1995; 8. Finland Ministry of the Environment Website, http://www.environment.fi/waterforecast; 9. GEMS website, www.gemstat.org; 10. Jansson *et al.*, 2000; 11. Kauppila and Koskiaho, 2003; 12. P. Kauppila, personal communication; 13. Keller, 1962; 14. Kempe *et al.*, 1991; 15. Koutaniemei, 1991; 16. Meybeck and Ragu, 1996; 17. Pitkånen, 1994; 18. Rand McNally, 1980; 19. UNESCO (WORRI), 1978; 20. UNESCO, 1971; 21. VH-Notat, personal communication; 22. (Swedish) Yearbook of Environmental Statistics 1986–1987

Estonia, Latvia, Lithuania and Poland

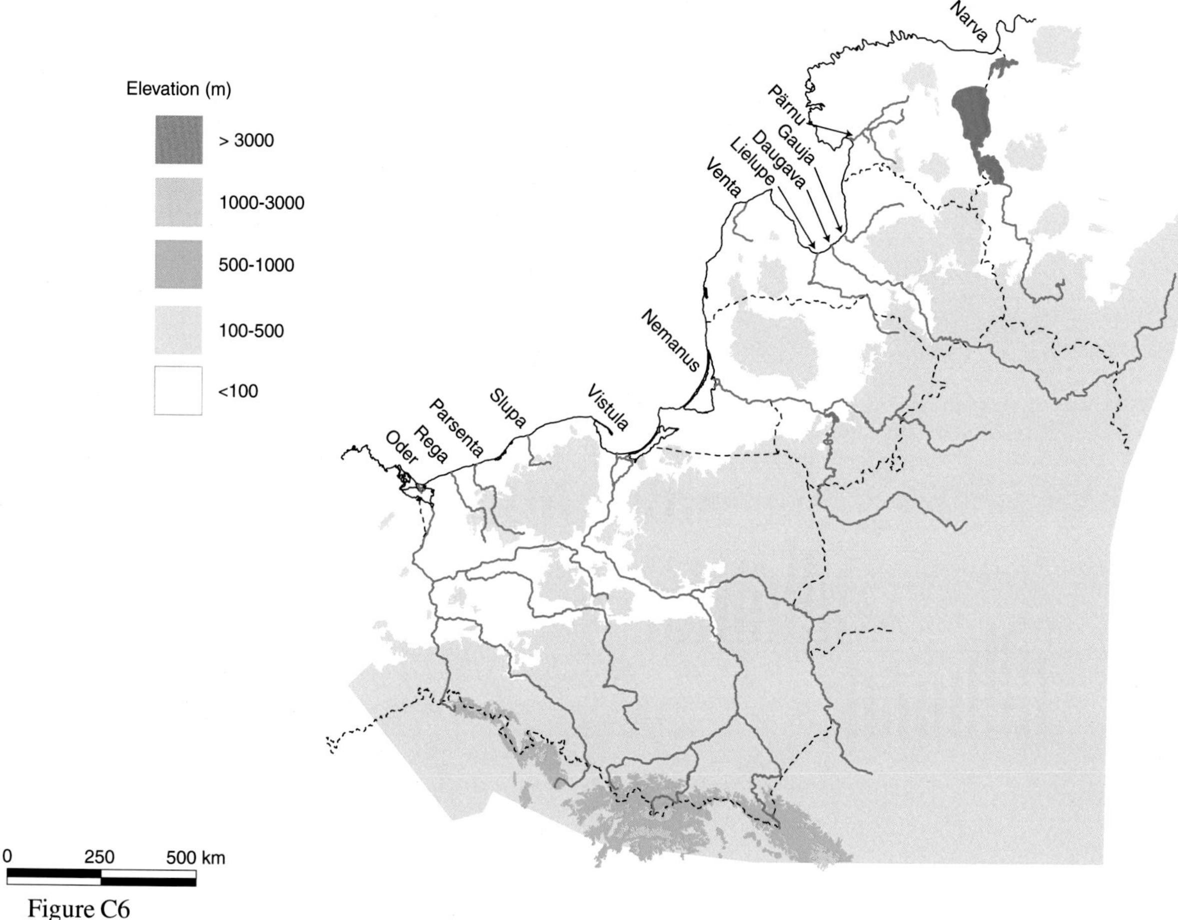

Figure C6

Estonia, Latvia, Lithuania and Poland

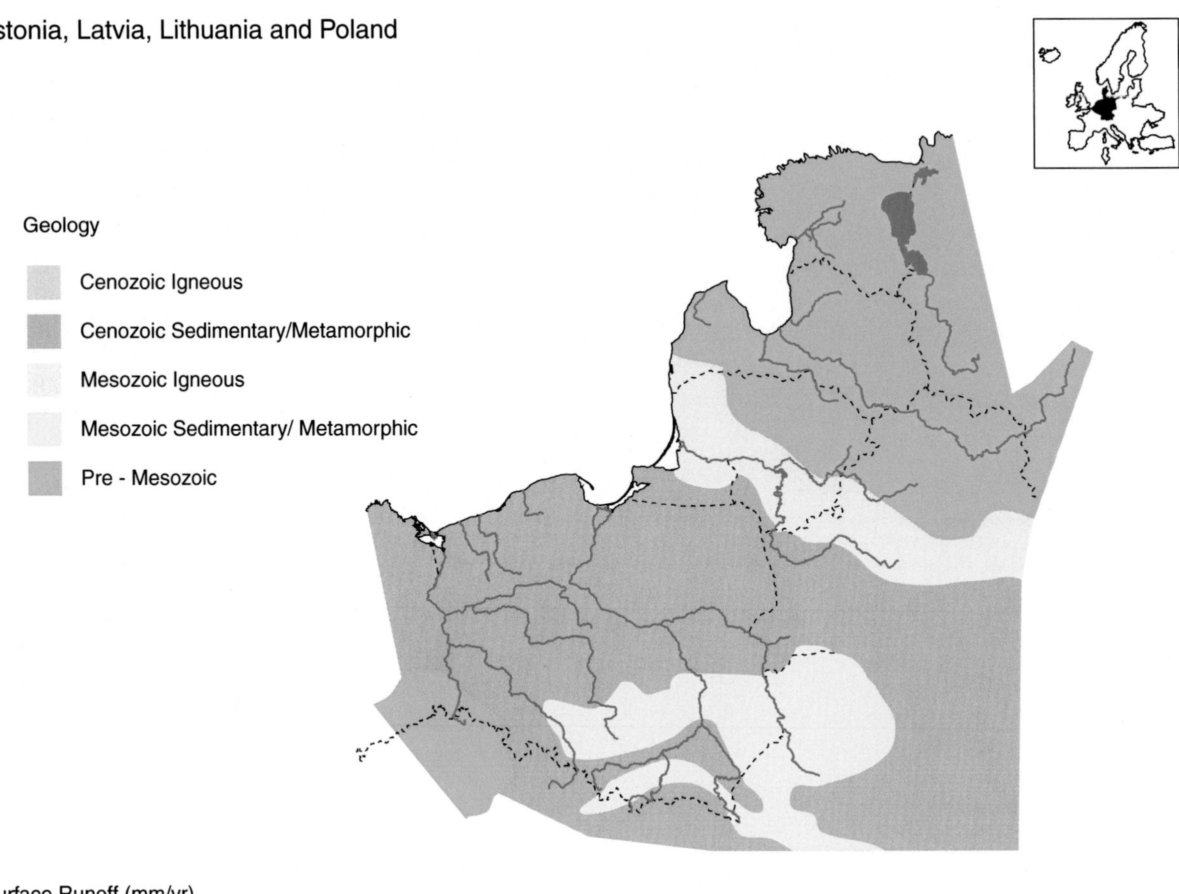

Surface Runoff (mm/yr)

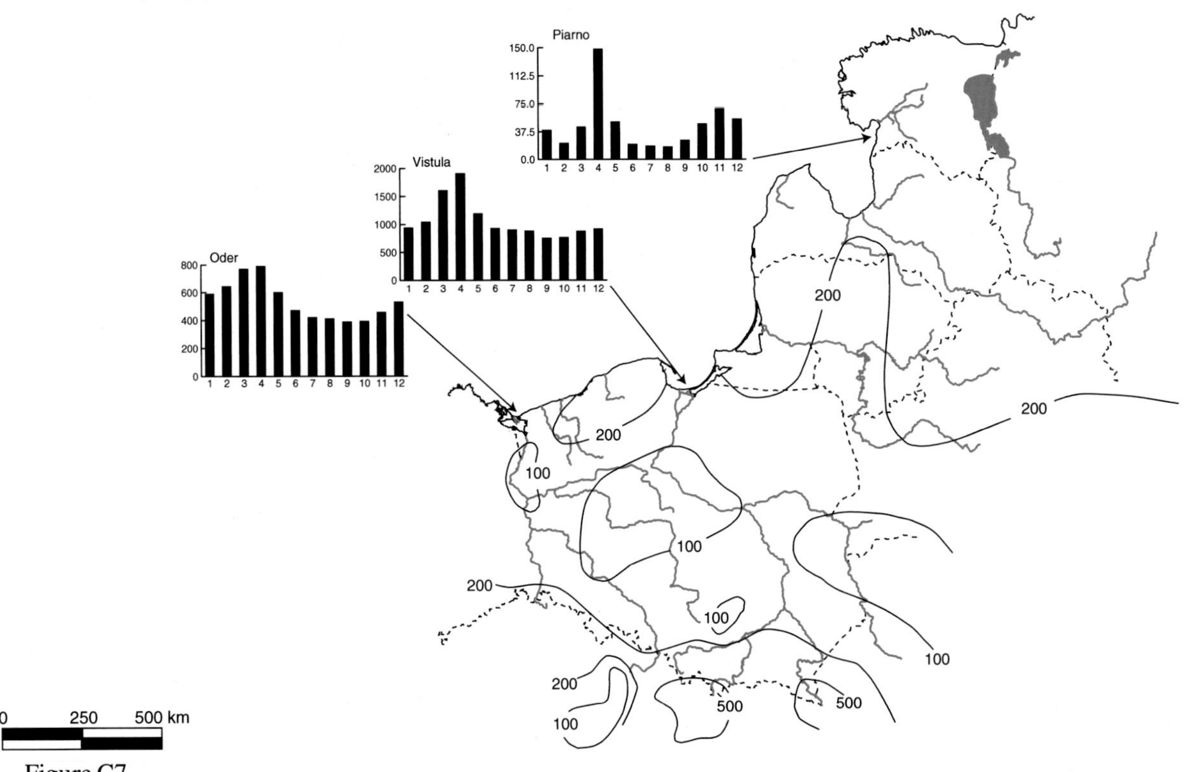

Figure C7

Table C3.

River name	Ocean	Area (10^3 km^2)	Length (km)	Max_elev (m)	Climate	Geology	Q (km^3/yr)	TSS (Mt/yr)	TDS (Mt/yr)	Refs.
Estonia										
Narva	G. of Finland	56	77	<100	SA-SA-S	PreMes	14	0.56		1, 3, 6, 8
Pärnu	Baltic Sea	6.9	140	<100	SA-H-S	PreMes	2.5			2, 4
Latvia										
Daugava	Baltic Sea	88	1000	220	SAr-H-S	PreMes	23	0.47	4.3	4, 6
Gauja	Baltic Sea	8.9	450	<100	SAr-H-S	PreMes	2.6		0.62	4, 6, 7
Lielupe	Baltic Sea	18	310	<100	SAr-H-S	PreMes	3		0.98	4, 6
Venta	Baltic Sea	8.3	200	<100	SAr-SA-S	PreMes	2	0.06	0.8	6, 7
Lithuania										
Nemanus	Baltic Sea	98	940	>100	SAr-SA-S	Mes S/M	20	0.66	7.7	6
Poland										
Oder	Baltic Sea	120	910	1600	SAr-SA-S	Cen S/M	16	0.35	8.8	6, 9
Parsenta	Baltic Sea	3	90	>100	SAr-H-S	Cen S/M	1	0.012		2
Rega	Baltic Sea	2.6	190	>100	SAr-H-S	Cen S/M	0.7	0.016		2, 7
Slupa	Baltic Sea	1.4	100	>100	SAr-H-S	Cen S/M	0.56	0.013		2
Vistula	Baltic Sea	200	1100	2500	SAr-SA-S	Cen S/M	33	1.8	19	5, 6

References:
1. Eurosion, 2004; 2. Global River Discharge Database, http://www.rivdis.sr.unh.edu/; 3. IAHS/UNESCO, 1974; 4. Laznik *et al.*, 1999; 5. Lisitzin, 1972; 6. Meybeck and Ragu, 1996; 7. Rand McNally, 1980; 8. Sovetskaya Entsiklopediya, 1989; 9. UNESCO (WORRI), 1978

Germany, Denmark, the Netherlands and Belgium

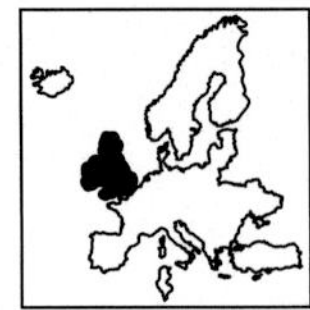

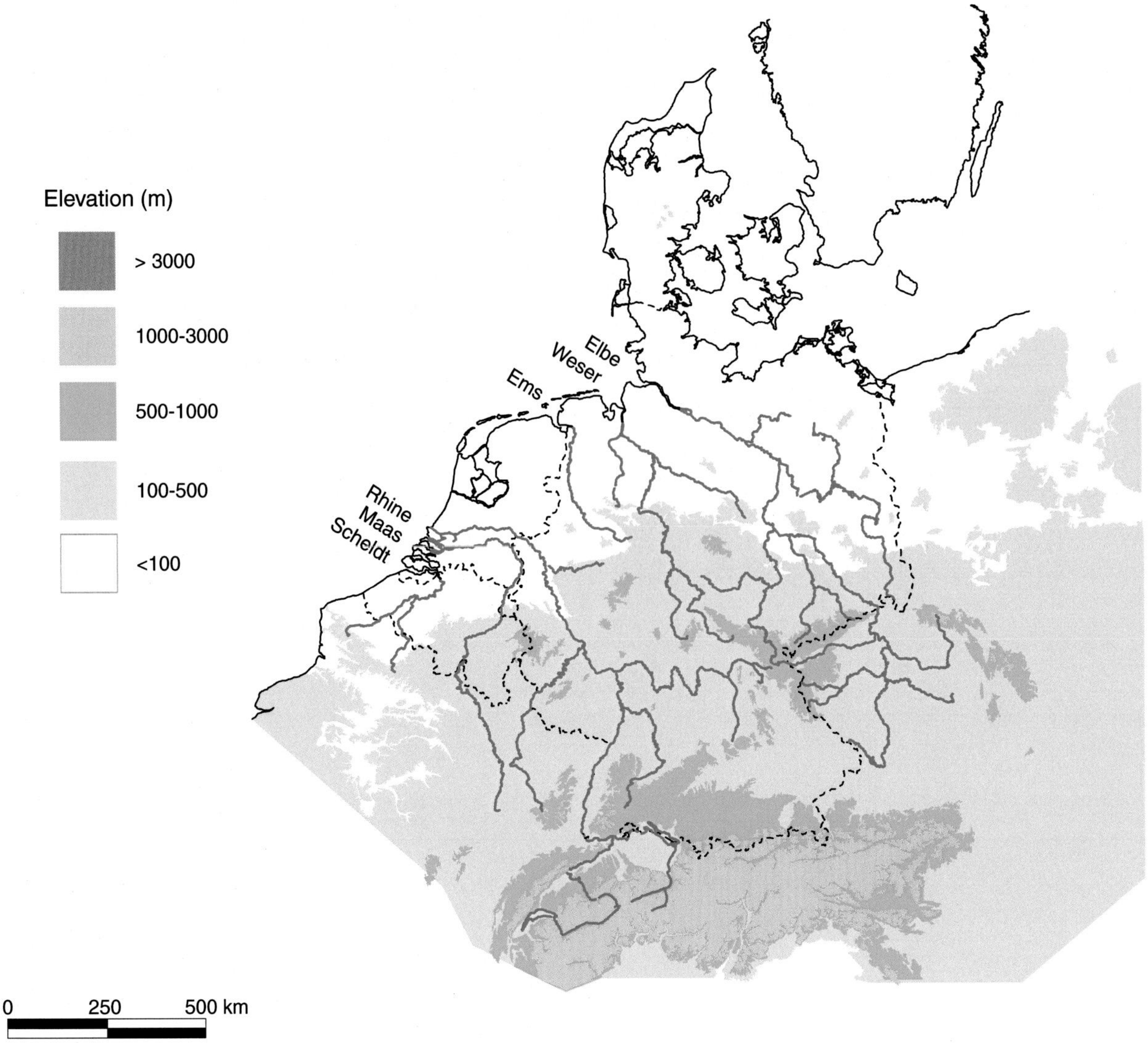

Elevation (m)

- > 3000
- 1000-3000
- 500-1000
- 100-500
- <100

Elbe
Weser
Ems
Rhine
Maas
Scheldt

0 250 500 km

Figure C8

Germany, Denmark, the Netherlands and Belgium

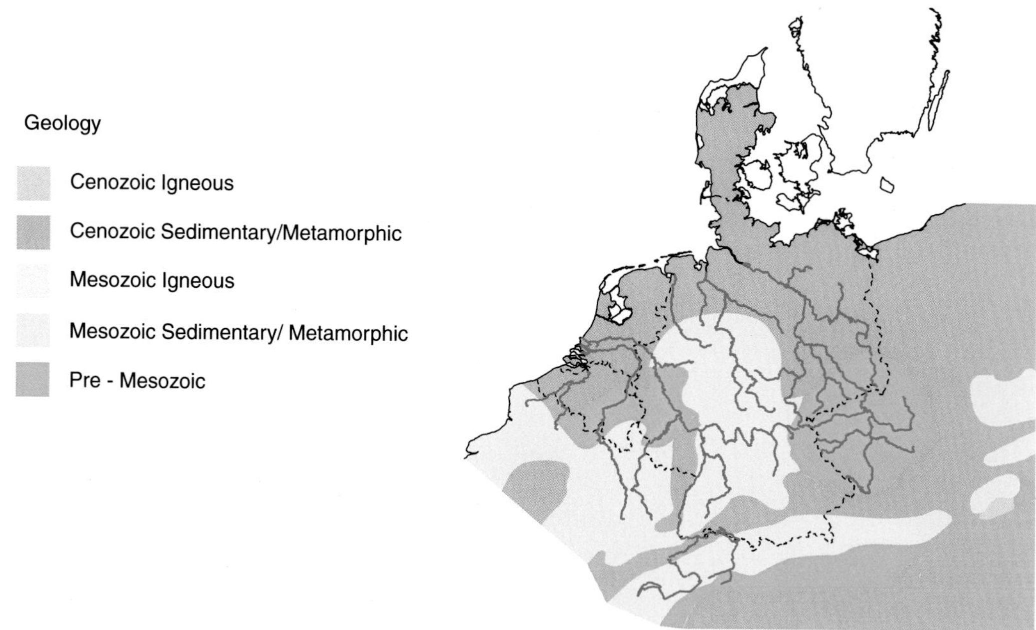

Surface Runoff (mm/yr)

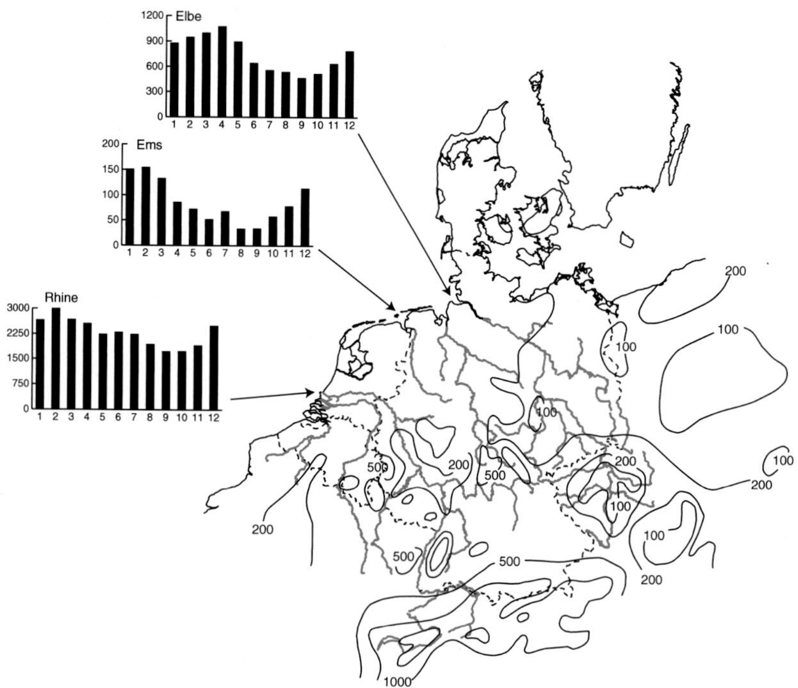

0 250 500 km

Figure C9

Table C4.

River name	Ocean	Area (10^3 km^2)	Length (km)	Max_elev (m)	Climate	Geology	Q (km^3/yr)	TSS (Mt/yr)	TDS (Mt/yr)	Refs.
Germany										
Elbe	North Sea	150	1100	>500	Te-SA-W	PreMes	24	0.84	14	7, 8, 9
Ems	North Sea	8.3	370	450	Te-H-W	PreMes	2.5		0.34	5, 7, 9, 12
Weser	North Sea	46	720	980	Te-SA-W	PreMes	11	0.33	23 (3.6)	1, 5, 7, 9, 12
Netherlands										
Maas	North Sea	36	920	>500	Te-SA-W	PreMes	10	0.7	1.7	1, 6, 9, 12
Rhine	North Sea	220	1400	1500	Te-H-C	PreMes	74	0.07	60	2, 3, 5, 6, 9, 12
Belgium										
Scheldt	North Sea	22	430	100	Te-H-C	PreMes	6	0.75	2.7	4, 5, 9, 10, 11

References:
1. Centre for Natural Resources, Energy and Transport (UN), 1978; 2. Eisma *et al.*, 1982; 3. Esser and Kohlmaier, 1991; 4. Fettweis *et al.*, 1998; 5. GEMS website, www.gemstat.org; 6. IAHS/UNESCO, 1974; 7. Kempe *et al.*, 1991; 8. Lisitzin, 1972; 9. Meybeck and Ragu, 1996; 10. Rand McNally, 1980; 11. Salomons and Mook, 1981; 12. UNESCO (WORRI), 1978

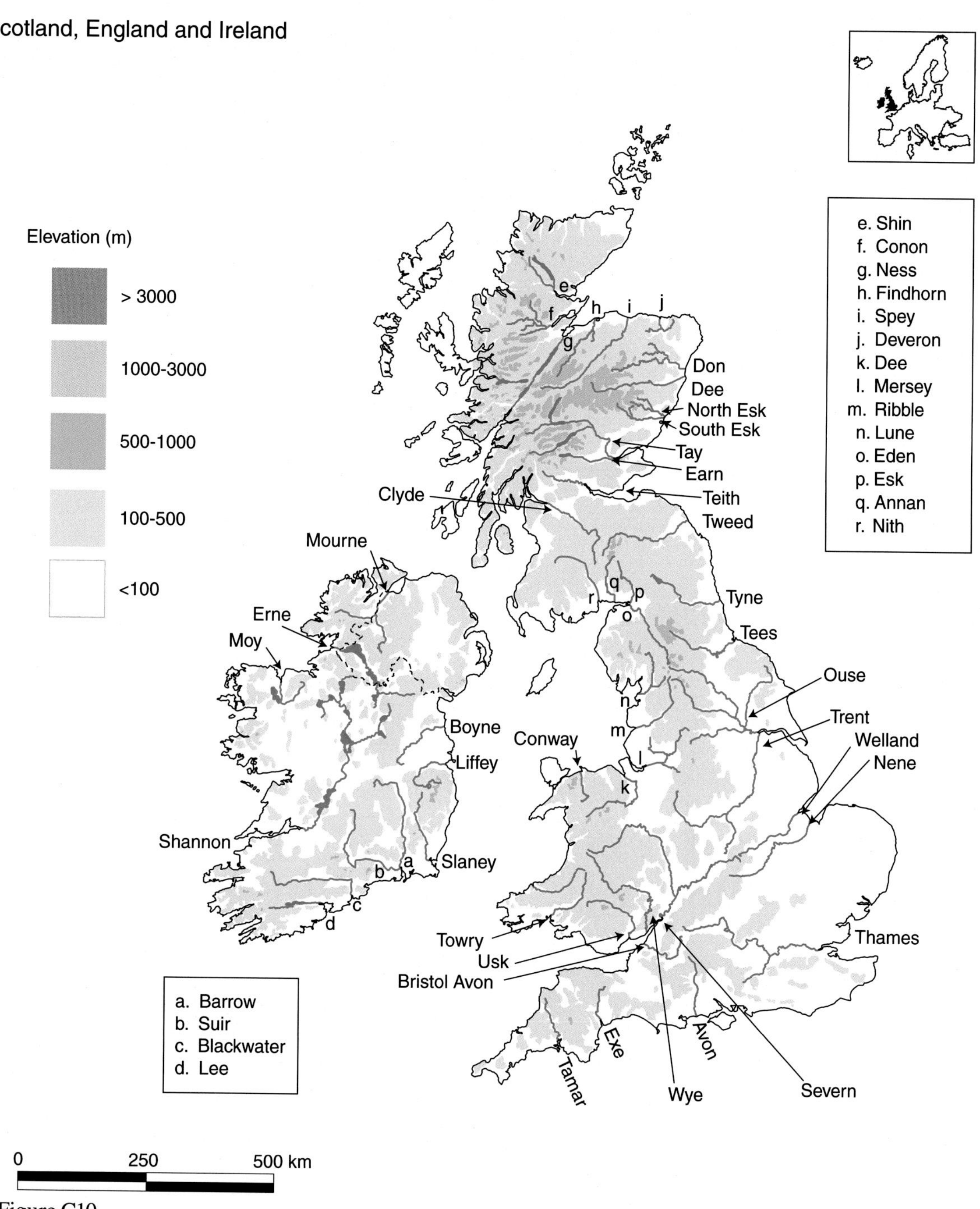

Scotland, England and Ireland
Elevation (m)
> 3000
1000-3000
500-1000
100-500
<100
e. Shin
f. Conon
g. Ness
h. Findhorn
i. Spey
j. Deveron
k. Dee
l. Mersey
m. Ribble
n. Lune
o. Eden
p. Esk
q. Annan
r. Nith
Don
Dee
North Esk
South Esk
Tay
Earn
Teith
Tweed
Clyde
Mourne
Erne
Moy
Tyne
Tees
Ouse
Trent
Welland
Nene
Boyne
Liffey
Conway
Shannon
Slaney
Towry
Usk
Bristol Avon
Thames
Exe
Avon
Tamar
Wye
Severn
a. Barrow
b. Suir
c. Blackwater
d. Lee
0
250
500 km

Figure C10

Scotland, England and Ireland

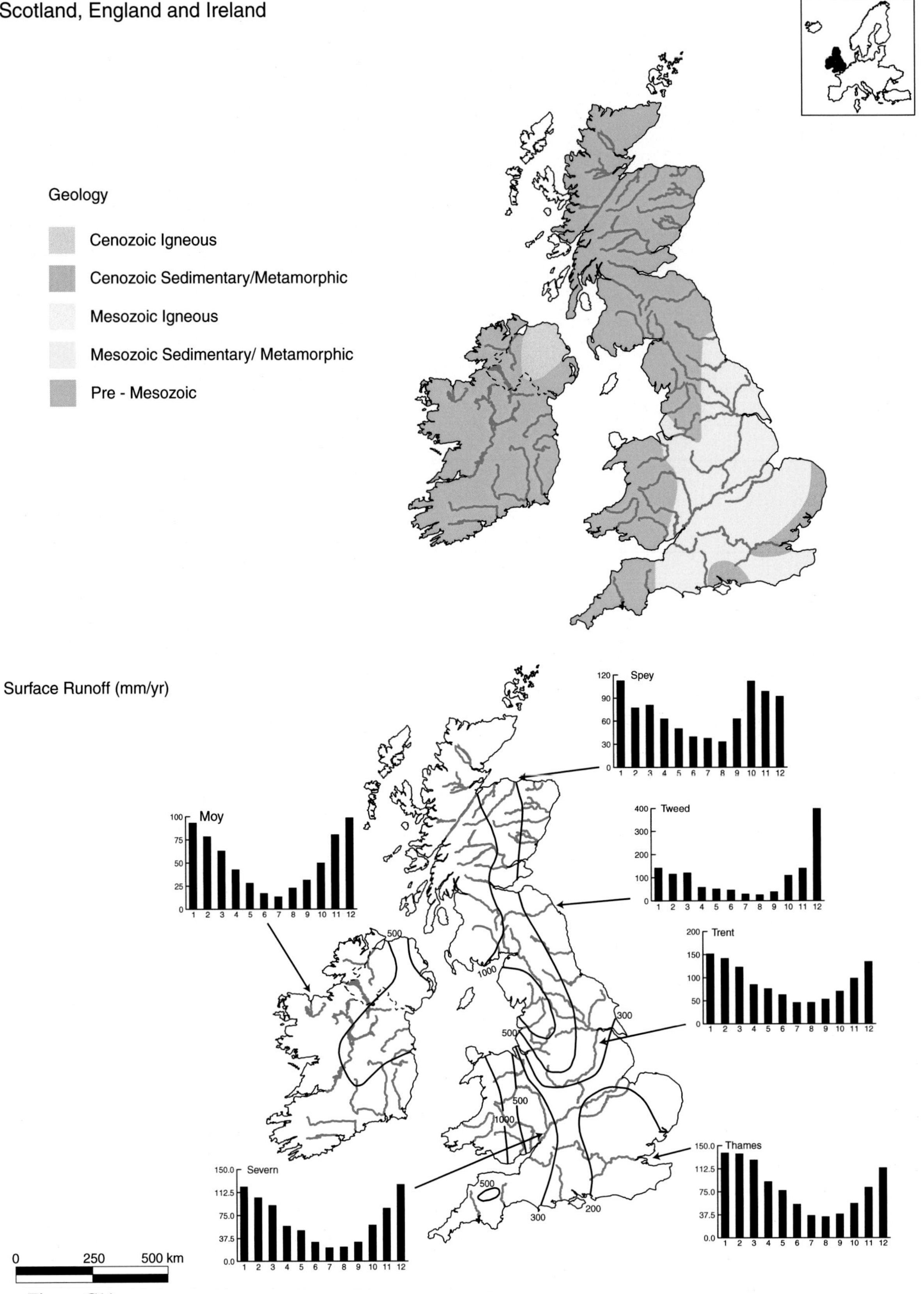

Figure C11

Table C5.

River name	Ocean	Area (10^3 km^2)	Length (km)	Max_elev (m)	Climate	Geology	Q (km^3/yr)	TSS (Mt/yr)	TDS (Mt/yr)	Refs.
Scotland										
Annan	Irish Sea	0.9	60	>500	Te-W-W	PreMes	0.9			6, 9
Clyde	North Sea	1.9	170	730	Te-H-W	PreMes	1.1	0.11		9, 13, 16, 17
Conon	North Sea	0.96	60	1000	Te-W-W	PreMes	1.5	0.006		19
Dee	North Sea	1.8	150	1300	Te-H-W	PreMes	1.5	0.03		8, 9, 14, 19
Deveron	North Sea	0.95	100	770	Te-W-W	PreMes	0.5	0.01		8, 9, 19
Don	North Sea	1.3	130	870	Te-W-W	PreMes		0.03		8, 9, 19
Earn	North Sea	0.78	90	980	Te-W-W	PreMes		0.06		8, 9, 19
Findhorn	North Sea	0.78	100	940	Te-W-W	PreMes	0.8	0.04		8, 9, 19
Ness	North Sea	1.8	50	1100	Te-H-W	PreMes	1.7	0.02		19
England										
Avon	English Channel	2.6	150	>100	Te-H-W	Mes S/M	0.62	0.42		1, 2, 13, 17
Bristol Avon	Irish Sea	2.3	120	>500	Te-H-W	Mes S/M	0.27	0.02	0.1	16
Conway	Irish Sea	0.6	45	1000	Te-H-W	PreMes				
Dee	Irish Sea	1	180	>100	Te-W-W	PreMes	1			6
Eden	Irish Sea	2.3	100	>200	Te-W-W	PreMes				8
Esk	Irish Sea	0.31	79	>500	Te-H-W	PreMes		0.18	0.016	1
Exe	English Channel	0.6	87	>200	Te-W-W	PreMes-Mes S/M	0.52	0.014	0.051	16
Lune	Irish Sea	0.8	90	>500	Te-H-W	PreMes				8
Mersey	Irish Sea	2	110	600	Te-H-W	Mes S/M	1.3			8, 10, 12
Nene	North Sea	1.6	140	120	Te-SA-W	Mes S/M	0.29	0.016		13, 17, 18
Ouse	North Sea	8.9	250	>100	Te-H-W	Mes S/M	2	0.11	0.28	2, 10, 12, 15, 16
Ribble	Irish Sea	1.6	160	>200	Te-W-W	PreMes	1.5			6, 8
Severn	Irish Sea	10	350	750	Te-H-W	Mes S/M	3.3	0.44	8.7	10, 13, 16, 17
Tamar	English Channel	1	100	>200	Te-H-W	PreMes				11
Tees	North Sea	2	100	>500	Te-H-W	Mes S/M	0.63			2, 12
Thames	North Sea	15	350	330	Te-SA-W	Mes S/M	2.8	0.04	0.82	4, 5, 7, 13, 17
Towry	Irish Sea	1	100	>200	Te-W-W	PreMes				
Trent	North Sea	9.5	290	640	Te-H-W	Mes S/M	2.6	0.08		2, 4, 5, 7, 13, 17
Tyne	North Sea	2.2	100	600	Te-H-W	Mes S/M	1.5	0.13	0.084	2, 13, 16, 17

Table C5. (*Continued*)

River name	Ocean	Area (10^3 km^2)	Length (km)	Max_elev (m)	Climate	Geology	Q (km^3/yr)	TSS (Mt/yr)	TDS (Mt/yr)	Refs.
Usk	North Sea	0.91	100	>200	Te-W-W	PreMes	0.95	0.042	0.12	8, 13, 16, 17
Welland	North Sea	0.53	110	>100	Te-SA-W	Mes S/M	0.11	0.007		18
Wye	North Sea	4.4	220	740	Te-H-W	Mes S/M	2.4	0.21	0.37	13, 16, 17
Ireland										
Barrow	Irish Sea	12	160	490	Te-H-W	PreMes	4.8	0.038		8, 19
Blackwater	Irish Sea	3.3	170	440	Te-H-W	PreMes	2.2	0.02		8, 19
Boyne	Irish Sea	3.3	110	140	Te-H-W	PreMes	1.1	0.011		8, 19
Erne	Atlantic (NE)	5.1	100	500	Te-H-W	PreMes	3	0.02		8, 19
Lee	Irish Sea	1.2	100	>500	Te-W-W	PreMes	1.3	0.01		8, 19
Liffey	Irish Sea	1.4	120	300	Te-H-W	PreMes	0.47	0.004		19
Mourne	Atlantic (NE)	2.9	110	670	Te-H-W	PreMes				3
Moy	Atlantic (NE)	2.1	100	510	Te-H-W	PreMes	1.6	0.01		3, 19
Shannon	Atlantic (NE)	23	320	570	Te-H-W	PreMes	8	0.64		8, 13, 19
Slaney	Irish Sea	1.8	120	930	Te-H-W	PreMes	1.1	0.006		3, 19
Suir	Irish Sea	3.6	180	450	Te-H-W	PreMes	2			3

References:

1. Collins, 1981; 2. Czaya, 1981; 3. Drainage Map of Ireland; 4. Eurosion, 2004; 5. GEMS website, www.gemstat.org; 6. GRDC website, www.gewex.org/grdc; 7. Meybeck, 1994; 8. Rand McNally, 1980; 9. Scottish Environment Protection Agency website, http://www.sepa.org.uk/; 10. Soulsby *et al.*, 2009; 11. Thornton and McManus, 1994; 12. UNESCO (WORRI), 1978; 13. UNESCO, 1971; 14. Vörösmarty *et al.*, 1996a; 15. Walling, 1999; 16. D. E. Walling, personal communication; 17. Willis, 1971 (cf. van der Leeden, 1975); 18. Wilmot and Collins, 1981; 19. J. Wilson, personal communication

France

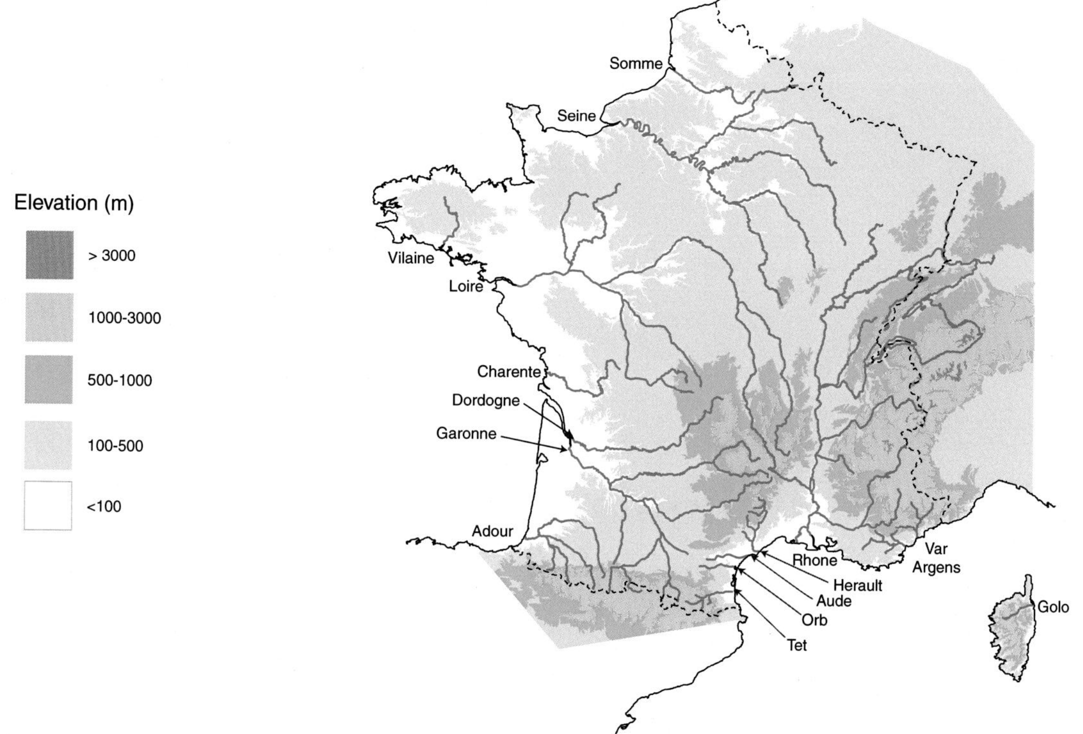

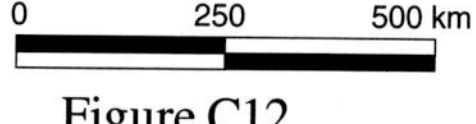

Figure C12

France

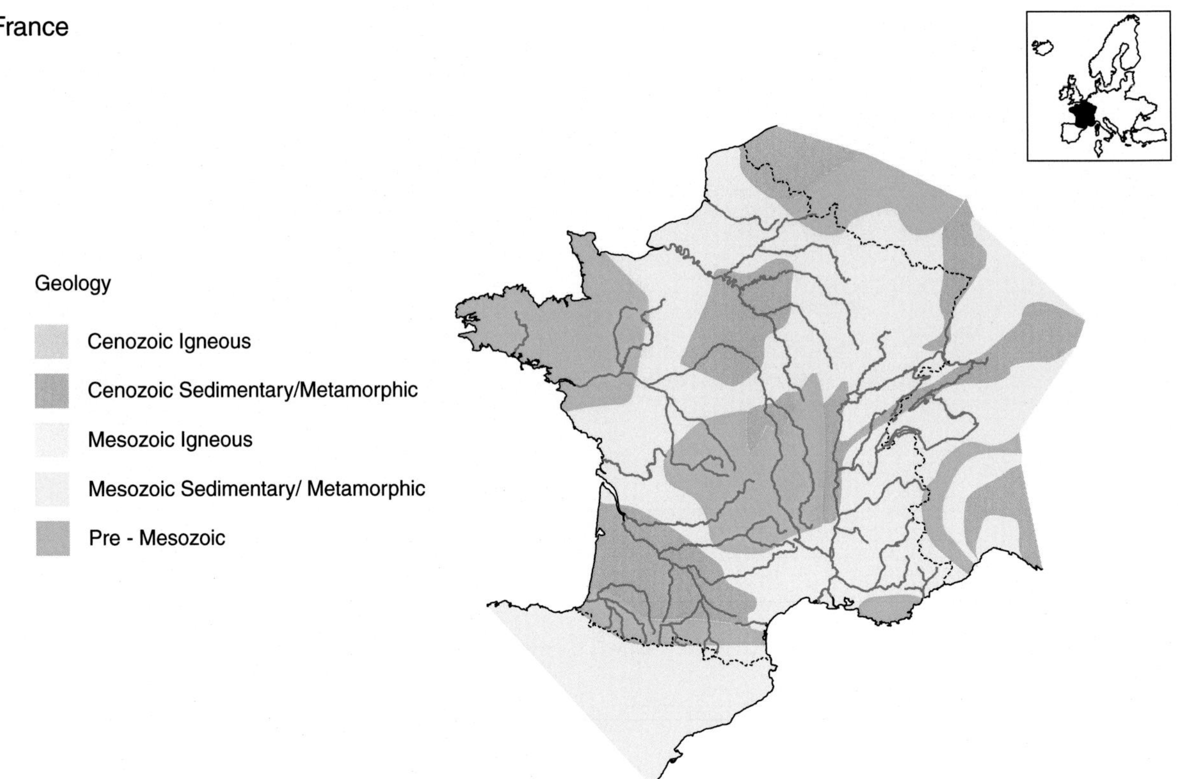

Surface Runoff (mm/yr)

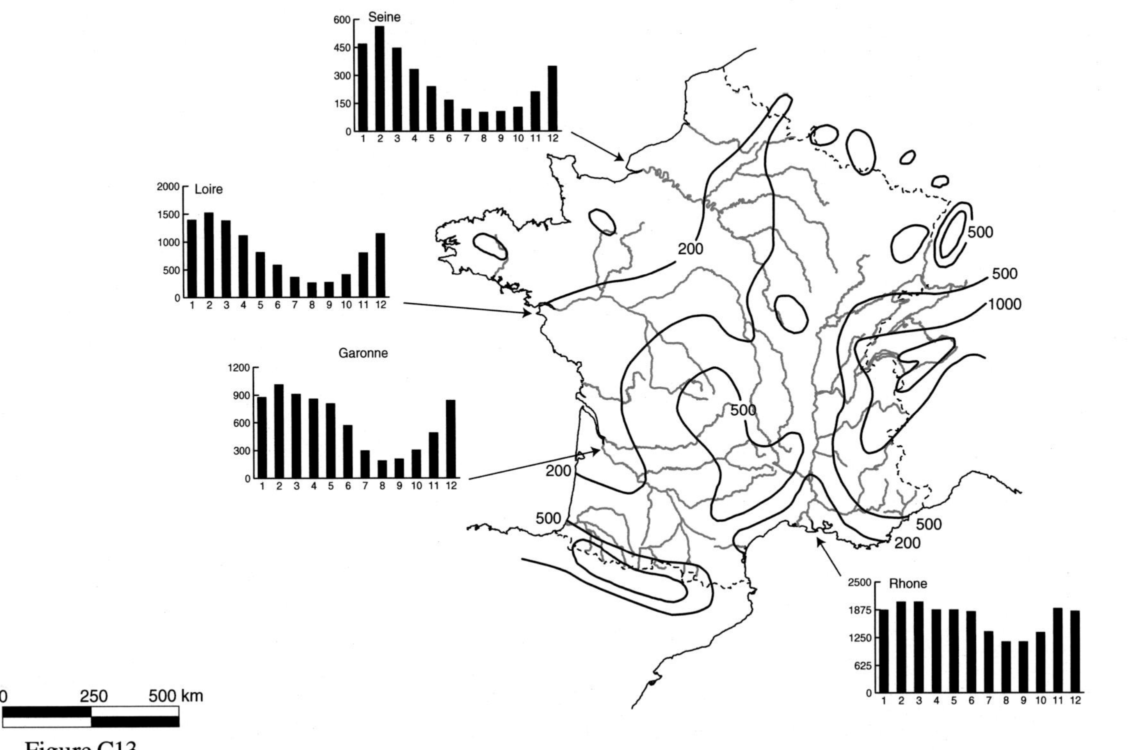

Figure C13

Table C6.

River name	Ocean	Area (10^3 km^2)	Length (km)	Max_elev (m)	Climate	Geology	Q (km^3/yr)	TSS (Mt/yr)	TDS (Mt/yr)	Refs.
France										
Mainland										
Adour	Atlantic (NE)	16	340	2800	Te-H-S	PreMes	11	0.29	1.9	8, 11, 12
Argens	Med. (W)	2.6	120	>1000	Te-H-W	PreMes	0.6	0.03		4
Aude	Med. (W)	5.9	220	>500	Te-H-S	Mes S/M	1.3	0.07		4, 9
Charente	Atlantic (NE)	9.1	350	>200	Te-H-W	PreMes	1.8	0.2		3, 11, 13, 14
Dordogne	Atlantic (NE)	24	470	1700	Te-H-W	PreMes	14			2, 3, 11, 13
Garonne	Atlantic (NE)	56	570	3400	Te-H-W	PreMes	21	1.1 (2.2)	4.5	2, 3, 4, 5, 7, 8, 10, 13
Herault	G. du Lion	2.9	160	1000	Te-H-W	Mes S/M	1.5	0.09		4, 9
Loire	Atlantic (NE)	120	1100	1900	Te-SA-W	PreMes	27	0.5	6.4	3, 5, 7, 8, 13
Orb	G. du Lion	1.8	100	>500	Te-H-W	Mes S/M	1.3	0.05		3, 4, 9, 13
Rhone	Med. (W)	96	1000	3600	Te-H-Sp	Mes S/M	54	6.2 (59)	17	3, 5, 7, 8, 9, 13
Seine	Atlantic (NE)	79	780	900	Te-SA-W	Mes S/M	13	0.7 (3.5)	7.7	3, 5, 7, 8, 13
Somme	Atlantic (NE)	5.5	240	>100	Te-SA-W	Mes S/M	0.85			3, 11, 13
Tet	Med. (W)	1.6	120	2400	Te-H-W	Mes S/M	0.3	0.5		6, 9
Var	Ligurian Sea	2.8	130	2500	Te-H-W	Mes S/M	1.3	1		1, 8, 15
Vilaine	Atlantic (NE)	11	230	150	Te-H-W	PreMes	2.5	0.2		11, 14
Corsica										
Golo	Med.	0.9	90	>1000	Te-H-S	Cen I	0.5	0.002		4

References:
1. Anthony and Julian, 1997; 2. Dauta *et al.*, 2009; 3. Direction du Gaz et de l'Electricite, 1966 (cf. van der Leeden, 1975); 4. European Environment Agency website, www.eea.europa.eu; 5. GEMS website, www.gemstat.org; 6. Guillén *et al.*, 2006; 7. Kempe, 1982; 8. Meybeck and Ragu, 1996; 9. Pont, 1997; 10. Probst, 1992; 11. Rand McNally, 1980; 12. Snoussi *et al.*, 1990; 13. UNESCO, 1967 (cf. van der Leeden, 1975); 14. Uriarte *et al.*, 2004; 15. Mulder *et al.*, 1998

Spain and Portugal

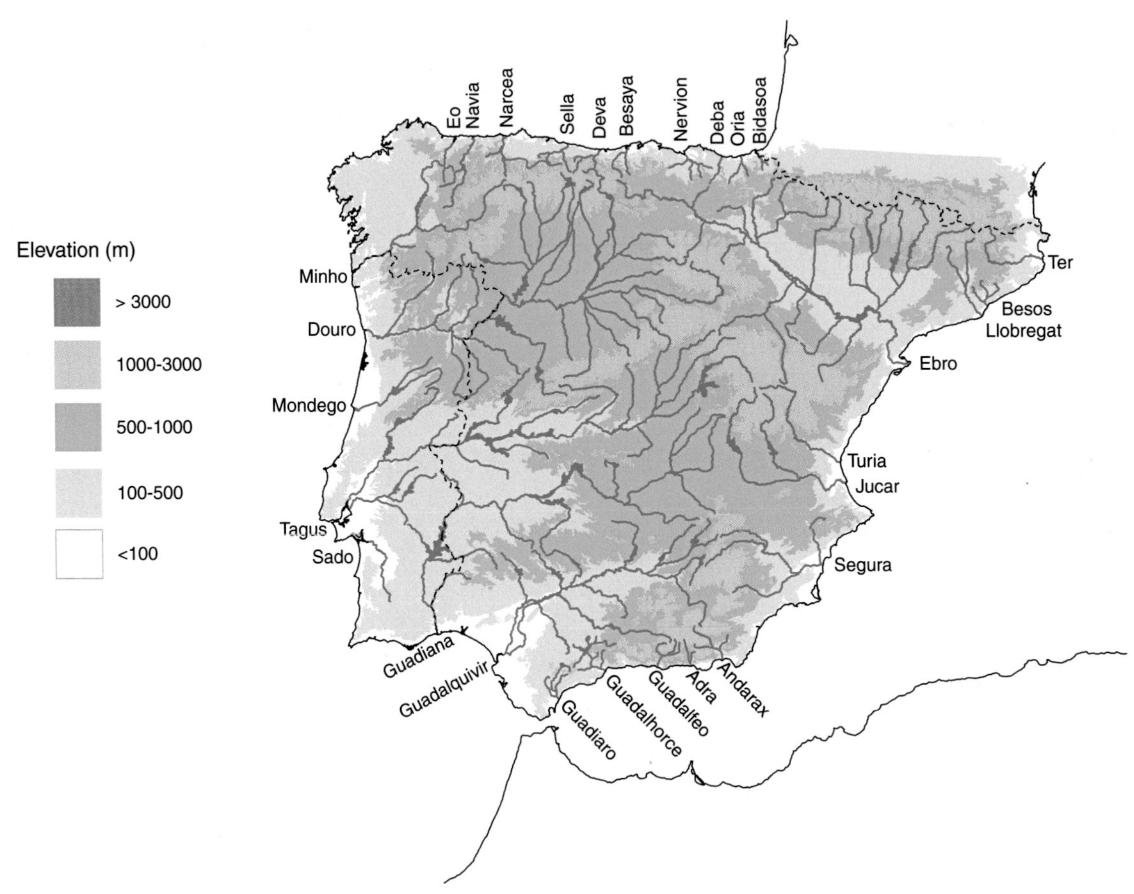

Figure C14

Spain and Portugal

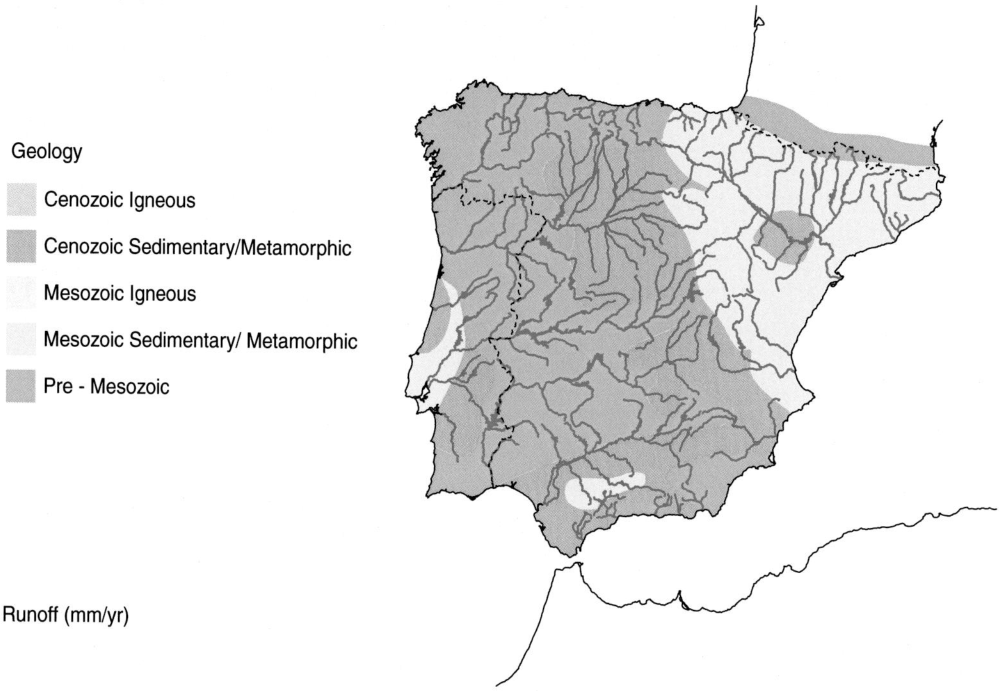

Geology

Cenozoic Igneous

Cenozoic Sedimentary/Metamorphic

Mesozoic Igneous

Mesozoic Sedimentary/ Metamorphic

Pre - Mesozoic

Surface Runoff (mm/yr)

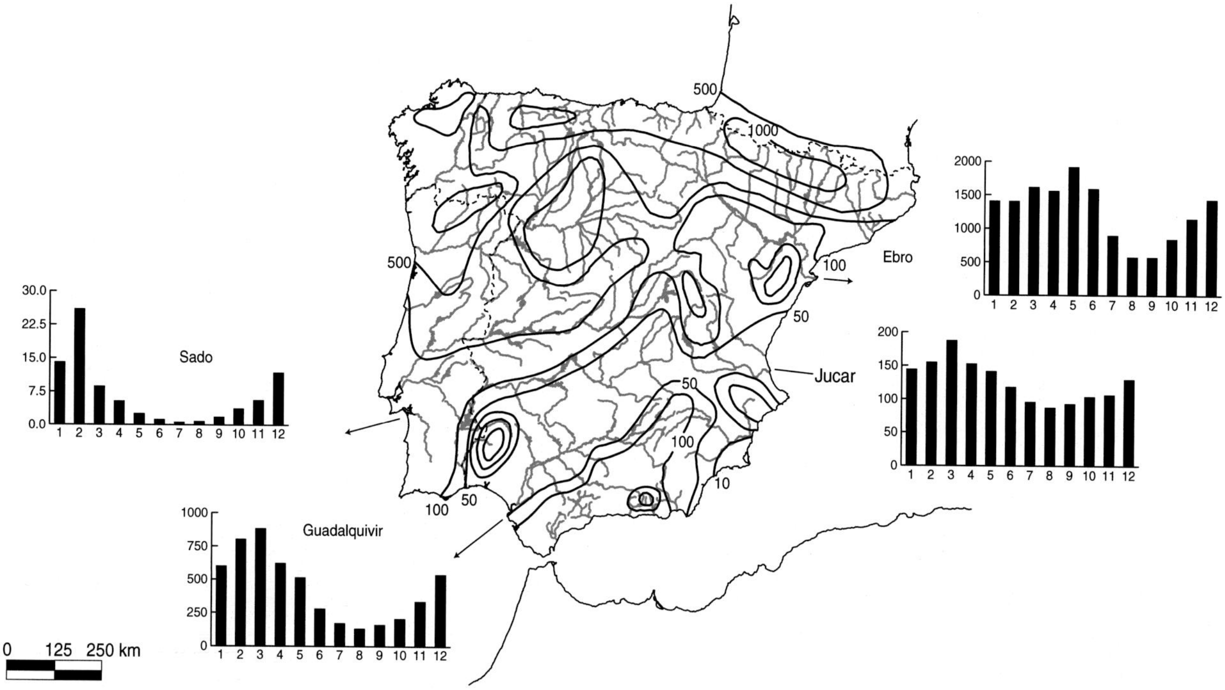

500
1000
500
100
50
50
100
10
100
50

Ebro
2000
1500
1000
500
0
1 2 3 4 5 6 7 8 9 10 11 12

Jucar
200
150
100
50
0
1 2 3 4 5 6 7 8 9 10 11 12

Sado
30.0
22.5
15.0
7.5
0.0
1 2 3 4 5 6 7 8 9 10 11 12

Guadalquivir
1000
750
500
250
0
1 2 3 4 5 6 7 8 9 10 11 12

0 125 250 km

Figure C15

Table C7.

River name	Ocean	Area (10^3 km^2)	Length (km)	Max_elev (m)	Climate	Geology	Q (km^3/yr)	TSS (Mt/yr)	TDS (Mt/yr)	Refs.
Spain										
Adra	Med. (W)	0.75	51	2700	Te-SA-W	Cen S/M	0.03	0.15		8
Andarax	Med. (W)	2.2	74	2500	Te-A-W	PreMes	0.01	0.18		8
Besaya	Atlantic (NE)	1		<1000	Te-W-W	PreMes	0.8	0.1		2, 13
Besos	Med. (W)	1	52	800	Te-SA-W	Mes S/M				9
Bidasoa	Bay of Biscay	0.71	66	780	Te-W-W	PreMes	0.9	0.15		13, 19
Deba	Bay of Biscay	0.5	62	750	Te-W-W	PreMes	0.5	0.03		13, 19
Deva	Atlantic (E)	1.2		<1000	Te-W-W	PreMes	1	0.13		2, 13
Ebro	Med. (W)	87	930	>2000	Te-SA-W	Cen S/M-Mes S/M	17 (50)	0.15 (18)	9	5, 11, 12, 15, 18
Eo	Bay of Biscay	0.93		>500	Te-W-W	PreMes				13
Guadalfeo	Med. (W)	1.3	72	3200	Te-SA-W	PreMes-Cen S/M	0.02	0.08		8
Guadalhorce	Med. (W)	3.2	154	1700	Te-SA-W	PreMes-Cen S/M	0.2	0.09		8
Guadalquivir	Atlantic (NE)	56	680	>1500	Te-SA-W	PreMes	2 (7.3)		5.9	3, 5, 11, 14, 17, 18
Guadiaro	Med. (W)	1.5	100	1600	Te-SA-W	Cen S/M	0.3	0.04		8
Jucar	Med. (W)	22	510	>500	Te-A-W	Mes S/M	1.2 (4.5)	0.8	1	11, 18
Llobregat	Med. (W)	5.2	170	1700	Te-SA-W	Mes S/M	0.69	0.07		9, 11, 14, 17
Narcea	Bay of Biscay	4.9	220	>1500	Te-H-W	PreMes	3.4	0.38		2, 13
Navia	Bay of Biscay	2.6	210	>1000	Te-H-W	PreMes	2.1	0.13		2, 13
Nervion	Atlantic (NE)	1.4		<1000	Te-W-W	PreMes	1.1	0.06		2
Oria	Bay of Biscay	0.86	78	750	Te-W-W	PreMes	0.9	0.07		13, 19
Segura	Med. (W)	19	340	1400	Te-A-W	PreMes	0.8 (3.1)	1.1		4, 5, 11, 16
Sella	Atlantic (NE)	1.3		<1000	Te-W-W	PreMes	1.1	0.03		2, 13
Ter	Med. (W)	3	210	2400	Te-H-W	Mes S/M	0.84			15
Turia	Med. (W)	6.4	240	>2000	Te-A-W	Mes S/M	0.46			14
Portugal										
Douro	Atlantic (NE)	98	930	2100	Te-H-W	PreMes	13 (20)	1.8	5.5	1, 10, 16
Guadiana	Atlantic (NE)	72	830	>500	Te-H-W	PreMes	9	0.07	4.6	5, 10, 11
Minho	Atlantic (NE)	20	260	>1500	Te-H-W	PreMes	13		0.63	11
Mondego	Atlantic (NE)	6.7	220	>1000	Te-H-W	PreMes	2.6			14, 16, 17
Sado	Atlantic (NE)	2.7	180	>200	STr-SA-W	Cen S/M	0.25			6, 14
Tagus	Atlantic (NE)	80	1000	>1500	Te-H-W	PreMes	9.6	0.4	3.1	5, 7, 10, 11, 17, 20

References:

1. Alt-Epping *et al.*, 2007; 2. Comisaria de Agua del Norte de España; 3. European Environment Agency website, www.eea.europa.eu; 4. Eurosion, 2004; 5. GEMS website, www.gemstat.org; 6. Global River Discharge Database, http://www.rivdis.sr.unh.edu/; 7. Jouanneau *et al.*, 1998; 8. Liquete *et al.*, 2005; 9. Liquete *et al.*, 2007; 10. Lugo, 1983; 11. Meybeck and Ragu, 1996; 12. Palanques *et al.*, 1990; 13. Prego *et al.*, 2008; 14. Rand McNally, 1980; 15. Sabater *et al.*, 1995; 16. Sabater *et al.*, 2009; 17. UNESCO (WORRI), 1978; 18. UNESCO, 1971; 19. Uriarte *et al.*, 2004; 20. Vörösmarty *et al.*, 1996a

Italy

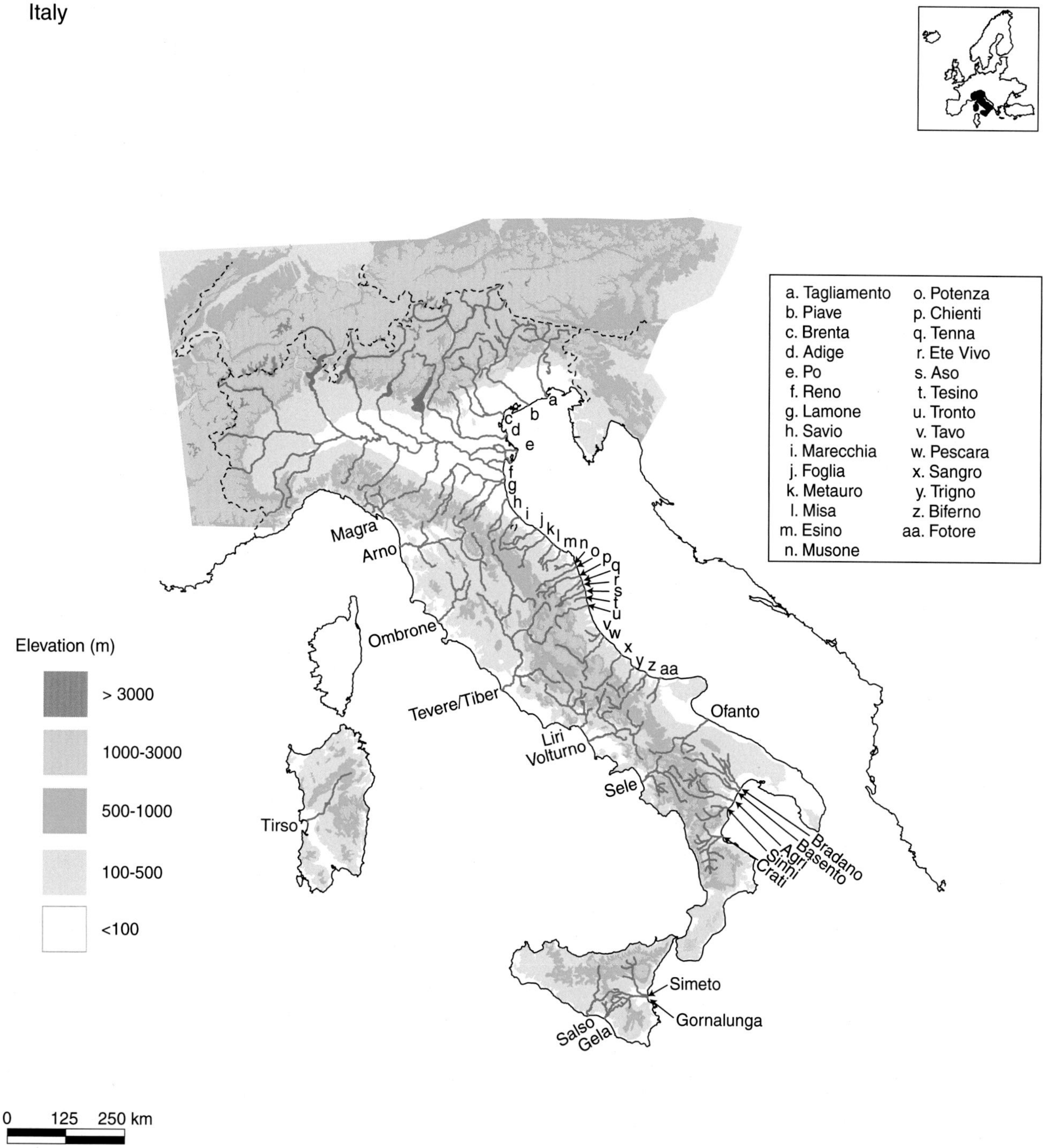

a. Tagliamento
b. Piave
c. Brenta
d. Adige
e. Po
f. Reno
g. Lamone
h. Savio
i. Marecchia
j. Foglia
k. Metauro
l. Misa
m. Esino
n. Musone
o. Potenza
p. Chienti
q. Tenna
r. Ete Vivo
s. Aso
t. Tesino
u. Tronto
v. Tavo
w. Pescara
x. Sangro
y. Trigno
z. Biferno
aa. Fotore

Magra
Arno
Ombrone
Tevere/Tiber
Liri
Volturno
Sele
Ofanto
Bradano
Basento
Agri
Sinni
Crati
Tirso
Simeto
Gornalunga
Salso
Gela

Elevation (m)

> 3000
1000-3000
500-1000
100-500
<100

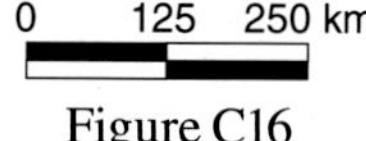

Figure C16

Italy

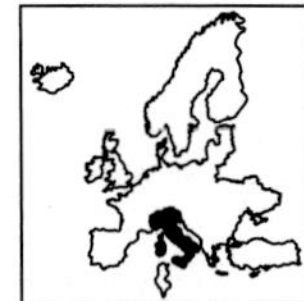

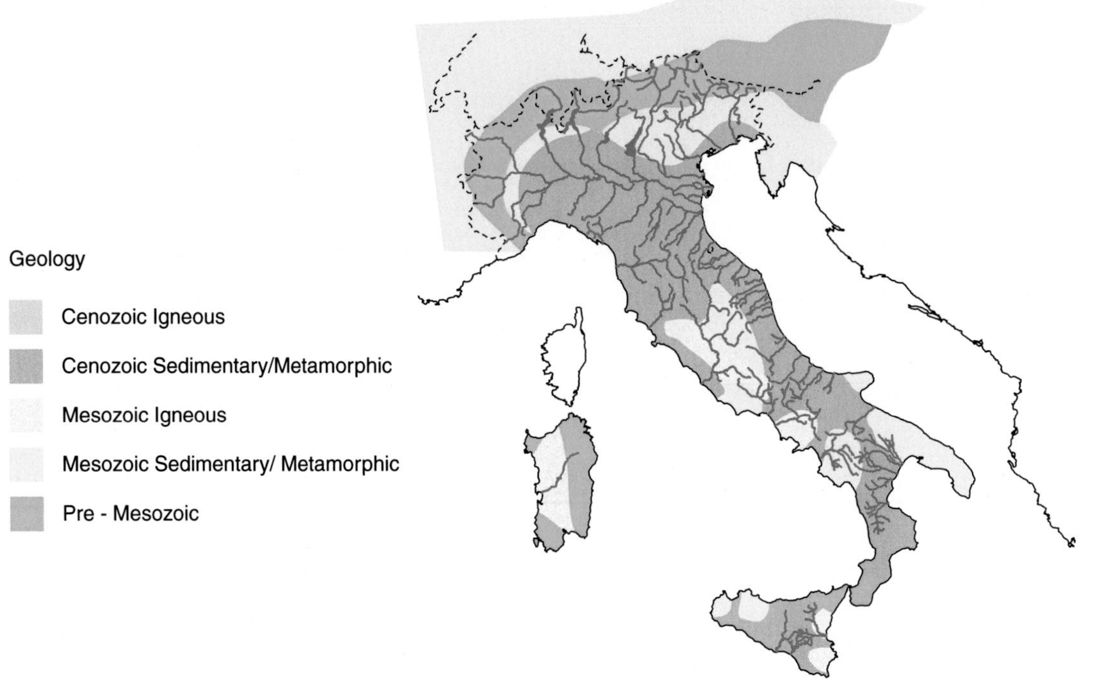

Surface Runoff (mm/yr)

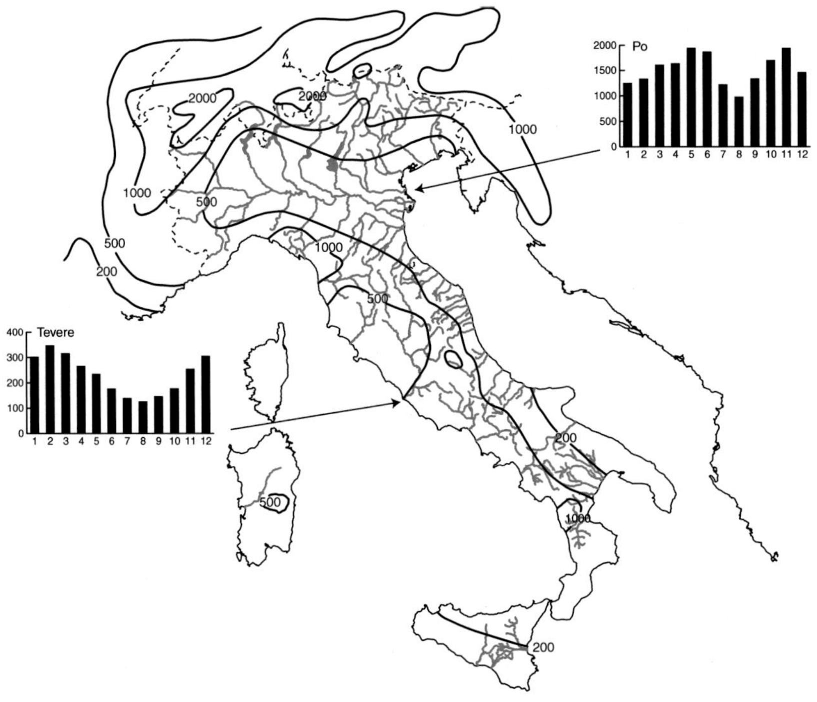

0 125 250 km

Figure C17

Table C8.

River name	Ocean	Area (10^3 km^2)	Length (km)	Max_elev (m)	Climate	Geology	Q (km^3/yr)	TSS (Mt/yr)	TDS (Mt/yr)	Refs.
Italy										
Mainland										
Adige	Adriatic Sea	12	410	2500	Te-H-W	Mes S/M	7.3	1.6	1.6	13, 14, 16, 21
Agri	G. of Tronto	0.28	110	2000	Te-W-W	Cen S/M	0.25	0.07		13, 17
Arno	Ligurian Sea	8.2	240	1700	Te-H-W	Cen S/M	1.8 (3.2)	2.2	1.4	11, 14, 16
Aso	Adriatic Sea	0.28	70	2500	Te-H-W	Cen S/M		0.18		1
Basento	G. of Tronto	1.4	25	1500	Te-H-W	Cen S/M				
Biferno	Adriatic Sea	1.3	95	1800	Te-H-W	Cen S/M	0.66	2.2		9, 13
Bradano	G. of Tronto	2.7	120	1100	Te-SA-W	Cen S/M	2	2.8		13
Brenta	Adriatic Sea	1.6	160	2100	Te-W-W	Mes S/M	2.3	0.19		13, 20
Chienti	Adriatic Sea	1.3	99	1600	Te-H-W	Cen S/M	0.3	0.56 (0.85)		1, 13, 19
Crati	G. of Tronto	2.4		1000	Te-H-W	Cen S/M	0.85	1.2		7, 13
Esino	Adriatic Sea	1.2	75	1400	Te-H-W	Cen S/M		0.9		1
Ete Vivo	Adriatic Sea	0.18	30	580	Te-H-W	Cen S/M		0.29		1
Foglia	Adriatic Sea	0.7	80	1500	Te-H-W	Cen S/M	0.25	1.4		1, 13
Fortore	Adriatic Sea	1.1		900	Te-H-W	Cen S/M	0.42	1.5		13
Lamone	Adriatic Sea	0.71	95	910	Te-H-W	Cen S/M	0.28	1.3		1, 13
Liri	Mediterranean	5	160	>1500	Te-H-W	Mes S/M				
Magra	Ligurian Sea	1.2		>500	Te-W-W	Mes S/M	1.3	0.5		13
Marecchia	Adriatic Sea	0.6		900	Te-H-W	Cen S/M	0.31	1.6		13
Metauro	Adriatic Sea	1.4	91	1600	Te-H-W	Cen S/M	0.3	0.55 (0.81)		1, 18, 19, 20
Misa	Adriatic Sea	0.38	35	790	Te-H-W	Cen S/M		0.47		1
Musone	Adriatic Sea	0.64	70	<100	Te-H-W	Cen S/M		1.1		1
Ofanto	Adriatic Sea	2.7	130	1100	Te-SA-W	Cen S/M	0.37	0.9		13, 17, 18
Ombrone	Mediterranean	3.2	160	1700	Te-H-W	Cen S/M	0.79	1.9 (10)		8, 13, 20
Pescara	Adriatic Sea	3.3	154	2700	Te-H-W	Cen S/M	0.9 (1.7)	1 (1.9)		13, 19, 20
Piave	Adriatic Sea	4.1	220	1900	Te-H-W	Mes S/M	3.2			18
Po	Adriatic Sea	74	680	4800	Te-H-W/Sp	Mes S/M-PreMes	46	10 (15)	17	1, 5, 6, 12, 13, 14, 16, 21
Potenza	Adriatic Sea	0.8	89	1500	Te-H-W	Cen S/M	0.2	0.35 (0.56)		1, 19
Reno	Adriatic Sea	3.4	210	1300	Te-H-W	Cen S/M	0.9	2.7		1, 18, 20
Sangro	Adriatic Sea	1.9	90	>200	Te-H-W	Cen S/M	0.75			10, 18

Table C8. (*Continued*)

River name	Ocean	Area (10^3 km^2)	Length (km)	Max_elev (m)	Climate	Geology	Q (km^3/yr)	TSS (Mt/yr)	TDS (Mt/yr)	Refs.
Savio	Adriatic Sea	0.67	90	1200	Te-H-W	Cen S/M	0.31	0.92		1, 2, 13, 14
Sele	Mediterranean	3.4	110	>1000	Te-H-W	Mes S/M				
Sinni	G. of Tronto	1.1	85	1000	Te-H-W	Cen S/M	0.65	2.5		13
Tagliamento	Adriatic Sea	3.6	170	>500	Te-H-W	Mes S/M	2.7			17, 18
Tavo	Adriatic Sea	0.25		>1000	Te-H-W	Cen S/M	0.06	0.04		13
Tenna	Adriatic Sea	0.49	70	2400	Te-H-W	Cen S/M		0.45		1
Tesino	Adriatic Sea	0.11	35	750	Te-H-W	Cen S/M		0.12		1
Tevere	Mediterranean	17	400	2500	Te-H-W	Mes S/M	7.4	0.33 (7.5)		1, 3, 14, 15, 20
Trigno	Adriatic Sea	1.2	80	>1000	Te-H-W	Cen S/M	0.1	0.42		13, 18
Tronto	Adriatic Sea	1.2	86	2300	Te-H-W	Cen S/M	0.3	0.5 (1)		13, 19, 20
Volturno	Mediterranean	5.5	170	2200	Te-H-W	Cen S/M	3.1	4.2		13, 14
Sardinia										
Tirso	Mediterranean	3.1	150	1800	Te-H-W	Cen S/M				4
Sicily										
Gela	Sicilian Channel	0.24		700	Te-SA-W	Cen S/M	0.02	0.13		13
Gornalunga	Sicilian Channel	0.23		600	Te-A-W	Cen S/M	0.005	0.03		13
Salso	Sicilian Channel	1.8	110	>1000	Te-SA-W	Cen S/M				17
Simeto	Ionian Sea	4.2	110	1700	Te-H-W	Cen	0.8	1 (3.5)		13

References:
1. Aquater, 1982; 2. Bartolini *et al.*, 1996; 3. Bellotti *et al.*, 1994; 4. Cattaneo, 1995; 5. Centre for Natural Resources, Energy and Transport (UN), 1978; 6. Correggiari *et al.*, 2005; 7. De Bartolo *et al.*, 2000; 8. Frangipane and Paris, 1994; 9. Global River Discharge Database, http://www.rivdis.sr.unh.edu/; 10. Gumiero *et al.*, 2009; 11. Holeman, 1968; 12. Hovius and Leeder, 1998; 13. IAHS/UNESCO, 1974; 14. Meybeck and Ragu, 1996; 15. Mikhailova *et al.*, 1998; 16. Pettine *et al.*, 1985; 17. Rand McNally, 1980; 18. Simeoni and Bondesan, 1997; 19. Syvitski and Kettner, 2007; 20. UNEP/MAP/MED_POL, 2003; 21. UNESCO (WORRI), 1978

Albania, Croatia and Greece

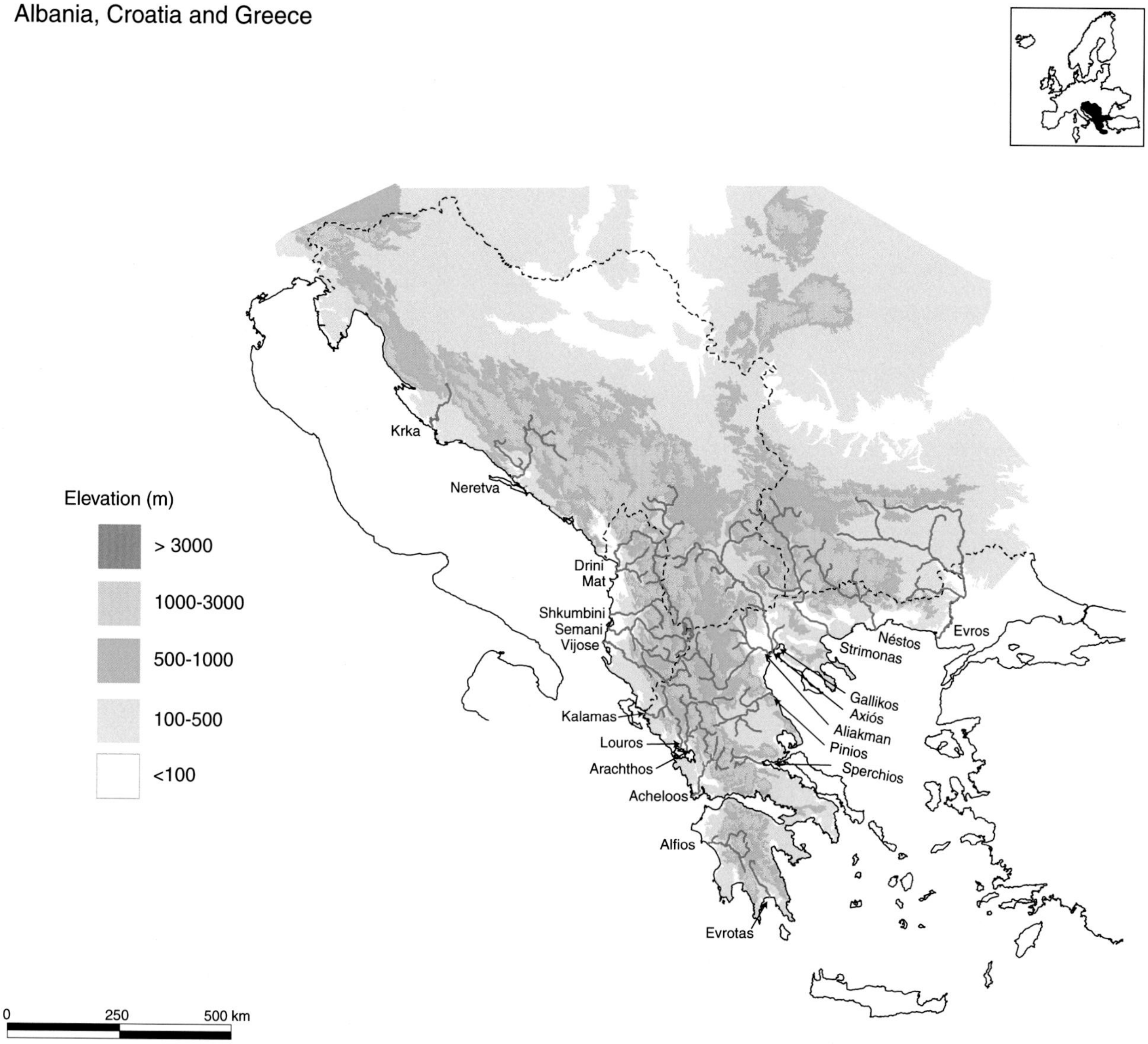

Krka
Neretva
Drini
Mat
Shkumbini
Semani
Vijose
Kalamas
Louros
Arachthos
Acheloos
Alfios
Evrotas
Néstos
Strimonas
Evros
Gallikos
Axiós
Aliakman
Pinios
Sperchios

Elevation (m)

> 3000
1000-3000
500-1000
100-500
<100

0 250 500 km

Figure C18

Albania, Croatia and Greece

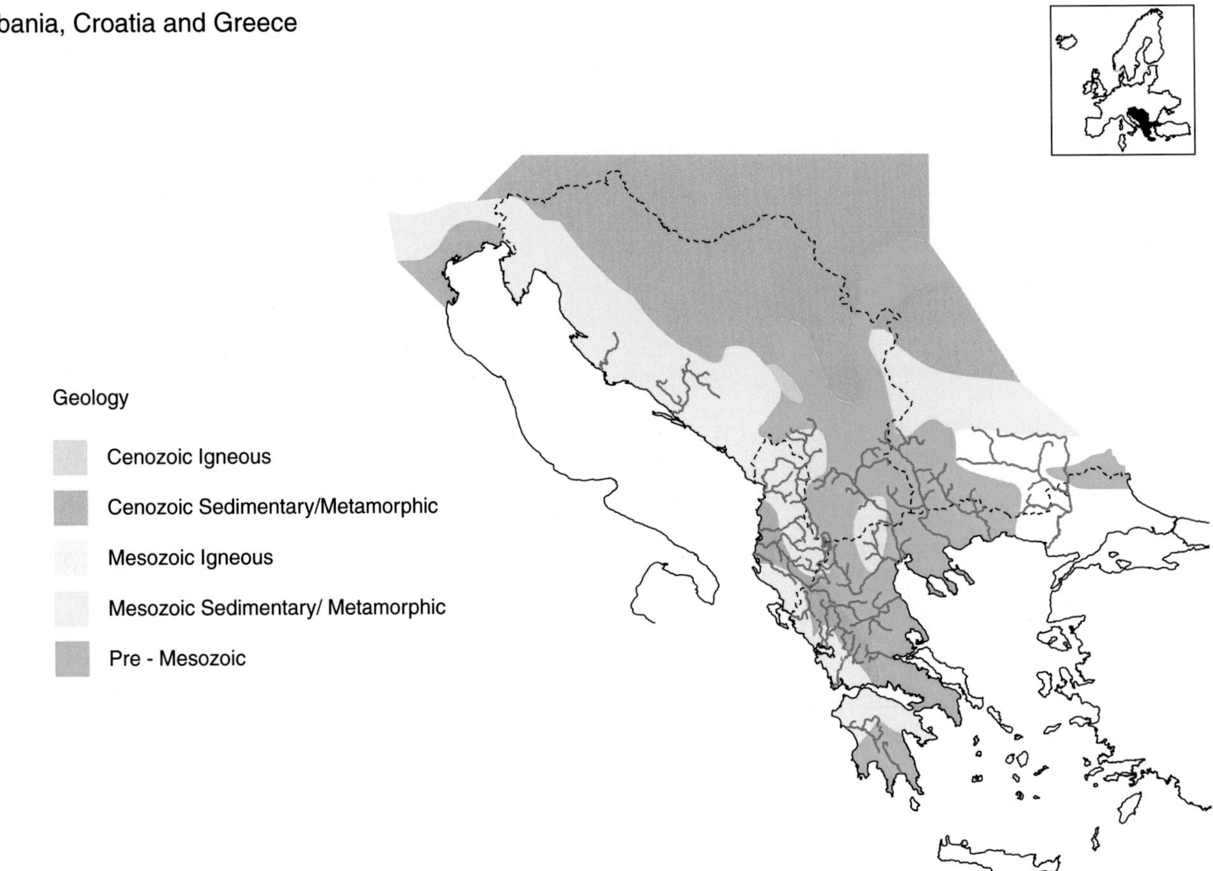

Surface Runoff (mm/yr)

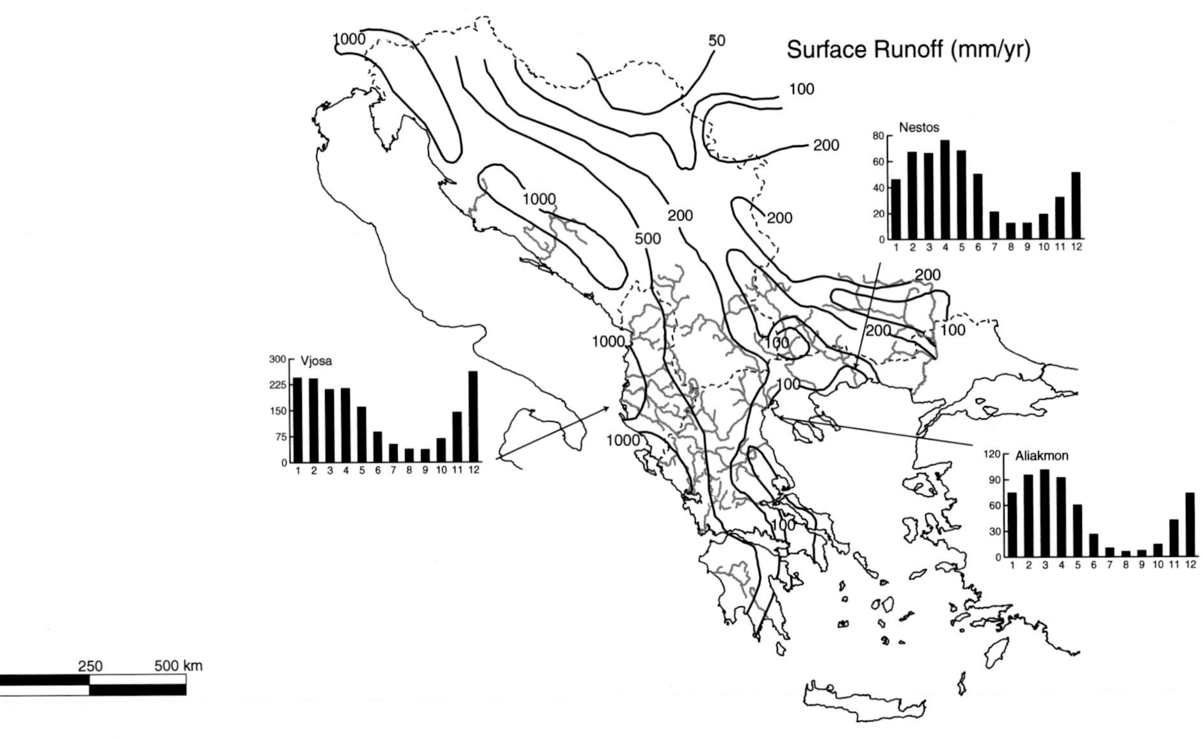

Figure C19

Table C9.

River name	Ocean	Area (10^3 km^2)	Length (km)	Max_elev (m)	Climate	Geology	Q (km^3/yr)	TSS (Mt/yr)	TDS (Mt/yr)	Refs.
Albania										
Drini	Adriatic Sea	20	280	2300	Te-W-W	Mes S/M	12	2.1 (16)		2, 4, 11, 13, 20, 23, 26
Mat	Adriatic Sea	2.3	110	>500	Te-W-W	Cen S/M	3.4	2.6		4, 13
Semani	Adriatic Sea	5.6	280	2500	Te-H-W	Mes S/M	2.9	16 (30)		4, 7, 13, 21, 22
Shkumbini	Adriatic Sea	2.4	180	2400	Te-H-W	Mes S/M	1.9	7.2		4, 8, 13, 21
Vijose	Adriatic Sea	6.8	260	2400	Te-W-W	Mes S/M	5.5	8.3 (29)		2, 4, 22, 23, 26
Croatia										
Krka	Adriatic Sea	2.2	70	>500	Te-W-W	Mes S/M				
Neretva	Adriatic Sea	13	220	>1000	Te-W-W	Mes S/M	12	14		5, 25
Greece										
Acheloos	Ionian Sea	5.4	190	2300	Te-W-W	Cen S/M	5.7	3.3	1.5	14, 19
Alfios	Ionian Sea	3.7	110	2200	Te-H-W	Cen S/M	1.7	3		12, 18
Aliakmon	Aegean Sea	9.5	310	2200	Te-SA-W	PreMes	3.1	4.4	1.2	10, 15, 18, 19, 20
Arachtos	Ionian Sea	1.9	100	2300	Te-W-W	Cen S/M	2.2	7.3		15, 19, 23
Axiós	Aegean Sea	24	310	2800	Te-SA-W	PreMes	4.9	4.7 (11)	1.7	5, 10, 15, 24
Évros	Aegean Sea	53	520	2900	Te-H-W	Mes S/M	6.8	8.5	2.6	1, 9, 10, 23
Evrotas	Aegean Sea	2.4	90	>2000	Te-H-W	Te-H-W	0.8			23
Gallikos	Aegean Sea	0.9	65	2200	Te-W-W	Cen S/M	1.2	0.004		17
Kalamas	Ionian Sea	1.8	100	>1000	Te-W-W	Mes S/M	6.8	1.9		6, 19, 23
Louros	Ionian Sea	0.8		1600	Te-W-W	Cen S/M	0.6	0.8	0.17	15
Néstos	Aegean Sea	6.2	210	2900	Te-H-W	Cen S/M	3.1	1	0.78	2, 3, 14
Pinios	Aegean Sea	11	220	1900	Te-SA-W	PreMes	3.2	4.4	1.6	10, 15, 20
Sperchios	Aegean Sea	1.8	70	2300	Te-H-W	PreMes	0.74	1.4	0.52	10, 15, 16
Strimonas	Aegean Sea	17	410	2700	Te-SA-W	Mes S/M	4.1	4	1.5	5, 20

References:

1. Artinyan *et al.*, 2008; 2. Center for Natural Resources, Energy and Transport (UN), 1978; 3. Chorafas, 1963 (cf. van der Leeden, 1975); 4. Ciavola *et al.*, 1999; 5. Eurosion, 2004; 6. GRDC website, www.gewex.org/grdc; 7. Holeman, 1968; 8. IAHS/UNESCO, 1974; 9. Kanellopoulos *et al.*, 2008; 10. Meybeck and Ragu, 1996; 11. Milliman and Meade, 1983; 12. Nicholas *et al.*, 1999; 13. Pano, 1992; 14. Poulos and Collins, 2002; 15. Poulos *et al.*, 1995; 16. Poulos *et al.*, 1996; 17. Poulos *et al.*, 2000; 18. Poulos *et al.*, 2002; 19. S. E. Poulos, personal communication; 20. Rand McNally, 1980; 21. Regional Activity Center for Environment Remote Sensing (RACERS), 1996; 22. Simeoni *et al.*, 1997; 23. Skoulikidis, 2009; 24. UNEP/MAP/MED_POL, 2003; 25. UNESCO (WORRI), 1978; 26. UNESCO, 1971

Bulgaria, Romania and Ukraine

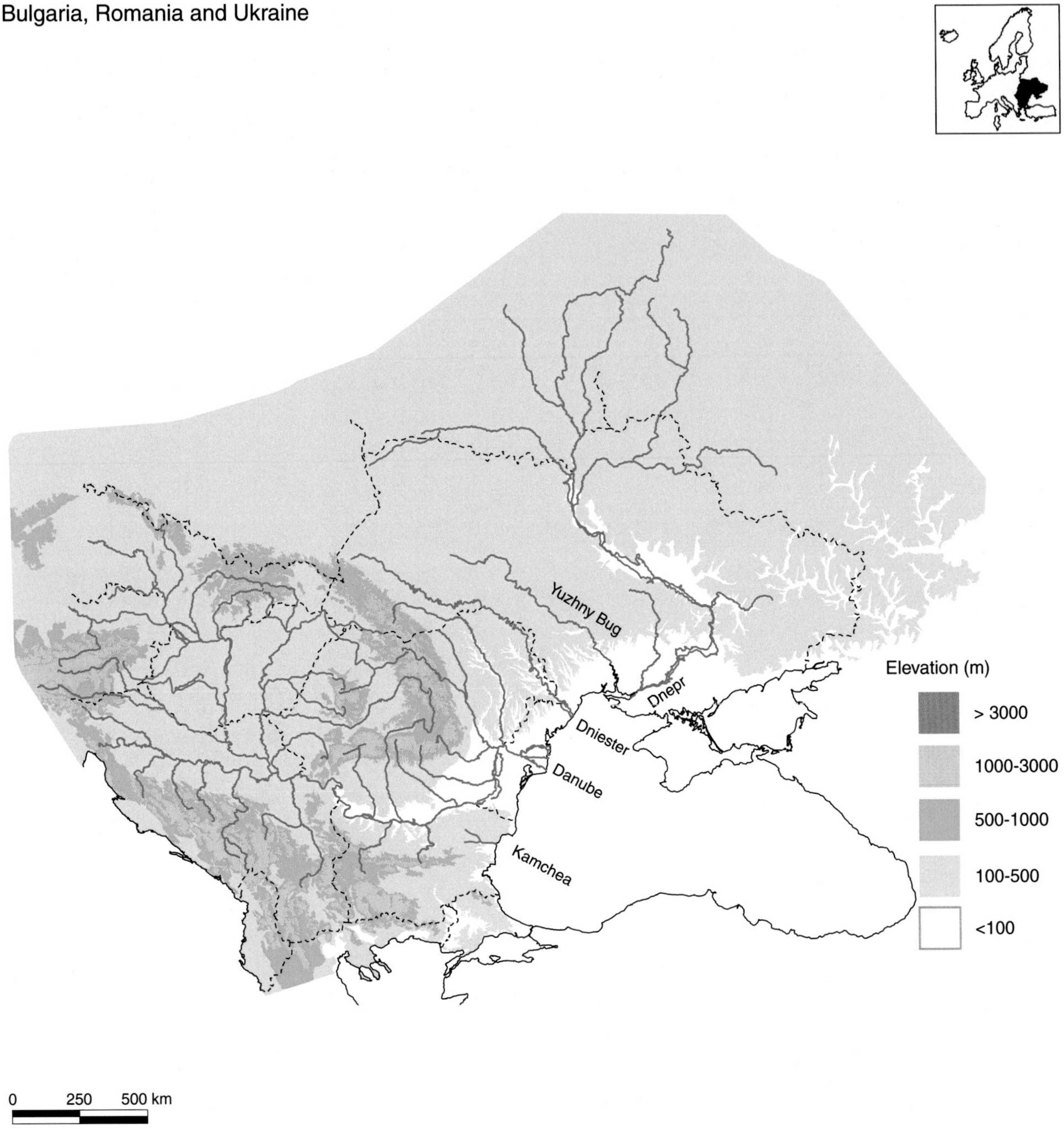

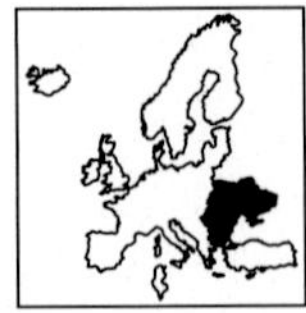

Yuzhny Bug
Dnepr
Dniester
Danube
Kamchea

Elevation (m)
> 3000
1000-3000
500-1000
100-500
<100

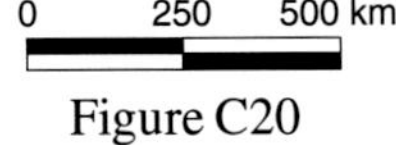

Figure C20

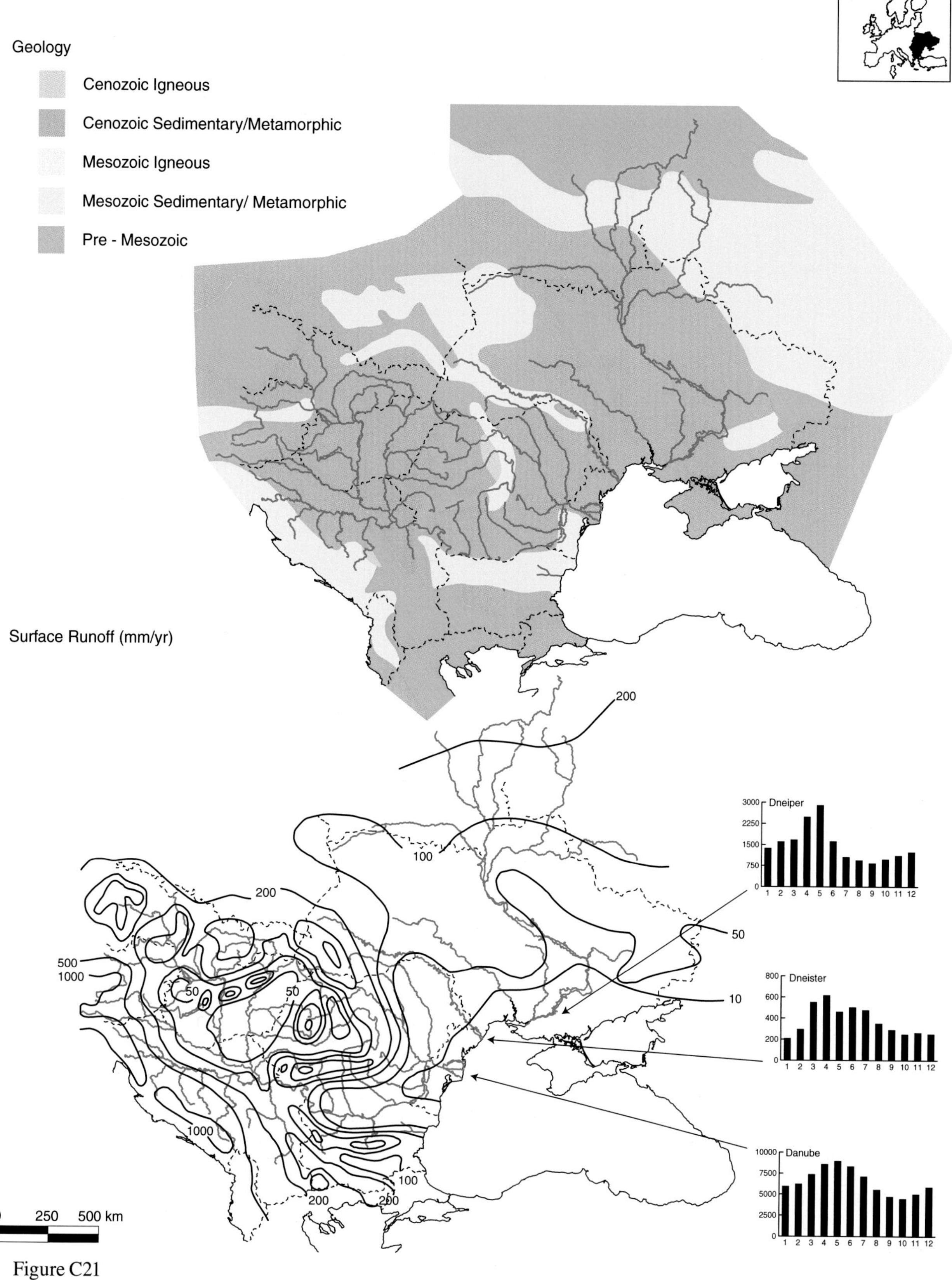
Bulgaria, Romania and Ukraine
Geology
Cenozoic Igneous
Cenozoic Sedimentary/Metamorphic
Mesozoic Igneous
Mesozoic Sedimentary/ Metamorphic
Pre - Mesozoic
Surface Runoff (mm/yr)
200
100
50
10
500
1000
50
50
1000
100
200
200
Dneiper
3000
2250
1500
750
0
1 2 3 4 5 6 7 8 9 10 11 12
Dneister
800
600
400
200
0
1 2 3 4 5 6 7 8 9 10 11 12
Danube
10000
7500
5000
2500
0
1 2 3 4 5 6 7 8 9 10 11 12
0 250 500 km

Figure C21

Table C10.

River name	Ocean	Area (10^3 km^2)	Length (km)	Max_elev (m)	Climate	Geology	Q (km^3/yr)	TSS (Mt/yr)	TDS (Mt/yr)	Refs.
Bulgaria										
Kamchea	Black Sea	5.3	165	>500	Te-SA-W	Mes S/M	0.3 (0.9)	0.46 (1.1)		3, 5
Romania										
Danube	Black Sea	820	2900	4100	Te-H-S	Cen S/M	210	42 (67)	80	2, 4, 6, 7, 8, 10
Ukraine										
Dnepr	Black Sea	510	2200	330	Te-A-S	PreMes-Mes S/M	43 (53)	2.3	15	1, 4, 6, 8, 9, 10
Dniester	Black Sea	72	1400	900	Te-SA-S	Cen-Mes-PreMes	9.3	0.49 (3)	6.1	2, 4, 8, 10
Yuzhny Bug	Black Sea	64	860	390	Te-A-S	Cen S/M-Mes S/M	2.8 (3.4)	0.2	1 (0.83)	6, 8, 10, 11

References:
1.Algan *et al.*, 1997; 2. GEMS website, www.gemstat.org; 3. GRDC website, www.gewex.org/grdc; 4. Hay, 1994; 5. Jaoshvili, 2002; 6. Kostianitsin, 1964; 7. Levashova *et al.*, 2004; 8. Meybeck and Ragu, 1996; 9. Skoulikidis, 2009; 10. Varga *et al.*, 1989; 11. Zhukinsky *et al.*, 1989

Appendix D Africa

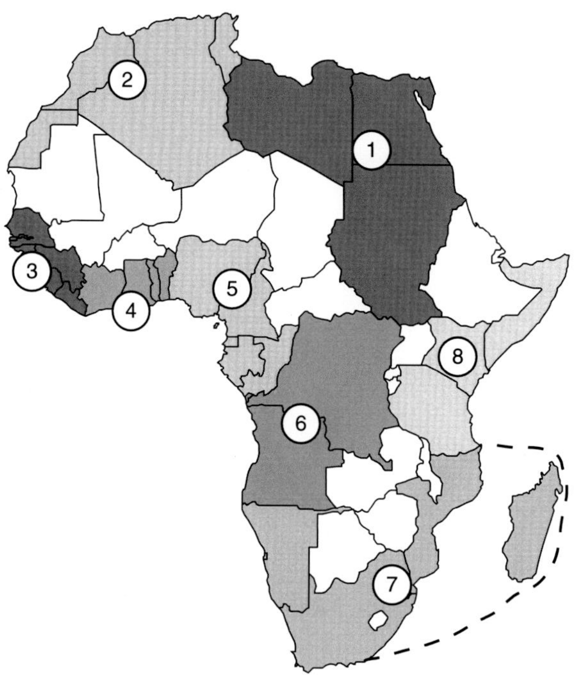

Figure D1

(1) Sudan, Egypt, Libya
(2) Tunisia, Algeria, Morocco, Western Sahara
(3) Senegal, Gambia, Guinea Bissau, Guinea, Sierra Leone and Liberia
(4) Ivory Coast, Ghana, Togo and Benin
(5) Nigeria, Cameroon, Equatorial Guinea, Gabon, Republic of Congo
(6) Democratic Republic of Congo and Angola
(7) Namibia, South Africa, Mozambique, Madagascar
(8) Tanzania, Kenya, Somalia

Sudan, Egypt and Lybia

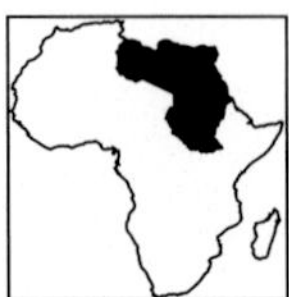

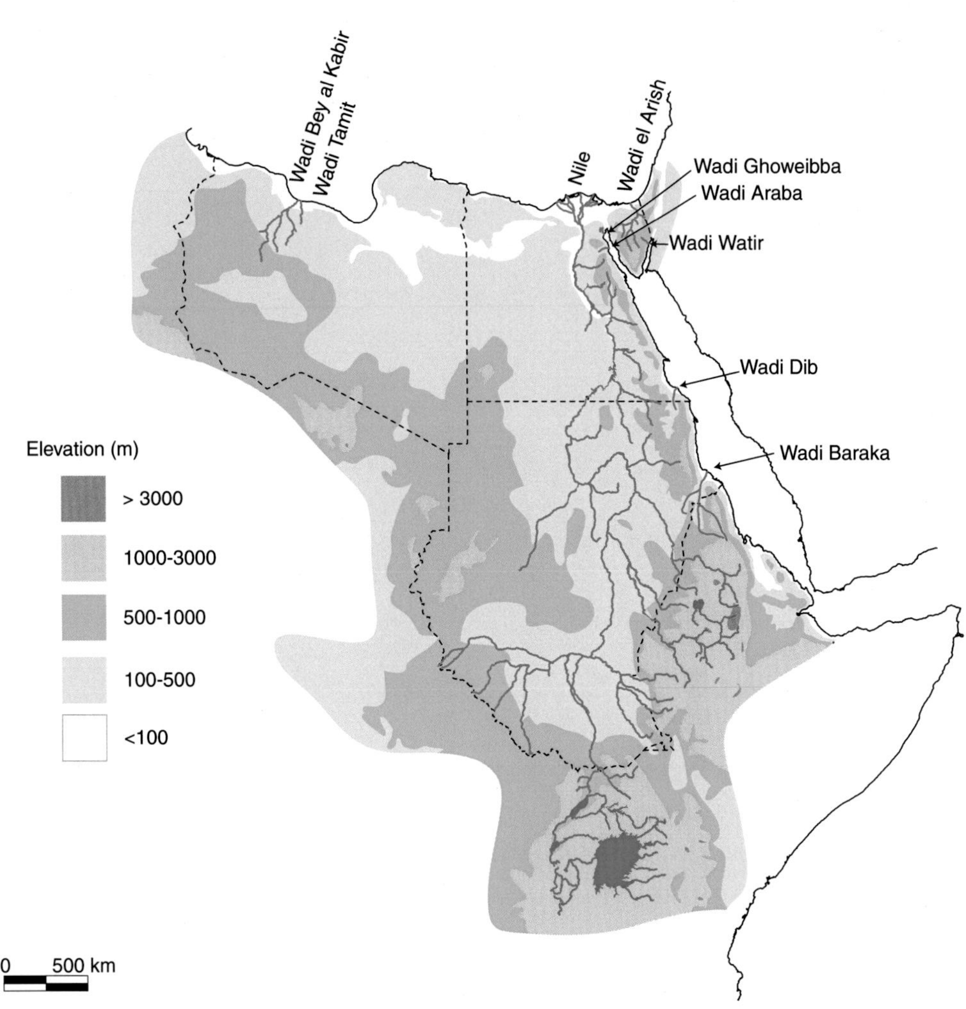

Figure D2

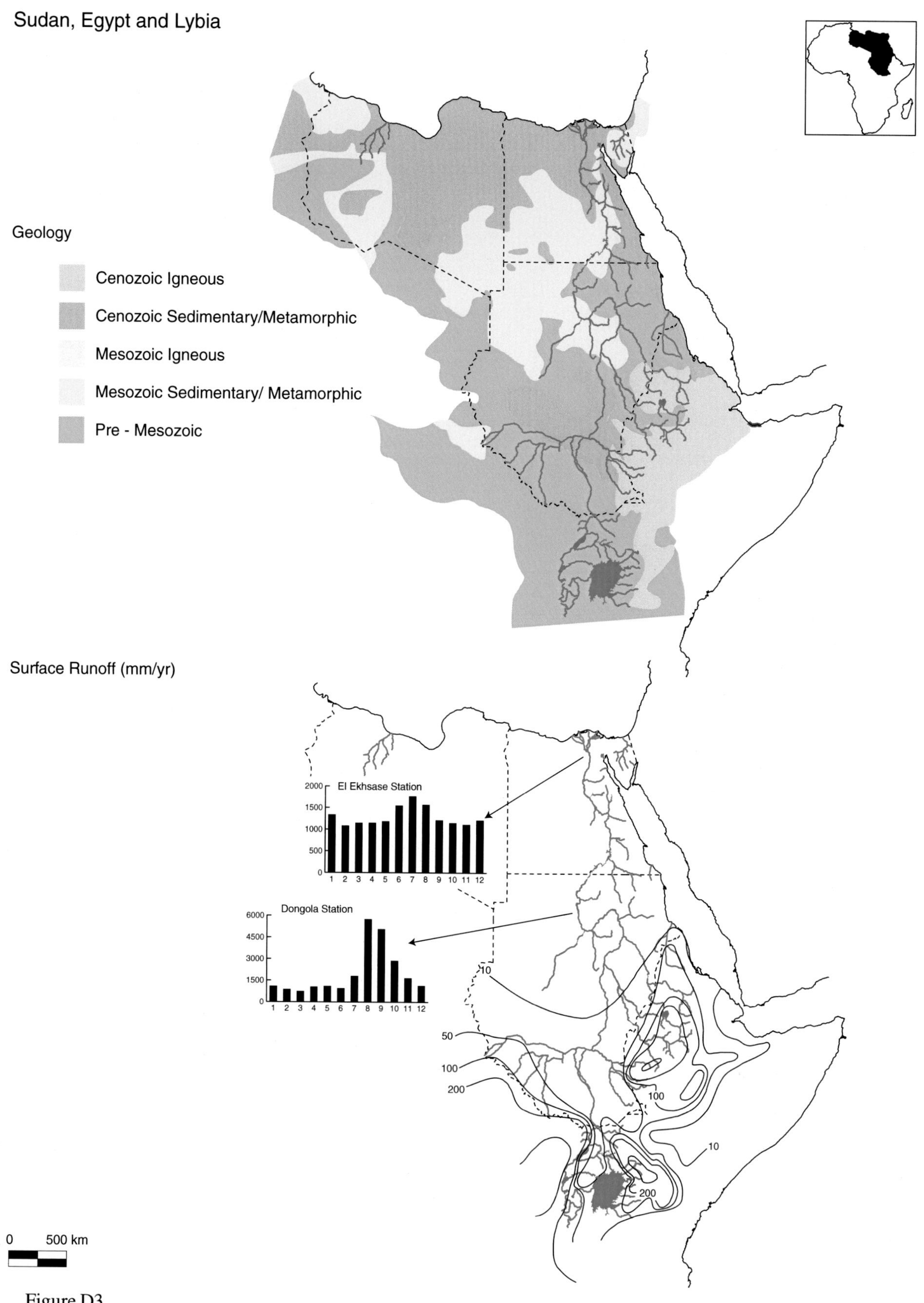
Sudan, Egypt and Lybia
Geology
Cenozoic Igneous
Cenozoic Sedimentary/Metamorphic
Mesozoic Igneous
Mesozoic Sedimentary/ Metamorphic
Pre - Mesozoic
Surface Runoff (mm/yr)
El Ekhsase Station
2000
1500
1000
500
0
1 2 3 4 5 6 7 8 9 10 11 12
Dongola Station
6000
4500
3000
1500
0
1 2 3 4 5 6 7 8 9 10 11 12
10
50
100
200
100
10
200
0 500 km

Figure D3

Table D1.

River name	Ocean	Area (10^3 km^2)	Length (km)	Max_elev (m)	Climate	Geology	Q (km^3/yr)	TSS (Mt/yr)	TDS (Mt/yr)	Refs.
Sudan										
Wadi Baraka	Red Sea	9	250	2500	Tr-A-D	PreMes				
Egypt										
Nile	Med. (E)	2900	6700	3800	STr-A-W	PreMes-Cen I	30 (80)	0.2 (120)	1.2 (6.1)	2, 3, 4, 5, 7, 8, 9
Wadi Araba	Red Sea	3.9	140	>1000	STr-A-D	PreMes				
Wadi Dib	Red Sea	3.9	160	>500	STr-A-D	PreMes				
Wadi el Arish	Med. (E)	19	140	1000	STr-A-D	Mes S/M				1
Wadi Ghoweibba	Red Sea	3.3	110	1200	STr-A-D	PreMes				
Wadi Watir	Red Sea	3.9	90	1500	STr-A-D	PreMes				6
Lybia										
Wadi Bey al Kabir	Med. (E)	36	390	>500	STr-A-D	Cen S/M				
Wadi Tamit	Med. (E)	18	220	>500	STr-A-D	Cen S/M				

References:
1. El-Etr *et al.*, 1999; 2. GEMS website, www.gemstat.org; 3. Global River Discharge Database, http://www.rivdis.sr.unh.edu/; 4. Meybeck and Ragu, 1996; 5. Probst, 1992; 6. Schick and Lekach, 1987; 7. Sestini, 1991; 8. Shahin, 2002; 9. UNEP/MAP/MED_POL, 2003

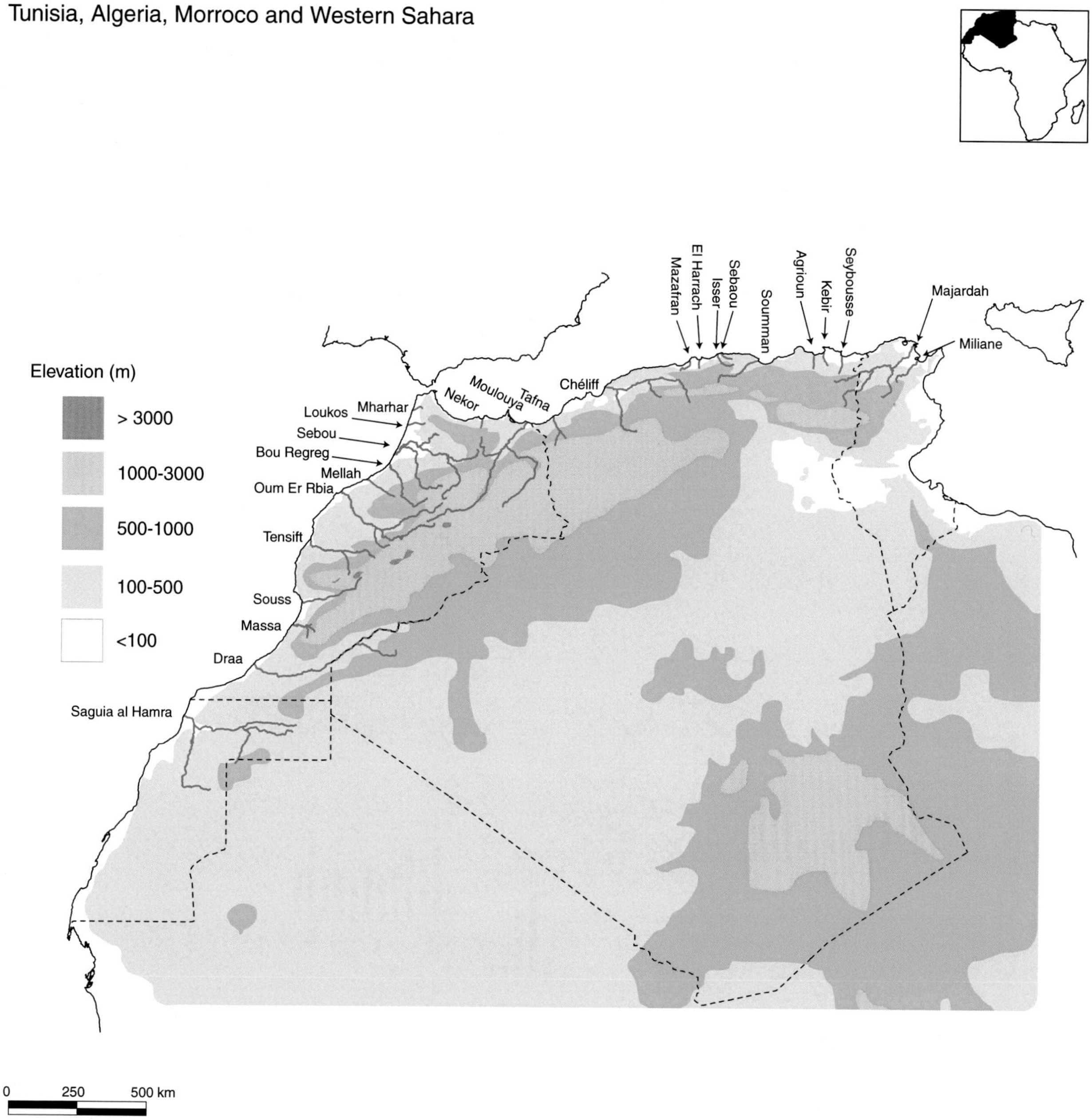

Tunisia, Algeria, Morroco and Western Sahara

Elevation (m)

> 3000

1000-3000

500-1000

100-500

<100

Mazafran
El Harrach
Isser
Sebaou
Soumman
Agrioun
Kebir
Seybousse
Majardah
Miliane
Chéliff
Tafna
Moulouya
Nekor
Mharhar
Loukos
Sebou
Bou Regreg
Mellah
Oum Er Rbia
Tensift
Souss
Massa
Draa
Saguia al Hamra

0 250 500 km

Figure D4

Tunisia, Algeria, Morroco, and Western Sahara

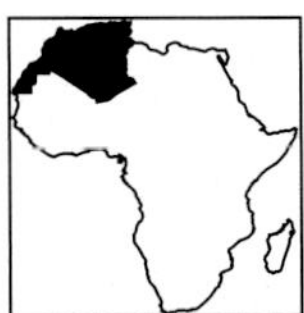

Geology

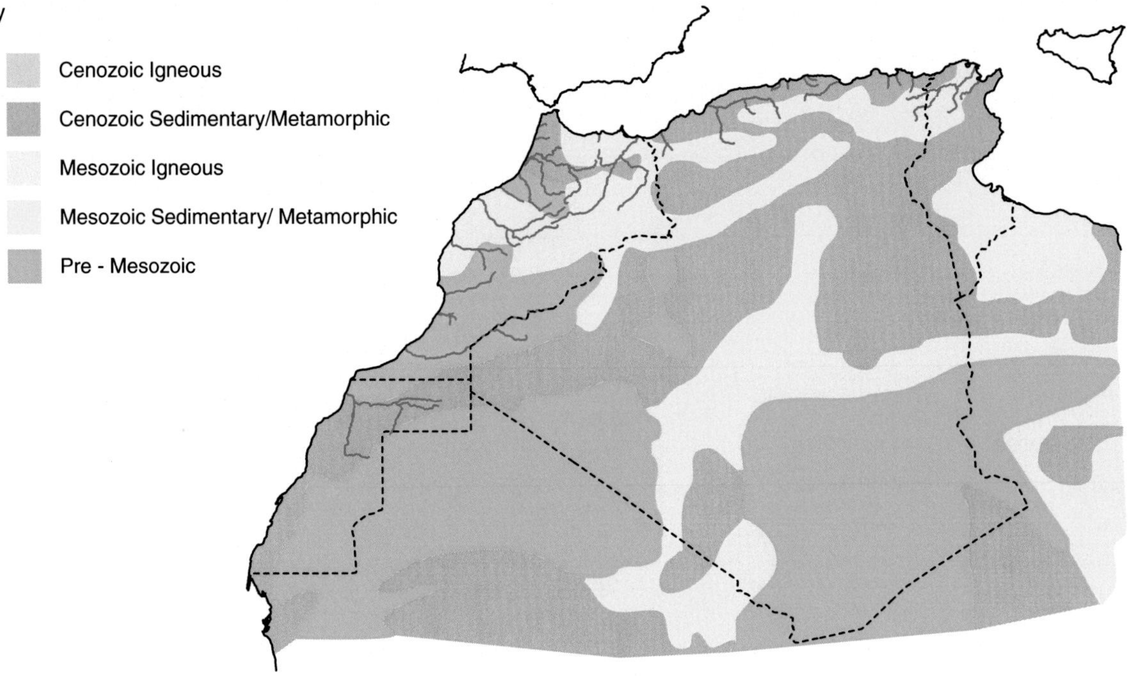

- Cenozoic Igneous
- Cenozoic Sedimentary/Metamorphic
- Mesozoic Igneous
- Mesozoic Sedimentary/ Metamorphic
- Pre - Mesozoic

Surface Runoff (mm/yr)

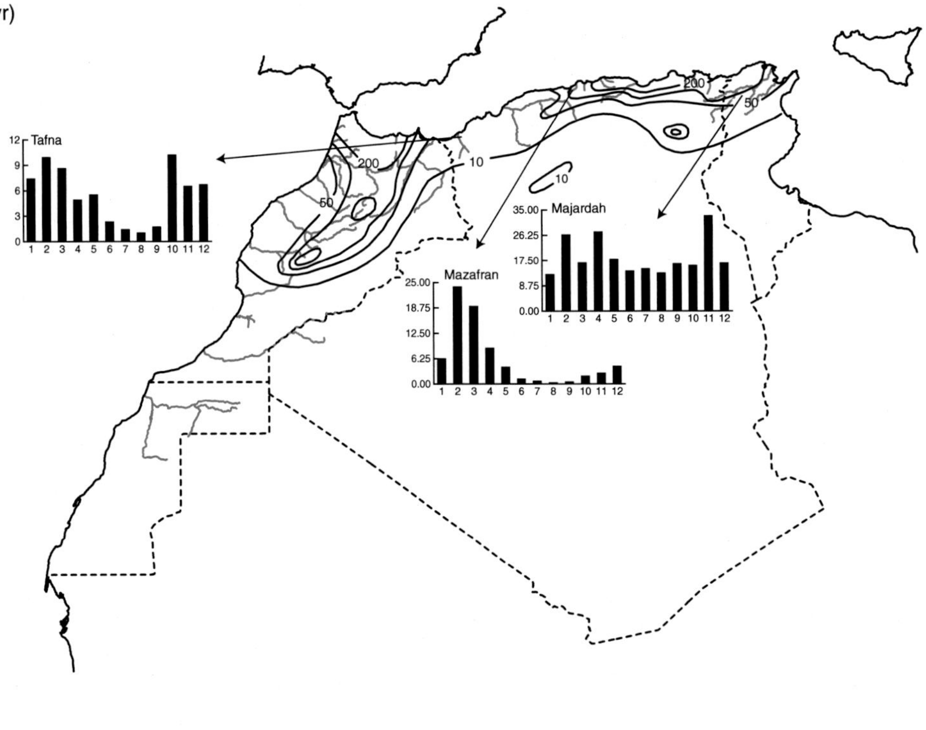

Tafna
12
9
6
3
0
1 2 3 4 5 6 7 8 9 10 11 12

200
50
10
10
200
50

Majardah
35.00
26.25
17.50
8.75
0.00
1 2 3 4 5 6 7 8 9 10 11 12

Mazafran
25.00
18.75
12.50
6.25
0.00
1 2 3 4 5 6 7 8 9 10 11 12

0 250 500 km

Figure D5

Table D2.

River name	Ocean	Area (10^3 km^2)	Length (km)	Max_elev (m)	Climate	Geology	Q (km^3/yr)	TSS (Mt/yr)	TDS (Mt/yr)	Refs.
Algeria										
Agrioun	Med. (W)	0.66	70	>1000	STr-H-W	Mes S/M	0.17	4.8		6
Chéliff	Med. (W)	44	720	1900	STr-A-W	Mes S/M	1.3	4		6, 7, 19
El Harrach	Med. (W)	0.39	60	>1000	STr-H-W	Mes S/M	0.13	0.63		18
Isser	Med. (W)	4.2	90	>500	STr-A-W	Mes S/M	0.36	8.3		3, 6
Kebir	Med. (W)	1.1	80	>1000	STr-SA-W	Mes S/M	0.23	0.22		6
Mazafran	Med. (W)	1.9	70	>500	STr-H-W	Mes S/M	0.44	3		6, 17
Sebaou	Med. (W)	2.5	80	>500	STr-H-W	Mes S/M	0.51	1.2		3, 6
Seybousse	Med. (W)	5.5	200	>500	STr-A-W	Mes S/M	0.43	1.2		6, 9
Soumman	Med. (W)	8.5	200	>500	STr-A-W	Mes S/M	0.79	4.1		6, 7, 9
Tafna	Med. (W)	8.8	80	>500	STr-A-W	Mes S/M	0.28	1		5, 9
Morroco										
Bou Regreg	Atlantic (NE)	9.8	180	>500	STr-A-W	PreMes	0.56	4.7		5, 9
Draa	Atlantic (NE)	114	1100	>1500	STr-A-W	PreMes	0.4 (0.8)	14		2, 5, 9
Loukos	Atlantic (NE)	1.8	190	>500	STr-H-W	Cen S/M	0.9	1.8		5, 9
Massa	Atlantic (NE)	3.8	90	>200	STr-A-W	PreMes	0.16	1.6		5
Mellah	Atlantic (NE)	1.8	80	>500	STr-A-W	PreMes	0.16	1		5
Mharhar	Atlantic (NE)	0.18	20	>500	STr-H-W	Cen S/M	0.06	0.21		5
Moulouya	Med. (W)	51	450	>1500	STr-A-W	Mes S/M	0.2 (1.3)	0.8 (12)		2, 4, 13, 16
Nekor	Med. (W)	0.79	40	>100	STr-H-W	Mes S/M	0.9	2.8		1
Oum Er Rbia	Atlantic (NE)	30	560	>1500	STr-SA-W	Mes S/M	3.3	6.6		4, 9
Sebou	Atlantic (NE)	37	500	>1000	STr-SA-W	Cen S/M	1.4 (4.4)	2 (37)	3.1	7, 10, 11, 12, 13, 14
Souss	Atlantic (NE)	16	230	2700	STr-A-W	Mes S/M	0.31	4.2	0.03	8, 12
Tensift	Atlantic (NE)	20	240	3600	STr-A-W	Mes S/M	0.91			
Tunisia										
Majardah	Med. (W)	22	370	1700	STr-A-W	Mes S/M	0.94	9.4		7, 9, 15
Miliane	Med. (W)	2	130	>200	STr-A-D	Mes S/M	0.02	0.9		9
Western Sahara										
Saguia al Hamra	Atlantic (NE)	68	210	>500	STr-A-D	Cen S/M				

References:

1. Boufous, 1982; 2. Combe, M., 1968, c.f. van der Leeden (1975); 2. FAO, 1997; 3. Global River Discharge Database, http://www.rivdis.sr.unh.edu/; 4. Heusch and Milles-Lacrois, 1971; 5. Lahlou, 1982; 6. Licitri and Normand, 1969; 7. Meybeck and Ragu, 1996; 8. Olivera F, Website, 6/7/99 https://ceprofs.civil.tamu.edu/folivera/UTexas/morocco/report.htm; 9. Rand McNally, 1980; 10. Shahin, 2002; 11. Shahin, 2007; 12. Snoussi *et al.*, 1990; 13. Snoussi *et al.*, 2002; 14. Snoussi, 1988; 15. Tiveront, 1960; 16. UNEP/MAP/MED_POL, 2003; 17. Vörösmarty *et al.*, 1996d; 18. Walling, 1985; 19. D. E. Walling, personal communication

Senegal, Gambia, Guinea Bissau,
Guinea, Sierra Leone, and Liberia

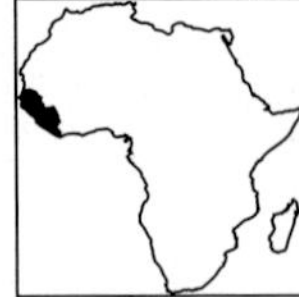

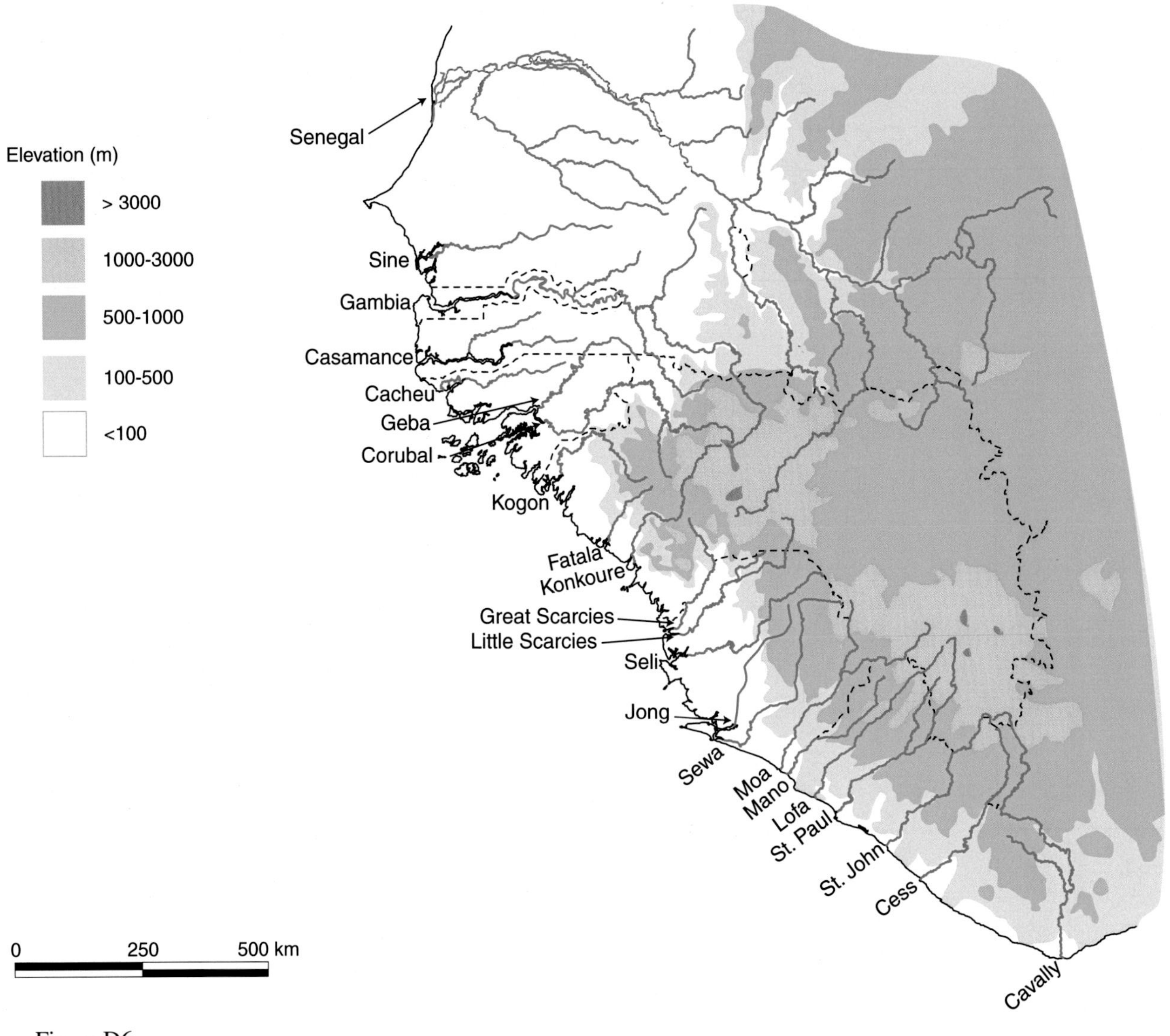

Elevation (m)

- > 3000
- 1000-3000
- 500-1000
- 100-500
- <100

Senegal
Sine
Gambia
Casamance
Cacheu
Geba
Corubal
Kogon
Fatala
Konkoure
Great Scarcies
Little Scarcies
Seli
Jong
Sewa
Moa
Mano
Lofa
St. Paul
St. John
Cess
Cavally

0 250 500 km

Figure D6

Senegal, Gambia, Guinea Bissau, Guinea, Sierra Leone, and Liberia

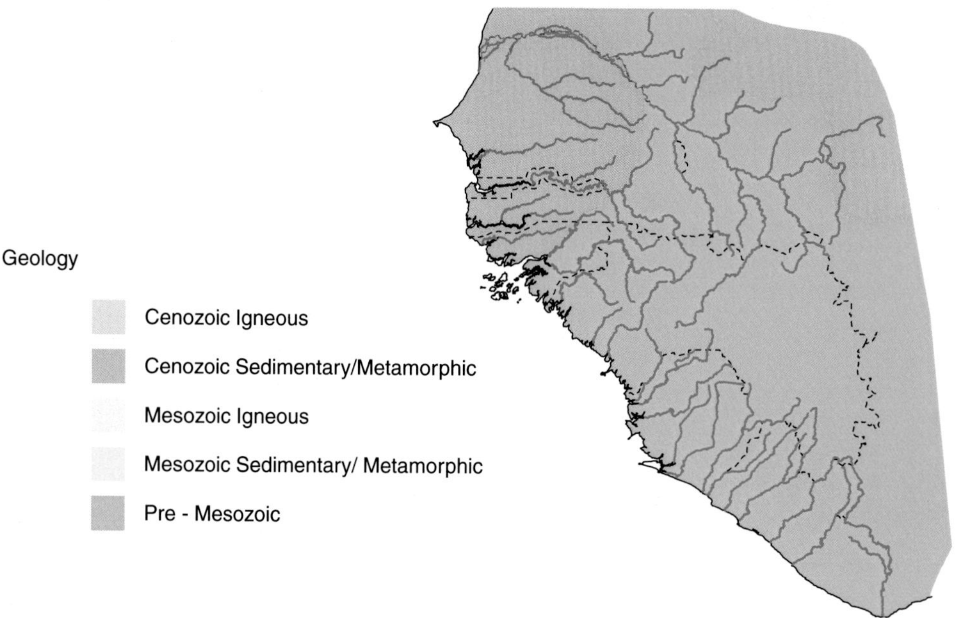

Surface Runoff (mm/yr)

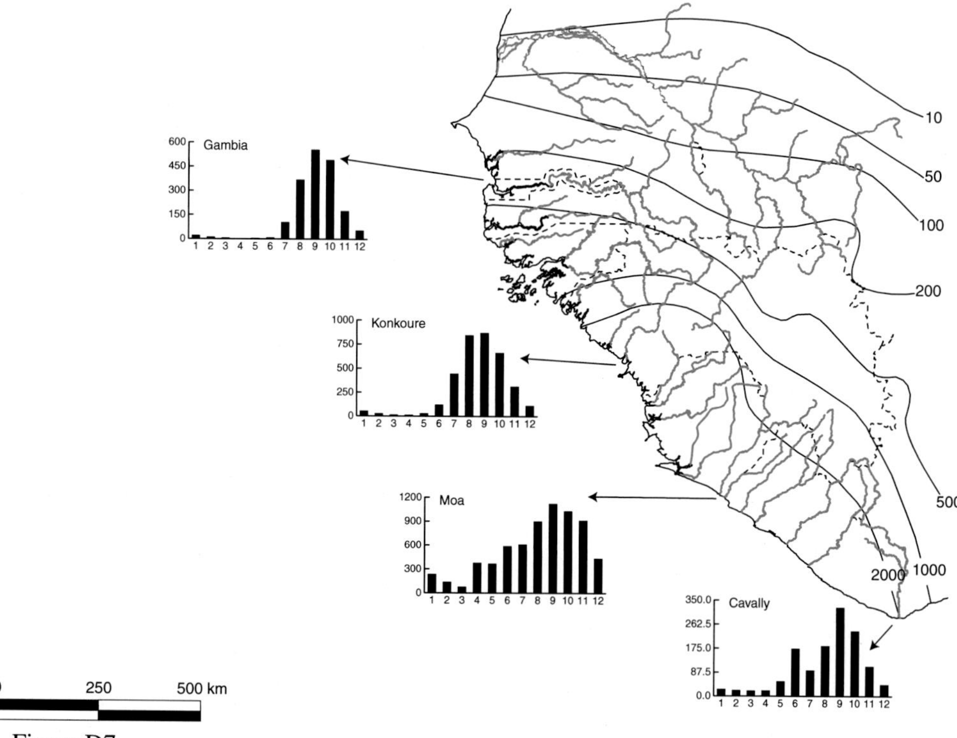

Figure D7

Table D3.

River name	Ocean	Area (10^3 km^2)	Length (km)	Max_elev (m)	Climate	Geology	Q (km^3/yr)	TSS (Mt/yr)	TDS (Mt/yr)	Refs.
Senegal										
Casamance	Atlantic (E)	37	320	57	Tr-A-S	Cen S/M	0.1			4, 12, 16
Senegal	Atlantic (E)	270	1600	1200	Tr-A-S	PreMes	22	3	0.6	2, 3, 10, 11, 13, 14, 16
Sine	Atlantic (E)		270	<100	Tr-A-S	Pre-Cen S/M				
Gambia										
Gambia	Atlantic (E)	77	1100	1100	Tr-A-S	Cen S/M	4.9	0.2	0.22	11, 14, 16
Guinea Bissau										
Cacheu	Atlantic (E)		190	40	Tr-H-S	Cen S/M				16
Corubal	Atlantic (E)	23	600	650	Tr-H-S	PreMes	13			8, 16
Geba	Atlantic (E)	14	340	<100	Te-H-W	Cen S/M				8
Guinea										
Fatala	Atlantic (E)		150	<100	Tr-W-S	PreMes				
Kogon	Atlantic (E)		210	<100	Tr-W-S	PreMes				
Konkoure	Atlantic (E)	16	360	900	Tr-H-S	PreMes	7 (21)	0.38	0.13	9, 11, 12, 15
Sierra Leone										
Great Scarcies	Atlantic (E)	8.5	260	>200	Tr-W-S	PreMes	1.8			1, 12
Jong	Atlantic (E)	4.5		>500		PreMes				
Little Scarcies	Atlantic (E)	15	280	>1000	Tr-W-S	PreMes				1, 16
Mano	Atlantic (E)	22	360	>200	Tr-H-S	PreMes	3.1			8, 15
Moa	Atlantic (E)	20	420	>500	Tr-W-S	PreMes	17.7			8, 17
Seli	Atlantic (E)	4	400	>500	Tr-W-S	PreMes	3.9			6, 15
Sewa	Atlantic (E)	19	380	490	Tr-W-S	PreMes	3.6			8, 15, 16
Liberia										
Cavally	Atlantic (E)	28	700	>200	Tr-W-S	PreMes	13	5.3	1	7, 8, 11
Cess	Atlantic (E)	11.5	380	>500	Tr-W-S	PreMes				8, 12, 16
Lofa	Atlantic (E)	9	350	900	Tr-W-S	PreMes				8, 16
St. John	Atlantic (E)	14	360	800	Tr-W-S	PreMes	4.5			4, 12, 16
St. Paul	Atlantic (E)	18	430	900	Tr-W-S	PreMes	12			5, 8, 15, 16

References:
1. Center for Natural Resources Energy and Transport (UN), 1978; 2. Gac and Kane, 1986; 3. GEMS website, www.gemstat.org; 4. Global River Discharge Database, http://www.rivdis.sr.unh.edu/; 5. GRDC website, www.gewex.org/grdc; 6. Gresswell and Huxley, 1966; 7. Guinea Current Large Marine Ecosystem, 2006; 8. Hugueny and Lévêque, 1994; 9. IAHS/UNESCO, 1974; 10. Martins and Probst, 1991; 11. Meybeck and Ragu, 1996; 12. Rand McNally, 1980; 13. Shahin, 2002; 14. UNESCO (WORRI), 1978; 15. UNESCO, 1995; 16. Vanden Bossche and Bernacsek, 1991; 17. Vörösmarty *et al.*, 1996d

Ivory Coast, Ghana, Togo and Benin

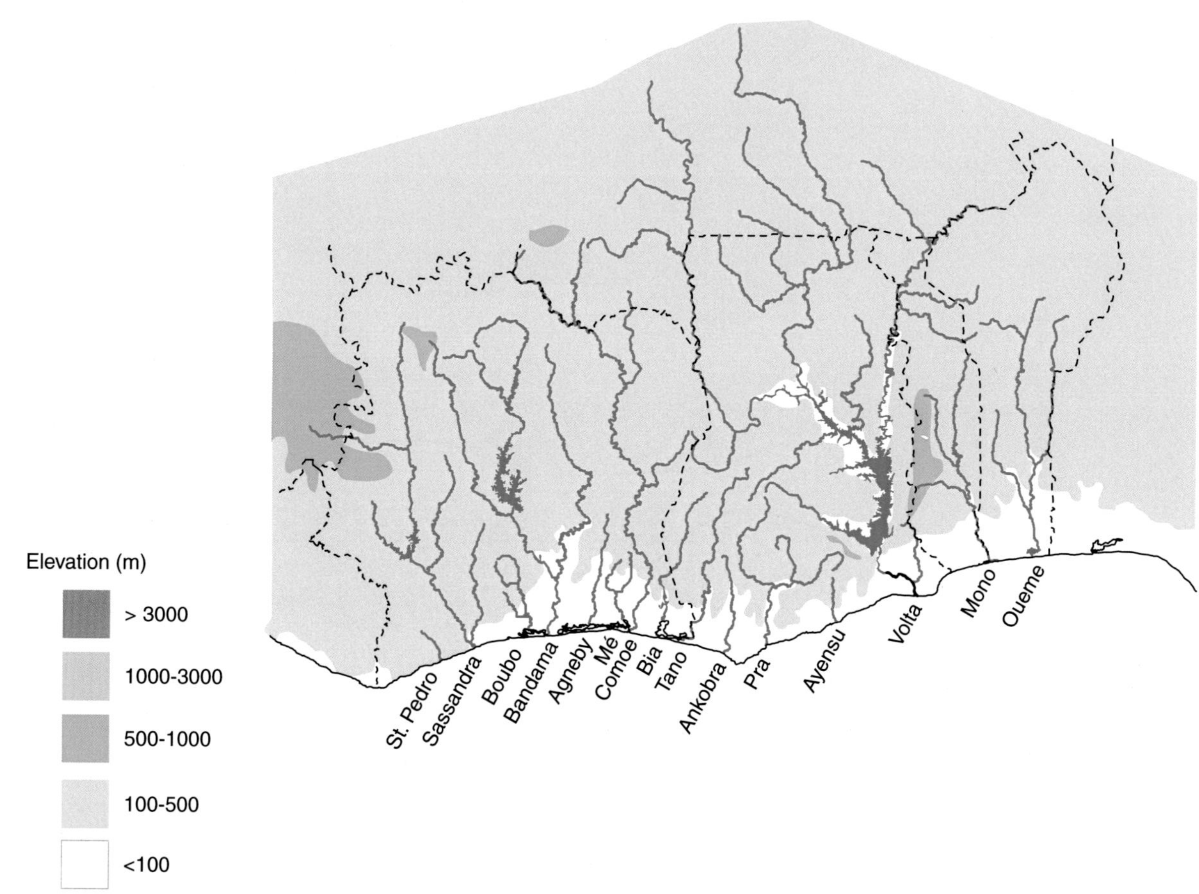

Figure D8

Ivory Coast, Ghana, Togo and Benin

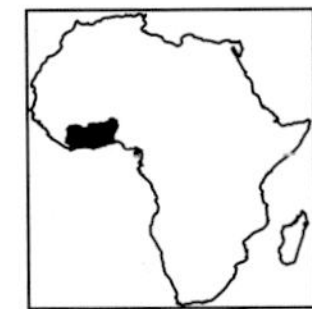

Geology

- Cenozoic Igneous
- Cenozoic Sedimentary/Metamorphic
- Mesozoic Igneous
- Mesozoic Sedimentary/ Metamorphic
- Pre - Mesozoic

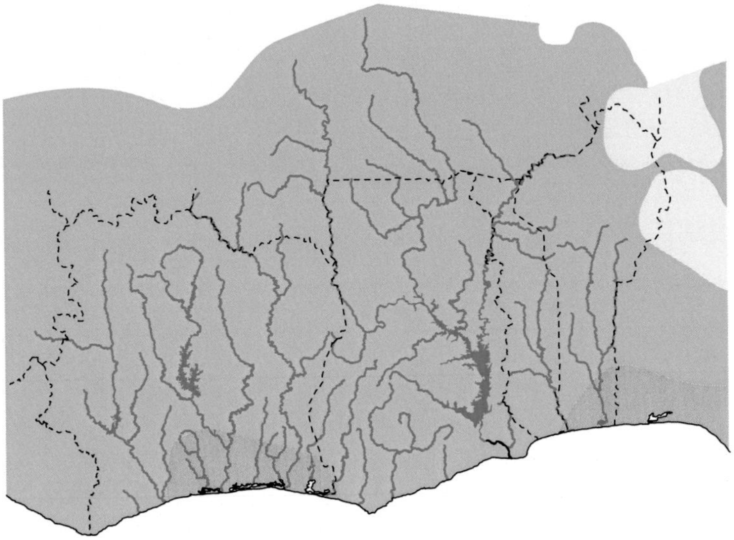

Surface Runoff (mm/yr)

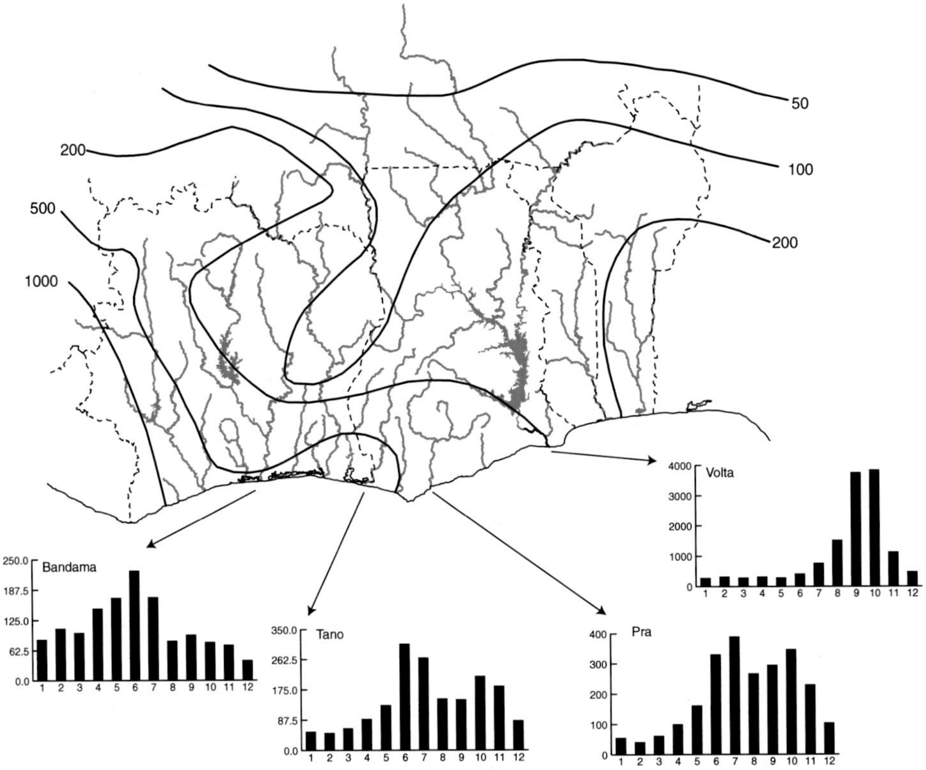

50
100
200
500
1000
200

Volta

Bandama

Tano

Pra

0 250 500 km

Figure D9

Table D4.

River name	Ocean	Area (10^3 km^2)	Length (km)	Max_elev (m)	Climate	Geology	Q (km^3/yr)	TSS (Mt/yr)	TDS (Mt/yr)	Refs.
Ivory Coast										
Agneby	G. of Guinea	8.9	210	<100	Tr-SA-S	PreMes	1.5	1.5		7, 9
Bandama	G. of Guinea	97	1050	1000	Tr-SA-S	PreMes	9.7	1.2 (7.2)	0.79	5, 7, 8, 9
Bia	G. of Guinea	9.5	300	310	Tr-SA-S	PreMes	2.4			3, 7, 16
Boubo	G. of Guinea	51	130	>100	Tr-H-S	PreMes				7
Comoe	G. of Guinea	78	1200	420	Tr-SA-S	PreMes	7.9	9	0.17	2, 5, 7, 8, 15, 18
Mé	G. of Guinea	4.3	140	>100	Tr-H-S	PreMes	1.5			7
St. Pedro	G. of Guinea	3.3	80	<100	Tr-W-S	PreMes		0.07		8, 9, 15
Sassandra	G. of Guinea	79	840	>500	Tr-SA-S	PreMes	10 (17)	2.9	1.3	17
Tano	G. of Guinea	16	620	430	Tr-H-S	PreMes	4.8	0.35		1, 4, 12, 16
Ghana										
Ankobra	G. of Guinea	6.2	190	330	Tr-SA-S	Cen S/M	0.8	1.8		2, 10, 13
Ayensu	G. of Guinea	1.7	100	>200	Tr-A-S	PreMes	0.07	0.15		2
Pra	G. of Guinea	38	450	550	Tr-SA-S	PreMes	6.2	2.4		5, 14, 16
Volta	G. of Guinea	400	1270	330	Tr-A-W	PreMes	40	1.6 (19)	2.1	1, 8, 11
Togo										
Mono	G. of Guinea	29	500	420	Tr-SA-W	PreMes	4.9	1.6		5, 14, 15, 16
Benin										
Oueme	G. of Guinea	50	700	600	Tr-SA-W	PreMes	5.7	2.4		5, 6, 9, 16

References:
1. Akrasi and Ayibotele, 1984; 2. Ayibotele and Tuffour-Darko, 1979; 3. Center for Natural Resources Energy and Transport (UN), 1978; 4. GEMS website, www.gemstat.org; 5. Guinea Current Large Marine Ecosystem, 2006; 6. Hugueny and Lévêque, 1994; 7. LeLoeuff *et al.*, 1993; 8. Meybeck and Ragu, 1996; 9. ORSTOM, 1969 (cf. van der Leeden, 1975); 10. Rand McNally, 1980; 11. Shahin, 2002; 12. UNESCO, 1971; 13. UNESCO, 1995; 14. UNESCO/UNEP, 1982; 15. UNESCO (WORRI), 1978; 16. Vanden Bossche and Bernacsek, 1991; 17. D. E. Walling, personal communication; 18. Welcomme, 1972 (cf. van der Leeden, 1975)

Nigeria, Cameroon, Equatorial Guinea, Gabon and Republic of the Congo

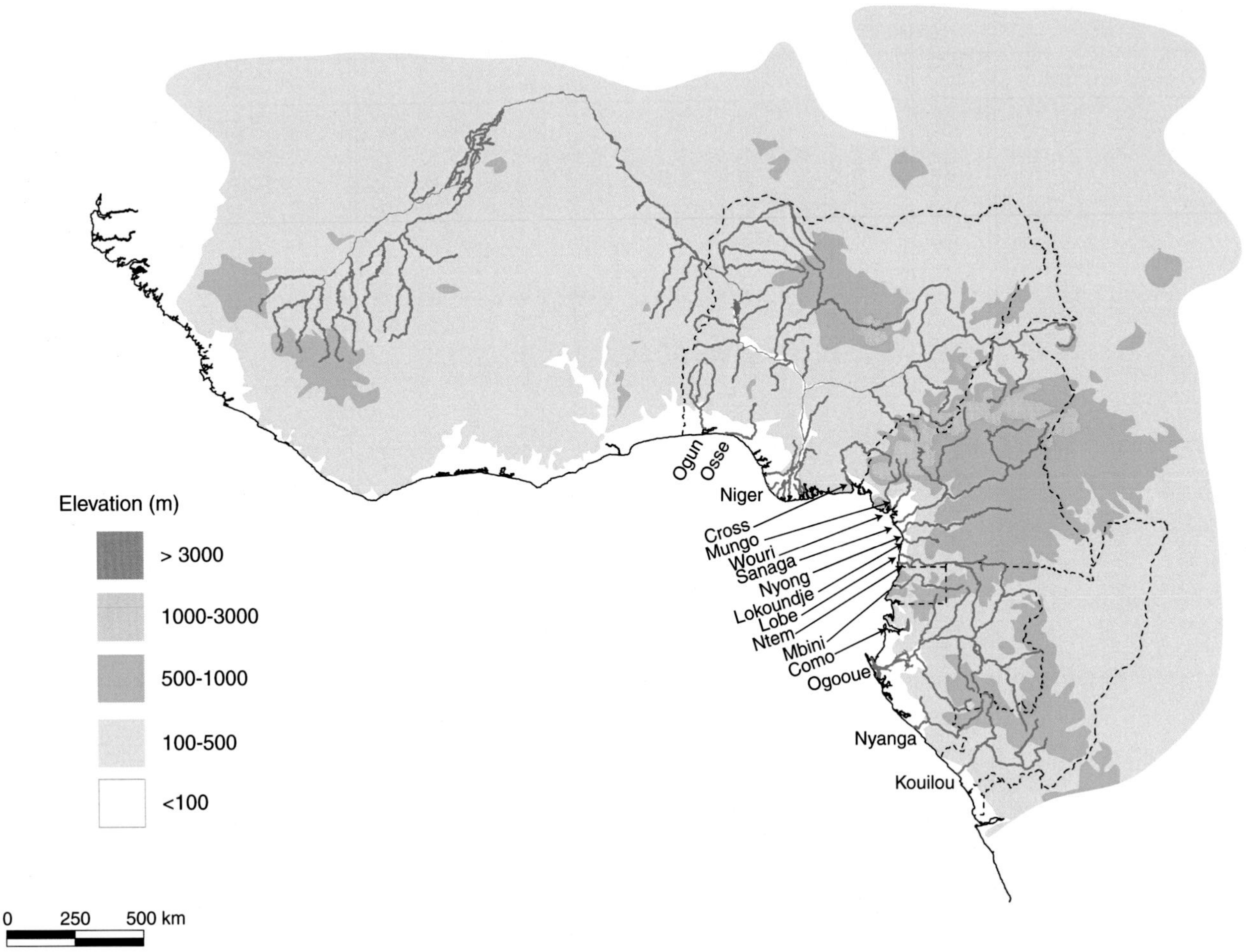

Ogun
Osse
Niger
Cross
Mungo
Wouri
Sanaga
Nyong
Lokoundje
Lobe
Ntem
Mbini
Como
Ogooue
Nyanga
Kouilou

Elevation (m)

> 3000

1000-3000

500-1000

100-500

<100

0 250 500 km

Figure D10

Nigeria, Cameroon, Equatorial Guinea, Gabon and Republic of the Congo

Geology

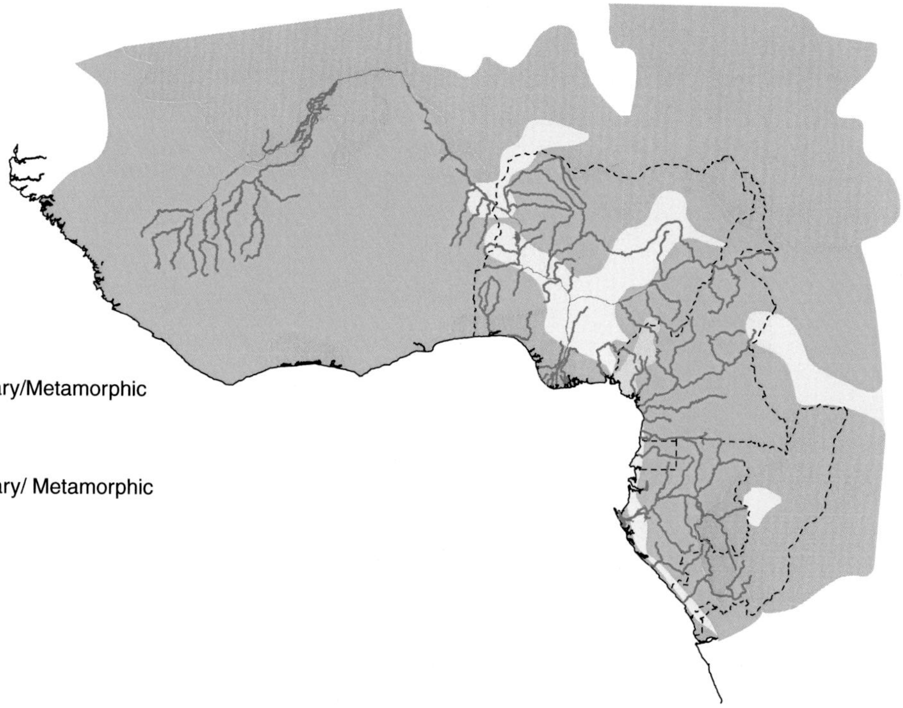

- Cenozoic Igneous
- Cenozoic Sedimentary/Metamorphic
- Mesozoic Igneous
- Mesozoic Sedimentary/ Metamorphic
- Pre - Mesozoic

Surface Runoff (mm/yr)

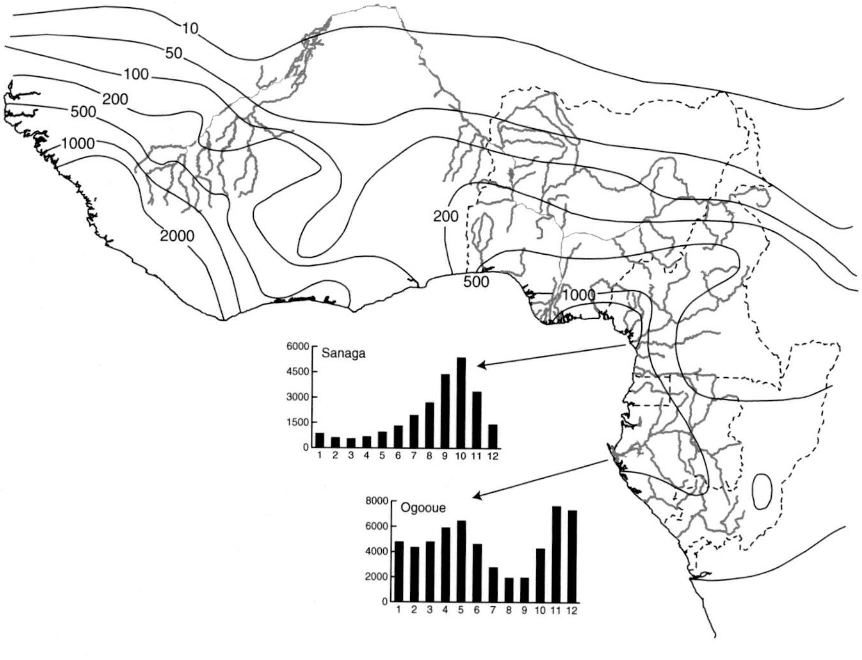

10
50
100
200
500
1000
2000
200
500
1000

Sanaga
6000
4500
3000
1500
0
1 2 3 4 5 6 7 8 9 10 11 12

Ogooue
8000
6000
4000
2000
0
1 2 3 4 5 6 7 8 9 10 11 12

0 250 500 km

Figure D11

Table D5.

River name	Ocean	Area (10^3 km^2)	Length (km)	Max_elev (m)	Climate	Geology	Q (km^3/yr)	TSS (Mt/yr)	TDS (Mt/yr)	Refs.
Nigeria										
Cross	G. of Guinea	60	480	2000	Tr-SA-W	PreMes	10 (52)	7.5		5, 10, 17
Niger	G. of Guinea	2200	4000	820	Tr-A-W	PreMes	160 (190)	40	11	3, 9, 12, 13, 14, 16, 20
Ogun	G. of Guinea	47	320	>200	Tr-H-W	PreMes		1.1		9, 17
Osse	G. of Guinea		200	>200	Tr-H-W	PreMes				
Cameroon										
Lobe	G. of Guinea	1.9	100	>500	Tr-W-S	PreMes	3.4			8, 9, 15, 21
Lokoundje	G. of Guinea	1.2	160	>500	Tr-W-S	PreMes	1.3			1, 15
Mungo	G. of Guinea	4.5	200	>500	Tr-W-S	Cen I	5.2			15
Ntem	G. of Guinea	33	390	>500	Tr-H-S	PreMes	8.7			1
Nyong	G. of Guinea	28	640	>500	Tr-H-S	PreMes	6.1	0.1		1, 15
Sanaga	G. of Guinea	130	860	>500	Tr-H-S	PreMes	65	6	2.6	1, 11, 13, 17, 19
Wouri	G. of Guinea	8.2	260	>500	Tr-W-S	Cen I	10			1, 6, 13, 15, 18
										1, 15
Equatorial Guinea										
Mbini	Atlantic (E)	16	400	>500	Tr-W-W	PreMes				2
Gabon										
Como	Atlantic (E)			>500	Tr-H-S	PreMes				
Nyanga	Atlantic (E)	26	390	>500	Tr-H-S	PreMes	16			13, 16
Ogooue	Atlantic (E)	200	920	880	Tr-H-S	PreMes-Cen S/M	150			1, 4, 12, 17, 20
Republic of the Congo										
Kouilou	Atlantic (E)	62	600	880	Tr-H-S	PreMes	35		0.63	7, 13

References:
1. Brummett and Teugels, 2004; 2. Center for Natural Resources Energy and Transport (UN), 1978; 3. Czaya, 1981; 4. Dai and Trenberth, 2002; 5. FAO, 1997; 6. Giresse *et al.*, 1998; 7. Global River Discharge Database, http://www.rivdis.sr.unh.edu/; 8. Guinea Current Large Marine Ecosystem, 2006; 9. Hugueny and Lévêque, 1994; 10. IAHS/UNESCO, 1974; 11. Lienou *et al.*, 2005; 12. Martins and Probst, 1991; 13. Meybeck and Ragu, 1996; 14. NEDECO, 1959; 15. ORSTOM, 1969 (cf. van der Leeden, 1975); 16. Probst, 1992; 17. Rand McNally, 1980; 18. Shahin, 2002; 19. UNESCO (WORRI), 1978; 20. van Blommestein/FAO, 1969 (cf. van der Leeden, 1975); 21. Vanden Bossche and Bernacsek, 1991

Democratic Republic of Congo and Angola

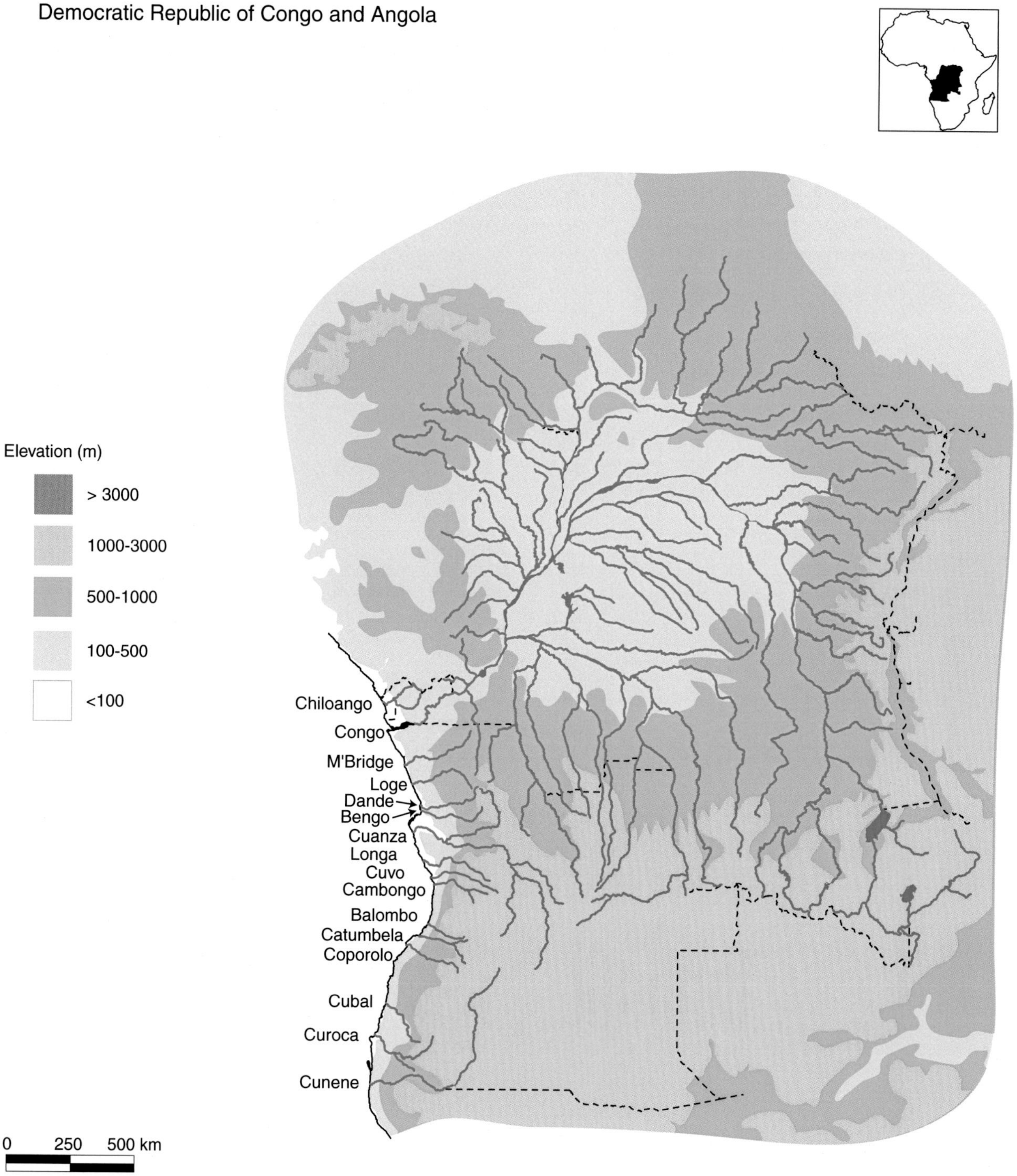

Elevation (m)

> 3000

1000-3000

500-1000

100-500

<100

Chiloango

Congo

M'Bridge

Loge

Dande

Bengo

Cuanza

Longa

Cuvo

Cambongo

Balombo

Catumbela

Coporolo

Cubal

Curoca

Cunene

0 250 500 km

Figure D12

Democratic Republic of Congo and Angola

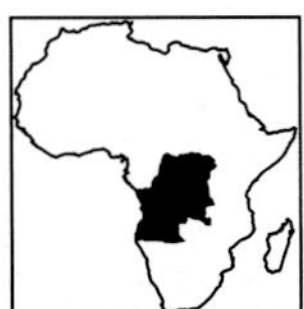

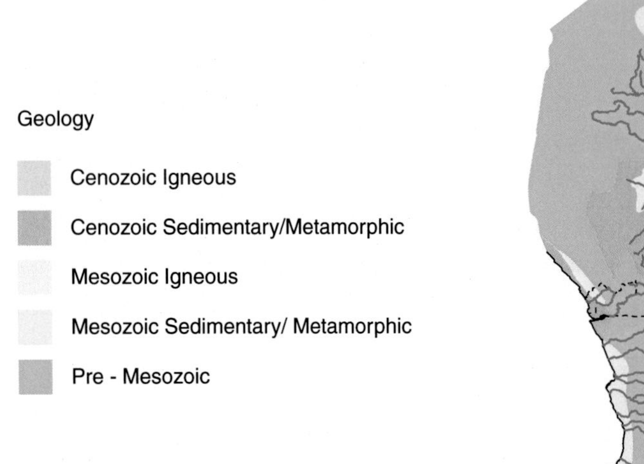

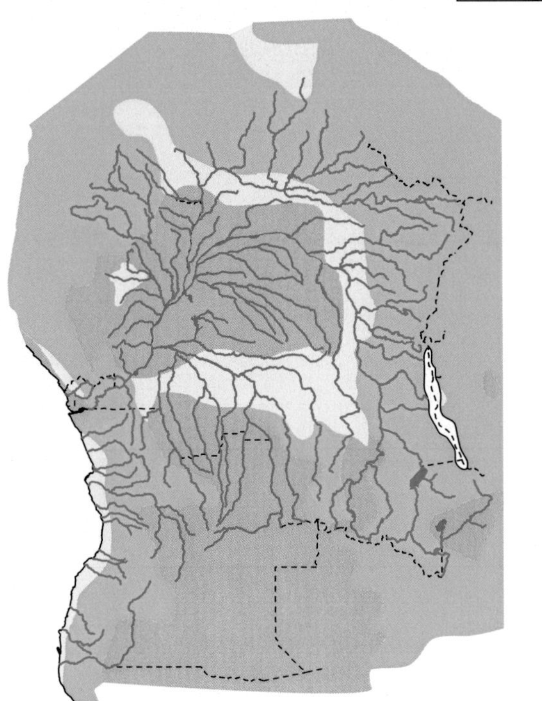

Surface Runoff (mm/yr)

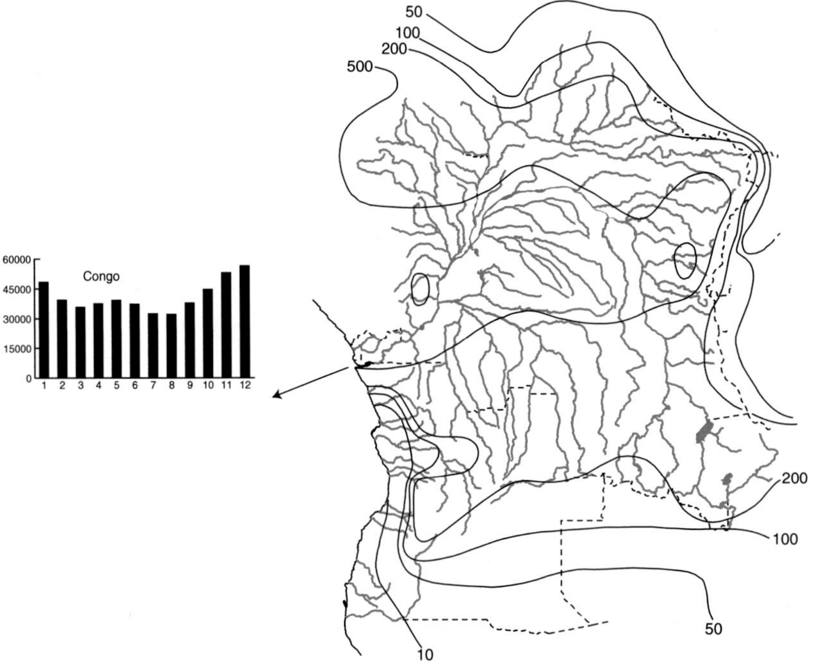

0 500 km

Figure D13

Table D6.

River name	Ocean	Area (10^3 km^2)	Length (km)	Max_elev (m)	Climate	Geology	Q (km^3/yr)	TSS (Mt/yr)	TDS (Mt/yr)	Refs.
Democratic Republic of Congo										
Congo	Atlantic (E)	3800	4700	1100	Tr-H-S	Cen S/M-Mes S/M	1300	43	56	2, 3, 4, 5, 8, 10, 12, 15
Angola										
Balombo	Atlantic (E)	8	200	>2000	STr-SA-S	PreMes				
Bengo	Atlantic (E)	9	300	1400	STr-SA-S	PreMes				7, 13, 14
Cambongo	Atlantic (E)	14	310	>1500	STr-SA-S	PreMes				
Catumbela	Atlantic (E)	9.2	220	1500	STr-A-S	PreMes				
Chiloango	Atlantic (E)	11	140	>500	STr-H-S	PreMes				1, 11
Coporolo	Atlantic (E)	15	220	1500	STr-SA-S	PreMes				
Cuanza	Atlantic (E)	120	960	1600	STr-SA-S	PreMes	26			7, 9, 11, 13, 14
Cubal	Atlantic (E)	7.7	200	>1000	STr-A-S	PreMes				
Cunene	Atlantic (E)	83	980	1800	STr-A-S	PreMes	6.8		0.35	6, 7, 9, 11, 13, 14
Curoca	Atlantic (E)	13	290	>1000	STr-SA-S	PreMes				
Cuvo	Atlantic (E)	10	300	>1000	STr-SA-S	PreMes				
Dande	Atlantic (E)	8.2	290	1300	STr-A-S	PreMes	1.5			14
Loge	Atlantic (E)	9.5	230	>1000	STr-SA-S	PreMes				
Longa	Atlantic (E)	18	310	1700	STr-SA-S	PreMes				
M'Bridge	Atlantic (E)	14	290	>1000	STr-H-S	PreMes				11

References:
1. Center for Natural Resources, Energy and Transport (UN), 1978; 2. Czaya, 1981; 3. Eisma *et al.*, 1978; 4. Esser and Kohlmaier, 1991; 5. GEMS website, www.gemstat.org; 6. Heyns, 2003; 7. IAHS/UNESCO, 1974; 8. Martins and Probst, 1991; 9. Meybeck and Ragu, 1996; 10. Probst, 1992; 11. Rand McNally, 1980; 12.; UNESCO (WORRI), 1978; 13. van der Leeden, 1975; 14. Vanden Bossche and Bernacsek, 1991; 15. Welcomme, 1972 (cf. van der Leeden, 1975)

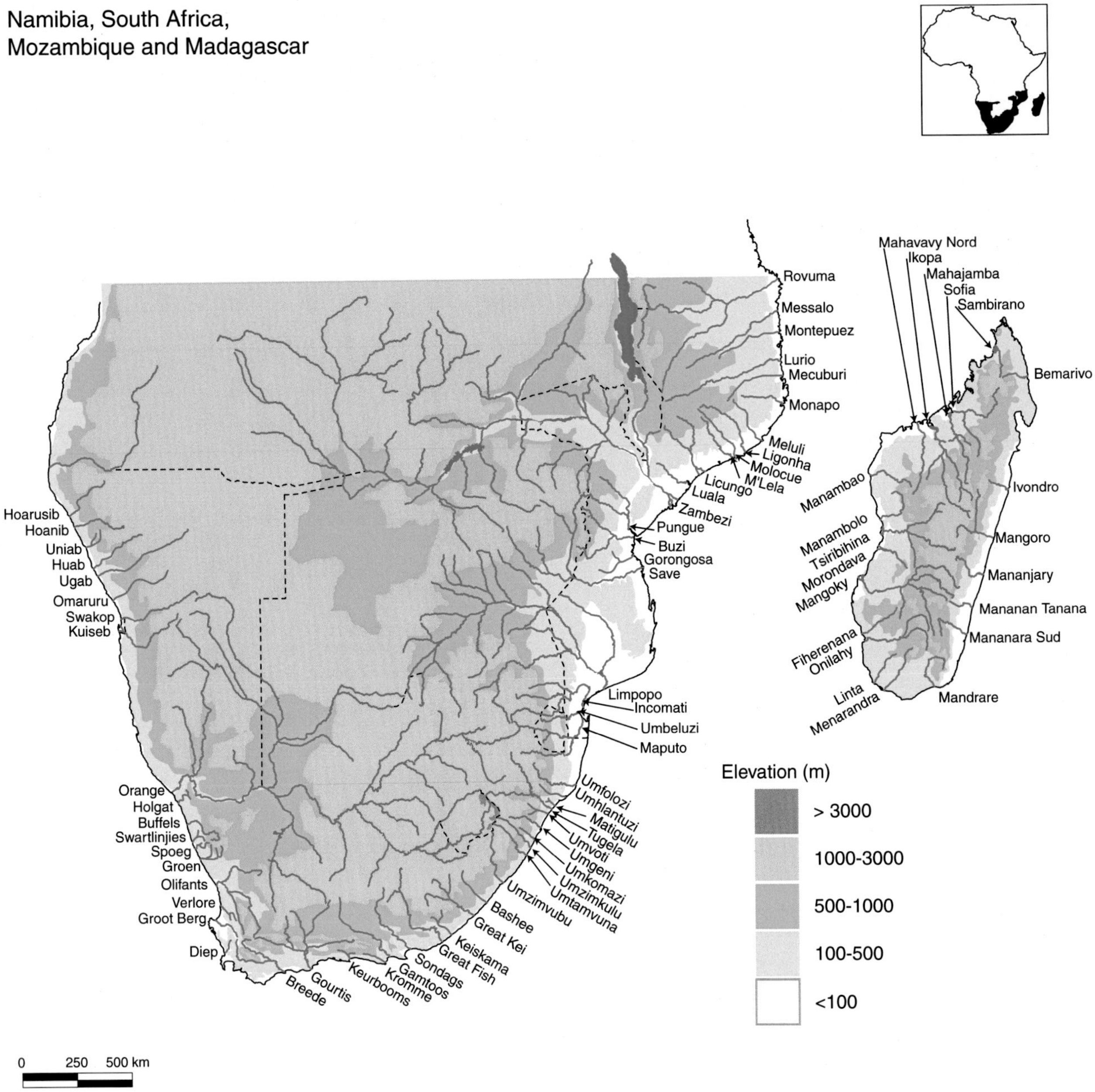

Namibia, South Africa,
Mozambique and Madagascar
Rovuma
Messalo
Montepuez
Lurio
Mecuburi
Monapo
Meluli
Ligonha
Molocue
M'Lela
Licungo
Luala
Zambezi
Pungue
Buzi
Gorongosa
Save
Limpopo
Incomati
Umbeluzi
Maputo
Hoarusib
Hoanib
Uniab
Huab
Ugab
Omaruru
Swakop
Kuiseb
Orange
Holgat
Buffels
Swartlinjies
Spoeg
Groen
Olifants
Verlore
Groot Berg
Diep
Breede
Gourtis
Keurbooms
Kromme
Gamtoos
Sondags
Great Fish
Keiskama
Great Kei
Bashee
Umzimvubu
Umtamvuna
Umzimkulu
Umkomazi
Umgeni
Umvoti
Tugela
Matigulu
Umhlantuzi
Umfolozi
Mahavavy Nord
Ikopa
Mahajamba
Sofia
Sambirano
Bemarivo
Ivondro
Mangoro
Mananjary
Mananan Tanana
Mananara Sud
Mandrare
Manambao
Manambolo
Tsiribihina
Morondava
Mangoky
Fiherenana
Onilahy
Linta
Menarandra
Elevation (m)
> 3000
1000-3000
500-1000
100-500
<100
0 250 500 km

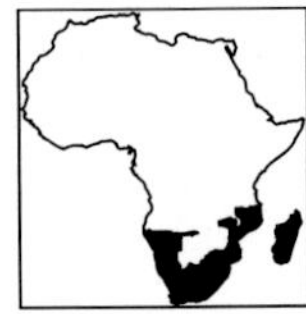

Figure D14

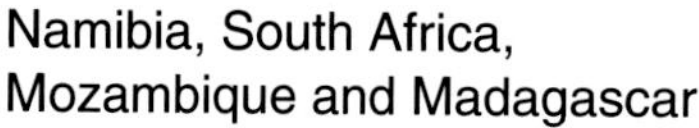

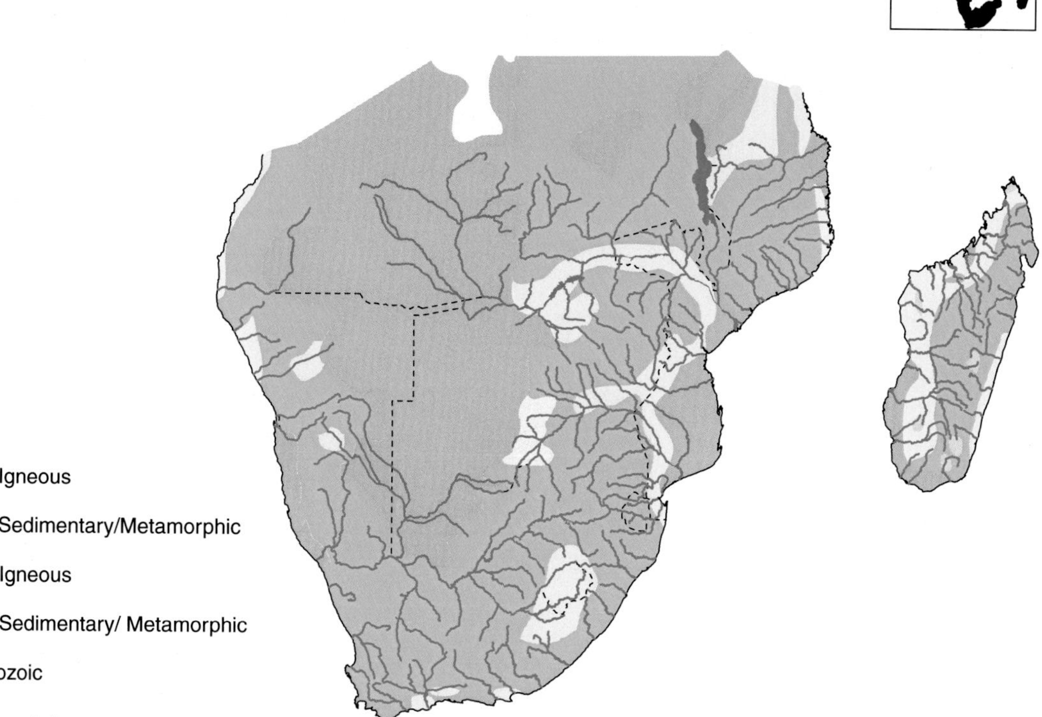

Geology

- Cenozoic Igneous
- Cenozoic Sedimentary/Metamorphic
- Mesozoic Igneous
- Mesozoic Sedimentary/ Metamorphic
- Pre - Mesozoic

Surface Runoff (mm/yr)

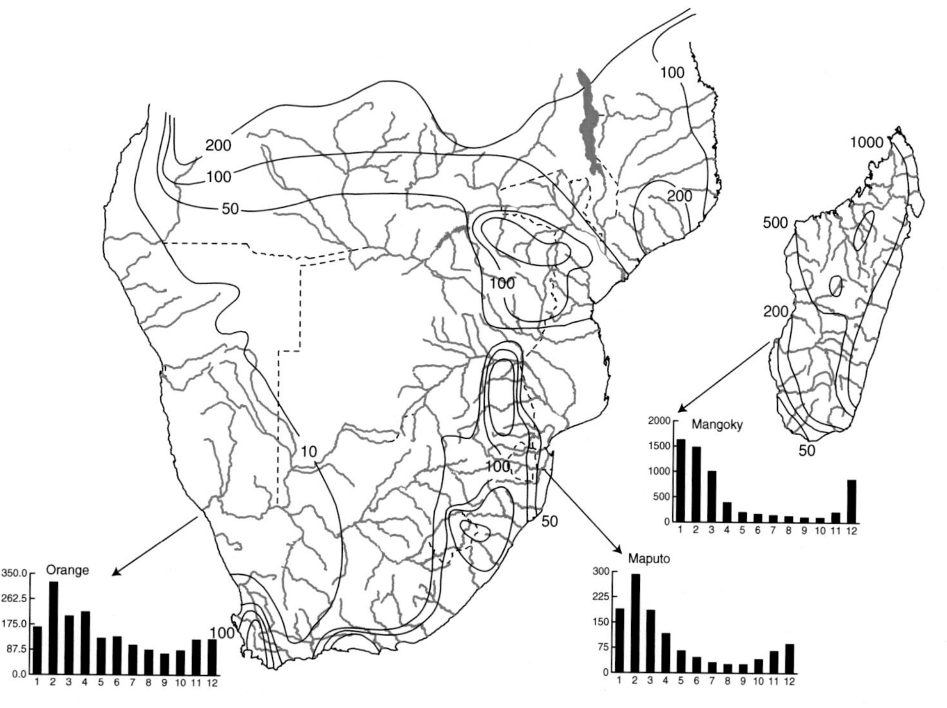

0 500 km

Figure D15

Table D7.

River name	Ocean	Area (10^3 km^2)	Length (km)	Max_elev (m)	Climate	Geology	Q (km^3/yr)	TSS (Mt/yr)	TDS (Mt/yr)	Refs.
South Africa										
Bashee	Indian (W)	6	180	2800	STr-SA-S	PreMes				
Breede	Indian (SW)	15	280	2300	STr-SA-S	PreMes	2			25
Buffels	Atlantic (SE)	9.7	260	1500	STr-A-D	PreMes	0.053			1
Diep	Atlantic (SE)	1.5	90	610	STr-H-S	PreMes	0.8			1
Gamtoos	Indian (SW)	34	640	2200	STr-A-S	PreMes	0.6			1, 25
Gourtis	Indian (SW)	45	310	1500	STr-A-D	PreMes	0.67			11, 25
Great Fish	Indian (SW)	30	640	2500	STr-A-D	PreMes	0.25			6, 11, 14, 25
Great Kei	Indian (SW)	20	220	2800	STr-A-S	PreMes	1.2			11, 14, 25
Groen	Atlantic (SE)	4.7	110	1200	STr-A-D	PreMes	0.034			1
Groot Berg	Atlantic (SE)	7.7	290	1500	Te-SA-S	PreMes	1			1
Holgat	Atlantic (SE)	1.6	100	1300	STr-A-D	PreMes	0.003			1
Keiskama	Indian (SW)	5.2	230	>1500	STr-SA-S	PreMes	0.2			
Keurbooms	Indian (SW)	1.1	85	1300	STr-SA-S	PreMes	0.18	0.2		1
Kromme	Indian (SW)	0.9	110	1100	STr-SA-S	PreMes	0.1	0.18		1, 17
Matigulu	Indian (SW)	1	96	760	STr-SA-S	PreMes	0.2	0.22		1
Olifants	Atlantic (SE)	55	285	1500	STr-SA-S	PreMes	1.2			1, 24
Orange	Atlantic (SE)	1000	2200	3000	STr-A-S	PreMes	4.5 (11)	17 (89)	2.8	1, 6, 7, 11, 13, 15, 16, 22, 30
Sondags	Indian (SW)	15	320	2400	STr-SA-S	PreMes	2			7, 14
Spoeg	Atlantic (SE)	1.6	110	1100	STr-A-D	PreMes	0.003			1
Swartlinjies	Atlantic (SE)	1.7	80	680	STr-A-D	PreMes	0.002			1
Tugela	Indian (SW)	29	400	3100	STr-SA-S	PreMes-Mes S/M	4.6	8.8		1, 24
Umfolozi	Indian (SW)	10	390	1600	Str-A-S	PreMes	0.89	2.4		1, 29
Umgeni	Indian (SW)	4.4	230	1800	STr-SA-S	PreMes	0.68	1.7		1, 19
Umhlantuzi	Indian (SW)	3.7	210	1300	STr-SA-S	PreMes	0.47	1.1		1
Umkomazi	Indian (SW)	4.3	300	2600	STr-SA-S	PreMes	1	1.6		1
Umtamvuna	Indian (SW)	1.5	160	1900	STr-SA-S	PreMes	0.3	0.43		1
Umvoti	Indian (SW)	2.8	200	1500	STr-SA-S	PreMes	0.47	0.81		1
Umzimkulu	Indian (SW)	6.7	330	>2000	STr-SA-S	Mes S/M				25
Umzimvubu	Indian (SW)	16	250	2800	STr-SA-S	Mes S/M	1.5	2.2		
Verlore	Atlantic (SE)	1.9	100	720	STr-SA-S	PreMes	0.38			1
Mozambique										
Buzi	Indian (SW)	29	360	1300	Tr-A-S	Mes S/M	1.4			11, 24, 26

Table D7. (*Continued*)

8. Global River Discharge Database, http://www.rivdis.sr.unh.edu/	Indian (SW)	13	230	200	Tr-SA-S	Cen S/M				12
Incomati	Indian (SW)	46	710	1800	Tr-A-S	PreMes	4.2			4, 11, 23, 24, 26
Licungo	Indian (SW)	28	340	2200	Tr-A-S	PreMes	1.2 (3.9)			4, 5, 11
Ligonha	Indian (SW)	16	290	>500	Tr-A-S	PreMes	0.82 (2.6)			4, 5, 11
Limpopo	Indian (SW)	410	1700	1700	Tr-A-S	PreMes	4.8 (26)	6 (33)	1.3 (6.2)	4, 5, 11, 13, 16, 20, 23, 24, 28
Luala	Indian (SW)	12	260	>1000	Tr-SA-S	PreMes				
Lurio	Indian (SW)	61	600	1300	Tr-SA-S	PreMes	7.3			4, 11, 26
Maputo	Indian (SW)	34	560	2300	Tr-A-S	PreMes	2.8			4, 24
Mecuburi	Indian (SW)	8.9	280	>200	Tr-A-S	PreMes	0.46			4
Meluli	Indian (SW)	9.7	290	>500	Tr-SA-S	PreMes	1.9			4
Messalo	Indian (SW)	24	530	>500	Tr-A-S	PreMes	1 (3)			4, 5
M'Lela	Indian (SW)		160	>500	Tr-SA-S	PreMes				
Molocue	Indian (SW)	6.5	350	>500	Tr-SA-S	PreMes	0.86			4
Monapo	Indian (SW)	8.8	240	>500	Tr-SA-S	PreMes	1			4
Montepuez	Indian (SW)	9.5	320	>500	Tr-A-S	PreMes	0.19			4
Pungue	Indian (SW)	29	300	2000	Tr-SA-S	Mes S/M-PreMes	3.1			4, 14, 26
Rovuma	Indian (SW)	160	640	1600	Tr-A-S	PreMes	2.2			4, 25, 26
Save	Indian (SW)	88	710	1700	Tr-SA-S	PreMes	5			4, 11, 22, 23, 26
Umbeluzi	Indian (SW)	5.6	200	1100	Tr-A-S	Mes S/M	0.31			4, 9, 25
Zambezi	Indian (SW)	1300	2600	1600	Tr-A-S	PreMes	100	9 (48)	11	2, 4, 6, 11, 13, 20, 21
Madagascar										
Bemarivo	Indian (SW)	6.5		>1000	Tr-W-S	PreMes	13			6
Fiherenana	Indian (SW)			1200	Tr-SA-S	Mes S/M				
Ikopa	Indian (SW)	30	520	1700	Tr-H-S	PreMes	19	15		3, 11, 24
Ivondro	Indian (SW)	2.8	160	>1000	Tr-W-S	PreMes	3.3			12
Linta	Indian (SW)	1.7		>500	Tr-SA-S	Mes S/M	0.33			7
Mahajamba	Indian (SW)	17	320	>1000	Tr-H-S	PreMes				14
Mahavavy Nord	Indian (SW)	3.2	320	1300	Tr-W-S	Mes S/M	4			24
Manambao	Indian (SW)	9	270	1300	Tr-H-S	Mes S/M				
Manambolo	Indian (SW)	12	210	1300	Tr-H-S	Mes S/M	3.7			

Table D7. (*Continued*)

River name	Ocean	Area (10^3 km^2)	Length (km)	Max_elev (m)	Climate	Geology	Q (km^3/yr)	TSS (Mt/yr)	TDS (Mt/yr)	Refs.
Mananan Tanana	Indian (SW)	6.5		>1000	Tr-H-S	PreMes	3			6
Mananara Sud	Indian (SW)	14	340	2000	Tr-H-S	PreMes	7.2			3, 11, 24
Mananjary	Indian (SW)	2.3	190	>1000	Tr-W-S	PreMes	3.1			13
Mandrare	Indian (SW)	12		1600	Tr-SA-S	PreMes	2			6, 11, 14
Mangoky	Indian (SW)	59	560	2000	Tr-H-S	PreMes	20	10		12, 24, 27
Mangoro	Indian (SW)	3.6	270	1300	Tr-H-S	PreMes	2.9			6, 14
Menarandra	Indian (SW)	5.3		>1000	Tr-SA-S	Mes S/M	1			7
Morondava	Indian (SW)	4.6	160	1100	Tr-H-S	Mes S/M	1.5	6.7		6, 27
Onilahy	Indian (SW)	28	340	1500	Tr-W-S	Mes S/M				3, 14
Sambirano	Indian (SW)	3		>1000	Tr-W-S	PreMes-Mes S/M	3.6			13
Sofia	Indian (SW)	4	350	1800	Tr-H-S	PreMes-Mes S/M	4			
Tsiribihina	Indian (SW)	45	460	1700	Tr-H-S	PreMes	31	12		3, 8, 14, 26
Namibia										
Hoanib	Atlantic (E)	18	250	>1500	STr-A-D	PreMes	0.01			7
Hoarusib	Atlantic (E)	16	300	>1500	STr-A-D	PreMes				
Huab	Atlantic (E)	18	250	>1000	STr-A-D	PreMes				
Kuiseb	Atlantic (E)	17	560	1500	STr-A-D	PreMes	0.001			7, 10, 19
Omaruru	Atlantic (E)	14		>500	STr-A-D	PreMes	0.02			
Swakop	Atlantic (E)	34	360	1500	STr-A-D	PreMes				
Ugab	Atlantic (E)	33	470	1500	STr-A-D	PreMes-Mes S/M	0.01			7
Uniab	Atlantic (E)	4.6	120	>500	STr-A-D	PreMes				18

References:
1. CSIR Reports; 2. Borchet and Kempe, 1985; 3. Chaperon *et al.*, 1993; 4. de Ataida, (1972). Service Hydrolique (cf. van der Leeder, 1975); 5. FAO, 1997; 6. GEMS website, www.gemstat.org; 7. GRDC website, www.gewex.org/grdc; 8. Hiscott, 2001; 9. Heyns, 2003; 10. Jacobsen *et al.*, 1999; 11. Meybeck and Ragu, 1996; 12. ORSTROM, 1969; 13. Probst, 1992; 14.Rand McNally, 1980; 15. Rooseboom and Harmse, 1979; 16. A. Rooseboom, personal communcation; 17. Scharler and Baird, 2003; 18. Scheepers and Rust, 1999; 19. Schulze *et al.*, 2004; 20. Shahin, 2002; 21. Syvitski, 1992; 22. UNESCO (WORRI), 1978; 23. UNESCO, 1985 (cf. Global River Discharge Database, http://www.rivdis.sr.unh.edu/;)24. UNESCO, 1995; 25. van der Leeden, 1975; 26. Vanden Bossche and Bernacsek, 1991; 27. D. E. Walling, personal communication; 28. Ward, 1980; 29. Watson, 1996; 30. Welcomme, 1972 (cf. van der Leeden, 1975)

Tanzania, Kenya, and Somalia

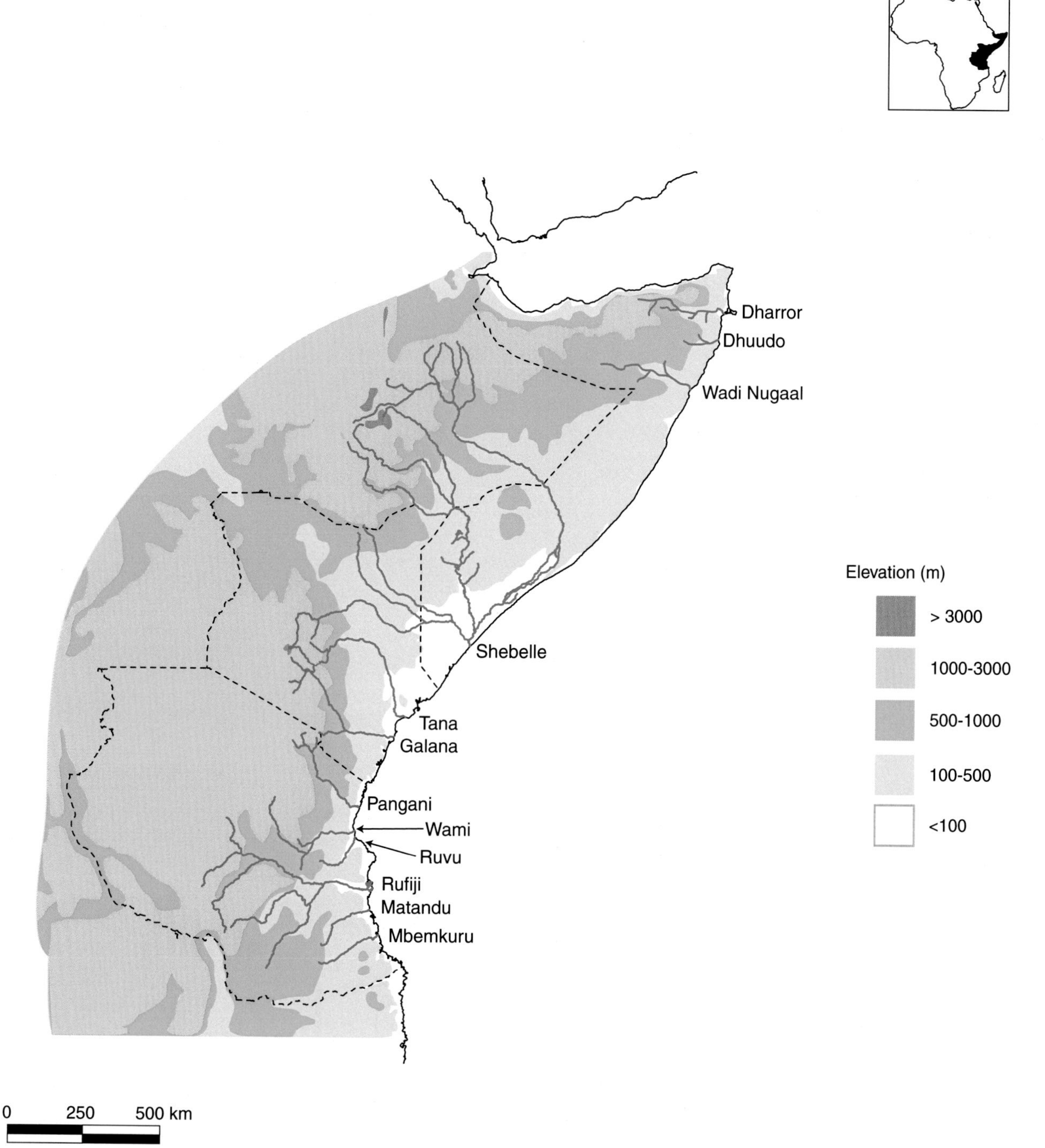

Dharror
Dhuudo
Wadi Nugaal
Shebelle
Tana
Galana
Pangani
Wami
Ruvu
Rufiji
Matandu
Mbemkuru

Elevation (m)

> 3000
1000-3000
500-1000
100-500
<100

0 250 500 km

Figure D16

Kenya, Somalia and Tanzania

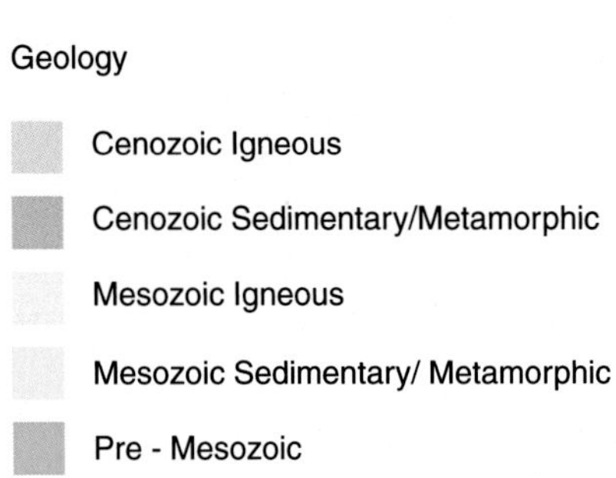

Surface Runoff (mm/yr)

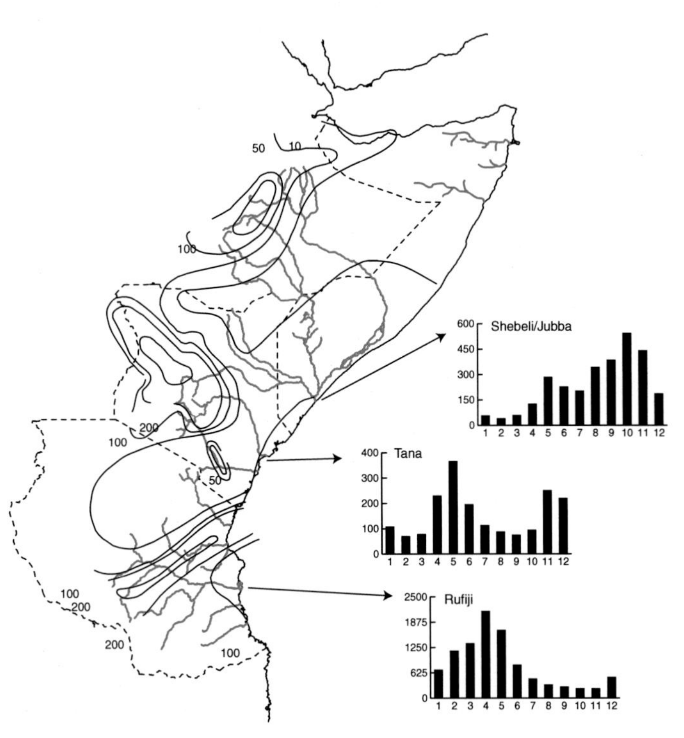

0 500 km

Figure D17

Table D8.

River name	Ocean	Area (10^3 km^2)	Length (km)	Max_elev (m)	Climate	Geology	Q (km^3/yr)	TSS (Mt/yr)	TDS (Mt/yr)	Refs.
Tanzania										
Matandu	Indian (W)	14	280	>500	Tr-SA-S	PreMes				
Mbemkuru	Indian (W)	13	270	>500	Tr-SA-S	PreMes				
Pangani	Indian (W)	29	480	5900	Tr-A-S	Cen I	0.95			3, 7
Rufiji	Indian (W)	180	1200	2900	Tr-A-S	Mes S/M	35	17 (26)	3	2, 3,7
Ruvu	Indian (W)	18	270	2600	Tr-SA-S	PreMes	2.2			6, 8
Wami	Indian (W)	46	490	1900	Tr-A-S	PreMes	2			3, 6, 7, 8
Kenya										
Galana	Indian (W)	40	560	1900	Tr-A-S	Cen I	1			3
Tana	Indian (W)	42	800	4000	Tr-SA-S	Cen I	4.7	32	0.65	3, 7
Somalia										
Dharror	Indian (W)	30		>1000	Tr-A-D	Cen S/M				
Dhuudo	Indian (W)	26	330	2000	Tr-A-W	Cen S/M				
Shebelle	Indian (W)	810	1600	>3000	Tr-A-W	Cen S/M	17		5.7	3
Wadi Nugaal	Indian (W)			>1000	Tr-A-D	Cen S/M				

References
1. T. Dunne, 1982, personal communication; 2. FAO, 1997; 3. Meybeck and Ragu, 1996; 4. Rand McNally, 1980; 5. Shahin, 2002; 6. *Tanzania Hydrological Yearbook*, 1967; 7. UNESCO, 1995; 8. Vanden Bossche and Bernacsek, 1991

Appendix E Eurasia

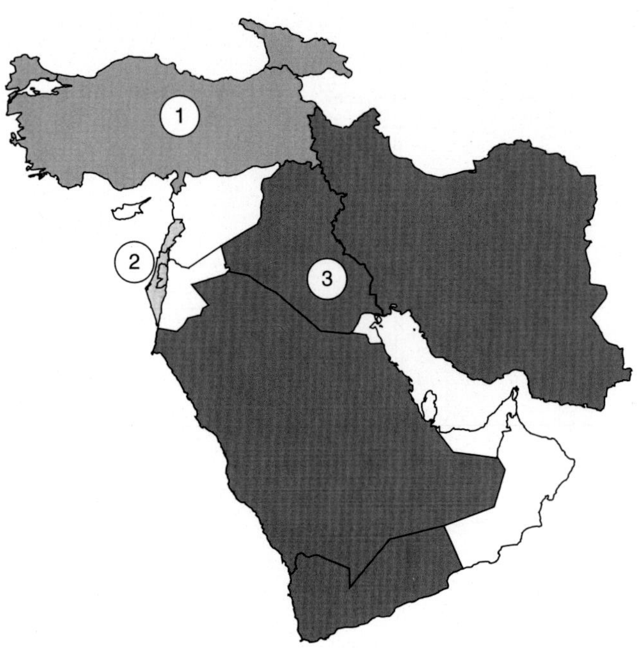

Figure E1

(1) Turkey and Georgia
(2) Lebanon and Israel
(3) Saudi Arabia, Yemen, Iraq, Iran

Turkey and Georgia

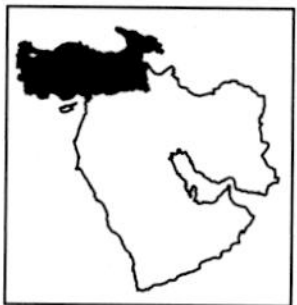

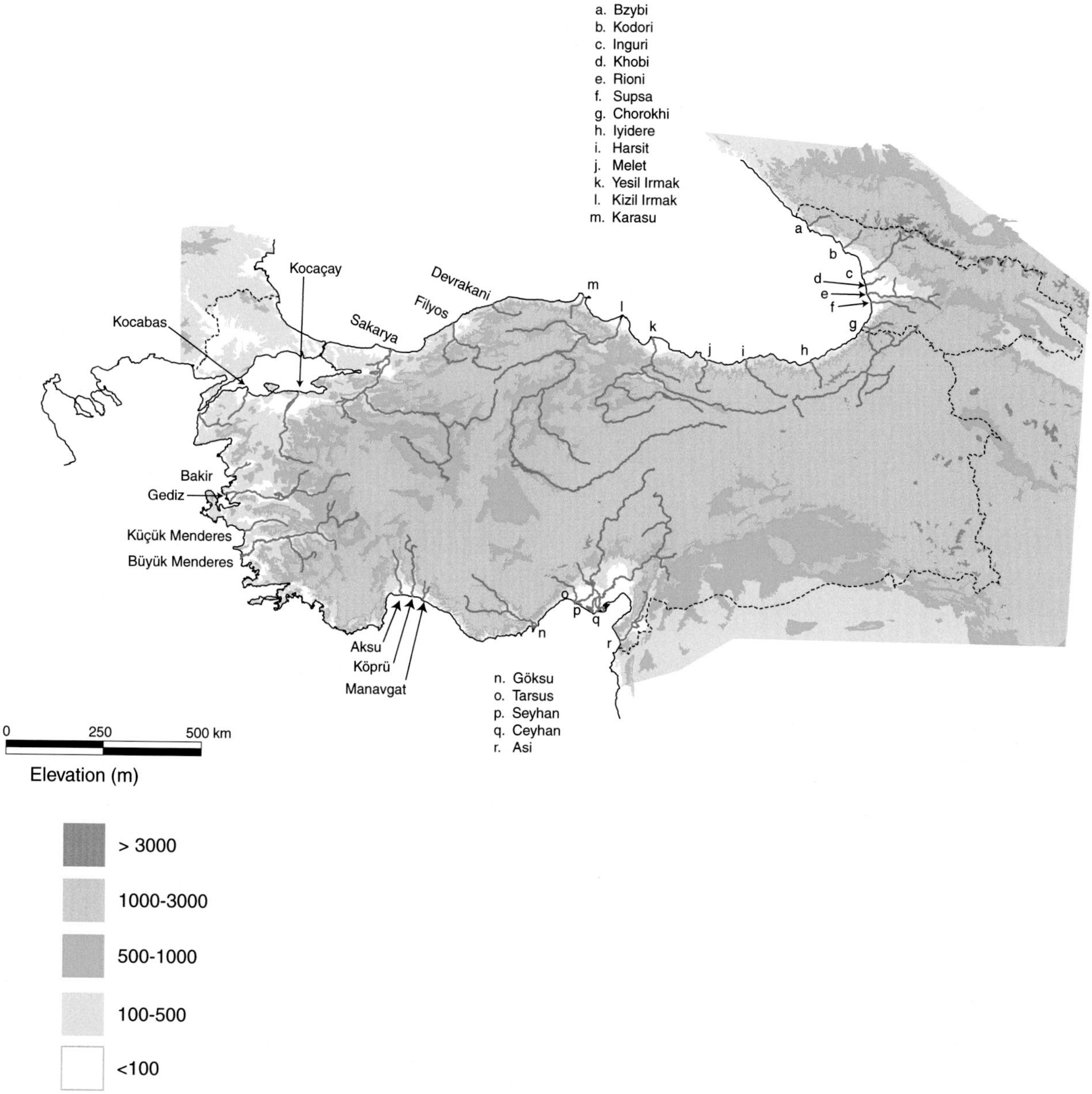

a. Bzybi
b. Kodori
c. Inguri
d. Khobi
e. Rioni
f. Supsa
g. Chorokhi
h. Iyidere
i. Harsit
j. Melet
k. Yesil Irmak
l. Kizil Irmak
m. Karasu

Kocaçay
Devrakani
Filyos
Sakarya
Kocabas
Bakir
Gediz
Küçük Menderes
Büyük Menderes
Aksu
Köprü
Manavgat

n. Göksu
o. Tarsus
p. Seyhan
q. Ceyhan
r. Asi

0 250 500 km

Elevation (m)

> 3000
1000-3000
500-1000
100-500
<100

Figure E2

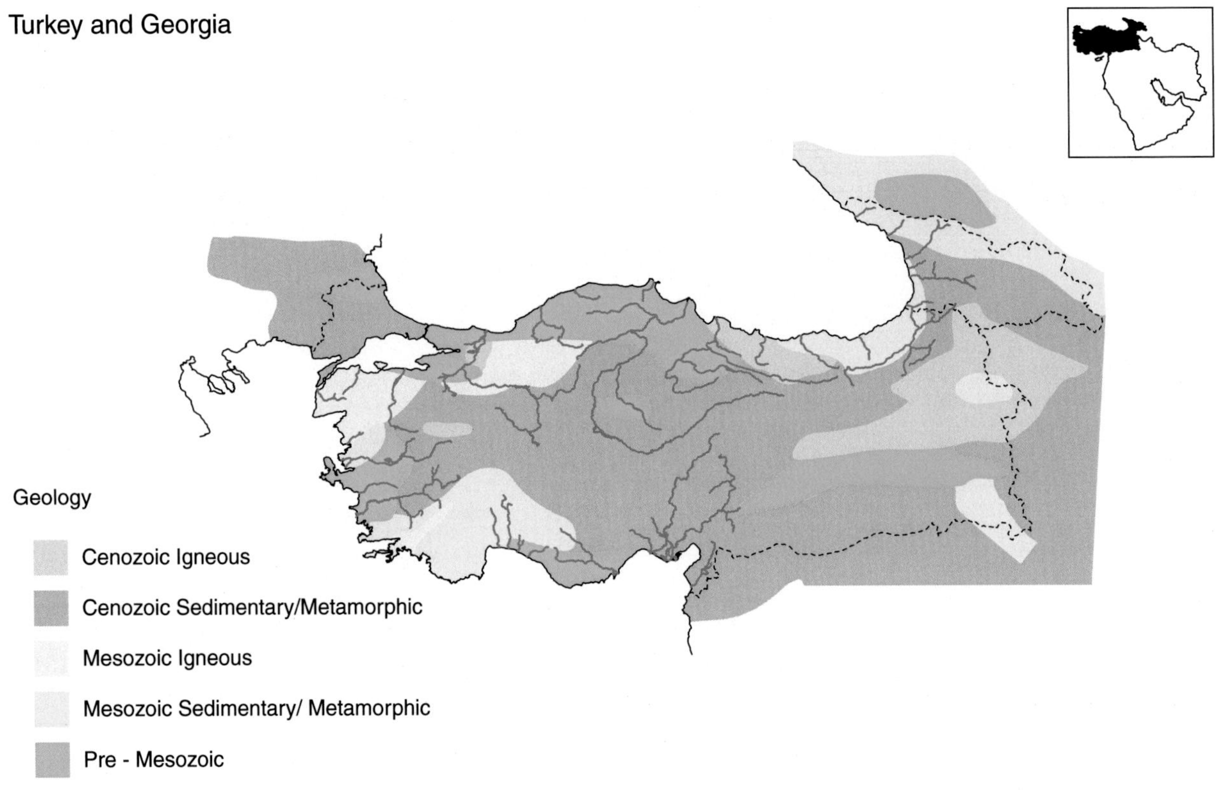
Turkey and Georgia
Geology
Cenozoic Igneous
Cenozoic Sedimentary/Metamorphic
Mesozoic Igneous
Mesozoic Sedimentary/ Metamorphic
Pre - Mesozoic

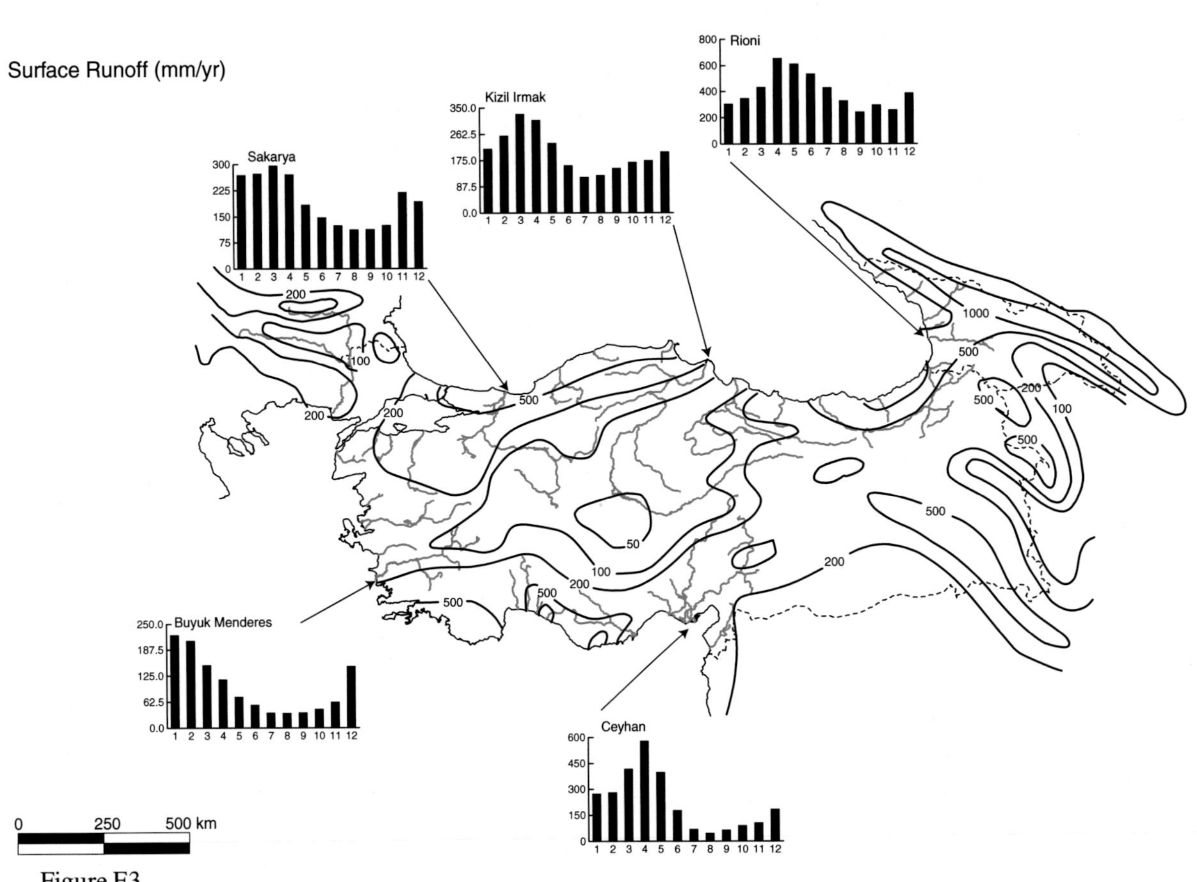
Surface Runoff (mm/yr)
Sakarya
Kizil Irmak
Rioni
Buyuk Menderes
Ceyhan
1000
500
200
100
50
0 250 500 km

Figure E3

Table E1.

River name	Ocean	Area (10^3 km^2)	Length (km)	Max_elev (m)	Climate	Geology	Q (km^3/yr)	TSS (Mt/yr)	TDS (Mt/yr)	Refs.
Turkey										
Aksu	Black Sea	20	120	2000	Te-H-W	Mes S/M	14			10
Asi	Med. (E)	23	500	3100	STr-SA-W	Cen S/M	2.7	0.36 (19)	1	10
Bakir	Aegean Sea	3.4	100	1400	Te-SA-W	Mes I				
Büyük Menderes	Aegean Sea	20	560	2000	Te-SA-W	Cen S/M-Mes S/M	4.7	0.78	3.4	10, 13, 14
Ceyhan	Med. (E)	21	380	2800	Te-H-Sp	Cen S/M	7	4.8 (5.5)	2	10, 13, 14
Devrekani	Black Sea	1.1		>1000	Te-H-W	PreMes	0.25	0.18		2, 7
Filyos	Black Sea	13	230	2400	Te-H-W	PreMes	3.1	0.1 (3.7)		2, 7, 10, 17
Gediz	Aegean Sea	18	350	1500	Te-SA-W	Cen S/M-Mes S/M	2.3	1.3	0.32	10, 13, 14, 15
Göksu	Med. (E)	10	180	2400	STr-H-W	Cen S/M-Mes S/M	2.5	2.5		10, 14
Harsit	Black Sea	2.6	160	3000	Te-H-W	Cen S/M	0.8	0.52		2, 8, 17
Iyidere	Black Sea	0.84		>3000	Te-H-W	Mes S/M	0.92	0.18		2, 3
Karasu	Sea of Marmara	2.9		>200	Te-H-W	PreMes	0.6	0.04		2, 6
Kizil Irmark	Black Sea	79	1400	>1500	Te-A-W	Cen S/M	7.6	0.44 (17)	5.5	2, 3, 7, 9, 10, 13, 17
Kocabas	Sea of Marmara	2.3	75	<1000	Te-SA-W	Mes I				
Kocaçay	Sea of Marmara	23		>1000	Te-SA-W	Cen S/M	4.4		1.2	10
Köprü	Med. (E)	2.8	130	3000	STr-H-W	Cen S/M				
Küçük Menderes	Aegean Sea	3.6	400	2100	Te-SA-W	PreMes	1	0.6		1, 3, 13, 15
Manavgat	Med. (E)	1.3	640	>1000	STr-H-W	Mes S/M	4.1		0.9	10, 14
Melet	Black Sea	1	170	>1500	Te-H-W	Cen	0.4	0.27		2, 7, 18
Sakarya	Black Sea	57	820	>1000	Te-SA-W	Cen S/M	3.6 (5.6)	3.8 (12)	2.9	2, 7, 9, 10, 13, 14, 17, 18
Seyhan	Med. (E)	22	510	3000	Te-H-W	Cen S/M	8	5.2	1.3	1, 10, 14
Tarsus	Med. (E)	1.4	100	2600	STr-SA-W	Cen S/M	0.1	0.13		10
Yesil Irmak	Black Sea	65	520	2600	Te-SA-W	Cen	7.2	6.2	1	7, 10, 13, 17
Georgia										
Bzybi	Black Sea	1.5	110	3000	Te-W-S	Mes S/M	3	0.48	0.23	5, 10, 16
Chorokhi	Black Sea	22	440	3100	Te-H-S	PreMes-Mes S/M	9	8.2		2, 4, 5, 7, 12, 16
Inguri	Black Sea	4.1	210	5000	Te-H-S	Cen S/M	1.6 (6)	0.13 (1.8)	0.51	5, 10, 16, 20
Khobi	Black Sea	1.3	95	3000	Te-W-S	Cen S/M	1.6	0.45 (2.7)		9
Kodori	Black Sea	2	80	2780	Te-W-S	Cen S/M	3.9	0.82	0.39	5, 10, 16, 19
Rioni	Black Sea	13	330	2600	Te-W-S	Cen S/M	13	6.9	2.8	5, 7, 10, 11, 16, 20
Supsa	Black Sea	1.1		>1000	Te-W-S	Cen S/M	1.5	0.25		9

References:
1. Akbulut *et al.*, 2009; 2. Algan *et al.*, 1997; 3. Cecen, Wasser (cf. van der Leeden, 1975); 4. Center for Natural Resources Energy and Transport (UN), 1978; 5. Dzhaoshvili, 1986; 6. GEMS website, www.gemstat.org; 7. Hay, 1994; 8. IAHS/UNESCO, 1974; 9. Jaoshvili, 2002; 10. Meybeck and Ragu, 1996; 11. Mikhailova and Dzhaoshvili, 1998; 12. Nace, 1970; 13. Ozturk, 1996; 14. X. Piper, personal communication; 15. Rand McNally, 1980; 16. Sovetskaya Entsiklopediya, 1989; 17. Tuncer *et al.*, 1998; 18. UNESCO (WORRI), 1978; 19. UNESCO, 1971; 20. Varga *et al.*, 1989

Lebanon and Israel/Palestine

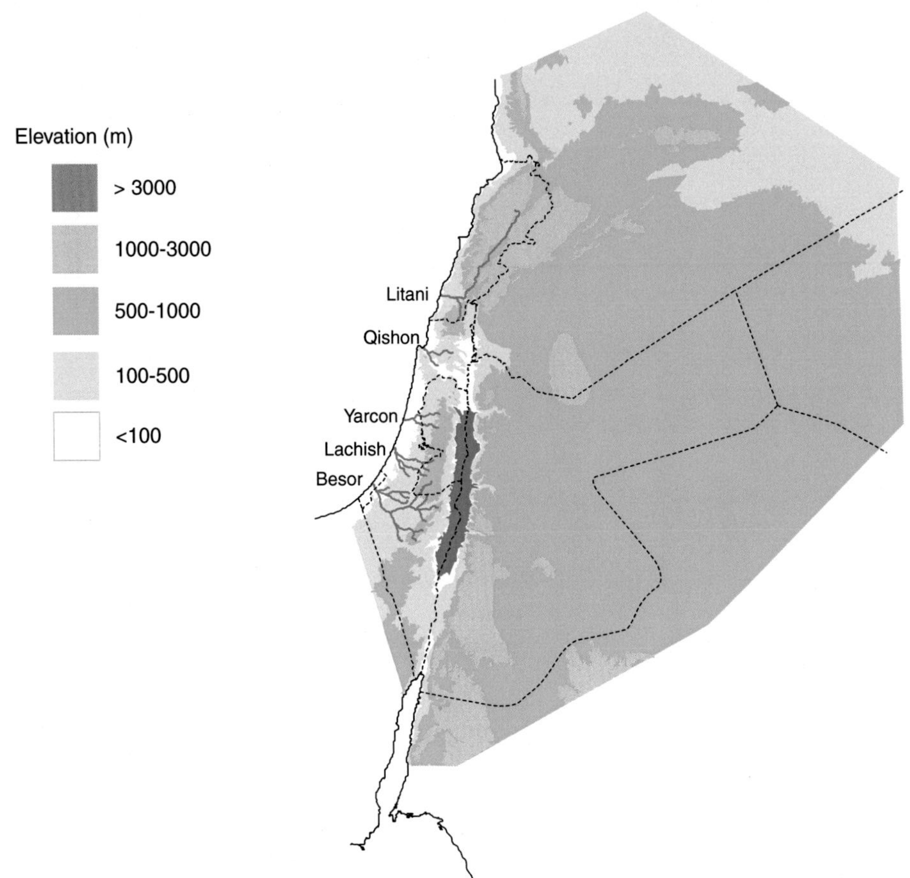

0 250 500 km

Figure E4

Lebanon and Israel/Palestine

Geology

Cenozoic Igneous

Cenozoic Sedimentary/Metamorphic

Mesozoic Igneous

Mesozoic Sedimentary/ Metamorphic

Pre - Mesozoic

Surface Runoff (mm/yr)

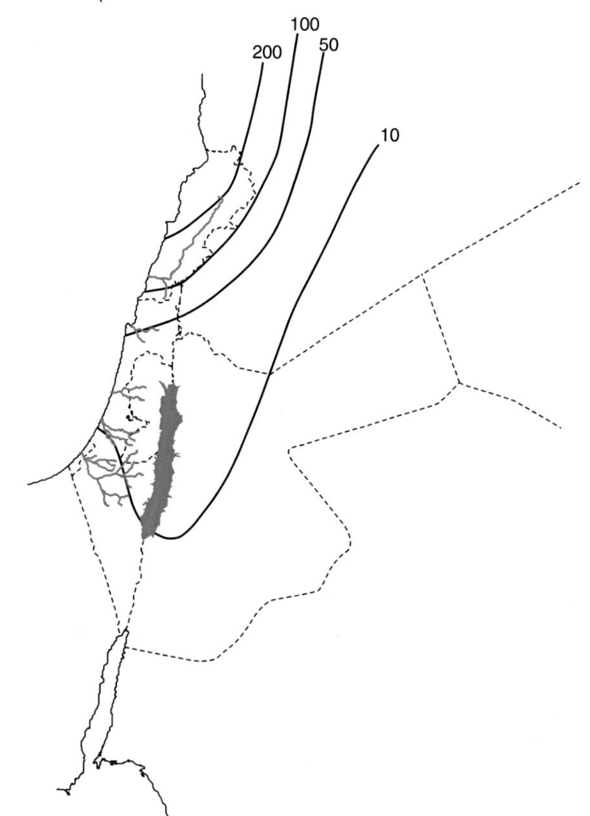

0 250 500 km

Figure E5

Table E2.

River name	Ocean	Area (10^3 km^2)	Length (km)	Max_elev (m)	Climate	Geology	Q (km^3/yr)	TSS (Mt/yr)	TDS (Mt/yr)	Ref.
Lebanon										
Litani	Med. (E)	2.5	140	>1000	STr-A-W	Mes S/M				1
Israel										
Besor	Med. (E)	3.7	110	>500	STr-A-W	Mes S/M-CenS/M	0.01			
Lachish	Med. (E)	1	70	>500	STr-A-W	Cen S/M				
Qishon	Med. (E)	1.1	70	>100	STr-A-W	Cen S/M				
Yarcon	Med. (E)	1.8	28	>600	STr-A-W	Mes S/M	0.03 (0.2)			

Reference:
1. Amery, 1993

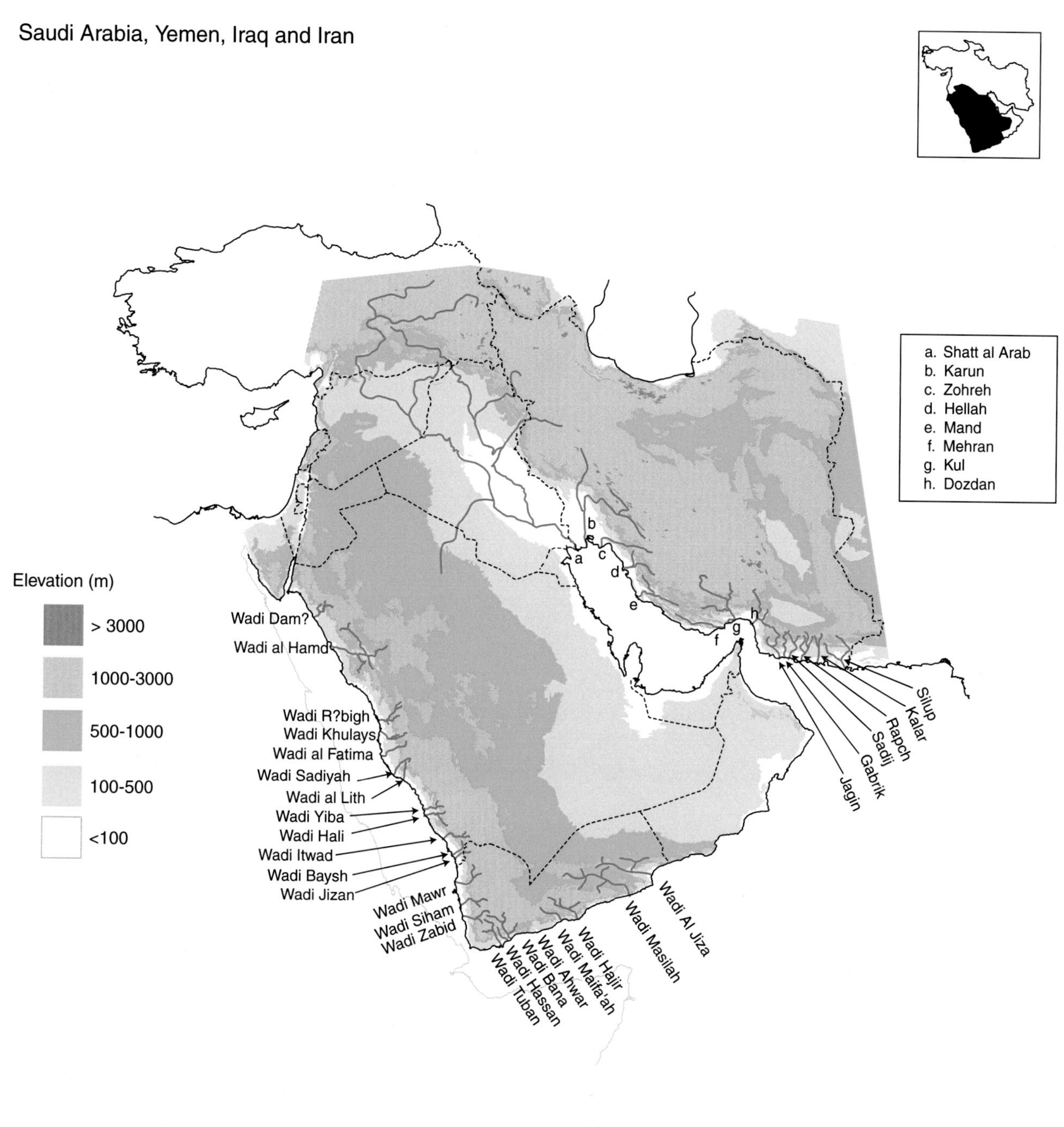

Saudi Arabia, Yemen, Iraq and Iran
Elevation (m)
> 3000
1000-3000
500-1000
100-500
<100
a. Shatt al Arab
b. Karun
c. Zohreh
d. Hellah
e. Mand
f. Mehran
g. Kul
h. Dozdan
a
b
c
d
e
f
g
h
Wadi Dam?
Wadi al Hamd
Wadi R?bigh
Wadi Khulays
Wadi al Fatima
Wadi Sadiyah
Wadi al Lith
Wadi Yiba
Wadi Hali
Wadi Itwad
Wadi Baysh
Wadi Jizan
Wadi Mawr
Wadi Siham
Wadi Zabid
Wadi Tuban
Wadi Hassan
Wadi Bana
Wadi Ahwar
Wadi Maifa'ah
Wadi Hajir
Wadi Masilah
Wadi Al Jiza
Silup
Kalar
Rapch
Sadij
Gabrik
Jagin
0
500
1000 km

Figure E6

Saudi Arabia, Yemen, Iraq and Iran

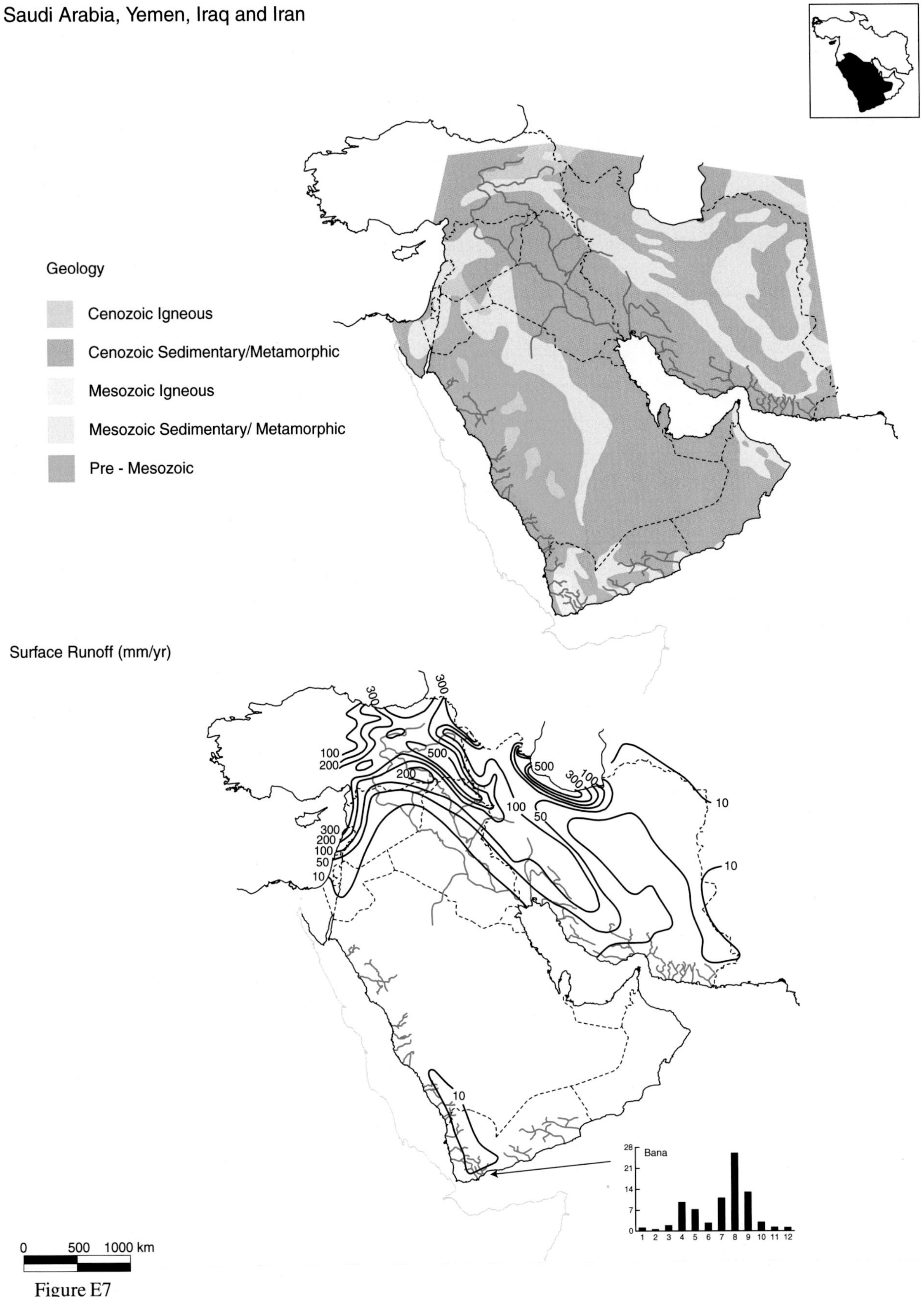

Figure E7

Table E3.

River name	Ocean	Area (10^3 km^2)	Length (km)	Max_elev (m)	Climate	Geology	Q (km^3/yr)	TSS (Mt/yr)	TDS (Mt/yr)	Refs.
Saudi Arabia										
Wadi al Fatima	Red Sea		180	>1000	STr-A-D	PreMes	0.04			5, 7, 9
Wadi al Hamd	Red Sea	62	380	>200	STr-A-D	PreMes				
Wadi al Lith	Red Sea	3	98	>1000	STr-A-D	Cen S/M	0.05			10, 12
Wadi Baysh	Red Sea	5.2	120	2100	STr-A-D	Cen S/M	0.2			10
Wadi Damā	Red Sea	5.5		1300	STr-A-D	PreMes	0.06			12
Wadi Hali	Red Sea	4.5	120	2500	STr-A-D	Cen S/M	0.09			10
Wadi Itwad	Red Sea	1.5	60	2000	STr-A-D	Cen S/M	0.003			10
Wadi Jizan	Red Sea	1.2	70	980	STr-A-D	Cen S/M	0.07			9, 10
Wadi Khulays	Red Sea	5.2		>1000	STr-A-D	PreMes	0.06			9
Wadi Rābigh	Red Sea	5		500	STr-A-D	PreMes	0.06			12
Wadi Sadiyah	Red Sea	1.4	110	>1000	STr-A-D	Cen S/M	0.001			10
Wadi Yiba	Red Sea	2.7	99	1500	STr-A-D	Cen S/M	0.02			10
Yemen										
Wadi Ahwar	G. of Aden	6.4		>560	STr-A-D	Mes I	0.07			9
Wadi al Jiza	G. of Aden	19	260	>1000	STr-A-D	Cen S/M	0.1			6
Wadi Bana	G. of Aden	6.4	200	2000	STr-A-D	PreMes	0.17			6, 8
Wadi Hajir	G. of Aden	8.3	170	>1000	STr-A-D	Mes S/M	0.05			8, 9
Wadi Hassan	G. of Aden	3.5		<500	STr-A-D	Mes I	0.04			9
Wadi Maifa'ah	G. of Aden	8.6		>1000	STr-A-D	Mes S/M	0.08			9
Wadi Masilah	G. of Aden	45	460	>1000	STr-A-D	Cen S/M				
Wadi Mawr	Red Sea	8	300	2000	STr-A-D	Mes S/M-Cen S/M	0.15			
Wadi Siham	Red Sea	4.9		>1000	STr-A-D	Mes I	0.1	2.8		6
Wadi Tuban	G. of Aden	6.5		<1000	STr-A-D	Mes I	0.1	3		1, 14
Wadi Zabid	Arabian Sea	4.6	250	1500	STr-A-D	Mes I	0.14			8, 9
Iraq										
Shatt el Arab	Persian Gulf	420	2500	>2500	Te-A-W	Mes S/M	46 (77)	(100)	18	4, 11
Iran										
Dozdan	Strt. of Hormuz	11	260	>1500	STr-A-W	Mes S/M	0.3			3

Table E3. (*Continued*)

River name	Ocean	Area (10^3 km^2)	Length (km)	Max_elev (m)	Climate	Geology	Q (km^3/yr)	TSS (Mt/yr)	TDS (Mt/yr)	Refs.
Gabrik	G. of Oman	4.2	160	>1500	STr-A-W	Mes S/M				
Hellah	Persian Gulf	4.8	210	>1000	STr-A-W	Mes S/M				
Jagin	G. of Oman	5.6	180	1500	STr-A-W	Mes S/M				
Kalar	G. of Oman	3.7	160	>1500	STr-A-W	Mes S/M				
Karun	Persian Gulf	61	850	>2000	STr-SA-W	Mes S/M	18			2, 7, 13
Kul	Strt. of Hormuz	35	400	>1000	STr-A-W	Cen S/M				
Mand	Persian Gulf	41	480	>2000	STr-A-W	Mes S/M				7
Mehran	Persian Gulf	14	410	>500	STr-A-W	Cen S/M				
Rapch	G. of Oman	6.2	160	>1000	STr-A-W	Mes S/M				
Sadij	G. of Oman	4.6	150	1500	STr-A-W	Mes S/M				
Silup	G. of Oman	6.1	13	500	STr-A-W	Cen S/M				
Zohreh	Persian Gulf	14	320	>2000	STr-A-W	Mes S/M				7

References:
1. FAO, 1997; 2. GEMS website, www.gemstat.org; 3. Global River Discharge Database, http://www.rivdis.sr.unh.edu/; 4. Meybeck and Ragu, 1996; 5. Ministry of Agriculture and Water, 1984; 6. Nouh, 2006; 7. Rand McNally, 1980; 8. Riggs, 1977; 9. Shahin, 2007; 10. Sorman and Abdulrazzak, 1987; 11. UNESCO (WORRI), 1978; 12. Vincent, 2008; 13. Vörösmarty *et al*., 1996e; 14. (http://www.fao.org/landwater laglw/sediment/default.asp)

Appendix F Asia

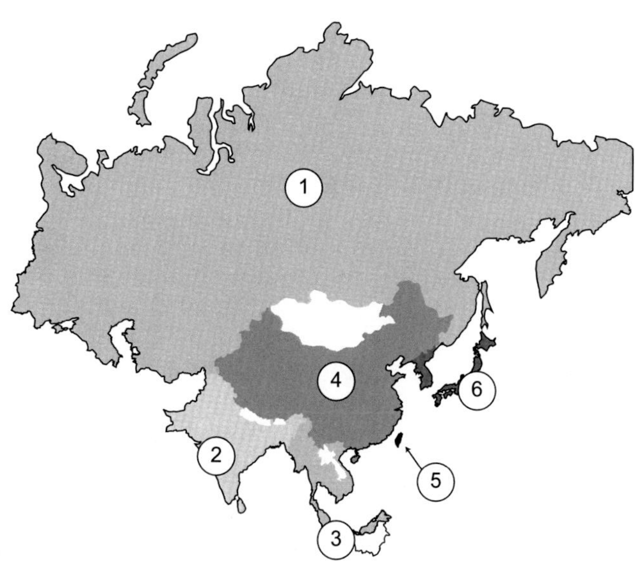

Figure F1

(1) Russia
(2) Pakistan, India, Sri Lanka, Bangladesh
(3) Burma, Brunei, Malaysia, Thailand, Vietnam
(4) China
(5) Taiwan
(6) Japan and Korea

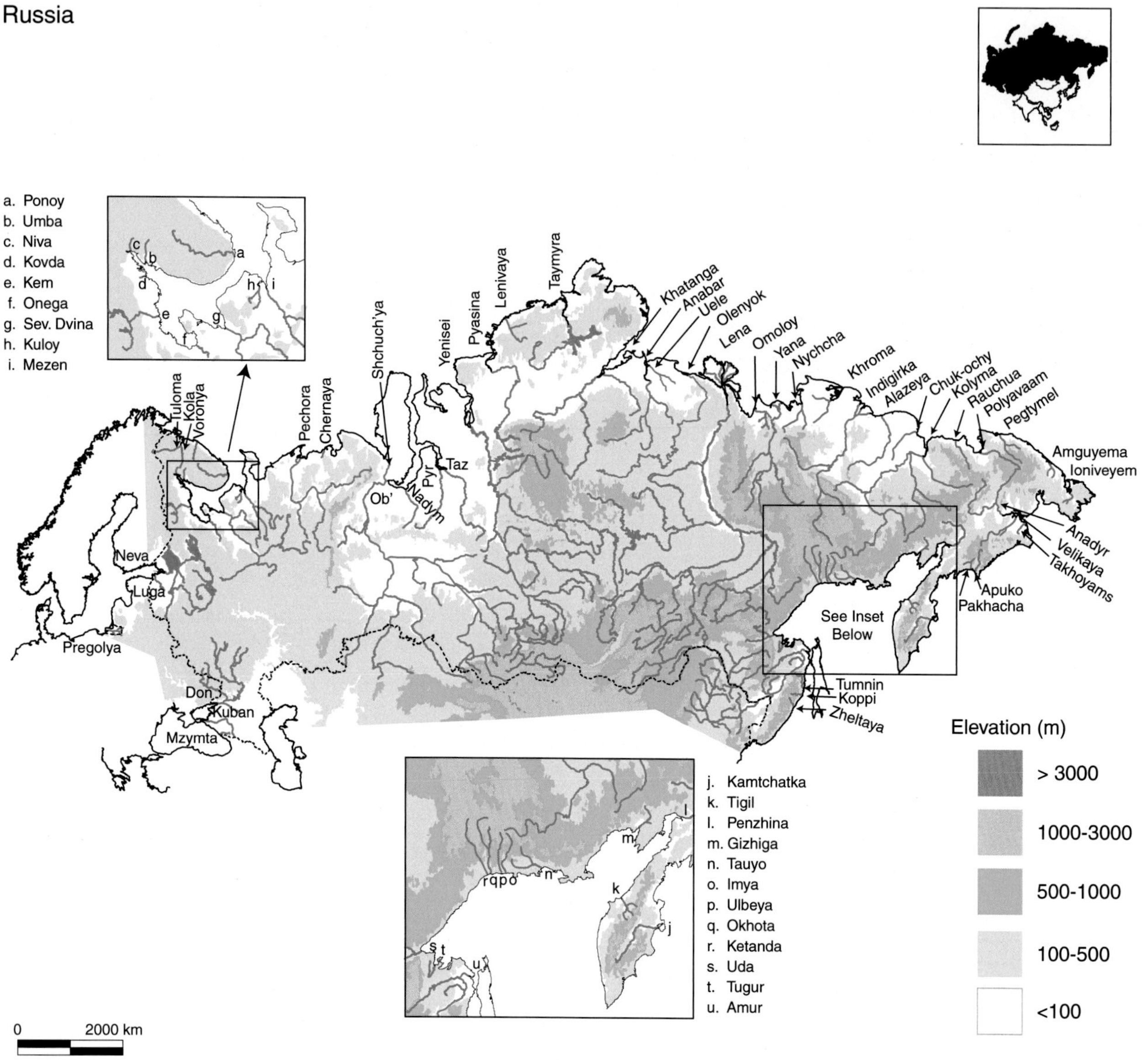

Russia

a. Ponoy
b. Umba
c. Niva
d. Kovda
e. Kem
f. Onega
g. Sev. Dvina
h. Kuloy
i. Mezen

Tuloma
Kola
Voronya
Pechora
Chernaya
Shchuch'ya
Yenisei
Pyasina
Lenivaya
Taymyra
Khatanga
Anabar
Uele
Olenyok
Lena
Omoloy
Yana
Nychcha
Khroma
Indigirka
Alazeya
Chuk-ochy
Kolyma
Rauchua
Polyavaam
Pegtymel
Amguyema
Ioniveyem
Anadyr
Velikaya
Takhoyams
Apuko
Pakhacha
See Inset Below
Tumnin
Koppi
Zheltaya
Taz
Pyr
Nadym
Ob'
Neva
Luga
Pregolya
Don
Kuban
Mzymta

j. Kamtchatka
k. Tigil
l. Penzhina
m. Gizhiga
n. Tauyo
o. Imya
p. Ulbeya
q. Okhota
r. Ketanda
s. Uda
t. Tugur
u. Amur

Elevation (m)
> 3000
1000-3000
500-1000
100-500
<100

0 2000 km

Figure F2

Russia

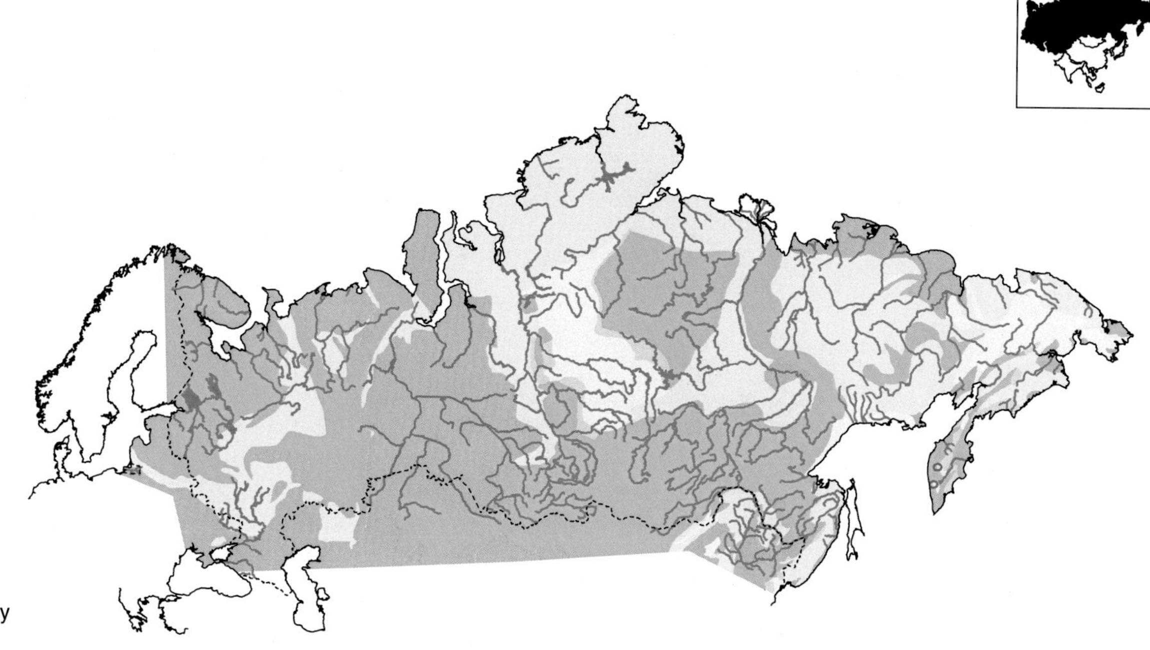

Geology

Cenozoic Igneous

Cenozoic Sedimentary/Metamorphic

Mesozoic Igneous

Mesozoic Sedimentary/ Metamorphic

Pre - Mesozoic

Surface Runoff (mm/yr)

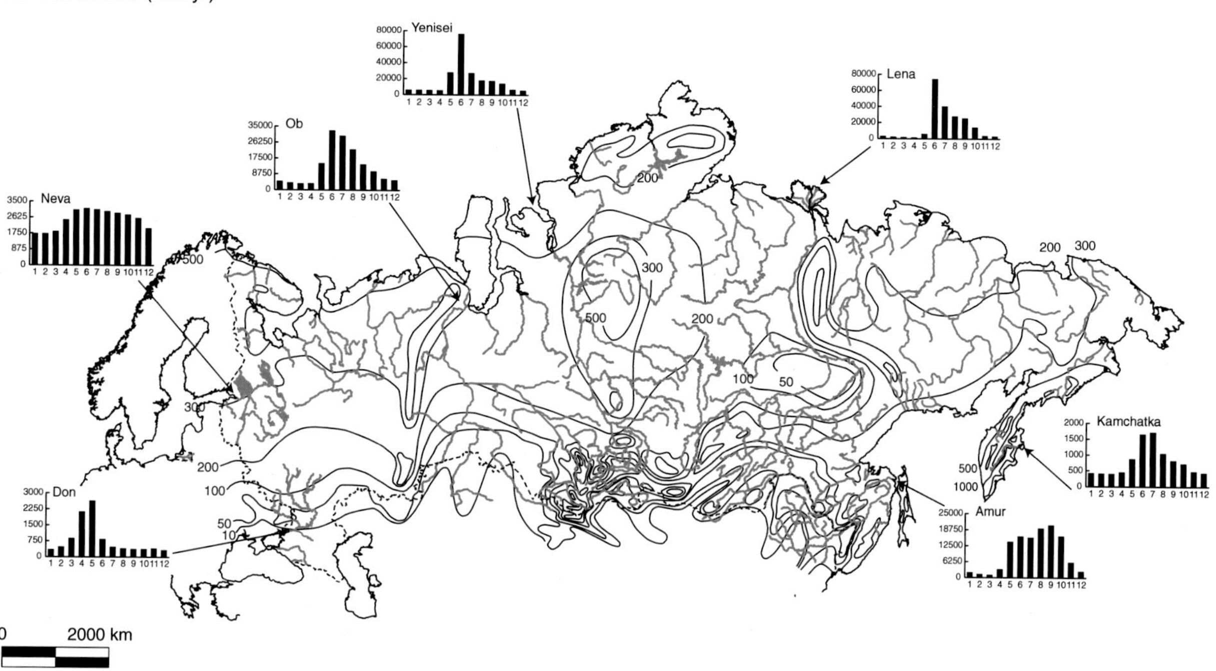

Figure F3

Table F1.

River name	Ocean	Area (10^3 km^2)	Length (km)	Max_elev (m)	Climate	Q (km^3/yr)	TSS (Mt/yr)	TDS (Mt/yr)	Refs.
Russia									
Alazeya	Siberian Sea	68	1600	>200	Ar-SA-S	1.5	0.1 (0.7)	0.19	12
Amguema	Arctic	30	500	2300	Ar-H-S	9.2	0.05	0.16	12
Amur	Sea of Okhotsk	1900	4400	>2000	Ar-SA-S	350	52	26	8, 11, 17, 23
Anabar	Laptev Sea	100	940	>500	Ar-SA-S	17	0.4	0.87	11, 12, 23
Anadyr	Bering Sea	200	1100	680	Ar-H-S	68	3.6	1.9	11, 13, 14, 23
Apuko	Bering Sea	9	210	2500	Ar-H-S				
Chernaya	Kara Sea	11	170	>100	Ar-H-S				
Chuk-ochy	Arctic	15	330	<100	Ar-H-S				
Don	Black Sea	420	1900	180	SAr-A-S	21 (29)	1.9 (6.3)	11 (12)	11, 13, 16
Gizhiga	Sea of Okhotsk	12	280	1500	Ar-H-S	4.9	0.33		17
Imya	Sea of Okhotsk	20	310	1600	Ar-H-S				
Indigirka	Siberian Sea	360	2000	>2000	Ar-SA-S	55	11	1.6	11, 13, 17, 23
Ioniveyem	Arctic	27	470	1200	Ar-H-S				
Kamchatka	Bering Sea	56	760	1200	Ar-H-S	33	3.1	3.4	11, 13, 17, 19
Kem	White Sea	28	190	270	Ar-H-S	8.7	0.07		7, 11, 13
Ketanda	Sea of Okhotsk	14	220	2000	Ar-H-S				
Khatanga	Laptev Sea	360	1600	>1500	Ar-H-S	85 (100)	1.7	7.9	11, 12, 17, 18, 23
Khroma	Laptev Sea	14	290	500	Ar-SA-S				
Kola	Barents Sea	38	83	>200	Ar-H-S	14		0.03	17
Kolyma	Siberian Sea	660	2100	2600	Ar-A-S	120	10	10	8, 11, 12, 13, 17, 19
Koppi	Sea of Japan	8	200	1500	SAr-H-S				
Kovda	White Sea	26	230	110	Ar-H-S	8.5			13, 17
Kuban	Black Sea	58	870	3000	Ar-SA-S	13	8.4		1
Kuloy	White Sea	19	350	>200	Ar-H-S	5.7			3, 13, 23
Lena	Laptev Sea	2500	4400	2500	Ar-SA-S	520	20	60	10, 11, 19
Lenivaya	Kara Sea	9	200	>200	Ar-SA-S				
Luga	Baltic Sea	13	350	>100	Ar-H-S	3		0.37	16, 19
Mezen	White Sea	78	970	480	Ar-H-S	27	0.9	3.5	11, 12, 17, 23
Mzymta	Black Sea	1		>3000	SAr-W-S	1.5	0.3		5
Nadym	Arctic	64	250	>100	Ar-H-S	18	0.7	0.14	11, 12, 21, 22
Neva	Baltic Sea	74	1100	<500	Ar-H-S	80	0.5 (2.9)	1.4	8, 11, 22
Niva	White Sea	13	40	>100	Ar-H-S	5	0.7		11, 23
Nychcha	Laptev Sea	15	310	500	Ar-SA-S				
Ob’	Kara Sea	3000	5400	3700	Ar-SA-S	390	16	34	11, 12, 17, 23
Okhota	Sea of Okhotsk	20	170	2000	Ar-H-S	5.2			4
Olenyok	Laptev Sea	220	2300	>500	Ar-SA-S	32	1.1	4.1	11, 17, 20
Omoloy	Laptev Sea	39	1600	>1500	Ar-SA-S	7	0.04		8, 11, 12, 22
Onega	White Sea	57	420	>200	Ar-H-S	16	0.25	3	11, 23

Table F1. (*Continued*)

Pakhacha	Bering Sea	31	400	>1500	Ar-H-S				
Pechora	Barents Sea	320	1800	680	Ar-H-S	130	4.4	14	8, 11, 12, 13, 17
Pegtymel	Arctic	19	270	>1500	Ar-SA-S				2
Penzhina	Sea of Okhotsk	73	710	>2000	Ar-H-S	25	1	0.83	4, 11, 17
Polyavaam	Arctic	15	370	>1500	Ar-SA-S		0.05		22
Ponoy	White Sea	16	430	<500	Ar-H-S	5.3			23
Pregolua	Baltic Sea	16	190	>200	Ar-SA-S	2.6			23
Pyasina	Kara Sea	180	820	>200	Ar-H-S	86	3.4		11, 12, 13
Pyr	Kara Sea	110	1000	170	Ar-H-S	28 (33)	0.7	1.3	11, 17, 23
Rauchua	Arctic	16	220	1700	Ar-H-S				
Severnaya Dvina	Arctic	360	740	>200	Ar-H-S	110	4.5	20	6, 14, 17
Shchuch'ya	Kara Sea	11	250	1000	Ar-SA-S				
Takhoyams	Bering Sea	13	230	>1500	Ar-H-S				
Tauyo	Sea of Okhotsk	25	290	1600	SAr-H-S	11	0.48	0.05	17
Taymyra	Kara Sea	120	380	500	Ar-H-S	38	0.58		17
Taz	Kara Sea	150	1400	>100	Ar-H-S	34 (48)	0.6	4.3	11, 15, 17
Tigil	Sea of Okhotsk	21	370	>1000	Ar-H-S				
Tugur	Pacific (NW)	12		>1500	Ar-H-S	6.3	0.15		17
Tuloma	Barents Sea	24	62	<500	Ar-H-S	7			13, 23
Tumnin	Sea of Japan	24	320	>1000	Ar-H-S	4.1			
Uda	Sea of Okhotsk	61	460	>1500	Ar-H-S	25	0.93	0.83	17, 23
Uele	Laptev Sea	19	250	100	Ar-SA-S				
Ulbeya	Sea of Okhotsk	22	310	>1500	Ar-SA-S				
Umba	White Sea	6.5	120	150	Ar-H-S	2.5			9
Velikaya	Bering Sea	10	170	>1500	Ar-H-S				
Voronya	Barents Sea	9.9	150	4500	Ar-H-S	3.2			17
Yana	Laptev Sea	240	870	>3000	Ar-SA-S	32	3	1.5	11, 19
Yenisei	Kara Sea	2600	4100	3500	Ar-SA-S	620	4.1 (13)	43	11, 17, 19
Zheltaya	Sea of Japan	8	170	1000	Ar-H-S				

References:
1. Algan *et al.*, 1997; 2. Arctic RIMS Website, rims.unh.edu/data.shtml; 3. Center for Natural Resources, Energy and Transport (UN), 1978; 4. Czaya, 1981; 5. Dzhaoshvili, 1986; 6. European Environment Agency Website, www.eea.europa.eu; 7. Eurosion, 2004; 8. GEMS website, www.gemstat.org; 9. Global River Discharge Database, http://www.rivdis.sr.unh.edu/; 10. Gordeev and Sidorov, 1993; 11. Gordeev *et al.*, 1996; 12. Gordeev, 2006; 13. IAHS/UNESCO, 1974; 14. Kimstach *et al.*, 1998; 15. Korotaev, 1991; 16. Lisitzin, 1972; 17. Meybeck and Ragu, 1996; 18. O'Grady and Syvitski, 2002; 19. Probst, 1992; 20. Rachold *et al.*, 1996; 21. Rand McNally, 1980; 22. Shiklomanov and Skakalsky, 1994; 23. Sovetskaya Entsiklopediya, 1989

India, Pakistan, Sri Lanka and Bangladesh

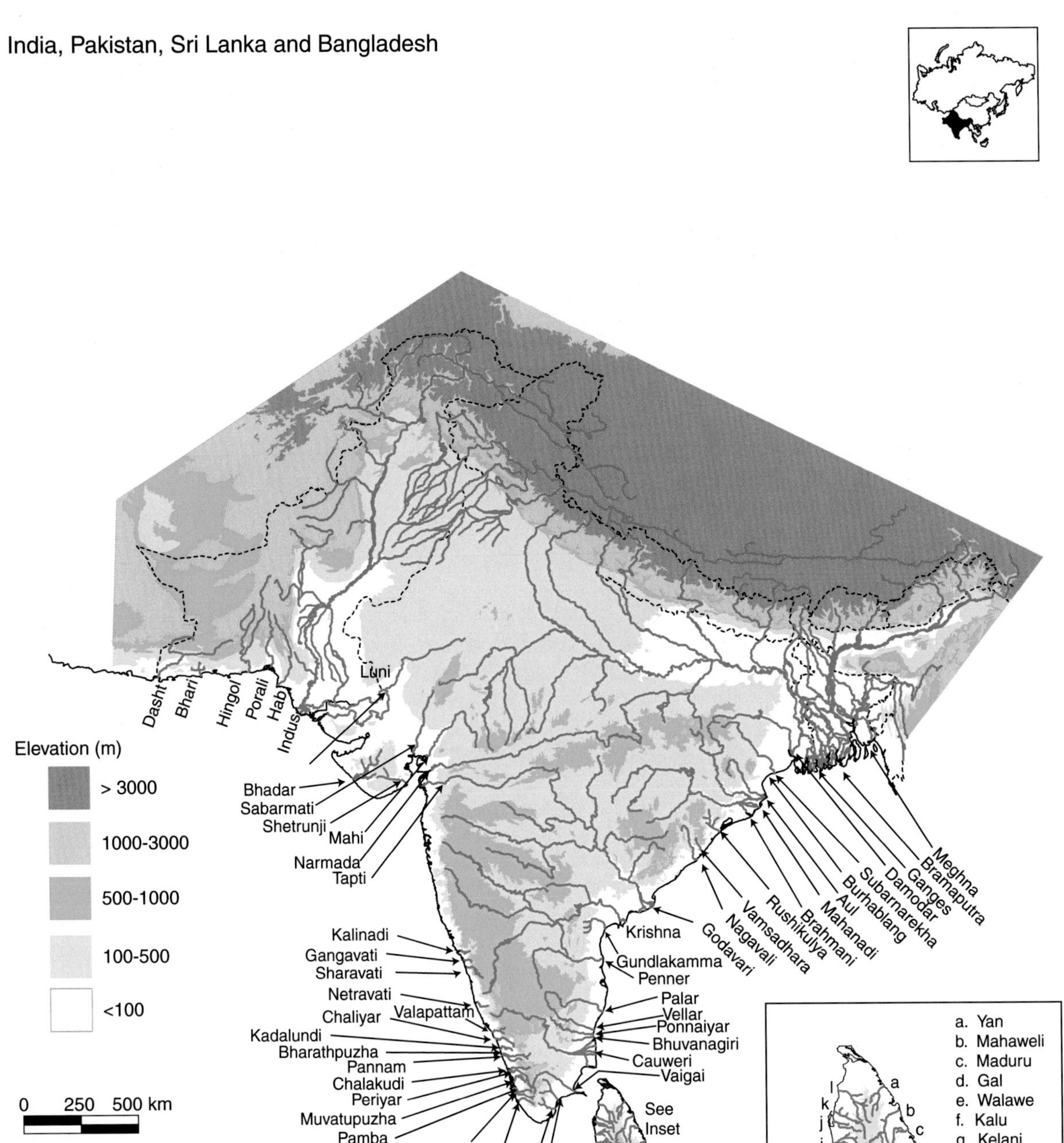

Figure F4

India, Pakistan, Sri Lanka and Bangladesh

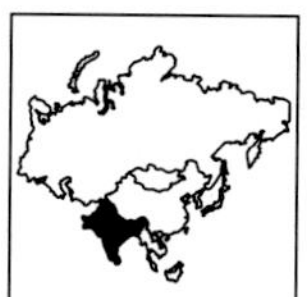

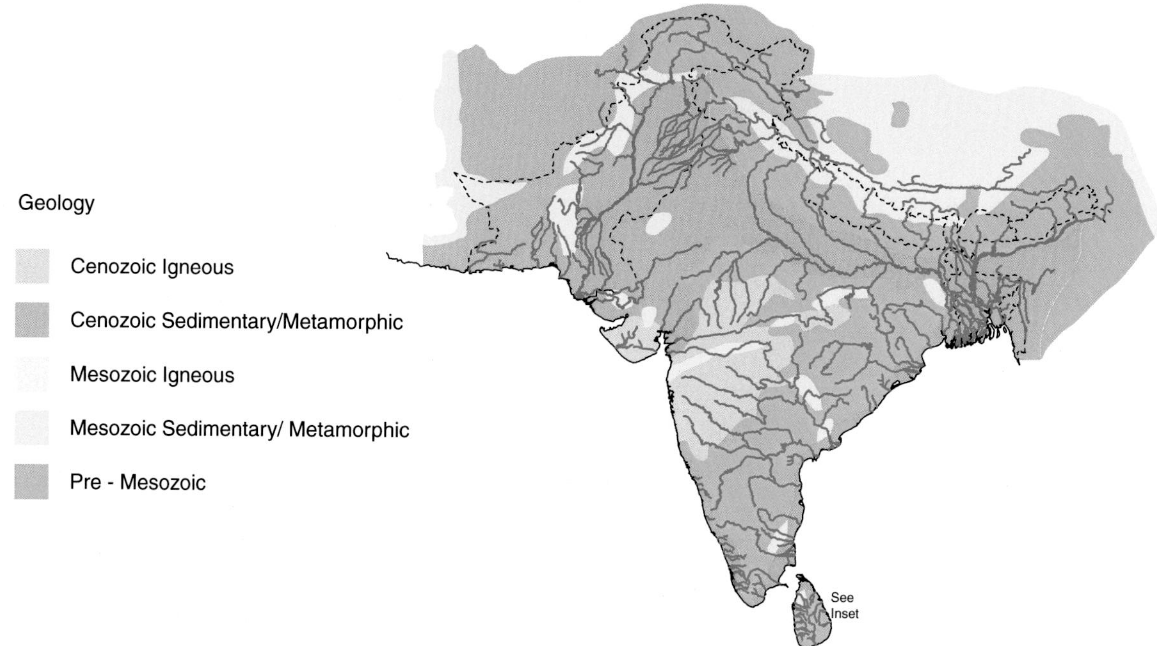

Surface Runoff (mm/yr)

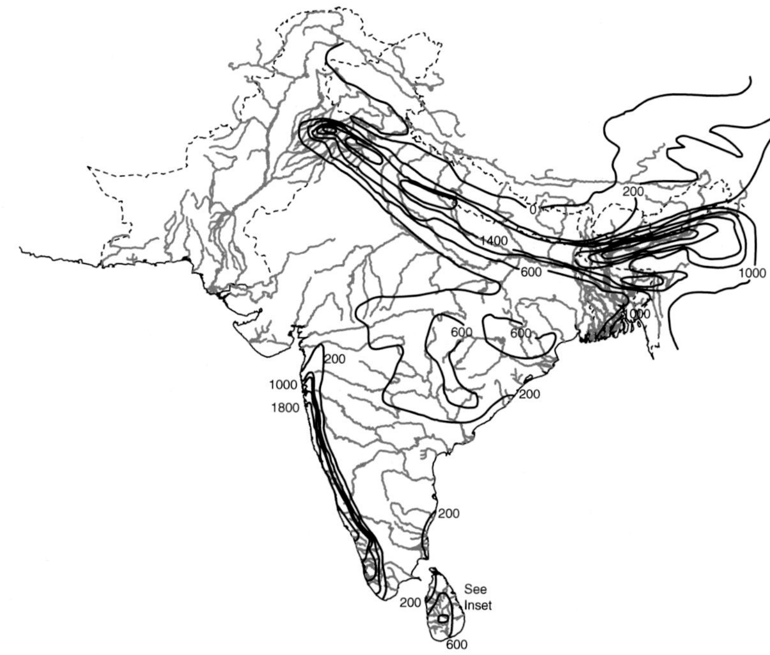

Figure F5

Table F2.

River name	Ocean	Area (10^3 km^2)	Length (km)	Max_elev (m)	Climate	Geology	Q (km^3/yr)	TSS (Mt/yr)	TDS (Mt/yr)	Refs.
Pakistan										
Bhari	Arabian Sea	4.2	100	1300	STr-A-S	Cen S/M				
Dasht	Arabian Sea	36	420	1600	STr-A-S	Cen S/M	0.04			3, 4, 37
Hab	Arabian Sea		400	>2000	STr-A-S	Mes S/M	0.1			3, 37
Hingol	Arabian Sea	32	560	2100	STr-A-S	Cen S/M	0.26			37
Indus	Arabian Sea	980	3200	7800	STr-A-S	Cen S/M	5 (90)	10 (250)	10	4, 16, 17, 18, 20, 34
Porali	Arabian Sea	4		>500	STr-A-S	Mes S/M	0.5			9, 37
Sri Lanka										
Aruvi Aru	Indian	3.2	160	>500	Tr-SA-Au	PreMes	0.6			27, 36
Deduru Oya	Indian	2.6	130	>200	Tr-H-Au	PreMes	1.6			27
Gal Oya	Indian	1.8	110	>500	Tr-H-Au	PreMes	1.2			27
Kala Oya	Indian	2.8	150	>200	Tr-SA-Au	PreMes	0.6			27
Kalu Ganga	Indian	2.7	130	>1000	Tr-W-Au	PreMes	7.9			27, 35, 36
Kelani Ganga	Indian	2.3	140	>1000	Tr-W-Au	PreMes	5.4			27, 35
Maduru Oya	Indian	1.5	200	500	Tr-H-Au	PreMes	0.8			27
Maha Oya	Indian	1.5	90	>1000	Tr-W-Au	PreMes	1.8			27
Mahaweli Ganga	Indian	11	330	>1000	Tr-W-Au	PreMes	11			23, 27, 35, 36
Mi Oya	Indian	1.5	100	>100	Tr-SA-Au	PreMes	0.3			
Walawe Ganga	Indian	2.4	130	>1000	Tr-W-Au	PreMes	2.2			23, 27, 35
Yan Oya	Indian	1.5	150	<200	Tr-H-Au	PreMes	0.3			27
Bangladesh										
Bramaputra	Bay of Bengal	670	2600	5500	STr-W-S	Mes S/M	630	540	63	6, 13, 16
Ganges	Bay of Bengal	980	2200	7000	STr-H-S	Mes-PreMes	490	520	91	6, 13, 16
Meghna	Bay of Bengal	80	900	<100	Tr-W-S	Mes S/M	150		8	2, 5, 7
India										
Achenkovil	Arabian Sea	1.5	130	1500	Tr-W-S	PreMes	1.5	0.058	0.06	1, 32, 39
Aul	Bay of Bengal	12	370	>500	Tr-H-S	PreMes				
Bhadar	Arabian Sea	16	220	200	Tr-A-S	Cen I				
Bharathpuzha	Arabian Sea	6.2	210	2000	Tr-W-S	PreMes	7	0.21	0.38	1, 14, 19, 28, 32
Bhuvanagiri	Bay of Bengal	8.1	180	>500	Tr-H-S	PreMes				
Brahmani	Bay of Bengal	52	480	600	Tr-H-S	PreMes	36	7 (20)	1.2	16, 22, 23
Burhabalang	Bay of Bengal	4.8	160	>200	Tr-H-S	PreMes				14
Cauweri	Bay of Bengal	88	760	1300	Tr-H-S	PreMes	21	0.4 (32)	8.3	11, 15, 16, 21, 23, 33, 34, 39
Chalakudi	Arabian Sea	1.7		1200	Tr-W-S	PreMes	1.6	0.07		19
Chaliyar	Arabian Sea	2.9	170	2100	Tr-W-S	PreMes	5.9	0.49	0.23	1, 19, 32, 33
Damodar	Bay of Bengal	22	550	600	Tr-H-S	PreMes	9.4	28	1.6	16, 23, 36
Gangavati	Arabian Sea	3.9	150	700	Tr-W-S	PreMes	5			24, 33
Godavari	Bay of Bengal	310	1400	1600	Tr-H-S	Cen I	92 (120)	47 (170)	20	3, 15, 16, 22, 25, 39
Gundlakamma	Bay of Bengal	8.5	210	600	Tr-A-S	PreMes	1.3	0.1		39

Table F2. (*Continued*)

Kadalundi	Arabian Sea	1.1	130	900	Tr-W-S	PreMes	1.1	0.6		39
Kalinadi	Arabian Sea	5.2	150	600	Tr-W-S	PreMes	6.5	0.7	0.6	19, 24, 30, 33
Kallada	Arabian Sea	1.7	120	1500	Tr-W-S	PreMes	3.4	0.09	0.17	1, 19, 32, 39
Krishna	Bay of Bengal	260	1300	1300	Tr-A-S	Cen I	12 (62)	1 (64)	22	8, 16, 23, 34, 38, 39
Luni	Arabian Sea	35	530	890	Tr-SA-S	Cen S/M				
Mahanadi	Bay of Bengal	140	900	440	Tr-H-S	PreMes	47 (54)	16 (61)	8.1	8, 12, 15, 16, 20, 22, 34, 39
Mahi	Arabian Sea	35	580	500	Tr-H-S	PreMes	4.6 (12)	2 (22)	2.8	8, 22, 29, 39
Muvatupuzha	Arabian Sea	1.5	120	1100	Tr-W-S	PreMes	3.6	0.037	0.12	1, 19, 32
Nagavali	Bay of Bengal	9.4	220	1600	Tr-H-S	PreMes	2.9	1		14, 16, 39
Narmada	Arabian Sea	99	1300	1100	Tr-SA-S	Cen I	23 (38)	15 (70)	13	8, 9, 10, 15, 22, 31, 34, 39
Netravati	Arabian Sea	4.2	100	<500	Tr-W-S	PreMes	4.6	1.4	1	24, 30, 33
Palar	Bay of Bengal	18	350	>500	Tr-SA-S	PreMes	2			16
Pamba	Arabian Sea	2.2	180	1600	Tr-W-S	PreMes	3.4	0.25	0.1	19, 32, 33
Pannam	Arabian Sea	5.4	400	>1000	Tr-W-S	PreMes	8.8			23, 24, 33
Penner	Bay of Bengal	55	560	1500	Tr-A-S	PreMes	2.4 (6.3)	1.8 (7)	1.1	14, 22, 23, 33, 35, 36, 39
Periyar	Arabian Sea	5.4	240	1800	Tr-W-S	PreMes	4.9	0.25	0.15	1, 8, 16, 19, 33, 36
Ponnaiyar	Bay of Bengal	16	400	900	Tr-A-S	PreMes	1.6	0.07		16, 39
Rushikulya	Bay of Bengal	7.7	150	1000	Tr-H-S	PreMes	2.1	1.2 (1.8)		14, 39
Sabarmati	Arabian Sea	21	420	760	Tr-A-S	Cen S/M	1 (1.4)	0.2 (4.6)	0.5	14, 22, 23, 39
Sharavati	Arabian Sea	2.2	120	700	Tr-W-S	PreMes	4.5			24, 33
Shetrunji	Arabian Sea	5.5	180	380	Tr-SA-S	PreMes	0.17	0.6		14, 39
Subarnarekha	Bay of Bengal	19	390	600	Tr-H-S	PreMes	10	3 (6)	1	8, 16, 30, 39
Tamraparni	Bay of Bengal	6	130	1400	Tr-W-S	PreMes	5.3	1	0.16	26
Tapti	Arabian Sea	65	720	750	Tr-SA-S	Cen I	9	20 (25)	3	8, 14, 22, 39
Vaigai	Bay of Bengal	7	260	>1000	Tr-H-S	PreMes				14
Vaippar	Bay of Bengal	5.3	130	>200	Tr-H-S	PreMes				14
Valapattanam	Bay of Bengal	1.9	110	900	Tr-W-S	PreMes	3.9	0.26		19
Vamsadhara	Bay of Bengal	11	380	600	Tr-H-S	PreMes	3.5	6.8	0.3	14, 16, 30
Vellar	Bay of Bengal	8.6	193	900	Tr-A-S	Cen S/M	0.85			

References:

1. Bajpajee, unpublished data; 2. Bangladesh Water Development Board, 1983; 3. Biksham and Subramanian, 1988; 4. Center for Natural Resources, Energy and Transport (UN), 1978; 5. Datta and Subramanian, 1997; 6. Esser and Kohlmaier, 1991; 7. FAP 24, 1994; 8. GEMS website, www.gemstat.org; 9. Global River Discharge Database, http://www.rivdis.sr.unh.edu/; 10. Gupta and Chakrapani, 2005; 11. Gupta *et al.*, 1999; 12. Harrison, 2000; 13. Hossain, 1991; 14. Jain *et al.*, 2007; 15. Law, 1968; 16. Meybeck and Ragu, 1996; 17. Milliman and Meade, 1983; 18. Milliman *et al.*, 1987; 19. Narayana, 2006; 20. Probst, 1992; 21. Ramanathan *et al.*, 1996; 22. Ramesh and Subramanian, 1993; 23. Rand McNally, 1980; 24. Rao, 1979; 25. Rao, 2006; 26. Ravichandran, 2003; 27. X. Silva, personal communication; 28. Y. Soman, personal communication; 29. Sridhar, 2007; 30. Subramanian, 1987; 31. Subramanian, 1993; 32. Subramanian, 2004; 33. V. Subramanian, personal communication; 34. UNESCO (WORRI), 1978; 35. UNESCO, 1971; 36. Vörösmarty *et al.*, 1996e; 37. World Bank, 2005; 38. Nageswara Rao *et al.*, 2010; 39. Gamage and Smakhtin, 2009. 39. H. Gupta, personal communication.

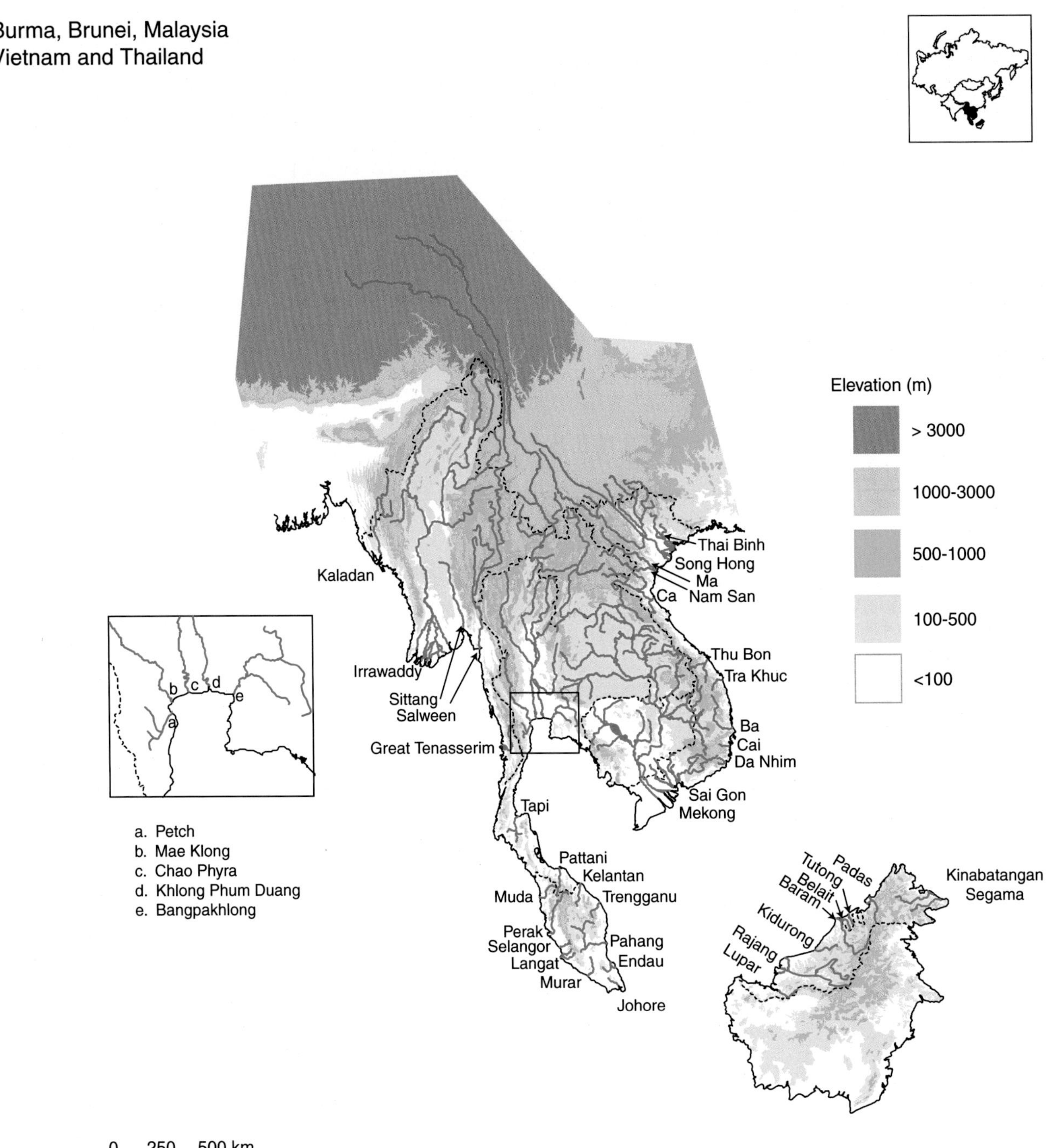
Burma, Brunei, Malaysia
Vietnam and Thailand
Elevation (m)
> 3000
1000-3000
500-1000
100-500
<100
Kaladan
Irrawaddy
Sittang
Salween
Great Tenasserim
Thai Binh
Song Hong
Ma
Ca
Nam San
Thu Bon
Tra Khuc
Ba
Cai
Da Nhim
Sai Gon
Mekong
Tapi
Pattani
Kelantan
Trengganu
Muda
Perak
Selangor
Langat
Murar
Pahang
Endau
Johore
Padas
Tutong
Belait
Baram
Kidurong
Rajang
Lupar
Kinabatangan
Segama
a
b
c
d
e
a. Petch
b. Mae Klong
c. Chao Phyra
d. Khlong Phum Duang
e. Bangpakhlong
0 250 500 km

Figure F6

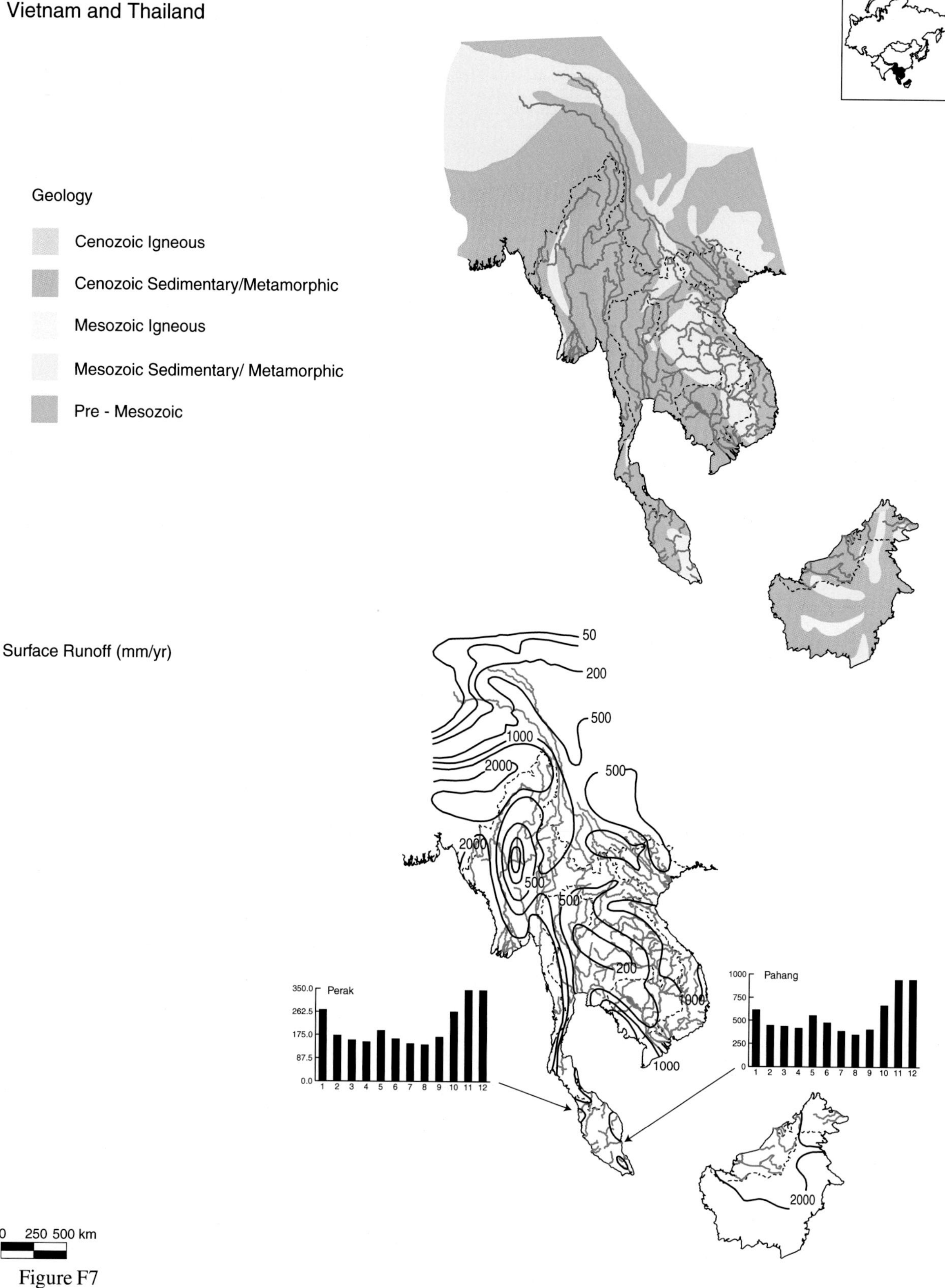
Burma, Cambodia, Malaysia
Vietnam and Thailand
Geology
Cenozoic Igneous
Cenozoic Sedimentary/Metamorphic
Mesozoic Igneous
Mesozoic Sedimentary/ Metamorphic
Pre - Mesozoic
Surface Runoff (mm/yr)
50
200
500
1000
2000
Perak
Pahang
0 250 500 km

Figure F7

Table F3.

River name	Ocean	Area (10^3 km^2)	Length (km)	Max_elev (m)	Climate	Geology	Q (km^3/yr)	TSS (Mt/yr)	TDS (Mt/yr)	Refs.
Burma										
Great Tenasserim	Andaman Sea	15	450	>1000	Tr-W-S	PreMes				21
Irrawaddy	Bay of Bengal	430	2300	>3000	Tr-W-S	PreMes	380 (430)	325 (360)	98	7, 10, 17, 22
Kaladan	Bay of Bengal	40	320	>1000	Tr-W-S	Cen S/M				2, 21
Salween	Bay of Bengal	270	2800	>2000	Tr-H-S	PreMes	210	180	65	8, 17, 21
Sittang	Andaman Sea	35	560	>1000	Tr-W-S	Cen S/M	50			21
Malaysia										
Peninsula										
Endau	South China Sea	4		>500	Tr-W-S	Mes S/M				15
Johore	Singapore Strt.	4.3	120	>500	Tr-H-S	Mes S/M	2.4			14
Kelantan	South China Sea	12	420	1900	Tr-W-S	PreMes	18	2.5	1.1	2, 8, 17, 21, 28
Kinabatangan	South China Sea	16	560	800	Tr-W-S	Mes S/M	20			3
Langat	Mallaca Strait	2.5	160	860	Tr-H-S	PreMes	1.8	0.6		2, 21
Muda	Mallaca Strait	7.4		>200	Tr-W-S	PreMes	3.6	0.09		6, 15
Murar	Mallaca Strait	3.2		<100	Tr-H-S	PreMes	1.7	0.07		6
Pahang	South China Sea	19	320	>1000	Tr-W-S	PreMes	18	3		2, 3, 8, 15, 21, 28
Perak	Mallaca Strait	13	270	1900	Tr-W-S	PreMes	12	0.9		13, 21, 28
Selangor	Mallaca Strait	3.2		1500	Tr-W-S	PreMes	3	0.09		4, 15
Trengganu	South China Sea	3.3	130	>500	Tr-W-S	PreMes	6.5	0.58		6, 8, 15, 29
Borneo										
Baram	South China Sea	23	630	2400	Tr-W-S	Mes S/M	44			
Kidurong	South China Sea	5.4		1000	Tr-W-S	Cen S/M				
Lupar	South China Sea	7	170	>500	Tr-W-S	Cen S/M				
Padas	South China Sea	5		2600	Tr-W-S	Cen S/M	7.1			9
Rajang	South China Sea	51	530	1800	Tr-W-S	Cen S/M	110	30		23, 24, 25
Segama	Celebes Sea	6		1600	Tr-W-S	Mes I				
Thailand										
Bangpakhlong	G. of Thailand	10	200	1300	Tr-H-S	PreMes	4.9		0.09	13, 17, 21
Chao Phraya	G. of Thailand	160	1200	>1000	Tr-H-S	PreMes	30	3 (30)	5.3	17, 18, 31
Khlong Phum Duang	G. of Thailand	3		< 100	Tr-W-S	Cen S/M	3.4	0.12		30
Mae Klong	G. of Thailand	31	520	2200	Tr-H-S	PreMes	13	8.1		5, 12, 21, 28
Pattani	G. of Thailand	4	210	>200	Tr-W-S	PreMes	3	0.35		6, 30
Petch	G. of Thailand	6		>1000	Tr-H-S	PreMes	1.4			5

Table F3. (*Continued*)

Tapi	G. of Thailand	12	160	>100	Tr-W-S	PreMes	17			21, 30
Vietnam										
Ba	South China Sea	14	390	1800	Tr-H-S	PreMes	9	1		19, 25
Ca	South China Sea	27	530	2700	Tr-W-S	PreMes	22	4	2.7	20
Cai	South China Sea	2	55	>200	Tr-W-S	PreMes	5			19
Da Nhim	South China Sea	8	270	>1000	Tr-H-S	PreMes				
Ma	South China Sea	28	540	2000	Tr-H-S	PreMes	17	3	2.2	20
Mekong	South China Sea	800	4800	5100	Tr-H-S	PreMes-Mes	550	110 (150)	65	1, 8, 16, 17, 19
Nam San	South China Sea	11	230	>1000	Tr-W-S	PreMes				
Sai Gon	South China Sea	44	590	2300	Tr-W-S	Cen S/M	42	3	2.25	20, 21
Song Hong	South China Sea	160	1100	3100	Tr-W-S	PreMes-Mes	120	50 (110)	20	20, 26, 27
Thai Binh	South China Sea	15	330	1300	Tr-H-S	PreMes	9	1	1.9	20
Thu-Bon	South China Sea	10	200	2600	Tr-W-S	PreMes	14	2	1.9	20, 25
Tra-Khuc	South China Sea	3	120	>500	Tr-H-S	PreMes				19
Brunei										
Belait	South China Sea	1.3		>1000	Tr-W-S	Cen S/M	1			11
Tutong	South China Sea	2.3		>1000	Tr-W-S	Cen S/M	4.4	0.25		11

References:

1. Borland, 1973; 2. Center for Natural Resources, Energy and Transport (UN), 1978; 3. Dai and Trenberth, 2002; 4. Douglas, 1968; 5. ECAFE, 1968 (cf. van der Leeden, 1975); 6. FAO Global River Sediment Yield Database, http://www.fao.org/nr/water/aquastat/sediment/; 7. Furuichi *et al.*, 2009; 8. GEMS website, www.gemstat.org; 9. Global River Discharge Database, http://www.rivdis.sr.unh.edu/; 10. Gordon, 1885; 11. Hiscott, 2001; 12. IAHS/UNESCO, 1974; 13. IOC/WESTPAC, 1991; 14. Jayawardena *et al.*, 1997; 15. Keat and Alias, 1982; 16. Lu and Siew, 2005; 17. Meybeck and Ragu, 1996; 18. Milliman and Meade, 1983; 19. Ngo, T-T 1967. Water for Peace (cf. van der Leeden, 1975); 20. Pham Van Ninh, personal communication; 21. Rand McNally, 1980; 22. Robinson *et al.*, 2007; 23. Staub and Gastaldo, 2003; 24. Staub *et al.*, 2000; 25. Takeuchi *et al.*, 1995; 26. Tanabe *et al.*, 2006; 27. Thanh *et al.*, 2004; 28. UNESCO (WORRI), 1978; 29. Vörösmarty *et al.*, 1996e; 30. Wattayakorn, personal communication; 31. Winterwerp *et al.*, 2005

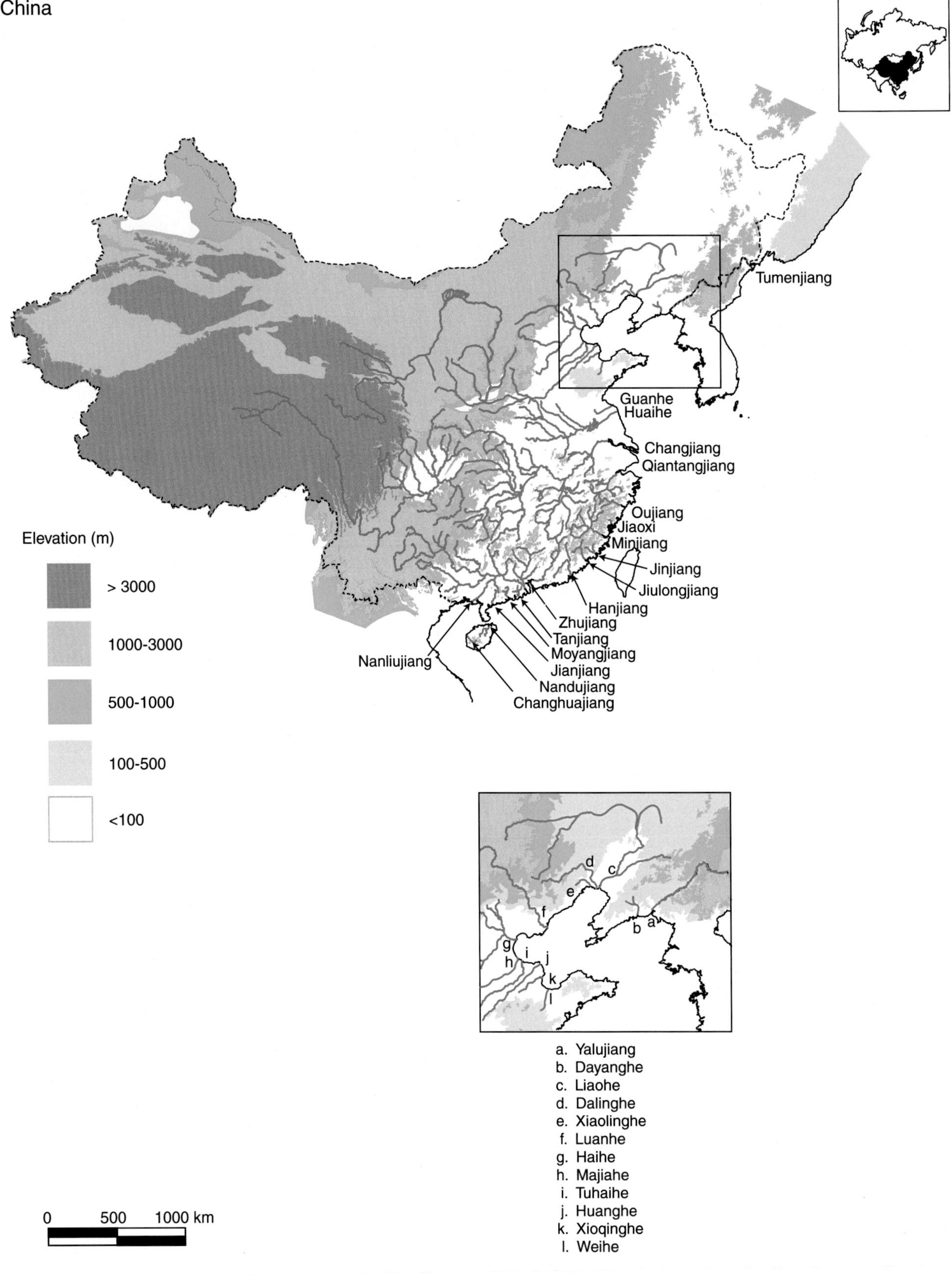
China
Tumenjiang
Guanhe
Huaihe
Changjiang
Qiantangjiang
Oujiang
Jiaoxi
Minjiang
Jinjiang
Jiulongjiang
Hanjiang
Zhujiang
Tanjiang
Moyangjiang
Jianjiang
Nandujiang
Changhuajiang
Nanliujiang
Elevation (m)
> 3000
1000-3000
500-1000
100-500
<100
0 500 1000 km
a
b
c
d
e
f
g
h
i
j
k
l
a. Yalujiang
b. Dayanghe
c. Liaohe
d. Dalinghe
e. Xiaolinghe
f. Luanhe
g. Haihe
h. Majiahe
i. Tuhaihe
j. Huanghe
k. Xioqinghe
l. Weihe

Figure F8

China

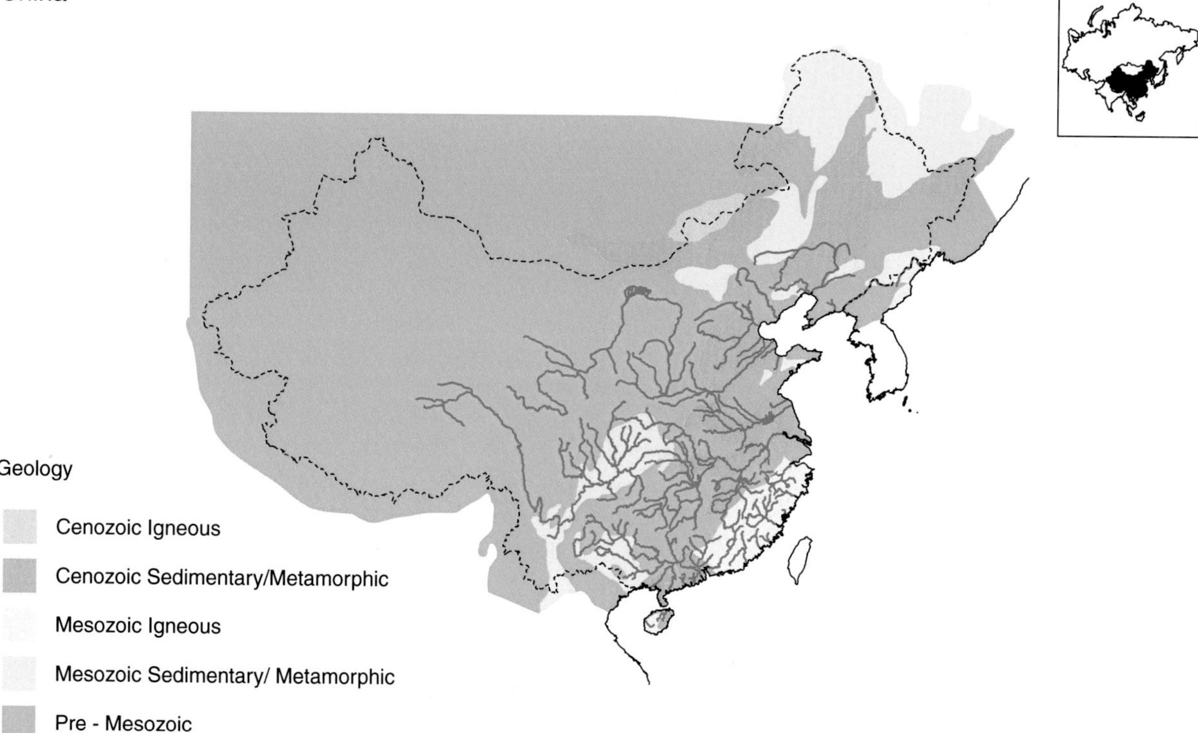
Geology
Cenozoic Igneous
Cenozoic Sedimentary/Metamorphic
Mesozoic Igneous
Mesozoic Sedimentary/ Metamorphic
Pre - Mesozoic

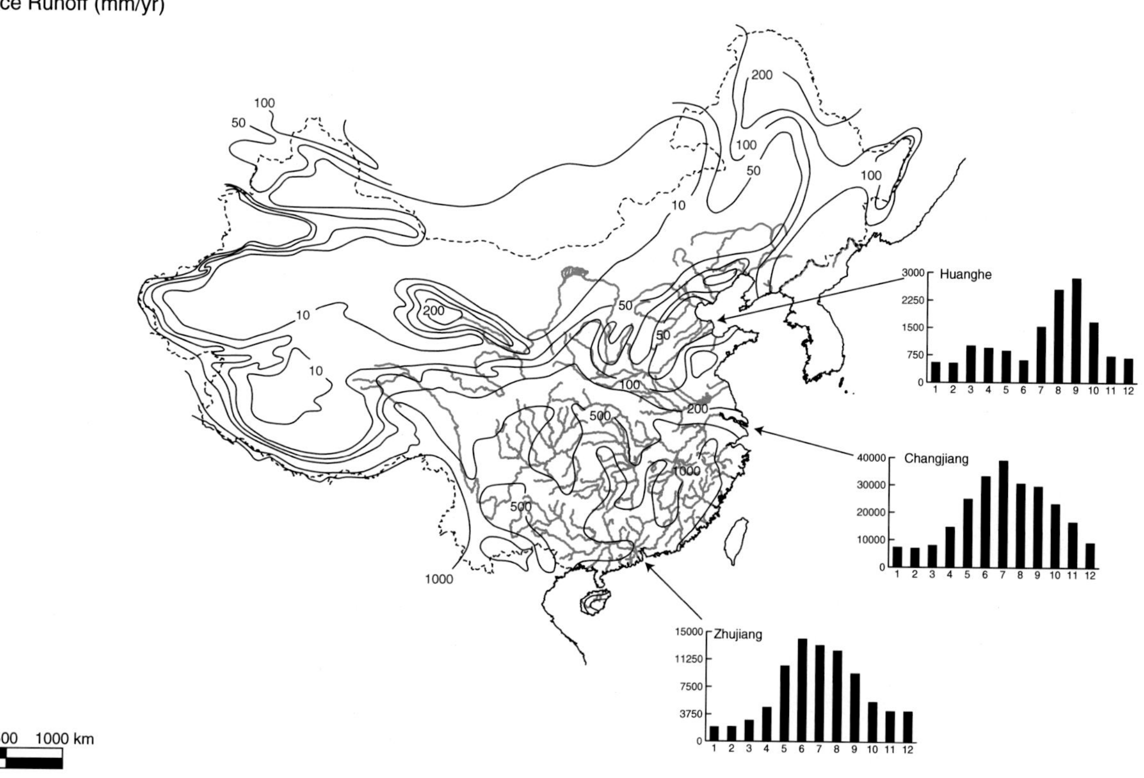
Surface Runoff (mm/yr)
100
50
200
10
500
1000
Huanghe
Changjiang
Zhujiang
0 500 1000 km

Figure F9

Table F4.

River name	Ocean	Area (10^3 km^2)	Length (km)	Max_elev (m)	Climate	Geology	*Q* (km^3/yr)	TSS (Mt/yr)	TDS (Mt/yr)	Refs.
China										
Mainland										
Changjiang	East China Sea	1800	6300	3200	Te-H-S	PreMes	900	470	180	2, 8, 16
Dalinghe	Yellow Sea	23		500	Te-A-S/Au	PreMes	2.1	36		2, 9
Dayanghe	Yellow Sea	6	190	440	Te-H-S/Au	Cen S/M	3			12
Guanhe	Yellow Sea	6.4		<100	Te-H-S	Cen S/M	3.4	0.7		
Haihe	Yellow Sea	320	1300	>1000	Te-SA-S	Cen S/M	27	0.007 (0.05)	5.3	1, 15
Hanjiang	South China Sea	30	430	400	Tr-H-S	PreMes	26	7.2 (24)		2, 5, 7
Huaihe	Yellow Sea	260	1100	1700	Te-A-S	Cen S/M	5.1 (22)	0.12 (14)	2.5	1, 2, 15
Huanghe	Yellow Sea	750	5500	3100	Te-A-S/Au	PreMes	15 (43)	150 (1100)	21	2, 10, 11, 15, 16
Jianjiang	South China Sea	9.5	170	1000	STr-H-S	PreMes	5.5	1.5		12
Jiaoxi	East China Sea	5.5	170	1400	STr-W-S	Mes I	6.9			12
Jinjiang	East China Sea	11	290	>500	Te-H-S	Mes I	5.1	2.4		2, 12
Jiulongjiang	South China Sea	15	260	790	Te-H-S	Mes I	8.3 (15)	3.1	1.3	2, 5, 7, 12
Liaohe	Yellow Sea	220	1400	1000	Te-A-S	Cen S/M	4.1 (31)	2.7 (39)	1.3	1, 2, 7, 9, 15
Luanhe	Yellow Sea	54	880	2300	Te-A-S/Au	PreMes	23	23	1.2	2, 7, 12, 15
Majiahe	Yellow Sea	8.7			Te-A-S/Au	Cen S/M	0.23	0.07		
Minjiang	East China Sea	61	580	>1000	STr-W-S	Mes I	58	2.4 (7.7)	2.4	1, 2, 5, 7, 10
Moyangjiang	South China Sea	6.1	200	460	Tr-W-S	Cen S/M	8.5	0.8		12
Nanliujiang	South China Sea	6.6	120	610	Tr-W-S	Cen S/M	5.1	1.1		12
Oujiang	East China Sea	18	390	1900	STr-W-S	Mes I	19	2.7		2, 4, 12
Qiantangjiang	East China Sea	42	490	1000	Te-W-S	PreMes	31	4.4	2.5	5, 7, 10, 15
Tanjiang	South China Sea	5.3	230	740	Tr-W-S	Cen S/M	6.3			12
Tuhaihe	Yellow Sea	19	440	>400	Te-A-S/Au	Cen S/M	1.2	0.5		12
Tumenjiang	Sea of Japan	33	540	>1000	Te-SA-S	Mes I	1.4			12
Weihe	Yellow Sea	6.4	250	>100	Te-SA-S/Au	Cen S/M				12
Xiaolinghe	Yellow Sea	5.5	180	220	Te-A-S	Cen S/M	0.22			12
Xiaoqinghe	Yellow Sea	12	240	>500	Te-H-S/Au	PreMes	8.1			12
Yalujiang	Yellow Sea	64	790	2400	Te-H-S	Cen S/M	25 (38)	4.8	2.1	3, 5, 12, 13, 15
Zhujiang	South China Sea	490	2200	2000	STr-H-S	PreMes	260	25 (80)	34 (58)	2, 6, 7, 9, 14, 15
Hainan Island										
Changhuajiang	South China Sea	5.1	220	1800	Tr-W-S	Mes I	3.8	0.08		
Nandujiang	South China Sea	6.6		200	Tr-W-S	PreMes	5.1	1.1		

References:
1. Dai *et al.*, 2009; 2. Dept. Water Conserv. & Electric Power, 1982; 3. Global River Discharge Database, http://www.rivdis.sr.unh.edu/; 4. Jayawardena *et al.*, 1997; 5. Liu *et al.*, 2001; 6. Luo *et al.*, 2006; 7. Meybeck and Ragu, 1996; 8. Probst, 1992; 9. Qian and Dai, 1980; 10. Wang *et al.*, 1998; 11. Wang *et al.*, 2007; 12. Xiong *et al.*, 1985; 13. Zhang *et al.*, 1998; 14. Zhang *et al.*, 2007; 15. Zhang, 1993; 16. Zhang, 1994

Taiwan, Republic of China

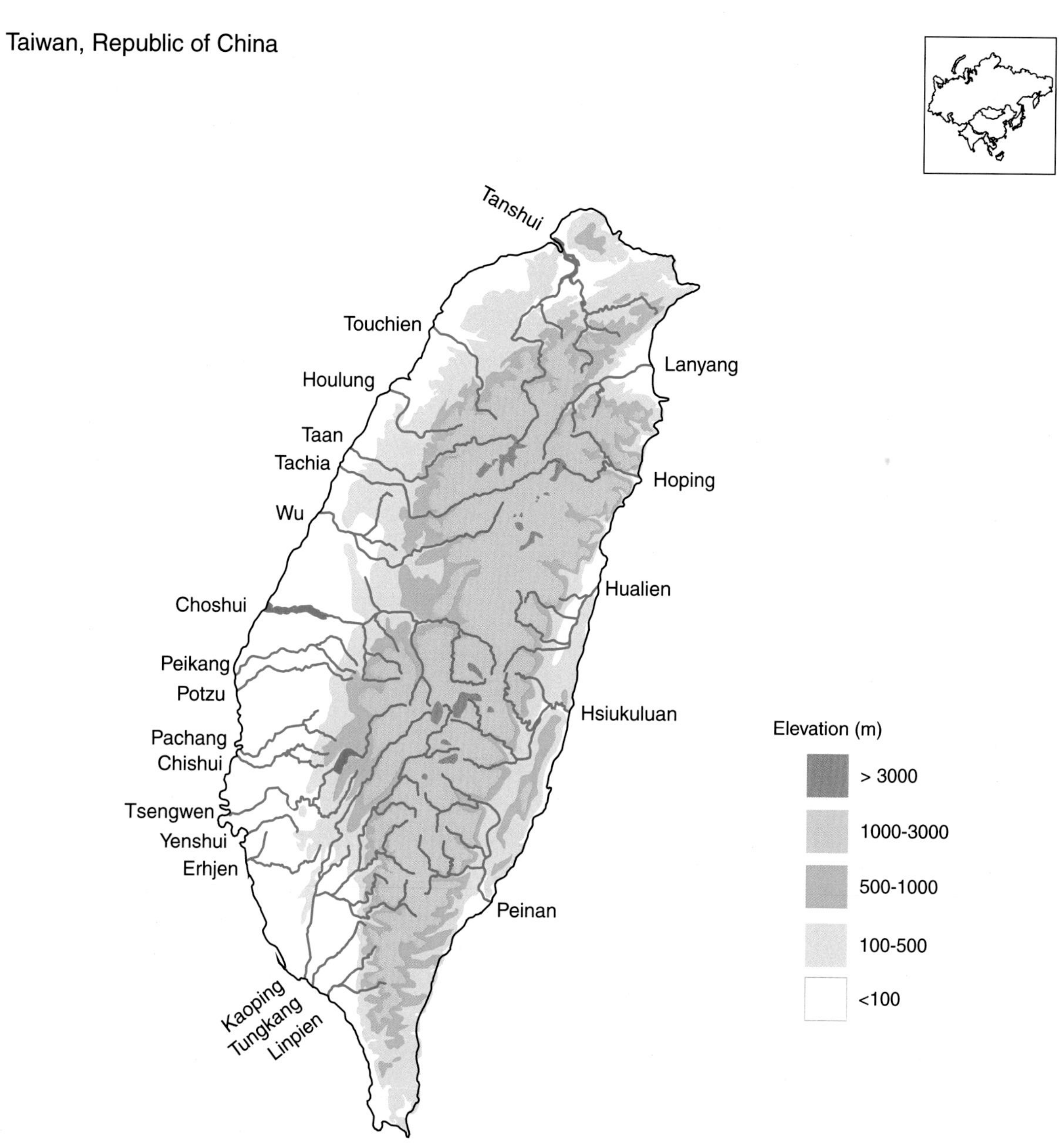

Figure F10

Taiwan, Republic of China

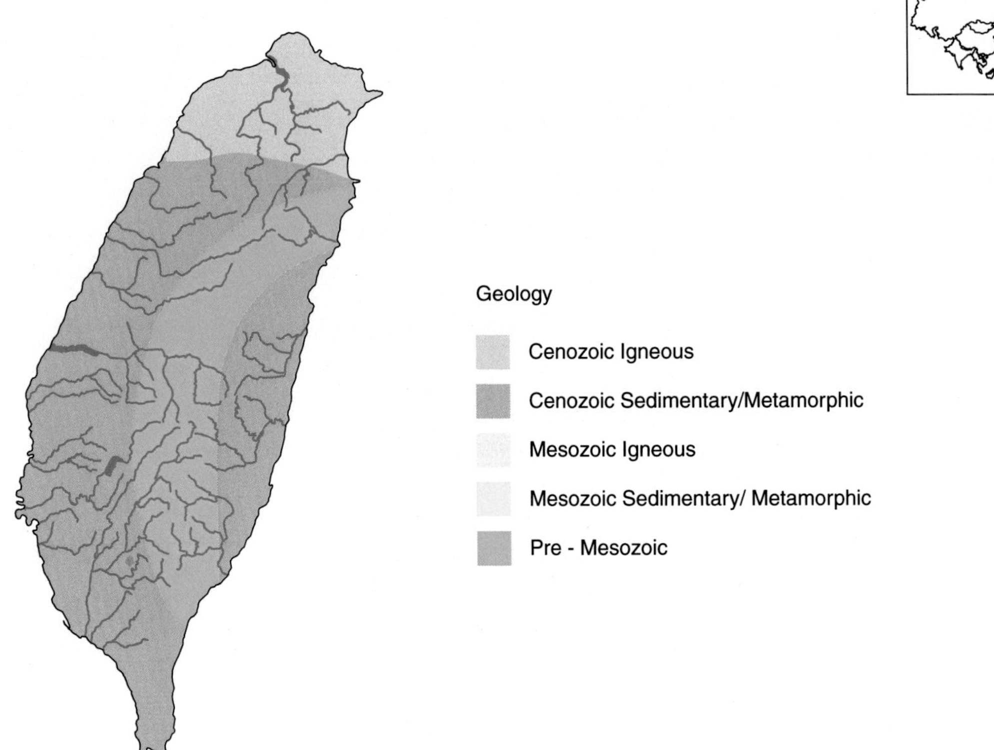

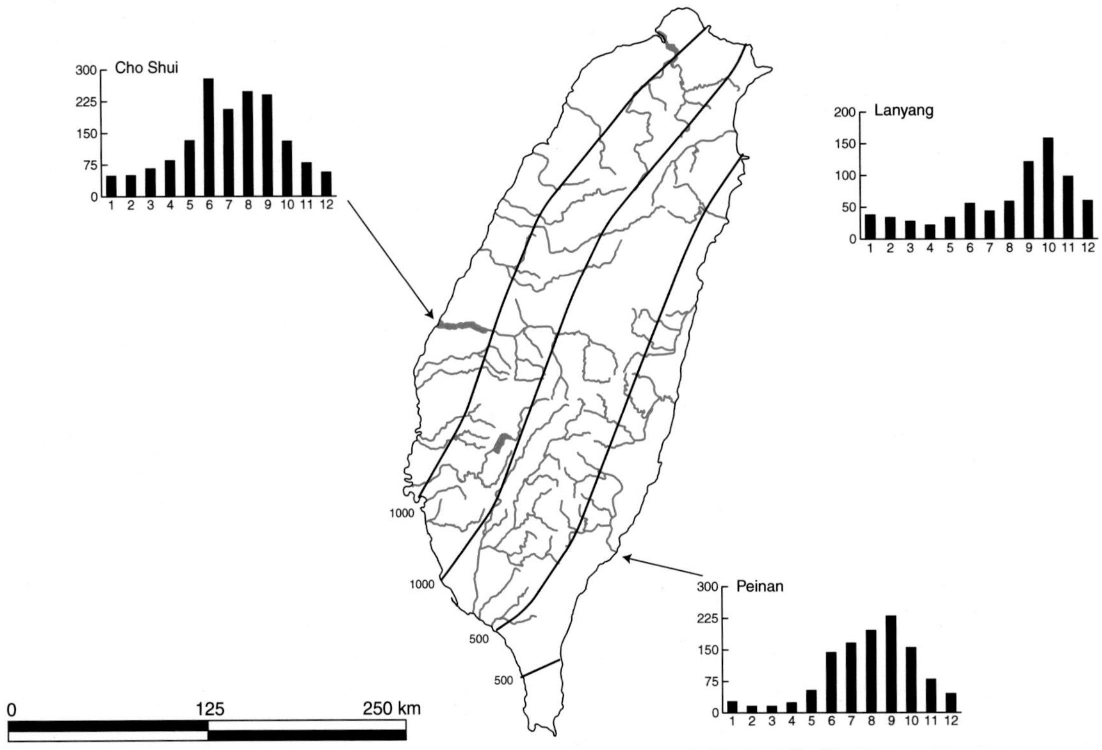

Figure F11

Table F5.

River name	Ocean	Area (10^3 km^2)	Length (km)	Max_elev (m)	Climate	Geology	Q (km^3/yr)	TSS (Mt/yr)	TDS (Mt/yr)	Refs.
Taiwan										
Chishui	Pacific (NW)	0.41	65	550	STr-W-S	Cen S/M	0.52	2.1		2, 4
Choshui	Taiwan Strait	3.1	190	3400	STr-W-S	PreMes-Cen S/M	4.3	40	1.6	1, 2, 3, 4
Erhjen	South China Sea	0.35	65	460	STr-W-S	Cen S/M	0.5	10	0.2	2, 4, 5
Hoping	Pacific (NW)	0.55	100	3700	STr-W-S	PreMes	1.2	16	0.6	1, 2, 5
Houlung	Taiwan Strait	0.47	58	2600	STr-W-S	Cen S/M	0.7	2.4	0.13	1, 2, 3, 4
Hsiukuluan	Pacific (NW)	1.8	81	2400	STr-W-S	PreMes-Cen S/M	3.9	12	1.3	1, 2, 3, 4
Hualien	Pacific (NW)	1.5	57	2300	STr-W-S	PreMes-Cen S/M	3.2	25	0.96	1, 2, 3, 4
Kaoping	South China Sea	3.3	170	4000	STr-W-S	Cen S/M	7.4	21	2.3	1, 2, 3, 4
Lanyang	Pacific (NW)	0.98	73	3500	STr-W-S	PreMes-Cen S/M	2.8	6.5	0.72	2, 3, 4
Linpien	South China Sea	0.34	42	2900	STr-W-S	Cen S/M	0.86	1.8	0.19	2, 3, 4
Pachang	South China Sea	0.47	80	1900	STr-W-S	Cen S/M	0.74	2.5	0.23	2, 3, 4
Peikang	South China Sea	0.64	82	520	STr-W-S	Cen S/M	0.8	1.4	0.29	1, 2, 3, 4
Peinan	Pacific (NW)	1.6	84	3700	STr-W-S	PreMes-Cen S/M	3	20	1	1, 2, 3, 4
Potzu	South China Sea	0.43	76	1400	STr-W-S	Cen S/M	0.55	0.83	0.21	2, 3, 4
Taan	Taiwan Strait	0.76	96	3300	STr-W-S	Cen S/M	1.1	4	0.27	1, 2, 3, 4
Tachia	Taiwan Strait	1.2	140	2600	STr-W-S	PreMes-Cen S/M	2.4	0.5	0.43	1, 2, 3, 4
Tanshui	Taiwan Strait	2.7	160	3500	STr-W-S	Cen	7	11	0.32	2, 4, 5
Touchien	Taiwan Strait	0.6	63	2200	STr-W-S	Cen S/M	0.8	1.1	0.17	1, 2, 3, 4
Tsengwen	South China Sea	1.2	140	2400	STr-W-S	PreMes-Cen S/M	1	12	0.49	1, 2, 3, 4
Tungkang	South China Sea	0.47	47	1100	STr-W-S	Cen S/M	1.1	5.2	0.28	2, 3, 4
Wu	Taiwan Strait	2	120	2600	STr-W-S	PreMes-Cen S/M	3.7	5.3	0.74	1, 2, 3, 4
Yenshui	South China Sea	0.22	87	140	STr-W-S	Cen S/M	0.3	2.2		2, 4

References:

1. Kao and Milliman, 2008; 2. Kao *et al.*, 2005; 3. Li, 1976; 4. Water Resources Planning Commission, 1984; 5. S. W. Lin, unpublished data.

Japan, North Korea and South Korea

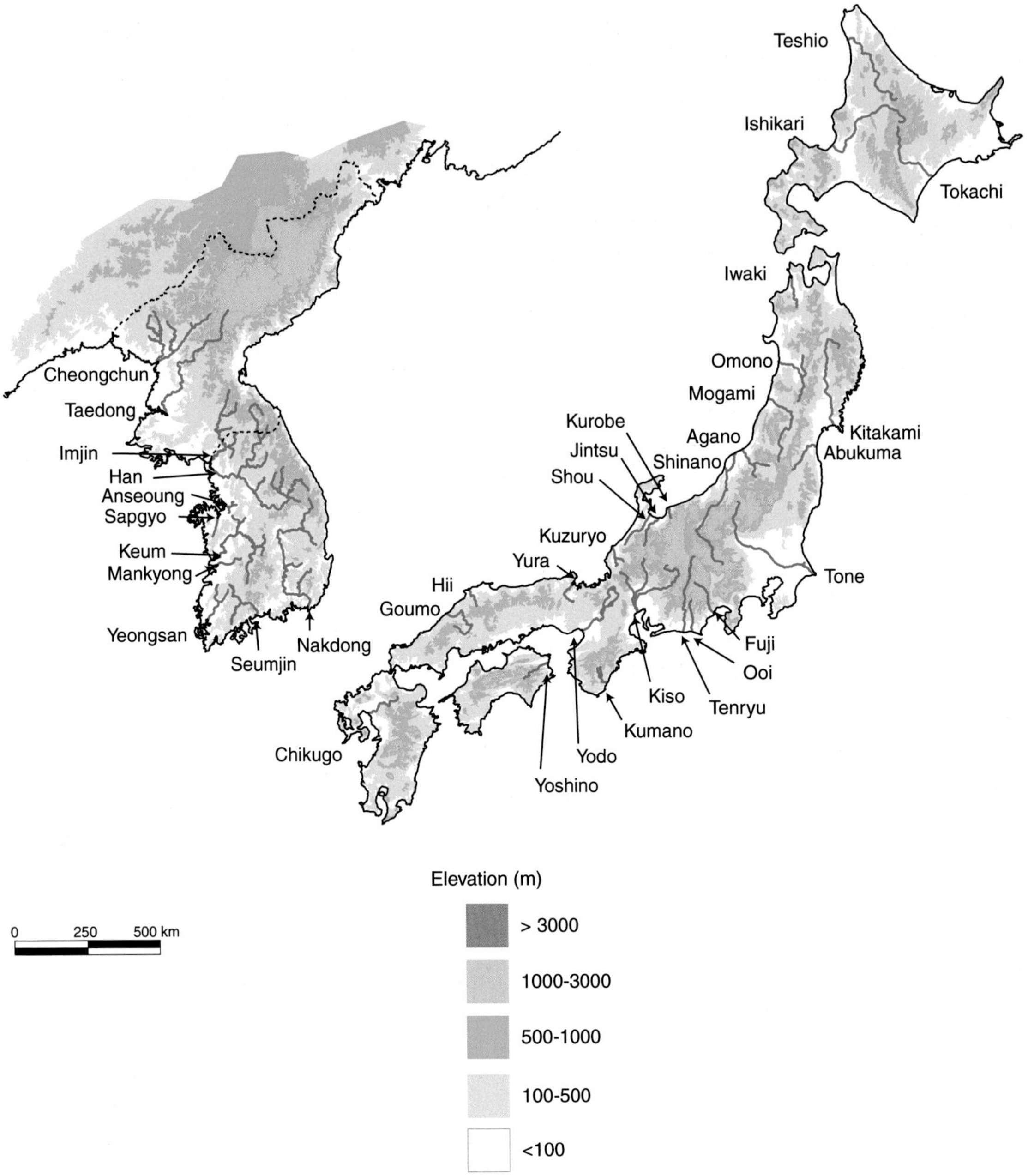

Teshio
Ishikari
Tokachi
Iwaki
Omono
Mogami
Kurobe
Agano
Kitakami
Jintsu
Shinano
Abukuma
Shou
Kuzuryo
Yura
Tone
Hii
Goumo
Fuji
Ooi
Kiso
Tenryu
Kumano
Yodo
Yoshino
Chikugo
Cheongchun
Taedong
Imjin
Han
Anseoung
Sapgyo
Keum
Mankyong
Yeongsan
Nakdong
Seumjin

0 250 500 km

Elevation (m)

> 3000
1000-3000
500-1000
100-500
<100

Figure F12

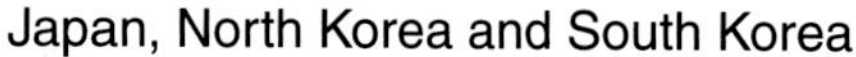

Japan, North Korea and South Korea

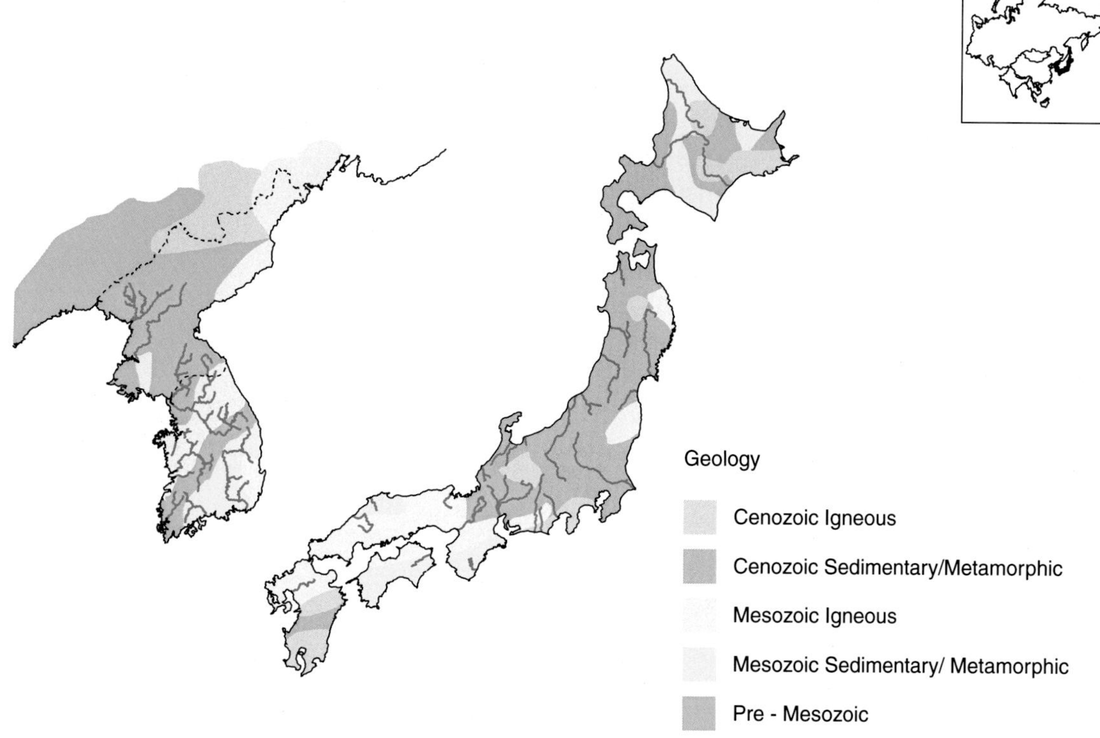

Surface Runoff (mm/yr)

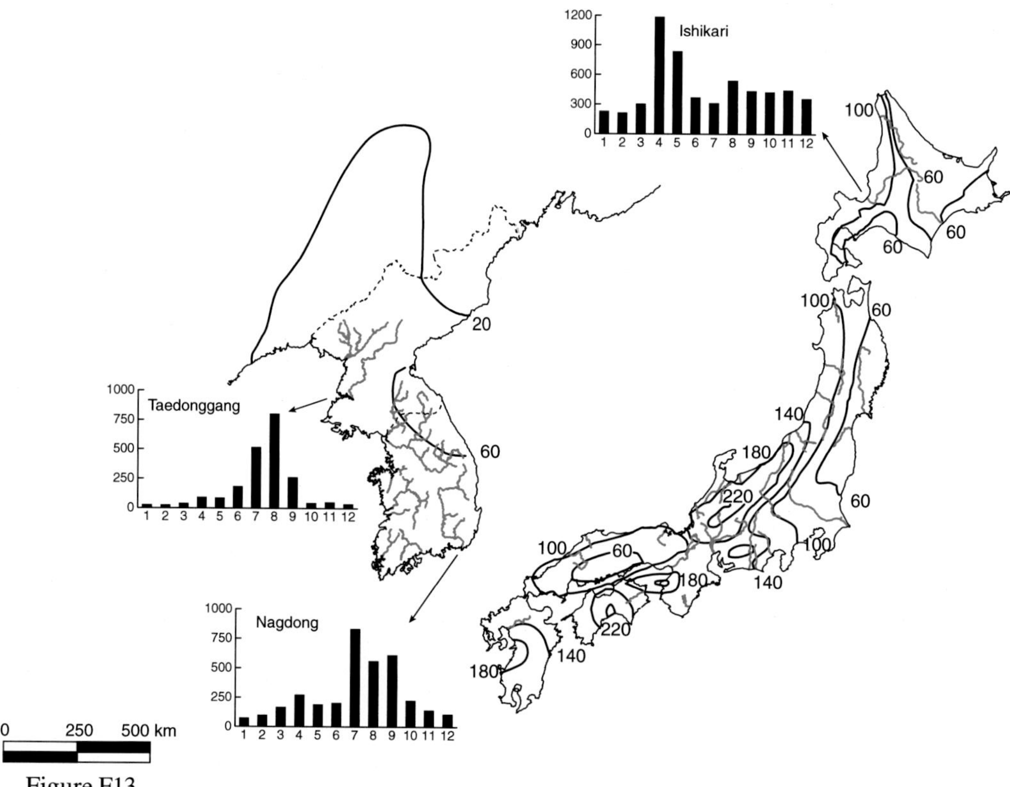

Figure F13

Table F6.

River name	Ocean	Area (10^3 km^2)	Length (km)	Max_elev (m)	Climate	Geology	Q (km^3/yr)	TSS (Mt/yr)	TDS (Mt/yr)	Refs.
Japan										
Hokkaido										
Ishikari	Sea of Japan	14	270	2300	Te-W-S	Cen S/M	16	1.8		5, 7, 11
Teshio	Sea of Japan	5.6	260	>1500	Te-W-S	PreMes	7		0.89	8, 11
Tokachi	Pacific (NW)	9	160	>1500	Te-H-S	PreMes	6.5			11
Honshu										
Abukuma	Pacific (NW)	5.4	240	>2000	Te-W-S	Cen S/M	3.5			11
Agano	Sea of Japan	7.7	210	>2500	Te-W-S	Cen S/M	10	0.9	0.58	4, 8, 11
Fuji	Pacific (NW)	3.9	130	3800	STr-H-S	Cen I	0.7			7
Goumo	Sea of Japan	3.9	240	>1500	Te-W-S	Mes I	3.7	0.053		3, 11, 13
Hii	Sea of Japan	2.1	150	>1500	STr-H-S	Mes S/M	1.1	0.45		8, 11
Iwaki	Sea of Japan	2.5	200	>1000	Te-W-S	Cen S/M	2.3	0.56		3, 11
Jintsu	Sea of Japan	2.7	240	>2000	Te-W-S	Cen S/M	4.7	0.024		3, 11
Kiso	Pacific (NW)	5	230	>2000	STr-W-S	PreMes	6.7	1.2		11
Kitakami	Pacific (NW)	10	250	>1500	Te-W-S	Cen S/M	7.9	1.1	0.78	8, 11
Kumano	Pacific (NW)	2.4	180	>1000	STr-W-S	Cen S/M	6	1.3		8, 11
Kurobe	Sea of Japan	0.68	86	>3000	Te-W-S	PreMes	1.8	0.24		11
Kuzuryo	Sea of Japan	2.9	230	>1500	Te-H-S	PreMes	2.2	0.039		3, 11
Mogami	Sea of Japan	7	230	2100	Te-W-S	Cen S/M	9.3	3.2	0.9	4, 6, 8, 11, 12
Omono	Sea of Japan	4.7	300	>1000	Te-W-S	Cen S/M	6.1	0.071		3, 11
Ooi	Pacific (NW)	1.3	170	3000	STr-W-S	Cen S/M	2	1.9	0.66	6, 8, 11
Shinano	Sea of Japan	12	370	>1500	Te-W-S	Cen S/M	14	1.6 (12)		11
Shou	Sea of Japan	1.2	110	>1500	Te-H-S	Cen S/M	0.7	1.1	1.1	11
Tenryu	Pacific (NW)	5.1	210	>2000	STr-W-S	Cen S/M	6	<5 (20)	0.43	8, 11, 12
Tone	Pacific (NW)	17	320	2500	Te-H-S	Cen S/M	6	3	0.88	8, 11
Yodo	Pacific (NW)	8.2	75	>1500	Te-W-S	PreMes	8.9	1.9	0.75	8, 11, 13
Yura	Sea of Japan	1.9	150	>1000	STr-W-S	PreMes	1.7	0.012		3, 11
Kyushu										
Chikugo	East China Sea	2.9	140	1800	STr-W-S	Mes I	3	1.8		6, 7, 8, 11

Table F6. (*Continued*)

Shikoku										
Yoshino	Pacific (NW)	3.2	190	1900	STr-W-S	Mes I	4.9			11, 12
Korea										
North										
Cheongchun	Yellow Sea	9.5	220	<1000	Te-H-S	PreMes				
Taedong	Yellow Sea	12	390	>1000	Te-H-S	PreMes	5.7			2
South										
Anseoung	Sea of Japan	1.7	76	< 100	Te-H-S	Mes I	0.88			3
Han	Yellow Sea	25	510	1300	Te-H-S	PreMes	17	10	1.2	1, 3, 8, 10, 13
Imjin	Yellow Sea	8.9	240	>1000	Te-H-S	PreMes				9
Keum	Yellow Sea	9.9	400	>500	Te-H-S	Mes I	4.2	2.2		3
Mankyong	Yellow Sea	1.6	81	>100	Te-H-S	PreMes	1.1	0.64		9
Nakdong	Tsushima Strait	24	530	>500	Te-H-S	Mes S/M	9.2	8.2		3, 13
Sapgyo	Sea of Japan	1.7	65	>100	Te-H-S	Mes I	1.2	0.13		3
Seumjin	Tsushima Strait	4.9	220	>400	Te-H-S	Mes S/M	3.1	2		3
Yeongsan	Yellow Sea	2.8	120	>400	Te-H-S	PreMes	1.5	0.7		3

References:

1. Center for Natural Resources, Energy and Transport (UN), 1978; 2. Global River Discharge Database, http://www.rivdis.sr.unh.edu/; 3. Hong *et al.*, 1997; 4. IAHS/UNESCO, 1974; 5. Jansen and Painter, 1974; 6. Jansen *et al.*, 1979; 7. Jayawardena *et al.*, 1995; 8. Meybeck and Ragu, 1996; 9. Republic of Korea Ministry of Construction and Transportation, 2007; 10. Ryu *et al.*, 2007; 11. Y. Saito, personal communication; 12. Takeuchi *et al.*, 1995; 13. Vörösmarty *et al.*, 1996e; 12. Y. Saito, personal communication.

Appendix G Oceania

Figure G1

(1) Philippines
(2) Indonesia and Papua New Guinea
(3) Australia
(4) New Zealand
(5) Fiji (see Fig. A8, Table A4)

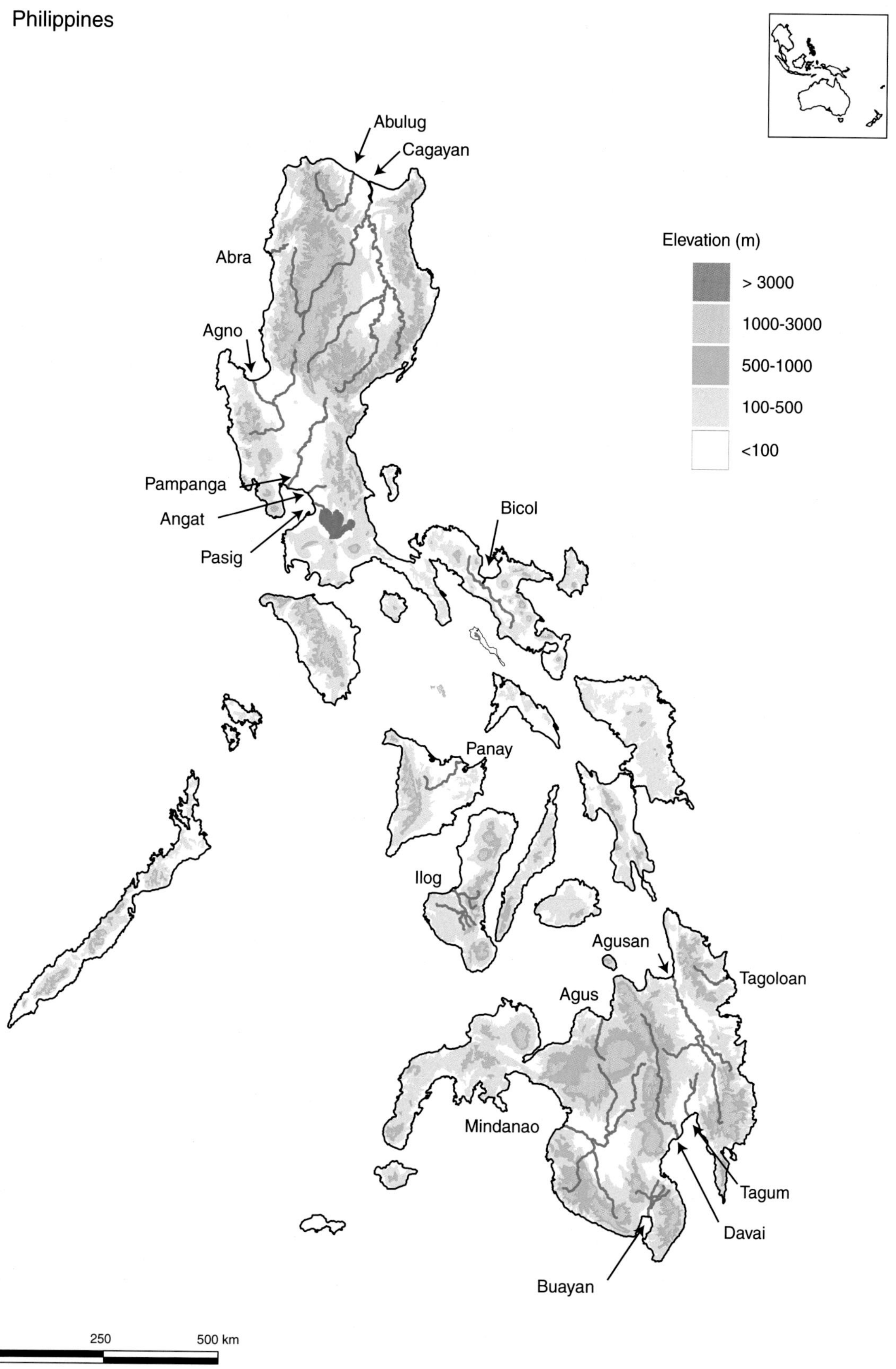
Philippines
Abulug
Cagayan
Abra
Agno
Pampanga
Angat
Pasig
Bicol
Elevation (m)
> 3000
1000-3000
500-1000
100-500
<100
Panay
Ilog
Agusan
Tagoloan
Agus
Mindanao
Tagum
Davai
Buayan
0
250
500 km

Figure G2

Philippines

Geology

Cenozoic Igneous

Cenozoic Sedimentary/Metamorphic

Mesozoic Igneous

Mesozoic Sedimentary/ Metamorphic

Pre - Mesozoic

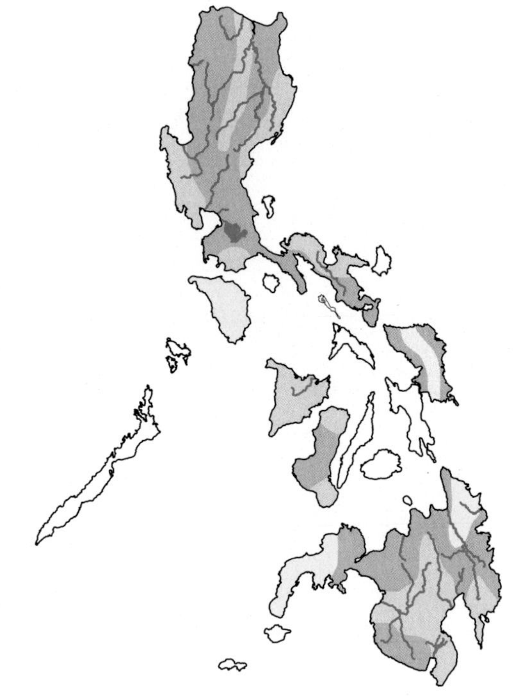

Surface Runoff (mm/yr)

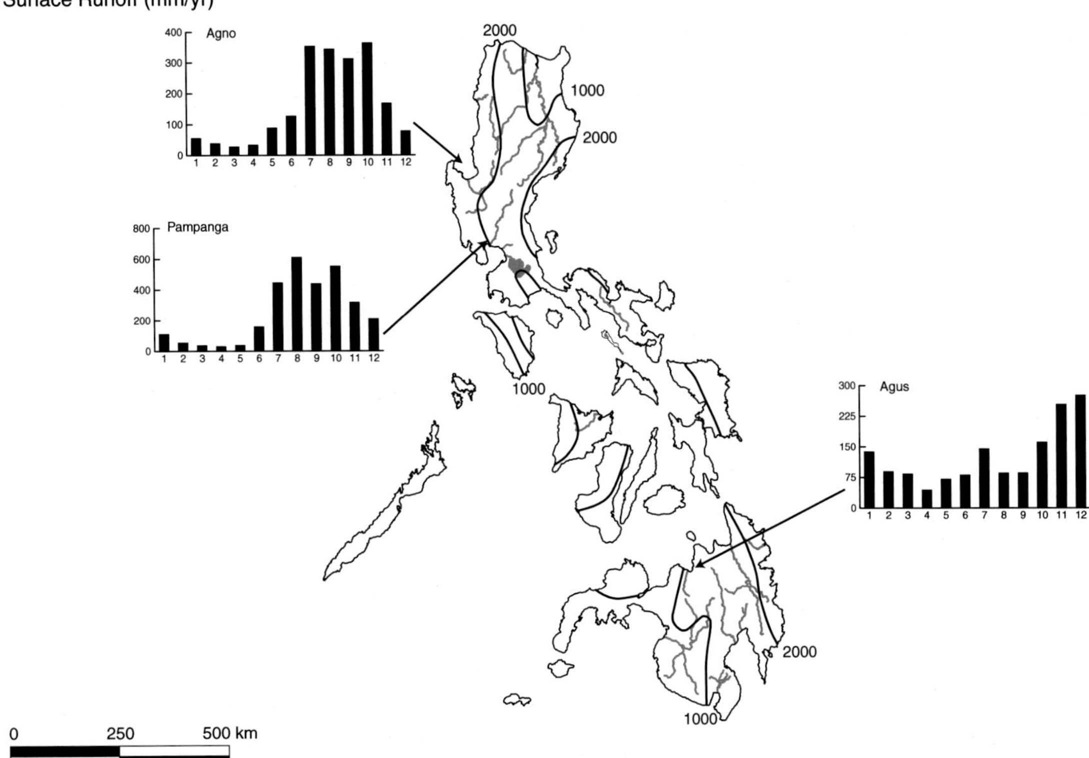

Figure G3

Table G1.

River name	Ocean	Area (10^3 km^2)	Length (km)	Max_elev (m)	Climate	Geology	Q (km^3/yr)	TSS (Mt/yr)	TDS (Mt/yr)	Refs.
Philippines										
Luzan										
Abra	South China Sea	6.2	88	2800	Tr-W-S	Cen S/M	13			1, 3, 5
Abulug	Pacific (W)	3.4	120	2500	Tr-W-S	Cen S/M	7.1			1, 3
Agno	South China Sea	6.9	270	3100	Tr-W-S	Cen S/M	6.7	5		1, 2, 3, 6
Angat	South China Sea	0.78		>500	Tr-W-S	Cen S/M		4.6		3, 6
Bicol	Pacific (W)	3.8	94	4400	Tr-W-S	Cen	5.1			1, 3
Cagayan	Pacific (W)	26	350	3100	Tr-W-S	Cen	54		2.8	1, 2, 3, 5
Pampanga	South China Sea	10	270	1200	Tr-W-S	Cen S/M	21	1.4		1, 2, 3, 5, 6
Pasig	South China Sea	0.6	25	>1000	STr-H-W	Cen	0.075	0.06		1, 3, 4, 5
Mindano										
Agus	Sulu Sea	1.6		>1000	Tr-W-S	Cen S/M	1.9			1, 3
Agusan	Sulu Sea	11	390	2800	Tr-W-S	Cen S/M	28			1, 2, 3, 5
Buayan	Celebes Sea	1.4		2300	Tr-W-S	Cen I	2.9			1, 3
Davai	Pacific (W)	1.6	150	>1000	Tr-W-S	Cen I	3.2			1, 3
Mindanao	Celebes Sea	23	320	3000	Tr-W-S	Cen	27			1, 3, 5
Tagoloan	Pacific (W)	1.7	70	1800	Tr-W-S	Mes S/M	4.3			1, 3
Tagum	Pacific (W)	3.1	80	1100	Tr-W-S	Cen S/M	6.1	1		4
Negros										
Ilog	Sulu Sea	2.2	124	>1000	Tr-W-S	Cen S/M				
Panay										
Panay	Sulu Sea	2.2	100	1800	Tr-W-S	Cen I	2.3			1, 3

References:
1. B. Gomez, personal communication; 2. Meybeck and Ragu, 1996; 3. National Water Resources Board, 1976; 4. Peucker-Ehrenbrink, 2009; 5. Rand McNally, 1980; 6. D. E. Walling, personal communication

Indonesia and Papua New Guinea

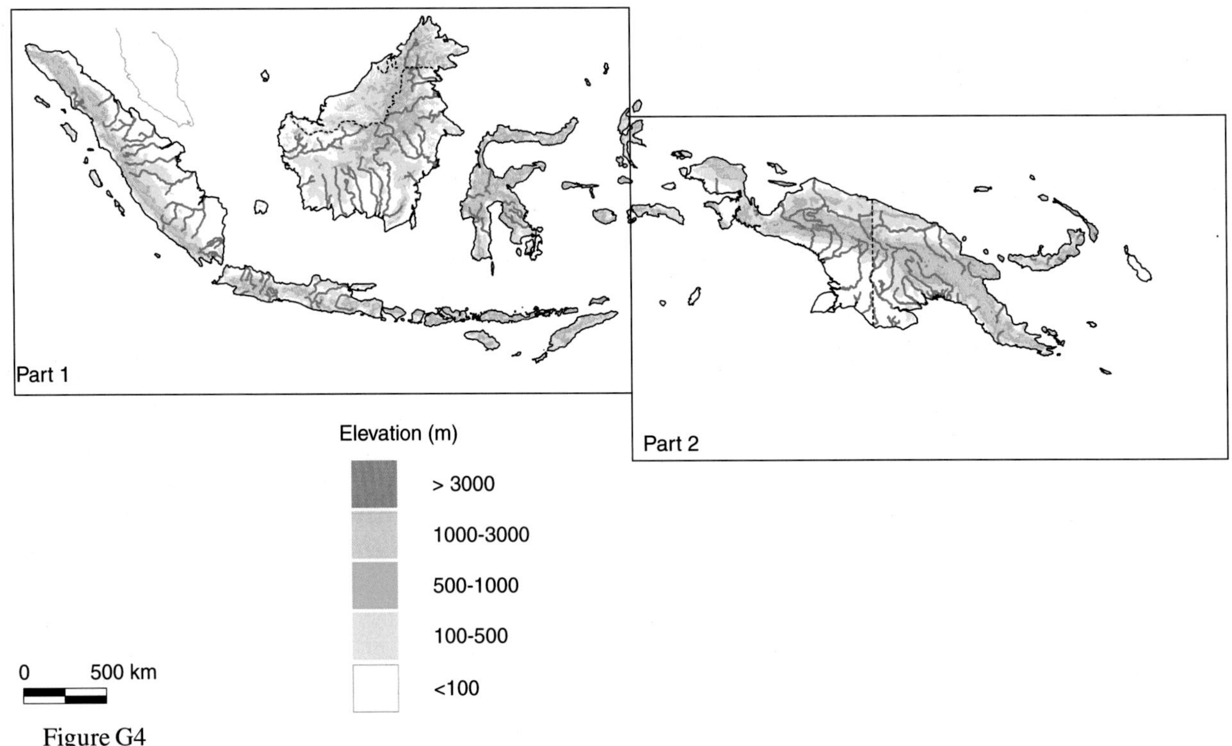

Figure G4

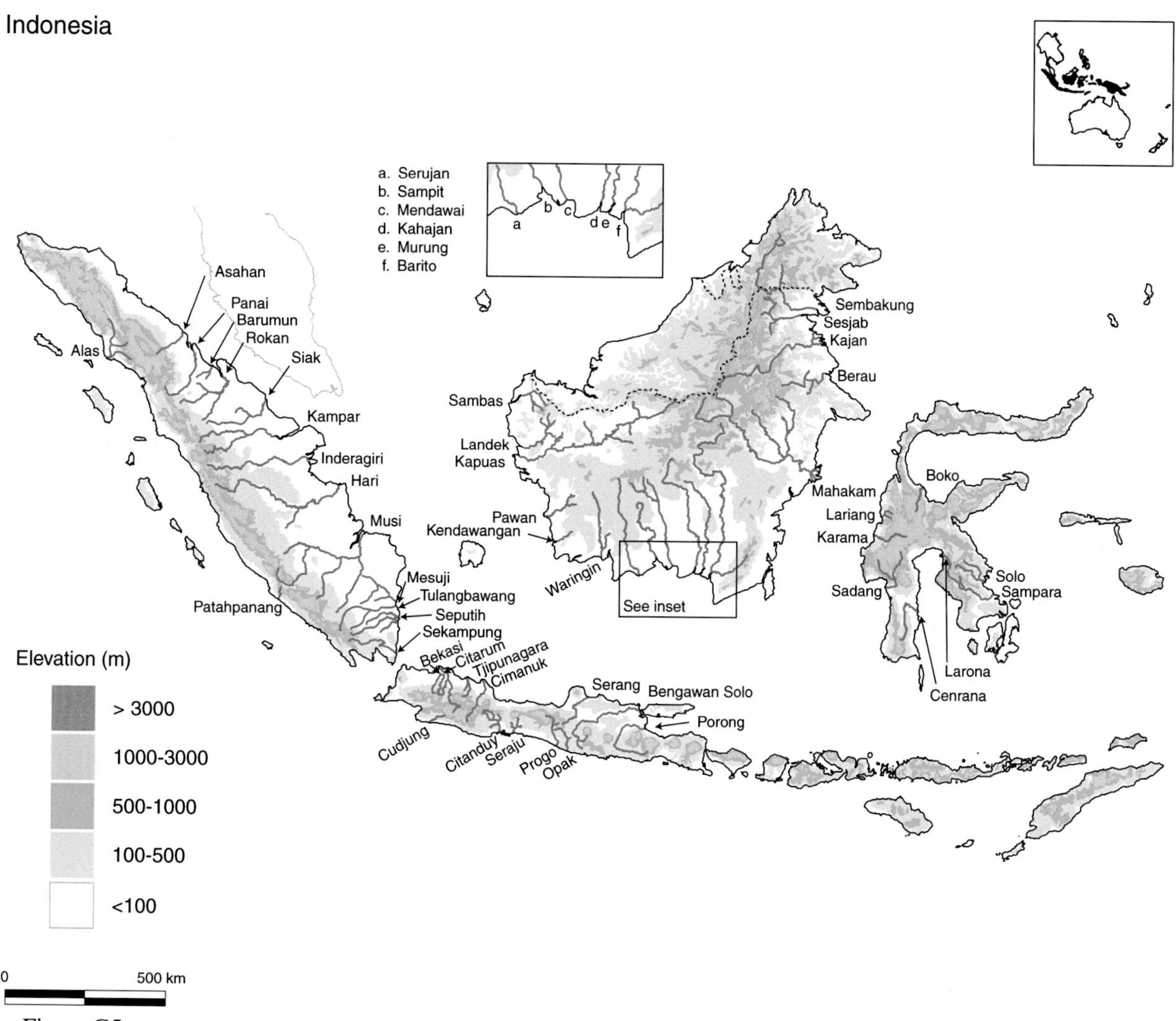
Indonesia
a. Serujan
b. Sampit
c. Mendawai
d. Kahajan
e. Murung
f. Barito
See inset
Asahan
Panai
Barumun
Rokan
Siak
Alas
Kampar
Inderagiri
Hari
Musi
Patahpanang
Mesuji
Tulangbawang
Seputih
Sekampung
Bekasi
Citarum
Tjipunagara
Cimanuk
Cudjung
Citanduy
Seraju
Progo
Opak
Serang
Bengawan Solo
Porong
Sambas
Landek
Kapuas
Pawan
Kendawangan
Waringin
Sembakung
Sesjab
Kajan
Berau
Mahakam
Lariang
Karama
Sadang
Boko
Solo
Sampara
Larona
Cenrana
Elevation (m)
> 3000
1000-3000
500-1000
100-500
<100
0
500 km

Figure G5

Indonesia

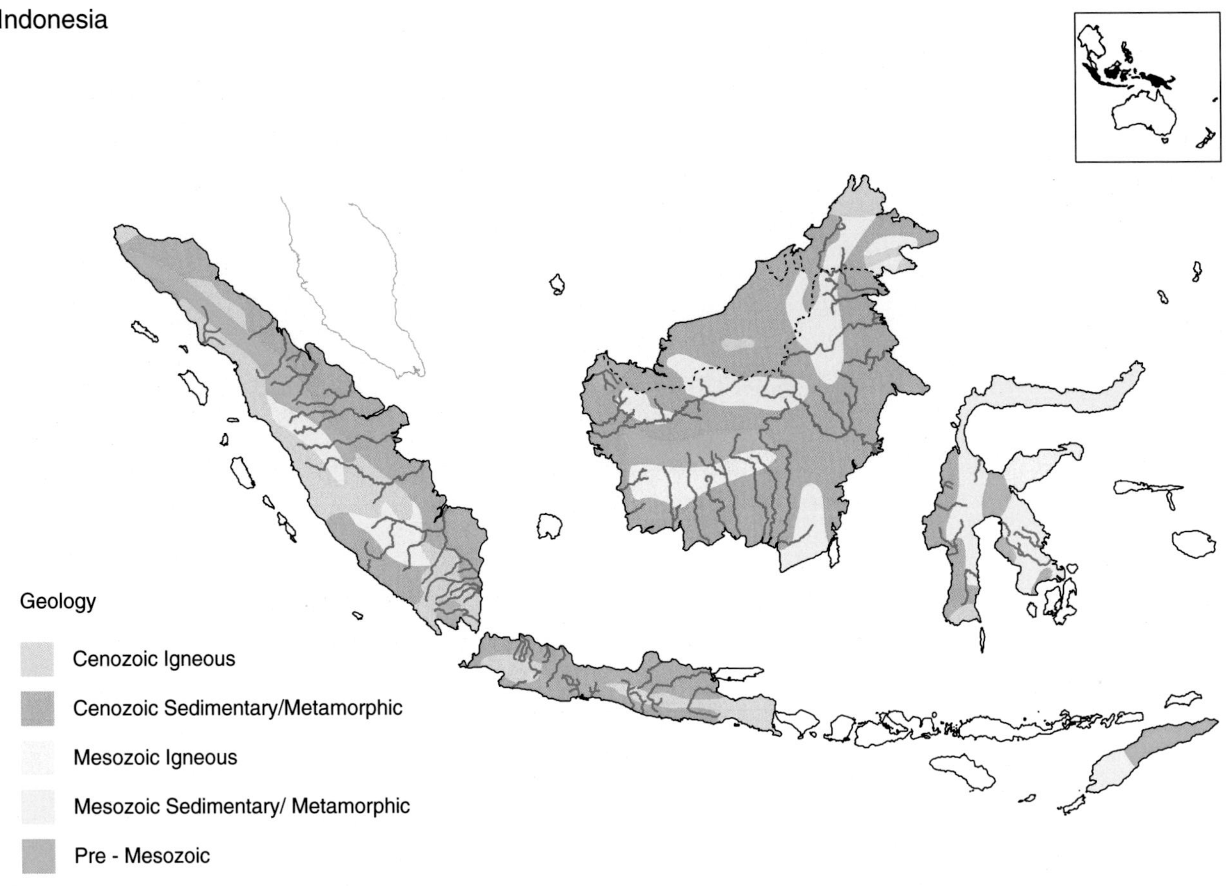

Surface Runoff

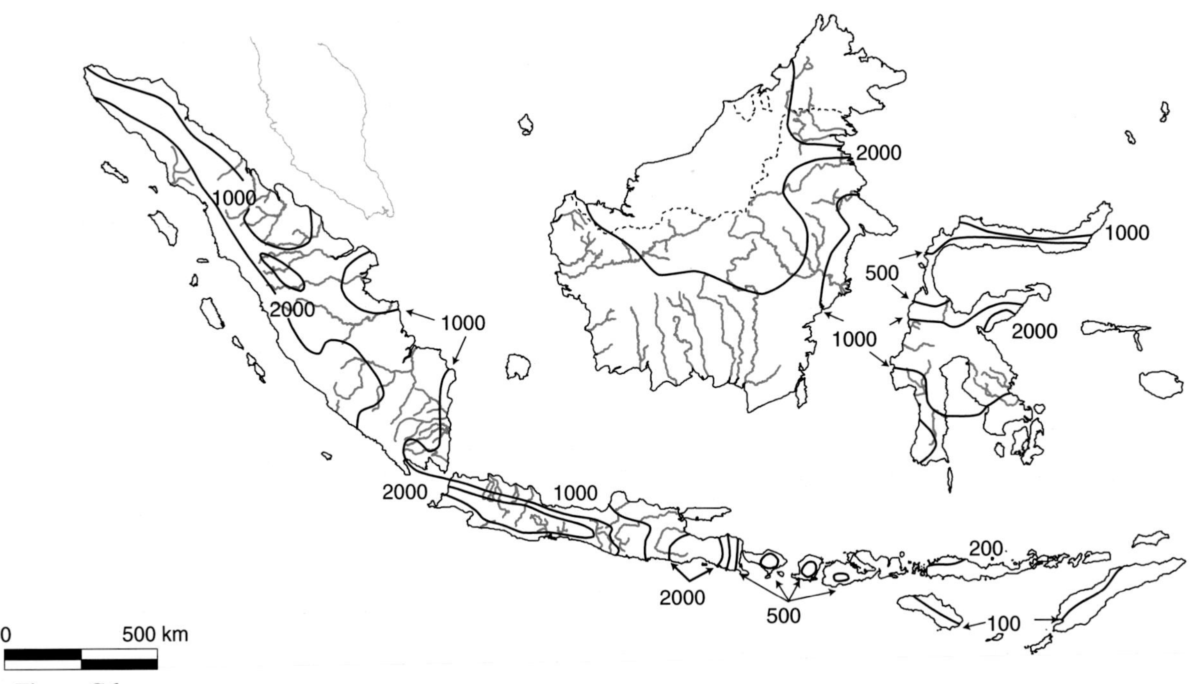

Figure G6

Irian Jaya and Papua New Guinea

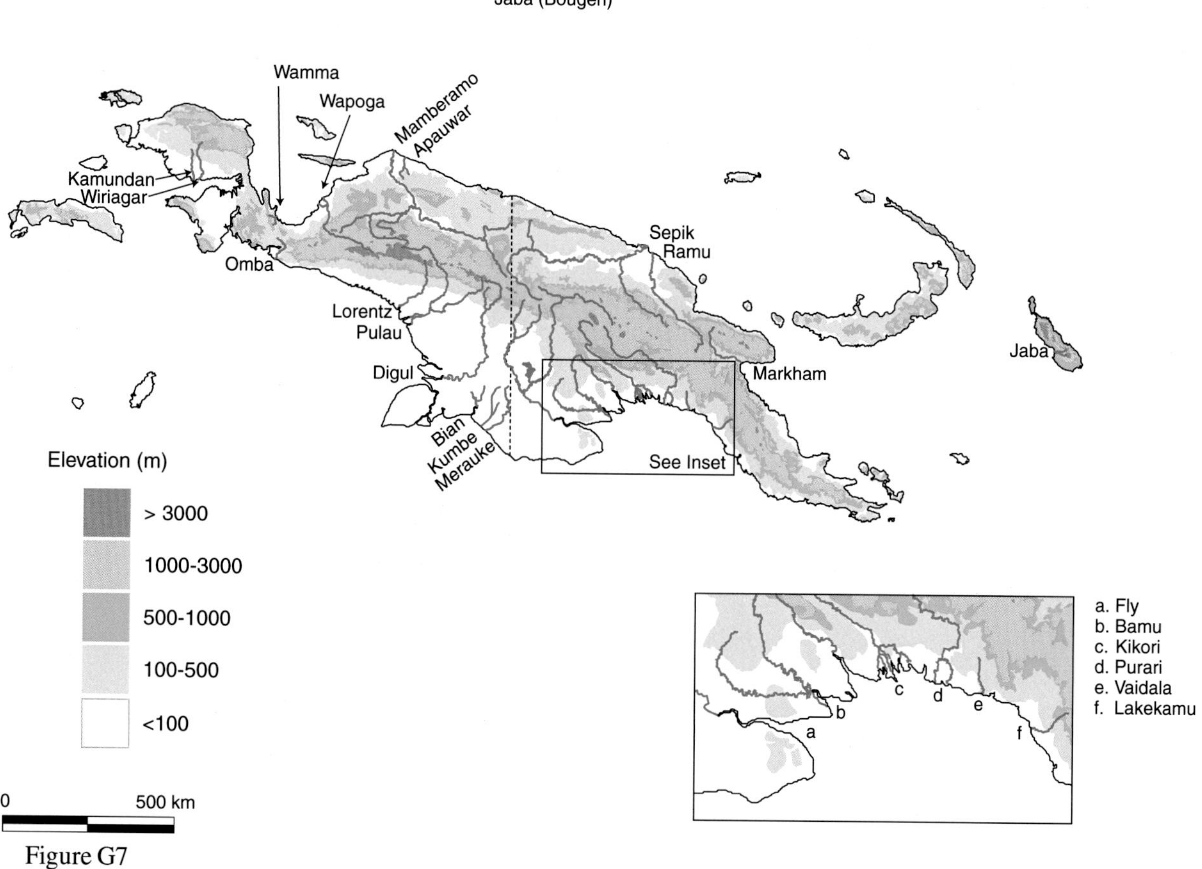

Figure G7

Irian Jaya and Papua New Guinea

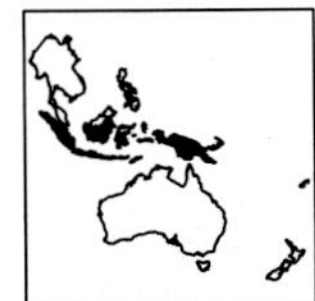

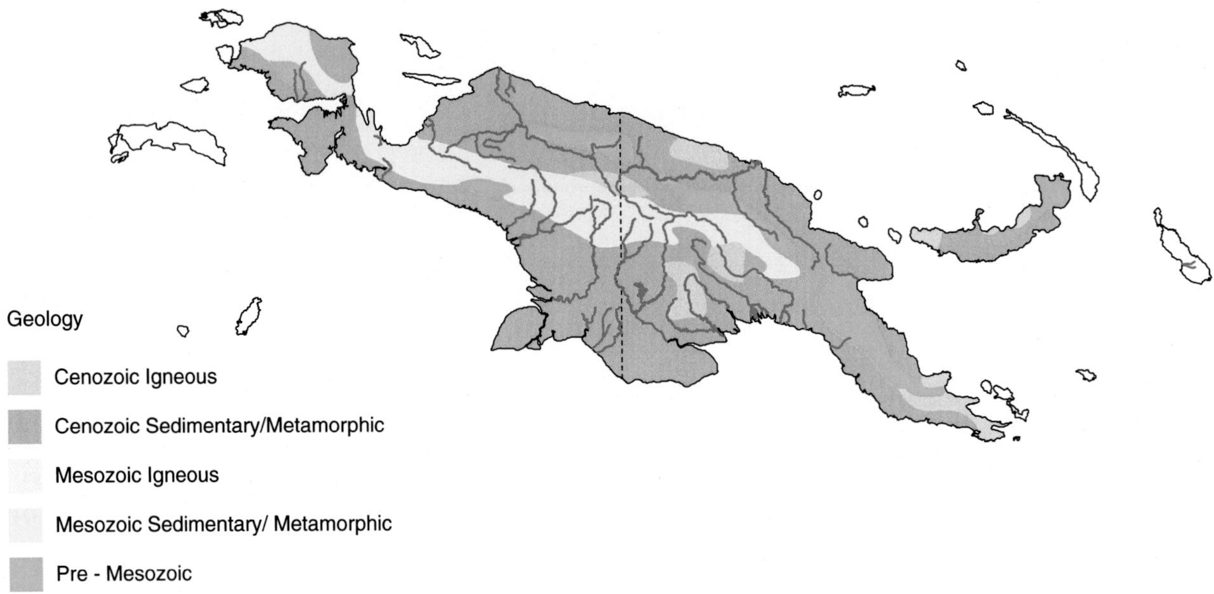

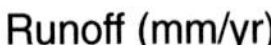

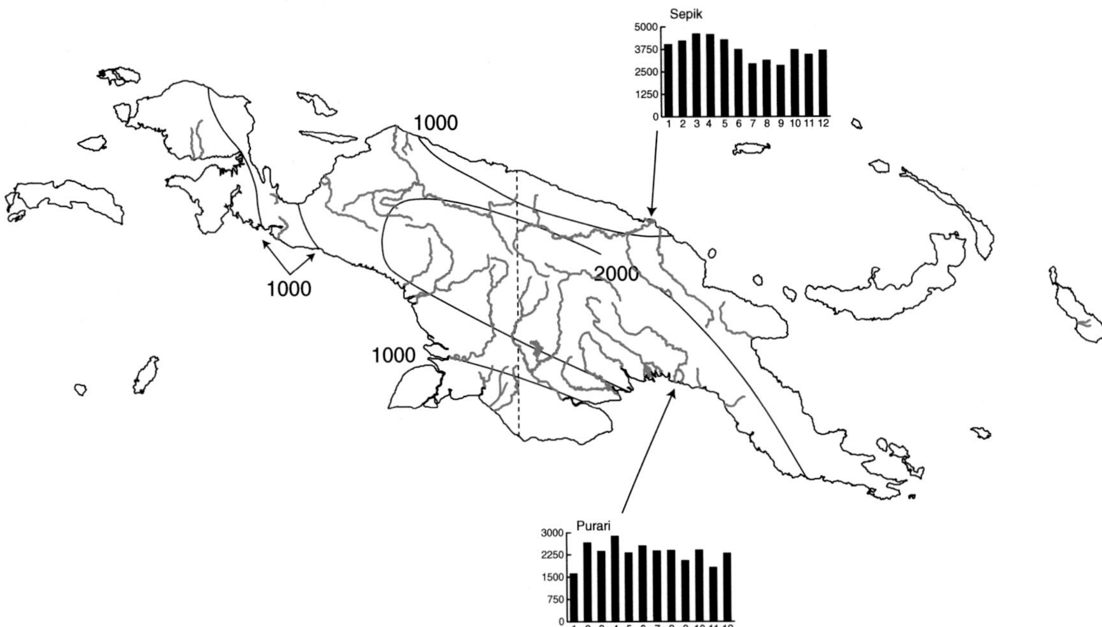

Figure G8

Table G2.

River name	Ocean	Area (10^3 km^2)	Length (km)	Max_elev (m)	Climate	Geology	Q (km^3/yr)	TSS (Mt/yr)	TDS (Mt/yr)	Refs.
Indonesia										
Borneo/Kalimantan										
Barito	Java Sea	57	880	1900	Tr-W-S	Mes S/M	87		4.2	5, 10
Berau	Celebes Sea	16	320	>1000	Tr-W-S	Cen S/M	12			15
Kahajan	Java Sea	26	410	1100	Tr-W-S	PreMes	22			15, 19
Kajan	Celebes Sea	37	460	2000	Tr-W-S	Cen S/M	39			15
Kapuas	Karimata Strait	82	1100	1800	Tr-W-S	Mes S/M	100			5, 10
Kendawangan	Karimata Strait	3	130	<100	Tr-W-S	Cen S/M				
Landek	Karimata Strait	10	200	1400	Tr-W-S	Mes S/M				
Mahakam	Celebes Sea	82	720	1700	STr-H-W	Mes S/M	87		3.6	10, 14
Mendawai	Java Sea	24	330	2300	Tr-W-S	PreMes	23			15
Murung	Java Sea	15	310	200	Tr-W-S	PreMes				
Pawan	Karimata Strait	13	240	1800	Tr-W-S	PreMes	38			15
Sambas	Karimata Strait	6.8	100	1600	Tr-W-S	Mes S/M				
Sampit	Java Sea	12	270	>500	Tr-W-S	Mes I	16			15
Sembakung	Celebes Sea	23	470	2400	Tr-W-S	Mes S/M				
Serujan	Java Sea	9.9	290	>1000	Tr-W-S	PreMes				
Sesjab	Celebes Sea	19	300	2400	Tr-W-S	Mes S/M	36			15
Waringin	Java Sea	13	350	>100	Tr-W-S	Mes I				
Celebes/Sulawesi										
Boko	Celebes Sea	5.3	170	3300	Tr-W-S	Mes S/M				
Cenrana	Java Sea	7.3	210	>1000	Tr-W-S	Cen S/M				
Karama	Celebes Sea	6	290	3100	Tr-W-S	Mes S/M				
Lariang	Makassar Strait	7.2	250	2900	Tr-W-S	Cen S/M				
Larona	Banda Sea	5		>500	Tr-W-S	Mes I				
Sadang	Karimata Strait	6.5	140	3400	Tr-W-S	Mes S/M				
Sampara	Banda Sea	8.1	190	2800	Tr-W-S	Mes S/M				
Solo	Banda Sea	6.2	130	1800	Tr-W-S	Mes S/M				
Java										
Bekasi	Java Sea	1.4	70	>1000	Tr-W-S	Cen S/M				5
Bengawan Solo	Java Sea	16	600	3300	Tr-W-S	Cen I	15	19	3	5, 9, 10, 17, 19
Cimanuk	Indian (E)	4.2	150	3100	Tr-W-S	Cen I	4.4	20	0.69	5, 7, 9, 10, 16, 20
Citanduy	Indian (E)	4.8	170	2800	Tr-W-S	Cen I	6.1	9.5	0.38	5, 8, 10, 20

Table G2. (*Continued*)

River name	Ocean	Area (10^3 km^2)	Length (km)	Max_elev (m)	Climate	Geology	Q (km^3/yr)	TSS (Mt/yr)	TDS (Mt/yr)	Refs.
Citarum	Java Sea	11	270	3000	Tr-W-S	Cen I	13		0.61	5, 10, 15, 17
Cudjung	Indian (E)	1.6	90	2400	Tr-W-S	Cen I	3.1			5
Opak	Indian (E)	1.7	50	2900	Tr-W-S	Cen I				5
Porong	Java Sea	11	320	3700	Tr-W-S	Cen I	12	6.2 (20)	3.3	1, 5, 10, 17, 19
Progo	Indian (E)	2.5	140	3400	Tr-W-S	Cen I	7.8		1.8	5, 8
Seraju	Indian (E)	3.7	110	2800	Tr-W-S	Cen I	5.2	9	0.54	3, 5, 10
Serang	Java Sea	5.7	150	2100	Tr-W-S	Cen		7.5		3, 5
Tjipunagara	Java Sea	1.5	75	2200	Tr-W-S	Cen S/M	3.4			5, 7
Sumatra										
Alas	Indian (E)	16	190	>1000	Tr-W-S	Cen I				
Asahan	Mallaca Strait	7.5	100	2500	Tr-W-S	Cen I	3			5, 8
Barumun	Karimata Strait	16	310	2400	Tr-W-S	Mes S/M				
Hari	Karimata Strait	50	490	2300	Tr-W-S	Cen I	46			15, 19
Inderagiri	South China Sea	22	500	2900	Tr-W-S	Cen I	18			15
Kampar	Mallaca Strait	36	300	1000	Tr-W-S	Cen I	33			15, 19
Mesuji	Java Sea	6.6	260	<100	Tr-W-S	Mes S/M				
Musi	Karimata Strait	61	750	1700	Tr-W-S	Mes S/M	80		3.5	10, 14, 18
Panai	Mallaca Strait	12	230	1900	Tr-W-S	Cen I				
Patahparang	Indian (E)	9.6	110	>500	Tr-W-S	Mes S/M				
Rokan	Mallaca Strait	16	280	1000	Tr-W-S	PreMes				
Sekampung	Java Sea	7	180	2100	Tr-W-S	Cen I				5
Seputih	Java Sea	7.1	180	>1000	Tr-W-S	Cen I				5
Siak	Mallaca Strait	16	130	>500	Tr-W-S	PreMes				
Tulangbawang	Java Sea	8.8	370	1500	Tr-W-S	Cen I				
West Papua										
Apauwar	Pacific (W)	4.7	140	<600	Tr-W-S	Mes S/M				
Bian	Arafura Sea	5.3		<100	Tr-W-S	Cen S/M				
Digul	Arafura Sea	25	520	>3000	Tr-W-S	Mes S/M				14, 19
Kamundan	Arafura Sea	6.1		>1000	Tr-W-S	PreMes				
Kumbe	Arafura Sea	6.3		<100	Tr-W-S	Mes S/M				
Lorentz	Arafura Sea	15	170	>3000	Tr-W-S	Mes S/M				
Mamberamo	Pacific (W)	81	800	>3000	Tr-W-S	Mes S/M	130			14, 19
Merauke	Arafura Sea	7.7		<100	Tr-W-S	Cen S/M				
Omba	Arafura Sea	5.7	64	>100	Tr-W-S	Mes S/M				

Table G2. (*Continued*)

Pulau	Arafura Sea	36		>3000	Tr-W-S	Mes S/M				
Wamma	Pacific (W)	5		>500	Tr-W-S	Mes S/M				
Wapoga	Pacific (W)	5.5		2500	Tr-W-S	Mes S/M				
Wiriagar	Seram Sea	5		2000	Tr-W-S	Mes S/M				
Papua New Guinea										
Mainland										
Bamu	G. of Papua	18	310	1000	Tr-W-S	Cen I				
Fly	G. of Papua	76	620	4000	Tr-W-S	Mes S/M	180	110 (80)	16	2, 4, 6, 9, 10, 11
Kikori	Pacific (W)	20	320	>2000	Tr-W-S	Mes S/M	50		7.1	10, 19
Lakekamu	G. of Papua	7		>1000	Tr-W-S	PreMes				
Markham	Pacific (W)	13	320	>1000	Tr-W-S	Mes S/M	15			
Purari	G. of Papua	33	630	3700	Tr-W-S	PreMes-Mes S/M	85	80	11	10, 11, 12, 13
Ramu	Pacific (W)	11	640	>4000	Tr-W-S	Cen S/M				8
Sepik	Pacific (W)	78	700	>2000	Tr-W-S	Mes S/M	120		13	10
Vaidala	G. of Papua	6	150	>3000	Tr-W-S	Cen S/M				
Bougainville										
Jaba	Pacific (SW)	0.46	50	2200	Tr-W-S	Mes S/M	1.3	26		21

References:
1. Aldrian *et al.*, 2008; 2. Dietrich *et al.*, 1999; 3. Douglas and Spencer, 1985; 4. GEMS website, www.gemstat.org; 5. Global River Discharge Database, http://www.rivdis.sr.unh.edu/; 6. Harris, 1991; 7. Hollerwoger, 1964; 8. Jayawardena *et al.*, 1997; 9. Markham and Day, 1994; 10. Meybeck and Ragu, 1996; 11. Pickup *et al.*, 1981; 12. Pickup, 1980; 13. Pickup, 1983; 14. Rand McNally, 1980; 15. Sea Around Us Website, www.seaaroundus.org; 16. Stevens, 1994; 17. Takeuchi *et al.*, 1995; 18. UN Water Resources Series #28, 1966 (cf. van der Leeden, 1975); 19. UNESCO (WORRI), 1978; 20. D. E. Walling, personal communication; 21. Wright *et al.*, 1980

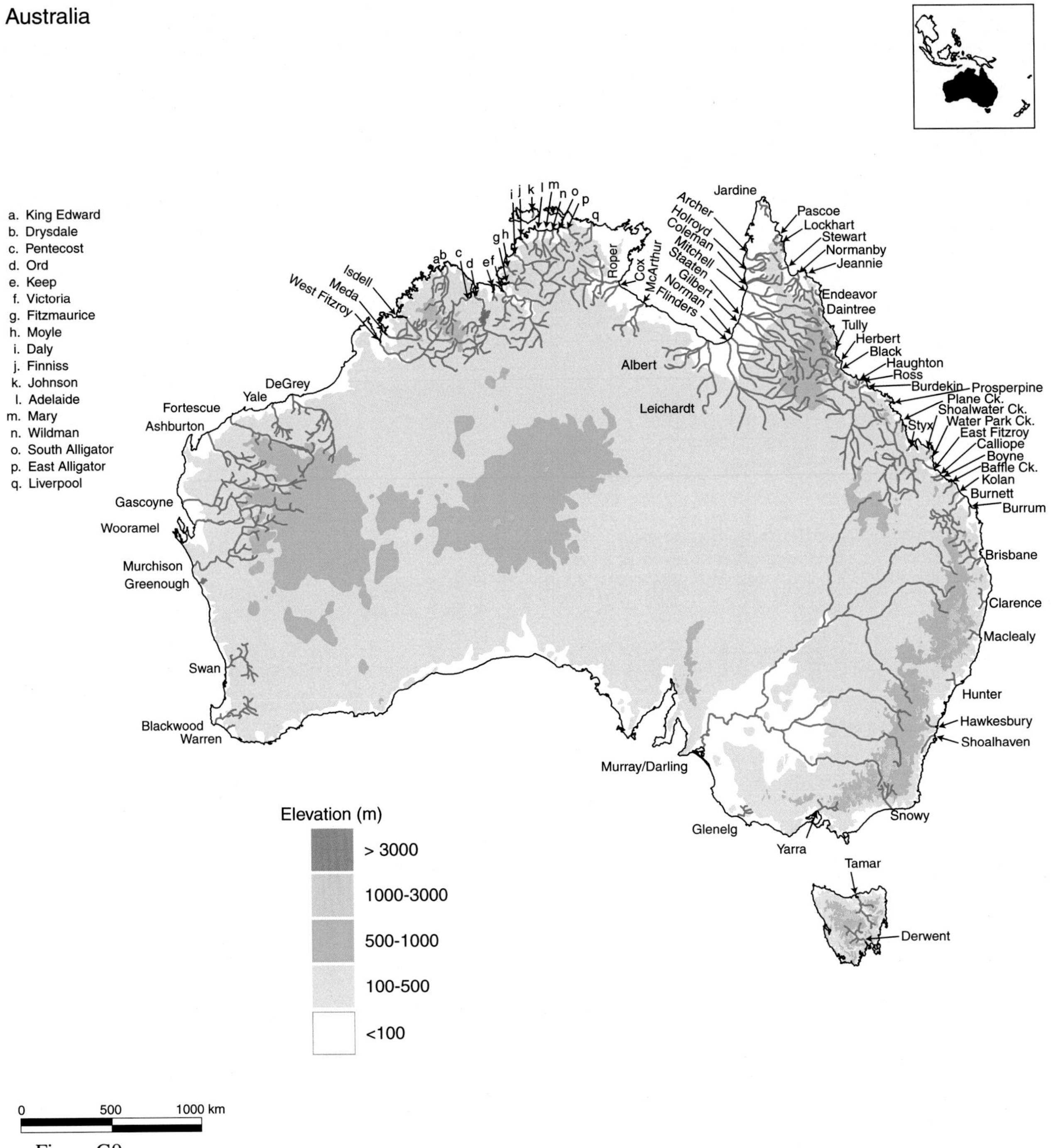

Australia

a. King Edward
b. Drysdale
c. Pentecost
d. Ord
e. Keep
f. Victoria
g. Fitzmaurice
h. Moyle
i. Daly
j. Finniss
k. Johnson
l. Adelaide
m. Mary
n. Wildman
o. South Alligator
p. East Alligator
q. Liverpool

Isdell
Meda
West Fitzroy
DeGrey
Yale
Fortescue
Ashburton
Gascoyne
Wooramel
Murchison
Greenough
Swan
Blackwood
Warren
Roper
Cox
McArthur
Albert
Leichardt
Jardine
Archer
Holroyd
Coleman
Mitchell
Staaten
Gilbert
Norman
Flinders
Pascoe
Lockhart
Stewart
Normanby
Jeannie
Endeavor
Daintree
Tully
Herbert
Black
Haughton
Ross
Burdekin
Prosperpine
Plane Ck.
Shoalwater Ck.
Styx
Water Park Ck.
East Fitzroy
Calliope
Boyne
Baffle Ck.
Kolan
Burnett
Burrum
Brisbane
Clarence
Maclealy
Hunter
Hawkesbury
Shoalhaven
Snowy
Murray/Darling
Glenelg
Yarra
Tamar
Derwent

Elevation (m)
> 3000
1000-3000
500-1000
100-500
<100

0 500 1000 km

Figure G9

Australia

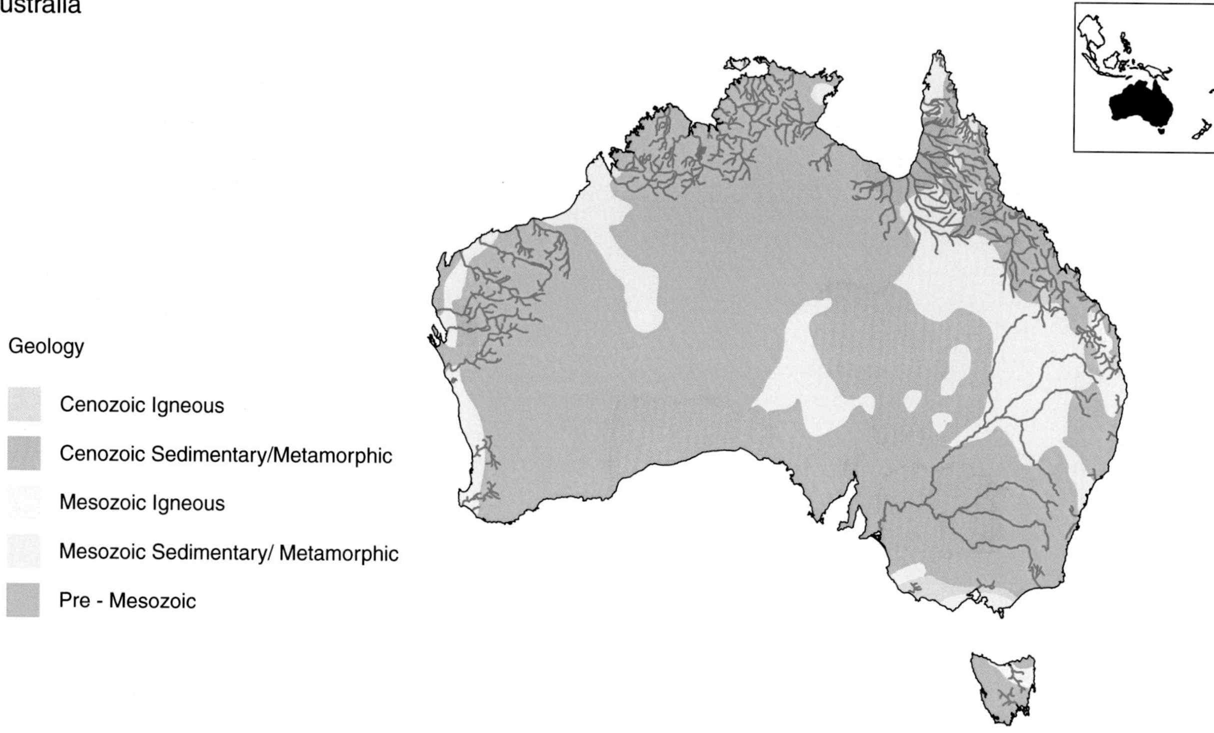

Surface Runoff (mm/yr)

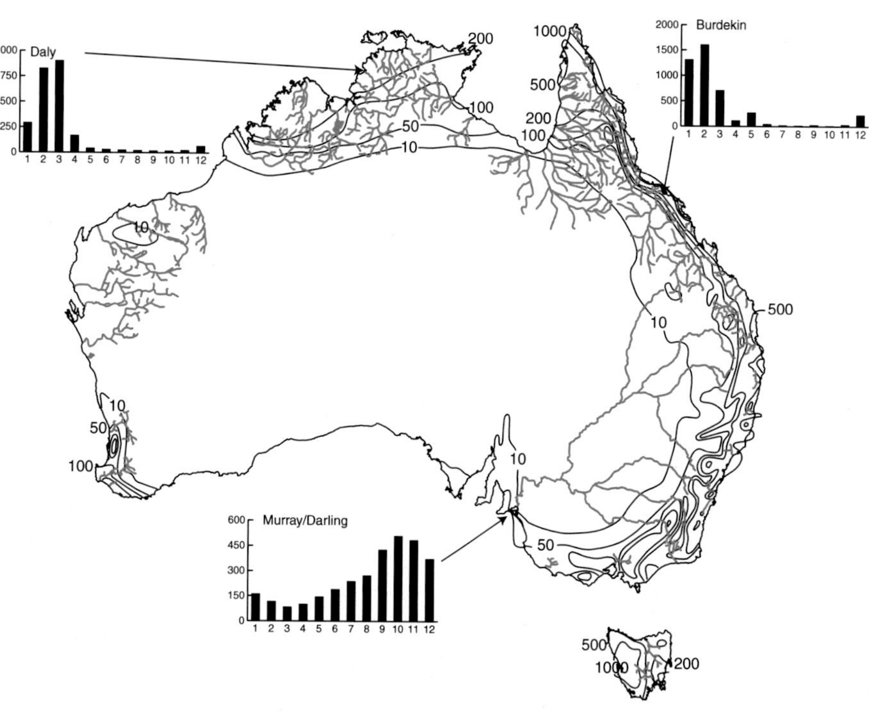

Figure G10

Table G3.

River name	Ocean	Area (10^3 km^2)	Length (km)	Max_elev (m)	Climate	Geology	Q (km^3/yr)	TSS (Mt/yr)	TDS (Mt/yr)	Refs.
Australia										
Adelaide	Arafura Sea	7.6	180	240	Tr-H-S	PreMes	2	0.5		10, 11, 12
Albert	G. of Carpentaria	34	320	300	Tr-A-S	PreMes				22
Archer	G. of Carpentaria	14	240	400	Tr-H-S	Mes S/M				
Ashburton	Indian (SE)	71	800	1000	STr-A-S	PreMes	0.6			8, 22
Baffle Creek	Pacific (SW)	3.9	100	770	STr-SA-S	PreMes	0.7	0.4		3
Black	Pacific (SW)	1.1	62	<500	Tr-H-S	PreMes	0.5	0.2		3
Blackwood	Indian (SE)	20	310	>200	STr-A-W	PreMes	0.6			23
Boyne	Pacific (SW)	2.5	110	670	STr-SA-S	PreMes	0.4	0.3		3
Brisbane	Tasman Sea	14	340	>1000	STr-H-W	Mes S/M	1.4	0.29 (0.18)		5
Burdekin	Pacific (SW)	130	710	980	STr-A-S	PreMes	10	3 (0.5)	2.4	3, 8, 17, 18, 22
Burnett	Pacific (SW)	33	400	1140	STr-A-S	PreMes	1.7	0.7		1, 18, 25
Burrum	Pacific (SW)	3.3	64	500	STr-SA-S	PreMes	0.7	0.3		3
Calliope	Pacific (SW)	2.3	100	300	STr-SA-S	PreMes	0.3	0.2		3
Clarence	Tasman Sea	17	400	1500	STr-SA-S	PreMes	4			1, 18, 22, 23, 25
Coleman	G. of Carpentaria	8.5	220	<200	Tr-H-S	Mes S/M				
Cox	G. of Carpentaria	12	200	500	Tr-A-S	PreMes				
Daintree	Pacific (SW)	2.1	86	1000	Tr-W-S	PreMes	3.6	1.2		3
Daly	Timor Sea	52	360	300	Tr-SA-S	PreMes	6.7			11
DeGrey	Indian (SE)	67	310	700	STr-A-S	PreMes	1			8, 22
Derwent	Bass Strait	9.2	170	1500	Te-H-W	PreMes	4.3	0.11		11, 25
Drysdale	Timor Sea	15	370	>500	Tr-H-S	PreMes	1.1			9
East Alligator	Arafura Sea	14	210	250	Tr-H-S	PreMes	6.9			11
East Fitzroy	Pacific (SW)	140	960	970	STr-A-S	PreMes	5.3	3 (0.3)	1.1	8, 11, 15, 17, 18, 24, 25
Endeavour	Pacific (SW)	2.2	45	1140	Tr-W-S	PreMes	1.8	0.75		3
Finniss	Timor Sea	9	33	<100	Tr-SA-S	PreMes		5.7		12
Fitzmaurice	Timor Sea	11	170	310	Tr-SA-S	PreMes	1.6			11, 25
Flinders	G. of Carpentaria	110	830	300	Tr-A-S	Mes S/M	0.5 (3)	9.3	0.3	11, 18
Fortescue	Indian (SE)	49	690	1000	STr-SA-S	PreMes	0.23			8
Gascoyne	Indian (SE)	79	760	700	STr-A-S	PreMes	0.6			18, 22
Gilbert	G. of Carpentaria	16	500	400	Tr-SA-S	Mes S/M	1.3			9
Glenelg	Grt. Austr.Bight	12	230	590	Te-A-W	Cen S/M	0.5			1, 2, 18

Table G3. (*Continued*)

Greenough	Indian (SE)	13	250	400	STr-A-W	Mes S/M				
Haughton	Pacific (SW)	3.6	100	1200	Tr-SA-S	PreMes	0.8	0.4		3
Hawkesbury	Tasman Sea	22	470	>1000	STr-SA-W	Mes S/M	2.8	1.5		2, 7, 16, 25, 26
Herbert	Pacific (SW)	10	340	1100	Tr-H-S	PreMes	4.9	1.6		3
Holroyd	G. of Carpentaria	7.7	220	200	Tr-H-S	Mes S/M				
Hunter	Tasman Sea	22	470	1200	STr-A-C	PreMes	1.8	11		1, 11, 22, 25
Isdell	Indian (SE)	7	200	900	Tr-SA-S	PreMes				
Jardine	G. of Carpentaria	3.3	150	<100	Tr-H-S	Mes S/M	2.2	0.76		11
Jeannie	Pacific (SW)	3.8	60	300	Tr-H-S	Mes S/M	2.4	1		11
Johnson	Arafura Sea	2.3	16	<100	Tr-H-S	PreMes	4.7	1.9		11
Keep	Timor Sea	12	140	400	Tr-A-S	PreMes	0.5			3
King Edward	Timor Sea	17	200	200	Tr-H-S	PreMes				20
Kolan	Pacific (SW)	3	150	2100	STr-SA-S	PreMes	0.5	0.3		3
Leichardt	G. of Carpentaria	31	450	500	Tr-A-S	Mes S/M				
Liverpool	Arafura Sea	7.5	200	500	Tr-H-S	PreMes				
Lockhart	Pacific (SW)	2.8	50	330	Tr-H-S	PreMes	1.6	0.6		3
Macleay	Tasman Sea	11	400	1600	STr-SA-S	PreMes	2	0.1 (1.7)		11, 22
Mary	Arafura Sea	9.6	270	700	Tr-H-S	PreMes	2.3	0.9		3, 11, 25
McArthur	G. of Carpentaria	10	250	>200	Tr-A-S	PreMes	0.7			8, 14
Meda	Indian (SE)	5.5	190	700	Tr-A-S	PreMes				
Mitchell	G. of Carpentaria	72	560	300	Tr-H-S	PreMes-Mes	12	0.43	2.1	11, 18, 25
Moyle	Timor Sea	7.5	100	250	Tr-A-S	PreMes	0.64			11
Murchison	Indian (SE)	82	780	>500	STr-A-S	PreMes	0.2			6
Murray–Darling	Grt. Austr.Bight	1100	3500	>1000	Te-A-S	PreMes-Mes	7.9 (24)	1	3.6	8, 11, 13, 18, 21
Norman	G. of Carpentaria	35	420	>200	Tr-A-S	Mes S/M				
Normanby	Pacific (SW)	14	260	1150	Tr-H-S	Mes S/M	5.9	2.7 (0.25)		3, 5, 17
Ord	Timor Sea	55	480	780	Tr-A-S	PreMes	5.1	20		11, 18, 22
Pascoe	Pacific (SW)	4.3	120	550	Tr-W-S	PreMes	4.2	1.2		3
Pentecost	Timor Sea	29	220	>1000	Tr-SA-S	PreMes	4.3			11
Plane Creek	Pacific (SW)	2.7		>1000	STr-H-S	PreMes	1.4	0.55		3
Prosperpine	Pacific (SW)	2.5		>500	STr-H-S	PreMes	1.4	0.1		4
Roper	G. of Carpentaria	75	500	>2000	Tr-A-S	PreMes	4			6
Ross	Pacific (SW)	1.8		<100	Tr-SA-S	Cen S/M	0.4	0.3		3
Shoalhaven	Tasman Sea	8	260	1400	STr-SA-W	PreMes	1.8	0.93		16, 19, 26
Shoalwater Creek	Pacific (SW)	3.7	40	850	STr-SA-S	PreMes	0.8	0.3		3

Table G3. (*Continued*)

River name	Ocean	Area (10^3 km^2)	Length (km)	Max_elev (m)	Climate	Geology	Q (km^3/yr)	TSS (Mt/yr)	TDS (Mt/yr)	Refs.
Snowy	Tasman Sea	12	430	2200	Te-SA-S	PreMes	1.7			1, 2, 8
South Alligator	Arafura Sea	12	220	200	Tr-H-S	PreMes	6.6	0.05		11, 12
Staaten	G. of Carpentaria	26	300	200	Tr-H-S	Cen S/M	17			
Stewart	Pacific (SW)	2.8	100	820	Tr-H-S	PreMes	1.6	0.5		3
Styx	Pacific (SW)	3.1	73	>100	STr-H-S	PreMes	0.8	0.3		3
Swan	Indian (SE)	120	390	470	STr-A-W	PreMes	0.88			24, 25
Tamar	Tasman Sea	12	64	1500	Te-H-W	PreMes	3.1	0.14		11, 22
Tully	Pacific (SW)	1.7	100	1100	Tr-W-S	PreMes	3.7	1.2		3
Victoria	Timor Sea	78	560	370	Tr-A-S	PreMes	5			11, 22, 25
Warren	Indian (SE)	14	200	400	STr-SA-W	PreMes				1
Water Park Creek	Pacific (SW)	1.9	30	700	STr-H-S	PreMes	0.7	0.3		3
West Fitzroy	Indian (SE)	86	520	910	Tr-A-S	PreMes	5.5			22
Wildman	Timor Sea	4.8	110	<100	Tr-SA-S	PreMes	0.8	0.3		11, 12
Wooramel	Indian (SE)	5.6	350	<500	STr-A-S	PreMes-CenS/M				
Yale	Indian (SE)	5.5	180	500	Str-A-S	PreMes				
Yarra	Bass Strait	4.1	180	1500	Te-SA-W	PreMes	1.1	0.15		11, 25

References:
1. Australian Water Resources Council, 1967 (cf. van der Leeden, 1975); 2. Australian Water Resources Council, 1978; 3. Belperio, 1979; 4. Belperio, 1983; 5. Bryce *et al.*, 1998; 6. Dai and Trenberth, 2002; 7. Erskine and Warner, 1999; 8. GEMS website, www.gemstat.org; 9. GRDC website, www.gewex.org/grdc; 10. Gresswell and Huxley, 1966; 11. Harris, 1991; 12. P. T. Harris, personal communication; 13. Jansen *et al.*, 1979; 14. Jones *et al.*, 2003; 15. Joo *et al.*, 2005; 16. Kjerfve *et al.*, 1992; 17. McKergow *et al.*, 2005; 18. Meybeck and Ragu, 1996; 19. NSW Department of Public Works, 1975; 20. Peucker-Ehrenbrink, 2009; 21. Probst, 1992; 22. Rand McNally, 1980; 23. Rodier and Roche, 1984; 24. UNESCO (WORRI), 1978; 25. Wasson *et al.*, 1996; 26. Wright *et al.*, 1980

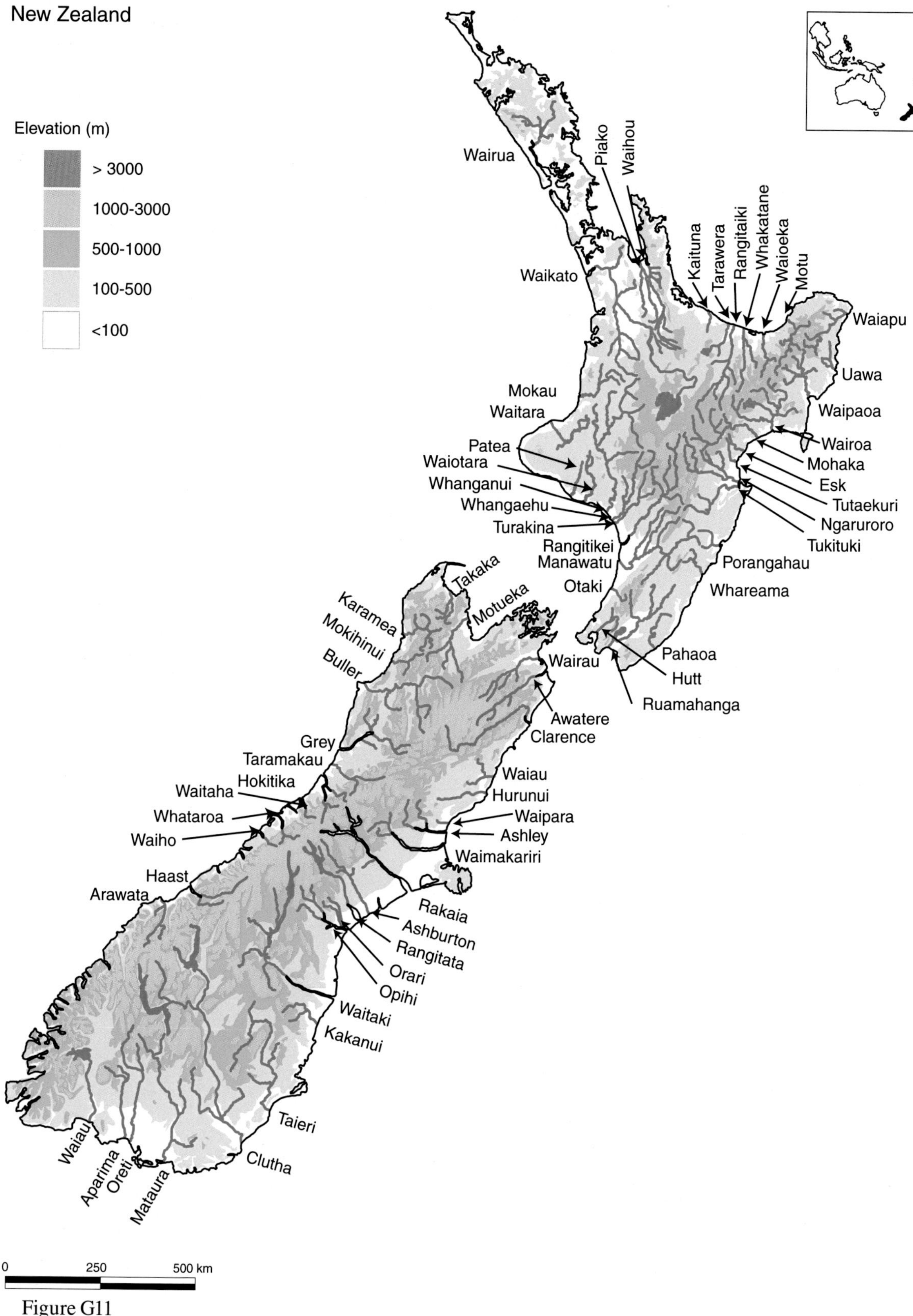
New Zealand
Elevation (m)
> 3000
1000-3000
500-1000
100-500
<100
Wairua
Piako
Waihou
Waikato
Kaituna
Tarawera
Rangitaiki
Whakatane
Waioeka
Motu
Waiapu
Uawa
Waipaoa
Wairoa
Mohaka
Esk
Tutaekuri
Ngaruroro
Tukituki
Porangahau
Whareama
Mokau
Waitara
Patea
Waiotara
Whanganui
Whangaehu
Turakina
Rangitikei
Manawatu
Otaki
Pahaoa
Hutt
Ruamahanga
Takaka
Motueka
Karamea
Mokihinui
Buller
Wairau
Awatere
Clarence
Grey
Taramakau
Hokitika
Waitaha
Whataroa
Waiho
Waiau
Hurunui
Waipara
Ashley
Waimakariri
Haast
Arawata
Rakaia
Ashburton
Rangitata
Orari
Opihi
Waitaki
Kakanui
Taieri
Clutha
Waiau
Aparima
Oreti
Mataura
0
250
500 km

Figure G11

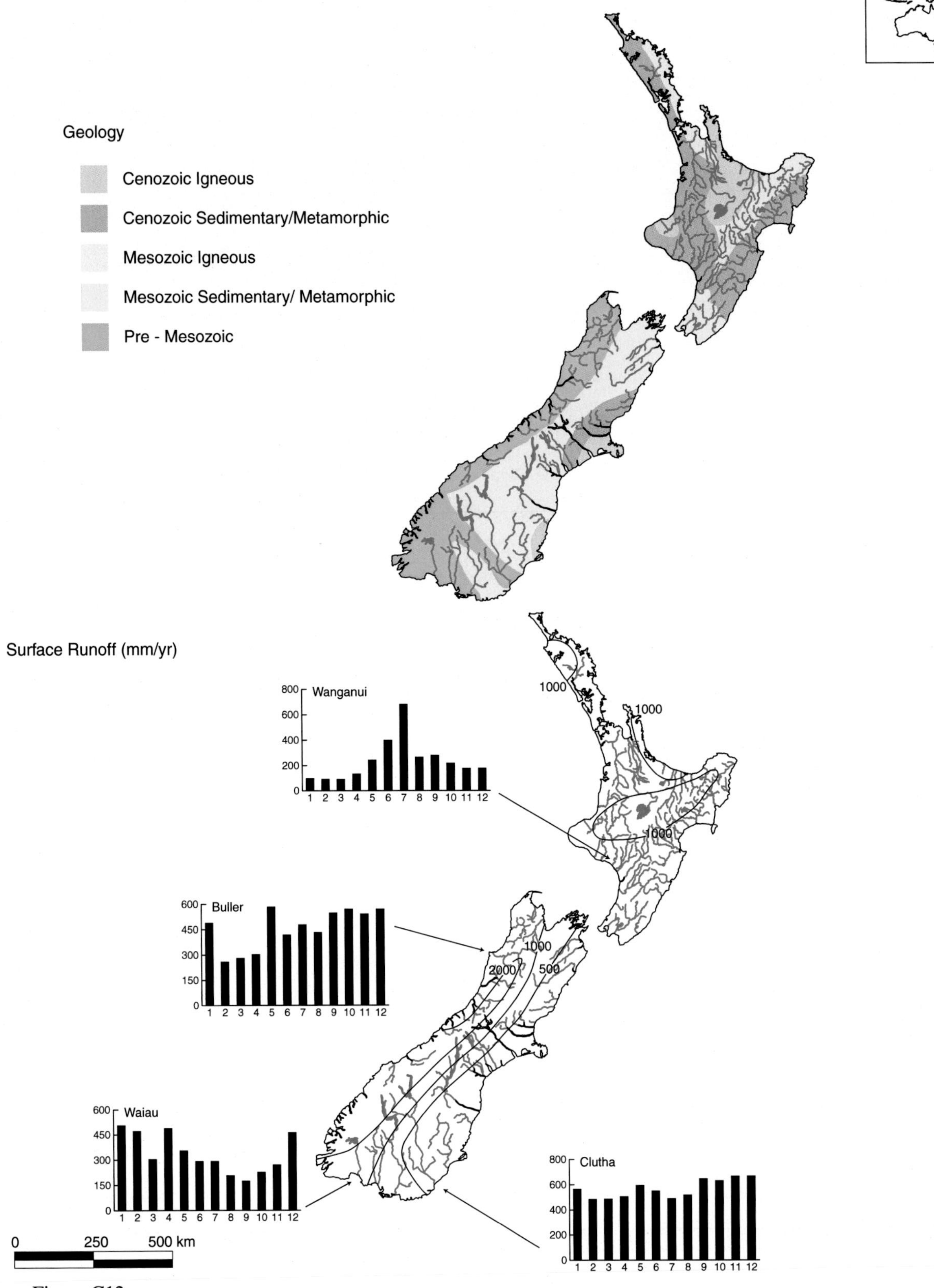

New Zealand
Geology
Cenozoic Igneous
Cenozoic Sedimentary/Metamorphic
Mesozoic Igneous
Mesozoic Sedimentary/ Metamorphic
Pre - Mesozoic
Surface Runoff (mm/yr)
Wanganui
Buller
Waiau
Clutha
1000
2000
500
0 250 500 km

Figure G12

Table G4.

River name	Ocean	Area (10^3 km^2)	Length (km)	Max_elev (m)	Climate	Geology	Q (km^3/yr)	TSS (Mt/yr)	TDS (Mt/yr)	Refs.
New Zealand										
North Island										
Esk	Pacific (SW)	0.25		>1000	STr-W-W	Mes S/M	1.7	0.035		6
Hutt	Pacific (SW)	0.43	56	>500	STr-W-W	Mes S/M	0.68	0.13		4, 6
Kaituna	Pacific (SW)	1.2	45	>500	STr-H-C	Cen S/M	0.7	0.04		1, 3, 6
Manawatu	Tasman Sea	6	180	>1000	STr-H-W	Cen S/M-Mes S/M	3	3.8		1, 3, 6
Mohaka	Pacific (SW)	2.4	110	1400	STr-W-W	Mes S/M	2.5	1.4		5, 7, 8
Mokau	Tasman Sea	1.5	120	1000	STr-H-C	Cen S/M	1.1	0.66		1, 3, 6
Motu	Pacific (SW)	1.4		>1000	STr-W-W	Mes S/M	2.8	3.5		5, 6, 7
Ngaruroro	Pacific (SW)	2.5	110	1700	STr-H-W	Mes S/M	1.3	1.3		5, 6, 7
Otaki	Tasman Sea	1.3	45	>1000	STr-H-W	Mes S/M	0.9	0.17		6
Pahaoa	Pacific (SW)	0.58		>500	STr-H-W	Cen S/M	0.3	0.44		6
Patea	Tasman Sea	1	100	>500	STr-H-W	Cen S/M	0.7	0.31		1, 3, 6
Piako	Pacific (SW)	1.5	100	>200	STr-H-C	Cen S/M		0.035		6
Porangahau	Pacific (SW)	0.84	45	>500	STr-H-W	Cen S/M	0.1	0.41		1, 3, 6
Rangitaiki	Pacific (SW)	2.3	180	>1000	STr-H-C	Cen I	1.7	0.08		5, 6, 12
Rangitikei	Tasman Sea	3.9	180	1000	STr-H-W	Cen S/M	2.4	1.1		6
Ruamahanga	Pacific (SW)	2.3	100	600	STr-W-W	Mes S/M	2.5	0.6		3, 5, 6
Tarawera	Pacific (SW)	1.1	65	>1000	STr-W-C	Cen I	1	0.07		1, 3, 6
Tukituki	Pacific (SW)	2.4	110	600	STr-H-W	Mes S/M	1.3	1		5, 6
Turakina	Tasman Sea	0.97		>1000	STr-H-W	Cen S/M	0.2	0.3		1, 3, 6
Tutaekuri	Pacific (SW)	0.84		>1000	STr-H-W	Mes S/M	0.5	0.21		1, 3, 6, 10
Uawa	Pacific (SW)	0.55	50	>500	Str-W-W	Mes S/M		5		6
Waiapu	Pacific (SW)	1.7	90	>2000	STr-W-W	Mes S/M	2.8	35		1, 3, 6
Waihou	Pacific (SW)	2	150	>500	STr-H-C	Cen S/M		0.16		1, 3, 6
Waikato	Tasman Sea	14	350	>1000	STr-W-C	Cen S/M	13	0.37	1.5	2, 6, 9, 10, 13
Waioeka	Pacific (SW)	0.66	65	>1000	STr-W-C	Cen I	1	0.7		5, 6
Waipaoa	Pacific (SW)	2.2	90	1200	STr-H-W	Mes S/M	1.8	15		3, 6
Wairoa	Pacific (SW)	3.7	100	2200	STr-H-W	Mes S/M	2.1	4.7		3, 6
Wairua	Tasman Sea	3.6		>1000	STr-H-C	Cen S/M	0.8	1.1		1, 3, 6
Waitara	Tasman Sea	1.1		>1000	STr-W-W	Cen I		0.97		1, 3, 6
Waitotara	Tasman Sea	1.2	100	>500	STr-H-W	Cen S/M		0.48		6
Whakatane	Pacific (SW)	1.8	90	600	STr-W-C	Cen I	1.8	0.69		5, 6, 12
Whangaehu	Tasman Sea	1.9	110	1100	STr-H-W	Cen S/M	1.2	0.69		6
Whanganui	Tasman Sea	7.1	230	2000	STr-W-W	Cen S/M	6.7	4.7		3, 5, 6, 10
Whareama	Pacific (SW)	0.4	50	500	STr-H-W	Cen S/M-Mes S/M	0.2	0.67		6
South Island										
Aparima	Foveaux Strait	1.6	100	>500	Te-W-C	PreMes		0.09		6
Arawata	Tasman Sea	0.93	68	>1000	Te-W-C	PreMes		7.2		6

Table G4. (*Continued*)

River name	Ocean	Area (10^3 km^2)	Length (km)	Max_elev (m)	Climate	Geology	Q (km^3/yr)	TSS (Mt/yr)	TDS (Mt/yr)	Refs.
Ashburton	Pacific (SW)	1.7	80	>3000	Te-W-C	Mes S/M	3	0.31		1, 3, 6
Ashley	Pacific (SW)	1.3	65	>1000	Te-W-C	Mes S/M		0.09		6
Awatere	Pacific (SW)	1.6	110	>1000	Te-H-C	Mes S/M	0.45	0.21		1, 3, 6
Buller	Tasman Sea	6.5	160	2400	Te-W-C	Mes S/M	13	2.7		4, 6, 11, 13
Clarence	Pacific (SW)	3.3	180	2100	Te-H-C	Mes S/M		0.65		6
Clutha	Pacific (SW)	21	340	>1000	Te-W-C	PreMes-Mes S/M	19	0.39 (2.9)	0.96	3, 4, 6, 10
Grey	Tasman Sea	3.9	120	1900	Te-W-C	PreMes-Mes S/M	11	2.1		5, 6, 12
Haast	Tasman Sea	0.93	100	3000	Te-W-C	Mes S/M	6	5.9		4, 6
Hokitika	Tasman Sea	0.35	65	1300	Te-W-C	Mes S/M	3.1	6.2		4, 6, 10
Hurunui	Pacific (SW)	0.74	130	1500	Te-W-C	Cen S/M	1.6	0.53		6
Kakanui	Pacific (SW)	0.89		>1000	Te-H-C	Mes S/M		0.11		6
Karamea	Tasman Sea	1.2	70	1000	Te-W-C	PreMes	3.2	0.15		4, 6, 12
Mataura	Foveaux Strait	5.4	190	>1000	Te-H-C	PreMes	2	0.69		1, 3, 6, 10
Mokihinui	Tasman Sea	0.75		>1000	Te-W-C	PreMes-Mes S/M	2.7	0.29		1, 4, 6, 10
Motueka	Tasman Sea	1.7	95	1800	Te-W-C	Cen S/M	1.8	0.35		6
Opihi	Pacific (SW)	2.4	75	>2000	Te-W-C	Mes S/M		0.16		6
Orari	Pacific (SW)	0.8		>2000	Te-W-C	Mes S/M		0.06		6
Oreti	Foveaux Strait	3.5	170	>1000	Te-W-C	PreMes		0.26		1, 3, 6, 10
Rakaia	Pacific (SW)	2.6	150	2600	Te-W-C	Mes S/M	6.2	4.2		4, 6
Rangitata	Pacific (SW)	1.8	110	>3000	Te-W-C	Mes S/M	3	1.6		1, 3, 6
Taieri	Pacific (SW)	5.7	200	>1000	Te-SA-C	Mes S/M	1	0.32		1, 3, 6, 10
Takaka	Tasman Sea	0.26	50	1700	Te-W-C	PreMes	0.46	0.07		6
Taramakau	Tasman Sea	1	70	1500	Te-W-C	Mes S/M	4.8	2.2		5, 6
Waiau	Foveaux Strait	8.2	180	>1500	Te-W-C	PreMes	10.6	0.78		6
Waiau	Pacific (SW)	2	100	2000	Te-W-C	Cen S/M	3	2.8	1.3	5, 6, 12, 13
Waiho	Tasman Sea	0.29	32	>1000	Te-W-C	PreMes		3.4		6
Waimakariri	Pacific (SW)	3.6	160	2300	Te-W-C	Mes S/M	3.8	3.1	0.19	1, 4, 6, 10
Waipara	Pacific (SW)	0.74	45	>1000	Te-W-C	Cen S/M		0.06		6
Wairau	Pacific (SW)	4.2	170	>2000	Te-W-C	Mes S/M	4	0.84		1, 3, 6
Waitaha	Tasman Sea	0.32	40	>1000	Te-W-C	PreMes		2.8		6
Waitaki	Pacific (SW)	12	150	1900	Te-H-C	Mes S/M	12	0.34 (1.1)	0.49	4, 6, 10, 13
Whataroa	Tasman Sea	0.59	51	>1000	Te-W-C	Cen S/M		4.8		6

References:
1. Duncan, 1992; 2. GEMS website, www.gemstat.org; 3. Griffiths and Glasby, 1985; 4. Griffiths, 1981; 5. Griffiths, 1982; 6. Hicks and Shankar, 2003; 7. Hicks *et al.*, 2000; 8. D. M. Hicks, personal communication; 9. Meybeck, 1994; 10. Rand McNally, 1980; 11. Takeuchi *et al.*, 1995; 12. UNESCO (WORRI), 1978; 13. UNESCO, 1971

References

Aalto, R., T. Dunne & J. L. Guyot (2006). Geomorphic control on Andean denudation rates. *J. Geol.*, **114**, 85–99.

Aalto, R. *et al.* (2003). Episodic sediment accumulation on Amazonian flood plains influenced by El Nino/Southern Oscillation. *Nature*, **425**, 493–497.

Abdul Rahmin, N. (1988). Water yield changes after forest conversion to agriculture land use in peninsular Malaysia. *J. Trop. For. Sci.*, **1**, 67–84.

Adam, J. C. *et al.* (2009). Implications of global climate change for snowmelt hydrology in the twenty-first century. *Hydrol. Proc.*, **23**, 962–972.

Adel, M. M. (2001). Effect on water resources from upstream water diversion in the Ganges Basin. *J. Environ. Qual.*, **30**, 356–368.

Ahnert, F. (1970). Functional relationships between denudation, relief, and uplift in large mid-latitude drainage basins. *Am. J. Sci.*, **268**, 243–263.

Akbulut, N. E. *et al.* (2009). Rivers of Turkey, in *Rivers of Europe*, eds. K. Tockner, C. T. Robinson & U. Uehlinger, Amsterdam: Elsevier, pp. 643–672.

Akrasi, S. A. & N. B. Ayibotele (1984). An appraisal of sediment transport measurement in Ghanian rivers. *IAHS Publ.*, **144**, 301–312.

Aldrian, E., C. T. A. Chen & S. Adi (2008). Spatial and seasonal dynamics of riverine carbon fluxes of the Brantas catchment in East Java. *J. Geophys. Res.*, **113** doi: 10.1029/2007JG000626Sep 2008.

Alexander, R. B. *et al.* (1996). *Data from Selected US Geological Survey National Stream Water Quality Monitoring Networks (WQN); USGS Digital Data Series DDS-37.*

Algan, O. *et al.* (1997). Riverine Fluxes into the Black and Marmara Seas. *CIESM Workshop Monographs*, **30**, 47–53.

Allen, G. P., D. Laurier & J. Thouvenin (1979). Etude sédimentologique du Delta de la Mahakam, in *Notes et Mémoires No. 15*, Paris: Compagnie Francaise des Petroles, p. 156.

Allen, P. A. & N. Hovius (2000). Sediment supply from landslide-dominated catchments: implications for basin-margin fans. *Basin Res.*, **10**, 19–35.

Alt-Epping, U. *et al.* (2007). Provenance of organic matter and nutrient conditions on a river-and upwelling influenced shelf: a case study from the Portuguese margin. *Mar. Geol.*, **243**, 169–179.

Alverson, K. D., R. S. Bradley & T. F. Pedersen (eds.) (2003). *Paleoclimate, Global Change and the Future,* Berlin: Springer-Verlag.

Amarasekera, K. N. *et al.* (1997). ENSO and the natural variability in the flow of tropical rivers. *J. Hydrol.*, **200**, 24–39.

Amery, H. A. (1993). The Litani River of Lebanon. *Geogr. Rev.*, **83**, 229–237.

Amoitte Suchet, P., J.-L. Probst & W. Ludwig (2003). Worldwide distribution of continental rock lithology: Implications for the atmospheric/soil CO2 uptake by continental weathering and alkalinity river transport to the oceans. *Global Biogeochem. Cycles*, **17**, 1–13.

Anchukaitis, K. J. *et al.* (2006). Forward modeling of regional scale tree-ring patterns in the southeastern United States and the recent influence of summer drought. *Geophys. Res. Lett*, **33**, L04705.

Anderson, S. P. (2005). Glaciers show direct linkage between erosion rate and chemical weathering fluxes. *Geomorph.*, **67**, 147–157.

Anderson, S. P., W. E. Dietrich & G. H. J. Brimhall (2002). Weathering profiles, mass-balance analysis, and rates of solute loss: linkages between weathering and erosion in a small steep catchment. *Geol. Soc. Amer. Bull.*, **114**, 1143–1158.

Anderson, S. P., J. I. Drever & N. F. Humphrey (1997). Chemical weathering in glacial environments. *Geology*, **25**, 399–402.

Andrews, E. D. *et al.* (2004). Influence of ENSO on flood frequency along the California coast. *J. Clim.*, **17**, 337–348.

Andrews, J. T. & J. P. M. Syvitski (1994). Sediment fluxes along high latitude glaciated continental margins: Northeast Canada and eastern Greenland, in *Global Sedimentary Geofluxes*, ed. W. W. Hay, National Academy of Sciences Press, pp. 99–115.

Anthony, E. J. & M. Julian (1997). The 1979 Var Delta landslide on the French Riviera; a retrospective analysis. *J. Coast. Res.*, **13**, 27–35.

Antweiler, R. C., D. A. Goolsby & H. E. Taylor (1995). Nutrients in the Mississippi River, in R. H. Meade (ed.), Contaminants in the Mississippi River, 1987–1992. *US Geological Survey Circular* **1133**, pp. 73–86.

Aquater (1982). *Regione Marche: Studio general per la difesa della costa primera fase: San Lorenzo in Campo.*

Arboleda, R. A. & M. L. Martinez (1992). Lahars in the Pasig–Potrero River system, in *Fire and mud: Eruptions and Lahars of Mount Pinatubo*, eds. C. G. Newhall & R. S. Punongbayan, Seattle: University of Washington Press.

Arnborg, L., H. J. Walker & J. Peippo (1967). Suspended load in the Colville River, Alaska, 1962. *Geogr. Ann.*, **49A**, 131–144.

Arnell, N. (1996). *Global Warming, River Flows and Water Resources*, Chichester: John Wiley and Sons, p. 224.

Artinyan, E. *et al.* (2008). Modeling the water budget and the riverflows of the Maritsa basin in Bulgaria. *Hydrol. Earth Syst. Sci.*, **12**, 475–521.

Asselman, N. E. M. & H. Middlelkoop (1995). Floodplain sedimentation: quantities, patterns and processes. Earth Surface Proc. Landforms. *Earth Surf. Processes Landforms*, **20**, 481–499.

Atwater, B. F. (1987). Status of glacial Lake Columbia during the last floods from glacial Lake Missoula. *Quat. Res.*, **27**, 182–201.

Audley-Charles, M. G., J. R. Curray & G. Evans (1977). Location of major deltas. *Geology*, **5**, 3341–3344.

Aulenbach, B. T., H. T. Buxton, W. A. Battaglin & R. H. Coupe (2007). Streamflow and Nutrient Fluxes of the Mississippi–Atchafalaya River Basin and Subbasins for the Period of Record Through 2005. USGS Open-file Report 2007–1080.

Australian Water Resources Council (1978). Variability of runoff in Australia, p. 44.

Awadallah, R. M. & M. E. Soltan (1995). Chemical survey of the Rile Nile from Aswan into the outlet. *J. Environ. Sci. Heal. A*, **30**, 1647–1658.

Axtmann, E. V. & R. F. Stallard (1995). Chemical weathering in the South Cascade Glacier basin, comparison of subglacial and extra-glacial weathering. *IAHS Publ.*, **228**, 431–438.

Ayibotele, N. B. & T. Tuffour-Darko (1979). Sediment loads in the southern rivers of Ghana, Water Resources Research Unit - CSIR, 44.

Baade, J. & R. Hesse (2008). An overlooked sediment trap in arid environments: ancient irrigation agriculture in the coastal desert of Peru. *IAHS Publ.* **325**, 375–382.

Bailey, R. C. (2005). Yukon River Basin, in *Rivers of North America*, eds. A. C. Benke & C. E. Cushing, Amsterdam: Elsevier, pp. 774–802.

Baker, P. A. *et al.* (2001). The history of South American tropical precipitation for the past 25,000 years. *Science*, **291**, 640–643.

Baker, V. R. (2002). High-energy megafloods: planetary settings and sedimentary dynamics. *Flood and Megaflood Processes and Deposits: Recent and Ancient Examples*, pp. 3–15.

Baker, V. R. & R. C. Bunker (1985). Cataclysmic late Pleistocene flooding from Glacial Lake Missoula: A review. *Quat. Sci. Rev.*, **4**, 1–41.

Baker, V. R. *et al.* (1992). Channels and valley networks, in *Mars*, eds. H. H. Kieffer, B. Jakosky & C. Snyder, Tucson: University of Arizona Press, pp. 493–522.

Baker, V. R. *et al.* (1995). Late Quaternary paleohydrology of arid and semi-arid regions, in *Global Continental Paleohydrology*, eds. K. J. Gregory, L. Starkel & V. R. Baker, Chichester: John Wiley & Sons, pp. 203–231.

Balamurugan, S. (1991). Tin mining and sediment supply in Peninsular Malaysia with special reference to the Kelang River basin. *The Environmentalist*, **11**, 281–291.

Balamurugun, S. Bergstrom & G. Lindstrom (1999). Floods in regulated rivers and physical planning in Sweden. Contribution to *ICOLD Workshop on Benefits and Concerns About Dams*, Antalya,Turkey, p. 11.

Baldwin, C. K. & U. Lall (1999). Seasonality of streamflow: the upper Mississippi River. *Water Resour. Res.*, **35**, 1143–1154.

Balogh-Brunstad, Z. *et al.* (2008). Chemical weathering and chemical denudation dynamics through ecosystem development and disturbance. *Global Biogeochem. Cycles*, **22** doi:10.1029/2007GB002957.

Bangladesh Water Development Board (1983). Basic consideration on the morphology and land accretion potentials in the estuary of the Lower Meghna River, 39 pp.

Bar-Matthews, M. *et al.* (2003). Sea–land oxygen isotopic relationships from planktonic foraminifera and speleothems in the Eastern Mediterranean region and their implication for paleorainfall during interglacial intervals. *Geochim. Cosmochim. Acta*, **67**, 3181–3199.

Bard, E., B. Hamelin & R. G. Fairbanks (1990). U-Th ages obtained by mass spectrometry in corals from Barbados: sea level during the past 130,000 years. *Nature*, **346**, 456–458.

Barnett, T. P. & D. W. Pierce (2008). When will Lake Mead go dry? *Water Resour. Res.*, **44**, W3201.

Barnett, T. P. *et al.* (2008). Human-induced changes in the hydrology of the western United States. *Science*, **319**, 1080.

Barrow, C. J. (1991). *Land Degradation: Development and Breakdown of Terrestrial Environments*, Cambridge: Cambridge University Press.

Bartolini, C., R. Caputo & M. Pieri (1996). Pliocene–Quaternary sedimentation in the northern Apennine foredeep and related denudation. *Geol. Mag.*, **133**, 255–273.

Baumgartner, A. & E. Reichel (1975). *The World Water Balance. Mean Annual Global, Continental and Maritime Precipitation, Evaporation and Run-off*, Amsterdam: Elsevier.

Bayley, S. E. *et al.* (1992). Effects of multiple fires on nutrient yields from streams draining boreal forest and fen watersheds: nitrogen and phosphorus. *Can. J. Fish. Aquatic Sci.*, **49**, 584–596.

Becker, A. *et al.* (2004). Responses of hydrological processes to environmental change at small catchment scales, in *Vegetation, Water, Humans and the Climate: A New Perspective on an Interactive System*, eds. P. Kabat *et al.*, Springer-Verlag, pp. 301–338.

Beckinsale, R. P. (1969). River regimes, in *Water, Earth and Man*, ed. R. J. Chorley, London: Methuen, pp. 176–192.

Begueria, S. (2006). Changes in land cover and shallow landslide activity: A case study in the Spanish Pyrenees. *Geomorph.*, **74**, 196–206.

Bellotti, P. *et al.* (1994). Sequence stratigraphy and depositional setting of the Tiber Delta – integration of high resolution seismics, well logs and archeological data. *J. Sed. Res.*, **B64**, 416–432.

Belperio, A. P. (1979). The combined use of washload and bed material load rating curves for the calculation of the total load, an example from the Burdekin River, Australia. *Catena*, **6**, 317–329.

Belperio, A. P. (1983). Terrigenous sedimentation in the central Great Barrier Reef lagoon; a model from the Burdekin region. *BMR J. Aust. Geol. Geoph.*, **8**, 179–190.

Beltaos, S. (1999). Climatic effects on the changing ice-breakup regime of the Saint John River, in *Proc. 10th Workshop on River Ice* Winnipeg, CA, pp. 252–264.

Beltaos, S. & B. C. Burrell (2002). Extreme ice jam floods along the Saint John River, New Bruswick, Canada. *IAHS Publ.* **271**, 9–14.

Benito, G. & J. E. O'Connor (2003). Number and size of last-glacial Missoula floods in the Columbia River valley between the Pasco Basin, Washington, and Portland, Oregon. *Geol. Soc. Amer. Bull.*, **115**, 624.

Berner, E. L. & R. A. Berner (1987). *The Global Water Cycle: Geochemistry and Environment*, Englewood Cliffs, NJ: Prentice-Hall.

Berner, R. A. (1994). GEOCARB II: a revised model of atmospheric CO_2 over Phanerozoic time. *Amer. J. Sci.*, **294**, 56–91.

Berner, R. A. (2004). *The Phanerozoic Carbon Cycle: CO_2 and O_2*, New York: Oxford University Press.

Berner, R. A. *et al.* (2003). Phanerozoic atmospheric oxygen. *Ann. Rev. Earth lPanet. Sci.*, 31.

Best, D. W. (1995). History of timber harvest in the Redwood Creek basin, northwestern California, USGS Prof. Paper 1454, C1-C7.

Beusen, A. H. W. *et al.* (2005). Estimation of global river transport of sediments and associated C, N, and P. *Global Biogeochem. Cycles*, **19**.

Bianchi, T. S. *et al.* (2008). Controlling hypoxia on the US Louisiana shelf: beyond the nutrient-centric view. *EOS*, **89**, 26.

Biksham, G. & V. Subramanian (1988). Sediment transport of the Godavari River basin and its controlling factors. *J. Hydrol.*, **101**, 275–290.

Billi, P. & M. Rinaldi (1997). Human impact on sediment yield and channel dynamics in the Arno River basin (central Italy). *IAHS Publ.* **245**, 301–311.

Binda, G. G., T. J. Day & J. P. M. Syvitski (1986). Terrestrial sediment transport into the marine environment of Canada, Environment Canada, p. 85.

Biswas, A. K. (1994). *International Waters of the Middle East: From Euphrates–Tigris to Nile*, Oxford University Press.

Bjørnsson, H. (1979). Glaciers in Iceland. *Jokull*, **29**, 74–79.

Bjørnsson, H. (1992). Jokulhlaups in Iceland: prediction characteristics and simulation. *Ann. Glaciol. Soc.*, **16**, 95–106.

Bjørnsson, H. (1995). Exursion Roadlog, in *AGS International Symposium on Glacial Erosion and Sedimentation* Rekkvjk, Iceland.

Blair, N. E. *et al.* (2003). The persistence of memory: the fate of ancient sedimentary organic carbon in a modern sedimentary system. *Geochim. Cosmochim. Acta*, **67**, 63–73.

Blong, R. J. (1991). The magnitude and frequency of large landslides in the Ok Tedi catchment. Report, Ok Tedi Mining, Ltd, Tabubil, Papua New Guinea.

Blum, M. D. & H. H. Roberts (2009). Drowning of the Mississippi Delta due to insufficient sediment supply and global sea-level rise. *Nat. Geosci.*, **2**, 488–491.

Bluth, G. J. S. & L. R. Kemp (1994). Lithologic and climatologic controls on river chemistry. *Geochim. Cosmochim. Acta*, **58**, 2341–2359.

Bobrovitskaya, N. N. (1996). Long-term variations in mean erosion and sediment yield from rivers of the former Soviet Union. *IAHS Publ.* **236**, 407–413.

Boesch, D. F. *et al.* (2009). Nutrient enrichment drives Gulf of Mexico hypoxia. *EOS*, **90**, 117–118.

Bogdanova, E. G. *et al.* (2002). A new model for bias correction of precipitation measurements, and its application to polar regions of Russia. *Russ. Meteorol. Hydrol.*, **10**, 68–94.

Bogen, J. (1996). Erosion and sediment yield in Norwegian Rivers. *IAHS Publ.* **236**, 73–84.

Bohannon, J. (2010). The Nile's Delta's shrinking future. *Science*,Bonell, M., M. M. Hufschmidt & J. S. Gladwell (eds.) (1993). *Hydrology and Water Management in the Humid Tropics,* Cambridge: Cambridge University Press.

Borchet, G. & S. Kempe (1985). A Zambezi Aqueduct, in *Transport of Carbon Minerals in Major World Rivers, pt. 3*, eds. E. T. Degens, S. Kempe & R. Herrera, Hamburg: Mitt. Geol.-Palaont. Inst. Univ., pp. 443–457.

Bork, H. R. *et al.* (1998). *Landschaftsentwicklung in Mitteleuropa*, Klett-Perthes Gotha.

Borland, W. M. (1973). *Pa Mong Phase II: Supplement to the Main Report (Hydraulics and Sediment Studies)*,US Bureau Recl.

Boufous, L. (1982). Definition des mesures contre l'envasement de la retenue sur l'Oued Nekur au Maroc, in *14th Congress CIBG,* Rio De Janeiro, pp. 11–20.

Bourgoin, L. M. *et al.* (2007). Temporal dynamics of water and sediment exchanges between the Curuai floodplain and the Amazon River, Brazil. *J. Hydrol.*, **335**, 140–156.

Bowling, L. C. & D. P. Lettenmaier (2001). The effects of forest roads and harvest on catchment hydrology in a mountainous maritime environment. *Water Sci. Appl.*, **2**, 145–164.

Brabets, T. P., B. Wang & R. H. Meade (2000). Environmental and hydrologic overview of the Yukon River basin, Alaska and Canada, US Geol. Survey, 106.

Braconnot, P. *et al.*, (2004). Evaluation of PMIP coupled ocean–atmosphere simulations of the mid-Holocene. In R. W. Battarbee, F. Gasse and C. E. Stickley (eds.), *Past Climate Variability through Europe and Africa. Developments in Paleoenvironmental Research, 6*. Springer, Dordrecht.

Brardinoni, F., M. A. Hassan & H. O. Slaymaker (2003). Complex mass wasting response of drainage basins to forest management in coastal British Columbia. *Geomorph.*, **49**, 109–124.

Bresson, L.-M. & C. V. Valentin (1994). Soil surface crust formation: contribution of micromorphology, in *Soil Micromorphology: Studies in Management and Genesis*, eds. A. J. Ringoses-Voase & G. S. Humphreys, Amsterdam: Elsevier, pp. 737–762.

Bretz, J. H. (1930). Lake Missoula and the Spokane flood. *Geol. Soc. Amer. Bull.*, **41**, 461–468.

Bretz, J. H. (1969). The Lake Missoula floods and the channeled scabland. *J. Geol.*, **77**, 505–543.

Bridge, J. S. (2003). *Rivers and Floodplains,* Wiley-Blackwell.

Brittain, J. E. *et al.* (2009). Arctic rivers, in *Rivers of Europe*, eds. K. Tockner, C. T. Robinson & U. Uehlinger, Amsterdam: Elsevier, pp. 337–379.

Broecker, W. S. (1989). Routing of meltwater from the Laurentide Ice Shelf during the younger Dryas cold episode. *Nature*, **341**, 318–321.

Brookfield, H. *et al.* (1992). Borneo and the Malay Peninsula, in *The Earth as Transformed by Human Action*, eds. B. L. Turner, II *et al.*, Cambridge: Cambridge University Press, pp. 495–512.

Brooks, D. (1999). *Bobos in Paradise,* New York: Simon and Schuster.

Brown, W. M. (1973). Streamflow, sediment, and turbidity in the Mad River basin, Humboldt and Trinity Counties, California. USGS Water-Resour. Invest. Rept. 73–36.

Brown, G. W. & J. T. Krygier (1971). Clear-cut logging and sediment production in the Oregon Coast Range. *Water Resour. Res.*, **7**.

Brown, W. M. & J. R. Ritter (1971). Sediment transport and turbidity in the Eel River basin, California, US Geol. Survey Open-File Rept., 67pp.

Brownlie, W. R. & B. D. Taylor (1981). Sediment management for the southern California mountains, coastal plains, and shoreline, Part C: Coastal sediment delivery by major rivers in southern California. California Institute of Technology, Environ. Qual. Lab Rep, 17-C.

Brummett, R. E. & G. G. Teugels (2004). Rivers of the Lower Guinean rainforest: Biogeography and sustainable exploitation, *FAO Regional Rept. 007,* 149–171.

Brunner, C. A. *et al.* (1999). Deep-sea sedimentary record of the late Wisconsin cataclysmic floods from the Columbia River. *Geology*, **27**, 463–466.

Brush, G. S. (2001). Natural and anthropogenic changes in Chesapeake Bay during the last 1000 years. *Hum. Ecol. Risk Assess*, **7**, 1283–1296.

Bryce, S., P. Larcombe & P. V. Ridd (1998). The relative importance of landward-directed tidal sediment transport versus freshwater flood events in the Normanby River estuary, Cape York Peninsula, Australia. *Mar. Geol.*, **149**, 55–78.

Budyko, M. I. (1974). *Climate and Life* (translated from Russian by D. H. Miller), San Diego: Academic Press.

Bue, C. D. (1970). Streamflow from the United States into the Atlantic Ocean during 1931–1960, U.S. Geol. Survey open-File Rept., 36 pp.

Burman, J.-O. (1983). Element transports in suspended and dissolved phases in the Kalix River. *Environ. Biogeochem. Ecol. Bull.*, **35**, 99–113.

Burns, D. A. *et al.*, (1998). Base cation concentrations in subsurface flow from a forested hillslope: The role of flushing frequency. *Water Resour. Res.*, **34**, 3535–3544.

Burn, D. A. et al. (2003). The geochemical evolution of riparian ground water in a forested piedmont catchment. *Ground Water,* **41**, 913–925.

Burt, T. P. (1992). The hydrology of headwater catchments, in *The Rivers Handbook*, eds. P. Calow & G. E. Petts, Oxford: Blackwell Science, pp. 3–28.

Burz, J. (1977). Suspended load discharge in the semiarid region of the northern Peru. *IAHS Publ.* **122**, 269–277.

Butler, D. R. (1995). *Zoogeomorphology – Animals as Geomorphic Agents,* Cambridge: Cambridge University Press.

Butler, D. R. (2006). Human-induced changes in animal populations and distributions, and the subsequent effects on fluvial systems. *Geomorph.*, **79**, 448–459.

Butler, D. R. & G. P. Malanson (2005). The geomorphic influences of beaver dams and failures of beaver dams. *Geomorph.*, **71**, 48–60.

Canadian National Committee (1972). *Discharge of Selected Rivers of Canada,* Ottawa: Canadian National Committee for the International Hydrological Decade.

Cañon, J., J. Gonzalvev & J. Valdes (2007). Precipitation in the Colorado River Basin and its low frequency associations with PDO and ENSO signals. *J. Hydrol.*, **333**, 252–264.

Canton, Y. *et al.* (2001). Hydrological and erosion response of a badlands system in semiarid SE Spain. *J. Hydrol.*, **252**, 65–84.

Carling, P. A. *et al.* (2002). Late Quaternary catastrophic flooding in the Altai Mountains of south-central Siberia: a synoptic overview and introduction to flood deposit sedimentology. *Flood and megaflood processes and deposits: recent and ancient examples*, Oxford, Blackwell Scientific, pp. 17–35.

Carter, J. L. & V. H. Resh (2005). Pacific coast rivers of the conterminous United States, in *Rivers of North America*, eds. A. C. Benke & C. E. Cushing, Amsterdam: Elsevier, pp. 540–589.

Carvalho, N. O. (2008). *Hidrossedimentologia Pratica* (2nd edn.), Rio de Janeiro: Editoria Interciencia.

Carvalho, N. O. & S. B. Cunha (1998). Estimata da carga solida do Rio Amazonas e seus principas tributarios para a foz e oceano: uma retrospectiva. *A Agua em Revista, CPRM*, **6**, 44–58.

Cattaneo, A. (1995). The rivers of Italy, in *River and Stream Ecosystems*, eds. C. E. Cushing, K. W. Cummins & G. W. Minshall, Amsterdam: Elsevier.

Cayan, D. R. & R. H. Webb (1992). El Niño/Southern Oscillation and streamflow in the western United States, in *El Niño. Historical and Paleoclimatic Aspects of the Southern Oscillation*, eds. H. F. Diaz & V. Markgraf, Cambridge: Cambridge University Press, pp. 58–68.

Center for Natural Resources, Energy and Transport (**UN**) (1978). Register of International Rivers, UN.

Cerdà, A. & T. Lasanta (2005). Long-term erosional responses after fire in the central Spanish Pyrenees 1. Water and sediment yield. *Catena*, **60**, 59–80.

Chang, J. & O. Slaymaker (2002). Frequency and spatial distribution of landslides in a mountainous drainage basin: Western Foothills, Taiwan. *Catena*, **46**, 285–307.

Chaperon, P., J. Danloux & L. Ferry (1993). *Rivers of Madagascar*, Monogr. Hydrol., ORSTOM, Paris.

Chatwin, S. C. & R. B. Smith (1992). Reducing soil erosion associated with forestry operations through integrated research: an example for coastal British Columbia, Canada. *IAHS Publ.* **209**, 377–385.

Chen, C. T. A. (2000). The Three Gorges Dam: reducing the upwelling and thus productivity in the East China Sea. *Geophys. Res. Lett.*, **27**, 381–383.

Chen, Y. (2009). Did the reservoir impoundment trigger the Wenchuar earthquake? *Science in China D. Earth Sci.* **52**, 431–433.

Chen, C. T. A., J. T. Liu & B. J. Tsuang (2004). Island-based catchment – The Taiwan example. *Reg. Environ. Change*, **4**, 39–48.

Chen, J., D. He & S. Cui (2003). The response of river water quality and quantity to the development of irrigated agriculture in the last 4 decades in the Yellow River Basin, China. *Water Resour. Res.*, **39**, 1047.

Chen, X. *et al.* (2001). Human impacts on the Changjiang (Yangtze) River basin, China, with special reference to the impacts on the dry season water discharges into the sea. *Geomorph.*, **41**, 111–123.

Chen, Z. *et al.* (2001). Yangtze River of China: historical analysis of discharge variability and sediment flux. *Geomorph.*, **41**, 77–91.

Chin, A. (2006). Urban transformation of river landscapes in a global context. *Geomorph.*, **79**, 460–487.

Choudhury, A. M. (1978). Bangladesh floods, cyclones and ENSO. IAEA-UNESCO Int. Centre for Theoretical Physics, 8.

Church, T. M. (1996). An underground route for the water cycle. *Nature*, **380**, 579–580.

Ciavola, P. *et al.* (1999). Relation between river dynamics and coastal changes in Albania: an assessment integrating satellite imagery with historical data. *Int. J. Remote Sens.*, **20**, 561–584.

Clague, J. J. *et al.* (2003). Paleomagnetic and tephra evidence for tens of Missoula floods in southern Washington. *Geology*, **31**, 247–250.

Clair, T. A., T. L. Pollock & J. M. Ehrman (1994). Exports of carbon and nitrogen from river basins in Canada's Atlantic Provinces. *Global Biogeochem. Cycles*, **8**, 441–450.

Clark, J. J. & P. R. Wilcock (2000). Effects of land-use change on channel morphology in northeastern Puerto Rico. *Geol. Soc. Amer. Bull.*, **112**, 1763–1777.

Clarke, F. W. (1924a). *The composition of the river and lake waters of the United States*, US Geol. Surv. Prof. Paper 135.

Clarke, F. W. (1924b). *The data of geochemistry*, US Geo. Surv. Bull. 770.

Clarke, G. K. C., W. H. Mathews & R. T. Pack (1984). Outburst floods from Glacial Lake Missoula. *Quat. Res.*, **22**, 289–299.

Claussen, M. & H.-J. Bolle (2004). The Sahara, in *Vegetation, Water, Humans and the Climate: A New Perspective on an Interactive System*, eds. P. Kabat, M. Claussen & P. A. Dirmeyer, Berlin: Springer-Verlag, pp. 42–44.

COHMAP Members (1988). Climatic changes of the last 18000 years: observations and model simulations. *Science*, **241**, 1043–1052.

Colin, C. *et al.* (2010). Impact of the East Asian monsoon rainfall changes on the erosion of the Mekong River basin over the past 25,000 yr. *Marine Geol.*, **271**, 84–92.

Colinvaux, P., J. E. de Oliveira & M. Patino (1999). *Amazon Pollen Manual and Atlas/Manual e Atlas Palinologico da Amazonia,* Amsterdam: Hardwood Academic Publishing.

Collins, B. D. & T. Dunne (1986). Erosion of tephra from the 1980 eruption of Mount St. Helens. *Geol. Soc. Amer. Bull.*, **97**, 896–905.

Collins, D. N. (1996). Sediment transport from glacierized basins in the Karakoram mountains. *IAHS Publ.* **236**, 85–96.

Collins, M. B. (1981). Sediment yield studies of headwater catchments in Sussex, S.E. England. *Earth Surf. Process. Landforms*, **6**, 517–539.

Collins, R. O. (2002). *The Nile,* Yale University Press.

Colman, S. M. & J. F. Bratton (2003). Anthropogenically induced changes in sediment and biogenic silica fluxes in Chesapeake Bay. *Geology*, **31**, 71–74.

Coltorti, M. (1997). Human impact in the Holocene fluvial and coastal evolution of the Marche region, Central Italy. *Catena*, **30**, 311–335.

Combe, M. (1968). *Resources en Eau du Maroc et Pourcetage d'Utilisation Etat des Connaissances*, Direction de l'Hydraulique.

Comision Nacional del Agua (1990). Datos hidrometricos de Mexico, 1937–1985. CD-ROM.

Conley, D. J. *et al.* (2000). The transport and retention of dissolved silicate by rivers in Sweden and Finland. *Limnol. Oceanogr.*, **45**, 1850–1853.

Cook, E. R. *et al.* (2004). Long-term aridity changes in the western United States. *Science*, **306**, 1015–1018.

Cook, E. R. *et al.* (2007). North American drought: reconstructions, causes, and consequences. *Earth Sci. Rev.*, **81**, 93–134.

Correggiari, A., A. Cattaneo & F. Trincardi (2005). The modern Po Delta system: Lobe switching and asymmetric prodelta growth. *Mar. Geol.*, **222**, 49–74.

Costa, J. E. & L. Schuster (1988). The formation and failure of natural dams. *Geol. Soc. Amer. Bull.*, **100**, 1054–1068.

Cowan, E. A., P. R. Carlson & R. D. Powell (1996). The marine record of the Russell Fiord outburst flood, Alaska, USA. *Ann. Glaciol.*, **22**, 194–199.

Craig, R. G. (1987). Dynamics of a Missoula flood, in *Catastrophic Flooding*, eds. L. Mayer & D. Nash, Boston: Allen & Unwin, pp. 305–332.

Crossland, C. J. *et al.* (eds.) (2005). *Coastal Fluxes in the Anthropocene*: Heidelberg: Springer-Verlag.

Crutzen, P. J. & E. F. Stoermer (2000). The "Anthropocene", in *IGBP Newsletter*, pp. 17–18.

Culp, J. M., T. D. Prowse & E. A. Luiker (2005). Mackenzie River Basin, in *Rivers of North America*, eds. A. C. Benke & C. E. Cushing, Amsterdam: Elsevier, 804–850.

Cunjak, R. A. & R. W. Newbury (2005). Atlantic coast rivers of Canada, in *Rivers of North America*, eds. A. C. Benke & C. E. Cushing, Amsterdam: Elsevier, 938–980.

Curtis, W. F., J. K. Culbertson & E. B. Chase (1973). Fluvial-Sediment Discharge to the Oceans from the Conterminous United States, US Geol. Surv. Circular, 17 pp.

Czaya, E. (1981). *Rivers of the World,* Van Nostrand Reinhold Company.

Dadson, S. J. *et al.* (2004). Earthquake-generated increase in sediment delivery from an active mountain belt. *Geology*, **32**, 733–736.

Dadson, S. J. *et al.* (2005). Hyperpycnal river flows from an active mountain belt. *J. Geophys. Res.*, **110**, f04016.

Dahm, D. H., R. J. Edwards & F. P. Gelwick (2005). Gulf Coast rivers of the Southwestern United States, in *Rivers of North America*, eds. A. C. Benke & C. E. Cushing, Amsterdam: Elsevier, 180–228.

Dai, A., K. E. Trenberth & T. T. Qian (2004). A global dataset of Palmer Drought Severity Index for 1870–2002: relationship with soil moisture and effects of surface warming. *J. Hydromet.*, **5**, 1117–1130.

Dai, A. G. & K. E. Trenberth (2002). Estimates of freshwater discharge from continents: latitudinal and seasonal variations. *J. Hydrometeorol.*, **3**, 660–687.

Dai, F. C. *et al.* (2005). The 1786 earthquake-triggered landslide dam and subsequent dam-break flood on the Dadu River, southwestern China. *Geomorphology*, **65**, 205–221.

Dai, S. B., S. L. Yang & A. M. Cai (2008). Impacts of dams on the sediment flux of the Pearl River, southern China. *Catena*, **76**, 36–43.

Dai, S. B., S. L. Yang & M. Li (2009). The sharp decrease in suspended sediment supply from China's rivers to the sea: anthropogenic and natural causes. *Hydrol. Sci.*, **54**, 135–146.

Datta, D. K. & V. Subramanian (1997). Texture and mineralogy of sediments from the Ganges-Brahmaputra-Meghna river system in the Bengal Basin, Bangladesh and their environmental implications. *Environ. Geol.*, **30**, 181–188.

Dauta, A. *et al.* (2009). The Adour–Garonne basin, in *Rivers of Europe*, eds. K. Tockner, C. T. Robinson & U. Uehlinger, Amsterdam: Elsevier, pp. 182–198.

De Bartolo, S. G., S. Gabriele **&** R. Gaudio (2000). Multifractal behaviour of river networks. *Hydrol. Earth Syst. Sci.*, **4**, 105–112.

de Vente, J. *et al.* (2007). The sediment delivery problem revisited. *Prog. Phys. Geog.*, **31**, 155–178.

de Villiers, M. (2000). *Water: The Fate of Our Most Precious Resource,* Boston: Houghton Mifflin Co.

de Wit, M. (1999). Modeling nutrient fluxes from source to river load: a macroscopic analysis applied to the Rhine and Elbe basins. *Hydrobiologia*, **410**, 123–130.

Dearing, J. A. & R. T. Jones (2003). Coupling temporal and spatial dimensions of global sediment flux through lake and marine sediment records. *Global Planet. Change*, **39**, 147–168.

Dedkov, A. P. & V. I. Mozzherin (1992). Erosion and sediment yield in mountain regions of the world. *IAHS Publ.* **209**, 29–36.

Degens, E. T. & D. A. Ross (1972). Chronology of the Black Sea over the last 25,000 years. *Chem. Geol.*, **10**, 1–16.

Degens, E. T., S. Kempe & J. E. Richey (eds.) (1991). *Biogeochemistry of major world rivers,* Chichester: John Wiley & Sons.

Dellapenna, J. W. (1996). Rivers as legal structures: the examples of the Jordan and the Nile. *Nat. Resour. J.*, **36**, 217–250.

Demissie, M. (1996). Patterns of erosion and sedimentation in the Illinois River basin. *IAHS Publ.* **236**, 483–490.

Depetris, P. J. **&** G. Irion (1996). Clay-size mineralogy of the suspend load from two Patagonian rivers: The Negro and The Colorado. *Actas del Trigésimo Congreso Geológico Argentino*, **3**, 275–280.

Depetris, P. J. & A. M. L. Lenardon (1982). Particulate and dissolved phases in the Parana River. *Mitt. Geol. -Palaont. Inst. University of Hamburg*, **52**, 385–395.

Depetris, P. J. & J. E. Paolini (1991). Biogeochemical aspects of South American rivers The Parana and the Orinoco, in *Biogeochemistry of Major World Rivers*, eds. E. T. Degens, S. Kempe & J. E. Richey, Chinchester: Wiley, pp. 105–126.

Depetris, P. J. *et al.* (1996). ENSO-controlled flooding in the Parana River (1904–1991). *Naturwissenschaften*, **83**, 127–129.

Depetris, P. J. *et al.* (2005). Biogeochemical output and typology of rivers draining Patagonia's Atlantic seaboard. *J. Coast. Res.*, **21**, 835–844.

Déry, S. J. & E. F. Wood (2004). Teleconnection between the Arctic Oscillation and Hudson Bay river discharge. *Geophys. Res. Lett.*, **31**, 18.

Deser, C. (2000). On the teleconnectivity of the "Arctic Oscillation". *Geophys. Res. Lett.*, **27**, 779–782.

Deser, C., A. S. Phillips & J. W. Hurrell (2004). Pacific interdecadal climate variability: Linkages between the tropics and North Pacific during boreal winter since 1900. *J. Climatol.*, **17**, 3109–3124.

Dettinger, M. D. & H. F. Diaz (2000). Global characteristics of stream flow seasonality and variability. *J. Hydrometeorol.*, **1**, 289–310.

Deverel, S. J. and Rojstaczer, S. (1996). Subsidence of agricultural lands in the Sacramento–San Joaquin Delta, California: role of aqueous and gaseous carbon fluxes. *Water. Reour. Res*, **32**, 2359–2367.

Dhanio, L. L. *et al.* (2008). Changes in sediment discharge after the collapse of Mount Bawakaraeng in south Sulawesi, Indonesia. *IAHS Publ.* **325**, 607–611.

Dietrich, W. E., G. Day & G. Parker (1999). The Fly River, Papua New Guinea: inferences about river dynamics, floodplain sedimentation and fate of sediment, in *Varieties of Fluvial Form*, eds. A. J. Miller & A. Gupta, Chichester, UK: John Wiley and Sons, pp. 345–376.

Dilley, M. & B. N. Heyman (1995). ENSO and disaster: droughts, floods and El Nino/Southern Oscillation warm events. *Disasters*, **19**, 181–193.

Dinehart, R. L. (1997). *Sediment transport at gauging stations near Mount St. Helens, Washington, 1980–90: data collection and analysis*, US Geological Survey., 105 pp.

Dixon, T. H. *et al.* (2006). Subsidence and flooding in New Orleans. *Nature*, **441**, 587–588.

Dole, R. & H. Stabler (1909). *Denudation*, US Geological Survey Water-Supply Paper 234, pp. 78–93.

Donnelly, J. P. *et al.* (2005). Catastrophic meltwater discharge down the Hudson Valley: a potential trigger for the Intra-Allerod cold period. *Geology*, **33**, 89–92.

Dorsey, R. J. (1988). Provenance evolution and unroofing history of a modern arc-continent collision: evidence from petrography of Plio-Pleistocene sandstones, eastern Taiwan. *J. Sediment. Petrol.*, **58**, 208–218.

Dosseto, A. *et al.* (2006). Time scale and conditions of weathering under tropical climate: study of the Amazon basin with U-series. *Geochim. Cosmochim. Acta*, **70**, 71–89.

Douglas, I. (1967). Man, vegetation, and the sediment yield of rivers. *Nature*, **215**, 925–928.

Douglas, I. (1968). Erosion in the Sungei Gombak catchment – Selangor Malaysia. *J. Tropic. Geog.*, **26**, 1–16.

Douglas, I. (1996). The impact of land-use changes, especially logging, shifting cultivation, mining and urbanization on sediment yields in humid tropical Southeast Asia: a review with special reference to Borneo. *IAHS Publ.* **236**, 463–471.

Douglas, I. & T. Spencer (1985). *Environmental change and tropical geomorphology,* Allen & Unwin Australia.

Dowidar, N. M. (1988). Effect of Aswan High Dam on the biological productivity of southeastern Mediterranean, in *Natural and Man-made Hazards*, eds. M. E. El-Sabh & T. S. Murty, Dordrecht: D. Reidel, pp. 477–498.

Drago, E. E. & M. L. Amsler (1988). Suspended sediment at a cross section of the Middle Parana River: concentrations, granulometry and influence of the main tributaries, in *Sediment Budgets*, IAHS Publ. **174**, 381–396.

Drake, D. E., R. Kolpack & P. J. Fischer (1972). Sediment transport on Santa Barbara–Oxnard shelf, Santa Barbara Channel, California, in *Shelf Sediment Transport: Processes and Pattern*, eds. D. J. P. Swift, D. B. Duane & O. H. Pilkey, Stroudsburg, PA: Dowdon, Hutchinson and Ross, pp. 307–332.

Drenzek, N. J. *et al.* (2009). A new look at old carbon in active margin sediments. *Geology*, **37**, 239.

Drever, J. I. (1994). The effect of landplants on weathering rates of silicate minerals. *Geochim. Cosmochim. Acta*, **58**, 2325–2333.

Duncan, M. J. (1992). Flow regimes of New Zealand rivers, in *Waters of New Zealand*, ed. M. P. Mosley, NZ Hydrol. Soc., pp. 13–27.

Dunne, T. (1978). Rates of chemical denudation of silicate rocks in tropical catchments. *Nature*, **274**, 244–246.

Dunne, T. (2001). Problems in measuring and modeling the influence of forest management on hydrologic and geomorphic processes, in *Land Use and Watersheds: Human Influence on Hydrology and Geomorpholgy in Urban and Forest Areas*, eds. M. S. Wigmosta & S. J. Burges, Washington, DC: American Geophysical Union, pp. 77–84.

Dunne, T. & W. E. Dietrich (1980). Experimental study of Horton overland flow on tropical hillslopes. *Z. Geomorphol. Suppl.*, **35**, 40–59.

Dunne, T. *et al.* (1998). Exchanges of sediment between the floodplain and channel of the Amazon River in Brazil. *Geol. Soc. Amer. Bull.*, **110**, 450–470.

Dupré, B. *et al.* (2003). Rivers, chemical weathering and Earth's climate. *C.R. Geosci.*, **335**, 1141–1160.

Dynesius, M. & C. Nilsson (1994). Fragmentation and flow regulation of river systems in the northern third of the world. *Science*, **266**, 753–762.

Dzhamalov, R. G. & T. I. Safronova (2002). On estimating chemical discharge into the world ocean with groundwater. *Water Resour. Res.*, **29**, 680–686.

Dzhaoshvili, W. (1986). *River sediments and formation of beaches on Georgia Black Sea shore (in Russian)*, Tbilisi: Sabchota Sakartvelo (cited in Meybeck and Ragu, 1997).

ECLAC (1990). *Latin America and the Caribbean: Inventory of Water Resources and Their Use,* Santiago, Chile: United Nations Economic Commission for Latin America and the Caribbean (ECLAC).

Edmond, J. M. & Y.-S. Huh (1997). Chemical weathering yields from basement and orogen terrains in hot and cold climates, in *Tectonic Uplift and Climate Change*, ed. W. F. Ruddiman, New York: Plenum Press, pp. 329–351.

Edmond, J. M. *et al.* (1995). The fluvial geochemistry and denudation rate of the Guayana Shield in Venezuela, Colombia, and Brazil. *Geochim. Cosmochim. Acta*, **59**, 3301–3325.

Edwards, T. K. & G. D. Glysson (1999). *Field methods for measurement of fluvial sediment, US Geological Survey Techniques of Water-Resource Investigation*, Book 3, chapter C2, 118.

Eisma, D., G. C. Cadee & R. Laane (1982). Supply of suspended matter and particulate and dissolved organic carbon from the Rhine to the coastal North Sea, in *Biogeochemistry of Major World Rivers, SCOPE-42*, eds. E. T. Degens, S. Kempe & J. E. Richey, Chichester: Wiley and Sons, pp. 297–322.

Eisma, D. L., L. Kalf & S. J. Van der Gaast (1978). Suspended matter in the Zaire estuary and the adjacent Atlantic Ocean. *Neth. J. Sea Res.*, **12**, 382–406.

Eittreim, S. L., R. Anima & A. J. Stevenson (2002). Seafloor geology of the Monterey Bay area continental shelf. *Mar. Geol.*, **181**, 3–34.

Ekholm, P. (1992). Reversibly adsorbed phosphorus in agriculturally loaded rivers in southern Finland. *Aqua Fennica*, **22**, 35–41.

El-Etr, H. A. *et al.* (1999). Regional study of the drainage basins of Sinai and eastern desert of Egypt, with a preliminary assessment of their flash flood potential. *Ann. Geo. Survey Egypt*, **22**, 335–356.

El-Sayed, R. (1993). *The River Nile: Geology, Hydrology and Utilization,* Oxford: Pergamon Press.

Eldridge, D. J. (1998). Trampling of microphytic crusts on calcareous soils, and its impact on erosion under rain-impacted flow. *Catena*, **33**, 221–239.

Emanuel, K. A. (1987). The dependence of hurricane intensity on climate. *Nature*, **326**, 483–485.

Emanual, K. A. (2005). Increasing destructiveness of tropical cyclones over the past 30 years. *Nature*, **436**, 686–688.

Embrey, S. S. & L. M. Frans (2003). *Surface-water quality of the Skokomish, Nooksack and Green-Duwamish rivers and Thornton Creek, Puget Sound basin, Washington, 1995–1998*, USGS Water Res. Invest. Rept. 4190.

Emile-Gray, J. *et al.* (2007). ENSO as a mediator for the solar influence on climate. *Paleoceanogr*, **22**, PA3210, doi:10.1029/2006PA1304, 2007.

Emiliani, C. *et al.* (1975). Paleoclimatological analysis of late Quaternary cores from the northeastern Gulf of Mexico. *Science*, **189**, 1083–1088.

Environment Canada (1984).

Enzel, Y. *et al.* (1999). High-resolution Holocene environmental changes in the Thar Desert northwestern India. *Science*, **284**, 125–128.

Erlingsson, U. L. F. (2008). A Jokulhlaup from a Laurentian-captured ice shelf to the Gulf of Mexico could have caused the Bølling warming. *Geog. Ann. Ser. A Phys. Geog.*, **90**, 125–140.

Ermini, L. & N. Casagli (2003). Prediction of the behavior of landslide dams using a geomorphological dimensionless index. *Earth Surf. Process. Landforms*, **28**, 31–47.

Erskine, W. D. & R. Warner (1999). Significance of river bank erosion as a sediment source in the altering of flood regimes of south-eastern Australia, in *Fluvial Processes and Environmental Change*, eds. A. G. Brown & T. A. Quine, New York: John Wiley and Sons, pp. 139–163.

Esser, G. & G. H. Kohlmaier (1991). Modeling terrestrial sources of nitrogen, phosphorus, sulfur and organic carbon to rivers, in *Biogeochemistry of Major World Rivers*, eds. E. T. Degens, S. Kempe & J. E. Richey, Chichester: Wiley, pp. 297–322.

Etchanchu, D. & J. L. Probst (1988). Evolution of the chemical composition of the Garrone River water during the period 1971–1984. *Hydrol. Sci. J.*, **33**, 243–256.

Eurosion (2004). *Living with coastal erosion in Europe; sediment and space for sustainabliity. Part II: Maps and Statistics*, DG Environment EC.

Fairbanks, R. G. (1989). A 17, 000-year glacio-eustatic sea level record: influence of glacial melting rates on the Younger Dryas event and deep-ocean circulation. *Nature*, **342**, 637–642.

Fang, X. *et al.* (2001). Changes in forest biomass carbon storage in China between 1949–1998. *Science*, **292**, 2320–2322.

Fanos, A. M. (1996). The impact of human activities on the erosion and accretion of the Nile delta coast. *J. Coast. Res.*, **11**, 821–833.

FAO (1997). Irrigation potential in Africa: a basin approach, Rome: FAO, Water Resources.

FAO (1999). State of world's forests, Rome: FAO, Forestry Department.

FAO/AGL (http://www.fao.org/landwater/aglw/sediment/default.asp).

FAP 24 (1994). Morphological Studies Phase I: Available Data and Characteristics, Government of Bangladesh.

Farnsworth, K. L. & J. D. Milliman (2003). Effects of climatic and anthropogenic change on small mountainous rivers: the Salinas River example. *Global Planet. Change*, **39**, 53–64.

Favier, V. *et al.* (2009). Interpreting discrepancies between discharge and precipitation in high-altitude area of Chile's Norte Chico region (26–32° S). *Water Resour. Res.*, **45**, W02424, doi:10.1029/2008WR006802.

Fekete, B. M., C. J. Vörösmarty & W. Grabs (1999). Global, Composite Runoff Fields Based on Observed River Discharge and Simulated Water Balances, Koblenz, Germany:WMO-Global Runoff Data Center Report #22.

Fekete, B. M., C. Vörösmarty & W. Grabs (2002). High-resolution fields of global runoff combining observed river discharge and simulated water balances. *Global Biogeochem. Cycles*, **16**, 1042, doi:10.1029/1999GB001254.

Fettweis, M., M. Sas & J. Monbaliu (1998). Seasonal, neap-spring and tidal variation of cohesive sediment concentration in the Scheldt estuary, Belgium. *Estuar. Coast. Shelf Sci.*, **47**, 21–36.

Fierro, P., Jr. & E. K. Nyer (2007). *The Water Encyclopedia (3rd edition)*, Boca Raton, Fla: CRC, Taylor and Francis.

Finlayson, B. L. & T. A. McMahon (1988). Australia vs. the world: a comparative analysis of streamflow characteristics, in *Fluvial Geomorphology of Australia*, ed. R. F. Warner, Sydney: Academic Press, pp. 17–40.

Fisk, H. N. (1947). Fine-grained alluvial deposits and their effects on Mississippi River activity, U.S. Army Corps of Engineers, Waterways Experiment Station, Vicksburg, MS, 82 pp.

Florsheim, J. L., Keller, E.A. and Best, D.W. (1991). Fluvial sediment transport in response to moderate storm flows following chaparral wildfire, Ventura County, southern California. *Geol. Soc. Amer. Bull.*, **103**, 504–511.

Foley, J. A. *et al.* (2002). El Nino-Southern oscillation and the climate, ecosystems and rivers of Amazonia. *Global Biogeochem. Cycles*, **16**, 1–17.

Foley, J. A. *et al.* (2005). Global consequences of land use. *Science*, **309**, 570–574.

Forbes, D. L. (1981). Babbage River Delta and lagoon: Hydrology and sedimentology of an Arctic estuarine system, Unpubl. Thesis, University of British Columbia, 554 p.

Foucart, A. & Stanley, D. J. (1989). Late Quaternary paleoclimatic oscillations in east Africa recorded by heavy minerals in the Nile Delta. *Nature* **229**, 44–46.

Fournier, F. (1949). Les factuer climatiques de l'erosion du sol. *Assoc. Geog. Francais Bull.*, **203**, 97–103.

Fournier, F. (1960). *Climat et Erosion,* Paris: Presses Universitaires de France.

Framji, K. K. & I. K. Mahajan (1969). *Irrigation and Drainage in the World; a Global Review*, 2nd edn. International Commission on Irrigation & Drainage. New Delhi. 2 v.

Francou, J. & J. Rodier (1967). Essai de classification des crues maximales observées dans le monde. *Cah. ORSTOM, sér. Hydrol*, **IV**.

Frangipane, A. & E. Paris (1994). Long-term variability of sediment transport in the Ombrone River Basin (Italy). *IAHS Publ.* **224**, 317–324.

Fraser, A. S., M. Meybeck & E. D. Ongley (1995). *Global Environment Monitoring System (GEMS): Water Quality of World River Basins*, UNEP Environment Library.

Frihy, O. E., E. A. Debes & W. R. El Sayed (2003). Processes reshaping the Nile delta promontories of Egypt: pre-and post-protection. *Geomorph.*, **53**, 263–279.

Fu, C. B. & H. L. Yuan (2001). A virtual numerical experiment to understand the impacts of recovering natural vegetation on summer climate and environmental conditions in East Asia. *Chin. Sci. Bull.*, **46**, 1199–1203.

Fuggle, R. & W. T. Smith (2000). Experience with Dams in Water and Energy Resource Development in the People's Republic of China, World Comm. Dams Country Review Paper.

Fuller, C. C. *et al.* (1990). Distribution and transport of sediment-bound metal contaminants in the Río Grande de Tárcoles, Costa Rica (Central America). *Water Res.*, **24**, 805–812.

Fuller, C. W. *et al.* (2003). Erosion rates for Taiwan mountain basins: new determinations from suspended sediment records and a stochastic model of their temporal variation. *J. Geol.*, **111**, 71–87.

Furuichi, T., Z. Win & R. J. Wasson (2009). Discharge and suspended sediment transport in the Ayeyarwady River, Myanmar: centennial and decadal changes. *Hydrol. Process.*, **23**, 1631–1641.

Gabet, E. J. & S. M. Mudd (2009). A theoretical model coupling chemical weathering rates with denudation rates. *Geology*, **37**, 151.

Gac, J. Y. & A. Kane (1986). Le fleuve Sénégal: II. Flux continentaux de matières dissoutes à l'embouchure. *Sciences Géol. Bull.*, **39**, 151–172.

Gagan, M. K. *et al.* (2004). Post-glacial evolution of the Indo-Pacific Warm Pool and El Niño-Southern Oscillation. *Quat. Int.*, **118**/119, 127–143.

Gaillardet, J. *et al.* (1997). Chemical and physical denudation in the Amazon River Basin. *Chem. Geol.*, **142**, 141–173.

Gaillardet, J. *et al.* (1999). Global silicate weathering and CO_2 consumption rates deduced from the chemistry of large rivers. *Chem. Geol.*, **159**, 3–30.

Galewsky, J. *et al.* (2006). Tropical cyclone triggering of sediment discharge in Taiwan. *J. Geophys. Res.*, **11** doi:10.1029/2005JF000428.

Galster, J. C. *et al.* (2006). Effects of urbanization on watershed hydrology: the scaling of discharge with drainage area. *Geology*, **34**, 713.

Galy, A. & C. France-Lanord (2001). Higher erosion rates in the Himalaya: geochemical constraints on riverine fluxes. *Geol. Soc. Amer. Bull.*, **29**, 23–26.

Gamage, N. and V. Smakhtin (2009). Do river deltas in east India retreat? A case of the Krishna Delta. Geomorph. **108**, 533–540.

Garbarino, J. R. *et al.* (1995). Heavy metals in the Mississippi River. *US Geol. Surv. Circular 1133*, pp. 53–72.

Garcia-Ruiz, J. M. (2010). The effects of land uses on soil erosion in Spain: a review. *Catena*, **81**, 1–11.

Gasse, F. (2000). Hydrological changes in the African tropics since the Last Glacial Maximum. *Quat. Sci. Rev.*, **19**, 189–211.

Gay, G. R. *et al.* (1998). Evolution of cutoffs across meander necks in Powder River, Montana, USA. *Earth Surf. Process. Landforms*, **23**.

Ge, S.-M. *et al.* (2009). Did the Zipingpu Reservoir trigger the 2008 Wenchuan earthquake? *Geophys. Res. Lett.*, **36**, L20315, doi:10.1029/2009GL040349.

GEMS (1983). GEMS/WATER Data Summary, Burlington, Ontario: WHO Collaborating Centre for Inland Waters.

Gerten, D. *et al.* (2008). Causes of change in 20th century global river discharge. *Geophys. Res. Lett.*, **35**, L20405.

Gibbs, R. (1967). The geochemistry of the Amazon River system. Part 1. The factors that control the salinity and composition and concentration of the suspended solids. *Geol. Soc. Amer. Bull.*, **78**, 1203–1232.

Gilbert, G. K. (1917). Hydraulic-mining debris in the Sierra Nevada: US, US Geological Survey Prof. Paper 105, 154 pp.

Gillett, N. P., T. D. Kell & P. D. Jones (2006). Regional climate impacts of the Southern Annular Mode. *Geophys. Res. Lett.*, **33**, L23704.

Gilluly, J. (1955). Geologic contrasts between continents and ocean basins. *Geol. Soc. Amer. Spec. Paper*, **62**, 7–18.

Giresse, P. A., A. Wiewiora & B. Lacka (1998). Processes of Holocene Ferromanganese-coated grains in the nearshore shelf of Cameroon. *J. Sediment. Res.*, **68**, 20–36.

Gleick, P. H. (1993). *Water in Crisis,* Oxford: Oxford University Press.

Gleick. P. H. (1999). *The World's Water, 1998/99*, Island Press, Washington D.C., 307pp.

Gleick, P. H. (2000a). The changing water paradigm. A look at twenty-first century water resources development. *Water Int.*, **25**, 127–138.

Gleick, P. H. (2000b). *The World's Water. 2000–2001,* Washington, DC: Island Press.

Gleick, P. H. (ed.) (2002). *The World's Water 2002–2003: The Biennial Report on Freshwater Resources,* Washington, DC: Island Press.

Gleick, P. H. *et al.* (2000). *Water: The Potential Consequences of Climate Variability and Change. A Report of the National Water Assessment Group, US Global Change Research Program,* US Geol. Surv. and Pac. Inst. Stud. in Devel., Environ., Secur., Oakland, California.

Gleick, P. H. *et al.* (2006). *The World's Water, 2006–2007*, Island Press, Washington D.C., 368pp.

Goldenberg, S. B. *et al.* (2001). The recent increase in Atlantic hurreane activity: causes and implications. *Science*, **293**, 474–479.

Goldsmith, S. T. *et al.* (2008). Extreme storm events, landscape denudation, and carbon sequestration: Typhoon Mindulle, Choshui River, Taiwan. *Geology*, **36**, 483–486.

Golosov, V. N. *et al.* (2008). Response of a small arable catchment sediment budget to introduction of soil conservation measures. *IAHS Publ.* **425**, 106–113.

Gong, G.-Y. & J.-X. Xu (1987). Environmental effects of human activities on rivers in the Huanghe-Huaihe-Haihe plain, China. *Geogr. Ann.*, **69**, 181–188.

Gonzalez-Hidalgo, J. C. *et al.* (2009). Contribution of the largest events to suspended sediment transport across the USA. Land Degrad. Develop., doi: 10.1002/ldr.897.

Gonzalez-Hidalgo, J. C. and J. L. Peña-Monné (2007). A review of daily erosion in Western Mediterranean areas. *Catena*, **71**, 193–199.

Goodbred, S. L. (2003). Response of the Ganges dispersal system to climate change: a source-to-sink view since the last interstade. *Sediment. Geol.*, **162**, 83–104.

Goodbred, S. L., Jr. & S. A. Kuehl (1998). Floodplain processes in the Bengal Basin and the storage of Ganges-Brahmaputra River sediment; an accretion study using 137Cs and 210Pb geochronology. *Sediment. Geol.*, **121**, 239–258.

Goodbred, S. L., Jr. & S. A. Kuehl (2000). The significance of large sediment supply, active tectonism, and eustasy on margin sequence development: Late Quaternary stratigraphy and evolution of the Ganges–Brahmaputra delta. *Sediment. Geol.*, **133**, 227–248.

Goolsby, D. A. (1994). Flux of herbicdes and nitrate from the Mississippi River to the Gulf of Mexico, in *Coastal Oceanographic Effects of Summer 1993 Mississippi River Flooiding*, ed. M. J. Dowigiallo, US National Oceanic and Atmospheric Administration, Special NOAA Report, 77 pp.

Goolsby, D. A. & W. A. Battaglin (2001). Long-term changes in concentrations and flux of nitrogen in the Mississippi River Basin, USA. *Hydrol. Process.*, **15**, 1209–1226.

Goolsby, D. A. & W. E. Pereira (1996). Pesticides in the Mississippi river. *US Geol. Surv. Circular* **1133**, 87–102.

Goolsby, D. A. *et al.* (1999). Flux and sources of nutrients in the Mississippi-Atchafalaya River Basin: Topic 3 report for the integrated assessment on hypoxia in the Gulf of Mexico. NOAA Coastal Ocean Program Decision Analysis Series No. 17, US Department of Commerce, National Oceanic and Atmospheric Administration, National Centers for Coastal Ocean Science, Silver Spring, Maryland, p. 130.

Gordeev, V. V. (2000). River input of water, sediment, major ions, nutrients and trace metals from Russian territory to the Arctic Ocean, in *The Freshwater Budget of the Arctic Ocean*, ed. E. L. Lewis, Dordrecht: Kluwer Academic Publishing, pp. 297–322.

Gordeev, V. V. (2006). Fluvial sediment flux to the Arctic Ocean. *Geomorph.*, **80**, 94–104.

Gordeev, V. V. & I. S. Sidorov (1993). Concentrations of major elements and their outflow into the Laptev Sea by the Lena River. *Marine Geochem.*, **43**, 33–45.

Gordeev, V. V. *et al.* (1996). A reassessment of the Eurasian River input of water, sediment, major ions and nutrients to the Arctic Ocean. *Amer. J. Sci.*, **296**, 664–691.

Gordon, R. (1885). The Irawadi River. *Royal Geogr. Soc. Proc*, **7**, 292–331.

Gorg, S. K. (1999). *River Water Disputes in India.* Laxmi Publ., New Delhi, 132pp.

Gorsline, D. S. (1996). Depositional events in Santa Monica Basin, California Borderland, over the past five centuries. *Sediment. Geol.*, **104**, 73–88.

Goudie, A. (2000). *The Human Impact on the Natural Environment*, MIT Press, Cambridge, MA, p. 511.

Goudie, A. S. (2006). Global warming and fluvial geomorphalogy. *Geomorph.*, **79**, 384–394.

Graf, W. L. (1999). Dam nation: a geographic census of American dams and their large-scale hydrologic impacts. *Water Resour. Res.*, **35**, 1305–1311.

Graf, W. L. (2006). Downstream hydrologic and geomorphic effects of large dams on American rivers. *Geomorph.*, **79**, 336–360.

GRDC (1995). First Interim Report on the Arctic River Database for the Arctic Climate System Study, Koblenz, Germany: Federal Institute of Hydrology.

Green, W. J. *et al.* (2005). Geochemical proceses in the Onyx River, Wright Valley, Antarctica: Major ions, nutrients, trace metals. *Geochim. Cosmochim. Acta*, **69**, 839–850.

Greene, R. S. B., C. J. Chartres & K. C. Hodgkinson (1990). The effects of fire on the soil in a degraded semi-arid woodland 1. Cryptogam cover and physical and micromorphological properties. *Aust. J. Soil Res.*, **28**, 755–777.

Gresswell, R. K. & A. Huxley (eds.) (1966). *Standard Encyclopedia of the World's Rivers and Lakes,* New York: G. P. Putnam's Sons.

Griffiths, G. A. (1981). Some suspended sediment yields from south island catchments, New Zealand. *Water Resour. Bull.*, **17**, 662–671.

Griffiths, G. A. (1982). Spatial and temporal variability in suspended sediment yields of North Island basins, New Zealand. *Water Resour. Bull.*, **18**, 575–584.

Griffiths, G. A. & G. P. Glasby (1985). Input of river-derived sediment to the New Zealand continental shelf: I. Mass. *Estuar. Coast. Shelf Sci.*, **21**, 773–787.

Griggs, G. B. *et al.* (1970). Deep-sea gravel from Cascadia Channel. *J. Geology* **78**, 611–619.

Grove, A. T. & O. Rackham (2001). *The nature of Mediterranean Europe. An Ecological History,* New Haven: Yale University Press.

Gudmundsson, M. T., F. Sigmundsson & H. Björnsson (1997). Ice-volcano interaction of the 1996 Gjálp subglacial eruption, Vatnajökull, Iceland. *Nature*, **389**, 954–957.

Guillén, J. & A. Palanques (1997). A historical perspective of the morphological evolution in the lower Ebro river. *Environ. Geol.*, **30**, 174–180.

Guillén, J. *et al.* (2006). Sediment dynamics during wet and dry storm events on the Têt inner shelf (SW Gulf of Lions). *Mar. Geol.*, **234**, 129–143.

Gumiero, B. *et al.* (2009). The Italian Rivers, in *Rivers of Europe*, eds. K. Tockner, C. T. Robinson & U. Uehlinger, Amsterdam: Elsevier, pp. 467–496.

Gunnell, Y. (1998). Present, past and potential denudation rates: is there a link? Tentative evidence from fission-track data, river sediment loads and terrain analyses in south Indian shield. *Geomorph.*, **25**, 135–153.

Gupta, A. (1988). Large floods as geomorphic events in the humid tropics, in *Flood Geomorphology*, eds. V. R. Baker,

R. C. Kochel & P. C. Patton, Chichester: John Wiley & Sons, pp. 301–315.

Gupta, A., V. S. Kale & S. N. Rajaguru (1999). The Narmanda River, India, through space and time, in *Varieties of Fluvial Form*, eds. A. J. Miller & A. Gupta, Chichester: John Wiley & Sons, pp. 113–143.

Gupta, H. & G. J. Chakrapani (2005). Temporal and spatial variations in water flow and sediment load in Narmada River Basin, India: natural and man-made factors. *Environ. Geol.*, **48**, 579–589.

Gurnell, A., D. Hannah & D. Lawler (1996). Suspended sediment yield from glacier basins. *IAHS Publ.* **236**, 97–104.

Gutierrez, F. & J. A. Dracup (2001). An analysis of the feasibility of long-range streamflow forecasting for Colombia using El Niño–Southern Oscillation indicators. *J. Hydrol.*, **246**, 181–196.

Guyot, J. L. *et al.* (1996). Dissolved solids and suspended sediment yields in the Rio Maderia basin from the Bolivian Andes to the Amazon. *IAHS Publ.* **236**, 55–63.

Habidin, A. (2005). Land subsidence in urban areas of Indonesia. *GIM International*, **19**.

Haines, A. T., B. L. Finlayson & T. A. McMahon (1988). A global classification of river regimes. *Appl. Geog.*, **8**, 255–272.

Hales, T. C. & J. J. Roering (2009). A frost "buzzsaw" mechanism for erosion of the eastern Southern Alps, New Zealand. *Geomorph.*, **107**, 241–253.

Hallet, B., L. Hunter & J. Bogen (1996). Rates of erosion and sediment evacuation by glaciers: a review of field data and their implications. *Global Planet. Change*, **12**, 213–235.

Han, Z. (2003). Groundwater resources protection and aquifer recovery in China. *Environ. Geol.*, **44**, 106–111.

Hanna, E. J. *et al.* (2009). Hydrologic response of the Greenland ice sheet: the role of oceanographic warming. *Hydrol. Process.*, **23**, 7–30.

Hanninen, J., I. Vuorinen & P. Hjelt (2000). Climatic factors in the Atlantic control the oceanographic and ecological changes in the Baltic Sea. *Limnol. Oceanogr.*, 703–710.

Harden, D. R. (1995). A comparison of flood-producing storms and their impacts in northwestern California. *U.S. Geological Survey Professional Paper 1454*, D1–D9.

Hare, S. R. (1996). Low frequency climate variability and salmon production, Unpublished PhD Thesis, University Washington.

Harr, R. D. *et al.* (1975) Changes in storm hydrographs after road building and clear-cutting in the Oregon Coast Range. *Water Resour. Res.*, **11**, 436–444.

Harris, P. T. (1991). Sedimentation at the junction of the Fly River in the northern Great Barrier Reef, in *Sustainable Development for Traditional Inhabitants of the Torres Strait Region*, eds. D. Lawrence & T. Cansvield-Smith, Townsville: Great Barrier Reef Marine Park Authority, pp. 59–85.

Harrison, C. G. A. (1994). Rates of continental erosion and mountain building. *Geol. Rundsch.*, **83**, 431–447.

Harrison, C. G. A. (2000). What factors control mechanical erosion rates? *Int. J. Earth Sci.*, **88**, 752–763.

Hartmann, J. N. *et al.* (2007). High riverine fluxes of dissolved silica from Japan: the influence of lithology. *Geophys. Res. Abs., EGU General Assembly 2008*, **9**, 00861.

Hartshorn , K. *et al.* (2002). Climate-driven bedrock incision in an active mountain belt. *Science*, **297**, 2036–2038.

Hasholt, B. (1996). Sediment transport in Greenland. *IAHS Publ.* **236**, 105–114.

Haston, L. & J. Michaelsen (1994). Long-term central coastal California precipitation variability and relationships to El Niño-Southern Oscillation. *J. Climate.*, **7**, 1373–1387.

Hawdon, A. A. *et al.* (2008). Hydrological recovery of rangeland following cattle exclusion. *IAHS Publ.* **325**, 532–539.

Hay, B. J. (1994). Sediment and water discharge rates of Turkish Black Sea rivers before and after hydropower dam construction. *Environ. Geol.*, **23**, 276–283.

Hay, W. W., J. L. Sloan & C. N. Wold (1988). Mass/age distribution and the global rate of sediment subduction. *J. Geophys. Res.*, **93**, 14 933–14 940.

Hayes, S. K. (1999). Low-flow sediment transport on the Pasig–Potrero alluvial fan, Mount Pinatubo, Philippines, Unpublished MSc Thesis, University of Washington, 73 pp.

Helama, S., J. Merilainen & H. Tuomenvirta (2009). Multicentennial megadrought in northern Europe coincided with a global El Nino-Southern Oscillation drought pattern during the Medieval Climate Anomaly. *Geology*, **37**, 175.

Herbert, T. D. *et al.* (2001). Collapse of the California current during glacial maxima linked to climate change on land. *Science*, **293**, 71–76.

Herschy, R. (2003). World catalog of maximum observed floods. *IAHS Publ.* **284**, 285.

Herschy, R. W. & R. W. Fairbridge (1998). *Encyclopedia of hydrology and water resources,* Kluwer Academic Publishers.

Hettler, J., G. Irion & B. Lehmann (1997). Environmental impact of mining waste disposal on a tropical lowland river system: a case study on the Ok Tedi Mine, Papua New Guinea. *Mineral. Deposita*, **32**, 280–291.

Heusch, B. & A. Milles-Lacroix (1971). Une method pour estimer l'ecoulement et l'erosion dans un bassin, Application au Magreb. *Mines et Geologie*, **33**, 21–39.

Hewawasam, T. *et al.* (2003). Increase of human over natural erosion rates in tropical highlands constrained by cosmogenic nuclides. *Geology*, **31**, 597–600.

Heyns, P. (2003). Water-resources management in Southern Africa, in *International Waters in Southern Africa*, ed. M. Nakayama, Tokyo: UN Press, p. 318.

Hickin, E. J. (1989). Contemporary Squamish River sediment flux to Howe Sound, British Columbia. *Can. J. Earth Sci.*, **26**, 1953–1963.

Hicks, D. M. & L. R. Basher (2008). The signature of an extreme erosion event on suspended sediment loads: Motueka River catchment, South Island, New Zealand. *IAHS Publ.* **325**, 184–191.

Hicks, D. M., B. Gomez & N. A. Trustrum (2000). Erosion thresholds and suspended sediment yields, Waipaoa River basin, New Zealand. *Water Resour. Res.*, **36**, 1129–1142.

Hicks, D. M., B. Gomez & N. A. Trustrum (2004). Event suspended sediment characteristics and the generation of hyperpycnal plumes at river mouths: East Coast continental margin, North Island, New Zealand. *J. Geol.*, **112**, 471–485.

Hicks, D. M., J. Hill & U. Shankar (1996). Variation of suspended sediment yields around New Zealand: the relative importance of rainfall and geology. *IAHS Publ.* **236**, 149–156.

Hicks, D. M. & U. Shankar (2003). Sediment from New Zealand rivers, NIWA Chart Misc. Series No. 79.

Hillel, D. (1994). *Rivers of Eden : The Struggle for Water and the Quest for Peace in the Middle East,* Oxford: Oxford University Press.

Hilton, R. G. *et al.* (2008). Tropical-cyclone-driven erosion of the terrestrial biosphere from mountains. *Nature Geoscience*, **1**, 759–762.

Hirabayashi, Y. *et al.* (2008). Global projections of changing risks of floods and droughts in a changing climate. *Hydrol. Sci.*, **53**, 754–772.

Hirsch, R. M. *et al.* (1990). The influence of man on hydrologic systems, in *The Geology of North America: Surface Water Hydrology*, eds. M. G. Wolman & H. C. Riggs, Boulder, Colorado: Geol. Soc. Amer., O-1, pp. 329–359.

Hiscott, R. N. (2001). Depositional sequences controlled by high rates of sediment supply, sea-level variations, and growth faulting: the Quaternary Baram Delta of northwestern Borneo. *Mar. Geol.*, **175**, 67–102.

Hodgkins, G. A. (2009). Streamflow changes in Alaska between the cool phase (1947–1976) and the warm phase (1977–2006) of the Pacific Decadal Oscillation: The influence of glaciers. *Water Resour. Res.*, **45**, W06502, doi:10.1029/2008WR007575.

Hodson, A. *et al.* (1998). Suspended sediment yield and transfer processes in a small High-Arctic glacier basin, Svalbard. *Hydrol. Process.*, **12**, 73–86.

Hoelzmann, P. *et al.* (1998). Mid-Holocene land-surface conditions in northern Africa and the Arabian peninsula: a data set for the analysis of biogeophysical feedbacks in the climate system. *Global Biogeochem. Cycles*, **12**, 35–51.

Hogan, D. L. & J. W. Schwab (1991). Meteorological conditions associated with hillslope failures on the Queen Charlotte Islands, B.C. Ministry of Forests, Land Management Report 73, p. 36.

Holdgate, M. *et al.* (1982). The world environment 1972–1982: a report, in *Natural Resources and the Environment Series-United Nations Environment Programme (Irlanda)* Dublin.

Holeman, J. N. (1968). Sediment yield of major rivers of the world. *Water Resour. Res.*, **4**, 737–747.

Holeman, J. N. (1980). Erosion rates in the US estimated by the Soil Conservation Service's inventory (abs). *EOS*, **61**, 954.

Holland, H. D. (1978). *The Geochemistry of the Atmosphere and Oceans*, New York: Wiley-Interscience.

Hollerwoger, F. (1964). The progress of the river deltas in Java, in *Scientific problem of the humid tropic zone deltas and their implications*. UNESCO, pp. 347–355.

Holmes, R. M. *et al.* (2000). Flux of nutrients from Russian rivers to the Arctic Ocean: Can we establish a baseline against which to judge future changes? *Water Resour. Res.*, **36**, 2309–2320.

Holmes, R. M. *et al.* (2001). Nutrient chemistry of the Ob' and Yenisey rivers, Siberia: results from June 2000 expedition and evaluation of long-term data sets. *Mar. Chem.*, **75**, 219–227.

Holmes, R. M. *et al.* (2002). A circumpolar perspective on fluvial sediment flux to the Arctic Ocean. *Global Biogeochem. Cycles*, **16**, 1098, doi:10.1029/2001GB001849.

Hong, G. H. *et al.* (1997). 210Pb-derived sediment accumulation rates in the southwestern East Sea (Sea of Japan). *Geo-Mar. Lett.*, **17**, 126–132.

Hooke, J. M. (2006). Human impacts on fluvial systems in the Mediterranean region. *Geomorph.*, **79**, 311–335.

Hooke, R. L. B. (2000). Toward a uniform theory of clastic sediment yield in fluvial systems. *Geol. Soc. Amer. Bull.*, **112**, 1778.

Horowitz, A. J. (2010). A quarter century of declining suspended sediment fluxes in the Mississippi River and the effect of the 1993 flood. *Hydrol. Process.*, **24**, 13–34.

Hossain, M. M. (1991). Total sediment load in the lower Ganges and Jumuna, Bangladesh. University of Engineering and Technology, 15 pp.

Hossain, F. *et al.* (2009). Have large dams altered extreme precipitation patterns? *EOS* **90**, 453–454.

Hovius, N. (1988). Control on sediment supply by large rivers, in *Relative Role of Eustacy, Climate and Tectonism in Continental Rocks. SEPM Special Publ. 59*, eds. K. W. Shanley & P. J. McCabe, pp. 3–16.

Hovius, N. & M. Leeder (1998). Clastic sediment supply to basins. *Basin Res.*, **10**, 1–5.

Hovius, N., C. P. Stark & P. A. Allen (1997). Sediment flux from a mountain belt derived by landslide mapping. *Geology*, **25**, 231–234.

Hovius, N. *et al.* (2000). Supply and removal of sediment in a landslide-dominated mountain belt: Central Range, Taiwan. *J. Geol.*, **108**, 73–89.

Howell, P. P. & J. A. Allan (1994). *The Nile: Sharing a Scarce Resource,* Cambridge: Cambridge University Press.

Hren, M. T. *et al.* (2007). Major ion chemistry of the Yarlung Tsangpo–Brahmaputra river: Chemical weathering, erosion, and CO_2 consumption in the southern Tibetan plateau and eastern syntaxis of the Himalaya. *Geochim. Cosmochim. Acta*, **71**, 2907–2935.

Hrieche, A., W. Najem & C. Bocquillon (2007). Hydrological impact simulations of climate change on Lebanese coastal rivers. *IAHS Publ.* **52**, 1119–1133.

Huang, M., L. Zhang & J. Gallichand (2003). Runoff responses to afforestation in a watershed of the Loess Plateau, China. *Hydrol. Process.*, **17**, 2599–2609.

Huber, A., A. Iroumé & J. Bathurst (2008). Effect of *Pinus radiata* plantations on water balance in Chile. *Hydrol. Proc.*, **22**, 142–148.

Hudson, P. F. (2003). Event sequence and sediment exhaustion in the lower Panuco Basin, Mexico. *Catena*, **52**, 57–76.

Hudson, P. F. & J. Mossa (1997). Suspended sediment transport effectiveness of three large impounded rivers, US Gulf Coastal Plain. *Environ. Geol.*, **32**, 263–273.

Hudson, P. F. *et al.* (2005). Rivers of Mexico, in *Rivers of North America*, eds. A. C. Benke & C. E. Cushing, Amsterdam: Elsevier, pp. 1030–1084.

Hughes, P. J., M. E. Sullivan & D. Yok (1991). Human-induced erosion in a highlands catchment in Papua New Guinea: the prehistoric and contemporary records. *Z. Geomorphol. Suppl.*, **83**, 227–239.

Hugueny, B. & C. Lévêque (1994). Freshwater fish zoogeography in West Africa: faunal similarities between river basins. *Environ. Biol. Fish.*, **39**, 365–380.

Huh, Y.-S. (2003). Chemical weathering and climate – a global experiment. A review. *Geoscience Journal (Korea)*, **7**, 277–288.

Huh, Y. *et al.* (1998). The fluvial geochemistry of the rivers of eastern Siberia. II. Tributaries of the Lena, Omoloy, Yana, Indigirka, Kolyma, and Anadyr draining the collisional/accretionary zone of the Verkhoyansk and Cherskiy ranges. *Geochim. Cosmochim. Acta*, **62**, 2053–2075.

Humborg, C. *et al.* (1997). Effect of Danube River dam on Black Sea biogeochemistry and ecosystem structure. *Nature*, **386**, 385–388.

Humborg, C. *et al.* (2000). Silica retention in river basins: far-reaching effects on biogeochemistry and aquatic food webs in coastal marine environments. *Ambio*, **29**, 45–50.

Hurrell, J. W. (1995). Decadal trends in the North Atlantic Oscillation: regional temperatures and precipitation. *Science*, **269**, 676–679.

Hurrell, J. W. *et al.* (eds.) (2003). *The North Atlantic Oscillation: Climatic Significance and Environmental Impact*. Amer. Geophys. Union Geophys. Monograph 134.

IAHS/UNESCO (1974). Gross sediment transport into the oceans, preliminary edition, UNESCO, 4 pp.

Imram, J. & J. P. M. Syvitski (2000). Impact of extreme river events on the coastal ocean. *Oceanography*, **13**, 85–92.

Inbar, M. (1992). Rates of fluvial erosion in basins with a Mediterranean type climate. *Catena*, **19**, 383–409.

Inbar, M., M. Tamir & L. Wittenberg (1998). Runoff and erosion processes after a forest fire in Mount Carmel, a Mediterranean area. *Geomorph.*, **24**, 17–33.

Inman, D. L. & S. A. Jenkins (1999). Climate change and the episodicity of sediment flux of small California rivers. *J. Geol.*, **107**, 251–270.

Inman, D. L. & C. E. Nordstrom (1971). On the tectonic and morphologic classification of coasts. *J. Geol.*, **79**, 1–21.

Inman, D. L., S. Jenkins & J. Wasyl (1998). Database for streamflow and sediment flux of California rivers, University of California San Diego, Scripps Inst. Oceangr. Ref. 98–9, 13 pp.

Instituto Geografico Nacional Mapa de Cuencas de Guatemala.

IOC/WESTPAC (1991). *Workshop on River Input of Nutrients to the Marine Environment in the Western Pacific*, Intergovernmental Oceanographic Commission.

Isdale, P. *et al.* (1998). Palaeohydrological variation in a tropical river catchment: a reconstruction using fluorescent bands in corals of the Great Barrier Reef, Australia. *Holocene*, **8**, 1–8.

Isik, S. *et al.* (2008). Effects of anthropogenic activities on the lower Sakarya River. *Catena*, **75**, 172–181.

Ismail, W. R. (1996). The role of tropical storms in the catchment sediment removal. *J. Bioscience*, **7**, 153–168.

Issar, A. (2003). *Climate Changes during the Holocene and their Impact on Hydrological Systems*, Cambridge: Cambridge University Press.

Issar, A. S. & N. Brown (eds.) (1998). *Water, Environment and Society in Times of Climatic Change*, Dordrecht: Kluwer Academic.

Istanbulluoglu, E. & R. L. Bras (2006). On the dynamics of soil moisture, vegetation, and erosion: Implications of climate variability and change. *Water Resour. Res.*, **42** doi:10.1029/20005WR004113.

Itambi, A. C. *et al.* (2009). Millennial-scale northwest African droughts related to Heinrich events and Dansgaard–Oeschger cycles: evidence in marine sediments from offshore Senegal. *Paleoceanogr.*, **24**, doi:10.1029/2007PA001570.

Iverson, R. M. (2000). Landslide triggering by rain infiltration. *Water Resour. Res.*, **36**, 1897–1910.

Jackson, J. K. *et al.* (2005). Atlantic coast rivers of the northeastern United States, in *Rivers of North America*, eds. A. C. Benke & C. E. Cushing, Amsterdam: Elsevier, pp. 20–71.

Jackson, M. & J. J. Roering (2009). Post-fire geomorphic response in steep, forested landscapes: Oregon Coast Range, USA. *Quat. Sci. Rev.*, **28**, 1131–1146.

Jacobson, P. J. *et al.* (1999). Transport, retention, and ecological significance of woody debris within a large ephemeral river. *J. N. Amer. Benthol. Soc.*, **18**, 429–444.

Jacobson, A. D. *et al.* (2003). Climatic and tectonic controls on chemical weathering in the New Zealand Southern Alps. *Geochim. Cosmochim. Acta*, **67**, 29–46.

Jaeschke, A. *et al.* (2007). Coupling of millennial-scale changes in sea surface temperature and precipitation off northeastern Brazil with high-latitude climate shifts during the last glacial period. *Paleoceanogr.*, **22**, PA4206, doi:10.1029/2006PA001391.

Jaffe, R. *et al.* (1995). Pollution effects of the Tuy River on the central Venezuelan coast: anthropogenic organic compounds and heavy metals in *Tivela mactroidea*. *Mar. Pollut. Bull.*, **30**, 820–825.

Jain, S. K., P. K. Agarwal & V. P. Singh (2007). *Hydrology and Water Resources of India*, Heidelberg: Springer-Verlag.

James, L. A. (1994). Channel changes wrought by gold mining: Northern Sierra Nevada, California, in *Effects of Human-Induced Changes on Hydrologic Systems*, eds. R. Marston & V. R. Hasfurther, Amer. Water Resour. Assoc., pp. 629–638.

Janda, R. J. & K. M. Nolan (1979). Stream sediment discharge in northwestern California, in *A Fieldtrip to Observe Natural and Management Related Erosion in Franciscan Terrain of Northern California; a Guidebook*. Boulder, CO: Geol. Soc. America, Cordillerian Section, pp. 1–27.

Janda, R. J. *et al.* (1981). Lahar movement, effects and deposits. *US Geological Survey Prof. Paper 1250*, 461–478.

Janda, R. J. *et al.* (1996). Assessment and response to lahar hazard around Mount Pinatubo, 1991 to 1993, in *Fire and Mud, Eruptions and Lahars of Mount Pinatubo, Philippines*, eds. C. G. Newhall & R. S. Punongbayan, Seattle: University of Washington Press, pp. 107–139.

Jansen, I. M. L. & R. B. Painter (1974). Predicting sediment yield from climate and topography. *J. Hydrol.*, **21**, 371–380.

Jansen, P. *et al.* (1979). *Principles of River Engineering,* London: Pitman.

Jansson, M. B. (1988). A global survey of sediment yield. *Geogr. Ann.*, **70**, 81–98.

Jansson, R. *et al.* (2000). Effects of river regulation on river-margin vegetation: a comparison of eight boreal rivers. *Ecol. Appl.*, **10**, 203–224.

Jaoshvili, S. (2002). *The rivers of the Black Sea.* European Environmental Commission Technical Report 71, 58.

Jarrett, R. D. & J. E. Costa (1986). Hydrology, geomorphology and dam-break modeling of the July 15, 1982 Lawn Lake Dam and Cascade Lake Dam failures, Larimer Country, Colorado, *US Geol. Surv. Prof. Paper 1369*, 78 pp.

Jayawardena, A. W., K. Takeuchi & B. Machubub (eds.) (1997). *Catalogue of Rivers for Southeast Asia and The Pacific, Vol. 2,* The UNESCO-IHP Regional Steering Committee (RSC) for Southeast Asia and the Pacific.

Jennerjahn, T. C. *et al.* (2010). The tropical Brazilian continental margin. In K. K. liu *et al.* (eds.), *Carbon and Nutrient Fluxes in Continental Margins.* IGBP Series, Springer-Verlag, Berlin, pp. 427–442.

Jobin, W. R. (1999). *Dams and Disease: Ecological Design and Health Impacts of Large Dams, Canals, and Irrigation Systems,* London: Taylor & Francis.

Johnson, H. M., IV & D. L. Belval (1998). Nutrient and suspended solid loads, yields and trends in the non-tidal part of five major river basins in Virginia, 1985–96, Richmond, VA. US Geol. Surv. Open-File Report, 36 pp.

Johnsson, M. J. & R. H. Meade (1990). Chemical weathering of fluvial sediments during alluvial storage; the Macuapanim Island point bar, Solimoes River, Brazil. *J. Sediment. Petrol.*, **60**, 827–842.

Johnsson, M. J., R. F. Stallard & N. Lundberg (1991). Controls on the composition of fluvial sands from a tropical weathering environment: sands of the Orinoco River drainage basin, Venezuela and Colombia. *Geol. Soc. Amer. Bull.*, **103**, 1622.

Jones, B. G., C. D. Woodroffe & G. R. Martin (2003). Deltas in the Gulf of Carpentaria, Australia: Forms, Processes and Products, in *Tropical Deltas of Southeast Asia - Sedimentology, Stratigraphy, and Petroleum Geology*, eds. F. H. Sidi *et al.*, Tulsa, OK: SEPM, 21–43.

Jones, J. A. (2000). Hydrologic processes and peak discharge response to forest removal, regrowth, and roads in 10 small experimental basins, western Cascades, Oregon. *Water Resour. Res.*, **36**, 2621–2642.

Jones, P. D., T. Jonsson & D. Wheeler (1997). Extension to the North Atlantic Oscillation using early instrumental pressure observations from Gibraltar and South-West Iceland. *Int. J. Climatol.*, **17**, 1433–1450.

Jonsson, P., A. Snorrason & S. Palsson (1998). Discharge and sediment transport in the jökulhlaup on Skeiarrsandur in November (abs.). *EOS*, **79**, S13.

Joo, M. *et al.* (2005). Estimation of long-term sediment loads in the Fitzroy catchment, Queensland, Australia, in *MODSIM 2005 International Congress on Modelling and Simulation: Advances and Applications for Management and Decision Making: Proceedings*, eds. A. Zerger & R. M. Argent, pp. 1161–1167.

Jouanneau, J. M. *et al.* (1998). Dispersal and deposition of suspended sediment on the shelf off the Tagus and Sado estuaries, SW Portugal. *Prog. Oceanogr.*, **42**, 233–257.

Judson, S. (1968). Erosion of the land, or what's happening to our continents? *Amer. J. Sci.*, **56**, 356–374.

Judson, S. & D. F. Ritter (1964). Rates of regional denudation in the U.S. *J. Geophys. Res.*, **69**, 3395–6401.

Jury, M. R. (2003). The coherent variability of African river flows: composite climate structure and the Atlantic circulation. *Water Sci. Appl.*, **29**, 1–10.

Kabat, P. *et al.* (eds.) (2003). *Vegetation, Water, Humans and the Climate,* Heidelberg: Springer-Verlag.

Kahya, E. & J. A. Dracup (1993). U.S. streamflow patterns in relation to the El Niño/Southern Oscillation. *Water Resour. Res.*, **29**, 2491–2503.

Kahya, E. & J. A. Dracup (1994). The relationships between US streamflow and La Nina events. *Water Resour. Res*, **30**, 2133–2141.

Kanellopoulos, T. D. *et al.* (2008). The influence of the Evros River on the recent sedimentation of the inner shelf of the NE Aegean Sea. *Environ. Geol.*, **53**, 1455–1464.

Kao, S. J. & K. K. Liu (2002). Exacerbation of erosion induced by human perturbation in a typical Oceania watershed: insight from 45 years of hydrological records from the Lanyang-Hsi River, northeastern Taiwan. *Global Biogeochem. Cycles*, **16**, 1016.

Kao, S., T. Lee & J. D. Milliman (2005). Calculating highly fluctuated suspended sediment fluxes from mountainous rivers in Taiwan. *Terr. Atmos. Ocean Sci.*, **16**, 653.

Kao, S. J. & J. D. Milliman (2008). Water and sediment discharge from small mountainous rivers, Taiwan: the roles of lithology, episodic events, and human activities. *J. Geol.*, **116**, 431–448.

Karabork, M. C., E. Kahya & A. U. Komuscu (2007). Analysis of Turkish precipitation data: homogeneity and the Southern Oscillation forcings on frequency distributions. *Hydrol. Process.*, **21**.

Kauppila, P. & J. Koskiaho (2003). Evaluation of annual loads of nutrients and suspended solids in Baltic rivers. *Nord. Hydrol.*, **34**, 203–220.

Kayano, M. T. & R. V. Andreoli (2007). Relations of South American summer rainfall interannual variations with the Pacific Decadal Oscillation. *Int. J. Climatol.*, **27**, 531–540.

Keat, T. S. & K. B. Alias (1982). Average annual and monthly surface water resources of penninsular Malaysia, 41 pp.

Keefer, D. K. (1994). The importance of earthquake-induced landslides to long-term slope erosion and slope-failure hazards in seismically active regions. *Geomorph.*, **10**, 265–284.

Kehew, A. E. & M. L. Lord (1987). Glacial-lake outbursts along the mid-continent margins of the Laurentide ice-sheet, in *Catastrophic Floods*, eds. L. Mayer & D. Nash, Boston: Allen and Unwin, pp. 95–120.

Keller, E. A., D. W. Valentine & D. R. Gibbs (1997). Hydrological response of small watersheds following the Southern California Painted Cave Fire of June 1990. *Hydrol. Process.*, **11**, 401–414.

Keller, R. (1962). *Gewasser und Wasserhaushalt des Festlandes,* Leipzig: Teubner.

Kelly, M. (2004). Florida river flow patterns and the Atlantic Multidecadal Oscillation. Draft Report, Ecologic Evaluation Section. Southwest Florida Water Management District, 80pp.

Kelly, E. F., O. A. Chadwick & T. E. Hilinski (1998). The effect of plants on mineral weathering. *Biogeochem.*, **42**, 21–53.

Kelly, M. H. & J. A. Gore (2008). Florida river flow patterns and the Atlantic Multidecadal Oscillation. *Regul. Rivers Res. Manage.*, **24**, 598–616.

Kelsey, H. M. (1978). Earthflows in Franciscan melange, Van Duzen River basin, California. *Geology*, **6**, 361–364.

Kelsey, H. M. (1980). A sediment budget and an analysis of geomorphic process in the Van Duzen River basin, north coastal California, 1941–1975: Summary. *Geol. Soc. Amer. Bull.*, **91**, 190–195.

Kelsey, H. M. *et al.* (1995). Geomorphic analysis of streamside landslides in the Redwood Creek basin, northwestern California, *US Geol. Surv. Prof. Paper 1454*, J1–J12.

Kempe, S. (1982). Long-term records of CO_2 pressure fluctuations in freshwaters, in *Transport of Carbon and Minerals in Major World Rivers*, ed. E. T. Degens, Hamburg: Mitt. Geol.-Palaont. Inst. SCOPE/UNEP, pp. 91–332.

Kempe, S. (1988). Freshwater carbon and the weathering cycle, in *Physical and Chemical Weathering in Geochemical Cycles*, eds. A. Lerman & M. Meybeck, Dordrecht: Kluwer Academic Publishing, pp. 197–223.

Kempe, S., M. Pettine & G. Cauwet (1991). Biogeochemistry of European Rivers, in *Biogeochemistry of Major World Rivers*, eds. E. T. Degens, S. Kempe & J. E. Richey, Chichester: John Wiley and Sons Ltd., pp. 169–211.

Keown, M. P., E. A. Dardeau Jr & E. M. Causey (1986). Historic trends in the sediment flow regime of the Mississippi River. *Water Resour. Res.*, **22**, 1555–1564.

Keppler, E. T., R. R. Ziemer & P. H. Cafferata (1994). Changes in soil moisture and pore pressure after harvesting a forested hillslope in northern California, in *Effects of Human-Induced Changes on Hydrologic Systems*, Amer. Water Resour. Assoc., pp. 205–214.

Kerr, A. (2000). A North Atlantic climate pacemaker for the centuries. *Science*, **288**, 1984–1986.

Kesel, R. H. (2003). Human modifications to the sediment regime of the Lower Mississippi River flood plain. *Geomorph.*, **56**, 325–334.

Kettner, A. J., B. Gomez & J. P. M.Syvitski (2008). Will human catalysts or climate change have a greater impact on the sediment load of the Waipaoa River in the 21st century? *IAHS Publ.* **325**, 425–431.

Kettner, A. J., B. Gomez & J. P. M. Syvitski (2007). Modeling suspended sediment discharge from the Waipaoa River system, New Zealand: the last 3000 years. *Water Resour. Res.*, **43**, W07411.

Kiage, L. M. & K. Liu (2006). Late Quaternary paleoenvironmental changes in East Africa: a review of multiproxy evidence from palynology, lake sediments, and associated records. *Prog. Phys. Geog.*, **30**, 633.

Kiely, G. (1999). Climate change in Ireland from precipitation and streamflow observations. *Adv. Water Resour.*, **23**, 141–151.

Kim, T. W. *et al.* (2006). Quantification of linkages between large-scale climatic patterns and precipitation in the Colorado River Basin. *J. Hydrol.*, **321**, 173–186.

Kim, W. *et al.* (2009). Is it feasible to build new land in the Mississippi River delta? EOS *(Transactions, Amer. Geophys. Union)*, **90**, 373–374.

Kimstach, V. A., M. Meybeck & E. Baroudy (1998). *A Water Quality Assessment of the Former Soviet Union,* London: GEMS/Routledge.

Kjerfve, B. J. *et al.* (1992). Modeling of the residual circulation in Broken Bay and the lower Hawkesbury River – NSW Australia. *J. Mar. Freshwater Res.*, **43**, 1339–1357.

Klee, G. A. (1991). *Conservation of Natural Resources,* Englewood Cliffs, NJ: Prentice-Hall.

Kliot, N. (1994). *Water Resources and Conflict in the Middle East,* London: Routledge.

Knox, J. C. (1995). Fluvial systems since 20 000 years BP, in *Global Continental Palaeohydrology*, eds. K. J. Gregory, L. Starkel & V. R. Baker, Chichester: John Wiley & Sons, pp. 87–108.

Knox, J. C. (2007). The Mississippi River system, in *Large Rivers. Geomorphology and Management*, ed. A. Gupta, Chichester: John Wiley & Sons, pp. 145–182.

Köppen–Geiger (1954). *Climate of the Earth* (map: 1:16 mio), Justus Perthes, Darmstadt, Germany.

Korotaev, V. N. (1991). *Geomorphology of River Deltas* (in Russian), Moscow University Publ. House.

Korup, O., M. J. McSaveney & T. R. H. Davies (2004). Sediment generation and delivery from large historic landslides in the Southern Alps, New Zealand. *Geomorph.*, **61**, 189–207.

Korup, O., A. L. Strom & J. T. Weidinger (2006). Fluvial response to large rock-slope failures: examples from the Himalayas, the Tien Shan, and the Southern Alps in New Zealand. *Geomorph.*, **78**, 3–21.

Korzoun, V. I. *et al.* (eds.) (1977). *Atlas of World Water Balance,* Paris: UNESCO.

Kosmas, C. S., N. Moustakas, N. G. Danalatos & N. Yassoglou (1996). The Sparta field site: I. The impacts of land use and management on soil properties and erosion. II. The effect of reduced moisture on soil properties and wheat production. In C. J. Brandt and J. B. Thornes (eds.) *Mediterranean*

Desertification and Land Use. John Wiley and Sons, Chichester, pp. 207–228.

Kostaschuk, R., J. Terry & R. Raj (2003). Suspended sediment transport during tropical-cyclone floods in Fiji. *Hydrol. Process.*, **17**, 1149–1164.

Kostianitsin, M. N. (1964). *Hydrology of the Dnieper and Southern Bug mouth area (in Russian)*, Moscow: Gidrometeoizdat.

Koutaniemi, L. (1991). Kasvihuone-Suomen vesivoima. *Terra: Suomen maantieteellisen seuran aikakauskirja; Geografiska sällskapets i Finland tidskrift*, **14**.

Kranck, K. & A. Ruffman (1981). Sedimentation in James Bay. *Nat. Can.*, **109**, 353–361.

Krasovskaia, I. *et al.* (1999). Dependence of the frequency and magnitude of extreme floods in Costa Rica on the Southern Oscillation Index. *IAHS Publ.*, **255**, 81–89.

Kravtsova, V. I., V. N. Mikhailova & N. A. Efremova (2009). Variation of hydrological regime, morphological structure, and landscape of Indus River delta (Pakistan) under the effect of large-scale water management measures. *Water Resources*, **36**(4), 465–379.

Krishnaswami, S. & S. K. Singh (2005). Chemical weathering in the river basins of the Himalaya, India. *Curr. Sci.*, **89**, 841–849.

Krishnaswamy, J., P. N. Halpin & D. D. Richter (2001). Dynamics of sediment discharge in relation to land-use and hydro-climatology in a humid tropical watershed in Costa Rica. *J. Hydrol.*, **253**, 91–109.

Kristmannsdottir, H., *et al.* (2002). Seasonal variation in the chemistry of glacial-fed rivers in Iceland. *IAHS Publ.* **127**, 223–229.

Kuenzi, W. D., O. H. Horst & R.V.McGehee (1979). Effect of volcanic activity on fluvial-deltaic sedimenation in a modern arc-trench gap, southwestern Guatemala. *Geol. Soc. Amer. Bull.*, **90**, 827–838.

Kuhle, M. (2002). Outlet glaciers of the Pleistocene (LGM) south Tibetan ice sheet between Cho Oyu and Shisha Pangma as potential sources of former megafloods, in *Flood and megaflood processes and deposits: recent and ancient examples*, Int. Assoc. Sedimentol. Spec. Publ. 32, pp. 291–302.

Kump, L. R., S. L. Brantley & M. A. Arthur (2000). Chemical weathering, atmospheric CO_2, and climate. *Ann. Rev. Earth Planet. Sic.*, **28**, 611–667.

Kurtzman, D. & B. R. Scanlon (2007). El Niño Southern Oscillation and Pacific Decadal Oscillation impacts on precipitation in southern and central US: evaluation of spatial distribution and predictions. *Water Resour. Res.*, **43**, W10427 doi:10.1029/20007WR005863.

Kutzbach, J. E. & P. J. Guetter (1986). The influence of changing orbital parameters and surface boundary conditions on climate simulations for the past 18,000 years. *J. Atmos. Sci.*, **43**, 1726–1759.

L'Vovich, M. I. (1974). *World Water Resources and Their Future* (Mysl'PH Moscow, Engl. Transl. 1979, AGU, Washington).

Labat, D. *et al.* (2004). Evidence for global runoff increase related to climate warming. *Adv. Water Res.*, **27**, 631–642.

Lacerda, L. D. & R. V. Marins (2002). River damming and changes in mangrove distribution. *ISME/Glomis Electronic Journal*, **2**, 1–4.

Lahlou, A. (1982). La degradation specifique des bassins versants et son impact cur l'envasement des barrages. *IAHS Publ.* **137**, 163–169.

Lahlou, A. (1996). Environmental and socio-economic impacts of erosion and sedimentation in North Africa. *IAHS Publ.* **236**, 491–500.

Lajoie, F. *et al.* (2007). Impacts of dams on monthly flow characteristics. The influence of watershed size and seasons. *J. Hydrol.*, **334**, 423–439.

Lamoureux, S. (2000). Five centuries of interannual sediment yield and rainfall-induced erosion in the Canadian high arctic recorded in lacustrine varves. *Water Resour. Res.*, **36**, 309–318.

Lancaster, S. T. & G. E. Grant (2006). Debris dams and the relief of headwater streams. *Geomorph.*, **82**, 84–97.

Lane, P. N. J. & G. J. Sheridan (2002). Impact of an unsealed forest road stream crossing: water quality and sediment sources. *Hydrol. Process.*, **16**, 2599–2612.

Lanfear, K. & R. M. Hirsch (1999). USGS study reveals a decline in long-term stream gauges. *EOS*, **80**, 605–607.

Lang, A. *et al.* (2003). Changes in sediment flux and storage within a fluvial system: some examples from the Rhine catchment. *Hydrol. Process.*, **17**, 3321–3334.

Langbein, W. B. & S. A. Schumm (1958). Yield of sediment in relation to mean annual precipitation. *Trans. Amer. Geophys. Union*, **39**, 1076–1084.

Langland, M. J., P. L. Lietman & S. A. Hoffman (1995). Synthesis of nutrient and sediment data for watersheds within the Chesapeake Bay drainage basin. US Geol. Surv. Open-File Report, 121 pp.

Larsen, G. & S. Asbjornsson (1995). Volume of tephra and rock debris deposited by the 1918 jokulhlaups on western Myrdalssandur, south Iceland (abs), in *IGS International Symposium on Glacial Erosion and Sedimentation, August 20–25, 1995.*

Larsen, M. C. & J. E. Parks (1997). How wide is a road? The association of roads and mass-wasting in a forested montane environment. *Earth Surf. Process. Landforms*, **22**, 835–848.

Larsen, M. C. & A. Simon (1993). A rainfall intensity-duration threshold for landslides in a humid-tropical environment Puerto Rico. *Geogr. Ann.*, **75A**, 13–23.

Larsen, R. L. & W. C. I. Pittman (1985). *The Bedrock Geology of the World*, New York: W.H. Freeman and Company, Inc.

Latif, M. & T. P. Barnett (1994). Causes of decadal climate variability over the North Pacific and North America. *Science*, **266**, 634.

Lavigne, F. & H. Suwa (2004). Contrasts between debris flows, hyperconcentrated flows and stream flows at a channel of Mount Semeru, East Java, Indonesia. *Geomorph.*, **61**, 41–58.

Lavigne, F. & J. C. Thouret (2003). Sediment transportation and deposition by rain-triggered lahars at Merapi Volcano, Central Java, Indonesia. *Geomorph.*, **49**, 45–69.

Law, B. C. (ed.) (1968). *Mountains and Rivers of India*, Calcutta: India National Committee for Geography.

Lawler, D. M. (1994). Recent changes in rates of suspended sediment transport in the Jokulsa a Solheimasandi glacial river, southern Iceland. *IAHS Publ.* **224**, 343–358.

Lawler, D. M. & L. J. Wright (1996). Sediment yield decline and climate change in southern Iceland. *IAHS Publ.* **236**, 415–425.

Laznik, M. *et al.* (1999). Riverine input of nutrients to the Gulf of Riga – temporal and spatial variation. *J. Marine Syst.*, **23**, 11–25.

Lecce, S. A. (2000). Spatial variations in the timing of annual floods in the southeastern United States. *J. Hydrol.*, **235**, 151–169.

Leeder, M. R., T. Harris & M. J. Kirkby (1998). Sediment supply and climate change; implications for basin stratigraphy. *Basin Res.*, **10**, 7–18.

Legates, D. R., H. F. Lins & G. J. McCabe (2005). Comments on "Evidence for global runoff increase related to climate warming" by Labat et al. *Adv. Water Res.*, **28**, 1310–1315.

Leifeste, D. K. (1974). Dissolved-solids discharge to the oceans from the conterminous United States. US Geological Survey Open-File Report, 8 pp.

Leithold, E. L., N. E. Blair & D. W. Perkey (2006). Geomorphic controls on the age of particulate organic carbon from small mountainous and upland rivers. *Global Biogeochem. Cycles*, **20**, GB3022, doi:10.1029/2005GB002677.

LeLoeuff, P., E. Marchall & J.- B. B. A. Kothias (1993). Environment et Resources aquatiques de Cote-d'Ivoire, ORSTOM, 586.

Leopold, L. B. (1994). *A View of the River. Cambridge*: Harvard University Press, 290 pp.

Levashova, E. A. *et al.* (2004). Natural and human-induced variations in water and sediment runoff in the Danube River mouth. *Water Resour.*, **31**, 235–246.

Lewis, J. *et al.* (2001). Impacts of logging on storm peak flows, flow volumes and suspended sediment loads in Casper Creek, California, in *Land Use and Watersheds: Human Influence on Hydrology and Geomorphology in Urban and Forest Areas*, eds. M. S. Wigmosta & S. J. Burges, Washington, DC: American Geophysical Union, pp. 85–126.

Li, Q. *et al.* (2007). Impacts of human activities on the sediment regime of the Yangtze River. *IAHS Publ.* **314**, 11–19.

Li, Y.-H. (1976). Denudation of Taiwan Island since the Pliocene Epoch. *Geology*, **4**, 105–107.

Li, Z. (1992). The effects of forest in controlling gully erosion. *IAHS Publ.* **209**, 429–437.

Licitri, R. & D. Normand (1969). Etudes générales des aires d'irrigation et d'assainissement agricloe en Algérie, dossier O.Sogreah/Mara.

Lienou, G. *et al.* (2005). Regimes of suspended sediment flux in Cameroon: review and synthesis for the main ecosystems; climatic diversity and anthropogenic activities. *Hydrol. Sci. J.*, **50**, 111–123.

Lins, H. F. (1997). Regional streamflow regimes and hydroclimatology of the United States. *Water Resour. Res.*, **33**, 1655–1668.

Lins, H. F. & J. R. Slack (1999). Streamflow trends in the United States. *Geophys. Res. Lett.*, **26**, 227–230.

Liquete, C. *et al.* (2005). Mediterranean river systems of Andalusia, southern Spain, and associated deltas: a source to sink approach. *Mar. Geol.*, **222**, 471–495.

Liquete, C. *et al.* (2007). Long-term development and current status of the Barcelona continental shelf: a source-to-sink approach. *Cont. Shelf Res.*, **27**, 1779–1800.

Liquete, C. *et al.* (2009). Sediment discharge of the rivers of Catalonia, NE Spain, and the influence of human impacts. *J. Hydrol.*, **366**, 76–88.

Lisitzin, A. P. (1972). *Sedimentation in the World Ocean*. Soc. Econ. Paleont. Mineral Spec. Publ. 17.

Litchfield, N. *et al.* (2008). The Waipaoa Sedimentary System: research review and future directions. *IAHS Publ.* **325**, 408–416.

Little, C., A. Lara, J. McPhee & R. Urrutia (2009). Revealing the impact of forest exotic planatations on water yield in large scale watersheds in South-Central Chile. *J. Hydrol.*, **374**, 162–170.

Liu, G.-W. & J. P. Wang (1999). A study of extreme floods in China for the past 100 years. *IAHS Publ.* **255**, 109–119.

Liu, J. P. & J. D. Milliman (2004). Reconsidering melt-water pulses 1A and 1B: Global impacts of rapid sea-level rise. *Oceanic and Coastal Research (China)*, **3**, 1283–1290.

Liu, S. *et al.* (2001). Regional variation of sediment load of Asian rivers flowing into the ocean. *Sci. China, Ser. B Chem.*, **44**, 23–32.

Livingstone, D. A. (1963). Chemical composition of rivers and lakes, US Geol. Surv. Prof. Paper 440-G, 64 pp.

Livingstone, I. (1991). Livestock management and "overgrazing" among pastoralists. *Ambio*, **20**, 80–85.

Lloret, J. *et al.* (2001). Fluctuations of landings and environmental conditions in the north-western Mediterranean Sea. *Fish. Oceanogr.*, **10**, 33–50.

Lombard, R. E. *et al.* (1981). Channel conditions in the Lower Toutle and Cowlitz rivers resulting from the mudflows of May 18, 1980, *US Geol. Surv. Circular 850-C*, 16 pp.

Long, B. F., F. Morissette & J. Lebel (1982). Etude du material particulaire en suspension et du material dissous des rivieres Romaine et Saint-Jean durant un cycle saisonnies, Hydro-Quebec, 199.

Long, Y.-Q. *et al.* (1994). Variability of sediment load and its impacts on the Yellow River. *IAHS Publ.* **224**, 431–436.

Lopes, C. & A. C. Mix (2009). Pleistocene megafloods in the northeast Pacific. *Geology*, **37**, 79.

López-Moreno, J. I. *et al.* (2007). Influence of the North Atlantic Oscillation on water resources in central Iberia: Precipitation, streamflow anomalies, and reservoir management strategies (DOI 10.1029/2007WR005864). *Water Resour. Res.*, **43**, 9411 doi:10.1029/2007WR005864.

López-Moreno, J. I. *et al.* (2009). Dam effects on droughts magnitude and duration in a transboundary basin: The Lower River Tagus, Spain and Portugal. *Water Resour. Res.*, **45**, W02405 doi:10.1029/2008WR007198.

Love, D. *et al.* (2008). Changing rainfall and discharge patterns in the northern Limpopo Basin, Zimbabwe (abs.).

Geophys. Res. Abs., EGU General Assembly 2008, SRef-ID:1607–7962/gra/EGU2008-A-07350.

Lu, X. & D. L. Higgitt (1998). Recent changes of sediment yield in the Upper Yangtze, China. *Environ. Management*, **22**, 697–709.

Lu, X. X. & R. Y. Siew (2005). Water discharge and sediment flux in the lower Mekong River. *Hydrol. Earth Syst. Sci.*, **2**, 2287–2325.

Lu, X. X. *et al.* (2007). Rapid channel incision of the lower Pearl River (China) since the 1990s. *Hydrol. Earth Syst. Sci.*, **4**, 2205–2227.

Luce, C. H. & T. A. Black (1999). Sediment production from forest roads in western Oregon. *Water Resour. Res.*, **35**, 2561–2570.

Luce, C. H. & T. A. Black (2001). Spatial and temporal patterns in erosion from forest roads, in *Land Use and Watersheds: Human Influence on Hydrology and Geomorphology in Urban and Forest Areas*, eds. M. S. Wigmosta & S. J. Burges, Amer. Geophys. Union, pp. 165–178.

Ludwig, W. & J. L. Probst (1996). A global modelling of the climatic, morphological, and lithological control of river sediment discharges to the oceans. *IAHS Publ.* **236**, 21–22.

Ludwig, W. & J. L. Probst (1998). River sediment discharge to the oceans; present-day controls and global budgets. *Amer. J. Sci.*, **298**, 265–295.

Lugo, A. E. (1983). Organic carbon export by waters of Spain, in *Transport of Carbon and Minerals in Major World Rivers Pt. 2*, eds. E. T. Degens, S. Kempe & H. Soliman, Hamburg: SCOPE/UNEP, pp. 267–279.

Lundberg, N. & R. J. Dorsey (1990). Rapid Quaternary emergence, uplift, and denudation of the Coastal Range, eastern Taiwan. *Geology*, **18**, 638.

Luo, X. J. *et al.* (2006). Polycyclic aromatic hydrocarbons in suspended particulate matter and sediments from the Pearl River Estuary and adjacent coastal areas, China. *Environ. Pollut.*, **139**, 9–20.

Lyons, W. B. *et al.* (2002). Organic carbon fluxes to the ocean from high-standing islands. *Geology*, **30**, 443–446.

Lyons, W. B. *et al.* (2005). Chemical weathering in high-sediment-yielding watersheds, New Zealand. *J. Geophys. Res.*, **110**, F1008, doi:1029/2003JF000088.

Macdonald, R. *et al.* (1998). A sediment and organic carbon budget for the Canadian Beaufort shelf. *Mar. Geol.*, **144**, 255–273.

Macias, J. L. *et al.* (2004). The 26 May 1982 breakout flows derived from failure of a volcanic dam at El Chichón, Chiapas, Mexico. *Geol. Soc. Amer. Bull.*, **116**, 233–246.

Macklin, M. G. *et al.* (2006). Past hydrological events reflected in the Holocene fluvial record of Europe. *Catena*, **66**, 145–154.

Madej, M. A. (1995). Changes in channel-stored sediment, Redwood Creek, northwestern Califonia, *US Geol. Surv. Prof. Paper 1454*, O1–O27.

Magilligan, F. J. & K. H. Nislow (2005). Changes in hydrologic regime by dams. *Geomorph.*, **71**, 61–78.

Magilligan, F. J., K. H. Nislow & B. E. Graber (2003). Scale-independent assessment of discharge reduction and riparian disconnectivity following flow regulation by dams. *Geology*, **31**, 569.

Maizels, J. (1989). Sedimentology and paleohydrology of Holocene flood deposits in front of a Jokulhlaup glacier, South Iceland, in *Floods: Hydrological, Sedimentological and Geomorphological Implications*, eds. K. Beven & P. Carling, Chichester: John Wiley & Sons, pp. 239–251.

Major, J. J. (2004). Posteruption suspended sediment transport at Mount St. Helens: decadal-scale relationships with landscape adjustments and river discharges. *J. Geophys. Res.*, **109**, F01002, doi:10.1029/2002JF000010, 2004.

Major, J. J. *et al.* (2000). Sediment yield following severe volcanic disturbance. A two-decade perspective from Mount St. Helens. *Geology*, **28**, 819–822.

Malmstrom, V. H. (1958). *A Regional Geography of Iceland*, Nat. Acad. Sci-Nat. Res. Council.

Manabe, S., P. C. D. Milly & R. Wetherald (2004). Simulated long-term changes in river discharge and soil moisture due to global warming. *Hydrol. Sci. J.*, **49**, 625–642.

Mantua, N. J. & S. R. Hare (2002). The Pacific Decadal Oscillation. *J. Oceanogr.*, **58**, 35–44.

Mantua, N. J. *et al.* (1997). A Pacific interdecadal climate oscillation with impacts on salmon production. *Bull. Am. Meteorol. Soc.*, **78**, 1069–1079.

Marden, M. & D. Rowan (1993). Protective value of vegetation on tertiary terrain before and during Cyclone Bola, East Coast, North Island, New Zealand. *N. Z. J. Forest. Sci.*, **23**, 255–263.

Marden, M. *et al.* (1992). A decade of earthflow research and interrelated studies in the North Island of New Zealand. *IAHS Publ.* **209**, 263–271.

Marden, M. *et al.* (2008). Last glacial aggradation and postglacial sediment production from the non-glacial Waipaoa and Waimata catchments, Hikurangi Margin, North Island, New Zealand. *Geomorph.*, **99**, 404–419.

Maria, A. *et al.* (2000). Source and dispersal of jokulhlaup sediments discharged to the sea following the 1996 Vatnajokull eruption. *Geol. Soc. Amer. Bull.*, **112**, 1507.

Mariño, M. G. (1992). Implications of climate change on the Ebro Delta, in *Climate Change and the Mediterranean*, eds. L. J. Jeftic, J. D. Milliman & G. Sestini, London: Edward Arnold, pp. 304–327.

Markgraf, V. (1993). Climatic history of Central and South America since 18,000 yr B.P.: comparison of pollen records and model simulations, in *Global Climate Since the Last Glacial Maximum*, eds. H. E. Wright, Jr., J. E. Kutzbach, T. Webb, III, *et al.*, Minneapolis, MN: University of Minnesota Press, pp. 357–385.

Markham, A. & G. Day (1994). Sediment transport in the Fly River system, Papua New Guinea. *IAHS Publ.* **22**, 233–239.

Marques, M. A. & E. Mora (1992). The influence of aspect on runoff and soil loss in Mediterranean burnt forest (Spain). *Catena*, **19**, 333–344.

Martins, O. & J.-L. Probst (1991). Biogeochemistry of major African rivers: Carbon and mineral transport, in *Biogeochemistry of Major World Rivers*, eds. E. T. Degens,

S. Kempe & J. E. Richey, John Wiley and Sons Ltd, pp. 121–156.

Massong, T. M. & D. R. Montgomery (2000). Influence of sediment supply, lithology, and wood debris on the distribution of bedrock and alluvial channels. *Geol. Soc. Amer. Bull.*, **112**, 591–599.

Matsuoka, A., T. Yamakoshi & K. Tamura (2008). Sediment yield from seismically-disturbed mountainous watersheds revealed by multi-temporal aerial LiDAR surveys. *IAHS Publ.* **325**, 208–216.

Mauget, S. A. (2003). Multidecadal regime shifts in US streamflow, precipitation, and temperature at the end of the twentieth century. *J. Climate*, **16**, 3905–3916.

McCabe, G. J., M. A. Palecki & J. L. Betancourt (2004). Pacific and Atlantic Ocean influences on multidecadal drought frequency in the United States. *Proc. Nat. Acad. Sci. USA*, **101**, 4136–4141.

McClelland, J. W. *et al.* (2004). Increasing river discharge in the Eurasian Arctic: consideration of dams, permafrost thaw and fires as potential agents of change. *J. Geophys. Res.*, **109**, doi:10.1029/2004JD004583.

McClelland, J. W. *et al.* (2006). A pan-arctic evaluation of changes in river discharge during the latter half of the 20th century. *Geophy. Res. Lett.*, **33**, L06715, doi:10.1029/2006GL025753.

McCully, P. (1996). *Silenced Rivers: The Ecology and Politics of Large Dams,* London: Zed Books.

McDowell, W. H., A. E. Lugo & A. James (1995). Export of nutrients and major ions from Caribbean catchments. *J. N. Amer. Benthol. Soc.*, **14**, 12–20.

McKergow, L. A. *et al.* (2005). Sources of sediment to the Great Barrier Reef World Heritage Area. *Mar. Poll. Bull.*, **51**, 200–211.

McLennan, S. M. (1993). Weathering and global denudation. *J. Geol.*, **101**, 295–303.

McMahon, T. A. (1979). Hydrological characteristics of arid zones. *IAHS Publ.* **128**, 105–123.

McMahon, T. A. *et al.* (1992). *Global Runoff: Continental Comparisons of Annual Flows and Peak Discharges*, Cremlingen-Destedt, West Germany: Catena Verlag.

McNamara, J. P., D. L. Kane & L. D. Hinzman (1998). An analysis of streamflow hydrology in the Kuparuk River Basin, Arctic Alaska: a nested watershed approach. *J. Hydrol.*, **206**, 39–57.

McNeill, J. R. (1992). *The Mountains of the Mediterranean World: An Environmental History,* Cambridge: Cambridge University Press.

McPhee, J. (1989). *The Control of Nature,* New York: Farrar Straus Giroux.

McPhee, J. (1994). *Assembling California,* New York: Farrar, Straus and Giroux.

MDBMC (1999). The salinity audit of the Murray–Darling Basin: a 100-year perspective, Canberra, Australia, 39.

Meade, R. H. (1969). Errors in using modern stream-load data to estimate natural rates of denudation. *Geol. Soc. Amer. Bull.*, **80**, 1265–1274.

Meade, R. H. (1994). Suspended sediments of the modern Amazon and Orinoco rivers. *Quat. Int.*, **21**, 29–39.

Meade, R. H. (1996). River-sediment inputs to major deltas, in *Sea-Level Rise and Coastal Subsidence*, eds. J. D. Milliman & B. U. Haq, Dordrecht, Boston: Kluwer Academic Publishers, pp. 63–85.

Meade, R. H. (2008). Transcontinental moving and storage: the Orinoco and Amazon Rivers transfer the Andes to the Atlantic. In *Large Rivers: Geomorphology and Management*, ed. A. Gupta, Chichester: John Wiley, pp. 45–63.

Meade, R. H., N. N. Bobrovitskaya & V. I. Babkin (2000). Suspended-sediment and fresh-water discharges In the Ob and Yenisey rivers, 1960–1988. *Int. J. Earth Sci.*, **89**.

Meade, R. H. *et al.* (1985). Storage and remobilization of suspended sediment in the lower Amazon River of Brazil. *Science*, **228**, 488–490.

Meade, R. H. & J. A. Moody (2008). Changes in the discharge of sediment through the Missouri–Mississippi River system, 1940–2007. *Proc.VIII Encontro Nacional de Engenharia de Sedimentos*, Campo Grande, Brazil, 2–8 November, 27 pp.

Meade, R. H. & J. A. Moody (2010). Causes for the decline of suspended-sediment discharge in the Mississippi River system, 1940–2007. *Hydrol. Process.*, **24**, 35–49.

Meade, R. H. & R. S. Parker (1985). Sediment in rivers of the United States, US Geol. Surv. Water Supply Paper 2275, pp. 49–60.

Meade, R. H., T. R. Yuzyk & T. J. Day (1990). Movement and storage of sediment in rivers of the United States and Canada, in *The Geology of North America: Surface Water Hydrology*, eds. M. G. Wolman & H. C. Riggs, Geological Society of America, O-1, pp. 255–280.

Meade, R. H. (ed.) (1995). Contaminants in the Mississippi River, 1987–92, *US Geol. Surv. Circular 1133.*

Megnounif, A. *et al.* (2007). Key processes influencing erosion and sediment transport in a semi-arid Mediterranean area: the Upper Tafna catchment, Algeria. *Hydrol. Sci. J.*, **52**, 1271–1284.

Meigs, A. *et al.* (2006). Ultra-rapid landscape response and sediment yield following glacier retreat, Icy Bay, southern Alaska. *Geomorph.*, **78**, 207–221.

Meigs, P. (1953). World distribution of arid and sem-arid homoclimates, in *Arid Zone Hydrology (UNESCO)*, pp. 203–209.

Meneghini, B., I. Simmonds & I. N. Smith (2007). Association between Australian rainfall and the southern Annular Mode. *Int. J. Climatol.*, **27**.

Mensing, S. A., J. Michaelsen & R. Byrne (1999). A 560-year record of Santa Ana fires reconstructed from charcoal deposited in the Santa Barbara Basin California. *Quat. Res.*, **51**, 295–305.

Mernild, S. H., G. E. Liston & B. Hasholt (2008). East Greenland freshwater runoff to the Greenland-Iceland-Norwegian Seas 1999–2004 and 2071–2100. *Hydrol. Process.*, **22**, 4571–4586.

Mertes, L. A. K. & J. A. Warrick (2001). Measuring flood output from 110 coastal watersheds in California with field measurements and SeaWiFS. *Geology*, **29**, 659.

Métivier, F. & Y. Gaudemer (1999). Stability of output fluxes of large rivers in South East Asia during the last 2 million years: implications on floodplain processes. *Basin Res.*, **11**, 293–303.

Meybeck, M. (1979). Concentration des eaux fluviales en éléments majeurs et apports en solution aux océans. *Rev. Geogr. Phys. Geol.*, **21**, 215–246.

Meybeck, M. (1988). How to establish and use world budgets of riverine materials, in *Physical and Chemical Weathering in Geochemical Cycles*, eds. A. Lerman & M. Meybeck, Dordrecht: Kluwer Acad. Publ., pp. 247–272.

Meybeck, M. (1994). Origin and variable composition of present day riverborne material, in *Material Fluxes on the Surface of the Earth*. Washington, DC: Nat. Acad. Press, pp. 61–73.

Meybeck, M. (1998). Surface water quality: global assessment and perspectives, in *Water: A Looming Crisis?*, ed. H. Zebidi, UNESCO, IHP-V, pp. 173–184.

Meybeck, M. (2002). Riverine quality at the Anthropocene: propositions for global space and time analysis, illustrated by the Seine River. *Aqua. Sci.-Res. Across Boundaries*, **64**, 376–393.

Meybeck, M. (2003). Global analysis of river systems: from Earth system controls to Anthropocene syndromes. *Phil. Trans. R. Soc. London, Ser. B*, **358**, 1935–1955.

Meybeck, M., D. V. Chapman & R. Helmer (1989). *Global Freshwater Quality. A First Assessment,* Oxford: Blackwell for GEMS/WHO/UNEP.

Meybeck, M., G. de Marsily & E. Fustec (eds.) (1998). *La Seine en Son Bassin,* Elsevier.

Meybeck, M. & A. Ragu (1996). River discharges to the oceans: an assessment of suspended solids, major ions and nutrients, GEMS/EAP, 245 pp.

Meybeck, M. & C. Vörösmarty (2005). Fluvial filtering of land-to-ocean fluxes: from natural Holocene variations to Anthropocene. *C.R. Geosci.*, **337**, 107–123.

Meyer, W. B. & B. L. Turner (eds.) (1994). *Changes in Land Use and Land Cover: a Global Perspective,* Cambridge: Cambridge University Press.

Meza, F. J. (2005). Variability of reference evapotranspiration and water demands. Association to ENSO in the Maipo river basin, Chile. *Global Planet. Change*, **47**, 212–220.

Mikhailova, M. (1998). The hydrological regime and dynamics of the drainage network of the mouth of the HuangHe River. *Water Res.*, **25**, 98–110.

Mikhailova, M. V. & S. V. Dzhaoshvili (1998). Hydrological and morphological processes in the mouth area of the Rioni River and their anthropogenic changes. *Water Res.*, **25**, 134–142.

Mikhailova, M. V. *et al.* (1998). The Tiber River delta and the hydrological and morphological features of its formation. *Water Res.*, **25**, 572–582.

Milliman, J. D. (1975). Upper continental margin sedimentation off Brazil – a synthesis. *Contrib. Sediment. Geol.*, **4**, 151–175.

Milliman, J. D. (1980). Sedimentation in the Fraser River and its estuary, southwestern British Columbia (Canada). *Estuar. Coast. Mar. Sci.*, **10**, 609–633.

Milliman, J. D. (1990). River discharge of water and sediment to the oceans: variations in space and time, in *Facets of Modern Biogeochemistry*, ed. S. K. V. Ittekkot, W. Michaelis, pp. 83–90.

Milliman, J. D. (1995). Sediment discharge to the ocean from small mountainous rivers: the New Guinea example. *Geo-Mar. Lett.*, **15**, 127–133.

Milliman, J. D., J. M. Broadus & F. Gable (1989). Environmental and economic impact of rising sea level and subsiding deltas: the Nile and Bengal examples. *Ambio*, **18**, 340–345.

Milliman, J. D., K. L. Farnsworth & C. S. Albertin (1999). Flux and fate of fluvial sediments leaving large islands in the East Indies. *J. Sea Res.*, **41**, 97–107.

Milliman, J. D. & B. U. Haq (eds.) (1996). *Sea-level Rise and Coastal Subsidence. Causes, Consequences, and Strategies,* Dordrecht: Kluwer Academic Press.

Milliman, J. D. & S. J. Kao (2005). Hyperpycnal discharge of fluvial sediment to the ocean: impact of Super-Typhoon Herb (1996) on Taiwanese rivers. *J. Geol.*, **113**, 503–516.

Milliman, J. D. *et al.* (2010). Recent trends in fluvial discharge to the Black Sea. *CIESM Workshop Monograph 39*, pp. 45–51.

Milliman, J. D. & R. H. Meade (1983). World-wide delivery of river sediment to the ocean. *J. Geol.*, **91**, 1–21.

Milliman, J. D., C. Rutkowski & M. Meybeck (1995). River discharge to the sea: a global river index (GLORI), LOICZ, 125 pp.

Milliman, J. D., G. Sestini & L. Jeftic (1992). The Mediterranean Sea and climate change – an overview, in *Climate Change and Sealevel Rise in the Mediterranean Basin: Implications for the Future*, eds. L. Jeftic, J. D. Milliman & G. Sestini, London: Edward Arnold Publ., pp. 1–14.

Milliman, J. D. & J. P. M. Syvitski (1992). Geomorphic/tectonic control of sediment discharged to the ocean: the importance of small mountainous rivers. *J. Geol.*, **100**, 525–544.

Milliman. D. J., G. S. Quraishee & M. A. A. Beg (1984). Sediment discharge from the Indus River to the ocean: past, present, and future, in B. U. Haq and J. D. Milliman (eds.), *Marine Geology and Oceanography of Arabian Sea and Coastal Pakistan*, Van Nostrand Reinhold, New York, pp. 65–70.

Milliman, J. D. *et al.* (1987). Man's Influence on the erosion and transport of sediment by Asian rivers: the Yellow River (Huanghe) example. *J. Geol.*, **95**, 751–762.

Milliman, J. D. *et al.* (1996). Catastrophic discharge of fluvial sediment to the ocean: evidence of Jokulhlaups events in the Alsek Sea Valley, southeast Alaska (USA). *IAHS Publ.* **236**, 367–379.

Milliman, J. D. *et al.* (2007). Short-term changes in seafloor character due to flood-derived hyperpycnal discharge: Typhoon Mindulle, Taiwan, July 2004. *Geology*, **35**, 779–782.

Milliman, J. D. *et al.* (2008). Climatic and anthropogenic factors affecting river discharge to the global ocean, 1951–2000. *Global Planet. Change*, **62**, 187–194.

Millot, R. *et al.* (2002). The global control of silicate weathering rates and the coupling with physical erosion: new insights from rivers of the Canadian Shield. *Earth Planet. Sci. Lett.*, **196**, 83–98.

Millot, R. *et al.* (2003). Northern latitude chemical weathering rates: clues from the MacKenzie River Basin, Canada. *Geochim. Cosmochim. Acta*, **67**, 1305–1329.

Milly, P. C. D., K. A. Dunne & A. V. Vecchia (2005). Global pattern of trends in streamflow and water availability in a changing climate. *Nature*, **438**.

Milly, P. C. D. *et al.* (2008). Stationarity is dead: Whither water management? *Science*, **319**, 573–574.

Milner, A. M., M. W. Osgood & K. R. Munkittrick (2005). Rivers of Arctic North America, in *Rivers of North America*, eds. A. C. Benke & C. E. Cushing, Amsterdam: Elsevier, pp. 902–937.

Ministry of Agriculture and Water (1984). *Water Atlas of Saudi Arabia*, Saudi Arabia.

Minobe, S. (1997). A 50–70 year climatic oscillation over the North Pacific and North America. *Geophys. Res. Lett.*, **24**, 683–686.Miyata, S. *et al.* (2009). Effects of forest floor coverage on overland flow and soil erosion on hillslopes in Japanese cypress plantation forests. *Water Resour. Res.*, **45**, W06402.

Mizuyama, T. & S. Kobashi (1996). Sediment yield and topographic change after major volcanic activity. *IAHS Publ.* **236**, 295–301.

Mohammad, A. G. & A. A. Ada (2010). The impact of vegetative cover type on runoff and soil erosion under different land uses. *Catena*, **81**, 97–103.

Molnar, P. (2001). Climate change, flooding in arid environments, and erosion rates. *Geology*, **29**, 1071–1074.

Molnar, P. & P. England (1990). Late Cenozoic uplift of mountain ranges and global climate change: chicken or egg? *Nature*, **346**, 29–34.

Molnia, B. F. & P. R. Carlson (1995). Glacial-marine sedimentation in Vitus Lake, Bering Glacier, Alaska, USA (abs). *IGS International Symposium on Glacial Erosion and Sedimentation, August 20–25, 1995.*

Montgomery, D. R. (1994). Road surface drainage, channel initiation, and slope instability. *Water Resour. Res.*, **30**, 1925–1932.

Montgomery, D. R. (2007). *Dirt: The erosion of civilizations,* University of California Press.

Montgomery, D. R. & M. T. Brandon (2002). Topographic controls on erosion rates in tectonically active mountain ranges. *Earth Planet. Sci. Lett.*, **201**, 481–489.

Montgomery, D. R. & J. M. Buffington (1997). Channel-reach morphology in mountain drainage basins. *Geol. Soc. Amer. Bull.*, **104**, 596–611.

Montgomery, D. R., M. S. Panfil & S. K. Hayes (1999). Channel-bed mobility response to extreme sediment loading at Mount Pinatubo. *Geology*, **27**, 271–274.

Montgomery, D. R. & H. Piegay (2003). Wood and rivers: interactions with channel morphology and processes. *Geomorph.*, **51**, 1–5.

Montgomery, D. R. *et al.* (2000). Forest clearing and regional landsliding. *Geology*, **28**, 311–314.

Moody, D. W., E. B. Chase & D. A. Aronson (1986). National Water Summary 1985 – Hydrologic Events and Surface Water Resources, U.S. Geol. Surv. Water Supply Paper 2300, 519 p.

Moody, J. A. & D. A. Martin (2009). Synthesis of sediment yields after wildland fire in different rainfall regimes in the western United States. *Int. J. Wildland Fire*, **18**, 96–115.

Moon, S. *et al.* (2007). Chemical weathering in the Hong (Red) River basin: rates of silicate weathering and their controlling factors. *Geochim. Cosmochim. Acta*, **71**, 1411–1430.

Mooney, H. A. & D. J. Parsons (1973). Structure and function of the California chaparral – an example from San Dimas. *Ecol. Stud.*, **7**, 83–112.

Moore, W. S. (1996). Large groundwater inputs to coastal waters revealed by 226-Ra enrichment. *Nature*, **380**, 612–614.

Moore, R. D. & S. M. Wondzell (2005). Physical hydrology and the effects of forest harvesting in the Pacific Northwest: a review. *J. Amer. Water Resour. Assoc.*, **41**, 763–784.

Mosley, M. P. (2000). Regional differences in the effects of El Niño and La Niña on low flows and floods. *Hydrol. Sci. J.*, **45**, 249–267.

Mount, J. F. (1995). *California Rivers and Streams: The Conflict Between Fluvial Process and Land Use,* Berkeley: University of California Press.

Mount, J. F. and R. Twiss (2005).Subsidence, sea level rise and seismicity in the Sacramento–San Joaquin Delta. *San Francisco Estuary and Watershed Science*, **3**(1).

Moy, C. M. *et al.* (2002). Variability of El Nino/Southern Oscillation activity at millennial timescales during the Holocene epoch. *Nature*, **420**, 162–165.

Mulder, T. & J. P. M. Syvitski (1995). Turbidity currents generate at river mouths during exceptional discharges to the world oceans. *J. Geol.*, **103**, 285–299.

Mulder, T. *et al.* (1998). The Var submarine sedimentary system: understanding Holocene sediment delivery processes and their importance to the geological record, in *Geological Processes on Continental Margins: Sedimentation, Mass Wasting and Stability*, eds. M. Stoker, D. Evans & A. Cramp, Geol. Soc., London, Spec. Publ. 129, pp. 145–166.

Mulder, T. *et al.* (2003). Hyperpycnal turbidity currents: initiation, behavior and related deposits. A review. *Mar. Pet. Geol.*, **20**, 861–882.

Mulholland, P. J. & J. A. Watts (1982). Transport of organic carbon to the oceans by the rivers of North America: a synthesis of existing data. *Tellus*, **34**, 176–186.

Murray, S. P. *et al.* (1982). Physical processes and sedimentation on a broad, shallow bank. *Estuar. Coast. Shelf Sci.*, **14**, 135.

Mutz, M. (2003). Hydraulic effects of wood in streams and rivers, in *The Ecology and Management of Wood in World Rivers*, eds. S. V. Gregory, K. L. Boyer & A. M. Gurnell. Amer. Fish. Soc., pp. 93–107.

NSW. Department of Public Works (1975). Shoalhaven River Entrance Study: Interim Report, 33 pp.

Nace, R. (1967). Water resources: a global problem with local roots. *Environ. Sci. Technol.*, **1**, 550–560.

Nace, R. L. (1970). Hydrological and related data programs in the United States of America, in *CENTO Seminar on Evaluation of water resources with scarce data*, Beirut, Lebanon: Middle East Devel. Div, Ministry Overseas Devel., pp. 87–101.

Nageswara Rao, K. *et al.* (2010). Impacts of sediment retention by dams on delta shoreline regression: evidences from the Krishna and Godavari deltas, India. *Earth Surf. Process. Landforms*, **35**, 817–827.

Naiman, R. J., C. A. Johnston & J. C. Kelley (1988). Alteration of North American streams by beaver. *BioScience*, **38**, 753–762.

Nakajima, X. (2006). Hyperpycnites deposited 700 km away from river mouths in the central Japan Sea. *J. Sed. Res.*, **76**, 60–73.

Narayana, A. C. (2006). Rainfall variability and its impact on the sediment discharge from the rivers of Kerala region, southwestern India. *J. Geol. Soc. India*, **68**.

National Board of Water and the Environment *Hydrological Yearbook*, Helsinki, Finland.

Neal, E. G., M. Todd Walter & C. Coffeen (2002). Linking the pacific decadal oscillation to seasonal stream discharge patterns in Southeast Alaska. *J. Hydrol.*, **263**, 188–197.

NEDECO (1959). *River Studies and Recommendations on Improvement of Niger and Benue, Amsterdam*: North Holland Publ., 1000 p.

NEDECO (1968). Suriname transportation study, Delft: The Netherlands, 293 pp.

NEDECO (1973). Rio Magdalena and Canal del Dique Survey Project. Technical Report, The Hague, The Netherlands Engineering Consultants.

Nelson, C. H. (1990). Estimated post-Messinian sediment supply and sedimentation rates on the Ebro continental margin, Spain. *Mar. Geol.*, **95**, 395–418.

New, M. *et al.* (2001). Precipitation measurements and trends in the twentieth century. *Int. J. Climatol.*, **21**, 1899–1922.

Newman, M., G. P. Compo & M. A. Alexander (2003). ENSO-forced variability of the Pacific decadal oscillation. *J. Climate*, **16**, 3853–3857.

Newson, M. (1997). *Land, Water and Development. Sustainable Management of River Basin Systems,* London: Routledge.

Nicholas, A. P. *et al.* (1999). Modeling and monitoring river response to environmental change: the impact of dam construction and alluvial gravel extraction on bank erosion rates in the lower Alfios Basin, Greece, in *Fluvial Processes and Environmental Change*, eds. A. G. Brown, T. A. Quine & T. Brown, Chichester: John Wiley and Sons.

Nilsson, C. *et al.* (2005). Fragmentation and flow regulation of the world's large river systems. *Science*, **308**, 405.

Nittrouer, J. A., M. A. Allison & R. Campanella (2008). Bedform transport rates for the lowermost Mississippi River. *J. Geophys. Res.*, **113**, F03004 doi:10.1029/2007JF000795.

Nixon, S. W. (2003). Replacing the Nile: are anthropogenic nutrients providing the fertility once brought to the Mediterranean by a great river? *Ambio*, **32**, 30–39.

NOAA (1985). *Gulf of Mexico Coastal and Ocean Zones Strategic Assessment: Data Atlas*, Washington, DC: Goverment Printing Office.

Nolan, K. M. & R. J. Janda (1995). Movement and sediment yield of two earthflows, northwestern California. *US Geol. Surv. Prof. Paper 1454*, F1–12.

Nolan, K. M., H. M. Kelsey & D. C. Marron (eds.) (1995). *Geomorphic processes and aquatic habitat in the Redwood Creek basin, northwestern California. US Geol. Surv. Prof. Paper 1454.*

Nolan, K. M., T. E. Lisle & H. N. Kelsey (1987). Bankful discharge and sediment transport in northwestern California. *IAHS Publ.* **165**, 439–449.

Nolan, K. M. & D. C. Marron (1995). History, causes, and significance of changes in the channel geometry of Redwood Creek, northwestern California. *US Geol. Surv. Prfo. Paper 1454*, N1–22.

Normark, W. R. & J. A. Reid (2003). Extensive deposits on the Pacific Plate from late Pleistocene North American glacial lake outbursts. *J. Geol.*, **111**, 617–637.

Normark, W. R. *et al.* (1997). Tectonism and turbidites in Escanaba Trough, southern Gorda Ridge (abs.). *EOS*, **78**, F630.

Nouh, M. (2006). Wadi flow in the Arabian Gulf states. *Hydrol. Process.*, **20**, 2393–2413.

Nunes, C. & J. I. Augé (eds.) (1999). *Land-use and Land-cover Change (LUCC) Implementation Strategy*: IGBP.

Nutalaya, P., R. N. Yong, T. Chumnankit & S. Buapeng (1996). Land subsidence in Bangkok during 1978–1988. In J. D. Milliman & B. U. Haq (eds.), *Sea-level Rise and Coastal Subsidence: Causes, Consequences and Strategies.* Kluwer Acad. Publ., Dordrecht, pp. 105–129.

O'Connor, J. E. & G. E. Grant (eds.) (2003). *A Peculiar River. Geology, Geomorphology, and Hydrology of the Deschutes River, Oregon.* Amer. Geophys. Union, 219 pp.

O'Connor, J. E., G. E. Grant & J. E. Costa (2002). The geology and geography of floods, in *Ancient Floods, Modern Hazards: Principles and Application of Paleoflood Hydrology.* Amer. Geophys. Union Water Sci. Appl., 5, pp. 359–385.

O'Grady, D. B. & J. P. M. Syvitski (2002). Large-scale morphology of Arctic continental slopes: the influence of sediment delivery on slope form, in *Glacier-influenced Sedimentation on High-latitude Continental Margins*, eds. J. A. Dowdeswell & C. O. Cofaigh, Geol. Soc. London Spec. Publ. 203, pp. 11–31.

Ogi, M. & Y. Tachibana (2006). Influence of the annual Arctic Oscillation on the negative correlation between Okhotsk Sea ice and Amur River discharge. *Geophys. Res. Lett.*, **33**.

Ohmori, H. (1983). Erosion rates and their relation to vegetation from the viewpoint of world-wide distribution. *Bull. Dept. Geogr., University of Tokyo*, pp. 77–91.

Oki, T. (1999). The global water cycle, in *Global Energy and Water Cycles*, eds. K. A. Browning & R. J. Gurney, Cambridge: Cambridge University Press, pp. 10–29.

Oki, T. & S. Kanae (2006). Global hydrological cycles and world water resources. *Science*, **313**, 1068–1072.

Olive, L. J. & W. A. Rieger (1986). Low Australilan sediment yields – a question of inefficient sediment delivery? *IAHS Publ.* **159**, 355–364.

Onda, Y., W. E. Dietrich & F. Booker (2008). Evolution of overland flow after a severe forest fire, Point Reyes, California. *Catena*, **72**, 13–20.

Opperman, J. J. *et al.* (2009). Sustainable floodplains through large-scale reconnection to rivers. *Science*, **326**, 1487–1488.

Owens, P. & O. Slaymaker (1992). Late Holocene sediment yields in small alpine and subalpine drainage basins, British Columbia. *IAHS Publ.* **36**, 216–219.

Ozturk, F. (1996). Suspended sediment yields of rivers in Turkey. *IAHS Publ.*, **236**, 65–71.

Pachur, H. -J. & S. Kropelin (1987). Wadi Howar: Paleoclimatic evidence from an extinct river system in the southeastern Sahara. *Science*, **237**, 298–300.

Page, M. J., L. M. Reid & I. H. Lynn (1999). Sediment production from Cyclone Bola landslides Waipaoa catchment. *J. Hydrol. N.Z.*, **38**, 289–308.

Page, M. & N. A. Trustrum (1997). A late Holocene lake sediment record of the erosion response to land use change in a steepland catchment. *Z. Geomorphol.*, **41**, 369–392.

Page, M., N. Trustrum & B. Gomez (2000). Implications of a century of anthropogenic erosion for future land use in the Gisborne–East Coast region of New Zealand. *N. Z. Geogr.*, **56**, 13–24.

Pahnke, K. *et al.* (2007). Eastern tropical Pacific hydrologic changes during the past 27,000 years from D/H ratios in alkenones. *Paleoceanogr.*, **22**, A4214 doi:10.1029/2007PA001468.

Palanques, A., F. Plana & A. Maldonado (1990). Recent influence of man on the Ebro margin sedimentation system, northwestern Mediterranean Sea. *Mar. Geol.*, **95**, 247–263.

Pano, N. (1992). *Dynamic a del litorale Albanese (sintesi delle conoscenze). Proc. 19th AIGI Mtg., G. Land. Publ, Genoa*, pp. 3–18.

Pareschi, M. T. *et al.* (2000). May 5, 1998, debris flows in circum-Vesuvian areas (southern Italy): insights for hazard assessment. *Geology*, **28**, 639–642.

Parrett, C., N. B. Melcher & R. W. J. James (1993). *Flood discharges in the upper Mississippi River basin, 1993*, U.S. Geol. Surv. Circular 1120-A.

Pasquini, A. I. & P. J. Depetris (2007). Discharge trends and flow dynamics of South American rivers draining the southern Atlantic seaboard: an overview. *J. Hydrol.*, **333**, 385–399.

Patchineelam, S. M., B. J. Kjerfve & L. R. Gardner (1999). A preliminary sediment budget for the Winyah Bay estuary, South Carolina, USA. *Mar. Geol.*, **162**, 133–144.

Pearce, A. J. & A. Watson (1986). Effects of earthquake induced landslides on sediment budget and transport over a 50-year period. *Geology 14*, **14**, 52–55.

Peel, M. C. & T. A. McMahon (2006). Continental runoff – a quality-controlled global runoff data set. *Nature*, e14-E14.

Peel, M. C. *et al.* (2001). Identification and explanation of continental differences in the variability of annual runoff. *J. Hydrol.*, **250**, 224–240.

Penman, H. L. (1963). Vegetation and hydrology, in *Bureau Soils Tech. Comm. 53* Commonwealth Agricul. Bureau (Great Britian), p. 124.

Pereira, H. C. (1973). *Land Use and Water Resources in Temperate and Tropical Climates,* Cambridge: Cambridge University Press.

Peterson, B. J. *et al.* (2002). Increasing river discharge to the Arctic Ocean. *Science*, **298**, 2171–2173.

Petrone, K. C. *et al.* (2007). The influence of fire and permafrost on sub-arctic stream chemistry during storms. *Hydrol. Process.*, **21**, 423–434.

Pettine, A. *et al.* (1985). Organic and trophic loads of major Italian rivers, in *Transport of Carbon and Minerals in Major World Rivers Pt. 4*, eds. E. T. Degens, S. Kempe & R. Herrera, Hamburg: SCOPE/UNEP, pp. 407–416.

Peucker-Ehrenbrink, B. (2009). Land2Sea database of river drainage basin sizes, annual water discharges, and suspended sediment fluxes. *Geochem. Geophys. Geosyst.*, **10**, Q06014 doi:10.1029/2008GC002356.

Peucker-Ehrenbrink, B. & M. W. Miller (2003). Quantitative bedrock geology of east and southeast Asia. *Geochemistry Geophysics Geosystems*, **5** doi:10.1029/2003GC000619.

Phien-wej, N., P. H. Giao & P. Nutalaya (2006). Land subsidence in Bangkok, Thailand. *Eng. Geol.*, **82**, 187–201.

Phillips, J. D. (1990). Relative importance of factors influencing fluvial soil loss at the global scale. *Amer. J. Sci.*, **290**, 547–568.

Pickup, G. (1980). Hydraulic and sediment modeling studies in environmental impact assesment of a major tropical dam project. *Earth Surf. Process.*, **5**, 61–75.

Pickup, G. (1983). Sedimentation processes in the Purari River upstream of the delta, in *The Purari – Tropical Environment of a high rainfall river basin*, ed. T. Petr, The Hague: Monographiae Biologicae, pp. 205–225.

Pickup, G., R. J. Higgins & R. F. Warner (1981). Erosion and sediment yield in Fly River drainage basins, Papua New Guinea. IAHS Publ. 132, pp. 438–456.

Piégay, H. *et al.* (2004). Contemporary changes in sediment yield in an alpine mountain basin due to afforestation (the upper Drôme in France). *Catena*, **55**, 183–212.

Pimentel, D. *et al.* (1995). Environmental and economic costs of soil erosion and conservation benefits. *Science*, **267**, 117–1123.

Pinet, P. & M. Souriau (1988). Continental erosion and large-scale relief. *Tectonics*, **7**, 563–582.

Pitkånen, H. (1994). Eutrophication of the Finnish coastal waters: origin, fate and effects of riverine nutrient fluxes. *Publ. Water Environ. Res. Inst. 1994.*

Pitlick, J. (1993). Response and recovery of a subalpine stream following a catastrophic flood. *Geol. Soc. Amer. Bull.*, **105**, 657–670.

Pont, D. (1997). Les debits solides du Rhone a proximité de son embouchure donnees récentes (1994–1995). *Rev. Geogr. Lyon*, **72**, 23–33.

Ponting, C. (1991). *A Green History of the World,* London: Penguin.

Porter, S. C. & Z. An (1995). Correlation between climate events in the North Atlantic and China during the last glaciation. *Nature*, **375**, 305–308.

Postel, S. & B. Richter (2003). *Rivers for Life. Managing Water for People and Nature,* Washington, DC: Island Press.

Postel, S. L., G. C. Daily & P. R. Ehrlich (1996). Human appropriation of renewable fresh water. *Science*, **271**, 785.

Potter, P. E. (1978). Significance and origin of big rivers. *J. Geol.*, **86**, 13–33.

Poulos, S. E. & M. B. Collins (2002). Fluviatile sediment fluxes to the Mediterranean Sea: a quantitative approach and the influence of dams. *Geol. Soc. London Spec. Publ.*, **191**, 227–245.

Poulos, S. E., V. Lykousis & M. B. Collins (1995). Late Quaternary evolution of Amvrakikos Gulf, western Greece. *Geo-Mar. Lett.*, **15**, 9–16.

Poulos, S. E., B. D. Collins & G. Evans (1996). Water sediment fluxes of Greek rivers, southeastern Alpine Europe: annual yields, seasonal variability, delta formation and human impact. *Z. Geomorphol.*, **40**, 243–261.

Poulos, S. E. *et al.* (2000). Thermaikos Gulf Coastal System, NW Aegean Sea: an overview of water/sediment fluxes in relation to air–land–ocean interactions and human activities. *J. Marine Syst.*, **25**, 47–76.

Poulos, S. E. *et al.* (2002). Sediment fluxes and the evolution of a riverine-supplied tectonically-active coastal system: Kyparissiakos Gulf, Ionian Sea (eastern Mediterranean). *Geological Society London Special Publications*, **191**, 247–266.

Prego, R., P. Boi & A. Cobelo -García (2008). The contribution of total suspended solids to the Bay of Biscay by Cantabrian Rivers (northern coast of the Iberian Peninsula). *J. Marine Syst.*, **72**, 342–349.

Probst, J. L. (1992). Geochemie et hydrologie de l'erosion continental. Mechanisms, bilan global actuel at fluctuations au cours des 500 millions d'annees. *Sci. Géol. Bull.*, **94**, 161.

Probst, J.-L. & P. Amiotte-Suchet (1992). Fluvial suspended sediment transport and mechanical erosion in the Maghreb (North Africa). *Hydrol. Sci. J.*, **37**, 621–637.

Prowse, T. D. (1993). Suspended sediment concentration during river ice breakup. *Can. J. Civil Eng.*, **20**, 872–875.

Pyne, S. J. (1991). *Burning Brush: A Fire History of Australia,* New York: Holt.

Qi, S. Z. & F. Luo (2007). Environmental degradation in the Yellow River delta, Shandong Province, China. *Ambio*, **36**, 610–611.

Qian, N. & D. Z. Dai (1980). The problems of river sedimentation and the present status of research in China, in *China Hydraul. Eng. Proc. Int. Symp. River Sedimentation* Beijing, China: Guanghua Press, pp. 3–39.

Quadrelli, R., V. Pavan & F. Molteni (2001). Wintertime variability of Mediterranean precipitation and its links with large-scale circulation anomalies. *Clim. Dyn.*, **17**, 457–466.

Rabalais, N. N. & R. E. Turner (2001). Hypoxia in the northern Gulf of Mexico: description, causes and change. *Coastal Hypoxia: Consequences for Living Resources and Ecosystems, Amer. Geophys. Union Coast. Est. Stud. Ser.* **58**, pp. 37–48.

Rabalais, N. N. *et al.* (1998). Consequences of the 1993 Mississippi River flood in the Gulf of Mexico. *River Res. Appl.*, **14**, 161–177.

Rabalais, N. N. *et al.* (2007). Hypoxia in the northern Gulf of Mexico: does the science support the plan to reduce, mitigate and control hypoxia? *Estuaries Coasts*, **30**, 753–772.

Rachold, V. *et al.* (1996). Sediment transport to the Laptev Sea – hydrology and geochemistry of the Lena River. *Polar Res.*, **15**, 183–196.

Rachold, V. *et al.* (2003). Modern terrigenous organic carbon input to the Arctic Ocean, in *The Organic Carbon Cycle in the Arctic Ocean: Present and Past*, eds. R. Stein & R. W. MacDonald, Berlin: Springer Verlag, pp. 33–55.

Rahaman, W. *et al.* (2009). Climate control on erosion distribution over the Himalaya during the past ~ 100 ka. *Geology*, **37**, 559–562.

Rajagopalan, B. *et al.* (2009). Water supply risk on the Colorado River: can management mitigate? *Water Res. Resear.*, **45**, W08201, doi:10.1029/2008WR007652.

Ramanathan, A. L., V. Subramanian & B. K. Das (1996). Sediment and heavy metal accumulation in the Cauvery Basin. *Environ. Geol.*, **27**, 155–163.

Ramankutty, N. & J. A. Foley (1999). Estimating historical changes in global land cover: Croplands from 1700 to 1992. *Global Biogeochem. Cycles*, **13**, 997–1027.

Ramesh, R. & V. Subramanian (1993). Geochemical characteristics of the major tropical rivers of India. *IAHS Publ.* **216**, 157–164.

Rand McNally (1980). *Encyclopedia of World Rivers,* London: Bison Books Limited.

Rao, K. L. (1979). *India's Water Wealth:* Orient Longman.

Rao, K. N. (2006). *Coastal Morphodynamics and Asymmetric Development of the Godavari Delta: Implications to Facies Architecture and Reservoir Heterogeneity,* J. Geol. Soc. India.

Ravichandran, S. (2003). Hydrological influences on the water quality trends in Tamiraparani Basin, South India. *Environ. Monit. Assess.*, **87**, 293–309.

Raymo, M. E. & W. F. Ruddiman (1992). Tectonic forcing of late Cenozoic climate. *Nature*, **359**, 117–122.

Raymond, M. B. (1999). Geochemistry of small mountainous rivers of Papua New Guinea: local observations and global implications, unpublished MSc Thesis, College of William and Mary.

Reed, L. A., C. S. Takita & G. Barton (1997). Loads and yields of nutrients and suspended sediment in the Susquehanna River basin, 1985–89, US Geol. Survey Circular, 17 pp.

Regional Activity Center for Environment Remote Sensing (RACERS) (1996). Monitoring of Coastal Evolution through Space Remote Sensing, Palermo.

Reid, L. M. & T. Dunne (1984). Sediment production from forest road surfaces. *Water Resour. Res.*, **20**, 1753–1761.

Reid, L. M. & T. Dunne (1996). *Rapid Evaluation of Sediment Budgets,* Reiskirchen: Catena Verlag.

Reid, L. M. & M. J. Page (2003). Magnitude and frequency of landsliding in a large New Zealand catchment. *Geomorph.*, **49**, 71–88.

Rein, B., A. Luckge & F. Sirocko (2004). A major Holocene ENSO anomaly during the Medieval period. *Geophys. Res. Lett.*, **31**, L17211 doi:10.1029/2004GL020161.

Rein, B. *et al.* (2005). El Niño variability off Peru during the last 20,000 years. *Paleoceanogr.*, **20**, PA4003 doi:10.1029/2004PA001099.

Reiners, P. W. *et al.* (2003). Coupled spatial variations in precipitation and long-term erosion rates across the Washington Cascades. *Nature*, **426**, 645–647.

Reisner, M. (1993). *Cadillac Desert,* New York, Penguin.

Ren, M.-E. (1992). Human impact on coastal landform and sedimentation – the Yellow River example. *GeoJournal*, **28**, 443–448.

Ren, M.-E. (1995). Anthropogenic effect on the flow and sediment of the Lower Yellow River and its bearing on the evolution of Yellow River delta, China. *GeoJournal*, **37**, 473–478.

Ren, M.-E. & J. D. Milliman (1996). Effect of sea-level rise and human activity on the Yangtze delta, China. In J. D. Milliman and B. U. Haq (eds.) *Sea-level Rise and Coastal Subsidence: Causes Consequences and Strategies.* Kluwer Acad. Publ., Dordrecht, pp. 205–214.

Ren, M.-E. & X.-M. Zhu (1994). Anthropogenic influences on changes in the sediment load of the Yellow River, China, during the Holocene. *Holocene*, **4**, 314–320.

Renssen, H. & J. M. Knoop (2000). A global river routing network for use in hydrological modeling. *J. Hydrol.*, **230**, 230–243.

Renwick, W. H. (1996). Continental-scale reservoir sedimentation patterns in the United States. *IAHS Publ.* **236**, 513–522.

Renwick, W. H. *et al.* (2005). The role of impoundments in the sediment budget of the conterminous United States. *Geomorph.*, **71**, 99–111.

Republic of Korea Ministry of Construction and Transportation (2007). *Water Resources in Korea,* Seoul: Water Resources Bureau, Ministry of Construction and Transportation.

Restrepo, J. D. & B. J. Kjerfve (2000a). Magdalena River: interannual variability (1975–1995) and revised water discharge and sediment load estimates. *J. Hydrol.*, **235**, 137–149.

Restrepo, J. D. & B. J. Kjerve (2000b). Water discharge and sediment load from the western slopes of the Colombian Andes, with focus on the Rio San Juan. *J. Geol.*, **108**, 17–33.

Restrepo, J. D. & S. A. López (2008). Morphodynamics of the Pacific and Caribbean deltas of Colombia, South America. *J. South Amer. Earth. Sci.*, **25**, 1–21.

Restrepo, J. D. & J. P. M. Syvitski (2006). Assessing the effect of natural controls and land use change on sediment yield in a major Andean Basin: the Magdalena drainage basin, Colombia. *Ambio*, **35**, 65–74.

Reuther, A. U. *et al.* (2006). Constraining the timing of the most recent cataclysmic flood event from ice-dammed lakes in the Russian Altai Mountains, Siberia, using cosmogenic in situ 10Be. *Geology*, **34**, 913–916.

Revelle, R. R. & P. E. Waggoner (1983). Effect of a carbon dioxide-induced climatic change on water supplies in the western United States, in *Carbon Dioxide Assessment Committee Changing Climate.* Washington, DC: National Academy of Science, pp. 419–432.

Reynoldson, T. B. *et al.* (2005). Fraser River Basin, in *Rivers of North America*, eds. A. C. Benke & C. E. Cushing, Amsterdam: Elsevier, pp. 696–732.

Richards, J. F. (1990). Land transformation, in *The Earth as Transformed by Human Action*, eds. B. L. Turner, II, *et al.*, Cambridge: Cambridge University Press, pp. 163–178.

Richardson, J. S. & A. M. Milner (2005). Pacific coast rivers of Canada and Alaska, in *Rivers of North America*, eds. A. C. Benke & C. E. Cushing, Amsterdam: Elsevier, pp. 734–773.

Riebe, C. S. *et al.* (2001a). Minimal climatic control on erosion rates in the Sierra Nevada, California. *Geology*, **29**, 447–450.

Riebe, C. S. *et al.* (2001b). Strong tectonic and weak climatic control of long-term chemical weathering rates. *Geology*, **29**, 511–514.

Riggs, H. C. (1977). A Brief Investigation of the Surface Water Hydrology of Yemen Arab Republic, US Geol. Survey Water Supply Paper, 37 pp.

Robertson, D. M. & E. D. Roerish (1999). Influence of various water quality sampling strategies on load estimates for small streams. *Water Resour Res.*, **35**, 3747–3759.

Robinson, R. A. J. *et al.* (2007). The Irrawaddy River sediment flux to the Indian Ocean: the original nineteenth-century data revisited. *J. Geol.*, **115**, 629–640.

Robinson, R. S. & M. J. Johnsson (1997). Chemical and Physical Weathering of Fluvial Sands in an Arctic Environment : Sands of the Sagavanirktok River, North Slope, Alaska. *J. Sediment. Res.*, **67**, 560–570.

Rodier, J. A. & M. Roche (1984). World catalog of maximum obvserved floods. *IAHS Publ.* **143**, 378.

Rondeau, B. *et al.* (2000). Budget and sources of suspended sediment transported in the St. Lawrence River, Canada. *Hydrol. Process.*, **14**, 21–36.

Rooseboom, A. & H. J. v. M. Harmse (1979). Changes in the sediment load of the Orange River during the period 1929–1969. IAHS Publ., pp. 459–470.

Ropelewski, C. F. & M. S. Halpert (1989). Precipitation patterns associated with the high index phase of the Southern Oscillation. *J. Climate*, **2**, 268–284.

Rosenberg, D. M. *et al.* (2005). Nelson and Churchill River Basins, in *Rivers of North America*, eds. A. C. Benke & C. E. Cushing, Amsterdam: Elsevier, pp. 852–901.

Rossignol-Strick, M. *et al.* (1982). After the deluge: Mediterranean stagnation and sapropel formation. *Nature*, **295**, 105–110.

Rothacher, J. (1970). Increases in water yield following clear-cut logging in the Pacific Northwest. *Water Resour. Res.*, **6**, 653–658.

Rothacher, J. (1973). Does harvest in west slope Douglas-fir increase peak flow in small forest streams? *US Forest Serv. Res. Paper PNW-163*, **13**.

Rovira, A., R. J. Batalla & M. Sala (2004). Fluvial sediment budget of a Mediterranean river: the lower Tordera (Catalan coastal ranges, NE Spain). *Catena*, **60**, 19–42.

Rozengurt, M. & I. Haydock (1993). Freshwater flow diversion and its implications for coastal zone ecosystems. *58th N. Amer. Wildlife Nat. Resour. Conf.*, pp. 287–295.

Rozin, U. & A. P. Schick (1996). Land use change, conservation measures and stream channel response in the Mediterranean/semiarid transition zone: Nahal Hoga, southern coastal plain, Israel. *IAHS Publ.* **236**, 417–444.

Russell, A. J. *et al.* (2006). Icelandic jökulhlaup impacts: implications for ice-sheet hydrology, sediment transfer and geomorphology. *Geomorph.*, **75**, 33–64.

Russell, J. M. & T. C. Johnson (2007). Little Ice Age drought in equatorial Africa: Intertropical Convergence Zone migrations and El Nino-Southern Oscillation variability. *Geology*, **35**, 21–24.

Russell, J. M., T. C. Johnson & M. R. Talbot (2003). A 725-yr cycle in the climate of central Africa during the late Holocene. *Geology*, **31**, 677–680.

Ryu, J. S., K. S. Lee & H. W. Chang (2007). Hydrogeochemical and isotopic investigations of the Han River basin, South Korea. *J. Hydrol.*, **345**, 50–60.

Sabater, F. *et al.* (1995). The Ter: a Mediterranean river case-study in Spain, in *Ecosystems of the World 22: River and Stream Ecosystems*, eds. C. E. Cushing, K. W. Cummins & G. W. Mishall, Amsterdam: Elsevier, pp. 419–438.

Sabater, S. *et al.* (2009). The Iberian Rivers, in *Rivers of Europe*, eds. K. Tockner, C. T. Robinson & U. Uehlinger, Amsterdam: Elsevier, pp. 113–149.

Saito, Y., Z. Yang & K. Hori (2001). The Huanghe (Yellow River) and Changjiang (Yangtze River) deltas: a review on their characteristics, evolution and sediment discharge during the Holocene. *Geomorph.*, **41**, 219–231.

Salomons, W. & W. G. Mook (1981). Field observations of the isotopic composition of particulate organic carbon in the southern North Sea and adjacent estuaries. *Mar. Geol.*, **41**, 11–20.

Sandler, A. & B. Herut (2000). Composition of clays along the continental shelf off Israel: contribution of the Nile versus local sources. *Mar. Geol.*, **167**, 339–354.

Savage, S. M. (1974). Mechanism of fire-induced water repellency in soil. *Soil Sci. Soc. Am. Proc.*, **38**, 653–657.

Savini, J. & J. C. Krammerer (1961). Urban growth and the water regime. US Geol. Surv. Water Supply Paper 1591A.

Scharler, U. M. & D. Baird (2003). The influence of catchment management on salinity, nutrient stochiometry and phytoplankton biomass of Eastern Cape estuaries, South Africa. *Estuar. Coast. Shelf Sci.*, **56**, 735–748.

Scheepers, A. C. T. & I. C. Rust (1999). The Uniab River Fan: an unusual alluvial fan on the hyper-arid Skeleton Coast, Namibia. *Varieties of Fluvial Form:* Chichester, John Wiley & Sons, pp. 273–294.

Schettini, C. A. F. (2002). Near-bed sediment transport in the Itajaí-Açu River estuary, southern Brazil, in *Fine sediment dynamics in the marine environment.* Elsevier, Amsterdam, eds. J. C. Winterwerp & C. S. Kranenburg, Amsterdam: Elsevier, pp. 499–512.

Scheumann, W. & M. Schiffler (eds.) (1998). *Water in the Middle East,* Berlin: Springer-Verlag.

Schick, A. P. (1988). Hydrologic aspects of floods in extreme arid environments, in *Flood Geomorphology*, eds. V. R. Baker, R. C. Kochel & P. C. Patton, Chichester: John Wiley & Sons, pp. 189–203.

Schick, A. P. & J. Lekach (1987). A high magnitude flood in the Sinai Desert, in *Catastrophic Floods*, eds. L. Mayer & D. Nash, Boston: Allen & Unwin, pp. 381–410.

Schimmelmann, A. *et al.* (1998). A large California flood and correlative global climatic events 400 years ago. *Quat. Res.*, **49**, 51–61.

Schmidt, K. M. & D. R. Montgomery (1995). Limits to relief. *Science*, **270**, 617–620.

Schulze, R. E. (2004). River basin responses to global change and anthropogenic impacts, in *Vegetation, Water, Humans and the Climate: A New Perspective on an Interactive System*, eds. P. Kabat *et al.*, New York: Springer Verlag, pp. 339–374.

Schulze, R. E. *et al.* (2004). Case study 3: Modelling the impacts of land-use and climate change on hydrological responses in the mixed underdeveloped/developed Mgeni catchment, South Africa, in *Vegetation, Water, Humans and the Climate: A New Perspective on an Interactive System.* IGBP, pp. 441–453.

Schumm, S. A. (1963). *The disparity between present rates of denudation and orogeny.* US Geol Sur. Prof. Paper 454-H.

Schumm, S. A. (1977). *The Fluvial System,* New York: Wiley-Interscience.

Schumm, S. A. (2005). *River Variability and Complexity,* Cambridge: Cambridge University Press.

Schumm, S. (2007). Rivers and humans: unintended consequences, in *Large Rivers: Geomorphology and Management*, ed. A. Gupta, John Wiley, pp. 517–533.

Schumm, S. A., J. F. Dumont & J. M. Holbrook (2002). *Active Tectonics and Alluvial Rivers,* Cambridge: Cambridge University Press.

Schumm, S. A. & R. F. Hadley (1961). *Progress in the application of landform analysis in studies of semiarid erosion*, US Geological Survey, 14 pp.

Schumm, S. A. & D. K. Rea (1995). Sediment yield from disturbed earth systems. *Geology*, **23**, 391.

Seager, R. *et al.* (2007). Blueprints for Medieval hydroclimate. *Quat. Sci. Rev.*, **26**, 2322–2336.

Serrano Suarez, B. E. (2004). The Sinú river delta on the northwestern Caribbean coast of Colombia: bay infilling associated with delta development. *J. South Amer. Earth. Sci.*, **16**, 623–631.

Sestini, G. (1991). The implications of climatic changes for the Nile Delta, in *Climatic Change and the Mediterranean: Environmental and Societal Impacts of Climatic Change and Sea-level Rise in the Mediterranean Region*, eds. L. Jeftic, J. D. Milliman & G. Sestini, Edward Arnold Publishers, p. 673.

Sestini, G. (1992). Implications of climatic changes for the Po delta and Venice lagoon, in *Climatic Change and the Mediterranean*, eds. L. Leftick & J. D. Milliman, London: Arnold, pp. 428–494.

Shabbar, A. (2006). The impact of El Nino-Southern Oscillation on the Canadian climate. *Adv. Geos.*, **6**, 149–153.

Shady, A. M. *et al.* (1996). *Management and Development of Major Rivers,* Oxford University Press.

Shahin, M. (2002). *Hydrology and Water Resources of Africa,* Dordrecht: Kluwer Academic Pub.

Shahin, M. (2007). *Water Resources and Hydrometeorology of the Arab Region,* Dordrecht: Kluwer Academic Pub.

Shakun, J. D. & J. Shaman (2009). Tropical origins of North and South Pacific decadal variability. *Geophys. Res. Lett.*, **36**, L19711, doi:10.1029/2009GL040113.

Shaman, J & E. Tziperman (2005). The effect of ENSO on Tibetan Plateau snow depth: A stationary wave teleconnection mechanism and implications for the south Asian monsoons. *J. Climate*, **18**, 2067–2079.

Sharma, K. D. (1997). Assessing the impact of overgrazing on soil erosion in arid regions at a range of spatial scales. *IAHS Publ.* **245**, 119–123.

Sherman, D. J., K. M. Barron & J. T. Ellis (2002). Retention of beach sands by dams and debris basins in southern California. *J. Coast. Res.*, **36**, 662–674.

Shiklomanov, A. I., R. B. Lammers & C. J. Vörösmarty (2002). Widespread decline in hydrological monitoring threatens pan-Arctic research. *EOS*, **83**, 13,16.

Shiklomanov, I. A. & J. C. Rodda (eds.) (2003). *World Water Resources at the Beginning of the 21st Century,* Cambridge: Cambridge University Press.

Shiklomanov, I. A. & B. G. Skakalsky (1994). Studying water, sediment and contaminant runoff of Siberian rivers: modern status and prospects. *Arctic Research of the United States*, **8**, 295–306.

Shiklomanov, A. I. *et al.* (2006). Cold region river discharge uncertainty – estimates from large Russian rivers. *J. Hydrol.*, **326**, 231–256.

Shukla, J., C. Nobre & P. Sellers (1990). Amazon deforestation and climate change. *Science*, **247**, 1322.

Siakeu, J. *et al.* (2004). Change in riverine suspended sediment concentration in central Japan in response to late 20th century human activities. *Catena*, **55**, 231–254.

Sidle, R. C., A. J. Pearce & C. L. O ' Loughlin (1985). *Hillslope Stability and Land Use,* American Geophysical Union Water Resource Monograph.

Sidle, R. C. & W.-M. Wu (2001). Evaluation of the temporal and spatial impacts of timber harvesting on landslide occurrence, in *Land Use and Watersheds: Human Influence on Hydrology and Geomorphology in Urban and Forest Areas*, eds. M. S. Wigmosta & S. J. Burges, American Geophysical Union, pp. 179–193.

Sidorchuk, A. Y. & V. N. Golosov (2003). Erosion and sedimentation on the Russian Plain, II: the history of erosion and sedimentation during the period of intensive agriculture. *Hydrol. Process.*, **17**, 3347–3358.

Siebert, S. *et al.* (2005). Development and validation of the global map of irrigation areas. *Hydrol. Earth Syst. Sci.*, **2**, 1299–1327.

Sigurdsson, H., S. Vikingsson & I. Kaldal (1998). Course of events of the jokulhlaup on Skeidararsandur in November, 1996. *EOS*, **79**, S13.

Silins, U. *et al.* (2008). Impacts of wildfire and post-fire salvage logging on sediment transfer in the Oldman watershed, Alberta, Canada. *IAHS Publ.* **325**, 510–515.

Simeoni, U. & M. Bondesan (1997). The role and responsibility of man in the evolution of the Italian Adriatic coast. *Bull. Inst. Oceanog., Monaco Spec.*, **18**, 111–132.

Simeoni, U., N. Pano & P. Ciavola (1997). The coastline of Albania: morphology, evolution and coastal management issues. *Bull. Inst. Oceanog., Monac Spec.*, **18**, 151–168.

Simmons, C. E. (1988). Sediment characteristics of North Carolina streams, US Geol. Surv. Open-File Report, 130 pp.

Singh, G. (1991). Environmental changes in southern Asia during the Holocene, in *Current Perspectives in Palynological Research*, ed. S. Chanda, New Dehli: Today and Tomorrow Printers & Publishers, pp. 277–296.

Sirocko, F. *et al.* (1991). Atmospheric summer circulation and coastal upwelling in the Arabian Sea during the Holocene and the last glaciation. *Quat. Res.*, **36**, 72–93.

Sisson, T. W., Vallance, J.W. and Pringle, P.T. (2001). Progress made in understanding Mount Rainer's hazards. *EOS*, **82**, 113–120.

Sistemas de Informacion, Geografica, S.A. de C.V. (2007). Proyecto Mexico Informacion Cartografica Digital: Mexico City.

Skoulikidis, N. T. (2009). The environmental state of rivers in the Balkans – a review within the DPSIR framework. *Sci. Tot. Environ.*, **407**, 2501–2516.

Sloan, J., J. R. Miller & N. Lancaster (2001). Response and recovery of the Eel River, California, and its tributaries to floods in 1955, 1964, and 1997. *Geomorph.*, **36**, 129–154.

Smith, L. C. *et al.* (2002). Geomorphic effectiveness, sandur development, and the pattern of landscape responses during jokulhlaups: Skeidararsandur, southeastern Iceland. *Geomorph.* **44**, 95–113.

Smith, S. E. & A. Abdel-Kader (1988). Coastal erosion along the Egyptian delta. *J. Coast. Res.*, **4**, 245–255.

Smith, S. V. *et al.* (2001). Budgets of soil erosion and deposition for sediments and sedimentary organic carbon across the conterminous United States. *Global Biogeochem. Cycles*, **15**, 697–707.

Smock, L. A., A. B. Wright & A. C. Benke (2005). Atlantic coast rivers of the southeastern United States, in *Rivers of North America*, eds. A. C. Benke & C. E. Cushing, Amsterdam: Elsevier, pp. 71–122.

Snorrason, Á. *et al.* (2002). November 1996 jokulhlaup on Skeioararsandur outwash plain, Iceland. *Int. Assoc. Sed.*, 55–66.

Snoussi, M. (1988). Nature, estimetion et comparison des flux de matierres issus des bassins versants de l'Adour (France), du Sebon, de l'Oum-Er-Rbia et du Souss (Maroc). Impact du climat sur les apports fluviatiles a l'Ocean, Bordeaux,

France: Memoire de l'Institut du Geologie du Bassin d'Aquitaine.

Snoussi, M., J. M. Jouanneau & C. Latouche (1990). Flux de matieres issues de bassins versants de zones semi-arid [Bassins du sebon et du sons Maroc], importance dans le bilan global des apports d'origine cintinentale pavenant a l'Ocean Mondial. *J. Afr. Earth Sci.*, **11**, 43–53.

Snoussi, M., S. Haïda & S. Imassi (2002). Effects of the construction of dams on the water and sediment fluxes of the Moulouya and the Sebou Rivers, Morocco. *Reg. Environ. Change*, **3**, 5–12.

Solomon, S. (ed.). *Fourth Assessment Report of the Intergovernmental Panel on Climate Change.* Cambridge University Press.

Sommerfield, C. K. & C. A. Nittrouer (1999). Modern accumulation rates and a sediment budget for the Eel shelf: a flood-dominated depositional environment. *Mar. Geol.*, **154**, 227–241.

Sorman, A. U. & M. J. Abdulrazzak (1987). Regional flood discharge analysis southwest region of the Kingdom of Saudi Arabia, in *Regional Flood Frequency Analysis*, ed. V. P. Singh, D. Reidel Publ. Co., pp. 11–25.

Soulsby, C., D. Tetzlaff & M. Hrachowitz (2009). Tracers and transit times: windows for viewing catchment scale storage? *Hydrol. Process.*, **23**, 3503 – 3507.

Souza, W. F. L. & B. Knoppers (2003). Fluxos de Água e Sedimentos a Costa Leste do Brasil: Relações Entre a Tipologia e as Pressões Antrópicas. *Geochim. Brasil.*, **17**, 57–74.

Souza, M. F. L. *et al.* (2003). Nutrient budgets and trophic state in a hypersaline coastal lagoon: Lagoa de Araruama, Brazil. *Estuar. Coast. Shelf Sci.*, **57**, 843–858.

Sovetskaya Entsiklopediya (1989). (In Russian.) Moscow.

Sridhar, A. (2007). A mid-late Holocene flood record from the alluvial reach of the Mahi River, Western India. *Catena*, **70**, 330–339.

Stahle, D. W. *et al.* (2009). Early 21st-century drought in Mexico. *EOS*, **90**, 89–90.

Stallard, R. F. (1985). River chemistry, geology, geomorphology, and soils in the Amazon and Orinoco basins, in *The Chemistry of Weathering*, ed. J. I. Drever, NATO, Kluwer Academic Press, pp. 293–316.

Stallard, R. F. (1995a). Relating chemical and physical erosion, in *Chemical weathering rates of silicate minerals*, eds. A. F. White & S. L. Brantley, Mineral. Soc. Amer., pp. 543–564.

Stallard, R. F. (1995b). Tectonic, environmental, and human aspects of weathering and erosion: a global review using a steady-state perspective. *Ann. Rev. Earth Pl. Sci.*, **12**, 11–39.

Stallard, R. F. (1998) Terrestrial sedimentation and the carbon cycle: coupling weathering and erosion to carbon burial. *Global Biogeochem. Cycles*, **12**, 231–257.

Stallard, R. F. & J. M. Edmond (1983). Geochemistry of the Amazon. 2. The influence of geology and weathering environment on the dissolved load. *J. Geophys. Res.*, **86**, 9844–9858.

Stanford, J. A. *et al.* (2005). Columbia River basin, in *Rivers of North America*, eds. A. C. Benke & C. E. Cushing, Amsterdam: Elsevier, pp. 590–653.

Stanley, D. J. (1996). Nile delta: extreme case of sediment entrapment on a delta plain and consequent coastal land loss. *Mar. Geol.*, **129**, 189–195.

Stanley, D. J. & A. G. Warne (1993). Nile Delta: recent geological evolution and human impact. *Science*, **260**, 628–634.

Starkel, L. (1972). The role of catastrophic rainfall in the shaping of the relief of the lower Himalaya (Darjeeling Hills). *Geogr. Pol.*, **21**, 103–147.

Staub, J. R. & R. A. Gastaldo (2003). Late Quaternary sedimentation and peat development in the Rajang River delta, Sarawak, East Malaysia, in *Tropical Deltas of Southeast Asia - Sedimentology, Stratigraphy, and Petroleum Geology*, eds. F. H. Sidi, D. Nummedal, P. Imbert, *et al.*, Tulsa, OK: SEPM, pp. 71–87.

Staub, J. R., H. L. Among & R. A. Gastaldo (2000). Seasonal sediment transport and deposition in the Rajang River delta, Sarawak, East Malaysia. *Sediment. Geol.*, **133**, 249–264.

Staubwasser, M. *et al.* (2002). South Asian monsoon climate change and radiocarbon in the Arabian Sea during early and middle Holocene. *Paleoceanogr.*, **17**, DOI 10.1029/2000PA000608.

Steinke, S. *et al.* (2006). On the influence of sea level and monsoon climate on the southern South China Sea freshwater budget over the last 22,000 years. *Quat. Sci. Rev.*, **25**, 1475–1488.

Steffen, W. *et al.* (2003). *Global Change and the Earth System*, Berlin: Springer-Verlag, 332 pp.

Stevens, M. A. (1994). The Citanduy, Indonesia – one tough river, in *The Variability of Large Alluvial Rivers*, eds. S. A. Schumm & B. R. Winkley, New York: ACE Press, pp. 201–219.

Stewart, B. W., R. C. Capo & O. A. Chadwick (2001). Effects of rainfall on weathering rate, base cation provenance, and Sr isotope composition of Hawaiian soils. *Geochim. Cosmochim. Acta*, **65**, 1087–1099.

Stock, J. & W. E. Dietrich (2003). Valley incision by debris flows: evidence of a topographic signature. *Water Resour. Res.*, **39**, 1089.

Stover, S. C. & D. R. Montgomery (2001). Channel change and flooding, Skokomish River, Washington. *J. Hydrol.*, **243**, 272–286.

Stow, D. W. & H. H. Chang (1987). Magnitude–frequency relationship of coastal sand delivery by a southern California stream. *Geo-Mar. Lett.*, **7**, 217–222.

Street-Perrott, F. A. & R. A. Perrott (1990). Abrupt climate fluctuations in the tropics: the influence of North Atlantic circulation. *Nature*, **343**, 607–612.

Subramanian, V. (1987). Environmental geochemistry of Indian river basins: a review. *J. Geol. Soc. India*, **29**, 205–220.

Subramanian, V. (1993). Sediment load of Indian rivers. *Current Science (Bangalore)*, **64**, 928–930.

Subramanian, V. (2001). *Water Quantity - Quality Perspectives in South Asia,* Surrey: Kingston Intern. Publ.

Subramanian, V. (ed.) (2004). *Dialogue on River Links and Diversions*: ENVIS Centre in Biogeochemistry, Jawaharlal Nehru University.

Summerfield, M. A. & N. J. Hulton (1994). Natural controls on fluvial denudation rates in major drainage. *J. Geophys. Res.*, **99**, 13 871–13 883.

Sun, Y. *et al.* (2007). How often will it rain? *J. Climate*, **20**, 4801–4818.

Sutcliffe, J. V. & Y. P. Parks (1999). The hydrology of the Nile. *IAHS Spec. Publ.* **45**.

Sutton, R. T. & D. L. R. Hodson (2005). Atlantic Ocean forcing of North American and European summer climate. *Science*, **309**, 115–118.

Suwa, X. & Y. Yamakoshi (1999).

Swanson, F. J., R. D. Harr & R. L. Fredriksen (1979). *Geomorphology and hydrology in the H. J. Andrews Experimental Forest, western Cascades*, Corvallis, Oregon: Oregon Forestry Service Lab, 19.

Swanson, K. M. *et al.* (2008). Sediment load and floodplain deposition rates: comparison of Strickland and Fly rivers, Papua New Guinea. *J. Geophys. Res.*, **113**, F01S02, doi:10.1029/2006JF000623, 2008.

Swantson, D. N., R. R. Siemer & R. J. Janda (1995). Rate and mechanics of progressive hillslope failure in the Redwood Creek basin, northwestern California, in *US Geol. Surv. Profess. Paper 1454*, E1–16.

(Swedish) Yearbook of Environmental Statistics, 1986–1987, Stockholm, Sweden.

Syed, F. S. *et al.* (2006). Effect of remote forcings on the winter precipitation of central southwest Asia part 1: observations. *Theor. Appl. Climatol.*, **86**, 147–160.

Syvitski, J. P. M. (1992).

Syvitski, J. P. M. & J. M. Alcott (1993). GRAIN2: predictions of particle size seaward of river mouths. *Comput. Geosci.*, **19**, 399–446.

Syvitski, J. P. & J. M. Alcott (1995). RIVER3: Simulation of river discharge and sediment transport. *Comput. Geosci.*, **21**, 89–151.

Syvitski, J. P. M. & G. E. Farrow (1983). Structures and processes in bayhead deltas: Knight and Butte Inlets, British Columbia. *Sediment. Geol.*, **36**, 217–244.

Syvitski, J. P. M. & A. J. Kettner (2007). On the flux of water and sediment into the Northern Adriatic Sea. *Cont. Shelf Res.*, **27**, 296–308.

Syvitski, J. P. M. & J. D. Milliman (2007). Geology, geography, and humans battle for dominance over the delivery of fluvial sediment to the coastal ocean. *J. Geol.*, **115**, 1–19.

Syvitski, J. P. M. & M. D. Morehead (1999). Estimating river-sediment discharge to the ocean: application to the Eel margin northern California. *Mar. Geol.*, **154**, 13–28.

Syvitski, J. P. M. & Y. Saito (2007). Morphodynamics of deltas under the influence of humans. *Global Planet. Change*, **57**, 261–282.

Syvitski, J. P. *et al.* (2000). Estimating fluvial sediment transport: The rating parameters. *Water Resour. Res.*, **36**, 2747–2760.

Syvitski, J. P. *et al.* (2003). Predicting the terrestrial flux of sediment to the global ocean: a planetary perspective. *Sediment. Geol.*, **162**, 5–24.

Syvitski, J. P. M. *et al.* (2005). Impact of humans on the flux of terrestrial sediment to the global coastal ocean. *Science*, **308**, 376.

Syvitski, J. P. M. *et al.* (2009).

Takeuchi, K., A. Jayawardena & Y. Takahasi (eds.) (1995). *Catalogue of Rivers for Southeast Asia and the Pacific, Vol. 1,* UNESCO-IHP Regional Steering Committee (RSC) for Southeast Asia and the Pacific.

Talwani, X. (1997).

Tanabe, S. *et al.* (2006). Holocene evolution of the Song Hong (Red River) delta system, northern Vietnam. *Sediment. Geol.*, **187**, 29–61.

Tanzania Hydrological Yearbook (1967). Port Washington, New York, Water Information Center, INC.

Teller, J. T. (1987). Proglacial lakes, in *North America and Adjacent Oceans during the Last Deglaciation*, eds. W. F. Ruddiman & H. E. Wright, Jr., Geo. Soc. Amer., pp. 39–69.

Teller, J. T. & X. Leverington (2004).

Teller, J. T. & L. H. Thorliefson (1987). Catastrophic flooding into the Great Lakes from Lake Agassiz, in *Catastrophic Flooding*, eds. L. Mayer & D. Nash, Boston: Allen & Unwin, pp. 121–138.

Templet, P. H. & K. J. Meyer-Arendt (1988). Louisiana wetland loss: a regional water management approach to the problem. *Environ. Manag.*, **12**, 181–192.

Terrain Analysis Center (1995a). *Water Resources Areal Appraisal, Belize,* Alexandria, VA: US Army Topographic Engineering Center.

Terrain Analysis Center (1995b). *Water Resources Areal Appraisal, Costa Rica,* Alexandria, VA: US Army Topographic Engineering Center.

Terrain Analysis Center (1995c). *Water Resources Areal Appraisal, El Salvador,* Alexandria, VA: US Army Topographic Engineering Center.

Terrain Analysis Center (1995d). *Water Resources Areal Appraisal, Guatemala,* Alexandria, VA: US Army Topographic Engineering Center.

Thanh, T. D. *et al.* (2004). Regimes of human and climate impacts on coastal changes in Vietnam. *Reg. Environ. Change*, **4**, 49–62.

Thieler, E. R. *et al.* (2007). A catastrophic meltwater flood event and the formation of the Hudson Shelf Valley. *Palaeogeog., Palaeoclimatol., Palaeoecol.*, **246**, 120–136.

Thomas, M. F. & M. B. Thorp (1995). Geomorphic response to rapid climatic and hydrologic change during the late Pleistocene and early Holocene in the humid and sub-humid tropics. *Quat. Sci. Rev.*, **14**, 193–207.

Thompson, C. J., I. Takken & J. Croke (2008). Hydrological and sedimentological connectivity of unsealed roads. *IAHS Publ.* **325**, 524–531.

Thompson, D. W. J. & J. M. Wallace (1998). The Arctic Oscillation signature in the wintertime geopotential height and temperature fields. *Geophys. Res. Lett.*, **25**, 1297–1300.

Thompson, D. W. J. & J. M. Wallace (2000). Annular modes in the extratropical circulation. Part I: Month-to-month variability. *J. Climate*, **13**, 1000–1016.

Thompson, D. W. J. & J. M. Wallace (2001). Regional climate impacts of the Northern Hemisphere annular mode. *Science*, **293**, 85–89.

Thoms, M. C. & F. Sheldon (2000). Water resource development and hydrological change in a large dryland river: the Barwon–Darling River, Australia. *J. Hydrol.*, **228**, 10–21.

Thornton, S. F. & J. McManus (1994). Application of organic and nitrogen stable isotope and C/N ratios as source indications of organic matter provenance in estuarine systems: evidence from the Tay Estuary, Scotland. *Estuar. Coast. Shelf Sci.*, **38**, 219–333.

Thorp, J. H., G. A. Lamberti & A. F. Casper (2005). St. Lawrence River Basin, in *Rivers of North America*, eds. A. C. Benke & C. E. Cushing, Amsterdam: Elsevier, pp. 982–1028.

The Times World Atlas (1999).

Tipper, E. T. *et al.* (2006). The short term climatic sensitivity of carbonate and silicate weathering fluxes: Insight from seasonal variations in river chemistry. *Geochim. Cosmochim. Acta*, **70**, 2737–2754.

Tiveront, J. (1960). Debit solide des cours d'eau en Algerie et en Tunisie. *IAHS Publ.* **53**, 26–42.

Tiwari, V. W., J. Wahr & S. Swenson (2009). Dwindling groundwater resources in northern Inda, from satellite gravity observations. *Geophys. Res. Lett.*, **36**, L18401, doi:10.1029/2009GL039401.

Todd, M. C. & R. Washington (2003). Climate variability in Central Equatorial African: evidence of extra-tropical influence. *CLIVAR Exchanges*, **27**, 4.

Tomasson, H. (1991). Glacifluvial transport and erosion, in *Arctic Hydrology: Present and Future Tasks*, ed. N. N. C. f. Hydrology, Norweg. Nat. Comm. Hydrol. Report. 223, pp. 27–36.

Tomasson, H. (1997). Catastrophic floods in Iceland. *IAHS Publ.* **271**, 121–126.

Tomasson, H., S. Palsson & P. Ingolfsson (1980). Comparison of sediment load transport in the Skeioara jokulhlaups in 1972 and 1976. *Jokull*, **30**, 21–32.

Tootle, G. A. & T. C. Piechota (2006). Relationships between Pacific and Atlantic ocean sea surface temperatures and US streamflow variability. *Water Resour. Res.*, **42**, W07441 doi:10.1029/2005WR004184.

Torab, M. & M. Azab (2007). Modern shoreline changes along the Nile Delta coast as an impact of the construction of the Aswan High Dam. *Geographia Technica*, **2**, 69–76.

Tranter, M. (2004). Gechemical weathering in glacial and proglacial environments, in *Treatise on Geochemistry, v.5*, eds. H. D. Holland & K. K. Turekian, Oxford: Pergamon Press, pp. 189–205.

Trefry, J. H. *et al.* (2003). Trace metals in sediments near offshore oil exploration and production sites in the Alaskan Arctic. *Environ. Geol.*, **45**, 149–160.

Trenberth, K. E. *et al.* (2007). Chapter 3: Observations: surface and atmospheric climate change, in *Climate Change 2007: The Physical Science Basis. Contribution of Working Group I to the Fourth Assessment Report of the Intergovernmental Panel on Climate Change*, eds. S. Solomon *et al.*, Cambridge: Cambridge University Press, pp. 235–336.

Trimble, S. W. (1977). The fallacy of stream equilibrium in contemporary denudation studies. *Amer. J. Sci.*, **277**, 876–887.

Trimble, S. W. (1983). A sediment budget for Coon Creek basin in the Driftless area, Wisconsin. *Amer. J. Sci.*, **283**, 454–474.

Trimble, S. W. (1999). Decreased rates of alluvial sediment storage in the Coon Creek Basin, Wisconsin, 1975–93. *Science*, **285**, 1244.

Trimble, S. W. & P. Crosson (2000). US soil erosion rates – myth and reality. *Science*, **289**, 248–250.

Trimble, S. W. & A. C. Mendel (1995). The cow as a geomorphic agent – a critical review. *Geomorph.*, **13**, 233–253.

Troeh, F. & L. M. Thompson (1993). *Soils and Soil Fertility*, New York: Oxford University Press.

Troeh, F. R., J. A. Hobbs & R. L. Donahue (1999). *Soil and Water Conservation*, Upper Saddle River, NJ: Prentice-Hall.

Trustrum, N. A. *et al.* (1999). Sediment production storage and output: the relative role of large magnitude events in steepland catchments. *Z. Geomorphol.*, **115**, 71–86.

Tudhope, A. W. *et al.* (2001). Variability in the El Niño-Southern Oscillation through a glacial–interglacial cycle. *Science*, **291**, 1511–1517.

Tuncer, G. *et al.* (1998). Land-based sources of pollution along the Black Sea coast of Turkey: concentrations and annual loads to the Black Sea. *Mar. Pollut. Bull.*, **36**, 409–423.

Turner, B. L. *et al.* (eds.) (1990). *The Earth as Transformed by Human Action, Cambridge:* Cambridge University Press.

Turner, R. E. & N. N. Rabalais (1991). Changes in Mississippi River water quality this century. *BioScience*, **41**, 140–147.

Uchida, T. *et al.* (2000). Sediment yield on a devastated hill in southern China; effects of microbiotic crust on erosion process. *Geomorph.*, **32**, 129–145.

Umbal, J. V. (1997). Five years of lahars at Pinatubo Volcano: declining but still potentially lethal hazards. *J. Geol. Soc. Philippines*, **52**, 35813.

UNEP/MAP/MED_POL (2003). Riverine Transport of Water, Sediments and Pollutants to the Mediterranean Sea, in *MAP Technical Reports Series No 141*, UNEP/MAP, Athens, p. 111.

UNESCO (1967). *Discharge of selected rivers of the world*, Paris: UNESCO

UNESCO (1969). *Discharge of Selected Rivers of the World*, Paris: UNESCO.

UNESCO (1971). *Discharge of selected rivers of the world.* Studies and Reports in Hydrology, 194 pp.

UNESCO (1995). *Discharge of Selected Rivers of Africa*, Paris: UNESCO.

UNESCO/UNEP (1982).

UNESCO (WORRI) (1978). World register of rivers discharging into the oceans, Unpubl. ms.

Urey, H. L. (1952). *The Planets: Their Origin and Development*, New Haven: Yale University Press.

Uriarte, A. *et al.* (2004). Sediment supply, transport and deposition: contemporary and Late Quaternary evolution, in

Oceanography and Marine Environment of the Basque Country, eds. A. Borja & M. B. Collins, Amsterdam: Elsevier Oceanography Series 70, pp. 97–132.

US Geological Survey (USGS) (1994). *Water Resources Data for Alaska, Water Year 1994,* USGS-WRD-AK-00–1.

van der Leeden, F. (1975). *Water Resources of the World: Selected Statistics,* New York: Water Information Center.

Van der Weijden, C. H. & J. J. Middelburg (1989). Hydrogeochemistry of the river Rhine: long term and seasonal variability, elemental budgets, base levels and pollution. *Water Res.*, **23**, 1247–1266.

Vanacker, V. *et al.* (2007). Restoring natural vegetation reverts mountain erosion to natural levels. *Geology,* **35**, 303–306.

Vanden Bossche, J.-P. & G. M. Bernacsek (1991). *Source book for the inland fishery resources of Africa,* CIFA Technical Paper 18.

Varga, S., S. Bruk & M. Babic-Mladenovic (1989). Sedimentation in the Danube and tributaries upstream from the iron Gates (Djerdap) Dam, in *Fourth International Symposium on River Sedimentation,* Beijing, China: Ocean Press, pp. 111–118.

Varis, O. & P. Vakkilainen (2001). China's 8 challenges to water resources management in the first quarter of the 21st century. *Geomorph.*, **41**, 93–104.

Vatne, G., B. *et al.* (1995). Glaciofluvial sediment delivery from two dynamically different polythermal glaciers, northern Spitsbergen (abs), in *IGS International Symposium on Glacial Erosion and Sedimentation.*

Vega, A. J., C.-H. Sui & K.- M. Lau (1998). Interannual to interdecadal variations of the regionalized surface climate of the United States and relationships to generalized flow parameters. *Phys. Geogr.*, **19**, 271–291.

Vera, C. & G. Silvestri (2009). Precipitation interannual variability in South America from the WCRP-CMIP3 multi-model dataset. *Clim. Dyn.*, **32**, 1003–1014.

Verdon, D. C. *et al.* (2004). Multidecadal variability of rainfall and streamflow: Eastern Australia. *Water Resour. Res.*, **40**, W10201 doi:10.1029/204WR003234.

Verschuren, D., K. R. Laird & B. F. Cumming (2000). Rainfall and drought in equatorial east Africa during the past 1,100 years. *Nature,* **403**, 410–414.

Vezzoli, G., E. Garzanti & S. Monguzzi (2004). Erosion in the Western Alps (Dora Baltea basin) 1. Quantifying sediment provenance. *Sediment. Geol.*, **171**, 227–246.

Vincent, P. (2008). *Saudi Arabia: an Environmental Overview,* London: Taylor and Francis.

Vink, R. J., H. Behrendt & W. Salomons (1999). Point and diffuse source analysis of heavy metals in the Elbe drainage area: comparing heavy metal emissions with transported river loads. *Hydrobiologia,* **410**, 307–314.

Visser, F. *et al.* (2002). Quantifying sediment sources in lowlying sugarcane land: a sediment budget approach. *IAHS Publ.* **276**, 169–175.

Viviroli, D. *et al.* (2007). Mountains of the world, water towers for humanity: Typology, mapping, and global significance. *Water Resour. Res.*, **43**, W07447, doi:10.1029/2006WR005653.

Vogt, C. (1997). Regional and temporal variations of mineral assemblages in Arctic Ocean sediments as climatic indicator during glacial/interglacial changes. *Rep. Polar Res,* **251**, 309.

von Huene, R. & D. W. Scholl (1991). Observations at convergent margins concerning sediment subduction, subduction erosion, and the growth of continental crust. *Rev. Geophys.*, 29.

von Rad, U. *et al.* (1999a). A 5000-yr record of climate change in varved sediments from the oxygen minimum zone off Pakistan, northeastern Arabian sea. *Quat. Res.*, **51**, 39–53.

von Rad, U. *et al.* (1999b). Multiple monsoon-controlled breakdown of oxygen-minimum conditions during the past 30000 years documented in laminated sediments off Pakistan. *Paleogeog. Paleoclimatol. Paleoceanogr.*, **152**, 129–161.

Voorhis, A. D. *et al.* (1983). The estuarine character of the Gulf of Nicoya, an embayment on the Pacific coast of Central America. *Hydrobiologia,* **99**, 225–237.

Vörösmarty, C. J., B. M. Fekete & B. A. Tucker (1996a). Global River Discharge Database (RivDIS v1.0), Vol. 3: Europe. International Hydrological Programme, UNESCO: Paris.

Vörösmarty, C. J., B. M. Fekete & B. A. Tucker (1996b). Global River Discharge Database (RivDIS v1.0), Vol. 4: North America. International Hydrological Programme, UNESCO: Paris.

Vörösmarty, C. J., B. M. Fekete & B. A. Tucker (1996c). Global River Discharge Database (RivDIS v1.0), Vol. 5: South America. International Hydrological Programme, UNESCO: Paris.

Vörösmarty, C. J., B. M. Fekete & B. A. Tucker (1996d). Global River Discharge Database (RivDIS v1.0), Vol. 1: Africa. International Hydrological Programme, UNESCO: Paris.

Vörösmarty, C. J., B. M. Fekete & B. A. Tucker (1996e). Global River Discharge Database (RivDIS v1.0), Vol. 2: Asia. International Hydrological Programme, UNESCO: Paris.

Vörösmarty, C. J. & M. Meybeck (2004). Responses of continental aquatic systems at the global scale: new paradigms, new methods, in *Vegetation, Water, Humans and the Cliamte,* eds. P. Kabat, M. Claussen, P. A. Dirmeyer, *et al.*, Heidelberg: Springer, pp. 375–413.

Vörösmarty, C. J. & D. Sahagian (2000). Anthropogenic disturbance of the terrestrial water cycle. *BioScience,* **50**, 753–765.

Vörösmarty, C. J. *et al.* (1997a). The storage and aging of continental runoff in large reservoir systems of the world. *Ambio,* **26**, 210–219.

Vörösmarty, C. J. *et al.* (1997b). The potential impact of neo-castorization of sediment transport by the global network of rivers. *IAHS Publ.*, **245**, 261–273.

Vörösmarty, C. J. *et al.* (2000). Geomorphometric attributes of the global system of rivers at 30-minute spatial resolution. *J. Hydrol.*, **237**, 17–39.

Vörösmarty, C. J. *et al.* (2001). Global water data: a newly endangered species. *Eos. Trans. AGU,* **82**, 54–58.

Vörösmarty, C. J. *et al.* (2003). Anthropogenic sediment retention: major global-scale impact from the population of registered impoundments. *Global Planet. Change*, **39**, 169–190.

Vörösmarty, C. J. *et al.* (2004). Humans transofrming the global water system. *EOS*, **85**, 509–512.

Waananen, A. O. (1969). Floods of January and February 1969 in central and southern California. US Geol. Surv. Open File Report, 223 pp.

Waananen, A. O., D. D. Harris & R. C. Williams (1970). Floods of December 1964 and January 1965 in the Far Western States. Part 2. Streamflow and Sediment Data. *US Geol. Surv. Water-Supply Paper 1866-B*, 861 pp.

Wahby, S. K. & N. F. Bishara (1981). The effect of the river Nile on Mediterranean water, before and after the construction of the high dam at Aswan, in *River Input to the Ocean Systems (RIOS)* UNESCO, pp. 311–318.

Wainwright, J. & J. B. Thornes (2004). *Environmental Issues in the Mediterranean. Processes and Perspectives from the Past and Present,* London: Routledge.

Waisanen, P. J. & N. B. Bliss (2002). Changes in population and agricultural land in conterminous United States counties, 1790 to 1997. *Global Biogeochem. Cycles*, **16**, 1137.

Waitt, R. B., Jr. *et al.* (1983). Eruption-triggered avalanche, flood, and lahar at Mount St. Helens – Effects of winter snowpack. *Science*, **221**, 1394–1396.

Wakatsuki, T. & A. Rasyidin (1992). Rates of weathering and soil formation. *Geoderma*, **52**, 251–263.

Wallace, I. G. (2009). Components of precipitation and temperature anomalies and change associated with modes of the Southern Hemisphere. *Int. J. Climatol.*, **29**, 809–826.

Walling, D. E. (1978). Reliability considerations in the evaluation and analysis of river loads. *Z. Geomorphol. Suppl.*, **29**, 29–42.

Walling, D. E. (1985). The sediment yields of African rivers. *IAHS Publ.* **144**, 265–283.

Walling, D. E. (1997). The response of sediment yields to environmental change. *IAHS Publ.* **245**, 77–89.

Walling, D. E. (1999). Linking land use erosion and sediment yields in river basins. *Hydrobiologia*, **410**, 223–240.

Walling, D. E. (2006). Human impact on land–ocean sediment transfer by the world's rivers. *Geomorph.*, **79**, 192–216.

Walling, D. E. (2008). The changing sediment loads of the world's rivers. *IAHS Publ.* **325**, 323–338.

Walling, D. E. & D. Fang (2003). Recent trends in the suspended sediment loads of the world's rivers. *Global Planet. Change*, **39**, 111–126.

Walling, D. E. & Q. He (1999). Improved models for estimating soil erosion rates from Cesium-137 measurements. *J. Environ. Qual.*, **28**, 611–622.

Walling, D. E. & B. W. Webb (1981). The reliability of suspended sediment load data. *IAHS Publ.* **133**, 177–194.

Walling, D. E. & B. W. Webb (1983). Patterns of sediment yield, in *Background to Paleohydrology*, ed. K. J. Gregory, Chichester: John Wiley & Sons, pp. 69–100.

Walling, D. E. & B. W. Webb (1988). The reliability of rating curve estimates of suspended sediment yield: some further comments. *IAHS Publ.* **174**, 337–350.

Walling, D. E. & B. W. Webb (1996). Erosion and sediment yield: a global overview. *IAHS Publ.* **236**, 3–19.

Walling, D. E., B. W. Webb & J. C. Woodward (1992). Monitoring suspended sediment concentration in discharge from regulated lakes in glacial deposits. *IAHS Publ.* **210**, 269–278.

Walsh, K. (2004). Tropical cyclones and climate change: unresolved issues. *Climate Res.*, **27**, 77–83.

Walsh, G. & Y. Vigneault (1986). Analysis of water quality of the rivers of the north shore of the Gulf of St. Lawrence in relation to acidification processes. *Rapp. Tech. Can. Sci. Halieut. Aquat.*, **1540**, 118.

Wang, H. *et al.* (2006). Interannual and seasonal variation of the Huanghe (Yellow River) water discharge over the past 50 years: connections to impacts from ENSO events and dams. *Global Planet. Change*, **50**, 212–225.

Wang, H. *et al.* (2007). Stepwise decreases of the Huanghe (Yellow River) sediment load (1950–2005): impacts of climate change and human activities. *Global Planet. Change*, **57**, 331–354.

Wang, J. J. & X. X. Lu (2008). Influence of the changing environment on sediment loads of the lower Mekong River. *IAHS Publ.* **325**, 612–615.

Wang, Y., M.-E. Ren & J. P. M. Syvitski (1998). Sediment transport and terrigenous fluxes, in *The Sea, v. 10, The Global Coastal Ocean: Processes and Methods*, eds. K. Brink & A. R. Robinson, Harvard University Press.

Wang, Z.-S. (1992). Meteorological conditions associated with severe regional debris flows in China. *IAHS Publ.* **209**, 325–328.

Wanner, X. *et al.* (2008). Mid- to late Holocene climate change: an overview, *Quat. Sci. Rev.*, **27**, 1791–1828.

Ward, G. M., P. M. Harris & A. K. Ward (2005). Gulf Coast Rivers of the Southeastern United States, in *Rivers of North America*, eds. A. C. Benke & C. E. Cushing, Amsterdam: Elsevier, 124–178.

Ward, P. R. B. (1980). Sediment transport and reservoir siltation formula for Zimbabwe-Rhodesia. *Die Siviele Ingenieur, Suid-Afrika*, 9–15.

Warrick, J. A. & K. L. Farnsworth (2009). Dispersal of river sediment in the Southern California Bight, in *Earth Science in the Urban Ocean: The Southern California Continental Borderland,* eds. H. J. Lee & W. R. Normark, Geol. Soc. Amer. Spec. Paper 454, pp. 53–68.

Warrick, J. A. & J. D. Milliman (2003). Hyperpycnal sediment discharge from semiarid southern California rivers: Implications for coastal sediment budgets. *Geology*, **31**, 781–784.

Warrick, J. A. *et al.* (2008). Rapid formation of hyperpycnal sediment gravity currents offshore of a semi-arid California river. *Cont. Shelf Res.*, **28**, 991–1009.

Wasson, R. J., L. L. Olive & C. J. Rosewall (1996). Rates of erosion and sediment transport in Australia. *IAHS Publ.* **236**, 139–148.

Water Resources Agency (1992). *Hydrologic Yearbook of Taiwan*, Republic of China,1991.

Water Resources Planning Commission (Taiwan) (1984). *Hydrological yearbook of Taiwan,* Republic of China: Ministry of Economic Affairs.

Watson, R. T. (ed.) (1996). *Climate Change 1995; Impacts, Adaptations and Mitigation, Cambridge:* Cambridge University Press.

Watterson, I. G. (2009). Components of precipitation and temperature anomalies and change associated with modes of the Southern Hemisphere. *Int. J. Climatol.*, **29**.

Waylen, P. R. (1995). Global hydrology in relation to palaeohydrological change, in *Global Palaeohydrology*, eds. K. J. Gregory & L. Starkel, Chichester: John Wiley, pp. 61–86.

Waylen, P. R. & C. N. Caviedes (1987). El Niño and annual floods in coastal Peru, in *Catastrophic Flooding*, eds. L. Mayer & D. Nash, Boston: Allen & Unwin, pp. 57–77.

Waythomas, C. F. & G. P. Williams (1988). Sediment yield and spurious correlation – toward a better portrayal of the annual suspended sediment load of rivers. *Geomorph.*, **1**, 309–316.

Wells, J. T. (1996). The lower Mississippi River delta, in *Sea-level Rise and Coastal Subsidence*, eds. J. D. Milliman & B. U. Haq, Dordrecht: Kluwer Acad. Publ., pp. 281–311.

Wells, J. T., S. J. Chinburg & J. M. Coleman (1984). The Atchafalaya River delta: generic analysis of delta development, Vicksburg, MS: Waterways Experiment Station, Hydraulics Lab, 89 pp.

Wells, J. T. & J. M. Coleman (1987). Wetland loss and the subdelta life cycle. *Estuar. Coast. Shelf Sci.*, **25**, 111–125.

Wheatcroft, R. A. & C. K. Sommerfield (2005). River sediment flux and shelf sediment accumulation rates on the Pacific Northwest margin. *Cont. Shelf Res.*, **25**, 311–332 10.1016/j.csr.2004.10.001.

Whitaker, A. C., H. Sato & H. Sugiyama (2008). Changing suspended sediment dynamics due to extreme flood events in a small pluvial-nival system in northern Japan. *IAHS Publ.* **325**, 192–199.

White, A. F. (2004). Natural weathering rates of silicate minerals, in *Treatise on Geochemistry, v. 5*, eds. H. D. Holland & K. K. Turekian, Oxford: Pergamon Press, pp. 133–168.

White, A. F. & A. E. Blum (1995). Effects of climate on chemical weathering in watersheds. *Geochim. Cosmochim. Acta*, **59**, 1729–1747.

White, G. F. (1988). The environmental effects of the High Dam at Aswan. *Environment*, **30**, 5–40.

White, S. M. (1992). The influence of tropical cyclones as soil eroding and sediment transporting events. An example from the Philippines. *IAHS Publ.* **192**, 259–269.

White, S. M. (1996). Erosion, sediment delivery and sediment yield patterns in the Philippines. *IAHS Publ.* **236**, 233–240.

Wilkinson, B. H. & B. J. McElroy (2007). The impact of humans on continental erosion and sedimentation. *Geol. Soc. Amer. Bull.*, **119**, 140–156.

Williams, M. (2003). *Deforesting the Earth: From Prehistory to Global Crisis,* University of Chicago Press.

Williams, M. A. J. (1985). Pleistocene aridity in tropical Africa, Australia and Asia, in *Environmental Change and Tropical Geomorphology*, eds. I. Douglas & T. Spencer, London: George Allen & Unwin Publ. Ltd, pp. 219–233.

Williams, M. R. & J. M. Melack (1997). Effects of prescribed burning and drought on the solute chemistry of mixed-conifer forest streams of the Sierra Nevada, California. *Biogeochem.*, **39**, 225–253.

Willis, R. (1971). In *Water Resources of the World: Selected Statistics,* ed. F. van der Leeden, New York: Water Information Center.

Wilmot, R. D. & M. B. Collins (1981). Contemporary fluvial sediment supply to the Wash. *Int. Assoc. Sedimentol. Spec. Publ.* **5**, 99–110.

Wilson, A. & K. T. Iseri (1969). River discharges to the sea from the shores of the conterminous Unites States, Alaska and Puerto Rico, US Geol. Survey Open-File Report, 2 pp.

Wilson, C. J. (1999). Effects of logging and fire on runoff and erosion on highly erodible grantic soils in Tasmania. *Water Resour. Res.*, **35**, 3531–3546.

Wilson, L. (1973). Variations in mean annual sediment yield as a function of mean annual precipitation. *Amer. J. Sci.*, **273**, 335–349.

Winchester, S. (2008). *The Man Who Loved China.* Harper Perennial, New York. 316 pp.

Winkley, B. R. (1994). Response of the Lower Mississippi River to flood control and navigation improvement, in *The Variability of Large Alluvial Rivers*, eds. S. A. Schumm & B. R. Winkley, New York: American Soc. Civil Eng., pp. 45–74.

Winterwerp, J. C., W. G. Borst & M. B. de Vries (2005). Pilot study on the erosion and rehabilitation of a mangrove mud coast. *J. Coast. Res.*, **55**, 223–230.

Wohl, E. (2006). Human impacts to mountain streams. *Geomorph.*, **79**, 217–248.

Wolfe, M. D. & J. W. Williams (1986). Rates of landsliding as impacted by timber management activities in northwestern California. *Bull. Assoc. Eng. Geol.*, **23**, 53–60.

Wolman, M. G. (1967). A cycle of sedimentation and erosion in urban river channels. *Geog. Ann. Ser. A Phys. Geog.*, pp. 385–395.

Wolman, M. G. *et al.* (1990). The riverscape, in *Surface Water Hydrology*, eds. M. G. Wolman & H. C. Riggs, Geol. Soc. Amer., pp. 281–328.

Wolman, M. G. & J. P. Miller (1960). Magnitude and frequency of forces in geomorphic processes. *J. Geol.*, **68**, 54–74.

Wolman, M. G. & A. P. Schick (1967). Effects of construction on fluvial sediment, urban and suburban areas of Maryland. *Water Resour. Res.*, **3**, 451–464.

Woodhouse, C. A., S. T. Gray & D. M. Meko (2006). Updated streamflow reconstructions for the Upper Colorado River basin. *Water Resour. Res.*, **42**, W05415 doi:10.1029/2005WR004455.

Woodroffe, C. D. *et al.* (2003). Mid-late Holocene El Niño variability in the equatorial Pacific from coral microatolls. *Geophys. Res. Lett.* **30**.

Woodward, J. C. (1995). Patterns of erosion and suspended sediment yield in Mediterranean river basins, in *Sediment and Water Quality in River Catchments*, eds. I. D. L. Foster, A.

M. Gurnell & B. W. Webb, Chichester: John Wiley & Sons Ltd, pp. 365–389.

Woodward, J. C. *et al.* (2007). The Nile: evolution, Quaternary river environments and material fluxes, in *Large Rivers: Geomorphology and Management*, ed. A. Gupta, John Wiley, pp. 261–292.

World Wildlife Fund (2007). (http://www.wwf.org)

Wright, K. *et al.* (1990). Logging effects on streamflow at Caspar Creek in northwest California. *Water Resour. Res.*, **26**, 1657–1667.

Wright, L. D., B. G. Thom & R. J. Higgins (1980). Wave influences on River-mouth depositional process: Examples from Australia and Papua New Guinea. *Est. Coast. Mar. Sci.*, **11**, 263–277.

Wright, S. A. & D. H. Schoellhamer (2004). Trends in the sediment yield of the Sacramento River, California, 1957–2001. *San Francisco Estuary and Watershed Science (online serial)*, **2** (2), article 2.

Wu, X. D. (1992). Dendroclimatic studies in China, in *Climate since AD 1500*, eds. R. S. Bradley & P. D. Jones, London: Routledge, pp. 432–445.

Xia, D. X., S. Y. Wu & Z. Yu (1993). Changes of the Yellow River since the last glacial age. *Mar. Geol and Quat. Geol.*, **13**, 83–88.

Xiong, Y., J.-Z. Zhang and E.-B. Liu (1985). The hydrology of China's rivers. *Geojournal*, **10**, 173–181.

Xu, J. (1994). A study of the accumulation rate of the lower Yellow River in the past 10,000 years. *IAHS Publ.* **224**, 421–429.

Xu, J. (2003). Sedimentation rates in the lower Yellow River over the past 2300 years as influenced by human activities and climate change. *Hydrol. Process.*, **17**, 3359–3371.

Xu, J. (2004). A study of anthropogenic seasonal rivers in China. *Catena*, **55**, 17–32.

Xu, K. H. & J. D. Milliman (2009). Seasonal variations of sediment discharge from the Yangtze River before and after impoundment of the Three Gorges Dam. *Geomorph.*, **104**, 276–283.

Xu, K. H. *et al.* (2006). Yangtze sediment decline partly from Three Gorges Dam. *EOS*, **87**, 185.

Xu, K. H. *et al.* (2007). Climatic and anthropogenic impacts on water and sediment discharges from the Yangtze River (Changjiang), 1950–2005, in *Large Rivers: Geomorphology and Management*, ed. A. Gupta, Chichester: John Wiley, pp. 609–626.

Xu, Z. X., Y. N. Chen & J. Y. Li (2004). Impact of climate change on water resources in the Tarim River basin. *Water Resour. Manag.*, **18**, 439–458.

Yang, Y.-h. & F. Tian (2009). Abrupt change of runoff and its major driving factors in Haihe River catchment. *China. J. Hydrol.*, **374**, 373–383.

Yang, S. L., J. Zhang & J. Zhu (2002). Impact of dams on Yangtze River and the influences of human activities. *J. Hydrol.*, **263**, 56–71.

Yang, S. L. *et al.* (2005). Impact of dams on Yangtze River sediment supply to the sea and delta intertidal wetland response. *J. Geophys. Res.*, **110**, F03006, doi:10.1029/2004JF000271.

Yang, S. L. *et al.* (2006). Drastic decrease in sediment supply from the Yangtze River and its challenge to coastal wetland management. *Geophys. Res. Lett.* **33**, DOI 10.1029/2005GL025507).

Yang, S. L. *et al.* (2011). 50,000 dams later: erosion of the Yangtze River and its delta. *Global Planet. Change,* **75**, 14–20.

Yang, Z. S. *et al.* (1998). The Yellow River's water and sediment discharge decreasing steadily. *EOS*, **79**, 589–592.

Yang, Z. S. *et al.* (2006). Dam impacts on the Changjiang (Yangtze) River sediment discharge to the sea: The past 55 years and after the Three Gorges Dam. *Water Resour. Res.*, **42**, W04407.

Yang, Z. *et al.* (2007). Influence of the Three Gorges Dam on downstream delivery of sediment and its environmental implications, Yangtze River. *Geophys. Res. Lett.*, **34**, L10401, doi:10.1029/2007GL029472.

Young, R. A. (1995). Coping with a severe sustained drought on the Colorado River: introduction and overview. *Water Resour. Res.*, **31**, 779–788.

Zachos, J. *et al.* (2001). Trends rhythms and abberations in global climate 65 Ma to present. *Science*, **292**, 686–693.

Zebidi, H. (ed.) (1998). *Water: a looming crisis?,* Paris: UNESCO.

Zekster, I. S. & R. G. Dzhamalov (1988). *Role of ground water in the hydrologic cycle and its continental water balance,* Paris: UNESCO IHP-II Project 2.3, 133 pp.

Zekster, I. S. & H. A. Loaiciga (1993). Groundwater fluxes in the global hydrologic cycle: past, present and future. *J. Hydrol.*, **144**, 405–427.

Zhang, J. (1994). Biogeochemistry of Trace Metals from Chinese River-Estuary systems: An overview. *Dept Marine Chem.*, Ocean Univ., Qingdao, 29 p.

Zhang, J. *et al.* (1992). Transport of heavy metals towards the China sea: a preliminary study and compaison. *Marine Geochem.*, **40**, 161–178.

Zhang, J. *et al.* (1994). Eco-social impact and chemical regimes of large Chinese rivers – a short discussion. *Water Res.*, **28**, 609–617.

Zhang, J. *et al.* (1998). Riverine sources and estuarine fates of particulate organic carbon from north China in late summer. *Estuar. Coast. Shelf Sci.*, **46**, 439–448.

Zhang, Q. *et al.* (2007). Possible influence of ENSO on annual maximum streamflow of the Yangtze River. *Chin. J. Hydrol.*, **333**, 265–274.

Zhang, R. & T. L. Delworth (2006). Impact of Atlantic multidecadal oscillations on India/Sahel rainfall and Atlantic hurricanes. *Geophys. Res. Lett.*, **33**, L17712 doi:10.1029/2006GL026267.

Zhang, Y. G. (1993). Transport of sediment and organic matter in freshwater systems and offshore if China. *Mitt. Geol. Geol-Paleont. Inst., Univ. of Hamburg*, **64**, 243–249.

Zhukinsky, V. N. *et al.* (1989). *Ecosystem of Dnieper-Bug estuary,* Kiev: Naukova Dumka.

Zuffa, G. G. *et al.* (2000). Turbidite megabeds in an oceanic rift valley recording jokulhlaups of late Pleistocene glacial lakes of the western United States. *J. Geol.*, **108**, 253–274.

Geographic Index: Bodies of Water

Geographic Index: Rivers

Geographic Index: Terrestrial

Topical Index

Printed in the United States
By Bookmasters